The Structure, Development and Evolution of Reptiles

SYMPOSIA OF THE ZOOLOGICAL SOCIETY OF LONDON
NUMBER 52

The Structure, Development and Evolution of Reptiles

A Festschrift in honour of Professor A.d'A. Bellairs on the occasion of his retirement

(The Proceedings of a Symposium held at The Zoological Society of London on 26 and 27 May 1983)

Edited by

MARK W. J. FERGUSON

Department of Anatomy, The Queen's University of Belfast, Belfast, Northern Ireland

Published for
THE ZOOLOGICAL SOCIETY OF LONDON
BY
ACADEMIC PRESS
1984

Wingate College Library

ACADEMIC PRESS INC. (LONDON) LTD
24/28 Oval Road, London NW1 7DX

United States Edition published by
ACADEMIC PRESS INC.
(Harcourt Brace Jovanovich, Inc.)
Orlando, Florida 32887

Copyright © 1984 by
THE ZOOLOGICAL SOCIETY OF LONDON
All rights reserved. No part of this book may be reproduced in any form by photostat, microfilm, or any other means, without written permission from the publishers

British Library Cataloguing in Publication Data
The structure, development and evolution of
 reptiles. – (Symposia of the Zoological Society
 of London, ISSN 0084-5612; no. 52)
 1. Reptiles
 I. Ferguson, Mark W. J. II. Zoological
 Society of London   III. Series
 597.9   QL665
 ISBN 0-12-613352-2

Typeset by Oxford Verbatim Limited,
and printed in Great Britain by
Page Bros (Norwich) Ltd

Contributors

ARNOLD, E. N., *Department of Zoology, British Museum (Natural History), Cromwell Road, London SW7 5BD, England* (p. 47)
AVERY, R. A., *Department of Zoology, University of Bristol, Woodland Road, Bristol BS8 1UG, England* (p. 407)
BELLAIRS, A.d'A., *Department of Anatomy, St Mary's Hospital Medical School, Paddington, London W2, England* (p. 665)
BENTON, M. J., *University Museum, University of Oxford, South Parks Road, Oxford OX1 3PW, England* (p. 575)
BRYANT, S. V., *Department of Developmental and Cell Biology, Developmental Biology Center, University of California, Irvine, California 92717, USA* (p. 177)
BUSTARD, H. R., *Airlie Brae, Alyth, Perthshire PH11 8AX, Scotland* (p. 385)
CHARIG, A. J., *Department of Palaeontology, British Museum (Natural History), Cromwell Road, London SW7 5BD, England* (p. 597)
COULSON, R. A., *Department of Biochemistry, Louisiana State University Medical Center, 1100 Florida Avenue, New Orleans, Louisiana 70119, USA* (p. 425)
FERGUSON, M. W. J., *Department of Anatomy, The Queen's University of Belfast, Medical Biology Centre, 97 Lisburn Road, Belfast BT9 7BL, Northern Ireland* (pp. 3, 223 and 275)
GANS, C., *Department of Zoology, Division of Biological Sciences, The University of Michigan, Natural Science Building, Ann Arbor, Michigan 48109, USA* (p. 13)
GRAHAM, E. E., *Laboratory for Developmental Biology, University of Southern California, Andrus Gerontology Center, University Park, MC-0191, Los Angeles, California 90089-0191, USA* (p. 275)
HALL, B. K., *Department of Biology, Life Sciences Centre, Dalhousie University, Halifax, Nova Scotia, B3H 4J1, Canada* (p. 155)
HONIG, L. S., *Laboratory for Developmental Biology, University of Southern California, Andrus Gerontology Center, University Park MC-0191, Los Angeles, California 90089-0191, USA.* Present address: *Department of Anatomy and Cell Biology, R-124, University of Miami Medical School, PO Box 016960, Miami, Florida 33101, USA* (p. 197)
JONES, K. W., *Institute of Animal Genetics, University of Edinburgh, King's Buildings, West Mains Road, Edinburgh EH9 3JN, Scotland* (p. 305)
LANCE, V., *Department of Medicine, Tulane University School of Medicine, 1430 Tulane Avenue, New Orleans, Louisiana 70112, USA* (p. 357)
LANDSMEER, J. M. F., *Anatomisch-Embryologisch Laboratorium, Universiteit Leiden, Wassenaarseweg 62, 2333 AL Leiden, The Netherlands* (p. 27)
LOVERIDGE, J. P., *Department of Zoology, University of Zimbabwe, PO Box MP 167, Mount Pleasant, Harare, Zimbabwe* (p. 443)

Contributors

MADERSON, P. F. A., *Department of Biology, Brooklyn College of the City University of New York, Bedford Avenue & Avenue H, Brooklyn, New York 11210, USA* (p. 111)

MUNEOKA, K., *Developmental Biology Center, University of California, Irvine, California 92717, USA* (p. 177)

NORMAN, D. B., *Department of Zoology and University Museum, University of Oxford, South Parks Road, Oxford OX1 3PS, England* (p. 521)

OSBORN, J. W., *Department of Oral Biology, Faculty of Dentistry, The University of Alberta, Edmonton, Alberta T6G 2N8, Canada* (p. 549)

PRESLEY, R., *Department of Anatomy, University College, Cardiff, PO Box 78, Cardiff CF1 1XL, Wales* (p. 127)

PRITCHARD, P. C. H., *Florida Audubon Society, 1101 Audubon Way, Maitland, Florida 32751, USA* (p. 87)

REID, R. E. H., *Department of Geology, The Queen's University of Belfast, Elmwood Avenue, Belfast BT7 1NN, Northern Ireland* (p. 629)

RIEPPEL, O., *Paläontologisches Institut und Museum der Universität Zurich, Künstlergasse 16, CH-8006 Zurich, Switzerland* (p. 503)

RUSSELL F. E., *Department of Pharmacology and Toxicology, College of Pharmacy, University of Arizona, Tucson, Arizona 85721, USA* (p. 469)

SAMUEL, N., *Laboratory for Developmental Biology, University of Southern California, Andrus Gerontology Center, University Park, MC-0191, Los Angeles, California 90089-0191, USA* (p. 275)

SLAVKIN, H. C., *Laboratory for Developmental Biology, University of Southern California, Andrus Gerontology Center, University Park, MC-0191, Los Angeles, California 90089-0191, USA* (p. 275)

SMITH, A. M. A., *Conservation Commission of the Northern Territory, PO Box 38496, Winnellie, Northern Territory 5789, Australia. Present address: Department of Population Biology, Research School of Biological Sciences, The Australian National University, PO Box 475, Canberra, ACT 2601, Australia* (p. 319)

SNEAD, M. L., *Laboratory for Developmental Biology, University of Southern California, Andrus Gerontology Center, University Park, MC-0191, Los Angeles, California 90089-0191, USA* (p. 275)

UNDERWOOD, G. L., *Department of Biological Sciences, City of London Polytechnic, Calcutta House Precinct, Old Castle Street, London E1 7NT, England* (p. 483)

WEBB, G. J. W., *School of Zoology, The University of New South Wales, PO Box 1, Kensington, New South Wales 2033, Australia* (p. 319)

ZEICHNER-DAVID, M., *Laboratory for Developmental Biology, University of Southern California, Andrus Gerontology Center, University Park, MC-0191, Los Angeles, California 90089-0191, USA* (p. 275)

Organizer and Chairmen of Sessions

ORGANIZER
M. W. J. FERGUSON, on behalf of the Zoological Society of London

CHAIRMEN OF SESSIONS
A. S. BREATHNACH, *Department of Anatomy and Cell Biology, St Mary's Hospital Medical School, University of London, Paddington, London W2, England*

M. R. K. LAMBERT, *British Herpetological Society, Centre for Overseas Pest Research, College House, Wrights Lane, London W8 5SJ, England*

T. S. WESTOLL, *Department of Geology, University of Newcastle-upon-Tyne, Newcastle-upon-Tyne NE1 7RU, England*

L. WOLPERT, FRS, *Department of Biology as Applied to Medicine, The Middlesex Hospital Medical School, Cleveland Street, London W1P 6DB, England*

Preface

This symposium on "The Structure, Development and Evolution of Reptiles" was held at the Zoological Society of London on 26 and 27 May, 1983, in honour of Professor A.d'A. Bellairs on the occasion of his retirement. The symposium was held under the auspices of The Zoological Society of London, The Anatomical Society of Great Britain and Ireland and The British Herpetological Society. The papers presented at the symposium are published here as a Festschrift for Professor Bellairs. Just as his interests in reptile biology are broad ranging so too are the subjects covered by this volume. All emphasize the inseparable relationships of form and function, evolution and development, ecology and physiology. The contributors are each outstanding experts in their particular field, and many are former students, friends or collaborators of Professor Bellairs. In part this determined the nature of the subjects covered, but the latter was also influenced by Professor Bellairs' personal interests and by areas of current investigation in reptile biology. This volume therefore represents a series of selected highlights in the fields of reptile morphology, development, physiological ecology and evolution. Papers are arranged within the volume in this order.

The symposium was also an attempt to bring together investigators who are using reptiles in some particular field, but who would not normally meet and interact given the somewhat restricted and specialized programmes of many scientific societies and meetings. This volume likewise contains papers which span a great breadth of biological interest. It is to be hoped not only that it contains something for every biologist but also that those who consult it will be stimulated to read the other contributions which are perhaps peripheral to their spheres of investigation. In some cases (e.g. limb development and interactions between therapsids and archosaurs during the Triassic) investigators with opposing views were asked to contribute. Controversy, criticism and scepticism are necessary parts of the scientific process and knowledge which is neither challenged nor changed tends to become sterile dogma.

Many of the papers will be of interest to those outside the reptile community. Thus, some emphasize areas where reptiles offer unique advantages as experimental animals, e.g. in studies of craniofacial and limb development and in investigations of metabolic rate, anaerobic glycolysis and sex-determining mechanisms. Others provide interesting insights into the evolution of mammalian (and avian) structures such as the ear, secondary cartilage, teeth and sex chromosomes. Yet others, e.g. those on the morphology of the anterior limb and on dinosaur bone histology, demonstrate the striking similarities between different vertebrates in their structural adaptations to function.

All papers contain new data which are often extensive, e.g. those on sex ratios and thermoregulation in crocodiles. Some, e.g. those on reproductive endocrinology, dinosaur bone and triassic dinosaurian scenarios also provide extensive reviews and critical appraisals of existing literature, whilst others, e.g. on the squamate epidermis, suggest new hypotheses on the basis of existing data. All highlight areas of active research and look forward with ideas and suggestions for future investigations.

I am grateful to the contributors for their participation and prompt delivery of manuscripts, to the Zoological Society of London and the Anatomical Society of Great Britain and Ireland for financial support, and especially to Miss Unity McDonnell, Administrative Assistant at the Zoological Society, for her invaluable help with the organization of the symposium and the editing of the manuscripts. Finally, I apologize to all those who volunteered papers for the symposium and this volume but who could not be accommodated owing to limitations of available time and space.

Belfast, March 1984   Mark W. J. Ferguson

Contents

Contributors v
Organizer and Chairmen of Sessions viii
Preface ix

INTRODUCTION

Professor Angus d'Albini Bellairs D.Sc. – An Appreciation

MARK W. J. FERGUSON

MORPHOLOGY

Slide-pushing – A Transitional Locomotor Method of Elongate Squamates

CARL GANS

Synopsis 13
Introduction 13
Observations 14
Analysis of the propulsive pattern 16
Reconstruction 21
How does the body twist? 21
Transitions 22
Acknowledgements 25
References 25

Morphology of the Anterior Limb in Relation to Sprawling Gait in *Varanus*

JOHAN M. F. LANDSMEER

Synopsis 27
Introduction 27
Global approach 28
Functional considerations 33
   General remarks 33
   Functional morphology of the anterior extremity 36
Conclusions 42
Post scriptum 43
Acknowledgements 43

References 43
Appendix – Key to abbreviations used in figures 44

Variation in the Cloacal and Hemipenial Muscles of Lizards and its Bearing on their Relationships
E. N. ARNOLD

Synopsis 47
Introduction 47
Generalized pattern of cloacal and hemipenial muscles
   in male lizards 48
      Non-muscular structures and ventral caudal muscles 48
      Muscles of the cloacal and hemipenial region 50
      Function 52
Deviations from the generalized pattern 52
Systematic account 56
      Lizards 56
      Amphisbaenians, snakes and *Sphenodon* 69
Cloacal muscles in female lizards 71
Muscle patterns and phylogeny 72
      Origin of the hemipenis and its muscles 72
      Polarity of muscle characters 72
      Distribution of derived states and the phylogeny of lizards ... 73
      Apparent homoplasies 78
Conclusion 79
Acknowledgements 79
References 79
Appendix I. Material examined 81
Appendix II. Synonymy of names applied to the muscles of the cloacal
   and hemipenial region of squamates 83

Piscivory in Turtles, and Evolution of the Long-necked Chelidae
PETER C. H. PRITCHARD

Synopsis 87
Introduction 87
Angling 88
Netting 89
Spear-fishing 90
      General 90
      Origin of the Trionychidae 93
      Chelidae 94
Origin of *Hydromedusa* and *Chelodina* 95
      General 95
      Skull characters 95
      Neural bones 98

Relations between *Hydromedusa* and *Chelus* 100
The anterior plastron in *Chelus*, *Hydromedusa* and *Chelodina* 102
Evolution of *Chelus* and *Hydromedusa* 105
References 109

The Squamate Epidermis: New Light Has Been Shed

P. F. A. MADERSON

Synopsis 111
Introduction 111
The anatomy of the squamate epidermis 112
The control of cyclic activity in the squamate epidermis 113
Cutaneous water loss (CWL) in squamate reptiles 116
The responses of biological epithelia to changes in aqueous flux 117
Direct testing of the hypothesis 120
Ecological and behavioral correlates 120
Summary and conclusions 122
Acknowledgements and dedication 123
References 124

Lizards, Mammals and the Primitive Tetrapod Tympanic Membrane

R. PRESLEY

Synopsis 127
The dilemma 127
Introduction 128
Materials and methods 129
Embryological observations 130
   Concord with classical observations 130
   Principal present findings 130
Discussion 136
   Historical perspective 136
   Implications of embryological observations 140
Conclusion and dedication 149
Acknowledgements 150
References 150
Appendix – Key to abbreviations used in figures 152

DEVELOPMENT

Developmental Processes Underlying the Evolution of Cartilage and Bone

BRIAN K. HALL

Synopsis 155
Introduction 155
Secondary cartilage – Do reptiles possess it? 156

The structure of reptilian cartilage and bone 162
  Bone 162
  Cartilage and secondary centres 163
  Indeterminate growth 164
  Sesamoids, periosteal and intratendinous ossification 165
  Metaplastic bone 165
  The cardiac skeleton 166
Limb development 167
Timing of inductive tissue interactions 168
Summary and future research 170
Acknowledgements 171
References 172

Regeneration and Development of Vertebrate Appendages
KEN MUNEOKA and SUSAN BRYANT

Synopsis 177
Introduction 177
The question of universality 178
Models for development and regeneration 182
  Progress zone/polarizing zone model 182
  Polar co-ordinate model 183
The experimental test 184
Discussion 189
Acknowledgements 193
References 193

Pattern Formation during Development of the Amniote Limb
LAWRENCE S. HONIG

Synopsis 197
Vertebrate limb development 197
The avian embryonic limb 200
  The proximodistal and dorsoventral axes 202
  The anteroposterior axis 203
Retinoic acid and limb development 207
  Effects on pattern regulation 207
  Role of retinoids at the molecular level 208
  Retinoids at the cell and tissue level 209
  Mechanisms of retinoic acid signalling in the limb 210
The reptilian embryonic limb 211
  Presence of the polarizing region 211
  Uniqueness of the polarizing region 215
Conclusions 217
Acknowledgements 218
References 218

Craniofacial Development in *Alligator mississippiensis*
MARK W. J. FERGUSON

Synopsis	223
Introduction	224
Normal development	226
Branchial arches	226
Nasal placodes, pits and processes	233
Primary palate	235
Secondary palate and tectoseptal processes	236
Tongue	245
Longitudinal studies	245
Species variations	249
Spontaneous malformations	251
Introduction	251
Techniques for aging adult alligators	252
Relationship between maternal age and spontaneous malformation rate	253
Teratology	255
Introduction	255
Reduced lower jaw	256
Facial clefting	259
Acknowledgements	263
References	265

Amelogenesis in Reptilia: Evolutionary Aspects of Enamel Gene Products
HAROLD C. SLAVKIN, MARGARITA ZEICHNER-DAVID, MALCOLM L. SNEAD, EDWARD E. GRAHAM, NELSON SAMUEL and MARK W. J. FERGUSON

Synopsis	275
Introduction	276
Survey of amelogenesis in selected vertebrate species	277
Developmental and morphological features	277
Biochemical features of enamel matrix proteins	281
Amelogenesis in the American alligator	282
Developmental and morphological features	282
Preliminary biochemical features	285
Survey of immunological determinants of enamel proteins during evolution	288
Mouse amelogenins	290
Rabbit enamelins	291
Indirect immunohistochemical localization of enamel protein antigens in selected vertebrate species	291
Micro-ELISA assay to compare enamel protein antigens in lower and higher vertebrates	293
Molecular investigations of enamel gene expression	294
Experimental strategy	295
Preliminary results of enamel gene expression	297
Summary	298

Acknowledgements 298
References 299

The Evolution of Sex Chromosomes and Chromosomal Inactivation in Reptiles and Mammals

K. W. JONES

Synopsis 305
Introduction 305
  Systems of sex determination 305
  Sex determination in snakes 307
W Chromosome heterochromatin and repeated DNA 308
X Chromosome inactivation 310
Chromosomal hijacking 311
Regulatory gene function by allelic exclusion 312
Accounting for the differences in chromosomal inactivation in reptiles,
  mammals and insects 314
References 315

PHYSIOLOGICAL ECOLOGY

Sex Ratio and Survivorship in the Australian Freshwater Crocodile *Crocodylus johnstoni*

GRAHAME J. W. WEBB and ANTHONY M. A. SMITH

Synopsis 319
Introduction 319
The McKinlay River *C. johnstoni* population: A selected review 321
  Study area and history of protection 321
  Population size and age structure 321
  The population sex ratio 321
  Post-hatching mortality and growth 322
  Movement and dispersal 322
  Reproduction and nesting 323
  Temperature-dependent sex determination 323
  Temperature-dependent survivorship 326
  Annual variation in the number and sex ratio of hatchlings 326
Methods 326
  Field nests 326
  Geographic variation in hatchling sex ratios and time of hatching .. 327
  Age structure analysis 328
Results 330
  Field nests 330
  Geographic variation 338
  Age structure analysis 343
Discussion 346
  Temperature-dependent sex determination 346
  Sex determination and survivorship in the field 348

Theoretical considerations 351
Acknowledgements 353
References 353

Endocrinology of Reproduction in Male Reptiles

VALENTINE LANCE

Synopsis 357
Introduction 357
Chelonia 359
Squamata 366
    Serpentes 367
    Lacertilia 370
Crocodilia 374
Summary and conclusions 377
Acknowledgements 378
References 378

Breeding the Gharial (*Gavialis gangeticus*): Captive Breeding a Key Conservation Strategy for Endangered Crocodilians

H. ROBERT BUSTARD

Synopsis 385
Introduction 385
The need for crocodilian captive breeding programmes 386
Initiation and operation of captive breeding programmes and rehabilitation 388
Design of the gharial breeding enclosure 390
    Relevance of gharial biology 391
    Ideal breeding conditions 392
    The breeding complex 392
    Water supply 394
    Food 396
    Summary and conclusions 396
The gharial captive breeding programme 397
    The breeding stock 397
    The 1980 breeding season 399
    The 1981 breeding season 401
    The 1982 breeding season 402
Gharial breeding in sanctuaries 404
Acknowledgements 405
References 405

Physiological Aspects of Lizard Growth: The Role of Thermoregulation

R. A. AVERY

Synopsis 407

Introduction 407
Materials and methods 409
Results 411
  Growth under constant conditions 411
  Growth dynamics: the role of thermoregulation 413
  Growth dynamics: effects of reduced food intake 417
Discussion 418
  The pattern of growth in *L. vivipara* 418
  Growth parameters 419
  Growth dynamics 420
  General and ecological considerations 422
Acknowledgements 423
References 423

How Metabolic Rate and Anaerobic Glycolysis Determine the Habits of Reptiles

R. A. COULSON

Synopsis 425
Introduction 425
Factors affecting metabolic rate 426
  Size, metabolic rate and blood flow 426
  Length of fast 431
  Reproducibility 431
  Temperature 431
  Feeding 431
Anaerobic glycolysis 433
  Cost of resynthesis of muscle glycogen 433
  Rate of resynthesis of muscle glycogen 434
Relationship between metabolic rate and various activities 434
  Diving 434
  Energy for sustained effort 437
  Basking 437
  Food requirements 437
  Speculation on the reason reptiles are cold-blooded 438
  Energy production in a homeothermic 100-ton dinosaur 439
Acknowledgements 440
References 440

Thermoregulation in the Nile Crocodile, *Crocodylus niloticus*

J. P. LOVERIDGE

Synopsis 443
Introduction 443
Material and methods 444
Results 446
  Thermoregulatory behaviour 446
  Body temperatures 448

Body temperature and thermoregulatory behaviour	452
Heating and cooling rates	456
The significance of gaping	458
Discussion	461
Thermoregulatory behaviour	461
Body temperatures	463
Heating and cooling rates	464
The significance of gaping	465
Acknowledgements	466
References	466

Snake Venoms
FINDLAY E. RUSSELL

Synopsis	469
Introduction	469
Discussion	470
Conclusions	479
References	479

EVOLUTION

Scleral Ossicles of Lizards: An Exercise in Character Analysis
GARTH UNDERWOOD

Synopsis	483
Introduction and sources of data	483
Scoring system	485
Outgroup comparison	488
Character analysis	489
Phylogenetic analysis of patterns	492
Taxonomic implications	494
Discussion of character analysis	496
Discussion of taxonomy and evolution	497
General discussion	500
References	501

Miniaturization of the Lizard Skull: Its Functional and Evolutionary Implications
OLIVIER RIEPPEL

Synopsis	503
Introduction	503
Why become small?	504
The reduction of skull diameter	505
The jaw adductor musculature	513

The origin of snakes ... 516
Acknowledgements ... 518
References ... 518
Appendix – Key to abbreviations used in figures ... 520

On the Cranial Morphology and Evolution of Ornithopod Dinosaurs
DAVID B. NORMAN

Synopsis ... 521
Introduction ... 521
General cranial architecture ... 522
    The 'fabrosauroid' ... 522
    The 'hypsilophodontoid' ... 525
    The 'iguanodontoid' ... 528
    The 'hadrosauroid' ... 532
Craniological observations on *Iguanodon* ... 535
    Cranial flexibility ... 535
Systematics of the Ornithopoda ... 541
    General observations and comments ... 542
Conclusions ... 544
Acknowledgements ... 544
References ... 545
Appendix – Key to abbreviations used in figures ... 547
Note added in proof ... 547

From Reptile to Mammal: Evolutionary Considerations of the Dentition with Emphasis on Tooth Attachment
J. W. OSBORN

Synopsis ... 549
Introduction ... 549
Sequence, shape and number ... 550
    Sequences of tooth initiation in reptiles ... 550
    Gradients of shape ... 553
    Tooth number ... 554
    Evolution of mammalian from reptilian dentitions ... 555
Tooth attachment ... 556
    Introduction ... 556
    Acrodont ankylosis ... 558
    Gomphosis ... 559
    Mixed ankylosis/gomphosis ... 562
    Evolution and (differentiation) clades ... 563
    Functional advantages ... 566
    Eruption ... 569
    Root morphology ... 569
    Alveolar bone ... 570
References ... 572

The Relationships and Early Evolution of the Diapsida
MICHAEL J. BENTON

Synopsis	575
Introduction	575
Classification of the Diapsida	576
Classification of the Permo-Triassic diapsids	580
Permo-Triassic diapsids and faunal evolution	581
Origin of the diapsids	581
Late Permian: The Lower Sakamena Formation, Malagasy	583
Early Triassic: the *Lystrosaurus* and *Cynognathus* Zones, South Africa	585
Middle Triassic: the Grenzbitumenzone, Switzerland	587
Late Triassic: the Lossiemouth Sandstone Formation, Scotland	588
Latest Triassic: the Knollenmergel, Germany	589
Late Triassic/Early Jurassic: the Bristol fissures, England	592
Conclusions	592
Acknowledgements	594
References	594

Competition Between Therapsids and Archosaurs During the Triassic Period: A Review and Synthesis of Current Theories
ALAN J. CHARIG

Synopsis	597
Introduction	597
Replacement of therapsids by archosaurs: the main characteristics	598
The nature of "scenarios"	601
Was the replacement competitive?	602
Explanations proposed	604
Anatomical explanations: Charig's/Thulborn's	605
Physiological explanations	607
Combination theories	622
Other explanations	623
Conclusions	624
Acknowledgements	625
References	626

The Histology of Dinosaurian Bone, and its Possible Bearing on Dinosaurian Physiology
R. E. H. REID

Synopsis	629
Introduction	629
Extent of data	630
The bone of dinosaurs	631
General	631
Compact bone	634
Cancellous bone	643

Compacted coarse-cancellous bone 649
Endosteal compact bone 650
Metaplastic bone 650
Pathological bone 651
Bone and dinosaurian physiology 651
General 651
Arguments for endothermy 653
Arguments against endothermy 654
Discussion 656
Acknowledgements 659
References 660
Notes added in proof 662

Closing Address: With comments on the organ of Jacobson and the evolution of Squamata, and on the intermandibular connection in Squamata

A. d'A. BELLAIRS

Introductory remarks 665
The organ of Jacobson and the evolution of the Squamata 666
The intermandibular connection in Squamata 674
Acknowledgements 680
References 680

Subject Index 685

Introduction

Professor Angus d'Albini Bellairs.

Professor Angus d'Albini Bellairs – An Appreciation

MARK W. J. FERGUSON

Department of Anatomy, The Queen's University of Belfast, Medical Biology Centre, 97 Lisburn Road, Belfast BT9 7BL

Angus d'Albini Bellairs was born on 11 January, 1918, and educated at Stowe School, Queens' College, Cambridge, and University College Hospital, London. Interested in natural history, especially in reptiles, for as long as he can remember, he was delighted to find that it was possible to read zoology at Cambridge in combination with the traditional medical subjects. At Cambridge his zoological interests became directed to comparative anatomy and he was lucky enough to meet several teachers who later became lifelong friends: in particular Dixon Boyd, Frank Goldby, Hugh Cott and Rex Parrington who, as Angus later wrote, "showed in his elegant and stimulating lectures on fossil reptiles the best that the formal discipline of a university course can offer".

When Angus graduated from Cambridge in 1939 the clouds of World War II were looming, and foreseeing that he would be better off as a medical officer than as a private in the infantry, he qualified in medicine in 1942. In the same year he was called up into the Royal Army Medical Corps and found the army surprisingly less unpleasant than he had expected – and in some ways more democratic than university life! He was posted as Medical Officer to the 4th Divisional Engineers, a unit for which he retains the greatest affection and whose annual reunion dinner he is still happy to attend. His unit went almost immediately to North Africa and took part in the final battles in Tunisia; it went briefly to Egypt, and then to Italy, participating in the fighting around Cassino. Angus was then suddenly posted across the world to a branch of the Army Biological Directorate (at that time headed in London by the distinguished biologists F. A. E. Crew and Lancelot Hogben) at the headquarters of General Slim's 14th ("forgotten") Army in Burma. This transfer, which carried promotion to Major, may possibly have been in the nature of a "kicking upstairs" (a commendable fate!). Everywhere Angus went he collected reptiles and there was a general rumour (which he may have encouraged) that anyone in his unit who reported sick with a reptile in a tin was bound to receive a day off duty! As would be expected this posting to the Far East afforded further opportunities for collection, and also fascinating human experiences in the course of duty. One of Angus' main projects in the 14th Army was a follow-up of non-commissioned soldier patients in Indian hospitals, which took him into many weird and wonderful medical institutions throughout the length and breadth of the sub-continent. He has pleasant memories of a delightfully alcoholic Colonel commanding a hospital, who carried out his ward rounds on a bicycle when he was not shooting imaginary tigers through the window of his hut. He also remembers the capricious medical documentation of the Indian Army; an aged sepoy diagnosed as suffering from senility might be temporarily downgraded to Category B for further treatment. In 1945 Angus was attached to the British Army Staff in Washington and by the strangest of military chances, Victory Day found him in Times Square, New York, where he participated in celebrations, possibly reminiscent of Mafeking Night in London.

Angus had long decided that his career after the war was not to be in clinical medicine, and it would have been logical for him to seek a post in a zoology department of one of the expanding British Universities. However, J. D. Boyd, then Professor of Anatomy at The London Hospital Medical College, persuaded him that anatomy, which at that time contained a stronger tradition of comparative work than it does today, might be a suitable milieu in which to combine his zoological interests with his medical experience, and offered him a Lectureship in his department. He is remembered with affection at "The London", where a delightful story about his participation in the early Monday morning anatomy lectures attended by nursing staff is still related. As can be imagined, these lectures were extremely unpopular with the anatomy teachers but after many months Angus was persuaded, against his will, that he would have to do his share. His first five lectures to the bewildered nurses were not "as scheduled" but rather consisted of a detailed exposition of the snake chondrocranium, whereupon the Nursing Supervisor requested that Dr Bellairs be removed from any further teaching commitment to nurses – much to his delight! During his time at The London Hospital Medical College Angus naturally paid his respects to Professor G. R. (later Sir Gavin) de Beer, the doyen of cranial morphologists, in the Anatomy Department of University College London. There he met a research student, Ruth Morgan, whom he subsequently married, and who is now herself Professor of Embryology in the same Anatomy Department at University College. They have a daughter, Vivien, recently qualified in medicine at Oxford.

Introduction

In 1951 J. D. Boyd moved to a Chair of Anatomy in Cambridge and managed to take Angus with him as a lecturer in anatomy. Unhappily, the transplantation was not a successful one so that in 1953 Angus obtained the Readership in Anatomy (later in Embryology) at St Mary's Hospital Medical School, London, where he has remained happily ever since. In 1970 he was awarded a personal Chair in Vertebrate Morphology, and very recently he was made an Emeritus Professor of the University of London.

Unlike some of his colleagues, Angus has the highest regard for that "vast acephalic monster" (as H. G. Wells termed it), the University of London which, if properly utilised, offers a range of facilities and talent throughout its various colleges which few other universities in Britain can match. Moreover, the capital contains two other venerable scientific institutions to which Angus feels deeply attached: the British Museum (Natural History) – despite its current exhibition policy, which he deplores – and the Zoological Society of London. His relationship with the Zoo has been particularly happy and he derives the greatest pleasure from his honorary appointment (which he has held since 1957) as Consulting Herpetologist to the Society. He has also been a member of the Society's Publications Committee (in effect, the editorial board of the Journal of Zoology) for many years. He was a founder member of the British Herpetological Society in 1947, and for some years Editor of its journal, the British Journal of Herpetology. Both this Society and the American Society of Ichthyologists and Herpetologists have elected him to Honorary Membership. Angus is also a member of the Anatomical Society and the Linnean Society and a Fellow of the Institute of Biology.

Angus Bellairs has made a number of scientific travels in post-war years. In 1953 he studied and collected reptiles in Algeria during the tenure of an award from King's College, Cambridge. Then in 1955 he obtained a Royal Society and Nuffield Foundation Commonwealth Scholarship to visit universities in South Africa. He enjoyed the zoology, especially the crocodile watching, but he found the politics depressing. In 1970 he was Visiting Professor of Zoology at the University of Kuwait, and as a person of the highest reputed moral integrity, was put in charge of the girl students' field excursions, teaching them to handle lizards and harmless snakes without fear – a considerable feat. In 1973 he took part in an expedition to the Galapagos Islands to study the behaviour of giant tortoises, and he is an enthusiastic visitor to the USA, the principal centre of modern herpetology. In the early seventies he was instrumental (with former teacher Hugh Cott, friend Robert Bustard, and others) in establishing the Crocodile Specialists' Group of the International Union for the Conservation of

Nature and Natural Resources – on which he served for some years.

Angus' 70 or so scientific publications (appended) have been mainly on reptiles, though he has also studied birds and mammals. Many of them have been concerned with the structure and development of the skull and its associated organs, a once predominant biological discipline, but today a somewhat recondite field. He is at present examining the skull in a fascinating group of burrowing reptiles, the amphisbaenians, in collaboration with Carl Gans of the University of Michigan. Indicative of Angus' scientific acumen is the fact that he and Professor Gans have published an important paper in *Nature* on the amphisbaenian skull, in this the year of his retirement. His studies on cranial morphology are embodied in a substantial monograph (Bellairs & Kamal, 1981) in the important serial publication *Biology of the Reptilia*. Other work has been concerned with reptilian teratology; Angus can claim to have discovered cleft palate of a type very similar to the human malformation in snakes, and has just finished what is probably the first anatomical description of cyclopia in turtles. He has also studied regeneration of the tail in lizards in collaboration with Susan V. Bryant, a former graduate student, now a professor in the University of California; they have recently prepared another large monograph on autotomy and regeneration for a forthcoming volume in the *Biology of the Reptilia*. Another former graduate student is Paul Maderson of Brooklyn College, New York, renowned for his studies on the reptilian skin.

Although Angus enjoys scientific collaboration with his friends, he likes to think of himself as something of a loner, pursuing a little known subject in his own way, and relatively independent of the vagaries of biological fashion and of grant awards. Indeed he sometimes refers to himself as a "scientific antique collector". He prepares most of his own material, types his own papers and illustrates them with his own drawings. He likes writing and has written or co-authored three books, two of which (*The Life of Reptiles*, Vols 1 & 2, Weidenfeld, 1969, and *Reptiles* (with J. Attridge), Hutchinson's University Library, 1975) have been translated into numerous foreign languages. In these books he has tried to provide a synthesis of the various kinds of knowledge available about reptiles, from physiology to behaviour, and likes to paraphrase Terence in saying, "Nothing reptilian is alien to me". The wide scope of this symposium is in part a reflection of Angus' broad interests in reptilian biology. Throughout his lifetime Angus Bellairs has always done his best both for herpetology, and for all those who serve therein.

Despite his somewhat unusual scientific interests for a teacher in a medical school, Angus likes to feel that he has also contributed to medical education and has done something to mitigate its deplorably narrow but congested curriculum. On first coming to St Mary's he designed, in collaboration with Professor Goldby, a type of introductory course in which some of the more general aspects of biology could be touched upon. He feels that every doctor – indeed every educated person – should be taught something about man's place in the Animal Kingdom and his effect on the world's ecosystem; and about animal behaviour, in particular the sexual and territorial imperatives. He believes that every medical school should contain at least one teacher who is primarily a naturalist.

Angus hopes to continue his scientific research and writing into his retirement, but he has many other interests including: all natural history, especially of the domestic cat which seems to show an admirable combination of friendliness and independence, antique collecting, military history, and modern fiction. He was very much impressed by Grahame Webb's crocodilian novel, *Numunwari* (Aurora Press Sydney, 1980) and Angus himself hopes to write a herpetological novel, partly based on academic life. However, he realises that writing a novel is very different from writing books on reptiles, and whether he can really do it remains to be seen . . .

PUBLICATIONS BY A. d'A. BELLAIRS AND COLLABORATORS

Bellairs, A. d'A. (1942). Observations on Jacobson's organ and its innervation in *Vipera berus*. *J. Anat.* **76**: 167–177.

Smith, M. & Bellairs, A. d'A. (1947). The head glands of snakes, with remarks on the evolution of the parotid gland and teeth of the Opisthoglypha. *J. Linn. Soc. (Zool.)* **41**: 351–368.

Bellairs, A. d'A. & Boyd, J. D. (1947). The lachrymal apparatus in lizards and snakes. I. The Brille, the orbital glands, lachrymal canaliculi and origin of the lachrymal duct. *Proc. zool. Soc. Lond.* **117**: 81–108.

Bellairs, A. d'A. (1948). The eyelids and spectacle in geckos. *Proc. zool. Soc. Lond.* **118**: 420–425.

Bellairs, A. d'A. (1949a). The anterior brain-case and interorbital septum of Sauropsida, with a consideration of the origin of snakes. *J. Linn. Soc. (Zool.)* **41**: 482–512.

Bellairs, A. d'A. (1949b). Observations on the snout of *Varanus*, and a comparison with that of other lizards and snakes. *J. Anat.* **83**: 116–146.

Bellairs, A. d'A. (1949c). Ectopic calcified embryo in viper. *Br. J. Herpet.* **1**: 55.

Bellairs, A. d'A. (1949d). Orbital cartilages in snakes. *Nature, Lond.* **163**: 106–107.

Bellairs, A. d'A. (1949e). Dr Edward Phelps Allis, junr [Obituary]. *Proc. Linn. Soc.* **161**: 243.

Bellairs, A. d'A. (1950a). Observations on the cranial anatomy of *Anniella*, and a comparison with that of other burrowing lizards. *Proc. zool. Soc. Lond.* **119**: 887–904.

Bellairs, A. d'A. (1950b). The limbs of snakes with special reference to the hind limb rudiments of *Trachyboa boulengeri*. *Br. J. Herpet.* **1**: 73–83.

Bellairs, A. d'A. & Boyd, J. D. (1950). The lachrymal apparatus in lizards and snakes. II. The anterior part of the lachrymal duct and its relationship with the palate and with the nasal and vomeronasal organs. *Proc. zool. Soc. Lond.* **120:** 269–310.

Bellairs, A. d'A. (1951). Observations on the incisive canaliculi and nasopalatine ducts. *Br. dent. J.* **91:** 281–291.

Bellairs, A. d'A. & Underwood, G. (1951). The origin of snakes. *Biol. Rev.* **26:** 193–237.

Bellairs, A. d'A. (1952). Albert Peacock, M.Sc., M.B., M.R.C.S. [Obituary]. *J. Anat.* **86:** 229.

Bellairs, A. d'A. (1953). The teaching of comparative dental anatomy to dental students. *Br. dent. J.* **95:** 100–103.

Bellairs, A. d'A. & Shute, C.C.D. (1953). Observations on the narial musculature of Crocodilia and its innervation from the sympathetic system. *J. Anat.* **87:** 367–378.

Shute, C. C. D. & Bellairs, A. d'A. (1953). The cochlear apparatus of Geckonidae and Pygopodidae and its bearing on the affinities of these groups of lizards. *Proc. zool. Soc. Lond.* **123:** 695–709.

Smith, M. A., Bellairs, A. d'A. & Miles, A. E. W. (1953). Observations on the premaxillary dentition of snakes with special reference to the egg-tooth. *J. Linn. Soc. (Zool.)* **42:** 260–268.

Bellairs, A. d'A. & Shute, C. C. D. (1954). Notes on the herpetology of an Algerian beach. *Copeia* **1954:** 224–226.

Lester, J. W. & Bellairs, A. d'A. (1954). Australian reptiles. *Zoo Life* **9:** 50–54.

Bellairs, A. d'A. (1955). Skull development in chick embryos after ablation of one eye. *Nature, Lond.* **176:** 658–659.

Shute, C. C. D. & Bellairs, A. d'A. (1955a). The external ear in Crocodilia. *Proc. zool. Soc. Lond.* **124:** 741–749.

Shute, C. C. D. & Bellairs, A. d'A. (1955b). A case of polydactyly in the Colugo, *Cynocephalus*. *J. Mammal.* **36:** 131–132.

Bellairs, R., Griffiths, I. & Bellairs, A. d'A. (1955). Placentation in the adder, *Vipera berus*. *Nature, Lond.* **176:** 657–658.

Fearnhead, R. W., Shute, C. C. D. & Bellairs, A. d'A. (1955). The temporo-mandibular joint of shrews. *Proc. zool. Soc. Lond.* **125:** 795–806.

Crompton, A. W. & Bellairs, A. d'A. (1956). The ictidosaurs: a link between reptiles and mammals. *J. Anat.* **90:** 571–572.

Bellairs, A. d'A. (1957). *Reptiles*. London: Hutchinson's University Library.

Bellairs, A. d'A. & Boyd, J. D. (1957). Anomalous cleft palate in snake embryos. *Proc. zool. Soc. Lond.* **129:** 525–539.

Bellairs, A. d'A. (1958). The early development of the interorbital septum and the fate of the anterior orbital cartilages in birds. *J. Embryol. exp. Morph.* **6:** 68–85.

Bellairs, A. d'A. (1959a). Reproduction in lizards and snakes. *New Biology* No. 30: 73–90.

Bellairs, A. d'A. (1959b). Malcolm Smith's contribution to herpetology. *Br. J. Herpet.* **2:** 138–141.

Bellairs, A. d'A. (1959c). Malcolm Arthur Smith [Obituary]. *Proc. Linn. Soc.* **170:** 130–131.

Bellairs, A. d'A. & Gamble, H. J. (1960). Cleft palate, microphthalmia and other anomalies in an embryo lizard (*Lacerta vivipara* Jacquin). *Br. J. Herpet.* **2:** 171–176.

Bellairs, A. d'A. & Miles, A. E. W. (1960). Apparent failure of tooth replacement in monitor lizards with remarks on loss of teeth in other reptiles. *Br. J. Herpet.* **2:** 189–194.

Bellairs, A. d'A. & Jenkin, C. R. (1960). The skeleton of birds. In *Biology and comparative physiology of birds*, **1:** 241–300. Marshall, A. J. (Ed.). New York and London: Academic Press.

Bellairs, A. d'A. & Miles, A. E. W. (1961). Apparent failure of tooth replacement in monitor lizards. Addendum. *Br. J. Herpet.* **3:** 14–15.

Maderson, P. F. A. & Bellairs, A. d'A. (1962). Culture methods as an aid to experiment on reptile embryos. *Nature, Lond.* **195:** 401–402.

Holder, L. A. & Bellairs, A. d'A. (1962). The use of reptiles in experimental embryology. *Br. J. Herpet.* **3:** 54–61.

Holder, L. A. & Bellairs, A. d'A. (1963). The use of reptiles in experimental embryology: note 2. *Br. J. Herpet.* **3:** 131–132.

Bellairs, A. d'A. (1964). Skeleton. In *A new dictionary of birds*: 753–760. Landsborough Thomson, A. (Ed.). London: Nelson.

Moffat, L. A. [formerly Holder] & Bellairs, A. d'A. (1964). The regenerative capacity of the tail in embryonic and post-natal lizards (*Lacerta vivipara* Jacquin). *J. Embryol. exp. Morph.* **12:** 769–786.

Bellairs, A. d'A. (1965). Cleft palate, microphthalmia and other malformations in embryos of lizards and snakes. *Proc. zool. Soc. Lond.* **114:** 239–251.

Poyntz, S. V. & Bellairs, A. d'A. (1965). Natural limb regeneration in *Lacerta vivipara*. *Br. J. Herpet.* **3:** 204–205.

Bellairs, A. [d'A.] & Carrington, R. (1966). *The world of reptiles*. London: Chatto & Windus.

Bryant, S. V. [formerly Poyntz] & Bellairs, A. d'A. (1967a). Tail regeneration in the lizards *Anguis fragilis* and *Lacerta dugesii*. *J. Linn. Soc. (Zool.)* **46:** 297–305.

Bryant, S. V. & Bellairs, A. d'A. (1976b). Amnio-allantoic constriction bands in lizard embryos and their effects on tail regeneration. *J. Zool., Lond.* **152:** 155–161.

Bryant, S. V., Breathnach, A. S. & Bellairs, A. d'A. (1967). Ultrastructure of the epidermis of the lizard (*Lacerta vivipara*) at the resting stage of the sloughing cycle. *J. Zool., Lond.* **152:** 209–219.

Hughes, A., Bryant, S. V. & Bellairs, A. d'A. (1967). Embryonic behaviour in the lizard, *Lacerta vivipara*. *J. Zool., Lond.* **153:** 139–152.

Rubin, L., Bellairs, A. d'A. & Bryant, S. V. (1967). Congenital malformations in snakes. *Br. J. Herpet.* **4:** 12–13.

Bellairs, A. d'A. & Bryant, S. V. (1968). Effects of amputation of limbs and digits of lacertid lizards. *Anat. Rec.* **161:** 489–496.

Bellairs, A. [d'A.] (1969). *The life of reptiles*. 2 vols. London: Weidenfeld & Nicolson. [Also published in 1970 by Universe Books, New York: also with the late H. W. Parker as *Les Amphibiens et Les Reptiles* and *Les Reptiles*. 2 vols. Lausanne: Edition Rencontre. Also other foreign editions.]

Bryant, S. V. & Bellairs, A. d'A. (1970). Development of regenerative capacity in the lizard, *Lacerta vivipara*. *Am. Zool.* **10:** 167–173.

Bellairs, A. d'A. (1971). The senses of crocodilians. In *Crocodiles*. Paper No. 19: 181–191. Proceedings of the First Working Meeting of Crocodile Specialists sponsored by the New York Zoological Society and organised by the Survival Service Commission, IUCN, at the Bronx Zoo, New York, 1971. Morges, Switzerland: International Union for Conservation of Nature and Natural Resources.

Bellairs, A. d'A. (1972). Comments on the evolution and affinities of snakes. In *Studies in vertebrate evolution*: 157–172. Joysey, K. A. & Kemp, T. S. (Eds). Edinburgh: Oliver & Boyd.

Sheppard, L. & Bellairs, A. d'A. (1972). The mechanism of autotomy in *Lacerta*. *Br. J. Herpet.* **4:** 276–286.

Bellairs, A. d'A. & Frazer, J. F. D. (1973). Revision of *The British Amphibians and Reptiles* by (the late) Malcolm Smith. 5th Edn. London: Collins (The New Naturalist).

Bellairs, A. d'A. (1974a). An expedition to the Galapagos. *St Mary's Gaz.* **80:** 13–15.

Bellairs, A. d'A. (1974b). Reptiles. In *The encyclopedia of wild life*: 92–113. London: Salamander Books; Castle Books.
Bellairs, A. d'A. & Attridge, J. (1975). *Reptiles*. 4th Edn. London: Hutchinson University Library.
Rodhouse, P., Barling, R. W. A., Clark, W. I. C., Kinmonth, A.-L., Mark, E. M., Roberts, D., Armitage, L. E., Austin, P. R., Baldwin, S. P., Bellairs, A. d'A. & Nightingale, P. J. (1975). The feeding and ranging behaviour of Galapagos giant tortoises (*Geochelone elephantopus*). The Cambridge and London University Galapagos Expeditions, 1972 and 1973. *J. Zool., Lond.* **176**: 297–310.
Ball, D. J. & Bellairs, A. d'A. (1976). Reptiles. In *The UFAW handbook on the care and management of laboratory animals*: 495–515. UFAW (Ed.) (5th Edn). Edinburgh and London: Churchill, Livingstone. [Also previous editions.]
Bellairs, A d'A. & Ball, D. J. (1976). Reptiles. *Symp. zool. Soc. Lond.* No. 40: 119–132.
Bellairs, A. d'A. & Cox, C. B. (Eds). (1976). *Morphology and biology of reptiles. Linn. Soc. Symp.* No. 3. London: Academic Press.
Halfpenny, G. & Bellairs, A. d'A. (1976). A black grass snake. *Br. J. Herpet.* **5**: 541–542.
Bellairs, A. d'A. (1977). *The naturalist in Britain. A social history*, by D. E. Allen. [Review.] *Oryx* **13**: 499–500.
Bellairs, A. d'A. (1977). The nose and Jacobson's organ in reptiles: a review. *Cotswold Wildl. Symp.* **1977**: 27–36.
Winchester, L. [formerly Sheppard] & Bellairs, A. d'A. (1977). Aspects of vertebral development in lizards and snakes. *J. Zool., Lond.* **181**: 495–525.
Martin, B. G. H. & Bellairs, A. d'A. (1977). The narial excrescence and pterygoid bulla of the gharial, *Gavialis gangeticus* (Crocodilia). *J. Zool., Lond.* **182**: 541–558.
Bellairs, A. d'A. (1979). The exhibition policy of the Natural History Museum. *Biologist* **26**: 162–164.
Bellairs, A. d'A. (1981a). *Cat behaviour. The predatory and social behaviour of domestic and wild cats*, by P. Leyhausen. [Review.] *J. nat. Hist.* **15**: 353–354.
Bellairs, A. d'A. (1981b). Congenital and developmental diseases. In *Diseases of the Reptilia*, **2**: 469–485. Cooper, J. E. & Jackson, O. F. (Eds). London: Academic Press.
Bellairs, A. d'A. & Kamal, A. M. (1981). The chondrocranium and the development of the skull in Recent reptiles. In *Biology of the Reptilia*, **11**: 1–263. Gans, C. & Parsons, T. S. (Eds). London: Academic Press.
Bellairs, A. d'A. (1983). Partial cyclopia and monorhinia in turtles. In *Advances in herpetology and evolutionary biology* (Essays in honor of Ernest E. Williams): 150–158. Rhodin, A. G. J. & Miyata, K. (Eds). Harvard: University Museum of Comparative Zoology.
Bellairs, A. d'A. & Gans, C. (1983). A reinterpretation of the amphisbaenian orbitosphenoid. *Nature, Lond.* **302**: 243–244.
Evans, S. E. & Bellairs, A. d'A. (In press). Histology of a triple tail regenerate in a gecko, *Hemidactylus persicus*. *Br. J. Herpet.* **6**: 319–322.
Bellairs, A. d'A. (1984). Closing address: With comments on the organ of Jacobson and the evolution of Squamata, and on the intermandibular connection in Squamata. *Symp. zool. Soc. Lond.* No. 52: 665–683
Bellairs, A. d'A. (In press). *The behavioural ecology of the Komodo monitor*, by W. Auffenberg. [Review.] *J. nat. Hist.*
Bellairs, A. d'A. (In press). *Biology of the Reptilia* **11** and **12**. Gans, C. & Pough, P. H. (Eds). [Review.] *J. nat. Hist.*
Bellairs, A. d'A. & Bryant, S. V. (In press). Autotomy and regeneration. In *Biology of the Reptilia* **15**. Gans, C., Billett, F. & Maderson, P. (Eds). New York: J. Wiley.

Morphology

Slide-pushing — A Transitional Locomotor Method of Elongate Squamates

CARL GANS

Division of Biological Sciences, The University of Michigan, Ann Arbor, Michigan 48109, USA

SYNOPSIS

Certain squamates manage to propel themselves fairly rapidly across planar surfaces using a slide-pushing pattern. They transmit propulsive forces to the substrate by sliding more of their total body mass posteriorly than anteriorly and by having the movements of the posterior portion of the trunk proceed more rapidly than the anterior ones. The reaction force induced by the posterior slippage propels the anterior portion of the trunk forward. The magnitude of the reaction forces generated and the forward velocity produced tend to be less than those exerted in regular undulation, using laterally placed resistance points. Control of force transmission, rather than the nature of muscular contraction, limits the effectiveness of slide-pushing. This locomotor pattern is intermediate between the terrestrial lateral undulation and undulatory swimming and has implications for evolutionary transitions in the pattern of locomotion of elongate squamates.

INTRODUCTION

Snakes and other limbless squamates ordinarily propel themselves by one of four locomotor patterns, commonly referred to as lateral undulation, concertina, sidewinding, and rectilinear movement (Gans, 1974), with saltation being a rare fifth pattern. In all but lateral undulation, the horizontal propulsive forces are transmitted across zones in which the trunk makes static contact with the substrate. Horizontal forces can only be transmitted across such static zones. As long as the horizontal propulsive forces are less than the static friction, stasis is maintained; if they exceed these, the stationary elements start to slip relative to each other.

In contrast, the trunk of a squamate moving by lateral undulation moves continuously relative to the substrate. The mass of the animal is supported by forces transmitted via the continuously moving ventral surface, and the propulsive forces are exerted via the sides of the trunk. During lateral undulation, a series of undulant curves passes from front to rear, whereas the posteriorly facing surfaces contact and push against irregularities in the substrate. The body slides past the sites through which the resistance forces are exerted. Sliding friction is induced wherever slippage occurs and is intrinsically disadvantageous because it opposes forward movement. Gray (1946) long ago formulated the rule that lateral undulation does not allow limbless squamates to traverse smooth surfaces lacking resistance points against which forces can be exerted; other locomotor patterns are used to cross these.

An example to which Gray's rule seemingly applies is readily observed when one disturbs a snake resting on the warm, smooth asphalt of a road after nightfall. Once disturbed, such an animal apparently undulates in place, with a rapidly curving movement that proceeds too fast to let the sides of the body transmit forces at specific sites. However, the animals apparently utilize an exception to Gray's rule, because the undulating snakes do progress. Their center of mass tends gradually to shift in the direction toward which their head points. What is the reason for this exception?

Analysis of a large number of films, taken in the field and laboratory and documenting situations in which snakes, elongate lizards and amphisbaenians were exposed to a variety of substrates at varied inclinations, suggests that there is an undescribed locomotor pattern, here referred to as slide-pushing. This distinct locomotor pattern involves an undulating movement, superficially similar to lateral undulation, that allows progression over smooth surfaces by a costly rapid undulation. The present report, dedicated to my long-time friend, Professor Angus d'A. Bellairs, characterizes this pattern. Slide-pushing is of interest because its effectiveness is influenced more by motor control than by morphology. Its recognition seems to aid the formulation of hypotheses about the transitional states between some of the previously defined locomotor patterns, and consequently, allows one to derive some insights into the evolution of snake motor systems.

OBSERVATIONS

This report is based primarily upon analyses of locomotor sequences recorded in Africa, southern Asia, and Australia, on 16 mm and Super-8 film, of a large number of reduced limbed and limbless squamates, including snakes of the families Colubridae and Viperidae, as well as of caecilians. [It has also become clear to me that I have often observed this "new" pattern, without recognizing its significance, in several North American species of snakes, including members of the genera *Agkistrodon, Nerodia* and *Thamnophis*, when surprised during early evening hours on the smooth and warm asphalt roads in the southern

United States.] The slide-pushing movements were most clearly noted in a specimen of *Echis carinatus*, filmed (1968 at Cape Coast, Ghana) while it was crossing a highly polished terrazzo floor, in a number of pygopodid lizards moving over perspex and, less obviously, in some Indian snakes. These low-friction substrates are all artificial and their natural equivalents seem to be rare.

The pattern was most frequently observed when animals that had been resting on very smooth surfaces were suddenly frightened or otherwise stressed into rapid escape movements while they were close enough to their preferred body temperature to sustain muscular efforts prolonged over some minutes. The pattern was seen only briefly in animals with body temperatures (ambient, checked against body temperatures *after* test runs) 10 or 15°C below their selected body temperatures. When the animals were tired or the disturbance was less severe, they shifted to, or would use variants of, concertina and might then progress more rapidly, although the body motion was slower.

In all these cases, the slide-pushing superficially appears as an undulant movement in which body waves rapidly move posteriorly along the trunk, but in which the squamate as a whole progresses forward relatively slowly. A view from above confirms that the waves of the trunk travel posteriorly relative to the ground rather than "standing" (in the sense of Gans, 1962, but not of Daan & Belterman, 1968).

Close examination of the films (Fig. 1), using various stop-action

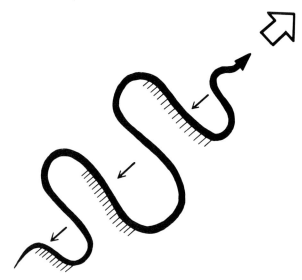

FIG. 1. View of a slide-pushing snake to indicate the zones in contact with the ground (hatched), the direction of travel of the center of mass (open arrow) and the direction of travel of the contact zones (filled arrows).

analysers, indicates: (a) that only the posteriorly facing portions of the curves tend to be in contact with the ground (the lateral-most portions of each loop being lifted out of contact); (b) that the contact zone often extends asymmetrically inward from the loop to just beyond the midline; (c) that the contact zones may be alternated, involving every other loop; and (d) that the trunk is twisted in the regions at which contact occurs (so that the median plane of the portion in contact forms a posteriorly acute angle with the substrate). The posteriorly facing portion of the animal then produces an angle, the edge of which slides posterolaterally across the substrate. The posteriorly moving contact zones may involve both sides of the animal, thus requiring reversing torsion of successive sections of the trunk. More often one sees that the contact is restricted to the portions of the loops that open to one side only.

ANALYSIS OF THE PROPULSIVE PATTERN

To move forward, an animal must exert backward forces against some portion of its environment. The center of mass will shift forward to the extent that the posteriorly directed force can (a) overcome the inertia of the mass of the animal to be propelled anteriorly, and (b) overcome the environmental resistance, i.e. the sliding friction, opposing forward progression. In slide-pushing locomotion, both anterior and posterior portions of the snake are slipping over the ground and both are in contact with it. Although forward progression is slow, the analysis is rendered difficult by the rapid body movements.

The mechanism by which the observed slide-pushing action propels the animals can best be understood by comparing and contrasting the movements of slide-pushing with the superficially similar lateral undulatory ones, with which they have previously been confused (Fig. 2). In an animal progressing by lateral undulation, both the head and tail move at a constant and essentially equal speed, although their direction of travel changes constantly as the snake moves along a curved path. As the path curves variably on land, so particular segments will accelerate and decelerate relative to the center of mass; however, the velocity of all of the parts is always positive (forward) relative to the ground. It is most important to note that an undulating animal retains the same contact points as it slides along; hence, these are fixed relative to the ground.

In lateral undulation, all supportive forces pass through the ventral surface and involve no muscular action. However, the lifting of the head or portions of the trunk, coincident with the traverse of rough terrain or for behavioral reasons, obviously involves a cantilevering action during which the axial musculature shifts the weight of the lifted portion

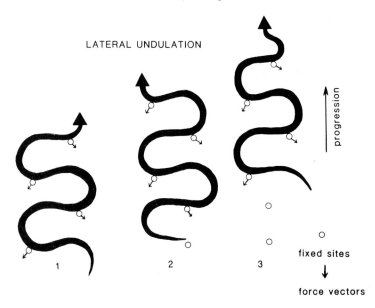

FIG. 2. Sequence of three stages in the progression of a snake by lateral undulation. Note that each step of the track is traversed by each portion of the animal.

toward the zone that remains in contact. In lateral undulation, the curves are forced into contact with variously spaced lateral resistance sites. Progression in such undulation requires at least three posteriorly directed force vectors (situations in which the snake pushes against a single point or two points can be shown to represent either transient states or to combine undulation with concertina; see Gans, 1974). However, the propulsive forces pushing against the resistance sites are generated entirely by axial muscles. As all the forces transmitted in lateral undulation pass essentially normal to the contact surfaces, friction is not used. Friction does produce a loss at sites at which portions of the body contact fixed places and either the contraction of axial muscles or the body mass loads the contact zone.

Slide-pushing differs from lateral undulation in several respects (Fig. 3). The parts of the animal move at different velocities, both relative to the center of mass and even more in relation to the ground. The head and neck tend to travel a less curved path, at a rate equivalent to that of the center of mass, whereas the posterior portions oscillate more widely and rapidly relative to the center. However, not all of the ventral surface of a slide-pushing animal need contact the ground. It is in the behaviour of contact zones that the slide-pushing differs most markedly from lateral undulation. The anterior sliding zone, positioned just beneath the head and neck, travels anteriorly at a relatively steady pace. How-

SLIDE PUSHING (slipping undulation)

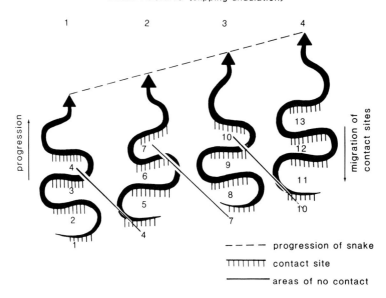

FIG. 3. Sequence of three stages in the progression of a snake by slide-pushing. Note that the zones in contact travel backwards much faster than the animal progresses forward.

ever, the most cranial of these posterior contact zones are formed anterior to the level of the center of mass and slide backward relative to the ground at several times the velocity of forward progression. Indeed, slide-pushing is one of the few modes of terrestrial locomotion in which the propulsive elements move posteriorly relative to the substrate. (A parallel mechanism is seen in a Pernambucan dance called the "frevo", in which the reaction forces to the action to the alternately backward sliding feet counterbalance the dancer, whose forward leaning posture is otherwise unstable.)

In slide-pushing, none of the propulsive forces is transmitted at right-angles to the locomotor surfaces; indeed, much of the propulsion occurs parallel to these surfaces. As the contacting surfaces move relative to each other, sliding friction is produced. The sliding friction of the portion of the body moving steadily anteriorly represents a loss that has to be overcome. However, the sliding friction generated in the more posterior portion of the trunk is utilized to propel the organism.

In an absolutely frictionless system, an undulating snake could not progress across a plane surface. Its center of mass would remain at a point or oscillate slightly along a transverse line, as the forces, and thus the movements, generated in both directions would be approximately equal. Even in the presence of some sliding friction, undirected undulations will not induce progression. Slide-pushing introduces asymmetry into the system by exerting differential forces at the anterior and posterior contact zones. The force generated in pure sliding friction, setting aside edge and area effects, of which more below, reflects the force that presses the two surfaces together times the coefficient of friction. The magnitude of the coefficient depends on the texture and surface properties of the areas in contact. Are there thus only two variables that may be modified to promote slide-pushing?

The force pressing the ventral surface against the ground is, to a large extent, a factor of the animal's mass. If only limited portions of the trunk are in contact with the ground, each contact zone will support the portion in contact and part of the adjacent bridged or cantilevered portion. The extent of contact is determined by the application of muscular action, i.e. by the formation of bridging curves and the elongation of the regions intermediate to the contact zones, tending to separate these (such elongation does not involve elongation of the animal, but the extension of an initially curved zone).

The influence of the coefficient of sliding friction depends upon directional anisotropy. As shown earlier (Gray & Lissmann, 1950), the ventral skin of squamates shows a strong directional component to the sliding friction (for all but the smoothest surfaces) with the coefficient for forward sliding lowest and that for posterior sliding much higher. While the coefficient for the lateral scales was considered equivalent, these were only loaded in the forward direction; in most reptiles, the coefficient for a lateral loading is likely to be equal to or higher than that for posterior sliding. There are no measurements of the transverse coefficient of the dorsolateral scales; however, their texture indicates that their coefficient should be higher than any of the others. This is, of course, critical, as the posterior curves slide laterally and the tilting of the trunk tends to bring these edges into contact with the ground. Thus the anterior portion of the trunk slides along the overlapping, low-friction ventral scales, whereas the more posterior sliding zones involve the sides of the trunk, with slippage at right-angles to the longitudinal axis of the animal.

The nature of the asymmetry in force generation should now be clear (Fig. 4). If we analyse the forces exerted at the bridge between the anteriormost major contact site and the first laterally placed one, we can calculate that the resistance to anterior slippage of the head region will be F_a (mass supported at the anterior zone) times the coefficient of sliding friction (0.4, of the ventral scales). The resistance to slippage in a posterior direction will be F_p (mass supported in the posterior zone, thus likely to be greater than F_a) times the coefficient of sliding friction (1.2, of the lateral scales). Thus, the resistance to slippage will be

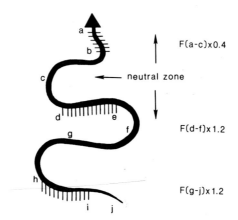

FIG. 4. Position of the snake at the start of a movement sequence, showing the forces resisting movement and with this the development of an anterior acceleration.

greater for the posterior portion of the trunk than for the anterior one, and the reaction forces produced will propel the animal forward.

Not only is the body twisted and the ventral skin bent to bring the edges of scales into posterior contact, but the deformation of the edge of the body into a more acutely angled wedge will increase contact with minor resistance sites. Whereas these by themselves are insufficient to block the slippage and to allow force transmission via a fixed site, their resistance will increase the resistance and reaction forces beyond those observed in pure sliding and simultaneously increase the asymmetry. The twisting action is important because the asymmetry of force generation is otherwise minor in slide-pushing, so that the magnitude of the reaction force is only a small fraction of that notable in lateral undulation.

Thus, the differential effect becomes more obvious. The change of the zone of progressive neutrality (or the site from which the curves either slide anteriorly or posteriorly) to a site anterior to the center of mass is a behavioral event that facilitates slide-pushing. The anterior excursion tends to be limited so that the head swings much less than the trunk. The anterior portions of the trunk and indeed their support zones will then progress more or less regularly, at most oscillating around the velocity of the center of mass, but always moving forward. The posterior portions of the animal, in contrast, will oscillate from a forward speed greater than that of the center of mass to a velocity that is slightly negative relative to that of the ground. Although these anteroposterior oscillations of the posterior trunk are minor, the lateral oscillations of each portion of the trunk involve major excursions with each wave.

RECONSTRUCTION

It is now possible to reconstruct the actual process of slide-pushing. Starting from the curved resting position at which the animal rests or toward which it contracts when disturbed, the head and neck are slid anteriorly by straightening of the anterior body. As this movement proceeds, the amount of force transmitted to the anterior portion of the trunk is sufficient to force the posterior portion of the body into slippage.

Though slippage then proceeds both forward and backward, the asymmetry noted in the pattern of sliding friction causes the center of mass to move forward. This movement reflects a gradual build-up of momentum that will quickly decay unless the sliding friction opposing forward movement is overcome by the formation of new contact sites by recurving of the neck (Fig. 3), thereby forming new bends and new contact zones that pass posteriorly along the ground. The backward sliding of these sites continues the generation of forward directed force. The torsional bending of the trunk at the contact sites both increases the coefficient of sliding friction and tends to contact any surface irregularities, thus increasing the force required to move the bends and the reaction force that will shift the center of mass forward.

The work involved in the bending of the body and its complex twisting must be considerable in view of the relatively slight nature of the asymmetry. The sliding friction is theoretically unaffected by velocity; however, the digging-in effect will increase with velocity. Thus the posteriorly travelling waves tend to be moved much more rapidly than the rate of forward progression. This induces a need for repeated curves that cannot be passed down the trunk at random intervals, but incorporate synchronized torsional twists.

HOW DOES THE BODY TWIST?

Films suggest that slide-pushing snakes twist the body up to some 30° from the center line. Because the observations of Mosauer (1931) indicate that there is essentially no capacity for intervertebral torsion between pairs of snake vertebrae, and because torsion occurs over short sections of the trunk (aggregating fewer than 40 vertebrae), it is unlikely that the twisting movement of the trunk proceeds by such intervertebral torsion. The observed torsion would seem to involve a kind of movement against which the vertebral column of snakes, with its characteristic double articular surfaces, is specifically protected. However, as indicated in earlier studies on sidewinding locomotion, effective torsion of the trunk may be achieved by a combination of the substantial capacity of dorsoventral (25°) and lateral (25°) flexibility between vertebrae; this level of flexibility is common in snakes. Combined dorsoventral and lateral flexure is likely involved in twisting the body at the contact regions. Even though the limited field of vidicon tubes makes cinefluoroscopic investigations of this mode of locomotion unlikely to be successful, the topic is ripe for some simple manipulative observations.

TRANSITIONS

Slide-pushing seems to belong among the group of intermediate locomotor patterns utilized by animals that are stressed by predators or traversing substrates not usually encountered. In some situations, slide-pushing progression may be little faster than that produced by concertina or a mixed concertina-undulation; however, the rapid movements involved in slide-pushing may have locomotion-independent benefits in predator avoidance. Although these intermediate patterns do not represent major components in terms of the period they occupy in the overall time budget, they may be quite important in allowing animals to survive unusual circumstances. Energetics would seem to be of secondary importance under such circumstances.

Slide-pushing has already been shown to differ substantially from lateral undulation in that the propulsive forces reach the substrate through the ventral rather than the lateral surface. We have also seen that this kind of undulation differs from lateral undulation in that the contact sites move posteriorly relative to the ground and do not remain in a fixed position. Unfortunately we lack any determination of the energetic costs of the different patterns of snake locomotion. The single, and often cited, study has been published only as an abstract (Chodrow & Taylor, 1973), and concerns a single species of snake crawling on a treadmill with an astroturf (artificial grass) tread surface and does not specify the method of propulsion. However, the relative inefficiency of slide-pushing relative to lateral undulation is suggested by noting that the former method requires many times' more oscillating movements per unit of progression than does the latter.

Slide-pushing is superficially similar to the movement of a swimming snake in that curves of the trunk travel posteriorly more rapidly than the entire animal travels forward. However, slide-pushing differs from swimming in that the latter receives its propulsion by accelerating portions of the surrounding medium and in that the reaction of this effort permits progression. Swimming is, thus, similar to lateral undulation in that all kinds of friction encountered by an animal are disadvantageous, rather than being used in force transmission. However, swimming and slide-pushing may be similar in terms of the Froude efficiency, which would seem to apply to both of them. This efficiency rises with the ratio between the absolute velocity of the animal and the wave speed, i.e. with the slip. This suggests that the absolute propulsive efficiency of slide-pushing increases as the ratio of backward velocity over forward velocity rises.

Slide-pushing does provide a condition transitional between an undulant movement and sidewinding much as predicted by Brain (1959, 1960), who suggested that the development of sidewinding would involve the addition to undulation of pushing loops primarily to one side (as already noted in the films of *Echis carinatus*). The shift to sidewinding would require the ability to exercise motor control over the forming loops, which would limit the posteriorly directed forces to a level at which posterior slippage was just avoided. A photograph of the track of a sidewinding rattlesnake traversing fine-grained mud shows the importance of this factor. One can see that the ventral scales slid for some distance before they stopped. The track also shows the effect of asymmetrical force with the backwards-pointing edge being much the deepest, as the scale edges had dug in and even deformed the track-edge. Selective bending of the trunk and more vertical application of force digs the edge into a compliant surface and keeps the animal from slipping posteriorly.

Of course, the observation of slide-pushing in animals, such as *Echis carinatus*, which are also effective lateral undulators and sidewinders (Gans & Mendelssohn, 1972), does not show that it is necessarily transitional in an evolutionary sense, but only that it is functionally intermediate. However, the method of progression does incorporate a strong behavioral component, as its effectiveness varies markedly with the site at which the backward pushing curves are formed, the rate at which the contact zones travel posteriorly, the extent to which the propulsive forces can be applied specifically to the intermediate sliding zones and the extent to which the trunk can here be twisted to bring the sides into contact. Changes in any of these behaviorally, rather than structurally, established aspects would generate a diversity of slightly different motor states and, with this, suitable raw material for natural selection.

The discovery of this method of progression allows a further refinement of the putative transitions between the generalized tetrapod locomotion and the various limbless ones (Table I). If data from field studies of skinks with various kinds of limb-reduction (Gans & Greer,

TABLE I

Transitions in the tetrapod/limbless sequence in squamates

State	Characteristics
I. Generalized tetrapod "lizard"	Trunk can bend Propulsive force generated by limbs Propulsive force transmitted by limbs Placement control mediated by limbs
II. Elongate tetrapod	Further flexibility of trunk Increase in vertebral number Propulsion by CONCERTINA of trunk Propulsive force transmitted by limbs Placement control mediated by limbs

III. Limb reduction and loss	Further flexibility of trunk
	Increase in motor control
	Propulsion by CONCERTINA at low speed
	   Force transmitted by limbs
	   Control mediated by limbs
	Propulsion by LATERAL UNDULATION at high speed
	   Force transmitted by sides of trunk
	   Limited control mediated by trunk

IV. Generalized limblessness	Flexibility of trunk Local motor control Animal has choice of locomotor repertoires involving differential bends LATERAL UNDULATION CONCERTINA
V. Specialized limbless condition (Major changes)	
A. Morphological shift	Liberation of integument from deeper muscles Development of costocutaneous muscles Control for symmetrical motor activation RECTILINEAR MOVEMENT
B. Control shifts	
1.	Local force application by rapidly travelling waves Torsion of the trunk SLIDE-PUSHING
2.	Force application controlled to avoid slippage of contact zone over ground Trunk placed down in straight line and lifted to new site SIDEWINDING
3.	Rapid, sequential, lift of trunk off ground, with marked vertical component SALTATION

1982) are incorporated, each of the major functional shifts, among "discrete" locomotor types, will now be bridged. Although the initial stages of the transition involved such obviously morphological modifications as multiplication of the segmentation and reduction of limb elements (Table I: I, II, III, V-A), most of the changes may be considered primarily behavioral as they involve motor control in the selection and placement of fixed sites, and in the use of friction (Table I: II, III, IV, V-B-1, 2, 3), rather than externally apparent structural shifts (Gans, 1975; Gans & Greer, 1982).

The present account of a previously undescribed method of locomotion by elongate squamates is presented here in the hope that it will generate further attention to the locomotor patterns of the Squamata. An understanding of behavioral as well as morphological diversity remains fundamental to an understanding of vertebrate adaptation. This makes it a double pleasure to dedicate this report to Professor Angus d'A. Bellairs, whose popular and technical writings have long provided an effective example of this viewpoint.

ACKNOWLEDGEMENTS

I thank the various students who helped discuss these issues and Barry Hughes, Allen Greer and other friends for support and aid in the field. Warren Porter kindly sent me a photograph of the track of a sidewinder traversing mud. The manuscript benefited from comments by David Carrier, Peter Pridmore, Paul Webb and George Zug. These studies on squamate locomotion have been supported by the U.S. National Science Foundation, most recently under NSF DEB 8121229.

REFERENCES

Brain, C. K. (1959). Sidewinding locomotion of the South West African adder *Bitis peringueyi*. *Bull. S. Afr. Mus. Ass.* **7:** 58–61.
Brain, C. K. (1960). Observations on the locomotion of the South west African adder *Bitis peringueyi* (Boulenger), with speculations on the origin of sidewinding. *Ann. Transv. Mus.* **24:** 19–24.
Chodrow, R. E. & Taylor, C. R. (1973). Energetic cost of limbless locomotion in snakes. *Fedn Proc. Fedn Am. Socs exp. Biol.* **32:** 422 Abs.
Daan, S. & Belterman, T. (1968). Lateral bending in locomotion of some lower tetrapods. I. and II. *Proc. K. ned. Akad. Wet.* (C) **71:** 245–266.
Gans, C. (1962). Terrestrial locomotion without limbs. *Am. Zool.* **2:** 167–182.
Gans, C. (1974). *Biomechanics: an approach to vertebrate biology.* Philadelphia: J. P. Lippincott. [Repr. Univ. Michigan Press.]
Gans, C. (1975). Tetrapod limblessness: Evolution and functional corollaries. *Am. Zool.* **15:** 455–467.

Gans, C. & Greer, A. (1982). Locomotor patterns in some crepuscular skinks of Australia. *Abstr. Meet. Soc. Stud. Amph. Rept.* **1982:** 72.

Gans, C. & Mendelssohn, H. (1972). Sidewinding and jumping progression of vipers. In *Toxins of animal and plant origin*, **1:** 17–38. De Vries, A. & Kochva, E. (Eds). London: Gordon & Breach Science Publ.

Gray, J. (1946). The mechanism of locomotion in snakes. *J. exp. Biol.* **23:** 101–120.

Gray, J. & Lissmann, H. (1950). The kinetics of locomotion of the grass-snake. *J. exp. Biol.* **26:** 354–367.

Mosauer, W. (1931). *The locomotion of snakes and its anatomical basis.* Ph.D. dissertation: University of Michigan, Ann Arbor.

Morphology of the Anterior Limb in Relation to Sprawling Gait in *Varanus*

JOHAN M. F. LANDSMEER

Department of Anatomy and Embryology, University of Leiden, Wassenaarseweg 62, 2333 AL Leiden, The Netherlands

SYNOPSIS

An account of the overall morphology of the anterior limb in *Varanus* is presented, followed by a functional approach to the articular chain, which takes into account the elementary demands of sprawling gait.

Sprawling gait is a three-dimensional mechanism which makes specific demands on the linkages of the articular chain (Rewcastle, 1980). The horizontal component of humeral sway is transmitted through the elbow joint as rotation of the forearm, provided that the elbow is flexed and the hand is fixed.

With regard to transmission and transformation of movements, the cubito-carpal mechanism is of major importance. The elbow, radius-radiale and ulna-ulnare-pisiform joints are parts of an interdependent system. A ligament system between radius, ulna, ulnare and radiale links these joints into one unit. This functional complex accounts for the fact that lateral rotation of radius and ulna is accompanied by a mutual approximation of these two bones.

Rotation of the radius and ulna, occurring in conjunction with horizontal humeral sway, is conveyed to the carpus and metacarpal rays. As soon as the range of movement of forearm-carpal joints has been fully employed, displacements within the hand – "rolling rotation of the rays" – come into play, including a change in the shape and internal condition of the hand. In this respect the oblique metacarpal-digital ligaments are important. Towards the end of power stroke the hand tends to narrow, whereas at the start of power stroke the metacarpals are widely set, the tensed metacarpo-digital bands providing a solid framework.

The asymmetry of the hand is emphasized. The function of the intrinsic muscles of the hand is discussed in relation to narrowing of the hand (contrahens) and to so-called claw retraction (superficial short flexor and lumbricals).

INTRODUCTION

Classic contributions to lacertilian morphology by Cuvier (1835) and Meckel (1828) are highlighted by Mivart (1867), Rüdinger (1868), Fürbringer (1900), Ribbing (1907) and Rabl (1916). Myology, osteology, innervation and brachial plexus were studied mainly from the specific viewpoint of comparative morphology.

An immense body of facts and features is made available in elaborate descriptions and superb illustrations. However, it seems that the morphology of the carpus and tarsus were not fully described because details, presumably considered unimportant in the comparative field, were often neglected. There are however some striking descriptions such as that of Born (1876: 4) on the radius-radiale joint: "Das r (Radiale) wendet, wie bei den übrigen Sauriern dem Radius einen Gelenkkopf zu, dem "Processus styloideus" desselben eine entsprechende Aushöhlung". More recently, Lécuru (1969) presented a detailed analysis of the lacertilian humerus, followed by a comparative and morphological study of equal thoroughness on the carpus in lepidosaurians (Renous-Lécuru, 1973). Recently Rewcastle (1980, 1981) presented a survey of sprawling gait with particular regard to its physiological and phylogenetic implications and with emphasis on posterior limb structure.

Functional approaches are valuable in the identification of structural coherences between bones, ligaments and muscles. In an analysis of the shoulder of *Varanus*, Haines (1952) paid due attention to the ligaments of this joint. Earlier Haines (1946), dealing with tetrapod locomotion, recognized pronation and supination of the forearm as a universal mechanism, not confined to a humanlike condition. Singular features of both elbow and forearm-carpal joints were revealed in quite a few reptilian and mammalian species, including *Varanus*. The functional significance of pronation and supination was described in *Chelonia* ". . . to adapt the hand to the ground in walking" (Haines, 1946: 3) which, however, does not pay full credit to the significance of pronation and supination in the sprawling gait of lizards. Haines (1946) was fully aware of the fact that the radius as well as the ulna was involved in forearm rotation. Earlier Vialleton (1924) had firmly denied such a mechanism; but Schaeffer (1941) had established rotation and crossing of tibia and fibula in Amphibia. The importance of functional concepts, in fossil material as well as in extant species, is demonstrated in studies by Watson (1917) and Jenkins (1971).

It is the purpose of this study to trace relationships between sprawling gait and the morphology of the articulated system of the anterior limb. Particular emphasis is laid on the joints – including their ligaments – as sites of transmission of movements.

GLOBAL APPROACH

A global morphological picture of the anterior limb in *Varanus* may serve as a background to explain some specific features and problems regarding function and structure of the anterior limb.

The muscular arrangement is presented in Figs 1, 2 and 3. The sites of attachments can be seen from the illustrations, or else current textbooks (Romer, 1970) or earlier literature (cited in the previous section) can be consulted. Trapezius, sternocleido-mastoideus, levator scapulae and serratus link the girdle to the trunk. Pectoralis major and latissimus lie between the girdle, trunk and arm. Proximally the humerus displays a lesser and greater tubercle, the latter carried on the marked deltopectoral crest. In addition to the muscles from which the crest derives its name, the coracohumeralis, a powerful muscle, taking origin from epi- and procoracoid, is also inserted into this crest. The coracohumeralis borders dorsoposteriorly on the scapulohumeralis. The latter, a conspicuous though small muscle from the coracoscapular membrane, is inserted into the dorsal aspect of the humerus. It passes underneath the scapulohumeral ligament which will be described later with the long head of triceps. The coracohumeralis partly covers the coracobrachialis which, as far as its short part is concerned, is inserted proximally into the ventral aspect of the humerus, the long portion descending down to the medial epicondyle. A powerful subscapularis taking origin from the inner side of the coracoid and a teres major from the posterior border of the scapula are inserted into the lesser tubercle. The entire open ventral surface serves as the area of origin of the brachialis, which inserts by one tendon with the one-headed biceps into the ulna, with a tendinous off-shoot into the radius.

The triceps is a most powerful muscle, inserted into the ulna with interposition of a patella-like olecranon. Its long tendon from the posterior border of the scapula is caught in a fibrous sling, explicitly mentioned by Mivart (1867), elaborately studied by Fürbringer (1900) and designated by Lécuru (1968) as the scapulohumeral ligament. This band, a sling around the long triceps tendon and firmly adherent to it, is attached to the posterior border of the scapula and finds its humeral attachment into the major tubercle close to the humeral head. Underneath this fibrous sling the tendon of the scapulohumeralis passes towards its attachment on the dorsal side of the humerus.

The forearm muscles are divided into a flexor and an extensor group. This division is clear-cut on the ventral side. Here the tendon of biceps and brachialis separates the entepicondylar flexors from the ectepicondylar extensors. However, on the dorsal side the extensor and flexor carpi ulnaris are linked by fascial bridges which serve also as areas of origin, as will be seen presently. The entepicondylar flexors are represented by pronator teres, flexor carpi radialis, flexor digitorum profundus s. longus and flexor carpi ulnaris. As regards their respective sites of insertion, the pronator teres is inserted all along the radius in a herringbone pattern, together with the most radial extensor, the

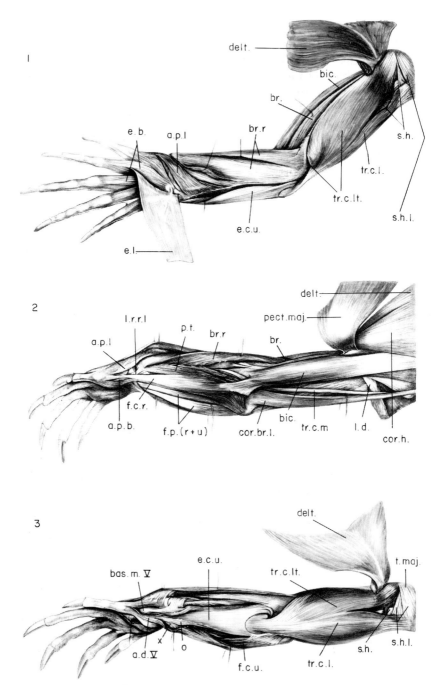

FIGS 1–3. Left arm *Varanus*. Extensor side of forearm, radial aspect of upper arm. A key to abbreviations used in the Figures is given on pp. 44–45. (2) Right arm *Varanus*. Flexor side of upper arm, radial aspect of forearm. (3) Left arm *Varanus*. Extensor aspect upper arm. Ulnar aspect forearm. 0 = insertion of e.c.u. into pisiform. x = insertion of f.c.u. into pisiform.

brachioradialis. The flexor carpi radialis is attached into the radiale and into the base of metacarpal I. The next muscle from the entepicondylar group is the flexor digitorum profundus s. longus. The proximal muscle body consists basically of three parts, a radial belly which originates as one mass with the flexor carpi radialis and pronator teres, an ulnar belly which does the same with the flexor carpi ulnaris and finally a third part which takes origin from the ventral side of the ulna. Another deep portion of this muscle will be dealt with in the description of the hand. Here a muscle like a quadratus plantae joins the tendon plate of the long flexor, approaching the deep aspect of this tendon from the side. The ulnar carpal flexor, in addition to its origin at the medial epicondyle, also receives fibres from a fascial bridge which runs from the tendon of the extensor carpi ulnaris around the triceps tendon to the medial epicondyle. The flexor carpi ulnaris is distally inserted into the pisiform bone with a superficial radiation into the flexor retinaculum. After removal of the flexor profundus a conspicuous muscular mass comes into view, the deep pronator. This muscle occupies the entire space between the ulna and radius running from the former bone to the latter in an oblique proximodistal direction, with a tendency to a more transverse course distally in the forearm. The ectepicondylar extensors comprise the brachioradialis, the extensor (digitorum) longus and the extensor carpi ulnaris. The brachioradialis is inserted all along the radius in a herringbone pattern with the pronator teres. The muscle, which very likely corresponds to the supinator, brachioradialis and the two radial extensors of the carpus, does not pass distally beyond the radius, although a fibrous band spans a bridge between the radius and radiale, running at the lateral aspect of the joint in an oblique palmar direction.

The extensor digitorum longus, also from the radial epicondyle of the humerus, ends in a tendinous plate inserting with three separate slips into the dorso-ulnar aspects of metacarpal bases II, III and IV. Between these slips bundles of extensor brevis take origin, for the most part from the dorsal aspects of the metacarpals.

The extensor carpi ulnaris has a strong tendon of origin which is hooked, as it were, around the radial border of the triceps tendon. This tendinous hook is part of a horseshoe-like sling around the triceps tendon. The ulnar part of the sling is less strong than the radial part, but nevertheless it serves as the area of origin of both the extensor and flexor carpi ulnaris. It is possible to produce a division between the two muscles by disrupting these fascial continuities. To sum up, the radial crus of this horseshoe is a powerful tendon which is inserted at the dorsal aspect of the humerus close to the elbow joint, near the radial epicondyle. The ulnar crus is thinner and is partly an extension of the medial triceps head. The extensor carpi ulnaris is inserted into the base of metacarpal V, with an equally powerful tendon into the pisiform.

The muscular system of the hand comprises the flexor digitorum sublimis, the intrinsic musculature, including lumbricals, contrahens and interossei, abductor digiti V and short abductor of the thumb. On the dorsal side the short digital extensors, the abductor pollicis longus and the extensor pollicis come into the picture. The flexor digitorum sublimis or brevis is an intriguing muscle. It originates from the pisiform and the flexor retinaculum, partly as a distal extension of the ulnar carpal flexor. From the muscular mass a small portion is isolated for the thumb. It is attached to metacarpal I and into the base of phalanx I. Tiny tendons emerge from the main muscle mass. They encircle the profundus tendon as paired reins. Each pair fuses into an unpaired median tendon on the dorsal side of the profundus and becomes inserted into the intermediate phalanges of the respective fingers. An intermediate phalanx is any phalanx between the first and terminal ones. Hence the flexor sublimis displays one pair of reins for the only intermediate phalanx of the three-phalanged index, two pairs for the four-phalanged medius, three pairs for the five-phalanged quartus and one pair for the three-phalanged quintus. Each unpaired tendon fuses with the capsular palmar plate proximal to the phalanx into which they are attached. Occasionally a tendon is represented by one unilateral slip only.

It may be appropriate now to pay attention to some remarkable muscles, viz. the paratendinous intravaginal flexors (Haines, 1950) or accessory lumbricals as they were termed by Rabl (1916). The true lumbricals will also be referred to. Ribbing (1907), on comparative arguments, combined both these muscle groups into the superficial short flexor. The tiny paratendinous intravaginal flexors are attached at their proximal ends to the radial side of the individual tendons of the long flexor of digits II–IV: they join the tendons of flexor brevis superficialis of the same digits. In the midhand, the lumbricals take origin from the long flexor distal to the site of its division into separate tendons for the individual fingers. The lumbricals are inserted into the base of the proximal phalanx of each finger.

Within the system of the intrinsics of the hand the following muscles can be distinguished: the abductor pollicis brevis, the abductor of digit V and a deep group of muscles which comprises both the contrahens and interossei (short, deep flexors). The contrahens is a muscular system radiating from its proximal point of origin on the ulnar side of the carpus into the bases of respective first phalanges of digits I–IV, extending along the ulnar side of these phalanges. A contrahens bundle to digit V emerges between the contrahens fascicles for the thumb. This muscular bundle is inserted on the radial side of the first phalanx of digit V (Fig. 4). Deep to the contrahens (and closely interwoven at least distally with the contrahens), the interossei take origin from the metacarpals and are inserted into the bases of the first phalanges. It is difficult, and probably unnecessary, to dissect the interosseous muscles in more detail but for one point, the relation of the interossei bundles to the system of digital bands.

Between the respective metacarpals, oblique longitudinal bands are found. At their proximal ends they are attached to a dorso-ulnar line on metacarpals I–IV. Each strong band runs toward the metacarpophalangeal joint of the next ulnar ray. The flat band becomes engaged in a capsular structure with palmar and lateral expansions and passes distally into a digital band on the radial side of the phalanges. A digital band runs from joint to joint on both lateral sides of the finger. In *Varanus* this system spans the proximal phalanx in digits II and V. In digits III and IV the second phalanx is also spanned by this system. In *Iguana* we found that this system spans the chains of phalanges down to the terminal phalanx. However, it may be that the radial band of each digit is the extension of a band from the next radial metacarpal; the ulnar band does not appear to be the distal extension of a band attached to a metacarpal – rather it appears to originate in the interosseous muscular system. As a result the symmetrical digital band displays asymmetrical proximal roots. On the radial side a direct fibrous continuation of the metacarpo-digital band runs into the digital band, while on the ulnar side the digital band takes origin from an interosseus (Fig. 5).

Finally, close to the metacarpal region, the oblique intermetacarpo-digital band has an offshoot to the adjacent metacarpo-phalangeal joint of the next radial ray. The system has only received passing mention in a few works, or else it has been illustrated without any allusion to its functional implications (Rabl, 1916; Haines, 1952; Rewcastle, 1980).

FUNCTIONAL CONSIDERATIONS

General Remarks

The sprawling gait of lacertilians imposes specific functional demands on the anterior limb. Rewcastle (1980) made an important contribution to the fundamental aspects of sprawling gait, emphasizing the three-dimensional nature of this gait, which poses particular demands on the joint systems:

"... there are few accounts of the form and function of the major limb joints. Such

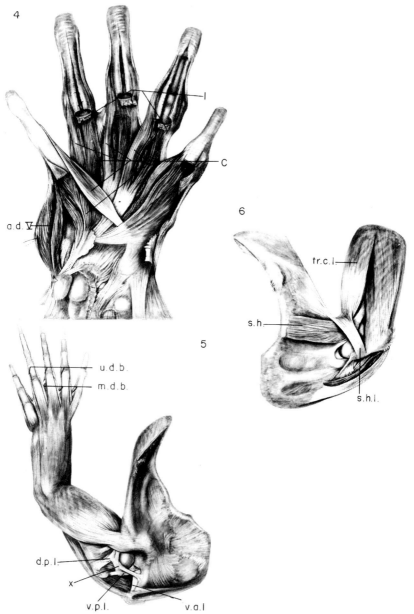

FIGS 4–6. (4) Palmar side of the right hand of *Varanus* after flexor profundus, including lumbricals, and flexor superficialis have been removed. The attachments of lumbricals at the bases of proximal phalanges II, III and IV are visible. The contrahens for digits I, II, III, IV and V originate from the ulnar carpalia (dist. carp. V and os centrale). Digit V receives an additional bundle emerging between contrahens bundles for the thumb. Insertion of contrahens bundles into the bases of the respective phalanges extending along the ulnar side of the first phalanges of I, II, III and IV and along the radial side of the first phalanx of V. (5) Left arm *Varanus*. Posterior dorsal view. Ligaments of the shoulder joint. x = ligament joining the two ligaments attached to the lesser tubercle. In the hand the metacarpo-digital bands are visible. These bands extend along the radial side of digits II, III and IV as the lateral digital bands. At the ulnar aspects of these same fingers the digital bands are extensions of interossei bundles. (6) Left shoulder *Varanus*. Anterior aspect. The scapulohumeral ligament in its relation to the long head of triceps and to scapulohumeralis. A key to abbreviations used in the figures is given on pp. 44–45.

is an unfortunate omission, since without arthrological data studies of lizard locomotion in particular, and reptile locomotion in general are hampered." (Rewcastle, 1980: 147.)

Support of the trunk by lateral horizontal humeri, resting on the vertical struts of the forearm, requires strong muscles on the ventral side of the shoulder joint (Romer, 1970). The elbow joints must be stable and the same applies to the joints in the wrist. Protraction and retraction of the upper arm, essential elements of the mechanism of locomotion, occur with a clear vertical component so that a rudder-like movement results (Haines, 1952). Adaptation of the shoulder joint to this motion needs careful analysis.

A major issue regarding sprawling gait can be explained in a simple way. Horizontal protraction and retraction of the forearm are transmitted to the forearm through the elbow joint. With the elbow bent at 90° the horizontal sway of the humerus or upper arm is transmitted to the forearm as a rotation of the latter, provided the hand keeps a stable position. Retraction of the humerus will be conveyed to the forearm and transformed into external rotation. Obviously retraction of the humerus, a major component of forward motion, occurs during power stroke. External rotation of the forearm with the hand fixed is identical to pronation of the hand when the humerus is fixed. Thus external rotation of the forearm can be produced actively by means of pronation. Powerful muscles such as pronator teres and pronator profundus sustain this movement. Retraction of the humerus is effected by the strong subscapularis. As soon as the range of movement of the pronation mechanism has expired, external rotation of the forearm, induced by humeral retraction, will be transferred to wrist and hand. How this movement will become manifest depends solely on the position of the hands with respect to the forearm. In the case of an extended, straight wrist, the hand will rotate laterally. This is not supination because supination, like pronation, can only occur as a result of articulation between the humerus, radius, ulna and carpus.

When these bones act as a unit there is no possibility of hand movement relative to the lower arm. Hence a change of position of the hand cannot be at once interpreted as pronation or supination. Neither, as we have seen, does a fixed position of the hand exclude pronation or supination. External rotation of the forearm induced by humeral retraction would immediately cause lateral rotation of the hand if some degree of pronation was not available. Indeed, rotation of the hand is unavoidable if the range of pronation has expired. So in order to use the full range of pronation it is of prime importance that the hand keeps contact with the ground.

Forward motion, sustained by retraction and pronation, will only occur if there is adequate grip of the hand on the ground. Claws and the surface condition of the palmar skin are no doubt important in this respect. On a slippery surface, when the hand cannot keep to the ground, pronation of the forearm would immediately lead to displacement of the hand. Which particular hand motion becomes manifest all depends on the position of the hand with respect to the forearm and with respect to the ground. With the forearm in a vertical position and the hand flat to the ground, so that the wrist is dorsiflexed, pronation of the forearm would bring about a medial rotation of the flat hand over a slippery surface. As soon as the range of pronation has expired, further rotation of the forearm (resulting from humeral retraction), would lead to lateral rotation of the hand over the surface. However, even when the surface provides frictional forces, a prime requisite for progression, there will always be a tendency for deformation within the hand as a result of pronation and/or lateral rotation. In the case of a tendency to inward or medial rotation of the hand, the rays of the hand will become bent laterally. In contrast a tendency towards outward or lateral rotation of the hand will result in the rays becoming bent medially.

It is hard to predict which tendency will be dominant in normal gait. However, it is unlikely that the two tendencies would be equal, for the position of each ray is definitely asymmetrical. The functional components of sprawling gait covered so far will now be considered more closely in connection with the structural features of the arm and hand.

Functional Morphology of the Anterior Extremity

First a brief survey of the more proximal joints. Structural and functional aspects of the shoulder joint of *Varanus* were studied by Haines (1952).

The condylarlike humeral head articulates in a saddle-shaped scapulocoracoidal socket. The joint is bridged by a conspicuous set of ligaments (Fig. 5). Opening of the capsule from the posterior or caudal side reveals two ligaments running anteriorly from the lesser tubercle, one dorsal to the humeral head, the other ventral (Fig. 5). It appears that the two ligaments are attached dorsally and ventrally (respectively) to the coracoscapular cavity in such a way that both contribute to the formation of a glenoidal labrum or rim around the anterior border of the articular surface. The two ligaments are connected to one another by a strong interligamentous bridge that runs in a vertical direction close to the lesser tubercle. Another ligament takes origin from the posterior border of the coracoid, heading anteriorly to the greater tubercle (Fig. 6). Consequently, this ligament and the ventral ligament from the lesser tubercle establish a ventral cruciform pair of ligaments, as recognized by Haines (1952). Also on the dorsal side of the joint a ligament runs between the scapula and greater tubercle, the scapulo-humeral ligament, which has been mentioned in connection with the insertion of the long head of triceps. This ligament becomes tensed by contraction of the long triceps head, and as a result it tends to raise the major tubercle to which it is attached. The latter motion becomes manifest as external or lateral axial rotation of the humerus which is definitely possible (though to a limited extent) owing to the condylar shape of the humerus. It is not impossible, therefore, that raising of the arm, as occurs in recovery stroke, would be accompanied by a limited external rotation of the humerus, which in turn would result in a more anterior placing of the hand, so as to start a new power stroke.

The posterior interligamentous bridge locks the humeral neck posteriorly, thus preventing the humeral head from luxation during anterior sway of the humerus. Elbow and carpal joints have been dealt with in detail in a recent paper (Landsmeer, 1983). Transmission of horizontal humeral sway into forearm rotation occurs by means of the distal humerus acting on the radius and ulna. As a result of the bicondylar shape of the distal humerus both the radius and ulna get involved in forearm rotation. Homonymous rotation of the radius and ulna requires specific adaptations in the elbow joint. The functional morphology of the elbow joint can easily be understood by keeping the humerus in a fixed position and letting the hand perform a tilt as in pronation or supination. It is seen that in pronation the radius flexes slightly with respect to the humerus and adducts towards the ulna. The obliquely placed radial condyle of the humerus and the anterior and posterior radio-ulnar ligaments are well adapted for this function (Fig. 7).

The axial rotation of the radius and ulna poses some problems at the forearm-carpus interface. The respective joints, the radius-radiale and the ulna-ulnare-pisiform, have to cope in some way with the longitudinal or axial rotation of the forearm bones. While the ulna-ulnare-pisiform joint figures as a ball-and-socket joint with a deformable socket composed of ulnare and pisiform, the radius-radiale is a complicated joint with eccentric properties. Born's description quoted earlier (Born, 1876: 4) is quite adequate. Figure 8 shows the radiale with its dorsal torus-like prominence, extending into a ventral-radial flange, the latter carrying a ventroradial prominence of the radius whilst the former borders on a dorsal concavity which articulates with the radiale torus.

A conspicuous ligament system links the radius, ulna, radiale and ulnare. What seems to be the essential component of this system consists of a meniscus-like ligament running along the ventral side of the radius-radiale joint, extending from the palmar prominence of the radius in a dorsal direction towards the dorso-ulnar ridge on the ulna.

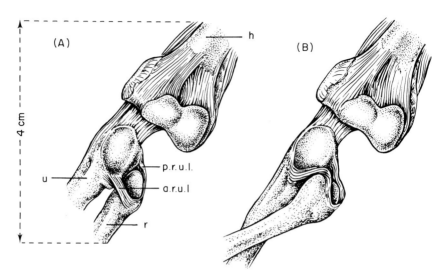

FIG. 7. Exploded view of the elbow joint in supination (A) and pronation (B). Notice condylar formation of the distal humerus (after Landsmeer, 1983).

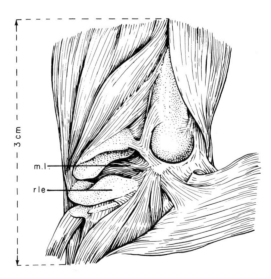

FIG. 8. Exploded view of forearm carpal joints. Left hand. Dorsal aspect. Hand palmarly flexed. The radius-radiale joint is opened to show the ligament system between radius, ulna, radiale and ulnare.

Axial rotation of the ulna and radius in the same direction (which may be designated as a homonymous rotation) will either tighten the ligament or slacken it, depending on the direction of rotation. The taut ligament tends to bring the radius and ulna towards one another (Fig. 9).

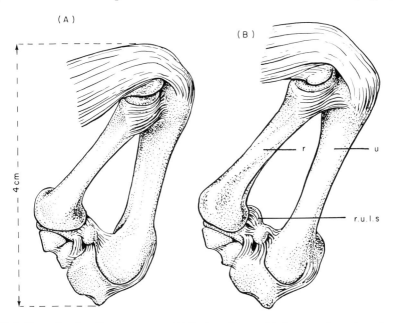

FIG. 9. The radio-ulnar ligament system. (A) Backward sway of the humerus. Lateral rotation of the radius and ulna, taut ligaments and radio-ulnar approximation should be noted. (B) Forward sway, with medial rotation of the radius and ulna and slackening of the ligament system. Left hand.

As soon as the range of motion in the radius-radiale and ulna-ulnare has expired, axial rotation of the radius and ulna will be transferred to radiale and ulnare and further to carpus, metacarpals and phalanges, provided the fingertips are kept in contact with the ground. In the rolling motion of the digital rays the intermetacarpal bands come into the picture. They appear to become tensed in internal or medial rotation and slackened during external or lateral rotation of the rays (Fig. 10). During lateral rotation the radius and ulna approach one another. This mutual approach of the forearm bones is necessarily transmitted to the carpus and metacarpals so that a narrowing of the hand is likely to occur with the rotation. The functional importance of narrowing and widening of the hand – the latter condition prevailing at the onset of power stroke, the former at the end of power stroke – will have to be elaborated further. However, it seems that at the start of power stroke a firm or tensed framework affords an optimal push off,

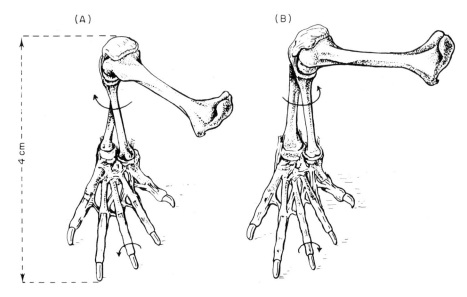

FIG. 10. An osseous ligamentous preparation of the right arm of *Varanus*. (A) Backward sway of the humerus imposes lateral rotation on the radius and ulna. After the range of movement at the forearm-carpal transition has expired, lateral rotation is conveyed to the rays. The metacarpo-digital bands become slack. (B) Forward sway leads eventually to medial rotation of the rays with tightening of the metacarpo-digital bands.

whereas at the end of power stroke and at the beginning of recovery stroke a pliable and narrow hand is more desirable. A significant component of this mechanism, which should be closely observed, can be traced along the following lines. External rotational twist of the forearm (the radius and ulna), provided the hand is fixed, leads to a mutual positioning of the humerus, radius, ulna and hand, which is identical to the position known as pronation. The forearm twist is transmitted to the very tips of the fingers (kept in contact with the ground), but eventually, with continued backward sway of the humerus, when the tolerance or range of movement expires the forearm and hand will have to follow the humeral sway. How the motion becomes manifest all depends on the angle of the elbow and wrist joints. Activity of the pronator teres, as is likely to occur during power stroke, could bring about a kind of burrowing or digging motion or else it could substantially sustain humeral retraction in the act of forward locomotion. It is not yet possible to relate activity of a specific muscle to a definite pattern of movement. A final point concerns the asymmetrical impact to which the hand is subjected during power stroke. This results in a bending of the phalanges to either the radial or the ulnar side. It is quite intriguing to note that the digital bands, which run at the lateral aspects of the phalangeal chain, display asymmetrical proximal roots (Landsmeer, 1981). The ulnar digital band in each finger represents the extension of an interosseous bundle – thus it can be actively powered. This would occur in a radial bending of the ray as the result of a lateral rotation or tendency to rotation of the hand. The latter movement tends to occur as the result of lateral rotation of the forearm, the sequel to humeral retraction.

A powerful contrahens system seems to support the narrowing of the hand which results from lateral rotation of the forearm towards the end of power stroke. It is obvious that many problems persist with respect to the significance of the intrinsic hand musculature. In this respect the roles of the superficial short flexor, the lumbricals and accessory lumbricals (intravaginal paratendinous flexors) are important. The accessory lumbricals are well developed in the opossum as are the lumbricals. There is reason to believe that both muscles are active in interphalangeal extension and quite specifically in retraction of the claw (Landsmeer, 1979). There is no reason to assume that this would not apply to *Varanus*. That the accessory lumbricals are poorly developed is of secondary importance in this context. However, the problem is of a greater complexity than merely the role of the lumbricals. It may be explained along the following lines:
– Synchronous activity of flexor profundus longus, flexor superficialis and the interossei is likely to occur in order to ensure a stable finger arch. Recently the co-operation of superficialis and the interossei in stabilizing the arch has been emphasized (Spoor, 1983).
– The lumbricals acting together with the extensor favour interphalangeal extension. Electromyographic studies in man and other primates (Long & Brown, 1964; Susman & Stern, 1980) support this assertion.
– If the same pattern is executed with an active flexor superficialis and active lumbricals, extension of the terminal claw-bearing lid is likely to precede extension of the other phalanges. The latter movement will take place as soon as the superficialis drops out.

Evidently, functional assessment of the superficialis can only be achieved by sophisticated experimental methods. It is crucial, however, to recognize the theoretical issues involved in this matter, in particular with regard to the phylogenetic origin of both superficialis and lumbricals (McMurrich, 1903; Ribbing, 1907; Haines, 1950). In the course of phylogeny this superficial flexor becomes more and more independent and also more powerful.

It seems legitimate to wonder whether a morphological change with respect to the superficial flexor is associated with a change in the functional setting of this muscle.

CONCLUSIONS

A major characteristic of sprawling gait concerns the transmission of movements through an articulated and angulated system in contact with the ground.

An important issue of gait, in general, concerns the transmission of forces throughout the articulated system. When the articulating components do not all move in the plane of progression, as happens in sprawling gait, transmission of movements from one part to the other involves especially the joints and ligaments. Rewcastle (1980) emphasized the three-dimensional nature of sprawling gait and also the asymmetrical function and structure of the foot and leg. Morphological features of crucial importance in relation to sprawling gait have been traced in the anterior limb of *Varanus*. The morphology of the shoulder and its functional aspects as elaborated by Haines (1952) is largely confirmed. The scapulohumeral ligament may act in conjunction with the triceps brachii. The elbow and the joints at the forearm-wrist interface (viz. the radius-radiale and ulna-ulnare-pisiform joints) must be considered as one mechanical system. The condylar shape of the distal humerus ensures transmission of horizontal sway to the forearm bones and vice versa. In pronation of the forearm in *Varanus*, the radius and ulna do not rotate with respect to one another. The movement can be described either as homonymous rotation of both the radius and ulna with respect to the carpus or else as flexion of the radius with respect to the humerus and adduction towards the ulna. Pronation or external rotation of the forearm involves a particular mechanism at the forearm-carpal interface. This mechanism may produce narrowing and broadening of the hand with lateral and medial rotation of the forearm with respect to the hand. The significance of the lumbricals (and accessory lumbricals) is discussed in relation to claw retraction and interphalangeal extension.

The superficial flexor comes into the picture with regard to stabilization of the arch as well as with regard to delaying interphalangeal extension in the case of extension of the claw-bearing lid. Comparative implications are discussed. The asymmetry of the finger rays, as expressed most clearly in the arrangement of the metacarpodigital bands, can be connected with sprawling gait, as can the arrangement of contrahens and interossei. The suggestion of a symmetrical arrangement of a contrahens or interosseous system – be it with respect to a definite finger, or with respect to the two sides of one finger – is shown to be fallacious.

POST SCRIPTUM

An important contribution to the shoulder mechanism in *Varanus* was published quite recently by Jenkins & Goslow (1983). This paper carries a wealth of information pertaining to electromyography of the shoulder muscles and more specifically to the tongue and groove joint between coracoid and sternum. The significance of this joint to increasing stride length of the anterior limb is stressed. Remarks worthy of consideration are made on muscle homology.

ACKNOWLEDGEMENTS

I wish to thank Dr Ruth St C. Gilmore, Queen's University of Belfast, for her critical reading of the manuscript and improving the English. Acknowledgements are also due to Mr Jan Tinkelenberg and Mr Henk Wetselaar (Department of Anatomy, University of Leiden) for the illustrations.

REFERENCES

Born, G. (1876). Zum Carpus und Tarsus der Saurier. *Morph. Jb.* **2:** 1–27.
Cuvier, G. (1835). *Leçons d'anatomie comparée*, **1**. Paris: Duméril, Laurillard et Duvernoy.
Fürbringer, M. (1900). Zur vergleichenden Anatomie des Brustschulterapparates und der Schultermuskeln. IV.T. *Jena Z. Naturw.* **34:** 215–718.
Haines, R. W. (1946). A revision of the movements of the forearm in tetrapods. *J. Anat.* **80:** 1–11.
Haines, R. W. (1950). The flexor muscles of the forearm and hand in lizards and mammals. *J. Anat.* **84:** 13–29.
Haines, R. W. (1952). The shoulder joint of lizards and the primitive reptilian shoulder mechanism. *J. Anat.* **86:** 412–422.
Jenkins, F. A. (1971). The postcranial skeleton of African cynodonts. *Bull. Peabody Mus. nat. Hist.* **36:** 1–216.
Jenkins, F. A. Jr & Goslow, G. E. Jr (1983). The functional anatomy of the shoulder of the savannah monitor lizard (*Varanus exanthematicus*). *J. Morph.* **175:** 195–216.
Landsmeer, J. M. F. (1979). The extensor assembly in two species of *Opossum*, *Philander opossum* and *Didelphis marsupialis*. *J. Morph.* **161:** 337–346.
Landsmeer, J. M. F. (1981). Digital morphology in *Varanus* and *Iguana*. *J. Morph.* **168:** 289–295.
Landsmeer, J. M. F. (1983). The mechanism of forearm rotation in *Varanus exanthematicus*. *J. Morph.* **175:** 119–130.
Lécuru, S. (1968). Myologie et innervation du membre antérieur des lacertiliens. *Mém. Mus. natn. Hist. nat., Paris* **48:** 127–215.
Lécuru, S. (1969). Etude morphologique de l'humérus des lacertiliens. *Annls Sci. nat.* (Zool.) (12) **11:** 515–558.

Long, C. & Brown, M. E. (1964). Electromyographic kinesiology of the hand: muscles moving the long finger. *J. Bone Joint Surg.* **46A:** 1683–1706.
McMurrich, J. P. (1903). The phylogeny of the forearm flexors. *Am. J. Anat.* **2:** 177–209.
Meckel, J. F. (1828). *System der vergleichenden Anatomie.* Halle: Renger.
Mivart, G. (1867). Notes on the myology of *Iguana tuberculata. Proc. zool. Soc. Lond.* **1867:** 776–797.
Rabl, C. (1916). Ueber die Muskeln und Nerven der Extremitäten von *Iguana tuberculata* Gray. *Anat. Hefte* **53** (160/161): 681–785.
Renous-Lécuru, S. (1973). Morphologie comparée du carpe chez les Lépidosauriens actuels (Rhynchocéphales, Lacertiliens, Amphisbéniens). *Gegenbaurs morph. Jb.* **119:** 727–766.
Rewcastle, S. C. (1980). Form and function in lacertilian knee and mesotarsal joints; a contribution to the analysis of sprawling locomotion. *J. Zool., Lond.* **191:** 147–170.
Rewcastle, S. C. (1981). Stance and gait in tetrapods: an evolutionary scenario. *Symp. zool. Soc. Lond.* No. 48: 239–267.
Ribbing, L. (1907). Die distale Armmuskulatur der Amphibien, Reptilien und Säugetiere. *Zool. Jb. (Anat.)* **23:** 587–682.
Romer, A. S. (1970). *The vertebrate body.* Philadelphia: W. B. Saunders Company.
Rüdinger, N. (1868). Die Muskeln der vordern Extremitäten der Reptilien und Vögel mit besonderer Rücksicht auf die analogen und homologen Muskeln bei den Säugethieren und dem Menschen. *Natuurk. Verh. Holland. Maatsch. Wet. Haarlem* **25:** 1–187.
Schaeffer, B. (1941). The morphological and functional evolution of the tarsus in amphibians and reptiles. *Bull. Am. Mus. nat. Hist.* **78:** 395–473.
Spoor, C. W. (1983). Balancing a force on the fingertip of a two-dimensional finger model without intrinsic muscles. *J. Biomech.* **7:** 497–504.
Susman, R. L. & Stern, J. T. (1980). EMG of the interosseous and lumbrical muscles in the chimpanzee (*Pan troglodytes*) hand during locomotion. *Am. J. Anat.* **157:** 389–397.
Vialleton, L. (1924). *Membres et ceintures des vertebrés tétrapodes. Critique morphologique du transformisme.* Paris: Librairie Octave Doin.
Watson, D. M. S. (1917). The evolution of the tetrapod shoulder girdle and fore-limb. *J. Anat.* **52:** 1–63.

APPENDIX – KEY TO ABBREVIATIONS USED IN FIGURES

a.d. V	abductor digiti V	e.b.	extensor digitorum brevis
a.p.b.	abductor pollicis brevis	e.c.u.	extensor carpi ulnaris
a.p.l.	abductor pollicis longus	e.l.	extensor digitorum longus
a.r.u.l.	anterior radio-ulnar	f.c.r.	flexor carpi radialis
	ligament	f.c.u.	flexor carpi ulnaris
bas.m. V	basis metacarpal V	f.p. (r + u)	flexor profundus radial
bic.	biceps		and ulnar head
br.	brachialis	h	humerus
br.r.	brachioradialis	l	lumbrical
c	contrahens	l.d.	latissimus dorsi
cor.br.l.	coracobrachialis longus	l.r.r.l.	lateral radius-radiale
cor.h.	coracohumeralis		ligament
delt.	deltoid	m.d.b.	metacarpo-digital band
d.p.l.	dorso-posterior ligament	m.l.	meniscus-like ligament

pect. maj.	pectoralis major	s.h.l.	scapulohumeral ligament
p.r.u.l.	posterior radio-ulnar ligament	tr. c.l.	triceps caput longum
		tr. c.lt.	triceps caput laterale
p.t.	pronator teres	tr. c.m.	triceps caput mediale
r	radius	t. maj.	tuberculum majus
rle	radiale	u	ulna
r.u.l.s.	radio-ulnar ligament system (sling)	u.d.b.	ulnar digital band
		v.a.l.	ventral anterior ligament
s.h.	scapulohumeralis	v.p.l.	ventral posterior ligament

Symp. zool. Soc. Lond. (1984) No. 52, 47–85

Variation in the Cloacal and Hemipenial Muscles of Lizards and its Bearing on their Relationships

E. N. ARNOLD

Department of Zoology, British Museum (Natural History), Cromwell Road, London SW7 5BD, England

SYNOPSIS

The muscles associated with the cloaca and hemipenis of lizards are described and variation among different taxonomic groups surveyed on the basis of over 180 species belonging to 135 genera. Functions of the muscles include maintenance of posture, movement of the cloaca and vent, for instance during evacuation, and eversion and retraction of the hemipenes. Female lizards often have a small structure in the same position as the hemipenis with similar musculature. In *Sphenodon*, muscles bordering the vent and anterior to it are arranged in a similar way to those of lizards, but there are no hemipenes or connected musculature. However, shallow, paired out-pouchings of the posterior wall of the cloaca, which each receive a posterior slip of muscle, may be similar to hemipenial precursors.

The likely primitive arrangement of cloacal and hemipenial muscles in the Squamata is estimated from comparison with *Sphenodon*, from the relative frequency of the different states and from their distribution in lizard groups for which phylogenies have been constructed on the basis of other characters. Derived states are then tentatively identified and provide *prima facie* evidence that the forms sharing them are related, although this naturally has to be assessed in the context of evidence from other sources. The following currently held hypotheses receive support: among iguanids, the holophyly of the sceloporines including *Phrynosoma* and the reality of various subgroups within the tropidurines; the relationship of chameleons to agamids, pygopods to geckoes, *Anniella* to anguids, lacertids to teiids, microteiids to macroteiids and advanced lacertids to each other. On the other hand xenosaurs lack features to be expected if they evolved from forms closely related to gerrhonotine anguids and neither lacertids nor amphisbaenians share derived states with microteiids. Although homoplasy certainly occurs in some characters, cloacal and hemipenial muscles are likely to repay further investigation in systematic and phylogenetic studies.

INTRODUCTION

The muscles of the cloaca and hemipenes of lizards and their relatives have received intermittent attention over a long period. Lereboullet (1851) described the situation in *Lacerta agilis* in considerable detail and *Phrynosoma coronatum* was examined less thoroughly by Sanders (1874);

Gadow (1882, 1886) discussed the muscles of several kinds of lizards and of *Sphenodon*, while Wöpke (1930) extended Lereboullet's work on *L. agilis*. D'Alton (1834), Hoffman (1890), Unterhössel (1902), Beuchelt (1936), Volsøe (1944) and Dowling & Savage (1960) have all dealt briefly with the hemipenial muscles of snakes. However, these structures have not been systematically compared between the various groups of lizards and their relatives and a preliminary survey is consequently presented here. Apart from its intrinsic interest, this was undertaken with the intention of increasing the number of characters available for judging relationships and a discussion of how the variation encountered correlates with currently held views of squamate phylogeny is therefore included.

Material examined is listed in Appendix I; unless otherwise indicated, statements are based entirely on this.

GENERALIZED PATTERN OF CLOACAL AND HEMIPENIAL MUSCLES IN MALE LIZARDS
(Figs 1–3)

The following description is a composite made up from the commonest conditions of the various muscles and other structures present in male lizards. As will be argued (p. 72), this arrangement may well be primitive for the Squamata. A variety of names have been used in the past for the structures concerned. Some of these are rejected because they are inappropriate, suggesting actions that the muscles concerned do not produce. In other cases very similar elements, scarcely separable in some taxa, were given very different names and these have been modified. Synonyms of the various muscles are given in Appendix II.

Non-muscular Structures and Ventral Caudal Muscles

The vent opens into the laterally expanded proctodeal area of the cloaca which becomes narrower anteriorly before passing fowards over the ischial symphysis. A rod-shaped hypoischium of usually calcified cartilage typically projects backwards from this. The area immediately anterior to the vent often contains substantial connective tissue which often encloses or supports a ventral gland (see for instance Disselhorst, 1904 and Gabé & Saint Girons, 1965). This is very variable in size and conformation and may be dumb-bell shaped or even divided into separate left and right sections. From the ischial tuberosity, the ilio-ischial ligament runs backwards and then upwards to join the ilium. An often poorly-defined extension of this ligament may run backwards to

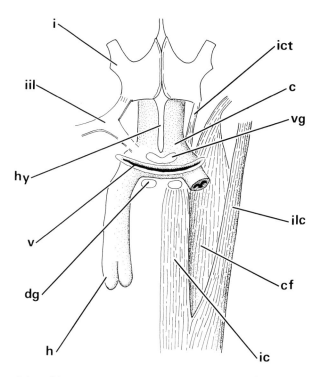

FIG. 1. Ventral view of the structures to which the cloacal and hemipenial muscles attach in a male lizard. c, cloaca; cf, caudifemoralis; dg, dorsal gland; h, hemipenis; hy, hypoischium; i, ilium; ic, pars ischiocaudalis of ilio-ischiocaudalis; ict, tendon of pars ischiocaudalis; iil, ilio-ischial ligament; ilc, pars iliocaudalis of ilio-ischiocaudalis; v, vent; vg, ventral gland.

merge with the connective tissue immediately in front of the vent. The retracted hemipenes lie behind the vent, forming diverticula of the cloaca, with their distal regions directed posteriorly; they project into the ventrolateral tail-base. The paired dorsal glands present in many forms are situated largely above the cloaca but often project downwards so that they are visible just behind the vent near the mid-line. Ventral tail muscles in the region of the hemipenis include the pars ischiocaudalis and pars iliocaudalis of the ilio-ischiocaudalis. The former lies superficially, close to the ventral mid-line, and passes forwards to insert on the ischial tuberosity via a tendon; the latter is placed laterally and runs to insert on the transverse process of the first caudal vertebra; posteriorly, these two elements merge. Between these muscles is situated the caudifemoralis which passes through the loop formed by the ilio-ischial ligament to insert on the femur and, via a tendon, on the tibia.

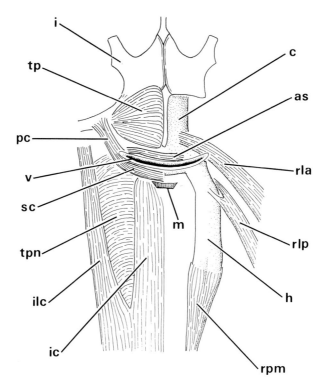

FIG. 2. Ventral view of the generalized pattern of cloacal and hemipenial muscles in a male lizard. The more superficial muscles are shown on the left. as, anterior fibres of sphincter cloacae; m, retractor medialis; pc, protractor commissurae; rla, retractor lateralis anterior; rlp, retractor lateralis posterior; rpm, retractor penis magnus; sc, posterior fibres of sphincter cloacae; tp, transversus perinei; tpn, transversus penis. Other abbreviations as in Fig. 1.

Muscles of the Cloacal and Hemipenial Region

(i) The transversus perinei runs more or less laterally from the hypoischium to insert mainly on the ilio-ischial ligament near its junction with the ischium.

(ii) The sphincter cloacae consists of transverse fibres lying anterior and posterior to the vent, the two sections often being more or less independent.

(iii) From the lateral extremity of the vent, muscle fibres called here the protractor commissurae run obliquely outwards and forwards to attach on the ilio-ischial ligament, usually posterior to the insertion of the transversus perinei. In fact, the protractor commissurae is confluent with the sphincter cloacae and frequently derives all or most of its fibres from the posterior section of that muscle.

(iv) The hemipenis is covered laterally and ventrally by the trans-

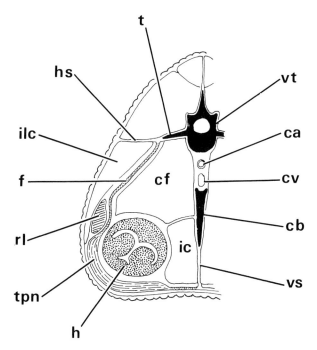

FIG. 3. Transverse section of the tail of a male lizard. ca, caudal artery; cb, chevron bone; cv, caudal vein; f, fascia joining transversus penis muscle to horizontal longitudinal septum; hs, horizontal longitudinal septum; rl, retractor lateralis; t, transverse process of vertebra; vs, vertical longitudinal septum; vt, vertebra. Other abbreviations as in Figs 1 and 2.

versus penis. This typically originates from a connective tissue fascia stemming from the more distal parts of the transverse processes of the caudal vertebrae or the horizontal longitudinal septum. It runs downwards and inwards to insert complexly on the superficial layers of the pars ischiocaudalis or, by a fascia, to the ventral mid-line.

(v) Two, often strap-shaped muscles, the retractores laterales, typically originate from the horizontal septum in the region of the transverse processes of the caudal vertebrae, or from fasciate connexions to this area. Their origins consequently lie between the upper parts of the pars iliocaudalis and the caudifemoralis. The retractor lateralis anterior runs downwards, inwards and forwards to insert just in front of the vent, either towards the mid-line or more laterally. Insertion is on to connective tissue attached to the ventral wall of the cloaca or joining the ilio-ischial ligament to the area in front of the vent. The retractor lateralis posterior inserts on the base of the hemipenis and the lateral dorsum of the posterior cloaca, there frequently being an often tendinous connexion to the ventral side of the hemipenial base and one or more to its dorsal surface and that of the cloaca.

(vi) The retractor medialis is unpaired and originates ventrally on a fascia attached to the skin around the mid-line, just behind the posterior fibres of the sphincter cloacae. It runs upwards and forwards, converging to insert on the dorsal surface of the cloaca via a median tendon.

(vii) The retractor penis magnus originates on one or more of the transverse processes of the caudal vertebrae and runs forwards to insert on the distal lobes of the uneverted hemipenis.

The small muscles described by Lereboullet (1851) as the dilatateur latéral and the dilatateur inférieur, which lie deep to the transversus perinei and close to the wall of the cloaca in *Lacerta agilis*, have not been investigated in this study.

Function

The muscle arrangement described above appears to be involved in a number of distinct functions: maintenance of posture, evacuation of cloacal contents and movement of the hemipenis. The transversus perinei probably helps maintain posture by drawing the ventral portion of the ilio-ischial ligament inwards, together with the proximal parts of the hind limb muscles originating from it. The posterior section of the sphincter cloacae and the confluent protractor commissurae may act antagonistically against the retractor lateralis anterior to keep the vent closed, so that relaxation would tend to open it. Evacuation, by reducing the volume of the cloaca, would appear to be mediated by the retractor medialis and perhaps the retractor lateralis posterior. At least the initial stages of hemipenial eversion seem likely to be partly effected by the transversus penis; anterior peristalsis in this would tend to move the hemipenis forwards (Beuchelt, 1936). Retraction is probably largely brought about by the retractor penis magnus, withdrawal of the base being completed by the retractor lateralis posterior.

DEVIATIONS FROM THE GENERALIZED PATTERN

In most squamates, the cloacal and hemipenial musculature differs from the generalized pattern in some features. The more obvious deviations are listed in Table I and their distribution shown in Table II. As is evident, most occur in only a minority of groups. Some of these are described more fully in the next section.

TABLE I
Deviations from the generalized squamate pattern of cloacal and hemipenial muscles

(1) Hypoischium strongly reduced or absent.
(2) Clearly defined posterior section of the transversus perinei present (Fig. 7).
(3) Posterior section of transversus perinei arched, with origin and insertion close together.
(4) Transversus perinei undivided but fibre pattern complex.
(5) Thin layers of muscle fibres diverging from lateroposterior border of abdominal muscles and from corner of vent to insert on skin superficial to transversus perinei.
(6) Anterior fibres of sphincter cloacae running on to tendon of retractor lateralis or confluent with its more posterior fibres (Fig. 7).
(7) A section of the protractor commissurae arising behind vent does not attach to the ilio-ischial ligament but instead terminates (a) on skin superficial to the transversus perinei, or (b) at margin of abdominal muscles (Fig. 6).
(8) Section of protractor commissurae arising in front of vent extends superficial to transversus perinei to insert at margin of abdominal muscles [Fig. 6(B)].
(9) Protractor commissurae runs forward superficial to transversus perinei to insert on ilio-ischial ligament.
(10) Protractor commissurae with some fibres extending deep to transversus perinei to insert on upper surface of ischium.
(11) Transversus penis arising lateral to the retractores laterales anteriorly and medial to them posteriorly.
(12) Transversus penis: (a) strongly reduced; (b) absent (Fig. 7).
(13) Fibres of apparent transversus penis run approximately longitudinally [Fig. 5(A)].
(14) Post-cloacal sacs present [Fig. 5(A)].
(15) Median post-cloacal bones present [Fig. 5(A)].
(16) Lateral post-cloacal sacs present [Fig. 5(A)].
(17) Cartilaginous body present just posterior to retracted hemipenis.
(18) Transverse muscle fibres arise on edge of anterior pars iliocaudalis and run towards mid-line [Fig. 5(A)].
(19) Dense fascia joins transversus penis dorsomedially to form hemipenial sheath [Fig. 4(B)].
(20) Transversus penis extends almost or entirely around hemipenial sheath [Fig. 4(D)].
(21) Accessory lateral, or medial, or both, sheath muscles present [Fig. 4(C)].
(22) Accessory dorsal sheath muscle present [Fig. 4(E)].
(23) Origins of retractores laterales on each side closely applied or fused [Fig. 5(B)].
(24) One or both retractores laterales originate approximately dorsal to hemipenes.
(25) Only a single retractor lateralis of uncertain identity present on each side (Fig. 7).
(26) Only a single retractor lateralis present approximately occupying the position of the retractor lateralis anterior.
(27) Retractor lateralis anterior short, not extending far behind level of vent.
(28) Retractor lateralis anterior attached to median cloacal outpushing behind vent.
(29) Retractor lateralis anterior fibres abruptly flexed just before their insertion in front of vent.
(30) Retractor lateralis anterior with fibres attached to lateral border of hemipenial base.
(31) Retractor lateralis anterior large, passing medial to the retractor lateralis posterior and originating well behind it.
(32) Retractor lateralis posterior completely divided into two sections.
(33) Retractor lateralis posterior with especially fleshy ventral insertion.
(34) Retractor lateralis posterior inserts only on lateroventral surface of hemipenial base.
(35) Retractor lateralis posterior substantially situated within hemipenial sheath.
(36) Hemipenis attached to apparent transversus penis by fasciate connexions [Fig. 5(B)].
(37) Penis retractor magnus often kinked when hemipenis retracted [Fig. 5(B)].

TABLE II
Distribution of states other than those found in the generalized squamate arrangement of cloacal and hemipenial muscles[a]

	1	2	3	4	5	6	7	8	9	10	11	12	13	14	15	16	17
Iguanidae																	
Iguanines																	
Basiliscines																	
Crotaphytines																	
Sceloporines																	
Tropidurines																	
Oplurines	+−																
Morunosaurs																	
Anoloids																	
Para-anoles																	
Anoles																	
Agamidae	+																
Chamaeleonidae	+−																
Gekkonidae	+																
Pygopodidae										+−							
Xenosauridae																	
Anguidae																	
Gerrhonotinae							b	+									
Anguinae	+						b	+									
Diploglossinae							a										
Anniellidae	+						b	+									
Helodermatidae																	
Lanthanotidae																	
Varanidae											+						
Teiidae																	
Tupinambini		+										b	+	+(−)	+(−)	+(−)	
Teiini		+										a	+	+	+	+	
Gymnophthalminae			+−									b					
Lacertidae		+							+−	+−							
Xantusiidae	+								+−			b					
Cordylidae												b				(+)	
Scincidae	+			+−	+−		a−					b					+

(continued)

	18	19	20	21	22	23	24	25	26	27	28	29	30	31	32	33	34	35	36	37
Iguanidae																				
Iguanines	+	+		+(−)		+−							+−							
Basiliscines	+	+																		
Crotaphytines	+	+																		
Sceloporines	+	+																		
Tropidurines	+	+	(+)		+−	(+)				(+)					+	(+)				
Oplurines	+	+								+−		+−								
Morunosaurs	+	+																		
Anoloids	+	+			(+)					(+)				(+)						
Para-anoles	+	+																		
Anoles	+	+	(+)			+−														
Agamidae	+	+				(+)				+(−)					(+)		+(−)	+		
Chamaeleonidae	+(−)	(+)								+	+						+	+		
Gekkonidae						+	+												+(−)	+(−)
Pygopodidae																			+	+
Xenosauridae																				
Anguidae																				
Gerrhonotinae						+														
Anguinae							+										+−			
Diploglossinae							+−							+						
Anniellidae						+	+													
Helodermatidae						+														
Lanthanotidae																				
Varanidae	+		(+)												+					
Teiidae																				
Tupinambini								+												
Teiini								+									+−			
Gymnophthalminae								+												
Lacertidae	+(−)												(+)							
Xantusiidae	+(−)					+														
Cordylidae						+			+(−)											
Scincidae	(+)																			

[a]Numbers refer to the various deviant states listed in Table I. + = state found throughout group; +− = state present or absent; brackets indicate that condition is only found in a small minority of forms in group; a, b indicate different states of same apparent transformation series.

SYSTEMATIC ACCOUNT

Only deviations from the generalized pattern are described in detail.

Lizards

(i) Iguanidae
The intrafamilial arrangement used here is that of R. E. Etheridge and K. De Queiroz (pers. comm.). Many iguanids conform quite closely to the generalized muscle pattern and this is approached in some members of nearly all the main groups in the family. However, iguanids differ consistently from the generalized pattern in possessing a fascia running dorsally and medially to the hemipenis which, with the transversus penis muscle, forms a sheath around the organ (Fig. 4(B)).

(a) Iguanines. *Dipsosaurus* and *Iguana* have the widespread iguanid condition. In many other members of this group, the retractores laterales are closely applied to each other, at least at their origins, and may be fused in this region (*Amblyrhynchus, Brachylophus, Cyclura, Sauromalus*); in some, the retractor lateralis anterior inserts partly along the outer edge of the base of the hemipenis. A number of iguanines possess accessory muscles arising on the inside of the anterior sheath and running forward onto the base of the hemipenis (Fig. 4(C)). *Brachylophus* and *Conolophus* have a single muscle on the medial side of the sheath whereas in *Cyclura* it is lateral; *Amblyrhynchus* and *Ctenosaura* have a muscle on both sides. These muscles overlap the transversus penis extensively.

(b) Basiliscines. *Basiliscus, Corythophanes* and *Laemanctus* all approach the widespread iguanid condition. The retractores laterales are substantially covered by the transversus penis.

(c) Crotaphytines. *Gambelia* is close to the widespread iguanid condition.

(d) Sceloporines including *Phrynosoma*. In all cases, the retractor lateralis posterior is divided into two separate sections. The more anterior of these lies largely above the hemipenis and is dorsoventrally flattened; it inserts on the dorsal confluence of the cloaca and hemipenis, just above the vent. The more posterior section is smaller and strap-shaped and inserts behind the anterior section on the lateral or ventral surfaces of the base of the hemipenis, or on both. The retractor lateralis anterior is often strap-like and extends nearly to the mid-line in front of the vent. In the area of insertion, this muscle is widened in many

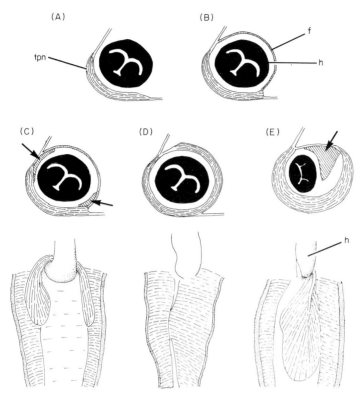

FIG. 4. (A)–(E): Transverse sections of the hemipenis and associated muscles. (A) Usual arrangement in lizards. (B) Fascia above hemipenis connects the two sides of transversus penis muscle forming a sheath, e.g. *Iguana*: (C) Small medial and lateral sheath muscles present, e.g. *Amblyrhynchus*: (D) Transversus penis muscle spreads around sheath fascia, e.g. *Ophryoessoides*: (E) Large dorsal accessory sheath muscle present, e.g. *Tropidurus*. The illustrations below (C), (D) and (E) show their respective conditions looking upwards into the opened sheath with the hemipenis turned forwards. f, fascia; h, hemipenis; tpn, transversus penis.

forms (*Callisaurus*, *Cophosaurus*, *Holbrookia*, *Phrynosoma*, *Sceloporus*, *Uma*, *Uta*) and in some the more anterior fibres are reflected outwards or posteriorly before insertion (*Callisaurus*, *Cophosaurus*, *Holbrookia*, *Phrynosoma*, *Uma*). In *Petrosaurus* the retractor lateralis anterior does not extend far posteriorly and is confined to the area in front of the vent. In the other genera both the retractores laterales originate from the area of the horizontal longitudinal septum, often via a fascia. The retractor lateralis posterior may be covered in part by the transversus penis.

(e) Tropidurines. *Leiocephalus* possesses the widespread iguanid condition. *Phymaturus* approaches this but the retractores laterales are closely applied and, in addition to its attachment at the dorsal confluence of the hemipenis and cloaca, the retractor lateralis posterior has a well defined fleshy insertion on the side of the base of the hemipenis (instead of one that is less clearly defined, tendinous or absent). In *Ctenoblepharis* and especially *Liolaemus*, this feature is more strongly developed and the retractor lateralis anterior does not extend backwards beyond the level of the vent, originating from a fascia immediately below the anterior caudifemoralis muscle.

In *Proctotretus* and *Stenocercus* the retractor lateralis posterior inserts only on the dorsal confluence of the hemipenis and cloaca, the retractor lateralis anterior is again short, not extending posterior to the level of the vent, and the transversus penis spreads round the median side of the hemipenial sheath so that the dorsomedial area of translucent fascia found in most iguanids is reduced in width. The condition in *Ophryoessoides* is similar but the transversus penis extends almost right round the sheath (Fig. 4(D)).

Tropidurus, *Plica*, *Strobilurus*, *Uracentron* and *Uranoscodon* all have an accessory sheath muscle. This is especially large in *Tropidurus torquatus* and *Uranoscodon* and lies in the dorsum of the hemipenial sheath (Fig. 4(E)). When big, it may fill a substantial portion of the anterior sheath cavity and is pinnate with an included ventral tendon. Insertion is on to the medial side of the base of the hemipenis. It differs from the sheath muscles found in some iguanines in being often larger and more dorsal in position and in not, or scarcely, overlapping the transversus penis. In other respects the cloacal and hemipenial musculature of these lizards is not unusual, except that the retractor lateralis anterior is often shortened and in *Tropidurus peruvianus* and *Uracentron* does not extend posterior to the vent.

(f) Oplurines. In the Madagascan iguanids, the muscles conform to the widespread iguanid condition. However, the retractor lateralis anterior, which approaches the mid-line, is not very thick and the narrow retractor lateralis posterior is closely bound to the hemipenial sheath although it does not lie within it. *Chalaradon* lacks a hypoischium.

(g) Morunosaurs. *Enyalioides* and *Hoplocercus* again have musculature of the widespread iguanid type.

(h) Anoloids. *Diplolaemus* possesses the widespread iguanid arrangement but in *Enyalius* the rector lateralis anterior does not extend far backwards and there is a dorsal accessory sheath muscle similar to that occurring in *Tropidurus* etc. *Polychrus* is singular among iguanids in having a very large retractor lateralis anterior incorporating a ventral tendon. It extends much further posteriorly along the tail than the retractor lateralis posterior and lies medial to this muscle. Insertion is into dense connective tissue in front of the vent that is attached to the distal part of the hypoischium.

(i) Para-anoles. *Urostrophus* has the widespread iguanid condition.

This is also found in *Anisolepis* and *Aptycholaemus*, except that the insertion of the retractor lateralis anterior is very swollen and the transversus perinei lies almost entirely anterior to it.

(j) Anoles. A few species of this very large assemblage, selected to cover most of the species series defined by Etheridge (1959), were examined. It appears that some deviation from the widespread iguanid condition is not uncommon. Thus the transversus penis extends right round the hemipenial sheath in *Anolis bimaculatus*, *A. biporcatus*, *A. chlorocyanus*, *A. chrysolepis*, *A. cuvieri*, *A. cybotes*, *A. nebulosus*, *A. sagrei* and *Chamaelinorops*. Fusion of the origins of the retractores laterales occurs in *A. chlorocyanus*, *A. coelestinus*, *A. cuvieri*, *A. equestris* and *A. richardi*; the insertion of the retractor lateralis anterior is especially large in *A. chlorocyanus*, *A. coelestinus*, *A. equestris* and *A. richardi*. Forms with the widespread iguanid condition include *A. gormani* and *A. valencienni*.

(ii) Agamidae

The sample of nine genera investigated suggests that agamids are more uniform than iguanids in their cloacal and hemipenial musculature. As in iguanids, a sheath is present (weak in *Gonocephalus*). The retractor lateralis anterior is short but may extend a brief distance posterior to the vent (in *Chlamydosaurus*, *Gonocephalus*, *Leiolepis*, *Physignathus* and especially *Uromastyx*); it may be quite plump, so that the portion of the transversus perinei superficial to it is comparatively thin (especially in *Leiolepis* and *Uromastyx*). The retractor lateralis posterior in *Leiolepis* is divided with the more anterior section going to the dorsum of the confluence of the hemipenis while the other part runs to the lateroventral surface of the base of the hemipenis. In other agamids examined, only the latter section of this muscle is present. A substantial portion of the retractor lateralis posterior is enclosed within the hemipenial sheath. In *Leipolepis* it only extends a short distance dorsal to this, being connected to the region of the horizontal septum by a fascia. In other forms it projects substantially beyond the sheath, often running upwards towards the septum between two layers of fascia.

(iii) Chamaeleonidae

In *Chamaeleo* the hypoischium is absent and the transversus perinei is small and simple. A hemipenial sheath is present although often rather thin. The paired retractor lateralis anterior muscles are closely applied on the mid-line and lie behind the transversus perinei. They are large but do not extend far posterior, although they do run backwards close to the mid-line to insert on the dorsum of a median cloacal outpushing just behind the vent. The posterior fibres of the sphincter cloacae may extend some way backwards on the mid-line and the transversus penis is often very thick and encloses much of the retractor lateralis posterior. The latter muscle is robust, arising from the area of the horizontal longitudinal septum and lying within the sheath; it inserts broadly on the lateroventral surface of the base of the hemipenis, although it may extend around the lateral edge of the hemipenis to the front of the confluence of the organ with the cloaca (*Chamaeleo africanus, C. dilepis*). In *Brookesia* the posterior median outpushing of the cloaca is more marked and the retractor lateralis posterior inserts on this rather than on the hemipenial base, having passed over the lateral and ventral surfaces of the hemipenis in a well-defined groove.

(iv) Gekkonidae (Fig. 5)

Geckoes depart substantially from the generalized lizard arrangement. The hypoischium is reduced or absent in the eublepharines and diplo-

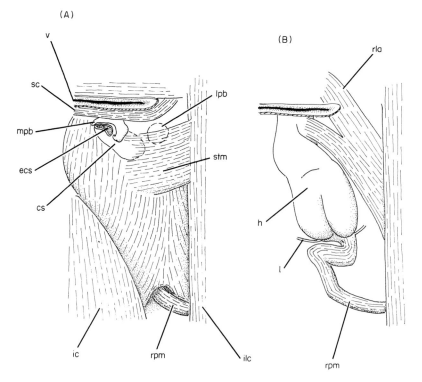

FIG. 5. Ventral views of cloacal and hemipenial muscles in geckoes. (A) Superficial muscles on right side. (B) Most superficial muscles removed. cs, post-cloacal sac; ecs, entrance to post-cloacal sac; ic, pars ischiocaudalis; ilc, pars iliocaudalis; l, fascia connecting hemipenis to transversus penis muscle; lpb, position of the lateral post-cloacal bone if present; mpb, medial post-cloacal bone; stm, superficial layer of transverse muscle. Other abbreviations as in Figs 1 and 2.

dactylines studied but not in the gekkonines (although Camp, 1923, indicates absence in some other members of this group). The transversus perinei is simple but some fibres may run to the anterior lip of the cloaca and it sometimes covers the insertion of the protractor commissurae. In some cases, the latter muscle has a section running forwards above the transversus perinei to the ischium. The fibres of what appears to be the transversus penis are arranged approximately longitudinally and not roughly transversely as in most other lizards. They run from behind the posterior cloacal lip to terminate on each side of the retractor penis magnus, attaching to the transverse processes of vertebrae laterally and in the region of their chevron bones medially (except in *Pristurus* where they reach the superficial mid-line). In some instances, the transversus penis itself is narrow and is replaced over the flanks of the retracted hemipenis by fascia that may extend dorsally to complete a weak hemipenial sheath (for instance in *Coleonyx* and *Teratoscincus*).

Most geckoes have paired post-cloacal sacs: epithelial diverticula just behind the vent that are superficial to the transversus penis and closely applied to it. Post-cloacal bones are also typically present just beneath the skin: a medial pair, each consisting of a transverse bar closely applied to the sphincter cloacae that laterally curves backwards around the entrance of the cloacal sac; a more lateral pair is also common but less frequently present, each bone often underlying and supporting one or more enlarged tubercles on the skin. The two bones on each side may be separate, fused together or joined by dense connective tissue. Kluge (1967, 1982) discusses the distribution of post-cloacal sacs and bones.

A superficial layer of transverse muscle often runs inwards from the lower edge of the anterior pars iliocaudalis to insert via a fascia on the skin towards the mid-line. Its anterior fibres frequently insert on the cloacal sac, if present, and sometimes on the medial post-cloacal bone as well; in cases where a lateral bone is also found, this lies superficial to the anterior fibres, at least in part. The superficial layer of transverse muscle is not always easily separated from the transversus penis and sometimes this seems to occupy its place. The present author could not locate the muscle running backwards from the end of the post-cloacal sac mentioned by Brongersma (1934) and Rieppel (1976).

The hemipenis and the attached retractor penis magnus are frequently, although not always, kinked in geckoes, with the hemipenial lobes flexed laterally, and the muscle medially and then laterally before running steeply upwards to its insertion on the transverse process of one of the caudal vertebrae. The hemipenis is often attached to the transversus penis muscle by fasciae. The retractores laterales are united at their origin; the smaller posterior section inserts on the lateral base of the hemipenis while the anterior part ends just in front of the vent, typically to the side, but near the mid-line in *Hoplodactylus*.

(v) Pygopodidae

The anatomy of the cloacal and hemipenial region is very like that of geckoes in *Pygopus*, the only genus investigated. The hypoischium is absent, transversus penis fibres run approximately longitudinally, postcloacal sacs and bones are present, fasciae attach the transversus penis muscle to the hemipenis and the retractor penis magnus muscle is kinked. However, a separate superficial layer of transverse muscle running from the edge of the pars iliocaudalis is absent and the two retractores laterales have separate origins situated dorsal to the hemipenis.

(vi) Xenosauridae

Xenosaurus is close to the generalized lizard condition with the retractores laterales having widely separated origins on the transverse processes of the caudal vertebrae. The retractor lateralis anterior does not reach the mid-line at its insertion.

(vii) Anguidae

(a) Gerrhonotines. The genera of this subfamily examined (*Abronia* and *Gerrhonotus*) deviate from the generalized lizard condition in having very extensive protractor commissurae muscles (Fig. 6): from the front of the vent, a band of fibres runs forward across the transversus perinei to the posterior termination of the superficial abdominal muscles. A parallel slip (broad in *Gerrhonotus*) originates behind the vent and lies lateral to this band; some additional fibres commencing posterior to the vent insert obliquely on the ilio-ischial ligament behind the insertion of the transversus perinei. The retractores laterales are closely applied at their origins in the region of the transverse processes of the caudal vertebrae and the retractor lateralis anterior inserts in front of the level of the vent but well away from the mid-line.

(b) Anguines. *Anguis* and *Ophisaurus* both have very reduced hindlimbs and pelves and lack a hypoischium. They possess an elaborate protractor commissurae system, like that found in gerrhonotines. The retractores laterales, which are quite short in *Anguis*, originate directly above the hemipenis (a common state in legless squamates) on the transverse processes of the caudal vertebrae where they are closely applied to each other. The retractor lateralis anterior does not reach the mid-line and the retractor posterior lacks a connection to the dorsal side of the hemipenial base in *Ophisaurus*, although it reaches the ventral surface in both genera.

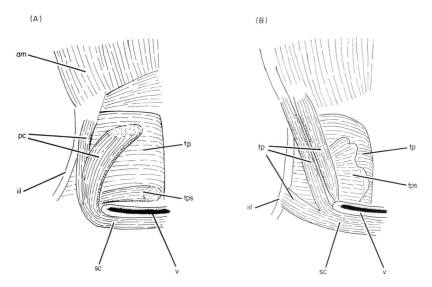

FIG. 6. Ventral view of superficial muscles in the vent region of anguid lizards (right side shown). (A) *Diploglossus stenurus*. (B) *Gerrhonotus coeruleus*. am, superficial abdominal musculature; iil, ilio-ischial ligament; pc, protractor commissurae; sc, sphincter cloacae; tp, transversus perinei; tps, section of transversus perinei attaching to skin; v, vent.

(c) Diploglossines. In *Diploglossus*, the tip of the hypoischium is dorsoventrally flattened and strongly expanded laterally. The transversus perinei is simple, but a few posterior fibres run from the ilio-ischial ligament to the skin near the mid-line just in front of the vent. The protractor commissurae is elaborate, as in anguids of other subfamilies, but differently arranged: it originates entirely behind the vent with one band running forwards to insert on the skin superficial to the anterior fibres of the transversus perinei; another band lies lateral and roughly parallel to this and runs to the ilio-ischial ligament. The retractor lateralis anterior inserts well away from the mid-line while the retractor lateralis posterior is much more slender and does not extend so far backwards, passing lateral to the former muscle. In *Ophiodes*, which has very reduced hind-limbs, the hypoischium is broad but very thin, especially near the mid-line. The protractor commissurae is similar to that in *Diploglossus*, as is the general arrangement of the retractores laterales. The retractor lateralis anterior is very large with a prominent ventral tendon incorporated in its more anterior parts and originates far behind the retractor lateralis posterior. This arrangement is reminiscent of that found in the iguanid *Polychrus* (p. 58). It is possible that in these forms, the muscle is involved in flexing the tail.

(viii) Anniellidae
The hypoischium is absent. The protractor commissurae muscles are similar to those found in the Anguinae, the inner band arising narrowly in front of the vent. The retractores laterales originate above the hemipenis as a single mass but their insertions are normal.

(ix) Helodermatidae
The musculature is like the generalized lizard type in most respects. An area of dense but cavernous connective tissue is present in front of the vent and behind the main bulk of the transversus perinei, although it is covered by a thin layer of this muscle. The retractores laterales are substantially fused and insert on the base of the hemipenis and the anterior wall of the vent.

(x) Lanthanotidae
The cloacal and hemipenial muscles are very close to the generalized condition.

(xi) Varanidae
The protractor commissurae is short but may include separate slips originating behind and in front of the vent. All species examined have a hemipenial sheath, although this is weak in *Varanus exanthematicus*. The transversus penis is well developed: anteriorly it arises medial to the pars iliocaudalis, often from a fascia, and covers most of the more anterior parts of the retractores laterales; more posteriorly it lies medial to these muscles, although it may arise from fascia that passes upwards on each side of them. In *V. dumerilii* and *V. gouldi* the transversus penis invades the sheath dorsally and in some species the anterior, more superficial section of this muscle can be quite easily separated from an inner layer which extends further posteriorly but overlaps it extensively. The retractores laterales are elongate and may arise from a fascia. The retractor lateralis anterior inserts well away from the mid-line and often lateral to the vent; it may terminate slightly anterior to this, or somewhat posterior (*V. dumerilii*, *V. gouldi*) and sometimes incorporates a ventral tendon (*V. dumerilii*, *V. gilleni*). The retractor lateralis posterior has two separate parts, one inserting anteriorly on the dorsum of the base of the hemipenis and the other laterally.

(xii) Teiidae
(a) Tupinambini [Fig. 7(A)]. Much dense connective tissue exists anterior to the vent, enclosing a well developed ventral gland which is laterally expanded and apparently divided at the mid-line in *Callopistes*

Cloacal and Hemipenial Muscles of Lizards

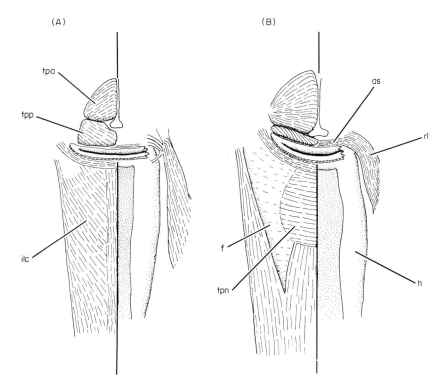

FIG. 7. Ventral views of the cloacal and hemipenial muscles of teiid lizards. (A) *Tupinambis nigropunctatus*. (B) *Teius teyou*. The more superficial muscles are shown on the left of each figure. as, anterior fibres of sphincter cloacae running into the retractor lateralis; f, fascia attached to transversus penis; h, hemipenis; ilc, fibres of pars iliocaudalis covering hemipenis ventrally; rl, retractor lateralis; tpa, transversus perinei – anterior section; tpn, transversus penis; tpp, transversus perinei – posterior section.

and *Tupinambis*. The transversus perinei has a very well differentiated posterior section with fibres running from the posterior mid-line of the dense connective tissue in front of the vent to insert on the ilio-ischial ligament behind the anterior section of the muscle. The transversus penis is absent but the hemipenis is covered ventrally by fibres of the pars iliocaudalis running backwards and downwards to contact the pars ischiocaudalis and the adjoining skin; anteriorly, some of these fibres arise from a fascia covering the retractor lateralis. Only one retractor lateralis is present, inserting by ligaments in the region of the hemipenial-cloacal confluence; it attaches to a posterior extension of the ilio-ischial ligament, to the cloacal wall anterior to the side of the vent, and to the ventral side of the very base of the hemipenis.

(b) Teiini [Fig. 7(B)]. Less connective tissue is present in front of the vent than in the Tupinambini, but a single or divided ventral gland is well developed in most forms, especially *Kentropyx* in which the tip of the hypoischium is strongly expanded laterally. The posterior section of the transversus perinei is quite well developed in *Teius*, small in *Ameiva*, *Cnemidophorus* and *Dicrodon* and apparently reduced to a small median patch of fibres in *Kentropyx*; its development correlates approximately inversely with that of the ventral gland. An apparent transversus penis is present but is restricted to the region close to the vent, near the mid-line; more laterally it attaches to a fascia which covers the anterior hemipenis. The retractor lateralis is again single and may originate from a fascia running from the outer parts of the transverse processes of the caudal vertebrae inwards across the ventral surface of the caudifemoralis muscle. The retractor lateralis inserts slightly anterior and lateral to the vent. Fibres of the anterior section of the sphincter cloacae lying in front of the vent run laterally into the posterior part of the retractor lateralis.

(c) Gymnophthalminae. The ventral gland is large and apparently always divided into left and right sections. The transversus perinei is more or less simple without a clear posterior section and the transversus penis is absent. A single retractor lateralis ends in a ligament which typically inserts by two branches to the very base of the hemipenis on its upper and lower surfaces. In most forms examined (but not in *Prionodactylus*), anterior fibres of the sphincter cloacae extend on to the ligament of the retractor lateralis and may occasionally contact the muscle itself (e.g. in *Anadia*).

It is uncertain whether the the retractor lateralis muscle of teiids is homologous to the anterior or posterior muscle of other lizards or whether it is a combination of both.

(xiii) Lacertidae

The transversus perinei has a well defined posterior section and the transversus penis is nearly always present (largely replaced by fascia in *Mesalina brevirostris*). In all forms examined except *Takydromus*, a well developed hemipenial sheath occurs. The retractores laterales arise more or less above the hemipenis, either on the dorsal wall of the sheath or from the vicinity of the horizontal longitudinal septum and transverse processes of the caudal vertebrae. The origins of the two muscles are closely applied and, although sometimes easily separated – for instance in *Latastia* – are often confluent. The insertion of the retractor lateralis anterior lies in front of the level of the vent. In *Takydromus* and *Lacerta vivipara* it ends very broadly and approaches the mid-line, while in some other *Lacerta* and *Nucras* species it extends as far as this but is narrow. In many other lacertids this muscle inserts more laterally, often via a tendon. When this occurs, the posterior section of the transversus perinei passing superficial to the termination of the retractor lateralis anterior has its origin and insertion close together, so that its fibres arch abruptly. This occurs in such forms as *Acanthodactylus*, many *Eremias sens. lat.*, *Ichnotropis*, *Latastia* and *Meroles*. In some cases the posterior section of the transversus perinei has its own ligamentous base separate from the ilio-ischial ligament. The retractor lateralis posterior usually runs onto the dorsum of the confluence of the hemipenis and cloaca where it often inserts via two ligamentous connexions. Typically, another branch or ligament from this muscle inserts on the side or ventral face of the base of the hemipenis.

(xiv) Xantusiidae

The hypoischium is absent and the transversus perinei is simple with a narrow insertion on the ilio-ischial ligament. The protractor commissurae arises behind the vent and in *Lepidophyma* and *Xantusia vigilis* the more medial fibres extend superficial to the transversus perinei to insert well forwards on the ilio-ischial ligament, while the more lateral ones pass deep to that muscle to reach the upper surface of the ischium. The protractor commissurae is not extended in this way in *X. henshawi*.

No transversus penis is present but transverse fibres often arise from the anterior edge of the pars iliocaudalis and run downwards and inwards to insert towards the mid-line (absent in *X. vigilis* where there is only a fascia). Contrary to Savage (1957), the present author found no trace of post-cloacal sacs, but, like Rieppel (1976), encountered a lateral post-cloacal bone in a male *X. vigilis* (BM 95.11.12.20). This is an irregular flat plate of bone surrounded at its margins by dense connective tissue and closely applied to the skin; it is also attached to a fascia which runs medial to the pars iliocaudalis and reaches the mid-line area. This lateral post-cloacal bone lies further behind the vent and more medially than is usual in geckoes and does not underlie one or more enlarged tubercles on the skin. In other *Xantusia* examined and in *Lepidophyma* no post-cloacal bone was encountered, although a patch of dense connective tissue occupied its place, just medial to where the muscle fibres running transversely from the edge of the pars iliocaudalis end.

The retractores laterales originate as a single mass with the retractor lateralis anterior inserting in front of the vent but to the side and the retractor lateralis posterior inserting on the lateral surface of the hemipenis well away from the base. There is a large piece of dense connective tissue partly covering the medial termination of the retracted hemipenis. It is very clearly defined, laterally expanded, and may be confluent across the mid-line with its equivalent on the other side. The anterodorsal surface is concave and conforms to the shape of the adjacent retracted hemipenial lobe, while the ventral face is closely applied to the skin. This structure is attached to the pars iliocaudalis laterally and posteriorly.

(xv) Cordylidae

The protractor commissurae may be lengthened in *Cordylus* and *Tetradactylus*, running forwards across the transversus perinei to terminate on connective tissue where this muscle inserts on the ilio-ischial ligament. The transversus penis is absent but fibres of the pars iliocaudalis may pass ventral to the hemipenis and pars ischiocaudalis to reach the mid-line (for instance in *Angolosaurus*, *Cordylus*, *Gerrhosaurus* and *Tetradactylus*). The retractores laterales are closely applied over much of their length, although they are often separable. They lie close to and parallel with the retractor penis magnus, which runs obliquely outwards and backwards to its origin, and may be bound to it by a fascia of varying strength. The retractor lateralis anterior inserts fleshily in front of the vent away from the mid-line and the retractor lateralis posterior attaches to the very base of the hemipenis, often with insertions above and below it.

(xvi) Scincidae

The hypoischium is absent or strongly reduced. In *Scincus* and *Eumeces algeriensis* the transversus perinei is simple with more or less transverse fibres, although posteriorly it becomes rather thin. The protractor commissurae originates behind the vent and runs forwards superficial to the transversus perinei to insert mainly on the skin below this muscle with a few fibres extending beyond it. In *Mabuya*, the muscles anterior to the vent are more complex. The superficial fibres in the posterior region of the transversus perinei run backwards from an aponeurosis and there is a thin layer of muscle arising from the lateroposterior border of the abdominal musculature and spreading backwards and inwards to insert on the skin superficial to the transversus perinei. Another thin layer of muscle originating from the corner of the vent passes forwards and inwards to insert in the same region, while the protractor commissurae runs to the ilio-ischial ligament. The arrangement in *Dasia* is rather similar, although the superficial posterior fibres of the transversus perinei do not run abruptly backwards. This area is also complex in *Leiolopisma* in which the hemipenis is surrounded by fascia. In all these forms there is only a single retractor lateralis mass which largely occupies the position of the retractor lateralis anterior: most fibres insert in front of the vent or to its lateral anterior wall; few if any fibres go to the base of the hemipenis. *Acontias* is generally similar to

Scincus or *Eumeces* but a substantial proportion of the retractor lateralis fibres insert on the hemipenis. The transversus penis is absent in all the skinks examined.

Amphisbaenians, Snakes and *Sphenodon*

(i) Amphisbaenians

In *Monopeltis* and *Diplometopon* the hypoischium is absent but muscles approach the generalized squamate condition: the transversus perinei is simple, the protractor commissurae arises behind the vent and runs to a ligamentous area posterior to the pelvic vestige: the transversus penis is thick and generally well developed and runs to the ventral mid-line area in *Monopeltis* but sweeps upwards onto the vertebrae in *Diplometopon*. The retractor lateralis anterior arises dorsal to the hemipenis and extends backwards as far as the area of origin of the retractor penis magnus. The retractor lateralis posterior arises laterally, close to the anterior origin of the retractor lateralis anterior. It inserts on the base of the hemipenis and also joins with the retractor lateralis anterior which then runs forwards to insert on the front wall of the vent.

(ii) Snakes

Snakes have not been surveyed in this study but, in some Caenophidia at least, there is a well developed transversus penis and two retractor muscles that are probably equivalent to the retractores laterales of other squamates. The one frequently called the retractor penis parvus is very similar to the retractor lateralis posterior. Like this it has its origin in the region of the horizontal longitudinal septum and inserts anteriorly on the proximal part of the retracted hemipenis: as in many elongate saurians, it lies largely above this organ. As described by Beuchelt (1936), the retractor penis basalis is attached to the very base of the hemipenis and runs to the ventral abdominal wall in front of the vent. In this it differs from the usual lizard condition of the retractor lateralis anterior where the muscle originates in the area of the horizontal longitudinal septum and does not involve the hemipenis at all. However, in some lizards (many Iguania), the retractor lateralis anterior is shortened and not connected to its usual site of origin and in some others (a few iguanines and lacertids for instance) a proportion of its fibres do insert on the hemipenial base. So, it is not impossible that the situation in snakes has arisen from the condition common in lizards.

(iii) *Sphenodon* (Fig. 8)

Contrary to Romer (1956) but as shown by Howes & Swinnerton (1901), a calcified hypoischium is present in *Sphenodon*. On each side,

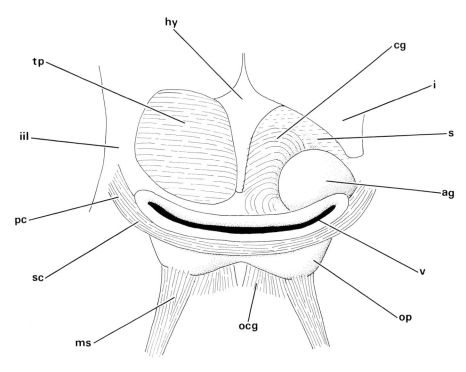

FIG. 8. Ventral view of cloacal region of *Sphenodon*. The transversus perinei has been removed and the deeper structures shown on the right. ag, 'anal' or scent gland; cg, compressor glandulae analis attaching to the hypoischium; hy, hypoischium; i, ischium; iil, ilio-ischial ligament; ms, muscle slip attached to posterior out-pouching of cloaca; ocg, posterior origin of compressor glandulae analis; op, posterior out-pouching of cloaca; pc, protractor commissurae; s, septum confluent with compressor glandulae analis; sc, sphincter cloacae; tp, transversus perinei; v, vent.

situated dorsal and slightly anterior to the lateral part of the vent, is a large globular 'anal' gland (Günther, 1868) which appears to produce scent (Gadow, 1886) and opens into the corner of the vent. No hemipenis is present but there is a distinct posterior out-pouching of the lower posterior wall of the cloaca on each side, just behind the scent gland, and dorsally its wall is smoothly continuous with that of this structure. The transversus perinei is simple, with fibres that run more or less laterally from the hypoischium to the ilio-ischial ligament with minor attachment to the ischium and the anterior wall of the vent. Sphincter cloacae fibres are not apparent in front of the vent but are well developed posteriorly, and laterally are confluent with the protractor commissurae which attaches on the ilio-ischial ligament behind the transversus perinei.

Muscle fibres arise posterior to the vent and sphincter cloacae and one relatively long slip, originating on the ventral skin of the tail base, runs inwards and forwards to insert on the upper hind surface of the posterior out-pouching of the wall of the cloaca. Fibres originating on the skin or associated fascia more anterior and more median to this slip run up and over the medial dorsum of the scent gland to curve downwards and inwards and insert on the edge of the hypoischium, along with other fibres arising on the mid-line of the posterior cloacal wall. This musculature is closely applied to the side of the posterior cloaca and is continuous laterally with a horizontal septum which runs from the side of the cloaca outwards beneath the anterior pars ischiocaudalis. Osawa (1898) called the muscle system the compressor glandulae analis. It has much in common with the retractor medialis of lizards in its area of origin and its association with the posterodorsal region of the cloaca, and together the right and left muscles may be homologous with the retractor medialis. They differ in not being confluent across the mid-line and in not converging forwards to insert via a tendon on the wall of the cloaca. Female *Sphenodon* are similar to males except that the scent gland and its associated musculature are smaller.

Without studies of living animals, it is uncertain how the scent gland, posterior out-pouching of the cloacal wall and their attached muscles function. One possibility is that, as the name given by Osawa (1898) suggests, the compressor glandulae analis and the septum confluent with it press down on the scent gland expelling its contents and perhaps everting its exit, so that it projects from the corner of the vent. If this were so, the posterior out-pouching may supply the necessary slack in the vent wall. Alternatively, or in addition, eversion of the lateral part of the vent might permit some shallow insertion into the cloaca of the opposite sex during copulation. The muscle slip attached to the posterior out-pouching could serve to retract this after eversion.

CLOACAL MUSCLES IN FEMALE LIZARDS

The cloacal muscles of female lizards have not been surveyed, but the few observations made suggest that their arrangement is often surprisingly like that found in males. There is frequently a narrow, posteriorly directed diverticulum at the corner of the vent in the same position as the hemipenis. Associated muscles originate from a thin fascia covering the caudifemoralis and extend medial to the pars iliocaudalis to the vicinity of the horizontal longitudinal septum. In some cases the muscles are short and arise just beneath the lateroventral skin of the tail base, but usually they originate more dorsally. In *Varanus griseus* the posterior diverticula are very well developed and, in one specimen (280mm snout – vent length, BM 1977.66) that was killed by its body being run over by a vehicle, they are about 7mm long and one is everted just like a hemipenis, although whether this occurs in untraumatized animals is unknown. In this species there are associated muscles corresponding to the retractor lateralis anterior, the retractor lateralis posterior (divided as in males) and the retractor penis magnus. Female *V. eremius* are similar. Among lacertids a diverticulum is frequent but the musculature is variable. There may be a retractor running to the diverticulum or the corner of the vent and a separate muscle corresponding to the retractor lateralis (e.g. in *Takydromus*, *Adolfus* and *Latastia*), or the latter may be absent (e.g. in *Acanthodactylus* and *Meroles*), or the two may form a single unit (e.g. in *Lacerta lepida*). In female *Callopistes* a diverticulum is present and there is only a single muscle corresponding to the retractores laterales, as in male teiids.

MUSCLE PATTERNS AND PHYLOGENY

Origin of the Hemipenis and its Muscles

The intromittent organs of squamates are singular among tetrapods in being paired and being turned inside out when not in use by a system of specialized retractor muscles. It is consequently of some interest to speculate about the origins of this system. Quite possibly, hemipenes have been derived from a condition similar to that found in *Sphenodon* (pp. 69–71), the apparent primitive sister taxon of the Squamata. Although its present function is uncertain, the paired posterior outpouching of the cloacal wall just behind the vent could be a hemipenial precursor. Like a hemipenis, each out-pouching is an apparently eversible diverticulum of the proctodeum (albeit a shallow one). It is also similar in having substantial amounts of cavernous tissue around its walls. In the same way, the slip of muscle inserting posteriorly on the out-pouching could represent an earlier, simpler stage in the evolution of the more elaborate retractor system found in squamates.

Polarity of Muscle Characters

What is likely to have been the primitive arrangement of cloacal and hemipenial muscles in squamates? Is the generalized condition close to this or are some of the minority muscle states listed in Table I involved? Three lines of evidence all point to the generalized condition being primitive. Firstly, *Sphenodon*, the apparent primitive sister taxon of the Squamata, is very close to the generalized condition in all the features that can be checked (hypoischium present, transversus perinei undifferentiated, greater part of sphincter cloacae situated behind vent, protractor commissurae simple, short and inserts entirely on the ilio-ischial ligament). The hemipenis and associated musculature are of course absent, so *Sphenodon* can give no indication of polarity for characters associated with these structures, Secondly, on average, widespread character states are more likely to be primitive than those which occur in only a few taxa (see for instance Arnold, 1981). From Table II it is clear that, although most lizard groups have one or more derived states, the majority of these are individually quite restricted in their distribution and, for nearly all characters, the state found in the generalized muscle arrangement is extremely widespread, being present in the great majority of lizard families. Finally, in cases where a reasonably convincing phylogeny of a group has been constructed using features other than cloacal and hemipenial muscles, the distribution of these states on the phylogeny often suggests that the generalized condition is likely to be primitive.

Distribution of Derived States and the Phylogeny of Lizards

If the states forming the generalized lizard condition are indeed primitive for the squamates, then the remaining minority states are likely to be derived conditions and, as such, form *prima facie* evidence that the forms sharing them are related. That is, that they have inherited the features concerned from a common ancestor not shared with taxa in which the features are absent. Whether such evidence is accepted depends of course on how well it agrees or conflicts with evidence from other character systems, for there is always the possibility that resemblance may be due not to common ancestry but to homoplasy (parallelism, convergence or reversal). In this section, the distribution of derived states of cloacal and hemipenial musculature will be compared with current hypotheses of squamate relationships based on other characters.

(i) Iguanidae

The restriction of lateral and medial accessory sheath muscles to the iguanines provides a little more evidence for the naturalness of this already well defined group. Because all combinations of two, two-state characters occur (no accessory muscles, lateral muscle only, medial muscle only, lateral and medial muscles both present), there must be some homoplasy in the evolution of these structures (Le Quesne, 1969). Their distribution should therefore be used with caution in any attempt to determine relationships within the family.

The holophyly of the sceloporines is supported by the occurrence in all of them of a completely divided retractor lateralis posterior, a state not known in other iguanids. This feature is present in *Phrynosoma*, a genus with imprecisely known relationships to the rest of the sceloporines (Etheridge, 1964) but now more certainly allocated to the group (R. E. Etheridge & K. De Queiroz, pers. comm). These workers place it close to the sand lizard assemblage (*Callisaurus*, *Cophosaurus*, *Holbrookia* and *Uma*) with which it shares reflection of the anterior fibres of the retractor lateralis anterior.

Among tropidurines, some derived features of the hemipenial musculature accord well with groupings tentatively proposed by R. E. Etheridge & K. De Queiroz (pers. comm.). Thus *Ctenoblepharis* and *Liolaemus* have a strongly developed fleshy insertion of the retractor lateralis posterior on to the sides of the hemipenis base and this is discernible, albeit weaker, in the apparently related *Phymaturus*. Members of the *Stenocercus- Ophryoessoides-Proctotretus* group are the only tropidurines known to have the transversus penis muscle spreading extensively into the medial and dorsal parts of the hemipenial sheath. *Tropidurus* shares a dorsal accessory sheath muscle with five other genera that are now believed to form a holophyletic group with it, although two of these were not clearly included in the tropidurines (*Strobilurus* and *Uranoscodon*; Etheridge, 1967).

Anisolepis and *Aptycholaemus*, two para-anoles which appear to be related on other grounds, share a distinctive, swollen insertion of the retractor lateralis anterior.

The deviations from the widespread iguanid condition encountered among anoles are not confined to either of the two main groups recognized by Etheridge (1959), the Alpha and Beta anoles. It is difficult to assess the significance of their distribution without more adequate sampling, but there are regularities. Thus, in the tentative scheme of relationships within the groups shown by Etheridge (1959: figs 10 & 11), a transversus penis extending right round the sheath is largely confined to the *cuvieri-bimaculatus-cristellatus* sequence of species series among the Alpha anoles, and to the *petersi-chrysolepis-nebuloides* sequence among the Betas. Fusion of the origins of the retractores laterales, on the other hand, occurs in most Alpha series except the *bimaculatus* and *cristellatus*, and in the *petersi* and *nebuloides* series of the Beta group. Enlargement of the insertion of the retractor lateralis anterior is found in these Beta anoles and also in the *latifrons-equestris-coelestinus-carolinensis* sequence among the Alphas.

(ii) Agamidae

As might be expected, the situation is reminiscent of that found in at least some Iguanidae, especially in such features as the presence of a hemipenial sheath and the usually abbreviated retractor lateralis anterior, although there is no general case for considering agamids to be especially closely related to the various, mainly tropidurine iguanids that possess the latter characteristic. The condition of the retractor lateralis posterior is distinctive in that much of it is usually enclosed in the hemipenial sheath and it inserts entirely on the lateroventral face of the base of the hemipenis, except in *Leiolepis*.

(iii) Chamaeleonidae
Although it has unique features, the general conformation of the cloacal and hemipenial region of chameleons is quite like that of other Iguania. The similarity of the retractor lateralis posterior to that of most agamids in its position of insertion tends to reinforce the case for the relationship of the two families, as does the inclusion of all or much of this muscle within the hemipenial sheath.

(iv) Gekkonidae and Pygopodidae
The extensive suite of derived features that they share adds weight to the view, raised by Boulenger (1885) and convincingly elaborated by Underwood (1957), that these two families are closely related.

(v) Anguidae, Anniellidae and Xenosauridae
The members of the Anguidae share an enlarged and complex protractor commissurae in which some fibres extend right across the transversus perinei. The rather more sophisticated arrangement found in gerrhonotines and anguines would support a closer relationship between these two sub-families than either have with the diploglossines. This would accord with the hypothesis of McDowell & Bogert (1954) of phylogeny within the family, but Rieppel (1980) argues that anguines and diploglossines are most closely related. *Anniella*, although given separate family status, is generally agreed to be clearly related to anguids and especially to anguines (McDowell & Bogert, 1954; Rieppel, 1980). The conformation of its protractor commissurae is in agreement with such a relationship. On the other hand, the authors cited suggest that xenosaurids arose from a stock similar to the gerrhonotine anguids and related to them, but the cloacal and hemipenial muscles of *Xenosaurus* lend no support to this view, for they exhibit the generalized lizard condition, which would fit better with their being the sister group of modern anguids as a whole plus *Anniella*.

(vi) Varanidae, Lanthanotidae and Helodermatidae
All the varanids examined have a characteristic combination of derived features in the hemipenial musculature but, as these are not shared with

the other two surviving platynotan families, which approach the generalized lizard condition, they provide no evidence of relationships among these groups.

(vii) Teiidae

Macroteiids share three synapomorphies in the cloacal and hemipenial musculature: the differentiated transversus perinei with a well defined posterior section, the reduced transversus penis and the presence of a single retractor lateralis of imprecisely known origin which inserts to the side of the vent. The ventral glands are also noticeably large, especially in the Teiini, although this feature has not been fully surveyed in other families. The two groupings within the macroteiids have characteristic derived features which support their reality: thus, Tupinambini have lost the transversus penis completely and the hemipenis is covered ventrally by fibres of the iliocaudalis, and Teiini have anterior fibres of the sphincter cloacae running into the posterior section of the retractor lateralis.

Microteiids, the Gymnophthalminae, have generally been regarded as most closely related to macroteiids but Presch (1983) suggests that they are not connected to the latter group and have most affinity with lacertids. As with most of the features that this author lists as characteristic of the Gymnophthalminae, there are no cloacal and hemipenial muscle characters that can be regarded as synapomorphies linking microteiids with the lacertids to the exclusion of the macroteiids. Instead, microteiids share derived features with at least some macroteiids including absence of the transversus penis, a single retractor lateralis of uncertain origin, anterior fibres of the sphincter cloacae usually running at least on to the tendinous insertion of this muscle and very large ventral glands.

(viii) Lacertidae

The presence of a well differentiated posterior section of the transversus perinei tends to support the often postulated relationship between lacertids and teiids. Within the lacertids, the absence of a hemipenial sheath only in the eastern Asian forms mainly assigned to *Takydromus* raises the possibility that the latter assemblage is sister group to the rest of the Lacertidae. The condition where the origin and insertion of the posterior section of the transversus perinei are close together and the fibres arched is best developed among genera which, on other grounds, may form a derived group within the family.

(ix) Xantusiidae

Xantusiids are often assigned to the Scincomorpha, for instance by

Camp (1923), but other workers, such as McDowell & Bogert (1954) suggest that they should be placed in the Gekkota. The absence of a hypoischium (shared with skinks) and of a transversus penis (shared with many teiids and with cordylids and skinks) could be seen as supporting the former hypothesis. But loss characters are weak indicators of possible relationships and absence of a hypoischium occurs elsewhere. The family does not possess the full suite of peculiarities that link geckoes with pygopods, but at least some xantusiids share certain derived features with a proportion of geckoes. Thus they all lack a hypoischium (absent in some geckoes), some fibres of the protractor commissurae may run deep to the transversus perinei to reach the ischium (as in some geckoes), there are often transverse muscle fibres running from the edge of the anterior pars iliocaudalis (present in many geckoes) and there may be a lateral post-cloacal bone (present in many geckoes). However, because none of these features is consistently present in both families, it is by no means certain that they have been inherited from a common ancestral form. Possibly more detailed information on their intrafamilial distribution would help form an opinion about this.

(x) Scincidae

Only a very small number of species in this large family were examined but the variation encountered suggests that differences in cloacal and hemipenial musculature could reflect relationship. Thus the complex arrangement of muscles anterior to the vent described on p. 68 was only encountered in lygosomines and not in more primitive forms like *Eumeces* and *Scincus*.

(xi) Amphisbaenians

As already noted, the cloacal and hemipenial muscles of amphisbaenians are quite similar to the generalized squamate condition and what derived features there are do not suggest credible relationships with any other group. For instance, a retractor lateralis posterior arising anterolateral to the origin of the retractor lateralis anterior has been found elsewhere only in the iguanid *Polychrus* and the Diploglossinae. Certainly there is nothing that indicates a relationship with all or part of the Teiidae, as suggested by Böhme (1981).

(xii) Snakes

As with the Amphisbaenia, there are no derived features that seem to suggest credible relationships with any other particular squamate group.

Apparent Homoplasies

In the previous section, emphasis has been laid on patterns of shared derived character states that conform to currently held hypotheses of relationship between lizard taxa. However, the distribution of a number of features does not entirely correspond to such accepted, or at least proposed, groupings. Among these are numbers 1, 19, 20, 22, 23, 24, 27, 31 and 32 in Table I. It seems probable that, in these cases, some resemblances are homoplasies arising from independent development of the feature concerned in different groups. In most instances, it is not possible at present to suggest factors that might have caused, or facilitated, such parallel or convergent development, but occasionally some speculation can be made. Thus, development of a hemipenial sheath (19), which has apparently occurred separately in the Iguania, varanids, lacertids and a minority of geckoes and skinks, may be a relatively simple evolutionary event since most lizards have considerable amounts of diffuse connective tissue around the retracted hemipenis which could presumably be quite readily elaborated into a distinct fascia. Origin of at least one of the retractores laterales dorsal to the hemipenis and associated retractor penis magnus (24) occurs in pygopods, anguines and *Anniella*, *Ophiodes*, lacertids, amphisbaenians and snakes. As the great majority of these forms are elongate and virtually limbless, it is probable that this habitus is a predisposing factor in the development of the condition. Another case involves development of an extended protractor commissurae of which all or part does not attach to the ilio-ischial ligament but instead runs superficial to the transversus perinei to reach the edge of the abdominal musculature or attach to the skin (7–9). This feature is found in anguids (and the related *Anniella*) and in skinks, both groups in which well developed osteoderms are usual. It could be argued that migration of the attachment of all or part of the protractor commissurae from the ilio-ischial ligament is only possible if there is a firm continuous surface along which it can move to successively more forward positions, so that it remains functional at all times. The stiff, osteoderm reinforced skin of the pre-anal region would provide this, filling the same role as the exoskeleton of many invertebrates to which muscles attach. The presence of a virtual exoskeleton in skinks may also have allowed the elaboration of superficial muscle layers in the pre-sacral region found in lygosomines.

CONCLUSION

As with the hemipenes themselves, variation in the cloacal and hemipenial muscles of lizards often seems to reflect the relationships of the forms concerned. Patterns of shared derived features frequently conform to hypotheses of phylogeny based on other sources of inference. The average rate of evolutionary change in this organ system appears to have been appropriate for providing information on relationships, being not so slow that the development of derived features is very rare nor so fast that members of higher taxa never share characteristic conditions inherited from the ancestor of the group concerned. Again, the wide variety of changes exhibited by the system means there is a greater possibility of these being unique, or at least rarely duplicated, so they are more likely to provide adequate markers of groups. Their performance as indicators of relationship in the present survey suggests that cloacal and hemipenial muscles merit further attention in taxonomic and phylogenetic studies.

Possible reasons why genital structure often provides useful information about relationships (see for instance Arnold, 1973, 1983) may also apply in part to the associated musculature. More information about the functional aspects of this region in squamates would be valuable, both for its inherent interest and in assessing polarity and the relative weight that should be given to characters in phylogeny reconstruction (Arnold, 1981). However, the field is difficult to investigate: reptile copulation is usually hard to observe in detail and invasive techniques, such as electromyography, seem more likely to interefere with normal function here than is the case with more regular activities like ventilation and ingestion.

ACKNOWLEDGEMENTS

I am most grateful to Professor R. E. Etheridge and Mr K. De Queiroz for generously allowing me to quote their as yet unpublished views on iguanid relationships.

REFERENCES

Alton, E. d' (1834). Beschreibung des Muskelsystems eines *Python bivittatus*. Von den Muskeln des Schwanzes und der Beckengegend. *Arch. Anat. Physiol.* **1834**: 528–541.
Arnold, E. N. (1973). The relationships of the Palaearctic lizards assigned to the genera

Lacerta, Algyroides and *Psammodromus* (Reptilia: Lacertidae). *Bull. Br. Mus. nat. Hist.* (Zool.) **25**: 289–366.
Arnold, E. N. (1981). Estimating phylogenies at low taxonomic levels. *Z. zool. Syst. Evolut.-Forsch.* **19**: 1–35.
Arnold, E. N. (1983). Osteology, genitalia and the relationships of *Acanthodactylus* (Reptilia: Lacertidae). *Bull. Br. Mus. nat. Hist.* (Zool.) **44**: 291–339.
Beuchelt, H. (1936). Bau, Funktion und Entwicklung der Begattungsorgane der männlichen Ringelnatter (*Natrix natrix* L.) und Kreuzotter (*Vipera berus* L.). *Gegenbaurs morph. Jb.* **78**: 445–516.
Böhme, W. (1981). *Handbuch der Reptilien und Amphibien Europas,* **1**. Wiesbaden: Akademische Verlagsgesellschaft.
Boulenger, G. A. (1885). *Catalogue of the lizards in the British Museum (Natural History),* (2nd Edn) **1**. London: British Museum (Natural History).
Brongersma, L. D. (1934). Contributions to Indo-Australian herpetology. *Zool. Meded., Leiden* **17**: 161–251.
Camp, C. L. (1923). Classification of the lizards. *Bull. Am. Mus. nat. Hist.* **48**: 289–481.
Disselhorst, R. (1904). Mannliche Geschlechtsorgane. Reptilien. In *Lehrbuch der vergleichenden mikroskopischen Anatomie der Wirbeltiere,* **4**: 60–89. Oppel, P. (Ed.). Jena.
Dowling, H. G. & Savage, J. M. (1960). A guide to the snake hemipenis: a survey of basic structure and systematic characteristics. *Zoologica, N. Y.* **45**: 17–28.
Etheridge, R. E. (1959). *The relationships of the anoles (Reptilia: Sauria: Iguanidae), an interpretation based on skeletal morphology.* Ph. D. Thesis, University of Michigan, USA.
Etheridge, R. E. (1964). The skeletal morphology and systematic relationships of sceloporine lizards. *Copeia* **1964**: 610–631.
Etheridge, R. E. (1967). Lizard caudal vertebrae. *Copeia* **1967**: 699–721.
Gabé, M. & Saint-Girons, H. (1965). Contribution à la morphologie comparée du cloaque et des glands épidermoides de la région cloacale chez les lépidosauriens. *Mém. Mus. natn. Hist. nat., Paris* (A) **33**: 150–292.
Gadow, H. (1882). Beiträge zur Myologie der hinteren Extremität der Reptilien. *Morph. Jb.* **7**: 329–466.
Gadow, H. (1886). Remarks on the cloaca and on the copulatory organs of the Amniota. *Phil. Trans. R. Soc.* (B) **178**: 5–35.
Günther, A. (1868). Contribution to the anatomy of *Hatteria* (*Rhynchocephalus*, Owen). *Phil. Trans. R. Soc.* **157**: 595–629.
Hoffman, C. K. (1890). Reptilien. *Bronn's Kl. Ordn. Tierreichs* **6**(3).
Howes, G. B. & Swinnerton, H. H. (1901). On the development of the skeleton of the Tuatara, *Sphenodon punctatus*: with remarks on the egg, on the hatching, and on the hatched young. *Trans. zool. Soc. Lond.* **16**: 1–86.
Kluge, A. G. (1967). Higher taxonomic categories of gekkonid lizards and their evolution. *Bull. Am. Mus. nat. Hist.* **135**: 1–59.
Kluge, A. G. (1982). Cloacal bones and sacs as evidence of gekkonid lizard relationships. *Herpetologica* **38**: 348–355.
Le Quesne, W. J. (1969). A method of selection of characters in numerical taxonomy. *Syst. Zool.* **18**: 201–205.
Lereboullet, A. (1851). Recherches sur l'anatomie des organes génitaux des animaux vertébrés. *Nova Acta Acad. Caesar. Leop. Carol.* **23**: 1–228.
McDowell, S. B. & Bogert, C. M. (1954). The systematic position of *Lanthanotus* and the affinities of the anguinomorphan lizards. *Bull. Am. Mus. nat. Hist.* **105**: 1–142.
Osawa, G. (1898). Beiträge zur Anatomie der *Hatteria punctata*. *Arch. mikrosk. Anat.* **51**: 481–691.

Presch, W. (1983). The lizard family Teiidae: is it a monophyletic group?. *Zool. J. Linn. Soc.* **77:** 189–197.
Rieppel, O. (1976). On the presence and function of postcloacal bones in the Lacertilia. *Monitore zool. ital.* **10:** 7–13.
Rieppel, O. (1980). The phylogeny of anguinomorph lizards. *Denkschr. schweiz. naturf. Ges.* **94:** 1–86.
Romer, A. S. (1956). *Osteology of the reptiles.* Chicago: University of Chicago Press.
Sanders, A. (1874). Notes on the myology of the *Phrynosoma coronatum. Proc. zool. Soc. Lond.* **1874:** 71–89.
Savage, J. M. (1957). Studies on the lizard family Xantusiidae. III. A new genus for *Xantusia riversiana* Cope, 1883. *Zoologica, N. Y.* **42:** 83–86.
Underwood, G. (1957). On lizards of the family Pygopodidae. A contribution to the morphology and phylogeny of the Squamata. *J. Morph.* **100:** 207–268.
Unterhössel, P. (1902). Morphologische Studien über Kloake und Phallus der Amnioten: die Eidechsen und Schlangen. *Gegenbaurs morph. Jb.* **30:** 541–580.
Volsøe, H. (1944). Structure and seasonal variation of the male reproductive organs of *Vipera berus* (L.). *Spolia zool. Mus. haun.* **5:** 1–157.
Wöpke, K. (1930). Die Kloake und die Begattungsorgane der männlichen Zauneidechse (*Lacerta agilis* L.). *Jena, Z. Naturw.* **65:** 275–318.

APPENDIX I. MATERIAL EXAMINED

If more than one male example of a species has been examined, the number is given in parentheses. All material is in the collection of the British Museum (Natural History).

Iguanidae. Iguanines: *Amblyrhynchus cristatus, Brachylophus fasciatus, B. vitiensis, Conolophus subcristatus, Ctenosaura similis, Cyclura nubila, Dipsosaurus dorsalis, Iguana iguana* (2), *Sauromalus ater.* Basiliscines: *Basiliscus vittatus, Corythophanes percarinatus, Laemanctus serratus.* Crotaphytines: *Gambelia wislizeni.* Sceloporines: *Callisaurus draconoides, Cophosaurus texanus, Holbrookia maculata, Petrosaurus thalassinus, Phrynosoma orbiculare, Sceloporus torquatus, Uma notata, Uta stansburiana.* Tropidurines: *Ctenoblepharis adspersus, Leiocephalus varius, Liolaemus multiformis* (2), *Ophryoessoides trachycephalus, Phymaturus pallumus, Plica umbra* (2), *Stenocercus roseiventris, S. simonsii, S. varius, Strobilurus torquatus, Tropidurus peruvianus, T. torquatus, Uracentron flaviceps, Uranoscodon superciliosus.* Oplurines: *Chalaradon madagascariensis, Oplurus cuvieri.* Morunosaurs: *Enyalioides heterolepis, Hoplocercus spinosus.* Anoloids: *Diplolaemus bibronii, Enyalius iheringii, Polychrus marmoratus.* Para-anoles: *Anisolepis grillii, Aptycholaemus longicaudus, Urostrophus torquatus.* Anoles: *Anolis biporcatus, A. bimaculatus, A. chlorocyanus, A. chrysolepis, A. coelistinus, A. cristellatus, A. cuvieri, A. cybotes, A. equestris, A. garmani, A. lemurinus, A. nebulosus, A. richardi, A. sagrei, Chamaelinorops wetmorei.*

Agamidae. *Agama flavimaculata, A. yemenensis, Amphibolurus muricatus, Calotes versicolor, Chlamydosaurus kingi, Gonocephalus modestus, Leiolepis belliana* (2), *Physignathus lesueurii, Uromastyx hardwicki, U. macfadyeni.*

Chamaeleonidae. *Brookesia brevicaudata, Chamaeleo africanus, C. dilepis, C. johnstonii.*

Gekkonidae. Eublepharines: *Coleonyx elegans, Eublepharis hardwickii, Hemitheconyx taylori.* Diplodactylines: *Hoplodactylus pacificus, Oedura marmorata, Rhacodactylus trachyrhynchus.* Gekkonines: *Cyrtodactylus louisiadensis, Gecko gekko, G. vittatus* (5), *Pristurus carteri, Ptyodactylus hasselquistii, Teratoscincus scincus.*

Pygopodidae. *Pygopus lepidopus.*

Xenosauridae. *Xenosaurus grandis.*

Anguidae. Gerrhonotines: *Abronia deppii, Gerrhonotus coeruleus* (2). Anguines: *Anguis fragilis, Ophisaurus gracilis.* Diploglossines: *Diploglossus monotropis, D. stenurus, Ophiodes intermedius.*

Anniellidae. *Anniella pulchra* (2).

Helodermatidae. *Heloderma horridum.*

Lanthanotidae. *Lanthanotus borneensis.*

Varanidae. *Varanus bengalensis, V. dumerilii, V. eremius, V. exanthematicus, V. gilleni, V. gouldii, V. griseus* (2).

Teiidae. Tupinambini: *Callopistes maculatus* (2), *Crocodilurus lacertinus, Tupinambis nigropunctatus* (2). Teiini: *Ameiva ameiva* (2), *Dicrodon guttulatum* (2), *Kentropyx calcaratus* (2), *Teius teyou* (2). Gymnophthalminae: *Anadia brevifrontalis, Aspidolaemus affinis, Cercosaura ocellata, Echinosaura horrida, Ecpleopus montius, Euspondylus maculatus, Gymnophthalmus pleii, Neusticurus bicarinatus, N. strangulatus, Prionodactylus manicatus, Proctoporus unicolor.*

Lacertidae. 40 species covering all recognized genera.

Xantusiidae. *Lepidophyma flavimaculatum* (2), *L. tuxtlae, Xenosaurus henshawi, X. vigilis* (3).

Cordylidae. *Angolosaurus skoogi, Cordylus cordylus, Gerrhosaurus flavigularis, G. nigrolineatus, Platysaurus intermedius, P. torquatus, Pseudocordylus subviridis* (2), *Tetradactylus africanus, Zonosaurus madagascariensis.*

Scincidae. *Acontias plumbeus, Dasia smaragdina, Eumeces algeriensis, Leiolopisma grande, Mabuya brevicollis, M. wrighti, Scincus scincus, S. mitranus* (2).

Amphisbaenians. Amphisbaenidae: *Monopeltis capensis* (2). Trogonophidae: *Diplometopon zarudnyi.*

Rhynchocephalians. *Sphenodon punctatus* (3).

APPENDIX II. SYNONYMY OF NAMES APPLIED TO THE MUSCLES OF THE CLOACAL AND HEMIPENIAL REGION OF SQUAMATES

Names used by other authors are listed under the ones used here. Abbreviations for the workers concerned, and the animals they describe, are listed below.

L	Lereboullet, 1851 (*Lacerta stirpium* = *L. agilis*)
S	Sanders, 1874 (*Phrynosoma coronatum*)
G	Gadow, 1882 (various lizards)
H	Hoffman, 1890 (snakes)
U	Unterhössel, 1902 (*Natrix natrix*)
W	Wöpke, 1930 (*Lacerta agilis*)
B	Beuchelt, 1936 (*Natrix natrix*)
V	Volsøe, 1944 (*Vipera berus*)
D & S	Dowling & Savage, 1960 (snakes)

Transversus perinei – Anterior Section

L	Premier releveur de la lèvre antérieure
S	Transversus perinei (part)
G	Transversus perinei
W	Relevator cloacalis anterior

Transversus perinei – Posterior Section

l	Second releveur de la lèvre antérieure
S	Transversus perinei (part)
G	5, ε
W	Relevator cloacalis posterior

Sphincter cloacae – Anterior and Posterior Sections

L	Constricteur des lèvres, sphincter des lèvres
S	Sphincter cloacae (posterior section only)
G	Sphincter cloacae (part), 2 (part), β (part)
H	Sphincter cloacae
W	Sphincter cloacalis

Protractor commissurae

L	Releveur de l'angle des lèvres

S	Sphincter cloacae (part)
G	Sphincter cloacae (part), 2 (part), β (part)
W	Relevator commissurae labii cloacalis

Transversus penis

L	Fourreau musculeux, muscle vaginien
H	Transversus penis
U	Bogenmuskeln, Bogenförmig verlaufenden muskeln
W	Vaginalis penis
B	Propulsoren
V	Propulsor muscle
D & S	Propulsor muscle

Retractor lateralis anterior

L	Releveur médian (de la lèvre anterieure)
S	Constrictor cloacae
G	4 (part), δ (part)
W	Relevator medialis cloacalis
B	Protractor penis basalis
V	Protractor penis basalis
D & S	Protractor penis basalis

Retractor lateralis posterior

L	Rétracteur lateral
S	Retractor cloacae
G	4 (part), δ (part)
U	Retractor phalli basalis
W	Retractor lateralis cloacalis
B	Retractor penis parvus
V	Retractor penis parvus
D & S	Retractor penis parvus

Retractor medialis (cloacae)

L	Rétracteur médian
W	Retractor medialis

Retractor penis magnus

L	Retracteur de la verge

G	3, γ
H	Retractor penis
U	Retractor phalli magnus
W	Retractor penis
B	Retractor penis magnus
V	Large retractor muscle
D & S	Retractor penis magnus

Symp. zool. Soc. Lond. (1984) No. 52, 87–110

Piscivory in Turtles, and Evolution of the Long-necked Chelidae

PETER C. H. PRITCHARD

Florida Audubon Society, 1101 Audubon Way,
Maitland, Florida 32751, USA

SYNOPSIS

Three different prey capture techniques utilized by turtles that feed upon live fish are compared with the human fishing techniques of "angling or luring", "netting or trawling", and "spear-fishing or harpooning". A single species, *Macroclemys temminckii*, uses the first technique and shows numerous behavioral and morphological specializations in addition to the unique lingual "lure". *Chelus fimbriatus* is assigned to the second group; its specializations and post-cranial parallels with *Macroclemys* are described. The third group includes members of several families – the emydid *Deirochelys*; the chelydrid *Chelydra*; the trionychid *Chitra*; the chelids *Chelodina* and *Hydromedusa*; and probably the Mesozoic turtle genera *Glyptops* and *Chitracephalus*. These turtles show common specializations in their elongate necks; narrow, elongate heads and jaws; enlarged hyoids; anteriorly-placed orbits; and, in the cryptodiran examples, specialized rib-heads to accommodate enlarged longissimus dorsi muscles, and plastra or bridges that can accommodate rapid visceral displacement, either by plastral reduction or kinesis, or by an increase in bridge depth.

The relationships of the two snake-necked chelid genera are analysed. The similarities in head and neck structure are considered to be the result of parallel evolution rather than phylogenetic relationship. *Hydromedusa*, on the basis of skull and shell structure, is considered closer to *Chelus* than to *Chelodina*. It is postulated that *Hydromedusa* and *Chelus* share a common ancestor and diverged morphologically as they became specialized for feeding upon live fish in clear and turbid water habitats, respectively. It is argued from analysis of the different patterns of anterior plastral and carapacial elongation that the neck elongation may have evolved independently in the three genera.

The progressive loss of neural bones in the chelids and certain other turtles is discussed; loss of neurals is correlated with regions of the carapace subject to asymmetrical stresses during feeding and locomotion.

INTRODUCTION

The evolution of the rounded, armored body form by the Testudines placed some major constraints upon feeding options. While other living reptiles – the lizards, snakes, and crocodilians – feed largely upon living and often quite agile prey, this has not been possible for most chelonians. Turtles and tortoises are not always slow, but they are rarely agile, and there are few creatures that could be expected to stay in one place while a tortoise ambled up and started biting pieces out of them.

Herbivory has been one option, and although an unusual one for reptiles (requiring profound modification of the digestive system) has been assumed by numerous turtles, including virtually all the terrestrial forms, many of the freshwater ones, and one major lineage of marine species. Other turtles have become omnivores, their food ranging from grass and fruits to carrion. A recurrent feeding strategy for turtles has also been durophagy; by developing powerful jaw muscles and specialized crushing surfaces on the jaws, turtles of several unrelated families subsist upon armored invertebrates even slower than themselves. Certain other turtles have developed surprising feeding niches. The giant leatherback turtle, for example, is highly specialized for feeding upon scyphomedusans, and certain South American pelomedusids, notably the more lentic members of the genus *Podocnemis*, have evolved means of filter feeding from the surface scum layer (Rhodin, Medem & Mittermeier, 1981).

Nevertheless, turtles have not been entirely unresponsive to the challenge of feeding upon agile prey, notably live fish. In fact, they have evolved techniques that parallel all three of the basic techniques that humans use for catching fish – "angling", "netting" and "spear-fishing".

ANGLING

The essence of "angling" is that the predator lies in ambush, attracts prey with a lure of some kind, and has a mechanism for capture of the prey once the lure is taken. Among turtles, only the alligator snapping turtle, *Macroclemys temminckii*, is a true angler, and it is largely the juveniles that practise this technique. The behavioural and anatomical adaptations shown by this species to make angling more effective are far-reaching. The "lure" itself is a bifurcate structure attached to the floor of the mouth, which can be manipulated directly by contraction of the musculus hypoglossoglossus and indirectly by a total of 10 paired muscles in the hyoid region (Spindel, 1980). Other physical modifications shown by this species include the unique pigmentation of the inside of the mouth, which is basically dark grey-brown with black spots, a highly procryptic pattern against which the pink lure and the white glottis are very conspicuous. The eyes of *Macroclemys* are lateral, presumably to enhance wide-angle rather than binocular vision, the neck is much shorter than that of the related *Chelydra*, the front part of the head, especially in juveniles, is distinctly narrowed, and the symphyseal parts of the jaws are strongly hooked. The anterior cephalic modifications presumably serve to facilitate rapid closure of the jaws and to impale fish that otherwise might escape, or which might be too large or strong to hold with unhooked jaws.

Behavioral adaptations of *Macroclemys* for "angling" are numerous. The young turtles are extremely sluggish, and have the habit of holding the mouth wide open and deliberately twitching the lure. Recent studies by Drummond & Gordon (1979) have also demonstrated that even neonates will angle with a great deal of apparent guile, usually opening the mouth only when fish are seen, carefully altering their position so that potential prey is presented with the clearest possible view of the "lure", and refraining from making "snaps" until a successful catch is virtually certain. Even so, this is not a particularly productive method of food acquisition. Large numbers of young *Macroclemys* may be kept with schools of small fish with only occasional captures taking place. The growth rate of young *Macroclemys* is very slow; and the subadults and adults derive much or all of their nutrition from other sources. The list of those "other sources" is lengthy and varied; it includes clams, acorns, carrion, parts of beavers and raccoons, sluggish fish such as *Amia*, and on occasion small alligators, which may be consumed entire (Pritchard, 1982a). Cheloniophagy is also common, and in some areas *Sternotherus minor* must be considered a staple rather than an incidental dietary item. Correlated with this diverse and frequently durophagous adult diet, the skull shows ontogenetic modifications, notably in the jaws which become shorter, broader, and less hooked.

NETTING

Netting, or trawling, is based upon the deployment of a device that will envelop a quantity of water, and extract the prey within that water by a filtration process. It has advantages over angling in that it does not depend upon co-operation of the prey, and more than one individual of the prey species may be taken at once. On the other hand, for turtles it requires rather profound anatomical adaptations, and within the Order this technique is only really developed by the South American matamata turtle, *Chelus fimbriatus*.

Netting, like trawling, requires that the prey approach the predator closely, and the crypsis demands of the two strategies are comparable. Indeed, *Chelus* shows some remarkable parallels with *Macroclemys* in general appearance, especially in the carapace which is similarly ridged, flattened and roughened, and in the limbs which are rather weak and poorly adapted for swimming.

Nevertheless, the superficial parallels end from the neck up; the head of *Chelus*, being modified for a completely different mode of feeding, is totally unlike that of *Macroclemys* except for the common feature of highly innervated, fleshy fringes and tassels on both sides of the head and the throat. These have a similar function of detection of the movements of prey near the head when turbid water reduces the role of vision.

Chelus is a pleurodire and engulfs its prey by means of a lateral swipe of the head and neck. The head is extraordinarily flattened, presumably to minimize water resistance during the fast lateral movement. The eyes are extremely small, reflecting not only the minimal role of vision in the activities of this turtle, but also the vulnerability of a larger eye to accidental damage, for example from fish spines, during the ingestion stroke, and probably also the difficulty of accommodating a large orbit in a skull whose crown-to-palate thickness is reduced to a few millimeters.

The rear of the skull of *Chelus*, on the other hand, is remarkable for its extreme width and the enormous size of the tympana, each tympanum being at least twice the diameter of the orbit. These modifications surely point to a major role of the auditory sense in prey acquisition, and the widely-spaced, highly sensitive tympana presumably not only detect the water movements caused by fish near the head but also give a good directional fix. Ingestion is facilitated by the enormous hyoid structure, whose extensive musculature has been well described by Poglayen-Neuwall (1966). The mouth is very wide, but the jaws are weak and serve purely as a valve, having no biting function.

SPEAR-FISHING

General

The third mode of chelonian piscivory depends upon great elongation of the neck to allow prey to be seized with a faster thrust than would be possible if the entire body had to be moved – a parallel to spear-fishing or harpooning. This can be done either when the turtle is standing on the bottom, or when it is in the water column, when the body of the turtle is stabilized by its own inertia during the "strike". Again drastic anatomical modifications are required. These include elongation of the head and neck to enhance the "reach"; enlargement of the mouth and hyoid, to promote effective prey seizure and ingestion; narrowing of the head for streamlining purposes; movement of the orbits to the front of the head to facilitate obtaining a binocular fix upon the prey before it is seized; enlargement of the musculi longissimi dorsi, responsible for the neck-thrusting movement; corresponding enlargement of the space for these muscles between the rib-heads and the pleural bones; and, in cryptodires, plastral or bridge modifications to accommodate the massive instantaneous changes in visceral volume as the strike is made. This mode of feeding may have occurred very early among certain turtles; all the necessary skull modifications are shown by *Glyptops plicatulus*, from the Jurassic of Wyoming.

The common snapping turtle, *Chelydra serpentina*, is the most familiar example of this specialization, and indeed it shows several of the above anatomical adaptations, including elongation of the neck; enlargement of the longissimus dorsi muscles with concomitant expansion of the area under the rib-heads; and reduction of the plastron to a cruciform condition to accommodate changes in visceral volume. However, *Chelydra* is not a specialized piscivore, but feeds upon a wide variety of animal and plant foods (Ernst & Barbour, 1972), and the famous "strike" is used in defense and aggression as well as in feeding.

Better examples among cryptodires include the emydid *Deirochelys*, which shows all the specializations mentioned above – orbital enlargement and anterior migration (Fig. 1); head narrowing; neck elongation;

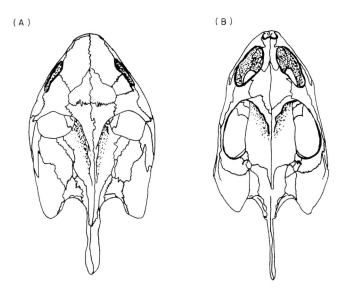

FIG. 1. Skull of a 'typical' emydid. *Chinemys reevesi* (A) compared with the skull of a piscivorous emydid, *Deirochelys reticularia* (B), in which the orbits have migrated anteriorly and the skull as a whole has assumed a streamlined form.

rib-head specialization; and a deepening of the bridge which increases the area of the inguinal pockets and thus facilitates accommodation of rapid visceral volume change. *Deirochelys* is in fact a crayfish-feeder rather than a piscivore, but the adaptations for seizing these similarly agile aquatic organisms are similar.

Among other cryptodires, the Asiatic trionychid *Chitra indica* is the best example of a specialized fish-grabber. This huge softshell, known to reach at least 108 kg (Nutaphand, 1979) and reputed to reach a carapace length of 6ft or 183 cm (Annandale & Shastri, 1914), shows striking cephalic specialization in the extreme narrowness of the skull and the location of the orbits virtually at the tip of the snout (Fig. 2). As with all trionychids, the neck is greatly elongated and the joint between the eighth cervical and the first dorsal vertebra is specialized so that these vertebrae can actually flex through a 180° angle and lie belly-to-belly when the neck is retracted. Visceral displacement in this and other softshells is accommodated by the highly flexible plastron. The habits of *Chitra* are very poorly known, but Smith (1931) reports that it will "suddenly shoot out its long neck with inconceivable rapidity and is

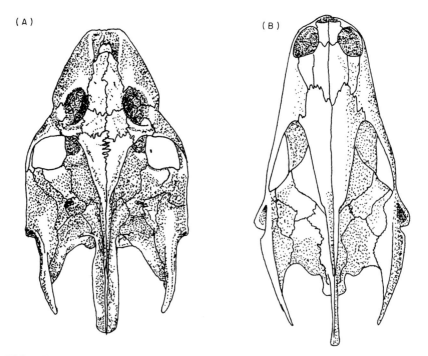

FIG. 2. Skull of a 'typical' trionychid, *Trionyx triunguis* (A) compared with the skull of a piscivorous species, *Chitra indica* (B), in which the orbits have migrated anteriorly and the skull as a whole has assumed a streamlined form.

capable of giving a very severe bite". Ms J. Vijaya of the Madras Snake Park Trust writes as follows (pers. comm.):

"*Chitra indica* seems very much a fish eater. We have a juvenile at the Crocodile Bank who keeps himself buried in sand most of the time. Just the tip of his snout and his eyes are exposed so it is almost impossible to find him. Just as a fish passes overhead he very quietly puts out his head and gulps it. Dr Moll once observing it said that while grabbing a particularly large fish it almost leapt out of its hiding place, throwing a fountain of sand while grabbing the fish".

Origin of the Trionychidae

The origins of the Trionychidae, of which *Chitra* is a member, remain somewhat mysterious. Many of the modern species, especially of the large genus *Trionyx*, are omnivorous, and it is hard to correlate such specialized body form with generalized feeding habits. However, it makes some sense if we consider the trionychids to have evolved originally as seizers of live fish. The extremely long, highly retractile neck accords with this role, and the shell accommodates the abrupt visceral displacement associated with rapid neck extension, partially by having a plastron that has considerable flexibility both along the mid-line and in the movable front lobe, and partially by loss of the peripheral bones and the horny scutes, which facilitates rapid volume changes for the shell as a whole. Unlike the other fish-seizing turtles, the trionychids seem to have become specialized for "chase and grab" rather than "stalk and grab" or even "lie in wait and grab", and the shell flatness and swimming prowess accord with this.

The unossified shell margin of the trionychids seems to be a liability that must have a compensating advantage. This advantage may lie partially in the facilitation of shuffling into mud or sand for crypsis, but it may also reflect the need to preserve both the long neck and extensive "reach" with a shell that does not have an anterior overhang and other impediments to streamlining. If the whole animal can become slightly flatter when the neck is suddenly fully extended, other means of accommodating visceral displacement become irrelevant. The vulnerability presented by the loss of the body armor is compensated by: (i) the very large body size of many species; (ii) retention of thick bone in the central parts of both carapace and plastron; (iii) aggressive disposition; (iv) extremely fast swimming ability; (v) development of secondary bony callosities in the shell surface; and (vi) completely retractile head and neck.

An interesting result of this hypothesis is that the small, humped, pond-dwelling *Lissemys* becomes the most advanced trionychid, rather than the most primitive as has been argued by Deraniyagala (1939) and others; the small bones in the shell margin of this genus would thus appear to be neomorphs rather than rudimentary peripherals. On the other hand, *Chitra*, normally considered to be highly advanced, becomes the most primitive softshell, retaining the fish-seizing mode of feeding as most trionychids became secondarily more generalized in diet (some even becoming durophagous). While this may seem a fanciful and unsubstantiated argument, it is noteworthy that the strange *Chitra* skull-form was uncannily anticipated by an extremely early turtle, the aptly-named *Chitracephalus dumonii*, from the Jurassic or Lower Cretaceous of Europe. While this genus has been attributed to the Mesozoic family Aperotemporalidae by Romer (1956), Gaffney (1979) examined the available material carefully and declared it to be *incertae sedis*, it not even being clear whether it was a cryptodire or a pleurodire. In the absence of reliable means of familial affiliation, one hypothesis is that *Chitracephalus* is an early trionychid, actually related to *Chitra*, and thus providing supporting evidence for the primitiveness of the latter genus.

The skull of *Chitra* is indeed amazingly similar to that of *Chitracephalus*, differing obviously only in its more parallel-sided form. However, although the plastron of *Chitracephalus* is reduced as is typical of modern trionychids, the forelimbs appear to have borne five claws, not three as in all known trionychids, and there was a complete though narrow ring of peripheral bones around the carapace. The specimen is thus indeed difficult to assign at the familial level, and R. Wood (pers. comm.) advises that the specimen has deteriorated since it was first described and is no longer possible to study in detail. The observations above were made from the photographs accompanying the original description of the species (Dollo, 1884).

Chelidae

The remaining species of the third group are the snake-necked turtles of the family *Chelidae*. There are two genera in this family that show the classical fish-seizing adaptations, namely *Hydromedusa*, with two species in South America, and *Chelodina* which includes two well-defined species groups, each with several species, in Australia and New Guinea. In both genera the neck is extremely long, and the head elongate and narrowed, with the eyes anterodorsally placed. However, since the thrust of the neck is sideways in these pleurodiran turtles rather than from deep within the shell when the strike is made, the shell specializations to accommodate enlarged longissimus dorsi and rapidly changing visceral volume are largely absent.

The two species groups of *Chelodina* have been discussed by Goode (1967); they are overdue for subgeneric or even generic recognition. In

Group B, piscivorous specializations are much more developed than in Group A; the neck is much longer and more powerful, the skull more elongate and flattened, and the plastron narrower.

Gaffney 1977) considered *Hydromedusa* and *Chelodina* to be closely related and together to form the Infratribe Hydromedusad. He considered the following characters to diagnose this Infratribe: a long, narrow, flat skull; reduced temporal roof covering; markedly narrow parietal area between the temporal fossae; narrow interorbital distance; and presence of a quadrate-basisphenoid contact. Gaffney included the Hydromedusad within the Subtribe Chelina, in which a third genus, *Chelus*, was included.

ORIGIN OF *HYDROMEDUSA* AND *CHELODINA*

General

The author's interpretation of the available evidence is that the assumption of a monophyletic origin for *Hydromedusa* and *Chelodina* is in error; that, although both are chelids, their fish-grabbing specializations are independently derived; and that *Hydromedusa* is more closely related to *Chelus* despite the striking superficial differences between these forms. Arguments for this are given below.

Skull characters

The overall similarity in skull shape between *Hydromedusa* and *Chelodina* – both show marked cranial flattening, narrowing and elongation, and anterior placement of the eyes – is a manifestation of characters common to all piscivorous turtles of the third group. Such characters do not appear to have taxonomic significance above the generic level. Similarly, the markedly narrow parietal area is a natural, possibly inevitable, consequence of the effects of overall narrowing of the skull without reduction of the adductor mandibulae muscles. Such narrowing is shown by certain members of the genus *Phrynops* (e.g. in the subgenus *Batrachemys*), but not by other species within that genus, so it appears to have little trenchant evolutionary significance.

On the other hand, the loss of the post-temporal arch, connecting the supraoccipital to the squamosal, in *Chelodina* and its retention in both *Hydromedusa* and *Chelus* (Fig. 3) is highly significant, and argues strongly that the specialized skulls of *Chelodina* and *Hydromedusa* evolved independently. Another peculiar feature of the skull of *Chelodina*, unique among the chelonians, is the presence of a single frontal bone; the

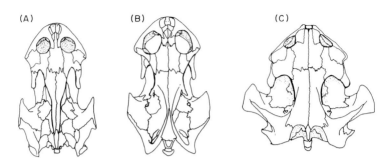

FIG. 3. Skulls of *Chelodina novaeguineae* (A), *Hydromedusa tectifera* (B) and *Chelus fimbriatus* (C). Note the single frontal and the absence of the post-temporal arches in *Chelodina*; the separation of the anterior and posterior parts of the frontals by the prefrontals in *Hydromedusa*; and the posteriorly expanded form and absence of nasal bones in the skull of *Chelus*.

frontals are paired in *Hydromedusa*, *Chelus*, and all other turtles. This is not an obviously adaptive character, and thus again emphasizes the phylogenetic uniqueness of *Chelodina*.

Chelus, by contrast, has lost the nasal bones which are present in all other chelids, but not in living turtles of other families. However, their loss is probably associated with the peculiar specialization of the snout of *Chelus*, in which the nostrils are placed at the tip of an elongate rostral prominence or tube. This tube is quite flexible and bends at the base easily on contact with obstacles. Such flexibility is presumably desirable, and facilitated by loss of the nasal bones. Why *Chelus* should also be unique among the chelids in having a single premaxillary bone is less obvious; but since the only other turtles in which this bone is single (the Trionychidae and Carettochelyidae) also have a flexible tubular rostral structure, there appears to be a functional connection of some kind.

Gaffney (1977) included the presence of a quadrate/basisphenoid contact in his list of synapomorphies for his Infratribe Hydromedusad. However, this contact is more probably a direct result of the narrowing of the posterior parts of the skull; in other chelids, these bones are separated by the prootic. An explanation for this might be that some bones have a predominantly "structural" role and others have a primarily "separative" role; or specifically that it is relatively important for the quadrate and basisphenoid to retain their overall shape even when the skull shape is modified, while it is less important for the intervening "separator" bone, the prootic, to maintain an unchanged shape. The lateral compression of the skull in the two genera has thus forced together two relatively unyielding bones, and the third bone that normally separates them has been squeezed posteriorly so that it now only partially separates them.

Two skull characters of *Hydromedusa* cited by Gaffney (1977) are not constant. In a skull of an adult female *H. tectifera* (PCHP 1625), the palatines and the vomer are in fairly broad contact (Fig. 4), and the prefrontals are separated by a long anterior process of the frontals. Gaffney's series of three skulls was apparently consistent in showing the opposite configuration.

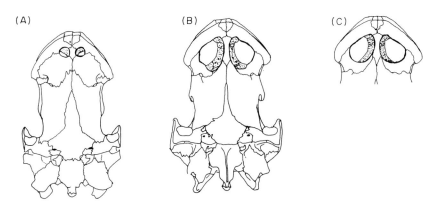

FIG. 4. Palatal view of the skull of *Chelodina novaeguineae* (A) and *Hydromedusa tectifera* (B). Figure on right (C) shows an alternative configuration, of the palatal bones of *H. tectifera* in which the vomer makes contact with the palatines. Note the greatly enlarged apertura narium interna of *Hydromedusa* and their small size in *Chelodina*.

The squamosal-supraoccipital contact in the temporal arch of *Hydromedusa* was considered by Gaffney (1977) to be a feature unique to this genus. However, these bones may be in contact in other chelids in which the parietals are strongly narrowed dorsally. For example, in *Phrynops* (*Batrachemys*) *nasutus wermuthi*, the parietals are extremely narrow, and the position of the parietal-squamosal suture is very variable, sometimes being near the middle point of the very narrow temporal arch, and in others being considerably displaced medially. In a series of three skulls of this form, one (PCHP 1060) has the parietal/squamosal suture displaced so far anteromedially that the squamosals also make contact with the supraoccipital. In a second specimen the separation is extremely narrow; in a third there is a moderate degree of separation.

The placement of the orbits of *Hydromedusa* is more like that of *Chelodina* than that of *Chelus*, but again this appears to be an adaptive parallelism. Both genera of snake-necked turtles sight their prey and strike directly at it, with the anterior part of the neck straight and thrusting the head directly towards the prey; the power for the thrust is generated in the posterior cervical area. Binocular vision and enlarged orbits are thus important in allowing the turtle to get an accurate "fix" on the direction and distance of the prey before a strike is made. Even so, despite the superficial similarity of the anterior part of the head in the two genera, the skull of *Hydromedusa* is unique in having exceptionally large apertura narium interna (Fig. 4). This feature essentially means that the floor of the orbit in *Hydromedusa*, instead of being bony as in other turtles, is composed of soft tissue. The development parallels that of anurans; it presumably represents an adaptation to the accommodation of very large eyeballs at the extreme front of a very flattened, wedge-shaped head. During blinking, the eyeballs are forced down so that they bulge into the oral cavity. *Chelodina* has not evolved such enlarged apertura, and thus may have greater constraints placed upon the degree to which the front of the skull can be flattened.

The presence of the apertura, on the other hand, would appear to place constraints upon the type of prey that could be consumed. Live fish that are flushed right past the front of the head into the throat present no problem; but biting or chewing spiny or brittle-shelled food organisms might result in fragments penetrating the orbits through the roof of the mouth.

In this connection it is of interest that *Hydromedusa* is reported by Freiberg (1971) to show a great appetite for snails. However, it does not consume them in the usual turtle fashion by crushing them between expanded jaw surfaces, but rather separates the operculum, places its snout into the opening and sucks out the soft morsel "con intenible fuerza". Freiberg had extensive familiarity with this species and presumably had watched this procedure himself. However the use of the verb "sorber" (to suck) is questionable, since the only sucking action possible in an organism lacking soft lips is the generation of an inrush of water into the oral cavity by hyoid depression. It is probable that the only way a turtle can remove the soft parts of a snail from its shell is by grasping the former in its mouth and thrusting the latter away with its front limbs.

Chelodina apparently eschews snails. Goode (1967) found that captive Australian chelids of all kinds "studiously ignored" snails even when he cracked the shells for them.

Neural bones

Gaffney (1977) admitted this character was a difficult one; it cut across his overall theory of chelid relationships, but he felt justified in ignoring it because acceptance of this single character required rejection of several others that were correlated.

Gaffney considered the neural bone character to divide the Chelidae into two groups, those with neurals (*Hydromedusa*, *Chelus*, *Chelodina* and

Phrynops) and those without (*Emydura, Elseya, Pseudemydura* and *Platemys*). However, the character may be more usefully defined as a *tendency* towards neural loss rather than complete attainment of that condition. Under this definition, only *Chelus* and *Hydromedusa* show no tendency towards neural loss, whereas neurals are unknown in *Emydura* and *Pseudemydura*; very rare in *Elseya* (Rhodin & Mittermeier, 1977); are usually absent in all species of *Chelodina* except one (*C. oblonga*, an otherwise typical species of *Chelodina*; Burbidge, Kirsch & Main, 1974); are almost invariably absent in *Platemys* but *P. platycephala* may occasionally have some neurals (A. Rhodin, pers. comm.); and are variably complete, reduced, or absent in *Phrynops*. This interpretation places *Hydromedusa* and *Chelus* together as a natural exclusive group, which the author believes is justified for reasons further elaborated below.

The possible significance of neural bone loss is worthy of discussion. Although the posterior pleural bones may meet medially by elimination of one or two neurals in several turtle families, only in the Pleurodira is there any tendency towards elimination of neural bones, and all the living pelomedusids have fairly complete series (but in *Pelusios sinuatus* the normal complement of eight may be reduced to five, though without size reduction of the individual neurals retained; Broadley, 1983). However, extinct pelomedusids (e.g. "*Podocnemis*" *venezuelensis* and *Eusarkia rotundiformis*) are known in which neurals are entirely lost. Increased neural width appears to correlate strongly with development of powerful longissimus dorsi muscles in turtles specialized for rapid neck extension (*Chelus, Chelydra, Deirochelys, Emydoidea*, etc.), in which the rib-heads arch away from the carapace to form "tunnels" to accommodate these muscles. However, the reasons for partial or total elimination of neurals, except perhaps as a shell-strenthening device (by elimination of the total number of bony elements), is more obscure. Conceivably, the thrusts of a sideways-bending neck in pleurodiran turtles produce asymmetrical tensions between the bands of muscle on each side of the vertebral column – a force that can be contained and stabilized by eliminating the neural bones and allowing the pleurals on opposite sides of the carapace to have a direct sutural connection.

This argument remains highly speculative. But it is worthy of note that in non-pleurodiran turtles, the only parts of the neural series that are subject to elimination – the extreme front and rear – are also the parts subject to asymmetrical shearing tensions transmitted to these areas by the limb girdles during locomotion. This point may be illustrated by consideration of the neural structure of the two largest cheloniid species, *Chelonia mydas* and *Caretta caretta*. The former has a symmetrical swimming action and terrestrial gait, and has a complete series of neural bones (Fig. 5). The latter has an alternating gait, and has the posterior pairs of pleural bones in contact by elimination of the posteriormost neurals. The loss of neurals is unlikely to make a contribution to the shell strength in adult turtles, in which analysis of shells of individuals smashed by automobiles shows no tendency for fractures to follow suture lines; but in hatchlings and juveniles fewer bones may make for a stronger shell.

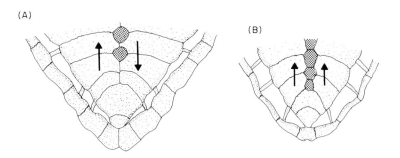

FIG. 5. Posterior part of the bony carapace of *Caretta caretta* (A) and *Chelonia mydas* (B). Note the medially contiguous pleurals and asymmetrical 'walking' stress in *Caretta*, and the pleurals separated by the neurals and the symmetrical 'walking' stress in *Chelonia*.

Relations between *Hydromedusa* and *Chelus*

The two South American genera *Hydromedusa* and *Chelus*, despite the bizarrely specialized features of the latter, share a number of noteworthy features. These include:

(i) Retention of a complete series of neural bones (unlike all other chelid genera).

(ii) Highly tuberculate carapace. This is retained throughout life in *Chelus*; it is most apparent in juvenile *Hydromedusa*, and more so in *H. tectifera* than in *H. maximiliani*. It is not shown by other chelids.

(iii) Interfemoral plastral concavity in males. In *Chelodina* there is no plastral concavity in males, and indeed the sexes may be so similar that even with a large series in hand sexing may be difficult. Cann (1978), for example, reports on a collection of 150 live *Chelodina longicollis*, individuals of which could only be sexed (as female) after they laid eggs.

The interfemoral plastral concavity in male *Hydromedusa* and *Chelus* has the obvious function of accommodation of the posterior vertebral tubercle of the carapace of the female during copulation. What is interesting, however, is that in adult female *Hydromedusa* this tubercle has disappeared, yet the concavity in the male plastron persists. A reasonable explanation may be that the carapacial tubercles are a fundamental and ancestral character in *Hydromedusa* that were once present in adult females as well as in juveniles. Adaptive pressures – possibly relating to streamlining – may have caused gradual loss of the tubercles in adults without corresponding loss of the male plastral concavity. This argues strongly in favor of the tuberculate carapace being a long-established character that may indeed link *Chelus* and *Hydromedusa* by phylogeny rather than convergence.

(iv) Long neck. This character has been independently derived in all three chelid genera in which it is present. Its obvious adaptive role lessens its value as a fundamental taxonomic character, but at least it serves to link, rather than separate, *Chelus* and *Hydromedusa*.

(v) Presence of temporal arches. The presence of these in *Chelus* and in *Hydromedusa* and their absence in *Chelodina* argues in favor of the former pair being a monophyletic group. The exact character of the arch is different in the two forms; in *Chelus* there is no contact between the squamosal and the supraoccipital, the arch being purely between the parietal and the squamosal. However, the unusual squamosal/supraoccipital suture of *Hydromedusa* may well result simply from the unique narrowing of the rear of the skull, forcing normally separated bones together just as the basisphenoid and quadrate were forced together in the ventral aspect.

(vi) Configuration of the intergular scute. This scute reaches the anterior margin of the plastron in *Hydromedusa* and in *Chelus*, and it is recessed from the anterior plastral margin in *Chelodina*. However, further elaboration is needed. In *Hydromedusa*, the intergular is extremely large, completely separates the gulars, and partially separates the humerals. The interhumeral seam may be highly reduced by intrusion of the intergular in *H. tectifera*, but as far as is known it is always present.

In *Chelus fimbriatus*, the intergular is extremely variable; it may partially separate the gulars, completely separate them, or partially separate the humerals as well. Moreover, in the fossil species *Chelus colombianus*, one of the available specimens had the gulars meeting in front of the intergular, as in *Chelodina*, while in another an extra pair of gulars was present. In other words, the scutation of the anterior plastral lobe of *Chelus* is extremely variable.

The condition of the intergular excluded from the front margin of the plastron but partially separating the gulars and pectorals and fully separating the humerals is very stable in *Chelodina*. However, in the type of *Chelodina intergularis* (a synonym of the "Group B" species *C. rugosa*) secondary reduction of the anterior plastral lobe has advanced to the point that the intergular has a narrow exposure on the front margin in the plastron.

The Anterior Plastron in *Chelus, Hydromedusa* and *Chelodina*

To interpret the unusual configurations of the anterior part of the plastron of the three chelid genera in question, it seems appropriate to postulate that in each case there has been a necessity to elongate or enlarge the front plastral lobe in order to provide some protection for the otherwise exposed neck. The mode of retraction in long-necked cryptodires, in which the posterior cervical vertebrae can be flexed through almost 180° relative to the anterior dorsal vertebrae and the neck folded so that the posteriormost point is actually in the rear half of the turtle's shell, is completely impossible in pleurodires. Consequently, protection must be afforded by increasing the length of the anterior lobe of the plastral and the anterior overhang of the carapace.

This can be achieved in different ways. In *Chelus* the anterior part of the plastron has simply lengthened and the anterior scutes have made accommodation to this on an almost individual basis – the intergular scute may or may not be lengthened from the primitive condition in which it is small, narrow, and only partially separates the gulars; and the gulars may once have met in front of the intergular (though only one fossil specimen is known in which this was the case), and then, for unknown reasons, the intergular once again reached the anterior plastral margin. On the other hand, the *Chelus colombianus* line in which this configuration was sometimes present may simply have become extinct – or the specimen may have been abnormal.

In *Hydromedusa* the anterior part of the plastron became substantially enlarged by considerable expansion of the entoplastron and the gular and humeral scutes as well as phenomenal enlargement of the intergular (Fig. 6). Similarly, the anterior part of the carapace is extended forward, with a considerable anterior extension of the nuchal bone and peripherals I and II on each side. The scutes made a unique accommodation to this; to avoid having a first vertebral scute of unconscionable length, the nuchal scute moved posteriorly, became excluded from the anterior carapace margin, greatly expanded laterally, and took the place of an additional vertebral scute (Fig. 6). The first marginal scutes then met anteriorly and, in *H. maximiliani*, so too did the anterior peripheral bones (Wood & Moody, 1976). It is as if the carapace achieved its additional anterior length by the areas on each side of the middle of the front of the carapace growing forwards and converging until they met.

In *Chelodina* the anterior part of the carapace shows no such peculiarities, the nuchal scute merely being somewhat broader than usual (Fig. 7). The anterior lobe of the plastron, however, became enlarged by

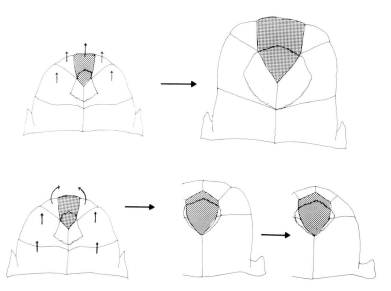

FIG. 6. Upper: modes of scute expansion in the evolution of the anterior plastral lobe of *Hydromedusa*; note the great enlargement of the entoplastron and the intergular scute. Lower: modes of scute expansion in the evolution of the anterior plastral lobe of *Chelodina*; note the anteroposterior elongation of the pectorals and the anterior migration of the gulars and humerals around the intergular scute. Figure on lower right shows the configuration in 'Group B' *Chelodina* in which the size of the anterior plastral lobe is secondarily reduced.

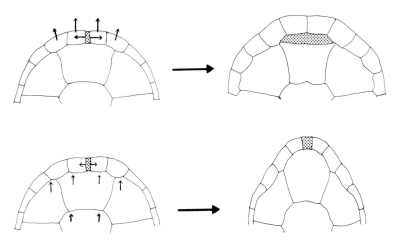

FIG. 7. Upper: mode of scute expansion in the evolution of the anterior part of the carapace of *Hydromedusa*; note the great lateral expansion of the nuchal scute and the anterior migration of marginals I and II. Lower: mode of scute expansion in the evolution of the anterior part of the carapace of *Chelodina*; note the modest lateral expansion and retention of the position of the nuchal scute, and the considerable elongation of the anterior vertebral and costal scutes.

the intergular expanding massively and the gular and humeral scutes expanding also so that the former finally met anteriorly (Fig. 6). Subsequently, the anterior lobe retreated in size in Group "B" *Chelodina*, leaving much of the neck apparently highly exposed to attack, and with the gulars and humerals small once again – but the intergular remaining huge and the gulars continuing to meet anteriorly. This is in sharp contrast to short-necked chelids in which the plastron had never gone through this expansion-contraction cycle.

The trenchant differences in the structure of the expanded anterior shell of *Chelodina*, *Chelus* and *Hydromedusa* suggest that the shell enlarged independently in these three genera, which also suggests that the long neck evolved independently.

In *Hydromedusa* and *Chelodina*, the dorsal scutes as a whole seem to have been destabilized; Zangerl (1969) comments upon the frequency of supernumerary scutes in *Hydromedusa*, and Luederwaldt (1926) gives examples; *Chelodina* also shows a high frequency of supernumerary scutes, especially extra vertebrals.

Four other characters shared by *Chelodina* and *Hydromedusa* and not shown by *Chelus* are: only four claws on each front foot; elongate eggs; a pattern of vermiculate or "marbled" superficial shell sculpturing in adults; and a broad posterior plastral lobe.

(i) The absence of the uppermost claw on the front foot may correlate with the narrow ventral aspect of the marginal scutes in the region of the axillary notch compared to other chelid genera. This has the result that, when the foot is retracted, the uppermost digit makes contact with soft skin rather than with the edge of the shell. The presence of a claw on this digit might cause chronic problems of scratching or abrasion of the skin in this area. A similar argument may account for reduction of claws in the trionychids.

(ii) Egg shape in chelids is very variable; although always brittle-shelled, the eggs may be spherical (*Chelus*, *Phrynops geoffroanus*, etc.), slightly ovate (*Batrachemys*), or elongate (*Hydromedusa*). They may be very small relative to the size of the turtle (*Chelodina novaeguineae*, etc.), or very large (*Platemys platycephala*). The size and shape of the egg can vary even within a species, depending upon female size and age (Goode, 1967). Egg size then appears to be more a function of mechanical and ecological considerations than of phylogenetic ones. All very large turtles (sea turtles, giant tortoises, large softshells, etc.) lay spherical eggs, all very small ones lay ovate eggs, and among turtles of intermediate size there is a tendency for eggs to be spherical if small (*Trionyx*, *Chelydra*), and elongate if large (*Rhinoclemmys*). Even within a genus (e.g. *Podocnemis*), the larger members may lay spherical eggs and the smaller ones elongate eggs.

The spherical eggs of *Chelus fimbriatus* are thus a probable correlate with the relatively large size of the species at maturity compared with *Hydromedusa* and *Chelodina*.

(iii) The vermiculate shell surface of *Chelodina* and *Hydromedusa* is found in other chelids (e.g. *Phrynops geoffroanus*). In *Chelus*, it may have been overridden as the unique deeply-incised carapacial sculpturing of that genus evolved.

(iv) The broad hind plastral lobe is in fact the normal condition in chelids. The hind lobe is very narrow in *Chelus* and in a few other forms, including a very large-headed undescribed *Batrachemys* from the Maracaibo Basin of Venezuela. The narrow plastron again probably has a visceral-displacement role in chelids with large heads and heavy neck musculature, though since the neck is curled laterally under the anterior carapace overhang in sidenecks rather than drawn deep within the shell when retracted, the reasons for the correlation are less obvious than they are in cryptodires.

Evolution of *Chelus* and *Hydromedusa*

From all available data, a reasonable postulate is that the common ancestor of *Chelus* and *Hydromedusa* was a late Mesozoic, short-necked, side-necked turtle with a tuberculate carapace, a complete set of neural bones, and narrow temporal arches. We can then argue that divergence between these genera arose as the population started to specialize in a predatory, fish-seizing mode. Populations in clear water, destined to become *Hydromedusa*, would continue to emphasize vision as the primary sense of prey location, and would be active animals with large eyes that could get a binocular "fix" on prey directly in front of the head. The neck would elongate (with the shell enlarging anteriorly to accommodate it), and the head would be highly streamlined, to facilitate lightning-like strikes. The force of the strike would be generated principally in the posterior cervical area, the anterior cervicals remaining relatively aligned behind the head. This reflects in the lateral processes in the cervical vertebrae of *Hydromedusa*, which are stronger posteriorly than anteriorly (Fig. 8).

On the other hand, murky water habitats might offer much more productive feeding opportunities if the problem of prey location could be overcome. *Chelus* overcame them. Prey location became auditory and tactile rather than visual. The tympana enlarged greatly, and became widely separated by widening of the posterior part of the skull so that sensitivity to vibrations and the direction from which they came would be enhanced. Prey detection was further enhanced by numerous highly innervated lappets of skin (Hartline, 1967), and by large veins in

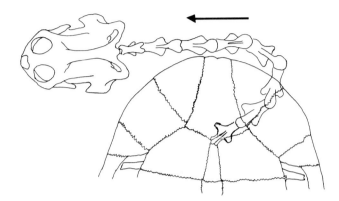

FIG. 8. Anterior axial skeleton of *Hydromedusa tectifera*, showing direction of predatory strike and relatively strong lateral processes on proximal cervical vertebrae.

the neck which possibly function in the fashion of the lateral line of fishes (Winokur, 1972). The turtle became highly sedentary in order to minimize water disturbance and self-generated vibrations confusing stimuli signalling the approach of prey.

Since prey cannot be pinpointed accurately in murky water even with these enhanced sensory capabilities, it became necessary to enlarge the mouth, the hyoid and the neck musculature in order to achieve a voluminous inrush of water that would carry all prey near the head into the throat. This made tight flexion of the neck difficult and a "side-swiping" rather than "frontal" attack became necessary (Fig. 9). The skull thus had to achieve overall flattening in order not to impede its lateral passage through the water. The eyes, no longer the primary prey-detection organs, became highly reduced, if for no other reason than to fit into the now extremely flattened anterior part of the head. The thrust of the "strike" being primarily with the anterior cervicals, it is these (in contrast to the condition in *Hydromedusa*) in which the lateral processes achieve their greatest development.

The tubular rostral structure of *Chelus*, unique in the Chelidae but also found in the long-necked soft-shelled turtles of the family Trionychidae, is probably an adaptation towards minimizing water disturbance during respiration – an important factor for a species whose mode of feeding depends upon the detection of agile prey that unknowingly approach close to the head.

The peculiar features of *Chelus* are all thus related to its feeding specializations, and if these are ignored a turtle very similar to *Hydromedusa* will result. On the other hand, several features of *Chelodina* suggest strongly that it represents an independent, parallel develop-

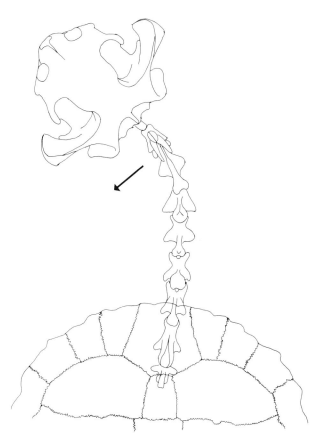

FIG. 9. Anterior axial skeleton of *Chelus fimbriatus*, showing direction of predatory strike and relatively strong lateral processes on distal cervical vertebrae.

ment that branched off the early chelid stem before *Chelus* and *Hydromedusa* diverged.

The fossil and geological evidence for this is scant but worthy of consideration. Wood & Moody (1976) mention a fossil nuchal bone from the early Eocene of southern Argentina that appears to be referable to *H. tectifera*, an impressively early find for a specimen similar to the modern form. The origins of the Chelidae remain unclear but it has been postulated elsewhere (Pritchard, 1982b, in press) that an extreme southern, possibly Antarctic, origin is reasonable for many reasons; that the continents of South America and Australia were colonized by chelids from the south in the Cretaceous; and that chelids, like most freshwater turtle families, have very poor powers of dispersal across oceanic barriers and generally require land bridges. This implies that the Australian and South American chelids have been separate since the end of the Cretaceous, and that the family never reached other continents (unlike the pelomedusids, which may have dispersed earlier when Laurasia and Gondwanaland were still united and intact, and which may have had a marine phase in their evolutionary history).

Maps showing the hypothesized forms of the continents at different stages of drift are given by Dietz & Holden (1970). These maps show Australia somewhat amorphously but with generally the same shape as today. However, Burbidge (1967), quoting Hill & Denmead (1960) and Glaessner & Parkin (1957), argues that Australia was divided into eastern and western subcontinents by seas throughout the Cretaceous, and that when these seas dried up in the Tertiary the resulting plain was devoid of fresh water and remained a barrier to freshwater turtle distribution ever since.

This would result in the ancestors of *Chelodina oblonga* (the most primitive species of *Chelodina*, retaining neural bones) and *Pseudemydura umbrina* (an extremely distinctive, possibly primitive chelid) having invaded south-western Australia from Antarctica independently from the invasion of eastern Australia by a similar early *Chelodina* stock as well as by the *Emydura/Elseya* lineage (and perhaps also by a *Pseudemydura* stock, long since extinct without trace). The eastern *Chelodina* stock, faced with a multiplicity of habitats, both tropical and temperate, diverged into two subgroups and many species. However, only one of these (*Chelodina steindachneri*) succeeded in crossing into northern Western Australia, assisted no doubt by coastal rivers and a highly developed capacity to withstand desiccation.

Nevertheless, the available fossils and geological arguments merely serve to illustrate the antiquity of the Chelidae as a family and its apparent divergence into existing genera by the end of the Cretaceous; they shed little light on the relative chronology of the divergence of these genera.

Several other authors have utilized varying techniques for classification of the chelid genera. Among the earliest was Kasper (1903), who confined his analysis to the relative lengths of the axis and atlas and the orientation of their zygapophyses. These characters are surely adaptive rather than fundamental, but it is still of interest that Kasper found that *Chelodina* and *Hydromedusa* fell into different groups (the latter allying with *Platemys*), while *Chelus* was placed in a third group with *Hydraspis* (= *Phrynops*).

Frair (1964), from serological studies, found that *Chelus* was more closely related to *Batrachemys* than to *Chelodina*, but did not study *Hydromedusa*. Burbidge, Kirsch & Main (1974) similarly found that genera within Australia were more closely related to each other than to

Platemys or *Mesoclemmys*. Bull & Legler (1980) examined karyotypes of all chelid genera, and obtained a uniform diploid number of 54 for *Chelodina*, 50 for *Chelus*, and 58 for *Hydromedusa*.

REFERENCES

Annandale, N. & Shastri, M. H. (1914). Relics of the worship of mud-turtles (Trionychidae) in India and Burma. *J. Asiat. Soc. Beng.* **10**: 131–138.

Broadley, D. G. (1983). Neural pattern – a neglected taxonomic character in the genus *Pelusios* Wagler (Pleurodira: Pelomedusidae). In *Advances in herpetology and evolutionary biology: Essays in honor of Ernest E. Williams*: 159–168. Rhodin, A. & Miyata, K. (Eds). Cambridge, Mass.: Mus. Comp. Zool., Harvard.

Bull, J. J. & Legler, J. M. (1980). Karyotypes of side-necked turtles (Testudines: Pleurodira). *Can. J. Zool.* **58**: 828–841.

Burbidge, A. A. (1967). *The biology of south-western Australian tortoises*. Ph.D. thesis: University of Western Australia, Nedlands, Australia.

Burbidge, A. A., Kirsch, J. A. W. & Main, A. R. (1974). Relationships within the Chelidae (Testudines: Pleurodira) of Australia and New Guinea. *Copeia* **1974**: 392–409.

Cann, J. (1978). *Tortoises of Australia*. Sydney: Angus & Robertson.

Deraniyagala, P. E. P. (1939). *The tetrapod reptiles of Ceylon*. Colombo: Colombo Museum.

Dietz, R. S. & Holden, J. C. (1970). The breakup of Pangaea. *Scient. Am.* Oct. 1970. [Reprinted in *Continents adrift, readings from Scientific American*: 102–113. San Francisco: W. H. Freeman & Co.]

Dollo, M. L. (1884). Première note sur les chéloniens de Bernissart. *Bull. Mus. r. Hist. nat. Belg.* **3**: 63–79.

Drummond, H. & Gordon, E. R. (1979). Luring in the neonate alligator snapping turtle (*Macroclemys temminckii*): description and experimental analysis. *Z. Tierpsychol.* **50**: 136–152.

Ernst, C. H. & Barbour, R. (1972). *Turtles of the United States*. Lexington, Kentucky: University Press.

Frair, W. (1964). Turtle family relationships as determined by serological tests. In *Taxonomic biochemistry and serology*: 535–544. Leone, C. A. (Ed.). New York: Ronald Press.

Freiberg, M. A. (1971). *El mundo de las tortugas*. Buenos Aires: Editorial Albatros.

Gaffney, E. S. (1977). The side-necked turtle family Chelidae: a theory of relationships using shared derived characters. *Am. Mus. Novit.* No. 2620: 1–28.

Gaffney, E. S. (1979). Comparative cranial morphology of recent and fossil turtles. *Bull. Am. Mus. nat. Hist.* **164**: 65–376.

Glaessner, M. F. & Parkin, L. W. (1957). The geology of South Australia. *J. geol. Soc. Aust.* **5**: 1–163.

Goode, J. (1967). *Freshwater tortoises of Australia and New Guinea (in the family Chelidae)*. Melbourne: Lansdowne Press.

Hartline, P. (1967). The unbelievable fringed turtle. *Int. Turt. Tort. Soc. J.* **1**: 24–29.

Hill, D. & Denmead, A. K. (1960). The geology of Queensland. *J. geol. Soc. Aust.* **2**: 1–474.

Kasper, A. (1903). Ueber den Atlas und Epistropheus bei den pleurodiren Schildkröten. *Arb. zool. Inst. Univ. Wien* **14**: 137–172.

Luederwaldt, H. (1926). Os Chelonios brasilieiros. *Revta Mus. paul.* **14**: 1–66.

Nutaphand, W. (1979). *The turtles of Thailand.* Bangkok: Mitbhadung Press.
Poglayen-Neuwall, I. (1966). Bemerkungen zur Morphologie und Innervation der Trigeminusmuskulatur von *Chelus fimbriatus* (Schneider). *Zool. Beitr.* **12**: 43–65.
Pritchard, P. C. H. (1982a). *The biology and status of the alligator snapping turtle (Macroclemys temminckii) with research and management recommendations.* Unpublished report to World Wildlife Fund (mimeo).
Pritchard, P. C. H. (1982b). Zoogeography of South American turtles. Abstract in *Proceedings, 25th Annual Meeting S.A.A.R. and 30th Annual Meeting, Herp. League*: 94. (Mimeo).
Pritchard, P. C. H. (In press). Turtles of Venezuela. *Soc. Study Amph. Rept.*
Rhodin, A. G. J., Medem, F. & Mittermeier, R. A. (1981). The occurrence of neustophagia among podocnemine turtles. *Br. J. Herpet.* **6**: 175–176.
Rhodin, A. G. J. & Mittermeier, R. A. (1977). Neural bones in chelid turtles from Australia and New Guinea. *Copeia* **1977**: 370–372.
Romer, A. S. (1956). *Osteology of the reptiles.* Chicago: University Press.
Smith, M. A. (1931). *Fauna of British India.* **1**. *Reptilia and Amphibia.* London: Taylor & Francis.
Spindel, E. L. (1980). *The fundamental mechanisms and histologic composition of the lingual appendage in the alligator snapping turtle,* Macroclemys temmincki *(Troost)* *(Testudines: Chelydridae).* M.S. thesis: Auburn University, Alabama, USA.
Winokur, R. M. (1972). *Cranial integumentary specializations of turtles.* Ph.D. thesis: University of Utah, USA.
Wood, R. C. & Moody, R. T. J. (1976). Unique arrangement of carapace bones in the South American chelid turtle *Hydromedusa maximiliani* (Mikan). *Zool. J. Linn. Soc.* **59**: 69–78.
Zangerl, R. (1969). The turtle shell. In *Biology of the Reptilia*, **1**: 311–339. Gans, C., Parsons, T. & Bellairs, A. (Eds). London and New York: Academic Press.

The Squamate Epidermis: New Light Has Been Shed

P. F. A. MADERSON

Biology Department, Brooklyn College of City University New York, Brooklyn, New York 11210, USA

SYNOPSIS

The unique pan-body synchrony of epidermal cell proliferation and differentiation which is the basis for squamate "skin-shedding" poses problems of interest to many biological disciplines. Whereas study of the control of the phenomenon has long centred on endogenous (mainly hormonal) factors, a new hypothesis is proposed suggesting an interaction between them and external environmental factors especially ambient humidity. The new model is suggested by data from recent studies of the relationship between the thyroid gland, environment and shedding frequency in snakes and lizards. Supportive data from other vertebrate embryonic and adult epithelia imply that the squamate epidermis may be a model system *par excellence* to further our understanding of the factors involved in epithelial tissue homeostasis.

INTRODUCTION

Squamate skin-shedding is arguably the oldest herpetological observation known to man, being, as it is, the basis for the caduceus, the symbol of Aesculapius and most of the medical societies in the world (Lillywhite & Maderson, 1982). Maurer (1892, 1895) wrote explicitly that the cellular basis of the phenomenon was so complex that it could only be understood by comparative study of sequential daily biopsies from an individual throughout a shedding cycle, but such an investigation was not reported until 70 years later (Maderson, 1965a). In spite of a plethora of histological, histochemical and ultrastructural studies which have appeared over the past 35 years we have as yet merely an understanding of basic cytologic data and two fundamental facts. First, periodic loss of "epidermal generations" is a lepidosaurian characteristic (Maderson, 1968; Baden & Maderson, 1970). Second, the associated pan-body synchronized patterns of cell proliferation and differentiation are unique when compared to the epidermis of other vertebrates, and indeed do not even have aspects analogous with any known metazoan epithelium except that the terminally differentiated cells are lost to the external environment.

A unifying hypothesis is presented here which explains much of the recorded data, further tenets of which are currently under investigation. The hypothesis derives from three fundamental, but as yet incompletely answered, questions. First, what controls cyclic epidermal cell proliferation and differentiation? Apparently this is a question for developmental biologists and endocrinologists. Second, how does periodic skin-shedding relate to water metabolism of the animals? Apparently this is a question for physiologists. Third, what is the function of the phenomenon in the life of the animals? Apparently this is a question for ecologists and naturalists. These questions are interrelated and an understanding of the system can only be achieved by an interdisciplinary approach.

A brief review of the current state of knowledge of the anatomy of the squamate integument from gross to microscopic levels is necessary.

THE ANATOMY OF THE SQUAMATE EPIDERMIS

All squamates have a scaled integument, i.e. the entire body and head is covered by a contiguous series of more or less overlapping scales (Maderson, 1965b) with outer epidermal and inner dermal components. Except for the cranial integument, or in those species where body scales are tuberculate, each unit has readily recognizable outer and inner surfaces and a hinge region (Maderson, 1965b). The relatively thicker cornified tissues on the outer surface make each scale stiff and inflexible, while the inner surface and hinge region have thinner cornified tissues and are therefore flexible, permitting considerable distensibility of the integument; this capacity is augmented by the relatively loose connection between the deep dermis and the muscle fascia. The precise form, shape and relative density of scales across the body surface varies within and among species and it is inferred that the gross form of the integument reflects the many different selective pressures associated with many different parameters in different environments (Lillywhite & Maderson, 1982: 405–407).

Periodically all squamates lose to the external environment a *complete, mature* epidermal generation. This complex unit consists of six terminally differentiated cell types named, from without inwards: the Oberhäutchen, β-layer, mesos layer, α-layer, lacunar tissue and clear layer. These six basic cell types are represented by single cell layers (the Oberhäutchen and clear layer) or by stratified squamous epithelia, which comprise varying numbers of layers of cells depending on the species, location on the body, or on an individual scale. The mature Oberhäutchen and β-layer form a syncytial tissue totally lacking plasma membranes and/or nuclear remnants, and contain predominantly feather (β-) type keratinaceous proteins. The cells of the rest of the shed generation contain only hair (α-) type keratinaceous protein, although there remains doubt as to the exact nature of structural proteins within individual cell populations (Maderson, Flaxman, Roth & Szabo, 1972; Maderson, in press).

Following a shed, the epidermis enters a resting phase which is defined as a period of time during which it consists of an incomplete outer epidermal generation lying atop a stratum germinativum. The resting phase may last anywhere from zero to "n" days and this variability is the basis for variation in shedding frequency (SF) from one individual or species to another (Maderson, Chiu & Phillips, 1970). During this phase of the cycle proliferative and differentiative activity within the system is minimal.

At the end of the resting phase, proliferative activity resumes in the germinal layer so completing the outer epidermal generation and, within the next 14 days, two series of events occur simultaneously. While the innermost components of the outer generation (lacunar tissue and clear layer) are becoming mature, most of a new inner epidermal generation is laid down and matures. Thus at the end of the 14-day "renewal phase" the older outer epidermation is shed and the erstwhile inner generation takes its place as the functional body surface; this cycle of events is repeated throughout the life of the animal.

The complex sequence of cytodifferentiative events associated with one complete shedding cycle is shown diagramatically in Fig. 1; references to the numerous histological and ultrastructural studies which have provided the data for the preceding account will be found in Landmann (1979), Maderson & Chiu (1981) and Maderson (in press).

The squamate epidermis may be conceptualized as a stratified squamous epithelium wherein, over time, some layers of daughter cells synthesize predominantly β-keratins followed by some which synthesize α-keratins. These cytologic data present two fundamental questions: (i) what controls shedding frequency, i.e. the periodicity of loss of mature generations from the body surface? and (ii) what controls the pathways of specific protein synthesis, notably β- to α-, α- back to β-, β- back to α- etc. etc.?

THE CONTROL OF CYCLIC ACTIVITY IN THE SQUAMATE EPIDERMIS

The history of studies concerning this topic began in the late 1920s, and the pioneering works of Drzewicki & Eggert (see references and discus-

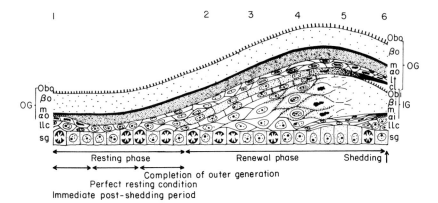

FIG. 1. Schematic representation of epidermal histogenesis throughout one shedding cycle in a squamate reptile. A shed is followed by a resting phase of variable length and the phase is subdivided into post-shedding, perfect resting and completion periods which describe the constitution and maturation of the outer epidermal generation. The period between the end of the resting phase and the next shed is the renewal phase; it is divided into five arbitrary stages which describe the maturation of the inner epidermal generation. Abbreviations on the figure are as follows. The constituent components of the outer epidermal generation (OG) are, from without inwards, the Oberhäutchen (Obo), the β-layer (βo), the mesos layer (m), and α-layer (αo), the lacunar tissue (lt) and the clear layer (cl); those of the inner epidermal generation (IG) are the Oberhäutchen (Obi), the β-layer (βi), the mesos layer (m) and the α-layer (αi). Above the germinal layer (sg) may be seen undifferentiated living cells (llc). Taken from Landmann (1975) by permission of the author and the Israel Journal of Zoology.

sion in Maderson, Chiu & Phillips, 1970) provided the foundation for the textbook maxim: "The thyroid controls skin-shedding in lower vertebrates". This not only proved to be, like most classic maxims, much more complex than previously imagined (Maderson, Chiu & Phillips, 1970; Maderson, in press) but also concealed an extraordinary paradox first revealed in the early 1930s. Whereas all data on lizards which have accumulated over the past 50 years imply that *hyperthyroidy* enhances shedding frequency (SF), Schaeffer (1933) reported that enhanced SF in snakes was associated with *hypothyroidy*. This paradox has proved extremely difficult to resolve since intact snakes do not shed sufficiently regularly in the laboratory to provide appropriate figures for statistical evaluation with experimental animals. Thus the majority of our information derives from studies on lizards, which are now reviewed.

While Maderson, Chiu & Phillips (1970) concluded that control of lizard SF resided within the pituitary-thyroid axis, experiments conducted from 1974 onwards have revealed that in Tokay geckos (*Gekko gecko*) (a) shedding occurs regularly, albeit at much lower frequency following hypophysectomy, and (b) particular SFs can be produced at will by a variety of combinations of endocrine interference and environmental temperature. These data have led to the conclusion that SF *in vivo* is not controlled by any particular hormonal milieu, but rather reflects metabolic rate which is affected directly or indirectly by a variety of hormones (Maderson & Chiu, 1981; Maderson, in press; Chiu, Maderson & Zucker, in preparation). This conclusion has been substantiated by measurements of oxygen consumption in Tokay geckos at various temperatures, under various hormonal milieux (Sham & Chiu, in preparation). All experiments indicate that increases or decreases in SF are effected by shortening or lengthening the resting phase, with the renewal phase always lasting approximately 14 days (Chiu & Maderson, 1980). This suggests that the capacity for cyclic cell proliferation and differentiation is genetically determined and is intrinsic to the integument, in fact to the epidermis alone, as has been shown by *in vitro* studies (Flaxman, Maderson, Szabo & Roth, 1968). This latter study implied the existence of an intra-epidermal feedback system, the mechanism of which remains enigmatic, controlling pathways of protein synthesis in emergent daughter cells. A final point which has emerged from all *in vivo* studies is that no agent(s) actively inhibiting germinal proliferation in lizards has (have) been identified.

Our knowledge of the control of snake SF is "simpler" yet still presents paradoxes. A series of publications by Dr K. W. Chiu of the Chinese University of Hong Kong, and his colleagues and students, have repeatedly confirmed Schaeffer's (1933) brief report that enhanced SF is associated with hypothyroidy. Only the thyroid gland affects SF; thyroidectomy always induces two to five consecutive epidermal cycles and the operated animals always die after having lost 20% of their pre-operative weight (Chiu, Leung & Maderson, 1983). From these experiments we conclude that thyroid gland secretion(s), and/or a metabolite(s) thereof, actively inhibit germinal proliferation and suppress SF. This conclusion is supported by the observation that administration of iodinated thyronines to thyroidectomized snakes restored the epidermal inhibition (Chiu, Leung *et al.* 1983). Since the thyroid gland does not influence oxidative metabolism in snakes (Chiu, Leung *et al.*, 1983) – one of many curious facts about thyroid physiology in these reptiles (Chiu, 1982) – on *a priori* grounds one would not have expected to see an effect of elevated temperature on the SF of intact snakes. This prediction has not been borne out and Maderson, Chiu, Contard & Bank (in preparation) have shown that moving intact garter snakes (*Thamnophis sirtalis*) from a constant temperature of 21°C to a constant temperature of 28°C induces a shed. A relationship between "thyroidectomy-induced multiple shedding" and "high-temperature induced single sheds" may be suggested by the work of Gupta, Thapliyal & Garg (1975) who reported that thyroidectomy produced disturbances of lipid constituents throughout the snakes' body. How could this relate to shedding?

CUTANEOUS WATER LOSS (CWL) IN SQUAMATE REPTILES

All terrestrial amniotes lose some water via respiration, excretion and defecation, and it was long assumed that their skin was "waterproof". Bentley & Schmidt-Nielsen (1966) refuted this dogma, and many subsequent studies revealed that CWL is a significant component of total evaporative water loss in reptiles (Lillywhite & Maderson, 1982). Only recently has study of reptilian CWL been pursued with the same rigor and with consideration of underlying theoretical implications as has been customary in studies of mammalian integument, especially that of man (Zucker, 1980). The bulk of available experimental evidence for reptiles implies that, as in mammals, the barrier to CWL is passive and lies in the epidermis (Zucker, 1980). The question therefore arises, what part or parts of squamate corneous tissues provide the primary barrier to CWL?

Landmann (1979, 1980a) using ultrastructural analysis, and Roberts & Lillywhite (1980, 1983) combining histochemical and physiological techniques, independently concluded that the primary barrier resided in the mesos layer, and depended upon the presence of extracellular lipids. These data permit the proposition of a general thesis applicable to "skin-shedding phenomena" in squamates. If the barrier to CWL is conceptualized as a passive membrane, application of derivatives of Fick's Laws of Diffusion (Zucker, 1980) leads to the biologically comprehensible relationship:

$$CWL = \frac{Permeability\ Constant\ (PC)}{Thickness\ (T)}$$

In principle, as Lillywhite & Maderson (1982: 400) have argued, two sympatric squamate species could show identical mean values for CWL with one species having a very thick mesos layer with a "poor" permeability constant (a qualitative faculty dependent on the macromolecular content and organization of extracellular lipids), and the other having a very thin mesos layer with an exceptionally "good" permeability constant. Further corollaries of the relationship are (a) if PC and/or T changed under constant environmental conditions, CWL would change, and (b) if environmental conditions (especially ambient humidity) changed, then CWL would change if PC and T remained constant.

While there are not yet data reporting SF under controlled ambient humidity conditions (Chiu & Maderson, 1980) those temperature effects which have been reported could be *indirect* via their effect on ambient humidity. From this premise we could postulate that while thyroidectomy-induced multiple sheds have a different etiology than those induced by a high temperature environment, both phenomena can be related to lipid metabolism. In short, it is proposed that integumentary germinal cells may proliferate not merely in response to genetically determined endogenous factors (whether intrinsic to the integument, or humoral) but may also respond directly to a change in net aqueous flux from the internal to the external environment. This is not an *ad hoc* thesis; there is a large body of experimental data from studies of terrestrial arthropods (Hadley, 1981), and at least two other vertebrate systems support it. Furthermore the postulate explains many anecdotal phenomena pertaining to squamate skin-shedding.

THE RESPONSES OF BIOLOGICAL EPITHELIA TO CHANGES IN AQUEOUS FLUX

As indicated in Fig. 2, there are four sets of experimental data which, *a priori*, support the thesis that germinal cells in epithelia respond to changes in aqueous flux.

The complexity of tissue interactions leading to the differentiated state of the amniote integument has been widely documented (Sawyer & Fallon, 1983). The chick chorion, an ectodermal derivative which has been extensively used as a model "competent, undetermined" embryonic tissue in a variety of studies (Sawyer, 1983), has long been known to show keratinous hyperplasia when its external milieu is changed (Moscona, 1959, 1960; Sawyer, 1978). Recent studies (W. M. O'Guinn & R. H. Sawyer, pers. comm.), have revealed that the hyperplastic tissue contains *all* the polypeptides and proteins (intact or subunits thereof?) detectable in the various regions of the adult integument. These data are surprising in view of the plenitude of studies showing that specific protein synthesis identifiable in such discrete structures as feathers, interfollicular epidermis, beak, claws and scutate and reticulate scales is the end-result of specific tissue interactions occurring at various times during embryogenesis (Sawyer, 1983). Normal chorionic tissue does not, by definition, produce such end products. The question therefore arises – what causes such a widespread pattern of transcription and translation of keratinaceous proteins in an *undetermined* tissue? We can propose the following explanation. In the face of a changed external environment, net

EXTERNAL ENVIRONMENT

PROVIDES PHYSIOLOGICAL CHALLENGE
WHICH SYSTEM MEETS (√), DOESN'T MEET
(X), OR UNKNOWN (?)

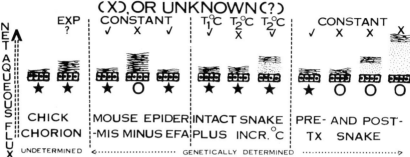

DOES (★) OR DOES NOT (O) PROVIDE
APPROPRIATE MOLECULAR PRECURSORS
FOR EPITHELIAL DIFFERENTIATION

INTERNAL ENVIRONMENT

FIG. 2. Diagrammatic representation of the data supporting the proposed hypothesis. The relationship between endogenous (e.g. mesenchymal influences, hormones, etc.) and extrinsic environmental factors producing net aqueous flux from the internal milieu to the external environment is illustrated. These influence germinal proliferation and daughter cell differentiation in a variety of vertebrate epithelia. For explanation see text pp. 117–120.

aqueous flux changes and chorionic cells proliferate. However, daughter cells leaving a germinal population have, by definition, left the pool of cycling cells and subsequently differentiate. In the absence of any *specific* inductive influence having impinged on the tissue, the daughter cells simply transcribe and translate everything in their "non-neural ectodermal genomic repertoire". This results in a stratified squamous epithelium which manifests thickness and *a* permeability constant although data on its physiological efficacy are lacking.

The mammalian stratum corneum presents a barrier to percutaneous water loss by virtue of extracellular lipid complexes originating from membrane coating granules which extrude their contents during the final phase of keratinization (Wertz & Downing, 1982). Menton (1968, 1970) maintained mice on a diet lacking essential fatty acids. Within a short period of time the interfollicular epidermis showed a spectacular hyperplasia resulting from greatly increased rates of germinal proliferation. However, the animals showed chronic symptoms of elevated cutaneous water loss and after three weeks of this diet drank almost continuously and were "soaking wet". In this instance, a determined organ system, in a *constant external environment*, responded to an enhanced "water-loss load" by elevating rates of mitotic activity (? an attempt to increase barrier thickness), and the daughter cells "tried" to differentiate appropriately. However, since the necessary molecular precursors were absent from the diet, the permeability constant of the corneous layer continued to diminish so that the experimental animals came to lack any degree of integumentary waterproofing. Addition of linoleic acid to their diet rapidly restored the epidermal tissues to their normal anatomical and physiological status.

As Lillywhite & Maderson (1982) have argued, barrier thickness and the permeability constraints of the appropriate epidermal components presumably reflect adaptations to the normal environment in which a reptilian species occurs. Hillman, Gorman & Thomas (1979) have shown that acclimation can influence rates of evaporative water loss for *Anolis* spp., and while this and similar studies either do not address CWL *per se*, or fail to differentiate between it and other possible pathways of water loss, certainly terrestrial arthropods show facultative adaptation to increased environmental water stress (Hadley, 1981). We offer the following explanation for "heat-induced shedding" in *Thamnophis sirtalis* (Maderson, Chiu, Contard *et al.*, in preparation). At temperature $T_1°C$ (Fig. 2) (probably representing a diurnal range of 14–24°C in nature) CWL may be "n" mg cm^{-2} h^{-1}, and imposes a physiological challenge that the mesos layer can meet adequately. Irrespective of the shedding history of a particular animal, its transfer to a constant environmental temperature of 28°C ($T_2°C$, Fig. 2) would expose it to an excessive water stress. Its epidermis would respond by producing a new generation, the mesos layer of which has an "improved" permeability constant, thus meeting the physiological challenge. That this response is not simply a reflection of "heat stress" disturbing the *milieu intérieur* in general is indicated by the following. The time taken for individuals to show a "heat-induced shed" varies considerably, presumably reflecting a degree of "excessive construction" (Gans, 1979). If an individual that shed after X days is then returned to 21°C for a similar period and subsequently returned to 28°C, it will not produce another heat-induced shed.

In the case of thyroidectomy-induced sheds (Chiu, Leung *et al.*, 1983) the animals, like Menton's essential fatty acid-deficient mice (Menton, 1968, 1970), remain in a constant environment. However, according to data of Gupta *et al.* (1975), they show a general disruption of lipid metabolism. Under these circumstances, CWL increases and the epidermis responds by producing new epidermal generations, but, since each mesos layer is "more defective" (in terms of its permeability constant) than the preceding, the animal eventually dies. The cause of death has not yet been determined directly; however, as intact animals kept under identical conditions and starved do not die, while *Ptyas korros* (Chiu, Maderson *et al.*, in preparation) and *Thamnophis sirtalis* (Maderson, Chiu, Contard *et al.*, in preparation) both die when the original weight decreases by 20+%, dehydration seems a likely cause.

DIRECT TESTING OF THE HYPOTHESIS

Since Zucker & Maderson (in preparation) have reported the extreme difficulty of getting accurate data on CWL for snakes, and most studies on arthropods have been pursued via biochemical analysis of the cuticular lipids (Hadley, 1981), Contard & Maderson (in preparation) used a quite different approach. They reasoned that if the mesos granules demonstrated by Landmann (1979, 1980a) were so obvious and readily demonstrable by transmission electronmicroscopy, their differentiation in thyroidectomized snakes should reveal striking differences between control and post-operative animals; this prediction has proved correct (Figs 3 & 4).

ECOLOGICAL AND BEHAVIORAL CORRELATES

Although "skin-shedding" is a well-known phenomenon, there are relatively few species in which it seems to occur regularly and/or frequently (Maderson, Chiu & Phillips, 1970a) and it seems possible that such behavior may represent a maintenance mechanism for specialized epidermal derivatives such as climbing foot-pads (Lillywhite & Maderson, 1982: 431). Surprisingly, species inhabiting xeric environments do not seem to be "regular shedders" (Maderson, Mayhew & Sprague, 1970; Maderson, unpublished data). *A priori* one might predict that the sophisticated thermoregulatory behaviors shown by many squamates would involve some feedback system monitoring CWL during solar exposure, with an associated mechanism controlling integumentary blood flow, but evidence for such is scanty (Lillywhite & Maderson, 1982: 430–433). Maderson (1965b) reviewed the anecdotal data concerning animals entering water before shedding and concluded that it seems to be an individual behaviour pattern which is in accord with our observations on "high-temperature induced shedding". Preliminary observations suggest that intact snakes kept at high tempera-

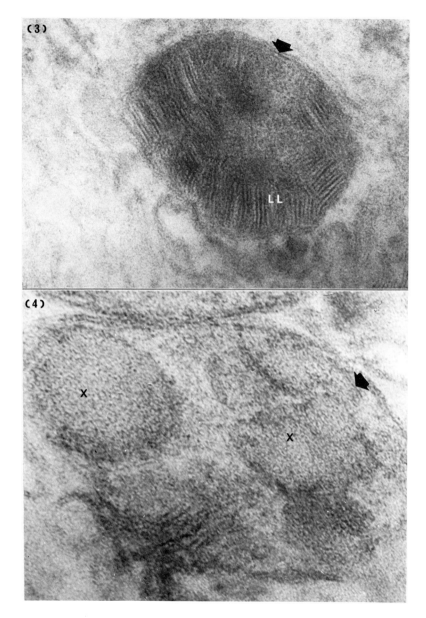

FIGS 3 & 4. (3) Electron-micrograph of an intracellular mesos granule during stage four of the shedding cycle of an intact garter snake (*Thamnophis sirtalis*). Note the precisely-stacked lipid lamellae (LL) lying within the membrane-bounded (arrow) organelle. Final magnification × 192,000. (4) Electron-micrograph of an intracellular mesos granule during stage four of the third post-operative shedding cycle of a thyroidectomized *T. sirtalis*. When compared with the normal organelle (Fig. 3), the absence of lipid lamellae is striking and the organelle is filled with amorphous material (X). Surrounding membranes are visible (arrow). Final magnification × 192,000. (Photographs by courtesy of Mr Paul Contard.)

tures, and thyroidectomized snakes, do tend to spend a good deal of time in their water dishes.

The fact that snakes tend to shed irregularly might reflect their unique locomotory and feeding behavior, both of which depend on exceptional flexibility and distensibility of the integument. The primary barrier to CWL, the mesos layer, consists of terminally differentiated keratinocytes with intercellular lipid lamellae. Such an architecture would be flexible, but would presumably be disrupted by mechanical stresses beyond a certain threshold. Biophysical calculations indicate that the permeability properties of a given membrane could be changed drastically by damage to 0.1% of its total surface (Zucker, 1980; Zucker & Maderson, 1980). Zucker & Maderson (in preparation) have argued that the extreme variability in values for CWL measured *in vivo* in chamber experiments might result from the excessive "bending" of the body: more consistent results are obtained from *in vitro* experiments (Maderson, Zucker & Roth, 1978). Thus exceptionally violent locomotory movements, e.g. escape from a predator, or ingestion of an unusually large prey item, could stress the integument beyond its normal mechanical limits, reduce barrier efficiency and precipitate a renewal phase. However, it should be noted that this postulate is inherently difficult to test since the "damage" to the mesos lipo-protein complexes could be in the order of Angstrom units and therefore not amenable to direct observation by available technology (Zucker & Maderson, 1980).

SUMMARY AND CONCLUSIONS

The working model for further study of skin-shedding centers around the functional premise that the unique flexibility of the squamate integument, especially that of snakes, causes special problems with reference to the maintenance of a barrier against CWL. It is emphasized that although the phenomenon of skin-shedding is so well-known in squamates, relatively few species are known to shed regularly. For those which do, e.g. gekkonids and anoline iguanids, there is little doubt that other selective pressures, notably maintenance of epidermal specializations, are of great importance.

Thyroid physiology of snakes presents many paradoxes by comparison with that of other vertebrates (Chiu, 1982; Chiu, Leung *et al.*, 1983) and the model demands further investigation of circulating titers of the gland's secretions in association with epidermal activity. However, it should be noted that attempts to monitor annual circulating thyroxine by radio-immune assay techniques in *Thamnophis* spp. suggest that levels are below the limits of technical resolution (P. Licht, pers. comm.). It remains to be determined whether this implies that the model cannot be tested or whether other iodinated thyronines (Chiu, Leung *et al.*, 1983) are the important factors.

The model does not fully explain the discrepancies that exist between thyroid activity and SF in snakes versus lizards. It may be that all the unique features of thyroid physiology in snakes are reflections of the special problems of integumentary maintenance in these animals. It is particularly interesting to note that Landmann (1980b) has documented the presence of zonulae occludentes encircling the outermost suprabasal cells lying beneath the corneous layers and representing another unique anatomical feature of snake epidermis. Do these membrane modifications help maintain a high relative humidity in the α-layer, thus making the mesos barrier more effective, and is it disturbance of their maintenance which precipitates the first renewal phase following thyroidectomy? These are important questions for future research. As the model depends on sustaining the *physical* composition of the mesos keratinocytes and their surrounding lipids, it would not be surprising to find that epidermal control mechanisms in snakes and those observable in lizards have evolved along divergent pathways.

The diversity of anatomical form characterizing the tetrapod integument does not detract from the fact that, like the integument of all terrestrial animals, it serves primarily to maintain the physiologic *milieu intérieur*. However, in addition to this primary function, all integuments manifest a diversity of secondary functions and functional adaptation is indubitably *the* common denominator in any discussion of integumentary diversity. In tetrapods, the ubiquitous epidermal-dermal interactions have facilitated such diversity to permit successful occupation of a range of ecological niches (Maderson, 1972). Inasmuch as understanding the *why* of integumentary diversity demands consideration of the entire organism in its natural habitat, so also does understanding the *how* of tissue homeostasis therein demand consideration of the variety of factors which influence cellular behaviour. The squamate integument represents *par excellence* a system which can only be understood by a synthesis of data from a broad spectrum of biological disciplines.

ACKNOWLEDGEMENTS AND DEDICATION

This essay is dedicated to my post-graduate mentor Professor A. d'A. Bellairs on the occasion of his retirement; I offer my thanks for the education I received from him and especially for his setting me to work

on a topic which has permitted me to explore so many of the diverse disciplines of contemporary zoology.
I wish to thank Dr C. Gans for commenting on an early draft of this manuscript. The original research upon which this review is based was supported by National Institutes of Health Grants CA-10844, CA-11536, NS-13924 and RR-07119, and PSC-BHE Grant ≠14056. Ms Judith Steinberg typed the manuscript.

REFERENCES

Baden, H. P. & Maderson, P. F. A. (1970). The morphological and biophysical identification of fibrous proteins in the amniote epidermis. *J. exp. Zool.* **174**: 225–232.

Bentley, P. J. & Schmidt-Nielsen, K. (1966). Cutaneous water loss in reptiles. *Science, N.Y.* **151**: 1547–1549.

Chiu, K. W. (1982). Thyroid function in squamate reptiles. In *Phylogenic aspects of thyroid hormone actions*: 107–122 (*Gunma Symp. Endocr.* No. 19). Inst. Endocr. Gunma Univ. (Ed.). Tokyo, Japan: Center for Academic Publishing.

Chiu, K. W., Leung, M. S. & Maderson, P. F. A. (1983). Thyroid and skin shedding in the Rat Snake (*Ptyas korros*). *J. exp. Zool.* **225**: 407–410.

Chiu, K. W. & Maderson, P. F. A. (1980). Observations on the interactions between thermal conditions and skin shedding frequency in the tokay (*Gekko gecko*). *J. Herpet.* **14**: 245–254.

Chiu, K. W., Maderson, P. F. A. & Zucker, A. H. (In prep.). *Interactions between thermal environments and hormones affecting skin-shedding frequency in the Tokay* (Gekko gecko) (*Gekkonidae, Lacertilia*).

Contard, P. & Maderson, P. F. A. (In prep.). *Studies of the functional anatomy of snake epidermis III. Ultrastructural observations following thyroidectomy.*

Flaxman, G. A., Maderson, P. F. A., Szabo, G. & Roth, S. I. (1968). Control of cell differentiation in lizard epidermis *in vitro. Devl Biol.* **18**: 354–374.

Gans, C. (1979). Momentarily excessive construction as the basis for protoadaptation. *Evolution, Lawrence, Kans*: **33**: 227–233.

Gupta, S. C., Thapliyal, J. P. & Garg, R. K. (1975). The effects of thyroid hormones on the chemical constituents of different tissues of the chequered water-snake *Natrix piscator*. *Gen. comp. Endocr.* **27**: 223–229.

Hadley, N. F. (1981). Cuticular lipids of terrestrial plants and arthropods: a comparison of their structure, composition and water-proofing function. *Biol. Rev.* **56**: 23–47.

Hillman, S., Gorman, G. C. & Thomas, R. (1979). Water loss in anolis lizards: evidence for acclimation and intraspecific differences along a habitat gradient. *Comp. Biochem. Physiol.* (A). **62**: 491–494.

Landmann, L. (1975). The sense organs in the skin of the head of squamates (Reptilia). *Israel J. Zool.* **24**: 99–135.

Landmann, L. (1979). Keratin formation and barrier mechanisms in the epidermis of *Natrix natrix* (Reptilia, Serpentes): an ultrastructural study. *J. Morph.* **162**: 93–126.

Landmann, L. (1980a). Lamellar granules in mammalian, avian and reptilian epidermis. *J. Ultr. Res.* **72**: 245–263.

Landmann, L. (1980b). Zonulae occludentes in the epidermis of the snake *Natrix natrix* L. *Experientia* **36**: 110–111.
Lillywhite, H. B. & Maderson, P. F. A. (1982). Skin structure and permeability. In *Biology of the Reptilia, physiology* C **12**: 397–442. Gans, C. & Pough, F. H. (Eds). London: Academic Press.
Maderson, P. F. A. (1965a). Histological changes in the epidermis of snakes during the sloughing cycle. *J. Zool., Lond.* **146**: 98–113.
Maderson, P. F. A. (1965b). The structure and development of the squamate epidermis. In *Biology of the skin and hair growth*: 129–153. Lyne, A. G. & Short, B. F. (Eds). Sydney: Angus & Robertson.
Maderson, P. F. A. (1968). Observations on the epidermis of the Tuatara (*Sphenodon punctatus*). *J. Anat.* **103**: 311–320.
Maderson, P. F. A. (1972). When? Why? and How?: Some speculations on the evolution of the vertebrate integument. *Am. Zool.* **12**: 159–171.
Maderson, P. F. A. (In press). Some developmental problems of the reptilian integument. In *Biology of the Reptilia: development*: 14. Gans, C., Billett, F. & Maderson, P. F. A. (Eds). New York: John Wiley & Sons.
Maderson, P. F. A. & Chiu, K. W. (1981). The effects of androgens on the glands of the tokay (*Gekko gecko*): modification of an hypothesis. *J. Morph.* **167**: 109–118.
Maderson, P. F. A., Chiu, K. W., Contard, P. & Bank, S. (In prep.) Studies on the functional anatomy of snake epidermis. II. The effects of temperature on shedding frequency.
Maderson, P. F. A., Chiu, K. W. & Phillips, J. G. (1970). Endocrine epidermal relationships in squamate reptiles. *Mem. Soc. Endocr.* **18**: 259–284.
Maderson, P. F. A., Flaxman, B. A., Roth, S. I. & Szabo, G. (1972). Ultrastructural contributions to the identification of cell types in the lizard epidermal generation. *J. Morph.* **36**: 191–210.
Maderson, P. F. A., Mayhew, W. W. & Sprague, G. (1970). Observations on the epidermis of desert-living iguanids. *J. Morph.* **130**: 25–36.
Maderson, P. F. A., Zucker, A. H. & Roth, S. I. (1978). Epidermal regeneration and percutaneous water loss following cellophane stripping of reptile epidermis. *J. exp. Zool.* **204**: 11–32.
Maurer, F. (1892). Hautsinnesorgane, Feder, und Haaranlagen und deren gegenseitige Beziehungen. *Morph. Jb.* **18**: 717–804.
Maurer, F. (1895). *Die Epidermis und ihre Abkommlinge*. Leipzig: Engelmann.
Menton, D. N. (1968). The effects of essential fatty acid deficiency on the skin of the mouse. *Am. J. Anat.* **122**: 337–355.
Menton, D. N. (1970). The effects of essential fatty acid deficiency on the fine structure of mouse skin. *J. Morph.* **132**: 181–206.
Moscona, A. (1959). Squamous metaplasia and keratinization of chorionic epithelium of the chick embryo in egg and in culture. *Devl Biol.* **1**: 1–23.
Moscona, A. (1960). Metaplastic changes in the chorio-allantoic membranes. *Transpl. Bull.* **26**: 120–124.
Roberts, J. B. & Lillywhite, H. B. (1980). Lipid permeability barrier in epidermis of reptiles. *Science, Wash.* **207**: 1077–1079.
Roberts, J. B. & Lillywhite, H. B. (1983). Lipids and the permeability of epidermis from snakes. *J. exp. Zool.* **228**: 1–10.
Sawyer, R. H. (1978). Keratogenic metaplasia of the avian chorionic epithelium: absence of the beta stratum which characterizes the epidermis of the avian scutellate scale. *J. exp. Zool.* **205**: 224–242.
Sawyer, R. H. (1983). The role of epithelial-mesenchymal interactions in regulating gene expression during avian scale morphogenesis. In *Epithelial-mesenchymal inter-*

actions in development: 115–146. Sawyer, R. H. & Fallon, J. F. (Eds). New York: Praeger Scientific.

Sawyer, R. H. & Fallon, J. F. (Eds) (1983). *Epithelial-mesenchymal interactions in development*. New York: Praeger Scientific.

Schaeffer, W. H. (1933). Hypophysectomy and thyroidectomy of snakes. *Proc. Soc. exp. Biol. Med.* **30**: 1363–1365.

Sham, S. K. & Chiu, K. W. (In prep.). Interaction of temperature, thyroid hormones, and metabolism in skin-shedding in Gekko gecko.

Wertz, P. W. & Downing, D. T. (1982). Glycolipids in mammalian epidermis: structure and function in the water barrier. *Science, Wash.* **217**: 1261–1262.

Zucker, A. (1980). Procedural and anatomical considerations of the determination of cutaneous water loss in squamate reptiles. *Copeia* **1980**: 425–439.

Zucker, A. H. & Maderson, P. F. A. (1980). Cutaneous water loss and the epidermal shedding cycle in the tokay (*Gekko gecko*) (Lacertilia, Reptilia). *Comp. Biochem. Physiol.* (A) **65**: 381–391.

Zucker, A. H. & Maderson, P. F. A. (In prep.). Studies on the functional anatomy of snake epidermis. I. Patterns of cutaneous water loss throughout the shedding cycle.

Lizards, Mammals and the Primitive Tetrapod Tympanic Membrane

R. PRESLEY

Department of Anatomy, University College, Box 78,
Cardiff CF1 1XL, UK

SYNOPSIS

The development of the tympanic region has been re-examined with special reference to lizards and mammals. The two forms differ greatly in the site of the external meatus and in the position and anatomy of the membrane during development and in the adult. In neither is the first pouch-first cleft (spiracular) contact utilized in the production of the definitive membrane. The meatal plate of mammals is unique and gives no indication of relationship with a post-quadrate membrane. It is supported by the tympanic bone blastema and in early stages by parts of the second arch cartilage represented by the tympanohyal and the element of Spence.

It is concluded that the considerable arguments presented by Gaupp in his classical studies for the view that there is no homology between the two forms of membrane and that they may have been independently evolved are fully upheld by the present findings. The contrary view, that the mammalian form evolved from a sauropsid-like stage in therapsids, is not supported by developmental evidence. No assumption of derivation from a common "primitive tetrapod" antecedent is suggested by the patterns of development. The architecture of the anlage of the tubo-tympanic recess is common to early stages of all tetrapods: repeated independent evolution of functional contacts with the surface is a reasonably parsimonious hypothesis as judged by developmental criteria.

The blastemata found in Recent mammals may be utilized to suggest a form of post-angular membrane in therapsids, supported posteriorly by the second arch cartilage and tensed by second arch muscle homologous with levator hyoidei rather than depressor mandibulae. Transmission from such a membrane need not have involved great inertial masses, if either liquid linked it to the round window, or the element of Spence acted to transmit directly to the stapes. But by anatomy and development such an element is not the homologue of the sauropsid extrastapes.

THE DILEMMA

"The morphology of the tympanum itself is not clear, as it is developed secondarily, and is apparently not strictly homologous in Anura, Sauropsida and Mammalia" (Parker, 1907).

"Rather should we consider the modern reptilian and mammalian plan as showing two divergent types derived from some intermediate plan of structure perhaps to be discovered among the Theromorpha" (Goodrich, 1916).

INTRODUCTION

The second of the two above views has prevailed. Until recently it seemed that among Recent forms tympanic membranes were so similar in form and function that it was futile to consider them independently evolved.

This situation is now quietly but dramatically altering. Allin (1975) proposed that cynodont reptiles had no tympanic membrane in the sauropsid position; instead this was a more mammal-like structure behind the angular notch. This has been accepted by others (Crompton & Parker, 1978; Kermack, Mussett & Rigney, 1981) while vigorously contested by Parrington (1979). The concept was extended more widely on comparative grounds (Lombard & Bolt, 1979). Primitive labyrinthodonts lack an otic notch and their stapes seems unsuitable for insertion into a tympanic membrane (Smithson, 1982). Therefore the hypothesis that there was no primitive tetrapod tympanic membrane and that those now extant may have evolved independently and are not strictly homologous structures (Tumarkin, 1955; Barry, 1963) must be carefully examined.

The magnum opus of Gaupp (1913) in favour of Reichert's theory of the homologies of the auditory ossicles is rightly acclaimed as a triumph of the methods of comparative anatomy. Ironically, an essential part of that meticulous work was the demonstration that the tympanic membranes of Anura, sauropsids and mammals are not homologous. Goodrich (1916) used developmental evidence effectively to convince the Western world of the incorrectness of that part of Gaupp's conclusions. By accepting that refutation, while also accepting the rest of Gaupp's work on the Reichert theory, comparative anatomists have elected to live with a paradox.

The swing back to Gaupp's view has been supported so far by evidence largely based on comparative anatomy of adult structure and on palaeontological reconstructions. The former can be tested by repeated observation and extension, but the latter method is very susceptible to the prevailing climate of opinion on what is probable in postulated anatomy, fashionably assessed in terms of "parsimony" of postulated evolutionary changes (with difficulties of methodology: Panchen, 1982). But the refutation of Gaupp was based mainly on embryological evidence, so clearly this must be examined again.

In this paper two of the principal examples used in the classical discussions, the lizard and the mammal, will be used to scrutinize developmental evidence for and against their homology. If these are not homologous by development the case for independent evolution of tympanic membranes is strengthened, and, by so much, it becomes reasonable to extend the hypothesis of independent evolution to other tetrapod groups. The embryological method can provide no direct information on the presence or absence of a tympanic membrane in "primitive tetrapods": here the fossil record is paramount. Embryology can, however, indicate what possibilities may be the most probable, for it enables comment upon whether developmental processes can be found among Recent groups which are, or are not, compatible with any anatomy postulated for fossil groups. If parsimony is a valid criterion, no adult structure should be favoured which is not found in Recent forms or is incompatible with the developmental pattern of any Recent form. For example, since no tetrapod develops a simple spiracular tympanic membrane, however tempting the idea of one may seem, it is not parsimonious to postulate that such a membrane was primitive to all tetrapods unless the fossil evidence gives overwhelming support for the concept. Conversely, in the presence of such palaeontological evidence it behoves the embryologist to consider the implications of the developmental pattern involved and how the associated anatomy, such as the courses of nerves and vessels, may have been arranged. In essence, no interpretation of the structure of an ancestor should involve the adoption of a uniquely specialized developmental pathway (i.e. one that has never been seen) incompatible with the observable organogenesis of the descendant groups.

MATERIALS AND METHODS

Principal observations and graphic reconstructions were made from serially-sectioned heads of embryos and adults of: *Lacerta vivipara*; *L. muralis*; *Gallus domesticus*; *Ornithorhynchus anatinus*; *Tachyglossus aculeatus*; *Didelphis virginiana*; *Mus musculus*; *Rattus norvegicus*; *Erinaceus europaeus*; *Sus scrofa*; *Tadarida mexicana*; *Felis domestica*; *Mustela domestica*; *Homo sapiens*.

For comparative purposes and to check observations in past publications reference was made to similar material from less full developmental series of: *Salmo trutta*; *Polypterus* sp.; *Amia calva*; *Rana pipiens*; *Triturus vulgaris*; *Lepidochelys olivacea*; *Crocodylus niloticus*; *Sphenodon punctata*; *Trichosurus vulpecula*; *Potamogale velox*; *Sorex araneus*; *Elephantulus myurus*; *Petrodromus tetradactylus*; *Myotis* sp.; *Hipposideros* sp.; *Pipistrellus pipistrellus*; *Balaenoptera* sp.; *Ovis aries*.

EMBRYOLOGICAL OBSERVATIONS

Concord with Classical Observations

The present paper is a review of past observations. Full descriptions will appear in a future account of the development of the mammalian tympanic cavity as a whole. Here the illustrations provided by Goodrich (1916) have been confirmed in all details and may be referred to. It is important, however, to note that those figures were reconstructed for mammals in a plane at right-angles to that used for his "sauropsids" and that in older mammalian embryos his "thick sections" were not extended ventrally beyond the skeletal elements and consequently omit the deeper part of the external auditory meatus and the meatal plate. Therefore those structures must be added mentally, noting that the lateral edge of the meatal plate must be sited at the level of, or medial to, the neck of the malleus. Similarly, Goodrich (1916) does not emphasise the auditory hillocks and orifice of the external auditory meatus. Reconstructions of these features at similar developmental stages and from similar aspects are presented for lizard and mammal in Fig. 1. The differences are striking. Their principal features will be presented in this section, while other incidental findings of the present study will be referred to later sections.

Principal Present Findings

Tympanic cavity
In early stages of lizard and mammal the tympanic cavity arises from part of the tubo-tympanic recess formed from pouches I and II which are expanded broadly through the head to make contact with the surface ectoderm of clefts I and II. This structure is common to all vertebrates. Thus the body of the tympanic cavity is homologous by development. In each case the contact at both clefts disappears: very early in mammals while the skeleton is mesenchymal; later in the lizard, after the start of chondrification (Goodrich, 1916). Thus in neither case is the tympanic membrane derived directly from pouch I meeting cleft I (*contra* Cartmill, 1975). In both cases the external auditory meatus is derived from more ventral ectoderm in the groove bridged by mesoderm connecting arches I and II.

External auditory meatus
The auricular hillocks were used to compare the site of the orifice of the external auditory meatus. Fixation artefacts cause difficulty in many

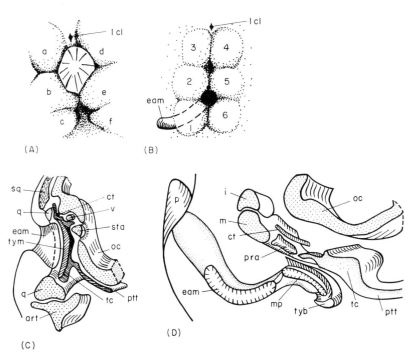

FIG. 1. Graphic reconstructions to show the developmental anatomy of the outer and middle ear regions in sauropsid [(A) and (C)] and mammal [(B) and (D)]. (A) The auditory hillocks of a chick (Stage 38, Hamilton, 1952). (B) The auditory hillocks of a pig embryo, 20 mm CRL. Note the different sites of the membrane blastemata. (C) The anterior part of the tympanic apparatus of a lizard, 6.2 mm Sn-par *Lacerta vivipara*. (D) The anterior part of the tympanic apparatus of a hedgehog, *Erinaceus europaeus*, 9 mm HL. Note contrasting external meati, position and attachment of membrane and position of the primary jaw-joints. Abbreviations: Hillocks in (A), (B) lettered/numbered after Lillie/His (see text). Otherwise see Appendix.

specimens (Wood-Jones & Wen, 1934) but Bouin-Hollande-fixed specimens of *Gallus* and *Mus* could be used to clarify and confirm the deep relations of these hillocks in sauropsids and mammals. In Fig. 1(A), (B) the surface anatomy is compared, using His's numbers (Keibel & Mall, 1910) for mammals and Lillie's letters (Hamilton, 1952) for the sauropsid condition.

The "spiracle" lies dorsal to hillocks a and d or 3 and 4; the groove between the first two arches runs ventrally from it between the two rows. In each case hillocks a or 3 lie entirely above the meatus; b or 2 are at the level of the jaw-joint; c and 1 lie below the axis of Meckel's cartilage. Over the stapes blastema lie d or 4; e or 5 are over the proximal part of Reichert's cartilage; f or 6 are extensions of the epipericardial ridge anterior to the cervical sinus. Thus the hillocks are a firm guide to deeper structures. In sauropsids the external auditory meatus or tympanic membrane lies between a, d, b and e; in mammals it lies more ventrally, between 1, 2, 5 and 6. These are not homologous positions.

The "meatus" of the lizard arises as a broad shallow opening in the groove, first recognizable well before the disappearance of the "spiracle". It deepens but little and becomes moulded over the supporting musculoskeletal elements as they mature.

The meatus is not visible in mammals until after the loss of the "spiracle" and the groove remains narrow and spaced by much mesenchyme from the skeletal elements. At the onset of chondrification a solid rod of epithelium extends in from the groove, eventually to end about two-thirds of the way from the surface to the mid-line. Its deep end broadens to form a solid flat epithelial meatal plate. Canalization of the epithelium occurs much closer to birth (eutherians) or well after birth (marsupials and monotremes). The dorsomedially facing aspect of the meatal plate then becomes the epithelium of the tympanic membrane. It covers mesenchyme between it and the tubo-tympanic recess which is the anlage of the fibrous layer of the membrane. At the onset of ossification this runs from the tympanic bone anteriorly (only latter surrounded by the bone) to the tympanohyal and more dorsal part of Reichert's cartilage posteriorly.

These aspects are shown in Fig. 1(C), (D) from which it can be seen that there can be no developmental homology of the meatus in the two forms. For mammals, Hammar (1902) made all of these observations in a neglected paper which was, however cited by Gaupp (1913).

Support of the tympanic membrane

In the lizard the anlage of the fibrous layer of the membrane is attached to the blastemata of quadrate, articular and depressor mandibulae muscle. The retro-articular process lies ventrally and the dorsal process of the quadrate curves upwards and backwards to the region of the paroccipital process (Figs 2(A), 3(A)). This accords with classical observations (Cords, 1909; Gaupp, 1913; Goodrich, 1916).

In mammals the anlage lies ventral to the axis of Meckel's cartilage. In early stages the manubrium is a weak blastema until the meatal plate forms (Fig. 2(B)), but the part of the blastema closest to the site of the plate is its most dense region. However, in no specimen has this blastema any clear connection to the second arch, nor is there evidence of a separate centre of chondrification for the manubrium, in agreement with classical studies and in contrast to reports of a second-arch origin for the manubrium (Findlay, 1944; Anson, Hanson & Richany, 1960; Jarvik, 1980). Failure here to observe such a phenomenon, described by others, does not disprove the second-arch hypothesis; but here one is

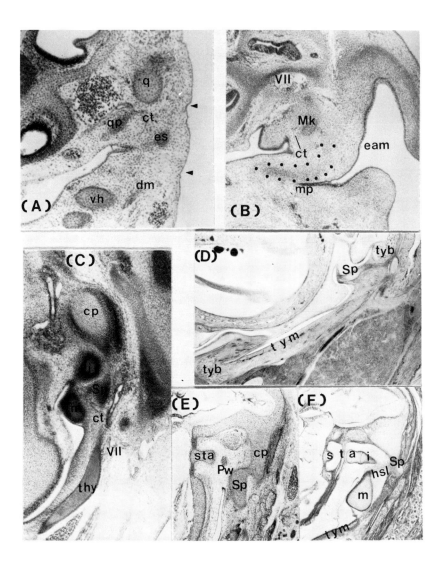

FIG. 2. Sections to show aspects of the developmental anatomy of the tympanic region of lizard and mammal. (A) *Lacerta vivipara* 4 mm HL. Site of ectoderm of membrane arrowed. (B) *Myotis* sp. (bat) 7.5 mm CRL. Note early meatal plate and blastema of manubrium mallei (inside dots) extending from Meckel's cartilage with its densest part near meatal plate. (C) *Hipposideros* sp. (bat) 13 mm CRL showing the element of Spence near stapedial process of incus and crossed ventrally by chorda tympani. (D) *Felis domesticus*, newborn, with a free (true) element of Spence below incus and with tympanic bone now providing attachment of fibrous layer of membrane. (E) *Myotis* sp. (bat) 24 mm CRL with the Spence-like element supporting membrane but fused to the crista parotica. (F) *Pipistrellus pipistrellus* 22.5 mm CRL with only the caudal part of the Spence-like element chondrified, fusing to the crista parotica, and the cranial part forming a fibrous hyostapedial ligament. Abbreviations: see Appendix.

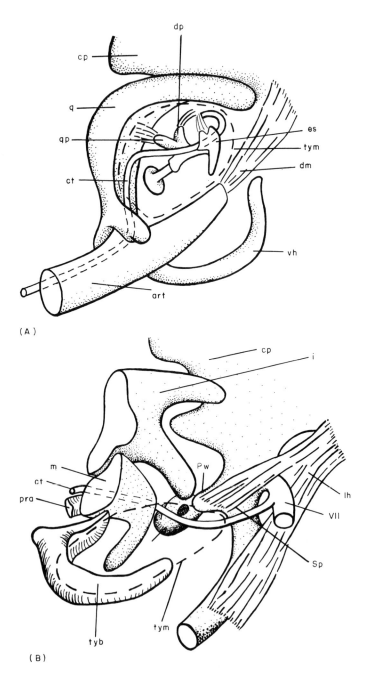

FIG. 3. Graphic reconstructions to show the anatomy of the skeletal elements, nerves, second arch muscles and membrane. (A) *Lacerta vivipara* 6.2 mm Sn-par. (B) A "generalized" mammal, proportions and positions from *Didelphis virginiana*, 14 mm CRL but including the second arch musculature of a 16.5 mm CRL *Ornithorhynchus anatinus*. Compare the blastemata of first and second arch derivatives and note the similarity of the "Paauw and Spence" component in the mammal to the quadrate process component in the lizard. Abbreviations: see Appendix.

obliged to prefer the classical view as consistent with the findings and simpler. There is here an important discrepancy between reports which may in future be settled by histochemical or experimental studies.

Expansion of meatal plate and tympanic cavities leads to the progressive enclosure of the manubrium within the definitive tympanic membrane. At no stage in mammals was a muscle blastema of second (or first) arch origin close to the manubrium.

The developing membrane bone of the tympanic produces a concave lateral edge which overlaps the anterior margin of the membrane anlage (Figs 1(D), 3(B)). Thus the plane of attachment of the membrane corresponds to that of the more medial part of the tympanic, and, before the horns of the latter have ossified, the lateral aspect of the perichondrium of Meckel's cartilage, from which it descends ventrally.

In later stages the tympanic extends round the membrane to support it posteriorly; prior to this, posterior support is provided by parts of the second arch cartilage: the tympanohyal and the element of Spence (Figs 2(C), 3(B)).

The tympanic membrane of the lizard having no such attachments, there can be no homology here.

The second visceral arch blastema

In each case in prechondrification stages the blastema of the second arch is dense and well-circumscribed, embedded in tracts of looser mesenchyme. In the blastema otic, dorsal, quadrate and hyoid processes can be seen (Figs 3(A), (B)). In the lizard an extrastapedial process with an insertion plate in the tympanic membrane is obvious, extending laterally from the site of dorsal and quadrate processes (Figs 2(A), 3(A)). No corresponding element is present in the mammal. The derivation of the adult structure of various lizards has been amply and accurately described (Cords, 1909; de Beer, 1937; Bellairs & Kamal, 1981) and has been judged to correspond sensibly with the hypothesis of the transformed hyomandibula (Westoll, 1943).

Descriptive accounts of mammals have emphasized only the anatomy of the parts of this blastema after chondrification; these vary greatly among mammals. Little attention has been given to that part of the blastema which runs forward from the tympanohyal to the region of the incudostapedial joint, lying above and medial to the chorda tympani nerve. The part of this in association with the insertion of the stapedius muscle chondrifies in some mammals (Cartilage of Paauw). In all mammals the blastema remains loose lateral to this element to give the break between stapes and the rest of the second arch. The next more lateral part of the blastema remains dense in all mammals but varies considerably in its fate. If remaining fibrous, it forms the hyostapedial ligament (Fig. 2(F)); chondrifying but separating from the tympanohyal, it is the element of Spence, which can lie well forward and thus ventral to the incudostapedial joint (Fig. 2(D)) or more posterolaterally to butt against the tympanic bone and participate in the attachment of the early tympanic membrane (Fig. 2(E), (F)). In this position it may remain fused with the tympanohyal and appear as a forward extension of the latter, or in other cases serve as the guiding plane for the formation of investing membrane bone: two forms of Chordafortsatz. The variability of these features in mammals has been extensively described (van der Klaauw, 1931).

The medial part of this blastema, by its anatomy, closeness to the incus and course of the chorda tympani may reasonably be homologized with the quadrate process of the second arch, though not possessing in detail the anatomy of that process in lizards. No part of the blastema projects into the tympanic membrane, or lies significantly outside the line of the chorda tympani or the plane of the second-arch cartilage proper. Thus there is no developmental or anatomical correspondence with the extrastapes of lizards.

The chorda tympani

The type chorda tympani is that of *Homo*; it can be defined functionally as conducting taste and secretomotor fibres between medulla and submandibular and lingual regions. In galliform birds (Goodrich, 1916) and in other vertebrates the possibility arises from present observations that the central part of this route can lie more caudally in the medulla, using the glossopharyngeal-vagus complex and "Jacobson's anastomosis" to reach the mandibular (and maxillary) nerves. In chick and crocodile this route is substantially larger than the atypical chorda tympani branches of the facial nerve. Thus it is possible that the anatomical chorda tympani may be replaced in evolution by Jacobson's anastomosis (or *vice versa*) or both routes may functionally co-exist (as, possibly, in man).

One must therefore agree with Gaupp (1913) that the chorda, while recognizable and with a characteristic route, may nevertheless be a poor guide to homologies unless carefully scrutinized with respect to the total anatomy of the tympanic region.

DISCUSSION

Historical Perspective

Lombard & Bolt (1979) give a full account of the history of this controversy, including discussion of much of the arguments of Gaupp

(1913) and Goodrich (1916). Therefore here the emphasis will be confined to the arguments affecting the tympanic membrane, rather than the homologies of bones.

Gaupp's case

Gaupp (1913) discusses the problem in his Section 10. This is brief, and it is clear that he was addressing a culture in which to take the anuran, sauropsid and mammalian tympanic membranes as not homologous was not new or eccentric. His points are set out briefly below.

(a) The anuran tympanic membrane alone is anatomically comparable with the selachian spiracle; it is clearly not homologous with the others.

(b) The chorda tympani lies wholly above the membrane in sauropsids (metachordal) but the membrane in mammals is part dorsal and part ventral to the nerve (amphichordal). However, the variability of the recesses of the tympanic cavity makes the nerve a poor guide to soft tissue homologies while being a reasonable guide for the skeletal elements, to which it shows reasonable constancy of relation.

(c) The sauropsid membrane lies behind the quadrate, above the jaw-joint and above the retro-articular process (supramandibular). In mammals it is not related to the incus (quadrate), lies below the body of the malleus and the axis of Meckel's cartilage (inframandibular) and, while the distal part of the manubrium mallei is within the membrane, the main attachment of the membrane is to membrane bone (tympanic, squamosal) which, despite its deep placing, corresponds morphologically to a plane superficial to that of the jaw cartilages. Therefore it is not homologous, by these features, with that of the sauropsid.

(d) In mammalian development there is no sign of relative movement of the components of the membrane which could suggest a previous sauropsid situation for it.

(e) The sauropsid membrane is superficial in the head and lies in a parasagittal plane; that of mammals lies deep in the head and is much more horizontal.

(f) The quadrate (incus) differs considerably between the two forms. Gaupp doubts whether that of mammals could readily be derived from one with sauropsid specializations.

(g) There being no sign of movement of the anlagen of malleus, incus and stapes relative to each other or the skull during mammalian development, and since they correspond in these respects to articular, quadrate and stapes in sauropsids, theories that the ossicles have moved substantially must be suspect. Therefore, since the membrane is indisputably in a different position, its displacement relative to the tympanic cavity must be postulated. Logically, such a postulate is an admission that there is no exact homological correspondence between the two forms of membrane.

(h) The configuration of quadrate, articular and stapes found in therapsids is conserved by the ossicles of mammals. No comment, perhaps wisely, is made on the tympanic membrane of fossils.

Goodrich's case

Goodrich (1916) was mainly concerned with resolving doubts about the homology of the tympanic cavity and chorda tympani of sauropsids and mammals, arising from (b) above. In summary, his points are set out below.

(a) The chorda tympani is post-trematic in both forms (except secondarily in galliform birds).

(b) The early embryological "spiracles" are homologous by development and anatomy in the two forms.

(c) In mammals malleus and incus "come to lie between stapes and tympanum". (This implicit movement is not seen in his figures where the blastemata mature *in situ*.) This "new" anatomy Goodrich attributes to the loss of the extrastapes.

(d) Chorda tympani has an "identical course" in the two forms. (Since it is clear from his figures that it has not, presumably the statement is to be restricted to the course with respect to skeletal elements).

(e) There is an otic notch in stegocephalians indicating the site of the primitive tympanic membrane. Anura show this membrane. The chorda is behind this membrane in Anura but in front of it in sauropsids and mammals because of "some relative shifting of parts".

(f) Gaupp (1913) "seems to me greatly to exaggerate the importance of a comparatively trivial difference" in his conclusion that the membranes are not homologous. The mammalian tympanic membrane "extends above and below the manubrium" . . . "If the manubrium of the malleus represents the posterior process of the articular" . . . "the difference between the two types is small".

This last position was restated (Goodrich, 1930): "For Gaupp's view that the tympanic membrane, situated dorsally to the retroarticular process and angular in modern Reptilia, and partly below this level in Mammalia, has been independently evolved in the two groups there seems to be no justification".

Differences between the two cases

Scrutiny of the two cases shows that in most matters of observation and interpretation Goodrich (1916) was in substantial agreement with

Gaupp (1913). Only the matter in (f) above represents major disagreement, and this fails to address the substantial arguments set out by Gaupp, summarized above, concerning the anatomy of the membrane. Apart from the omission of the anlage of the mammalian tympanic membrane in later stages, Goodrich's (1916) figures substantiate each of Gaupp's principal points. However, Goodrich's stylized drawings show the manubrium directed caudally, unlike his careful cameralucida reconstructions. One must conclude that Goodrich missed Gaupp's points, if not the mammalian tympanic membrane itself, in his considerations, and used his masterly style of writing to convince many others.

Further case for the "primitive tetrapod membrane"

Little further attention was paid to the problem of comparative development of the membrane. This is ironic, since the accounts of Cords (1909) for lizards and Hammar (1902) for humans so obviously substantiate Gaupp (1913) that loss of this fact, with its implicit difficulties for a simple evolutionary story, can only make sense if related to a more recent urge to diminish the challenging complexities of observed nature by simplistic generalization.

Instead the contribution of palaeontology became paramount. Relevant were the concept of the stegocephalian tympanum (Watson, 1926), the increasing evidence of close therapsid-mammal relationships (Broom, 1932), and the masterly derivation of Westoll (1943) of all tetrapod configurations from the five-processed hyomandibular of crossopterygians. These arguments were amplified (Westoll, 1945) by the suggestion that the cartilages of Paauw and Spence might represent the "lost" extrastapes of mammals and corroborated by Parrington (1955, 1979) finding in therapsids evidence of dorsal and extrastapedial processes of the stapes and the evolution of a groove on the squamosal. In advanced cynodonts this groove ended in a sharply incurved portion interpreted as the attachment of a tympanic membrane just behind the quadrate, with the groove occupied by a mammal-like external auditory meatus.

Nevertheless, some difficulties remained in assuming that cynodonts possessed lizard-like tympanic membranes.

(a) It was necessary to assume a depressor mandibulae muscle homologous with the sauropsid homonym, inserting into a retroarticular process and supporting the membrane. But where this process lay, and whether it was the homologue of the manubrium mallei was disputed (Kermack, Mussett & Rigney, 1973; cf. Parrington, 1978).

(b) The fate of the dorsal process of the second arch in therapsids was neglected, though it appears as a prominent styloid process in mammals (not seen in cynodonts). There was, however, some evidence that a hyoid apparatus lay close to the stapes (Barry, 1968).

(c) Westoll's hypothesis (1943) and its refinements (Shute, 1956; Hopson, 1966) imply the extinction of the post-quadrate membrane by loss or migration after development of the definitive membrane in its post-angular position. Gaupp (1913) posed the developmental case against this, set out in (d) and (g) on p. 37, but this was overlooked.

Thus, until the challenge of Allin (1975), and Lombard & Bolt (1979), the concept of a primitive amniote, if not tetrapod, tympanic membrane, though almost universally accepted, did not rest on exhaustively tested hypotheses. However, the challenge has been based on palaeontological and comparative grounds and the developmental evidence which was crucial to the rejection of Gaupp's (1913) view has been little discussed.

Two of Allin's (1975) main points were that direct articulation of stapes with quadrate rather than a post-quadrate membrane is primitive for therapsids, and that the squamosal groove may not have been for an external auditory meatus but for a depressor mandibulae muscle similar to that of chelonians. He postulated a *de novo* origin of the membrane in the angular notch. That the membrane was so sited in advanced synapsids has been widely accepted (Crompton & Jenkins, 1979; Kermack *et al.*, 1981; Kemp, 1982). These authors, *per contra* Allin (1975), accept, at least tacitly, that the meatus lay on the squamosal. Lombard & Bolt (1979) threw doubt upon the concept of the primitive tetrapod membrane from which all others were derived, arguing that comparative evidence indicated Gaupp's view to be equally as favoured as Goodrich's. Since then, Carroll (1980) and Smithson (1982) have questioned whether the earliest tetrapods show clear evidence of the possession of any tympanic membrane.

Implications of Embryological Observations

Spiracular tympanic membrane

Contact between ectoderm and endoderm at the site of the first pouch is common in the early stages of vertebrates. However, such contact is devoid of mesoderm and cannot, unmodified, become vascularized and persist as a membrane. Therefore the postulate that the ancestral tetrapod had such a membrane implies greater embryological complexity than at first sight, and should be regarded with more suspicion than heretofore.

Tubo-tympanic recess

In similar early stages in all vertebrates a broad spread of first and second pouch across the head is found. It is therefore reasonable to postulate a tympanic cavity derived by retention of all or part of this recess in any desired tetrapod group. Tympanic membranes could be established from such cavities where the overlying mesoderm could specialize to give fibrovascular support to an accessible area of ectoderm. The mechanism of this in lizards differs from that of mammals, and therefore anatomical and developmental criteria of homology fail to relate the two forms. It is most parsimonious to postulate independent evolution of the two forms of development. Also, since this was the classical ground on which the homologies were based, it becomes more reasonable to advocate independent acquisition of membranes in other Recent and fossil groups. A legion of possibilities is thus created, and therefore hypotheses in relation to fossil specimens must be carefully tested by seeking clear evidence of skeletal specializations for the support of the edges of the membranes: in the absence of this, however necessary for an evolutionary story, such a membrane must be regarded as speculative.

External auditory meatus

The development of the meatus in lizard and mammal is clearly different, and the anatomy of the site differs with respect to the auricular hillocks. It is therefore parsimonious to assume that the two forms are not homologous in these respects. This does not rule out the hypothesis that the ancestors of mammals utilized ectoderm from the sauropsid position to produce a membrane at some stage in their evolution, but this is not indicated by the development of any Recent mammal studied and will therefore require strong support from the fossil record. The squamosal groove (Parrington, 1955, 1978, 1979) in therapsids seems reasonably to correspond to the line of the meatus of Recent mammals, and, *per contra* Allin (1975) there is no sign of a potential depressor mandibulae blastema in any mammal in this position. This favours Parrington's view (1979) but with the qualification that in all Recent mammals the meatal tissue runs far beyond the skeletal markers in the post-quadrate position to a membrane in the tympanic (angular) position, so that the part of Allin's (1975) interpretation involving such a membrane in therapsids is quite parsimonious.

Position of the mammalian tympanic membrane

Nothing emerges from the present study to counter Gaupp's list of arguments on pp. 137–138 showing that the mammalian membrane is in a position morphologically dissimilar from that of sauropsids. The only attachment to the first arch cartilage is below its axis and where the manubrium extends by chondrification between its extending layers. This manubrium shows no sign of affording attachment for second arch musculature, throwing doubt on its homology with the retro-articular process of sauropsids and indeed therapsids.

In mustelids there is a slip of second arch retractor muscle, which is associated in development with the blastema of the posterior belly of digastric; but this inserts into and above the lateral process on the "neck" of the malleus, not into the manubrium. Such a muscle is absent in monotremes, marsupials and most placentals, and is perhaps best considered as a specialization, where it occurs, rather than as a vestige of a depressor mandibulae. But if it is regarded as such a vestige, then it argues against the homology of the manubrium with the retro-articular process in support of Parrington (1978) and *per contra* Kermack *et al.* (1973, 1981).

Detrahens mandibulae muscle

This unique muscle of monotremes runs from the paroccipital region to the posteroventral aspect of the mandible and has on occasions been compared to a sauropsid depressor mandibulae. The present study confirms that it is innervated by the mandibular nerve (Lubosch, 1906) and is of first arch origin. In a newly-hatched platypus its orientation suggests that initially it pulls anterior to the axis of the primary jaw-joint (i.e. malleo-incudal) and is thus at first an adductor mandibulae. In monotremes allometric growth allows the dentary-squamosal joint to move dorsally and posteriorly with respect to the attachments of detrahens, converting its action to that of a retractor and very weak depressor of the mandible. Thus it can be no homologue of the depressor of sauropsids.

The course of the external auditory meatus of monotremes with respect to cranial bones, mandible and first and second arch musculature in early stages is very similar to that of other mammals, and affords no support to the idea (Hopson, 1966; Parrington, 1974) that here there has been independent evolution amongst mammals. It should be noted that present-day monotremes show considerable post-hatching specialization of the architecture of their jaw apparatus and it is risky to take their adult structure as any sort of guide to the possible structure of primitive mammals.

Tympanic meatal plate

The meatal plate of mammals has no parallel in sauropsids. The fact that the tympanic bone overlaps it, even in early stages, is of interest in that if Recent mammals retain any correspondence in their tympanic with the angular of therapsids, then it should be considered possible that the plane of the therapsid membrane lay within the notch, overlain by the reflected lamina and close to the plane of the keel. Previous reconstructions (Allin, 1975; Kermack *et al.*, 1981) imply attachment to the reflected lamina itself.

Chorda tympani and supporting cartilages

If it is reasonable to use the course of the chorda tympani as a guide to the homologies of skeletal elements (Gaupp, 1913; Goodrich, 1916; pp. 135–136), then it provides a guide to the identity of the quadrate process of the second arch in the two forms. The line of this blastema in the lizard is easy to see and there is general agreement on its identity. It is less widely appreciated that a similar blastema can be seen in early stages of mammalian development.

In mammals the chorda passes outside the tympanohyal (always in early stages, though later migrating through, in e.g. *Trichosurus* and *Macropus*), then along the Chordafortsatz anlage on its lateroventral aspect to reach the medial side of the malleus below the primary jaw-joint. This part of the second arch blastema thus corresponds to the quadrate process. Prior to birth, the tympanic membrane lies deeply and horizontally, at first attached to the second arch blastema in this region. Further extension of the meatus and tympanic cavity to enclose the chorda below Shrapnell's membrane or pars flaccida occurs very late.

The deeper portion of this blastema, inside the chorda, is separated by a mesenchymatous break from the outer part after the earliest stages; its dense part is associated with the insertion of the stapedius muscle and when chondrified represents the element of Paauw. To be parsimonious in response to this universal pattern in mammals, any skeletal apophysis found in this place on the stapes of fossils should be regarded as an indication of the insertion of a homologue of stapedius, rather than for an extrastapes (*per contra* Parrington, 1979). The gap between this and the outer part of the second arch blastema in mammals must be regarded as the developmental equivalent of the vertical subdivision of the "primitive hyomandibula" required in Westoll's (1943) treatment of the evolution of a styloid process. Its effect is to leave mobility between the incudostapedial joint and the more distal Chordafortsatz or element of Spence. But it lies internal to the line of the chorda tympani and is very dissimilar in its relation to the gap between the stapedial elements and the hyoid elements of sauropsids: there is no anatomical or developmental support for homology here.

However, the mammalian blastema shows reasonable homologues of the otic process (stapes), quadrate process (element of Spence), dorsal process (styloid) and hyoid process (stylohyoid) implicit in Westoll's analysis. Lacking is the fifth process, an extrastapedial element running from the line of the otic process to an insertion plate in the tympanic membrane: this sauropsid feature shows no obvious homologue in the development of mammals.

Nevertheless, it is clear in early development that the membrane is supported by the element of Spence and the tympanohyal. The almost complete enclosure of the membrane by the tympanic which occurs later in mammals clearly could not have been complete in therapsids, in which the angular moved with the jaw. If these animals had a membrane in the post-angular position (Allin, 1975; Crompton & Parker, 1978; Kermack et al., 1981) some other mechanism must have been present for its posterior attachment. Mammalian development shows no sign of a depressor mandibulae muscle. If there was some homologue of the manubrium mallei attached to the angular (e.g. Kermack et al., 1981), then it is strange that it has left little trace on fossils, and the membrane would lie in a position anatomically dissimilar to that of mammals in being wholly anterior to the primary jaw-joint. It is also unclear how such a membrane, not connected directly to the stapes, could transmit its vibrations to the inner ear without reducing its detection efficiency almost to that of bone conduction because of the inertial mass of the lower jaw to which it was attached. The possibility of fluid transmission from such a membrane will, however, be mentioned below.

The attachment of the tympanic to the skull must be an advanced feature in mammalian evolution. It is only found, with great variety of form and function (Fleischer, 1978), in mammals and presumably can only evolve when the post-dentary elements have lost functional connection with the lower jaw. Only at this stage is there clear evidence of the existence of a manubrium mallei: a feature which arises quite late in skeletal maturation. Thus it is reasonable to hypothesize that the tympanic "ring" and the manubrium are co-members of an advanced complex adaptation late in therapsid evolution.

It follows that extrapolation back to therapsids using the information from mammalian development suggests strongly that the second arch equivalents of the tympanohyal and the element of Spence should be considered as posterior supports for the membrane. There is some fossil evidence that the hyoid portion of the second arch was closely associated with the stapes in therapsids (Barry, 1968). In large cynodonts (Presley, 1980) there is an ovoid facet on the paroccipital process just behind the meatal groove on the squamosal which corresponds exactly with the attachment of the second arch blastema to the crista parotica in Recent mammals. It must also be considered whether the levator hyoidei blastema in its present position could not have been present in therapsids, when it could have acted as a retractor of this part of the second arch and perhaps kept the membrane tense when the jaw was depressed. Such a mechanism would need a joint with the paroccipital process, which could account for the loss of the element during fossilization.

In view of the above considerations, the embryologist must respond to the challenge presented by fossils. Two models of the therapsid tympanic apparatus will therefore be presented which minimize changes in developmental pathways during the reptile-mammal transition.

Evolution of lizard tympanic membrane

That the squamate tympanic membrane may not have been evolved from a "primitive tetrapod" type has been well argued by Carroll (1980). It must be for herpetologists to determine at what stage a membrane evolved and to trace the phylogenetic implications of the very great diversity of auditory mechanisms in reptiles as a whole (Wever, 1978). Here it is reiterated that in lizard development there is no compelling evidence to suppose that evolution from a "spiracular" type was involved. The ectoderm of the membrane is from a more ventral location, though not from a site homologous with that of mammals. The site seems similar among reptile groups, but it must be remembered that squamates, birds, crocodiles and chelonians have middle and inner ears which show very great differences of form and function.

Evolution of the therapsid membrane

Given that there is nothing in mammalian development to support the notion of evolution from either a "spiracular" or a sauropsid form, it is very reasonable to propose a membrane evolving *de novo* in the post-angular position. In this connection it is tempting to wonder whether the appearance of the angular notch in pelycosaurs, with its retention as a unique feature of therapsids, might not have been associated with the appearance of an external auditory meatus of equally unique developmental pattern. This leads to the possibility of air vibrating in a space supported laterally by the reflected lamina of the angular, the receptor surface being the medial face of this cavity, homologous with the mammalian tympanic membrane by phylogenetic descent. From the argument on p. 141 it follows that no embryological case can be offered against the accounts of Westoll (1943), Shute (1956), Hopson (1966) or Allin (1975) as to which part of the tubo-tympanic recess might reach the membrane, though parsimony favours its development in all forms from an inframandibular site (cf. Shute, 1956). Nor, since the adductores mandibulae are not, as a group, inserted on the post-dentary bones in mammals, can the embryologist comment on the use of the reflected lamina as a muscular apophysis (Parrington, 1978; Kemp, 1982), though if the membrane were, as suggested above, in the plane of the keel of the angular this suggestion does not exclude the possibility that both features of function could have been present together.

The mass and inertia of the jaw elements and stapes has been a conceptual difficulty in visualizing transmission (Hotton, 1959). However, a membrane itself in this position with the vibratory properties of the mammalian type, given a tensing mechanism, should be as capable of resonance as any in Recent forms. It is important to recognize that the sensitivity of the ear is determined primarily by the properties of the inner ear (Manley, 1972) and that these can never be known for any fossil. By the same token, reasoning based on calculation of transducer ratios (Kermack *et al*., 1981) is suspect because the impedance offered by the components depends not only on the unit impedance of the fluid media but to a much greater extent on the configuration, size and resilience of the soft-tissues which confine those fluids. In fossils there is no reasonable chance of determining these.

Transmission from the membrane by liquid

Sound transmission to the inner ear by liquid is not unknown, e.g. in urodeles (Schmalhausen, 1968). In mammalian development the tympanic cavity is fluid-filled till birth, yet it is popularly felt to be reasonable to consider the benefits to the prenatal human of playing music, and it is certain that the prenatal mammal is subject to the vibrations from maternal respiration, borborygmi, cardiovascular sound and probably vocalization. From the onset of chondrification synovial cavities are present in the interossicular joints, and clearly from the moment of birth sound is meaningful to many mammals. Thus it may be supposed that a functioning auditory system can mature in liquid.

The cynodont fenestra rotunda is small and faces posteriorly towards the exit of the jugular foramen (Estes, 1961). In mammals this aperture is larger, more laterally placed and subdivided by the processus recessus (de Beer, 1937) into true fenestra, related to air in the tympanic cavity, and the small perilymphatic duct which remains related to liquids. The platypus lacks this subdivision, and even in the most mature specimen in this study (295 mm CR) the small, posteriorly-facing fenestra is spaced by 1 cm of mesenchyme from the tympanic air-space, which does not extend behind the stapes. In echidna the processus recessus develops late and is not anatomically homologous with that of other mammals (Kuhn, 1971). These findings hint at the possibility that the development of the mammalian fenestra in its typical form was late in evolution, perhaps concomitant with the evolution of the free ossicular complex, fixed tympanic, and (p. 144) possibly the manubrium.

It follows that consideration might be given to the possibility that a membrane in the therapsid post-angular position with an area sufficient to give an adequate transducer ratio at the small fenestra, might have transmitted vibrations through liquid, with the much more massive jaw-associated ossicular system serving as an inertial reference and pressure release element. This is untestable, and is offered only to draw attention to this neglected aspect of otic morphology. The fluid might have been secretions of the tympanic epithelium, or venous blood which, in many reconstructions of therapsids (Kemp, 1982), is accepted as converging in large extracranial veins on this region to join the jugular system.

The merit of this idea is that it allows the hypothetical development of therapsids so equipped to embrace only those features of mammalian development prior to the fixation of the tympanic to the skull, without hypothetical additions such as an extrastapes and without the inertial problem posed by the attachment of the membrane to the lower jaw. From such a stage the advance to true mammals would involve only fixation of the tympanic to the skull, freeing of the ossicular chain to reduce inertia, and changes in the fenestra rotunda – the expanding air-space giving new resonant qualities with possible adaptive advantages (Webster, 1962). Since the central receptors of the inner ear would still be recording standing waves in the labyrinth, the route change of primary input need not have been a severe problem.

Spence's element as an extrastapes

Findlay's hypothesis (1944) of the role of the second arch in tympanic evolution has been generally neglected. It contains the idea that the element of Spence may be the vestige of a connection between the stapes and the membrane. In detail, he suggested that the insertion plate of this element became the manubrium and lost its continuity with the element of Spence. The appearance of the manubrium, prominent in all mammals and unimpressive at best in all known therapsids, is a major problem of morphology. Here the interpretation of development favours a first-arch origin of the manubrium, but Findlay's specimens have not been examined. Nevertheless, the problem remains that in producing a mammal from an advanced therapsid both a manubrium and a styloid process must be accounted for.

In Fig. 4 an entirely hypothetical reconstruction is presented in

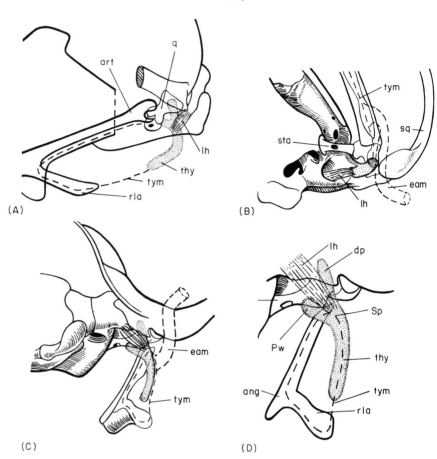

FIG. 4. Reconstructions of *Morganucodon* after Kermack *et al.* (1981) but including a post-angular tympanic membrane supported posteriorly by a second arch element based on the blastemata of mammals. In (A), (B) and (C) the position of meatus, membrane, second-arch element and muscle are indicated from lateral, ventral and posterior respectively. In (D) more detail of the element is shown and it is drawn as if it fitted dorsally into the deep pit present here in *Morganucodon*. Abbreviations: see Appendix.

which the vibrating skeletal element is the second arch (tympanohyal) supporting the rear of the membrane, linked by the element of Spence to the outer part of the stapes and kept in tension by muscles corresponding to the second arch levator hyoidei. This is closest to the very young monotreme configuration, where a fascicle lies dorsal to the chorda and runs to the Spence-like part of the second arch, anatomically somewhat reminiscent of the stapedius, but later degenerates. There is no true stapedius muscle in adult monotremes, but this is probably secondary to the very specialized pattern of the ossicles.

In this reconstruction the element is speculatively placed on a reconstruction of *Morganucodon*, with the dorsal process in the fossa for the "great quadrate ligament" (Kermack *et al.*, 1981). In cynodonts such a reconstruction would place the joint on the paroccipital process just behind the quadrate. In this reconstruction the course of the external meatus conforms to that of Parrington (1979) except that it continues ventrally past the quadrate to a post-angular membrane.

This hypothesis is again untestable, but it does utilize elements found in mammals in approximately this configuration in early development. It does not resolve the problem of the manubrium and styloid process, assuming for the latter that a loose connection was present between the tympanohyal and the hyoid suspensorium which did not impede vibration. Nor, if it could work, does it account for an adaptive advantage later in evolution for the development of the three-ossicle mechanism. Thus it is presented with diffidence, mainly to emphasize that there are many neglected and unresolved problems in our analysis of this region.

CONCLUSION AND DEDICATION

No case is found here for regarding the tympanic membrane of lizard and mammal as homologous by development or anatomy. There is some development in common for the tympanic cavity, but none for the ectodermal component. The case for common inheritance from a "primitive tetrapod" form is thereby diminished, and this is consistent with the current doubts about that structure having ever existed. Thus the arguments of Gaupp (1913) and Lombard & Bolt (1979) are upheld on developmental grounds.

Mammalian development here shows many unique features, and it can be argued that the appearance of the angular notch in therapsids may signify the start of the evolution of a specialized auditory system whose Recent form appeared only after an intermediate structure present in cynodonts and their more advanced non-mammalian descendants. Two new models of that intermediate form have been suggested, based on the need to minimize change in the developmental elements involved. These and others require criticism using meticulous anatomy, developmental studies, careful scrutiny of the fossil record and physiological experiments (especially on neonates and pre-natal animals).

All these disciplines, close to the heart of Angus Bellairs, must be put together as a form of morphological science recognizable to his hero, Gaupp, if our subject is to have a strong and healthy future evolution.

ACKNOWLEDGEMENTS

I wish to express my gratitude to Miss C. Hemington for the drawings, to Mr P. F. Hire for photographic work, and to Messrs D. Norman, D. Powell and M. K. Henderson and Mrs S. K. Singhrao for technical assistance with the specimens under grants from the Science and Engineering Research Council.

I am especially grateful to Professor A. d'A. Bellairs for generous gifts and loans of material without which this work would have been impossible, and to Dr H. L. H. H. Green for the loan of monotreme material.

REFERENCES

Allin, E. F. (1975). Evolution of the mammalian middle ear. *J. Morph.* **147**: 403–438.
Anson, B. J., Hanson, J. S. & Richany, S. F. (1960). Early embryology of the auditory ossicles and associated structures in relation to certain anomalies observed clinically. *Ann. Otol. Rhinol. Lar.* **69**: 427–448.
Barry, T. H. (1963). On the variable occurrence of the tympanum in Recent and fossil tetrapods. *S. Afr. J. Sci.* **59**: 160–175.
Barry, T. H. (1968). Sound conduction in the fossil anomodont *Lystrosaurus*. *Ann. S. Afr. Mus.* **50**: 275–281.
Beer, G. R. de (1937). *Development of the vertebrate skull.* Oxford: Clarendon Press.
Bellairs, A. d'A. & Kamal, A. M. (1981). The chondrocranium and the development of the skull in Recent reptiles. In *Biology of the Reptilia*, **11**: 1–263. Gans, C. & Parsons, T. S. (Eds). London: Academic Press.
Broom, R. (1932). *The mammal-like reptiles of South Africa and the origin of mammals.* London: Witherby.
Carroll, R. L. (1980). The hyomandibular as a supporting element in the skull of primitive tetrapods. In *The terrestrial environment and the origin of land vertebrates*: 293–317. Panchen, A. L. (Ed.). London: Academic Press.
Cartmill, M. (1975). Strepsirhine basicranial structures and the affinities of the Cheirogaleidae. In *Phylogeny of the primates*: 313–354. Luckett, W. P. & Szalay, F. S. (Eds). New York: Plenum Press.
Cords, E. (1909). Die Entwickelung der Paukenhöhle von *Lacerta agilis*. *Anat. Hefte* **38**: 219–319.
Crompton, A. W. & Jenkins, F. A. (1979). Origin of mammals. In *Mesozoic mammals*: 74–90. Lillegraven, J. A., Kielan-Jaworowska, Z. & Clemens, W. A. (Eds). Berkeley: University of California Press.
Crompton, A. W. & Parker, P. (1978). Evolution of the mammalian masticatory apparatus. *Am. Scient.* **66**: 192–201.
Estes, R. (1961). Cranial anatomy of the cynodont reptile *Thrinaxodon liorhinus*. *Bull. Mus. comp. Zool. Harv.* **125**: 165–180.
Findlay, G. H. (1944). The development of the auditory ossicles in the elephant shrew, the tenrec and the golden mole. *Proc. zool. Soc. Lond.* **114**: 91–99.
Fleischer, G. (1978). Evolutionary principles of the mammalian middle ear. *Adv. Anat. Embryol. Cell Biol.* **55**(5): 1–70.

Gaupp, E. (1913). Die Reichertsche Theorie (Hammer-, Amboss und Kieferfrage). *Arch. Anat. Physiol. (Suppl.)* **1912**: 1–416.
Goodrich, E. S. (1916). The chorda tympani and middle ear in reptiles, birds and mammals. *Q. Jl microsc. Sci.* **61**: 137–156.
Goodrich, E. S. (1930). *Studies on the structure and development of vertebrates.* London: Macmillan.
Hamilton, H. L. (1952). *Lillie's development of the chick* (3rd Edn). New York: Henry Holt & Co.
Hammar, J. A. (1902). Studien über die Entwicklung des Vorderdarms und einiger angrenzendend Organe I. *Arch. mikrosk. Anat.Entw..Mech.* **59**: 471–628.
Hopson, J. A. (1966). The origin of the mammalian middle ear. *Am. Zool.* **6**: 437–450.
Hotton, N. (1959). The pelycosaur tympanum and early evolution of the middle ear. *Evolution, Lancaster, Pa.* **13**: 99–121.
Jarvik, E. (1980). *Basic structure and evolution of vertebrates*, 2. London: Academic Press.
Keibel, F. & Mall, F. P. (1910). *Manual of human embryology.* Philadelphia: J. B. Lippincott Co.
Kemp, T. S. (1982). *Mammal-like reptiles and the origin of mammals.* London: Academic Press.
Kermack, K. A., Mussett, F. & Rigney, H. W. (1973). The lower jaw of *Morganucodon*. *Zool. J. Linn. Soc.* **53**: 87–175.
Kermack, K. A., Mussett, F. & Rigney, H. W. (1981). The skull of *Morganucodon*. *Zool. J. Linn. Soc.* **71**: 1–158.
Klaauw, C. J. van der (1931). The auditory bulla in some fossil mammals. *Bull. Am. Mus. nat. Hist.* **62**: 1–352.
Kuhn, H.-J. (1971). Die Entwicklung und Morphologie des Schadels von *Tachyglossus aculeatus*. *Abh. senckenb. naturforsch. Ges.***528**: 1–224.
Lombard, R. E. & Bolt, J. R. (1979). Evolution of the tetrapod ear: an analysis and reinterpretation. *Biol. J. Linn. Soc.* **11**: 19–76.
Lubosch, W. (1906). Über das Kiefergelenk der Monotremen. *Jena Z. Med. Naturw.* **41**: 549–606.
Manley, G. A. (1972). A review of some current concepts of the functional evolution of the ear in terrestrial vertebrates. *Evolution, Lawrence, Kans.* **26**: 608–621.
Panchen, A. L. (1982). The use of parsimony in testing phylogenetic hypotheses. *Zool. J. Linn. Soc.* **74**: 305–328.
Parker, W. N. (1907). *Comparative anatomy of vertebrates.* [Adapted from the German of Wiedersheim, R.] London: Macmillan.
Parrington, F. R. (1955). On the cranial anatomy of some gorgonopsids and the synapsid middle ear. *Proc. zool. Soc. Lond.* **125**: 1–40.
Parrington, F. R. (1974). The problem of the origin of monotremes. *J. nat. Hist.* **8**: 421–426.
Parrington, F. R. (1978). A further account of the Triassic mammals. *Phil. Trans. R. Soc.* (B) **282**: 177–204.
Parrington, F. R. (1979). The evolution of the mammalian middle and outer ears: a personal review. *Biol. Rev.* **54**: 369–387.
Presley, R. (1980). The braincase in Recent and Mesozoic therapsids. *Mem. Soc. géol. Fr.* (NS) **139**: 159–162.
Schmalhausen, I. I. (1968). *The origin of terrestrial vertebrates.* London: Academic Press.
Shute, C. C. D. (1956). The evolution of the mammalian ear drum. *J. Anat.* **90**: 261–281.
Smithson, T. R. (1982). The cranial morphology of *Greererpeton burkemorani* Romer (Amphibia: Temnospondyli). *Zool. J. Linn Soc.* **76**: 29–90.

Tumarkin, A. (1955). On the evolution of the auditory conducting apparatus. A new theory on functional considerations. *Evolution, Lancaster, Pa.* **9**: 119–140 & 193–216.
Watson, D. M. S. (1926). The evolution and origin of the amphibia. *Phil. Trans. R. Soc.* (B) **214**: 189–257.
Webster, D. B. (1962). A function of the enlarged middle-ear cavities of the kangaroo-rat, *Dipodomys*. *Physiol. Zool.* **35**: 248–255.
Westoll, T. S. (1943). The hyomandibular of *Eusthenopteron* and the tetrapod middle ear. *Proc. R. Soc.* (B) **131**: 393–414.
Westoll, T. S. (1945). The mammalian middle ear. *Nature, Lond.* **155**: 114.
Wever, E. G. (1978). *The reptile ear*. Princeton: University Press.
Wood-Jones. F. & Wen I-Chuan (1934). The development of the external ear. *J. Anat.* **68**: 525–533.

APPENDIX – KEY TO ABBREVIATIONS USED IN FIGURES

ang	angular keel	Pw	element of Paauw (site of)
art	articular	q	quadrate
cp	crista parotica	qp	quadrate process
ct	chorda tympani	rla	angular reflected lamina
dm	depressor mandibulae	Sp	element of Spence (site of)
dp	dorsal process	sq	squamosal
eam	external auditory meatus	sta	stapes
es	extrastapes	tc	tympanic cavity
hsl	hyostapedial ligament	thy	tympanohyal
i	incus	tyb	tympanic bone
lh	levator hyoidei	tym	tympanic membrane
m	malleus	v	lateral head vein
Mk	Meckel's cartilage	vh	ventrohyal
mp	meatal tympanic plate	1 cl	site of first cleft
oc	otic capsule	VII	facial nerve
p	pinna	CRL	crown-rump length (curved)
pra	pre-articular	HL	snout-occiput length (curved)
ptt	pharyngo-tympanic tube	Sn-par	snout-parietal length (straight)

Development

Developmental Processes Underlying the Evolution of Cartilage and Bone

BRIAN K. HALL

*Department of Biology, Life Sciences Centre,
Dalhousie University, Halifax, Nova Scotia, Canada B3H 4J1*

SYNOPSIS

The aim of this chapter is to examine the stage of skeletal differentiation found in the Reptilia, with especial emphasis on underlying developmental processes. It begins with secondary cartilage, which is found in the sutures, articulations and fractures of dermal bone of birds and mammals, where it differentiates from otherwise osteogenic periosteal cells as a response to localized mechanical stimulation. The author's own studies on the Australian tiger snake, *Notechis scutatus*, combined with a literature search, indicate that reptiles lack secondary cartilage. Either their periosteal cells are unable to synthesize cartilage-specific products or they cannot respond to the mechanical stimulation which evokes such synthesis. Distinctive features of reptilian bone and cartilage are then considered and it is concluded: (a) that bone is highly adaptive to local needs in forming the full range of vertebrate bony tissue types; (b) that the significant amounts of intratendinous and metaplastic bone found in reptiles are a consequence of the limited osteogenic ability of reptilian periostea; (c) that the reptiles were the first vertebrates to organize their cartilages as hyaline, fibro, etc. and (d) that the diversity of methods which reptiles use to form secondary centres makes them an ideal group in which to study the mechanism of both the resistance which uncalcified cartilage shows to vascular invasion and of indeterminate growth. Ectopic cartilage and bone are well developed in reptiles as illustrated by sesamoids and cardiac cartilage. The most distinctive new process seen in limb development in the Reptilia is programmed cell death, a process which is used in sculpting the digits and in removing the apical ectodermal ridge and somitic cell processes during regression of the limb buds of limbless reptiles. Finally, reference is made to the timing of inductive tissue interactions in the induction of Meckel's cartilage and to the possible importance of shifts in timing of developmental processes in skeletal evolution.

INTRODUCTION

This chapter will present several examples to illustrate the developmental processes which have underlain the evolution of cartilage and bone. As far as is posible examples will be restricted to reptiles although occasionally, when data on reptiles are not available, examples from other groups, largely birds, will be used. This is therefore *not* a comprehensive overview of the evolution of either the skeleton or its tissues.

It *will* try to point out where data from reptiles would be especially useful in furthering our knowledge of the evolution of the vertebrate skeletal tissues. But first, the contribution to a symposium on the structure, development and evolution of reptiles, from an author who has worked almost exclusively with avian embryos and occasionally with foetal mice, should be legitimized by a report of the author's only study on reptiles. These data, although not previously published, have been referred to in the literature. They concerned a search for secondary cartilage in the skull of an almost full-term embryonic Australian tiger snake (*Notechis scutatus*), a member of the subfamily Elapinae in the Family Elapidae, or Colubridae as it was at the time of the study. These data will be preceded by some background on the distribution and circumstances surrounding the origin of secondary cartilage. That forms the first section of this chapter, which then goes on to examine (a) any distinctive features of reptilian bone and cartilage, (b) the developmental processes which produce those tissues, and (c) the evolution of those processes, and finally to summarize some problems for future research on the development and evolution of reptilian cartilage and bone.

SECONDARY CARTILAGE – DO REPTILES POSSESS IT?

Secondary cartilage, sometimes known as adventitious cartilage, arises on dermal bones *after* the commencement of intramembraneous ossification. In this sense it is the opposite of cartilage which precedes the development of replacement bones, which cartilage forms the primary skeleton. Secondary cartilage is also histologically distinctive, having very little extracellular matrix. DNA synthesis and cell division continue *after* secondary chondrocytes are embedded in extracellular matrix, so that its growth is interstitial as well as appositional – another difference from primary cartilage. While primary cartilage differentiates in the absence of embryonic movement, as seen, for example, in paralysed embryos, or *in vitro*, secondary cartilage *only* differentiates in response to mechanical stimulation, and often from progenitor cells which would otherwise have differentiated into oesteoblasts and formed bone. For reviews of these attributes of secondary cartilage see Schaffer (1930), Murray (1963), Hall (1970, 1978, 1979), Durkin (1972), Beresford (1981) and Vinkka (1982).

Secondary cartilage forms in the sutures between dermal bones of the skulls of birds and mammals (Murray, 1963; Hall, 1970). Because secondary cartilage is mechanically induced and because the snake skull is highly kinetic, one would expect to find many sites of secondary chondrogenesis in the snake skull. For a third-year undergraduate Zoology project, P. D. F. Murray set the author the task of sectioning and examining the skull of a single nearly full-term tiger snake skull. The aim was to document the distribution of secondary cartilage. From this study came the present author's only contribution to reptilian biology. In the discussion of his 1963 paper Murray notes: "The presence of adventitious cartilage at so many sites on the kinetic skull of the chick . . . also caused one to expect its occurrence in the highly kinetic skull of snakes; but a study by Mr Brian Hall, a student in this Department, of the skull of a nearly full-term foetus of the viviparous snake *Notechis scutatus* revealed no sign of any adventitious cartilage at all" (Murray, 1963: 412) – an ignominious beginning to a scientific career!

Twenty articulations in this skull were examined. Those between endochondral bones consisted of a core of primary cartilage surrounded by a thick sheath of collagen fibres (Fig. 1); those between dermal bones consisted of collagen fibres alone (Fig. 2), as did all of those between dermal and endochondral bones with the exception of the articulation between the prootic and the parasphenoid (Fig. 3). The prootic is an endochondral bone. The parasphenoid is a dermal bone but one which fuses with an endochondral bone, the basisphenoid. Furthermore the trabecular cartilages lie along its surface (Bellairs & Kamal, 1981). This is a notoriously difficult region of the snake skull (Kesteven, 1940). Bellairs & Kamal (1981: 154) understate this when they comment "the mode of ossification of the skull base in snakes remains somewhat obscure". It was not possible to tell from the single-age tiger snake examined whether the cartilage at the parasphenoid-prootic articulation was secondary or not. This being so, the tiger snake must be said to lack secondary cartilage. Several possibilities exist; either (a) the tiger snake never has secondary cartilage, (b) secondary cartilage arises in younger embryos and disappears (by resorption or metaplasia to bone) by the late embryonic stage or (c) it arises after birth. Secondary cartilages in the embryonic chick and in the rat arise at quite varied times during development (Murray, 1963; Vinkka, 1982). Angular secondary cartilage arises at 16.5 days of gestation in the rat, while secondary cartilage in the mandible posterior to the molars arises at 13 days postnatally. The lifespan of these cartilages also varied tremendously (Vinkka, 1982). Clearly, one has to be cautious in drawing conclusions from one specimen at one age. However, the present author knows of only two other reports of secondary cartilage in reptiles. Beresford (1981, and pers. comm.) cites Fuchs (1909) as reporting cartilage on the pterygoid of *Lacerta vivipara*. That at the pterygoid-basipterygoid articulation might well be secondary. Cartilage at the

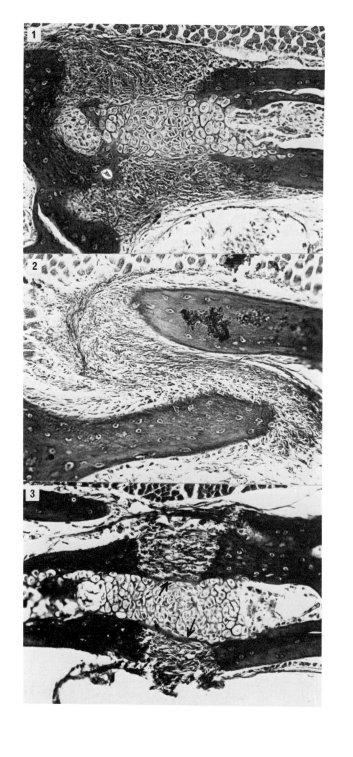

quadrate-pterygoid articulation would be primary. Beresford concludes: "Whether reptiles have secondary cartilage on their membrane bones needs thorough study". The other report is in a letter which Professor Angus Bellairs wrote to the present author on 12 February 1965, in response to an enquiry as to the presence of secondary cartilage in snakes. He said:

> "I have indeed noticed some secondary cartilage in the developing skull of a few snakes but I do not think I have reported this anywhere in print. The place I can recall having seen it most clearly is in the palatine bones of embryos of the anaconda. I rather think I saw it also in the pterygoids and/or ectopterygoids but cannot now be certain and have not got the material to hand. I would suggest however that these dermal palate bones might be the most likely source of cartilage. It is always possible, however, that bits of cartilage might represent isolated portions of the chondrocranium such as a posterior maxillary process which had got embedded in the dermal bones. I have not observed secondary cartilage in lizard skulls but again a critical examination of late embryos might show the same kind of thing."

Clearly, there is a need for a thorough study of a series of embryonic and postnatal stages of a number of different species to determine whether reptiles can and do produce secondary cartilage.

Why would this be important? Such studies would tell us whether the acquisition of the ability of periosteal tissues on dermal bones to respond to mechanical stimulation by forming secondary cartilage – a property exhibited by birds and mammals – arose independently in these two groups comparatively late in vertebrate evolution, or whether it is a more widespread phenomenon having its origin in the reptiles or the common ancestor of reptiles, birds and mammals. Amphibians do not form secondary cartilage. Secondary chondrogenesis is sufficiently different in birds and mammals (Durkin, 1972; Hall, 1978) that it could well have arisen independently. Presence of secondary cartilage is already being used as one of the characters to relate birds and mammals in schemes of tetrapod classification (Gardiner, 1982).

Mechanisms of repair of fractured dermal bones must also differ between reptiles on the one hand and birds and mammals on the other. Fractured dermal bones in birds repair by depositing a callus of

---

FIGS 1–3. (1) The articulation between the supraoccipital and the prootic, two bones which develop by endochondral ossification. The articular tissue consists of a core of primary cartilage bordered by thick bundles of collagen. Haematoxylin, alcian blue and chlorantine fast red. × 360. (2) The articulation between the parietal and the postorbital, two dermal bones which develop by intramembraneous ossification. Note complete absence of cartilage at the articulation. Haematoxylin, alcian blue and chlorantine fast red. × 360. (3) The articulation between the parasphenoid (a dermal bone) and the prootic (an endochondral bone, left). Note the primary cartilage at the articulation, the fibrous bundles uniting the two bones and the layer of perichondrial bone (arrow) bridging the articulation (arrow). Haematoxylin, alcian blue and chlorantine fast red. × 360.

secondary cartilage (Hall & Jacobson, 1975 and see Fig. 4), finally dispelling the notion that bones which lack a cartilage model cannot form cartilage during repair. Whether fractured dermal bones in reptiles form secondary callus cartilage is not known. The one piece of literature known to the author is an abstract by Nissenbaum (1971). He does not mention any formation of secondary cartilage. It is known that fractured amphibian dermal bones do *not* form cartilage (Goss & Stagg, 1958; Goss, 1983). A study of repair of fractured reptilian dermal bones would provide valuable insights into the evolution of this particular developmental process and the time when vertebrates first gained the ability to interpose a cartilaginous phase in this repair process. Nature has performed one experiment for us, viz. the splitting of the maxilla into two in the bolyerine snake *Casarea dussumieri* (Frazzetta, 1970). When such additional articulations form between dermal bones in *avian* skulls, secondary cartilage develops at the articulating surface (Bock, 1960; Bock & Morioka, 1968). Unfortunately only one specimen of *C. dussumieri* was available to Frazzetta for examination and no data on the nature of the articulating tissue are available. Even an X-ray would

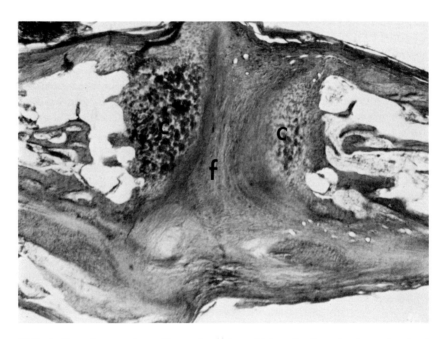

FIG. 4. Extensive secondary callus cartilage (c) seen 14 days after fracturing the quadratojugal (a dermal bone) of a 2-week-old chick (*Gallus domesticus*). A thick pad of fibrous tissue separates the two areas of cartilage (f). Haematoxylin, alcian blue and chlorantine fast red. × 80.

provide a first clue to whether this snake, like birds, has responded to the mechanical environment of this new joint by forming a secondary cartilage.

A further implicit reason for looking for secondary cartilage in reptiles and/or amphibians is to determine when the pluripotential periosteal cells evolved. What was entailed in that evolution is fundamental to our knowledge of the nature of the evolution of the developmental processes underlying tissue and morphological change during evolution. The background is as follows. It is known that the periosteal cells which form secondary cartilage on avian and mammalian dermal bones can also differentiate as osteoblasts and deposit bone. This happens when these cells are deprived of the mechanical environment required to initiate secondary chondrogenesis (Hall, 1967, 1968, 1979; Meikle, 1973; Thorogood, 1979). There is strong circumstantial evidence that the same cells form the bone as would have formed the cartilage (Hall, 1967, 1968, 1979; Meikle, 1973; Thorogood, 1979), i.e. that each cell is bipotential for the two differentiative fates of osteogenesis and secondary chondrogenesis. Clonal cell culture will be required to confirm this bipotentiality and Peter Thorogood and the present author have initiated an experimental programme using clonal culture of periosteal cells. Reptiles could lack secondary cartilage for one of two reasons: (i) the periosteal cells on their dermal bones are unipotential, containing only the genetic information for differentiating as osteoblasts; or (ii) these cells are bipotential but their environment (be it mechanical, hormonal, ionic or inductive) is inappropriate for activating their latent potential for secondary chondrogenesis. These same two possibilities apply to the lack of cartilage in fracture repair. Experiments could be readily designed to differentiate between these two alternatives. Acquisition of such data is crucial to our understanding of the evolution of developmental processes. If (i) is correct and these reptilian cells are unipotential, then the evolution of secondary cartilage in birds and mammals has involved the appearance of genes for components specific to chondrogenesis within these periosteal cells [type II (cartilage-type) collagen, glycosaminoglycans capable of forming aggregates with hyaluronic acid, cartilage-type proteoglycans, etc.]. If (ii) is correct, then evolution of secondary cartilage has involved changes in the environment of these cells or in their reception of that environment. By analogy to embryonic induction, it is the choice between evolution of the inducer and evolution of the responding tissue (see Hall, 1983a, b for a detailed discussion). Even one set of good data would greatly expand our knowledge of the evolution of the developmental processes which control cell and tissue diversity, especially given current controversies over the correlation between genomic and morphological evolution (Schopf, Raup, Gould & Simberloff, 1975; Bickham, 1981).

In summary, a close study of reptilian dermal bones during normal development and fracture repair would tell us when the ability of periosteal cells to respond to mechanical stimulation by forming secondary cartilage arose and what the basis for that evolution was. Given that secondary cartilage is restricted to endothermic tetrapods it might also be worth examining dinosaur skulls as a contribution to the now lukewarm arguments concerning hot- or cold-blooded dinosaurs.

THE STRUCTURE OF REPTILIAN CARTILAGE AND BONE

Bone

Our knowledge of the histology of reptilian cartilage and bone far exceeds our knowledge of the processes which produce these tissues. Enlow & Brown (1957, 1958) and Enlow (1969) provide the most detailed summaries of the types of bone found in reptiles. Enlow (1969) lists nine types – haversian, primary vascular, avascular, acellular, lamellar and non-lamellar, laminar, plexiform, compacted coarse-cancellous, periosteal and endosteal. He comments on the similarity between the bone structure of fossil amphibians and reptiles, on the correlation between bone type and growth rate, on the similarity in bone structure between dinosaurs and many mammals [a topic covered by Reid in this volume (p. 629), and beautifully illustrated by the scanning electron micrographs of osteocytes from 80 My old dinosaur bone produced by Pawlicki, 1978], and on the significance of growth rings and their interpretation (see also Castanet & Cheylan, 1979). Some of the biochemical and histochemical studies already carried out on fossil skeletal tissues provide invaluable information on the capabilities of these tissues, e.g. their basic amino acid composition (Biltz & Pellegrino, 1969; Matsumura, 1972) and the presence of glycosaminoglycans in dinosaur bone (Pawlicki, 1977).

The type of bony tissue found in a particular location is a function of many factors – shape of the bone and its rate of growth, growth of the entire body, remodelling, life span, muscle attachments, vascularization and body size (see also the chapters by Avery, Pritchard and Reid in this volume). Orthogenetic lines of phylogeny do not exist (see also Moss, 1964, and Hall, 1975, for developments of this view). Reptiles shown no single advance in bone structure over the Amphibia, nor can a single sequence leading to mammals be identified. Bone histology is opportunistic. What changed from reptiles to mammals was not bone structure but association, positions and function of skeletal elements – the new jaw articulation of the mammals, apical growth of the mammalian dentary replacing sutural growth of the many bones in the reptilian jaw, jointed tooth attachment in mammals replacing ankylosis in reptiles. [On the latter see the review by Osborn in this volume (p. 549) and the papers by Berkovitz & Sloan (1979) and Savitsky (1981) for examples of reptiles with socketed or hinged teeth.] Developmental studies, especially those involving functional units rather than single components (the jaw or skull rather than single mandibular or cranial bones), and those examining epigenetic interactions between components, become of paramount importance in the identification of mechanisms which govern bone structure and its adaptiveness during vertebrate evolution. Rieppel's studies reported in this volume (p. 503) examine miniaturization of the lizard skull using such techniques.

Cartilage and Secondary Centres

Reptilian cartilage has been less well studied. Haines (1969), Van Sickle & Kincaid (1978) and Moss & Moss-Salentijn (1983) provide surveys. The latter point to the reptiles as a transitional group in the evolution of cartilaginous tissues. Fish and amphibians have chondroid, pseudocartilage and vesicular cartilage (see Beresford, 1981, for a recent review of these). Reptiles, like birds and mammals, have hyaline cartilage, fibrocartilage, epiphyseal cartilage, etc. Although reptiles can differentiate these "higher" tissue types they do not organize them as birds and mammals do, i.e. they have the differentiative but not the morphogenetic phase. For example, the hypertrophic chondrocytes are not organized into the regular cellular columns seen in mammalian epiphyseal cartilages. The fate of these epiphyseal cartilages and of the secondary centres which develop within them is also quite variable within the few reptiles studied (Moss & Moss-Salentijn, 1983). In small lizards, a secondary centre of calcified cartilage forms to remain as a permanent structure. In *Sphenodon*, ossification of the secondary centre of calcified cartilage is from the diaphysis, i.e. from below, and not from an independent epiphyseal ossification centre as it is in most mammals (Haines, 1939). In most lizards, as in small mammals, ossification of epiphyseal cartilage is from the epiphyseal perichondrium-periosteum, i.e. laterally. Such epiphyses lack vascular cartilage canals (see Haines, 1942, 1969, for reviews). Varanid lizards, in common with most mammals and some birds, have epiphyses with such cartilage canals. Interestingly, among lizards, varanids have a very "mammalian" physiology: relatively high basal metabolic rate and reasonable capacity for aerobic metabolism. Also, while ectotherms, they do maintain a relatively high body temperature and are very active. So like the mammalian is ossification in these lizards that Haines had to use the presence of nucleated erythrocytes to distinguish the two! The cartilage canals either invade the epiphyseal cartilage or are incorporated into uncalcified channels in the growing cartilage providing access both to nutrients and to osteoprogenitor cells. The secondary centre of ossification develops from the inside of the cartilage outwards.

Several interesting developmental problems emerge. What determines whether vascular canals will invade a cartilage? Size is one factor, for small cartilages do not develop cartilage canals. Calcification must be another, for it is now well established that uncalcified cartilage is not invaded by blood vessels because such cartilage produces an anti-angiogenic factor (Kuettner & Pauli, 1983). As calcification takes place this factor stops being produced, or is neutralized, either by being broken down or by diffusing away. Varanid lizards are able to leave uncalcified channels around their canals while calcifying the remainder of the epiphyseal cartilage. Such epiphyses remain open throughout life so that growth never ceases (see below). What of the development of the avascular bone of modern lizards and snakes? Is it also related to production of anti-angiogenic factors? Given the interest in such factors for their use in slowing the growth and spread of tumours (by preventing the tumour from being vascularized) and that uncalcified cartilage is the major source of such factors, reptiles may play a significant role in this exciting area of research.

Indeterminate Growth

It is often noted that growth of many reptiles is indeterminate, a pattern of growth which is correlated with the structure of the epiphyses, especially the presence or absence of a secondary centre of ossification (Goss, 1978; Hinchliffe & Johnson, 1983). Such indeterminate growth is often associated with delayed ageing or with absence of signs of senescence (Goss, 1974). If ageing and growth are mutually exclusive, the former only commencing when growth ceases, and if determinate growth is set by using up all the cells of the epiphyseal growth centre, then how is indeterminate growth allowed? Is the primary control because chondroprogenitor cells can be continually produced, i.e. a time-independent mitotic clock? Is it because these cartilages actively resist vascular invasion and replacement by secondary ossification centres, or is it that, as in *Varanus*, vascular invasion is localized within uncalcified cartilage canals? Experiments on cell kinetics as outlined by Kember (1983) and on resistance to vascular invasion as outlined by

Kuettner & Pauli (1983) will have to be performed on reptile epiphyseal cartilage in order to understand the controls involved.

Sesamoids, Periosteal and Intratendinous Ossification

Sesamoid bones are commonly found in reptiles (Haines, 1969). Most arise by ossification within tendon rather than by periosteal ossification (Haines, 1942; Barnett & Lewis, 1958; Moss, 1969). The stimulus seems to be the local tension which is set up when tendons form along lines of stress (Haines & Mohiuddin, 1968). That cells within reptilian tendons can respond to stress by becoming osteoblasts when periosteal cells on dermal bones have not acquired the ability to respond to mechanical factors by forming secondary chondroblasts, nicely illustrates the independence of the evolution of mechanisms of cellular differentiation between individual skeletal elements in the same organism.

Intratendinous calcification or ossification is also the rule in the development of reptilian dermal skeletal elements. Only rarely is the periosteum involved (Moss, 1969). The fully formed osteogenic and chondrogenic periosteum is clearly a late feature in vertebrate evolution. Pathological conditions involving the periosteum do seem to have arisen early. Dinosaurs show osteopetrosis (Campbell, 1966), osteoarthritis (also seen in the marine plesiosaurs so perhaps not induced by mechanical injury resulting from the weight which dinosaurs carried around), and infections of bone such as periostitis and osteomyelitis (see Wells, 1973, for further information and charming anecdotes concerning tumours in crinoids and the like). Intratendinous ossifications are also seen as spines associated with the dorsal axial muscle masses of ornithischian dinosaurs. As with the evolution of secondary cartilage, we can ask what had to evolve to produce predominant periosteal ossification – the genes for synthesis of type I collagen, osteocalcin, bone-specific alkaline phosphatase and the other molecules specific to bone, or the epigenetic control required to activate present, but latent, genes?

Metaplastic Bone

Perhaps as a corollary of the limited osteogenic ability of the reptilian periosteum, reptiles exhibit much metaplastic bone formation as admirably reviewed by Haines (1969) and Beresford (1981). Cartilage undergoes a direct transformation into bone as chondroblasts become osteoblasts and cartilaginous extracellular matrix becomes osseous. Histologists and pathologists working with mammalian skeletons find metaplasia an unlikely concept (Willis, 1962); workers with reptilian skeletons find it the norm. This need for osteogenesis by metaplastic transformation of cartilage explains the need for the extensive cartilaginous cones seen in the long bones of crocodiles, turtles and birds. We now need molecular and biochemical studies to complement the existing histological analyses.

The Cardiac Skeleton

The ability to form ectopic cartilage and bones is certainly a reptilian property. Sesamoids and other intratendinous bones have already been mentioned. The cardiac skeleton is a further example.

There have been a number of sporadic reports of cartilage or bone in the hearts of various reptiles. Torres (1917) observed cartilage at the base of the ventricles in the septum intermedium in all specimens of several species of snakes (*Lachesis neuwiedii, Crotalus terrificus, Lachesis alternatus, L. lanceolatus, Oxirhopus* spp.) and cites Favaro (1903) as reporting cartilage at the same site in *Tropidonotus natrix*. The size of these cardiac cartilages increased with the size and age of the snakes. Hueper (1939) also reported both cartilage and bone in the hearts and aortae of turtles, alligators and crocodiles. Torres (1917) regarded these cartilages as examples of protocartilage or parenchymatous cartilage. In so doing he paralleled both the terminology and the philosophy of Schaffer (1930) who believed that cartilage which occurred outside the skeleton could not be true cartilage. However, when an element is present as a constitutive one, i.e. every individual has one, it seems reasonable to regard it as a normal part of the skeleton for that species. To determine whether these are true cartilages or some intermediate tissue will require a more extensive histological and biochemical examination than the studies performed by Torres (1917) and Hueper (1939). Torres, arguing from phylogenetic analogy, maintained (1917: 445) that these cartilages were homologues of genuine cartilage . . . "chondromucoid substance repeats in the phylogenesis the evolution of the cartilaginous tissue of the individual, for corresponding to the same point where we found this vesicular tissue, there exists in animals of a higher class of development a tissue which is typically cartilaginous". Beresford (1981) reported instances of normal cartilage, osteosarcoma, chondrosarcoma, myxoma and malignant mesenchymomas in the hearts of various mammals. The hearts of dogs and man normally contain a fibrocartilage, while hyaline cartilage is a normal element in the aorta of the rat. Such cartilage has been shown to form as cells begin to synthesize and deposit chondroitin sulphate in response to mechanical stimulation associated with blood flow (Rodbard, 1970) or after mechanical injury such as cardiac ligature (Beresford, 1981). Cardiac muscle itself acts as an inducer of both periosteal and metaplastic cartilage and bone during muscle regeneration (Zacks & Sheff, 1982). Again we see that one region of the body, the heart, can respond to mechanical stimulation by initiating chondrogenesis, while other regions – periostea – cannot.

LIMB DEVELOPMENT

Although there is neither time nor space to review adequately the processes involved in limb development within the reptiles, some summary comments will be made.

Like all vertebrates, reptilian limb buds possess an apical ectodermal ridge (AER) which governs the proximodistal sequence of limb skeletogenesis (Milaire, 1957; Goel & Mathur, 1977). Even limbless lizards start development with limb buds complete with AERs (Hall, 1978).

The patterns of condensation formation, chondrogenesis and periochondria formation are typical of those seen in birds and mammals (Mathur & Goel, 1976; Ede, 1977; Mathur, 1979). Very few skeletal condensations either fuse or are lost during limb development (Mathur & Goel, 1976). The approach of localizing condensations by uptake of S^{35} as developed by Hinchliffe using avian embryos (Hinchliffe, 1977; Hinchliffe & Johnson, 1983) is yielding vital information on the homology of skeletal elements and on the basic form and evolution of the pentadactyl limb. Such studies should be extended to the reptilian limb. The molecular basis of such patterns is currently under study (Honig, this volume, p. 197).

The interaction between somites and the limb, now known to be so important in the avian and mammalian limb (Hall, 1978; Kenny-Mobbs & Hall, 1982), operates within the reptilian limb bud as well (Vasse, 1974, 1977). Defects in this interaction are a major cause of limb bud regression in limbless lizards (Raynaud, 1977; Hall, 1978). Phylogenetically the process of limb reduction in lizards is a gradual one, with various intermediate stages being preserved (Lande, 1978). Continuation of somite-limb interactions beyond the normal time of limb-bud regression can result in the formation of atavistic limbs (Hall, in press).

In birds and mammals, but not in amphibians, separation of the digits from one another involves a genetically programmed interdigital cell death. Similar cell death is seen in turtles and lizards (Pieau & Raynaud, 1976; Goel & Mathur, 1977). Fallon & Cameron (1977) argue that this developmental process arose with the evolution of the amniotes. It is also precocious cell death which removes the apical ectodermal ridge and somitic processes in limbless reptiles. It may also be important in the reduction in digits seen in various vertebrate lineages (Hinchliffe & Johnson, 1980; Hall, 1984) including reptiles (Russell & Rewcastle, 1979).

Much work has been done on the regeneration of reptilian digits, limbs and tails, much of it by Angus Bellairs and his former Ph.D. student Susan Bryant (Bryant & Bellairs, 1967a, b; Bellairs & Bryant, 1968; Bryant & Bellairs, 1970; Bryant & Wozny, 1974). Goss (1969) and Muneoka & Bryant (this volume, p. 177) review these studies.

Paedomorphosis, or the retention of embryonic or juvenile structures in the adult is also seen in reptiles as in the extensively cartilaginous adult skeleton of the marine turtle *Dermochelys coriacea* (Rhodin, Ogden & Conlogue, 1981).

In summary, the fundamental processes established in lower vertebrates persist in the reptiles. The major new development is in limb pattern formation, especially in the appearance of programmed cell death. The major loss is also associated with cell death in the regression of limb buds in limbless reptiles.

TIMING OF INDUCTIVE TISSUE INTERACTIONS

This overview of developmental processes and the evolution of reptilian cartilage and bone will touch on one further topic. This is an area where no data for reptiles are available but where data would be very valuable. It concerns the induction of Meckel's cartilage.

In all vertebrates examined so far Meckel's cartilage, the single support of each lower jaw, develops from cells which arose in the embryonic neural crest (Hall, 1980). Such cells migrate to the future craniofacial region to initiate the formation of the mandibular arches. The differentiation of these cells as chondroblasts requires that they undergo an inductive interaction with an embryonic epithelium (see Hall, 1980, 1983a, c, d for literature). However, the epithelium with which these cells interact varies from class to class within the vertebrates, as is summarized in Fig. 5. In birds the interaction is with cranial ectoderm *before*, or very early during, neural crest cell migration. In amphibians – both urodeles and anurans – the interaction is with pharyngeal endoderm, *during* neural crest cell migration. (The same is true for the induction of cyclostome branchial cartilages, which although derived from neural crest cells are not homologous with Meckel's cartilage. They are included on Fig. 5 for completeness.) In mammals, specifically the mouse, the interaction is with mandibular epithelium, *after* neural crest cell migration is complete. What of reptiles? We do not know. The only circumstantial evidence available is a personal com-

Skeletal Development and Evolution

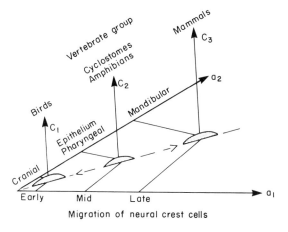

FIG. 5. The time when neural crest-derived cells are induced to form Meckel's cartilage in birds (C_1), amphibians (C_2) and mammals (C_3) or branchial cartilage in the cyclostomes (C_2) is shown as a function of both migration of neural crest cells (a_1, early, mid or late migration) and of the inductively-active epithelium (a_2, cranial, pharyngeal, mandibular).

munication from Dr M. W. J. Ferguson, the Editor of this volume. He found that organ-cultured mandibular arches from 10-day-old embryonic American alligators (*Alligator mississippiensis*) formed Meckel's cartilage even when the mandibular epithelium was accidentally removed. The implication is either that the epithelial interaction had already occurred or that an inductive interaction was not required. The former is more likely but why is it important to know? The reason is that the data available from the other vertebrates indicate that developmental processes can themselves evolve. Requirement for an epithelium as an inducer remains as a constant feature while the timing of the interaction and the particular epithelium varies. As Fig. 5 indicates, the temporal variation can be to advance, or to retard, the timing of the induction. These are just the type of data required to support the hypothesis that the developmental processes underlying heterochrony are a mechanism for morphological change during evolution (Gould, 1977; Hall, 1983b, 1984). Furthermore, such data would greatly enhance our knowledge of the processes involved in the transition of Meckel's cartilage from an element in the reptilian lower jaw to its role as a middle ear element in mammals. It is impossible to make this shift if neural crest-derived cells have already been induced to differentiate as jaw-bones and Meckel's cartilage. Acquisition of data of the type shown in Fig. 5 is a major way of understanding how evolution can act by modifying developmental processes (see also Maderson, 1975; Starck, 1979; Reif, 1982; and the papers cited in Mayr & Provine, 1980; Scudder &

Reveal, 1981; and Bonner, 1982, for further developments of this view).

SUMMARY AND FUTURE RESEARCH

This has been a very brief overview of the develomental processes which typify the stage of skeletal differentiation attained by the Reptilia. It has concentrated on just a few aspects. Ironically, there were often aspects for which we *lacked* data on any reptiles – presence or absence of secondary cartilage, process of repair of fractured dermal bones, the cellular mechanisms which allow indeterminate growth to occur, how Meckel's cartilage is induced, etc.

In preparing this review it has become clear that the periostea of reptilian dermal bones are not as highly differentiated as those of birds and mammals. This is most dramatically seen in the inability of reptilian periosteal cells to differentiate as chondroblasts and to deposit secondary cartilage, an inability shared with the Amphibia. Admittedly, the data base is not extensive. Detailed studies on ontogenetic series, both embryonic and post-hatching, are needed to confirm this lack of secondary chondrogenesis. Also required are experimental studies on the repair of fractured dermal bones, either in the mandible, or in the skull. Biochemical studies are needed to determine whether such periosteal cells can synthesize cartilage-specific products such as type-II collagen, cartilage-specific proteoglycan, etc. If they *cannot*, then lack of secondary chondrogenesis reflects absence of the parts of the genome responsible for the production of these products. If they *can*, then failure to chondrify must have another explanation which relates either to inability of periosteal cells to respond to the mechanical stimulation known to evoke secondary cartilage in birds and mammals, or to inadequate threshold levels of such stimulation. Cellular and neuromuscular physiologists could contribute the necessary data. In either case, such studies will provide invaluable information on how skeletal differentiation evolves either within the differentiating periosteal cells and/or through altered epigenetic tissue interactions.

The periostea on reptilian endochondral bones *appear* to have far less osteogenic potential than do similar periostea on avian or mammalian endochondral bones. Is this true and is it why intratendinous and metaplastic ossification are such common processes in reptiles? Are the persistent cartilaginous cores of the long bones of many reptiles an adaptation to the need for metaplasia necessitated by the limited periosteal osteogenesis? Can we relate this limitation with the inability of periostea of reptilian dermal bones to form secondary cartilage, and so conclude that reptilian periostea lack the pluripotentiality seen in avian and mammalian periostea? With the limited data at hand a reasonable working hypothesis is that the periostea of reptilian dermal bones consist of a population of unipotential osteogenic cells, and that many of the cells in the periostea of endochondral bones are not osteogenic. Again the question arises, why? Do these cells lack the ability to produce bone-specific products (type-II collagen, osteocalcin, bone-specific alkaline phosphatase) or are they just quiescent, lacking only the appropriate epigenetic activation?

The existence and variety of secondary centres (either of calcified cartilage or of bone) within reptilian epiphyses is now well described as is their correlation with determinate or indeterminate growth. Why are some epiphyses vascularized and some not? How can uncalcified channels be maintained within otherwise fully calcified epiphyses? Is proliferative activity of persistent epiphyses unlimited in reptiles which show indeterminate growth? Are reptile epiphyses a good model system for the mode of action of the cartilage-derived anti-angiogenic factors which have been shown to slow tumour growth? All these are unexplored areas for future research.

The appearance of programmed cell death in the reptilian limb bud appears to have been a major new development in vertebrate limb morphogenesis. Not only does it enable digits to be effectively separated from one another, it also provides a mechanism for the *failure* of limb development which is such a necessary corollary of adaptations to burrowing and to the locomotory patterns of the snakes. Colonization of major new habitats was facilitated by this new developmental process.

Last is the question of how and when inductive tissue interactions control initiation of the differentiation of Meckel's cartilage in any reptile. Similar information for other vertebrates has allowed us to discuss the evolution of a developmental process. In many respects, reptiles are a pivotal group for future studies in this area, standing, as they do, at the crossover between the "lower" and the "higher" vertebrates. The stage is set for the future Angus Bellairs to tell us what developmental processes have underlain the evolution of reptilian cartilage and bone.

ACKNOWLEDGEMENTS

I thank the Natural Sciences and Engineering Research Council of Canada for their continued support of my research programme on the developmental biology of the skeleton. Jim Hanken provided many stimulating discussions and critical comments on this manuscript. Bill

Beresford provided valuable information on the early German literature on secondary cartilage.

REFERENCES

Barnett, C. H. & Lewis, O. J. (1958). The evolution of some traction epiphyses in birds and mammals. *J. Anat., Lond.* **92**: 593–601.
Bellairs, A. d'A. & Bryant, S. V. (1968). Effects of amputation of limbs and digits of Lacertid lizards. *Anat. Rec.* **161**: 489–496.
Bellairs, A. d'A. & Kamal, A. M. (1981). The chondrocranium and the development of the skull in recent reptiles. In *Biology of the Reptilia* **11**, *Morphology (F)*: 1–264. Gans, C. & Parsons, T. S. (Eds). New York: Academic Press.
Beresford, W. A. (1981). *Chondroid bone, secondary cartilage and metaplasia*. Munich and Baltimore: Urban & Schwarzenberg.
Berkovitz, B. K. B. & Sloan, P. (1979). Attachment of tissues of the teeth in *Caiman sclerops* (Crocodilia). *J. Zool., Lond.* **187**: 179–194.
Bickham, J. W. (1981). Two hundred-million-year-old chromosomes: deceleration of the rate of karyotypic evolution. *Science, Wash.* **212**: 1291–1293.
Biltz, R. M. & Pellegrino, E. O. (1969). The chemical anatomy of bone. 1. A comparative study of bone composition in sixteen vertebrates. *J. Bone Jt Surg.* **51**(A): 456–466.
Bock, W. J. (1960). Secondary articulations of the avian mandible. *Auk* **77**: 19–55.
Bock, W. J. & Morioka, H. (1968). The ectethmoid-mandibular articulation in some Meliphagidae (Aves). *Am. Zool.* **8**: 808.
Bonner, J. T. (1982). *Evolution and development*. Berlin: Springer-Verlag.
Bryant, S. V. & Bellairs, A. d'A. (1967a). Tail regeneration in the lizards *Anguis fragilis* and *Lacerta dugesii*. *J. Linn. Soc. Lond.* **46**: 297–305.
Bryant, S. V. & Bellairs, A. d'A. (1967b). Amnio-allantoic constriction bands in lizard embryos and their effects on tail regeneration. *J. Zool., Lond.* **152**: 155–162.
Bryant, S. V. & Bellairs, A. d'A. (1970). Development of regenerative ability in the lizard, *Lacerta vivipara*. *Am. Zool.* **10**: 167–173.
Bryant, S. V. & Wozny, K. J. (1974). Stimulation of limb regeneration in the lizard *Xantusia vigilis* by means of ependymal implants. *J. exp. Zool.* **189**: 339–352.
Campbell, J. G. (1966). A dinosaur bone lesion resembling avian osteopetrosis and some remarks on the mode of development of the lesions. *J. R. microsc. Soc.* **85**: 163–174.
Castanet, J. & Cheylan, M. (1979). Les marques de croissance des os et des écailles comme indicateur de l'age chez *Testudo hermanni* et *Testudo gracea* (Reptilia, Chelonia, Testudinidae). *Can. J. Zool.* **57**: 1649–1665.
Durkin, J. F. (1972). Secondary cartilage: a misnomer? *Am. J. Orthod.* **62**:15–41.
Ede, D. A. (1977). Relations between ontogeny and phylogeny of the vertebrate limb suggested by studies on the talpid[3] mutant of the chick. *Colloques int. Cent. natn. Rech. scient.* **266**: 410–411.
Enlow, D. H. (1969). The bone of reptiles. In *Biology of the Reptilia*, **1**: 45–80. Gans, C. (Ed.). London: Academic Press.
Enlow, D. H. & Brown, S. O. (1957). A comparative histological study of fossil and recent bone tissues. Part II. *Tex. J. Sci.* **9**: 186–214.
Enlow, D. H. & Brown, S. O. (1958). A comparative histological study of fossil and recent bone tissues. Part III. *Tex. J. Sci.* **10**: 187–230.

Fallon, J. F. & Cameron, J. A. (1977). Interdigital cell death during limb development of the turtle and lizard with an interpretation of evolutionary significance. *J. Embryol. exp. Morph.* **40**: 285–289.
Frazzetta, T. H. (1970). From hopeful monster to bolyerine snakes? *Am. Nat.* **104**: 55–72.
Fuchs, H. (1909). Über Knorpelbilding in Deckknochen nebst Untersuchungen und Betrachtungen uber Gehörknöchelchen, Kiefer und Kiefergelenk der Wirbeltiere. *Arch. Anat. Physiol. Anat.* (Suppl.) 1909: 1–256.
Gardiner, B. G. (1982). Tetrapod classification. *Zool. J. Linn. Soc.* **74**: 207–232.
Goel, S. C. & Mathur, J. K. (1977). Morphogenesis in reptilian limbs. In *Vertebrate limb and somite morphogenesis*: 387–404. Ede, D. A., Hinchliffe, J. R. & Balls, M. (Eds). London: Cambridge University Press.
Goss, R. J. (1969). *Principles of regeneration.* New York: Academic Press.
Goss, R. J. (1974). Aging and growth. *Persp. Biol. Med.* **17**: 485–494.
Goss, R. J. (1978). *The physiology of growth.* New York: Academic Press.
Goss, R. J. (1983). Chondrogenesis in regenerating systems. In *Cartilage*, **3**: 267–307. Hall, B. K. (Ed.). New York: Academic Press.
Goss, R. J. & Stagg, M. W. (1958). Regeneration of the lower jaw in adult newts. *J. Morph.* **102**: 289–309.
Gould, S. J. (1977). *Ontogeny and phylogeny.* Cambridge, Mass.: Belknap Press of Harvard University Press.
Haines, R. W. (1939). The structure of the epiphyses in *Sphenodon* and the primitive form of secondary centres. *J. Anat., Lond.* **74**: 80–90.
Haines, R. W. (1942). The evolution of epiphyses and of endochondral bone. *Biol. Rev.* **17**: 267–292.
Haines, R. W. (1969). Epiphyses and sesamoids. In *Biology of the Reptilia*, **1**: 81–116. Gans, C. (Ed.). London: Academic Press.
Haines, R. W. & Mohiuddin, A. (1968). Metaplastic bone. *J. Anat., Lond.* **103**: 527–538.
Hall, B. K. (1967). The formation of adventitious cartilage by membrane bones under the influence of mechanical stimulation applied *in vitro. Life Sci.* **6**: 663–667.
Hall, B. K. (1968). *In vitro* studies on the mechanical evocation of adventitious cartilage in the chick. *J. exp. Zool.* **168**: 283–306.
Hall, B. K. (1970). Cellular differentiation in skeletal tissues. *Biol. Rev.* **45**: 455–484.
Hall, B. K. (1975). Evolutionary consequences of skeletal development. *Am. Zool.* **15**: 329–350.
Hall, B. K. (1978). *Developmental and cellular skeletal biology.* New York: Academic Press.
Hall, B. K. (1979). Selective proliferation and accumulation of chondroprogenitor cells as the mode of action of biomechanical factors during secondary chondrogenesis. *Teratology* **20**: 81–92.
Hall, B. K. (1980). Chondrogenesis and osteogenesis in cranial neural crest cells. In *Current research trends in prenatal craniofacial development*: 47–63. Pratt, R. M. & Christiansen, R. L. (Eds). New York: Elsevier, North Holland.
Hall, B. K. (1983a). Embryogenesis: cell-tissue interactions. In *Skeletal research – an experimental approach*, **2**: 53–87. Simmons, D. J. & Kunin, A. S. (Eds). New York: Academic Press.
Hall, B. K. (1983b). Epigenetic control in development and evolution. *Symp. Br. Soc. devel. Biol.* No. 6: 353–379. London: Cambridge University Press.
Hall, B. K. (1983c). Tissue interactions and chondrogenesis. In *Cartilage*, **2** 187–222. Hall, B. K. (Ed.). New York: Academic Press.
Hall, B. K. (1983d). Epithelial-mesenchymal interactions in cartilage and bone

development. In *Epithelial-mesenchymal interactions in development*: 189–214. Sawyer, R. H. & Fallon, J. F. (Eds). New York: Praeger Press.
Hall, B. K. (1984). Developmental processes and underlying heterochrony as an evolutionary process. *Can. J. Zool.* **62**: 1–6.
Hall, B. K. (In press). Develomental mechanisms underlying the formation of atavisms. *Biol. Rev.*
Hall, B. K. & Jacobson, H. N. (1975). The repair of fractured membrane bones in the newly hatched chick. *Anat. Rec.* **181**: 55–70.
Hinchliffe, J. R. (1977). The chondrogenic pattern in chick limb morphogenesis: a problem of development and evolution. In *Vertebrate limb and somite morphogenesis*: 293–310. Ede, D. A., Hinchliffe, J. R. & Balls, M. (Eds). London: Cambridge University Press.
Hinchliffe, J. R. & Johnson, D. R. (1980). *The development of the vertebrate limb.* London: Oxford University Press.
Hinchliffe, J. R. & Johnson, D. R. (1983). Growth of cartilage. In *Cartilage*, **2**: 255–296. Hall, B. K. (Ed.). New York: Academic Press.
Hueper, W. C. (1939). Cartilaginous foci in the hearts of white rats and of mice. *Archs Path.* **27**: 466–468.
Kember, N. F. (1983). Cell kinetics of cartilage. In *Cartilage*, **1**: 149–180. Hall, B. K. (Ed.). New York: Academic Press.
Kenny-Mobbs, T. P. & Hall, B. K. (1982). Muscle differentiation in cultures and chorioallantoic grafts of pre-bud wing territories. In *Limb development and regeneration, Part B*: 323–332. Kelley, R. O., Goetinck, P. F. & MacCabe, J. A. (Eds). New York: Alan R. Liss Inc.
Kesteven, H. L. (1940). The osteogenesis of the base of the saurian cranium and a search for the parasphenoid bone. *Proc. Linn. Soc. N.S.W.* **65**: 447–467.
Kuettner, K. E. & Pauli, B. U. (1983). Vascularity of cartilage. In *Cartilage*, **1**: 281–312. Hall, B. K. (Ed.). New York: Academic Press.
Lande, R. (1978). Evolutionary mechanisms of limb loss in tetrapods. *Evolution, Lawrence, Kans.* **32**: 73–92.
Maderson, P. F. A. (1975). Embryonic tissue interactions as the basis for morphological change in evolution. *Am. Zool.* **15**: 315–328.
Mathur, J. K. (1979). Histogenesis of cartilage and bone in humerus and femur of lizard *Calotes versicolor. Ind. J. exp. Biol.* **17**: 533–537.
Mathur, J. K. & Goel, S. C. (1976). Pattern of chondrogenesis and calcification in the developing limb of the lizard *Calotes versicolor. J. Morph.* **149**: 401–420.
Matsumura, T. (1972). Relationship between amino acid composition and differentiation of collagen. *Int. J. Biochem.* **3**: 265–274.
Mayr, E. & Provine, W. B. (1980). *The evolutionary synthesis. Perspectives on the unification of biology.* Cambridge, Mass.: Harvard University Press.
Meikle, M. C. (1973). In vivo transplantation of the mandibular joint of the rat: an autoradiographic investigation into cellular changes at the condyle. *Archs oral Biol.* **18**: 1011–1020.
Milaire, J. (1957). Contribution à la connaissance morphologique et cytochimique des bourgeons de membres chez quelques reptiles. *Archs Biol.* **68**: 429–512.
Moss, M. L. (1964). The phylogeny of mineralized tissues. *Int. Rev. gen. exp. Zool.* **1**: 297–331.
Moss, M. L. (1969). Comparative histology of dermal sclerifications in reptiles. *Act Anat.* **73**: 510–533.
Moss, M. L. & Moss-Salentijn, L. (1983). Vertebrate cartilages. In *Cartilage*, **1**: 1–30. Hall, B. K. (Ed.). New York: Academic Press.
Murray, P. D. F. (1963). Adventitious (secondary) cartilage in the chick embryo and

the development of certain bones and articulations in the chick skull. *Aust. J. Zool.* **11**: 368–430.

Nissenbaum, A. (1971). Preliminary observations on mandibular regenerations in *Anolis carolinensis*. *Am. Zool.* **11**: 707.

Pawlicki, R. (1977). Histochemical reactions for mucopolysaccharides in dinosaur bone studied on epon-embedded and methacrylate-embedded semithin sections as well as on isolated osteocytes and ground sections of bone. *Acta Histochem.* **58**: 75–78.

Pawlicki, R. (1978). Morphological differentiation of the fossil dinosaur bone cells: light, transmission electron and scanning electron microscopic studies. *Acta Anat.* **100**: 411–418.

Pieau, C. & Raynaud, A. (1976). Dégénérescence cellulaire dans la crête apical de l'ébauche du membre de la Turtue mauresque (*Testudo graeca* L. Chelonian). *C. r. Séanc. Acad. Sci., Paris.* **282**: 1797–1800.

Raynaud, A. (1977). Somites and early morphogeneses in reptile limbs. In *Vertebrate limb and somite morphogenesis*: 373–386. Ede, D. A., Hinchliffe, J. R. & Balls, M. (Eds). London: Cambridge University Press.

Reif, W.-E. (1982). Evolution of dermal skeleton and dentition in Vertebrates: the odontode regulation theory. *Evol. Biol.* **15**: 287–368.

Rhodin, A. G. F., Ogden, J. A. & Conlogue, G. J. (1981). Chondro-osseous morphology of *Dermochelys coriacea*, a marine reptile with mammalian skeletal features. *Nature, Lond.* **290**: 244–246.

Rodbard, S. (1970). Negative feedback mechanisms in the architecture and function of the connective and cardiovascular tissue. *Persp. Biol. Med.* **13**: 507–527.

Russell, A. P. & Rewcastle, S. C. (1979). Digital reduction in *Sitana* (Reptilia: Agamidae) and the dual roles of the fifth metatarsal in lizards. *Can. J. Zool.* **57**: 1129–1135.

Savitzky, A. H. (1981). Hinged teeth in snakes: an adaptation for swallowing hard-bodied prey. *Science, Wash.* **212**: 346–349.

Schaffer, J. (1930). Die Stutzgewebe. In *Handbuch der mikroskopischen Anatomie des Menschen*, **2**(2): 338–350. von Mollendorff, W. (Ed.). Berlin: Julius Springer.

Schopf, T. J. M., Raup, D. M., Gould, S. J. & Simberloff, D. S. (1975). Genomic versus morphologic rates of evolution: influence of morphologic complexity. *Paleobiologica* **1**: 63–70.

Scudder, G. E. & Reveal, J. L. (1981). *Evolution today*. Pittsburg: Carnegie-Mellon University.

Starck, D. (1979). *Vergleichende Anatomie der Wirbeltiere auf evolutionsbiologischer Grundlage*, **2**. Berlin: Springer-Verlag.

Thorogood, P. V. (1979). In vitro studies on skeletogenic potential of membrane bone periosteal cells. *J. Embryol. exp. Morph.* **54**: 185–207.

Torres, A. de L. (1917). On the cartilaginous tissue of the heart of *Ophidia*. *Anat. Rec.* **13**: 443–445.

Van Sickle, D. C. & Kincaid, S. A. (1978). Comparative arthrology. In *The joints and synovial fluid*, **1**: 1–48. Sokoloff, L. (Ed.). New York: Academic Press.

Vasse, J. (1974). Etudes expérimentales sur le rôle des somites au cours des premiers stades du développement du membre antérieur chez l'embryon du Chélonien, *Emys orbicularis* L. *J. Embryol. exp. Morph.* **32**: 417–430.

Vasse, J. (1977). Etude expérimentale sur les premiers stades du développement du membre antérieur chez l'embryon du Chélonien *Emys orbicularis* L.: détermination en mosaique et régulation. *J. Embryol. exp. Morph.* **42**: 135–148.

Vinkka, H. (1982). Secondary cartilages in the facial skeleton of the rat. *Proc. Finn. Dent. Soc.* **78** (Suppl. 7): 1–137.

Wells, C. (1973). The paleopathology of bone disease. *Practitioner* **210**: 384–391.

Willis, R. A. (1962). *The borderland of embryology and pathology.* London: Butterworths Press.

Zacks, S. I. & Sheff, M. F. (1982). Periosteal and metaplastic bone formation in mouse minced muscle regeneration. *Lab. Invest.* **46**: 405–412.

Regeneration and Development of Vertebrate Appendages

KEN MUNEOKA and SUSAN BRYANT

Developmental Biology Center and the Department of Developmental and Cell Biology, University of California, Irvine, California 92717, USA

SYNOPSIS

The relationship between limb development and limb regeneration is discussed with regard to the mechanisms by which pattern is established during outgrowth. We present results from grafting experiments where the interaction between cells from the developing limb bud and the regenerating limb blastema result in the production of supernumerary limbs. Furthermore, substantial contribution from both developing and regenerating cells to all supernumerary limbs analysed was shown using the triploid cell marker in the axolotl. From these data we conclude that both the developing and regenerating limb utilize similar patterning mechanisms during outgrowth. This conclusion is discussed in terms of patterning models for developing and regenerating limbs and it is proposed that the rules of the polar co-ordinate model can explain the behaviour of cells during limb development as well as limb regeneration. The possibilities of inducing limb regeneration in higher vertebrates are discussed in view of our findings.

INTRODUCTION

Despite our enormous capacity as humans to manipulate natural phenomena towards our own ends, we have been singularly unsuccessful to date in exerting any directed influence over our own development. Hence, while we can patch up failing, overworked parts with artificial replacements, we have been unable to call forth innate developmental mechanisms in the services of such highly desirable goals as increasing longevity, bringing cancerous cells under control, improving mental function and repairing or regenerating lost or damaged parts. While there are no immediate prospects for accomplishing any of these goals, regeneration of appendages following loss or damage can be studied in other, closely related vertebrates such as amphibians and reptiles. We would like to begin by considering the relationship, if any, between the prospects for human limb regeneration and the ability of urodeles to repeatedly form new limbs after amputation. This will involve an examination of a number of issues, the first of which is whether embryonic leg development in humans and embryonic leg development in urodeles involve similar mechanisms. Next, we must consider whether in fact leg development and leg regeneration in urodeles and humans are the same, and if the mechanisms of leg development and leg regeneration are the same, are there special features of human limbs which might preclude their regeneration? Finally, is it feasible to consider that the legs of higher vertebrates might be stimulated to regrow? In this paper we specifically focus on the first two issues and conclude that we are now at a stage where the remaining issues may be addressed in a meaningful way.

The study of the mechanisms underlying these processes involves the study of pattern formation, a field which is still in a relatively immature state. Still, we have learned enough about the limb patterning mechanisms during urodele limb regeneration to believe that it is at the level of pattern formation that we can most fruitfully approach the problem of mammalian limb regeneration. Each tissue of the mammalian limb has the ability to grow, repair, and differentiate independently of another, but in response to limb amputation, skin, muscle, and bone do not respond in a co-ordinated way by dedifferentiation and blastema formation to bring about regeneration of a spatially co-ordinated structure. Mammalian limb cells can be caused to dedifferentiate and grow after amputation (Neufeld, 1980; Smith, 1981) but the resulting regenerate does not possess a pattern which even closely resembles the amputated limb. Hence, the first and most basic issue we must address is the nature of the pattern-forming mechanisms in developing and regenerating vertebrate limbs.

THE QUESTION OF UNIVERSALITY

The issues raised above relate to the question of the universality of patterning mechanisms in the limbs of different vertebrates, both during development and during regeneration. Wolpert (1969, 1971) has suggested that patterning mechanisms may be universal, but it is important for us to determine the degree of similarity which exists from case to case in order to accurately assess the likelihood of universal mechanisms. Relevant to this consideration is the degree of similarity in structure between the limbs of different vertebrates. Even a superficial examination of a variety of vertebrate limbs leads to the conclusion that all are in fact very similar. All tetrapods, with the exception of those showing secondary reduction or loss (e.g. snakes, whales) have limbs which consist of three basic parts: stylopodium, zeugopodium, and autopodium, arranged in a proximal to distal sequence. There is no variation in the number of skeletal elements present in the stylopodium, almost none in the zeugopodium and very little in the autopodium. The upper limit of the number of digits among living vertebrates is six. Most tetrapods have five digits, and some have less than five. The most striking conclusion that can be drawn about the vertebrate limb is that despite the wide range of jobs for which limbs are adapted – flying, swimming, digging, walking, climbing – all have the same basic morphology. The simplest hypothesis is that all vertebrate limbs develop using the same mechanisms. In fact, during their embryonic development, all vertebrate limbs appear remarkably similar, passing through stages in which the limbs of one species are barely distinguishable from those of another.

We must next consider experimental evidence which might bear upon the issue of universality. Transplantation experiments show that posterior limb bud cells from a wide variety of different vertebrates are just as effective as posterior limb bud cells from the chick in inducing supernumerary limb parts to form when they are grafted to the anterior of the chick limb bud (Balcuns, Gasseling & Saunders, 1970; MacCabe & Parker, 1976; Tickle, Shellswell, Crawley & Wolpert, 1976; Fallon & Crosby, 1977). It does not matter if the posterior limb bud cells come from a chick, a human, or a turtle, all are able to cause the chick limb bud to respond by forming supernumerary limb structures. The assumption has been made, and is fundamental to the study of pattern formation, that when cells are transplanted to a new location, their response indicates something about the positional information they possess. Since the response is the same for cells from different animals it follows that all the transplanted cells share the same positional information and that the patterning mechanisms utilized by the cells of all the vertebrate limbs which have been tested are the same. Even though not all vertebrate limb buds have been subjected to this test, it seems reasonable to infer that all vertebrate limbs share common patterning mechanisms in the establishment of the limb pattern during development.

The next issue of interest is whether all regenerating limbs utilize the same patterning mechanisms during limb regeneraton. This question is a little more difficult to answer since not many vertebrate species have the ability to regenerate their limbs. However, among those animals that do regenerate their limbs, there is good evidence that cells from both forelimbs and hind-limbs of the same species share common patterning mechanisms during regeneration (Pescitelli & Stocum, 1980; Stocum, 1980a, 1980b; Rollman-Dinsmore & Bryant, 1982). A few transplants between the limbs of different regenerating species have been performed. These studies show that the transplanted cells behave in a manner typical of transplants within a species, and hence lend support to the concept that the regenerating limbs of different species utilize similar patterning mechanisms (Pescitelli & Stocum, 1980; Stocum, 1982).

The third and crucial issue is whether limb regeneration involves the re-utilization of the same patterning mechanisms which were used during the initial development of the limb. In other words, can the concept of universality be extended to cover both developing and regenerating vertebrate limbs? In fact, there are several reasons for thinking that similar mechanisms might be responsible for both limb development and limb regeneration. Probably the most immediate is that any limb which undergoes regeneration has at some previous time utilized developmental mechanisms to form in the embryo, and it seems unlikely that a new set of patterning mechanisms would evolve specifically for regeneration. Furthermore, the developing limb bud and the regenerating limb blastema progress through a number of morphologically similar stages during limb outgrowth. Both possess an apical epidermal thickening at the distal tip of the bud (apical ectodermal ridge) or blastema (apical cap), and in both, this apical thickening plays an important role in the outgrowth of the limb (for review see Hinchliffe & Johnson, 1980 – limb bud; Wallace, 1981 – blastema). In both regenerating and developing limbs differentiation of the skeleton and the musculature occurs in a proximal to distal sequence. Further, in animals which can regenerate, the anterior to posterior sequence of differentiation is similar for both limb development and limb regeneration. Both developing and regenerating limbs display a cell lineage restriction between cells which form muscle and those which form the other mesodermal tissues of the limb, i.e. cartilage, dermis, and connective tissue of the muscle (Namenwirth, 1974, and Lheureux, 1983 – regenerating limb; Christ, Jacob & Jacob, 1977, and Chevallier, Kieny & Mauger, 1977 – developing limb). Finally, in experiments designed to yield information about the patterning mechanisms of limbs there are many cases where similar responses are observed following similar experiments performed on the limb bud and blastema. All limb buds (*Xenopus, Rana* – Maden, 1981; *Ambystoma* – Maden & Goodwin, 1980; chick – Summerbell, Lewis & Wolpert, 1973) and blastemas (*Notophthalmus* – Iten & Bryant, 1975; *Ambystoma* – Stocum, 1975) which have been tested, respond identically to proximal-to-distal shifts by yielding limbs with a serial duplication in the proximal-distal axis. Contralateral grafts which appose anterior and posterior limb cells result in the production of supernumerary structures anterior and/or posterior to the graft in all buds (*Xenopus, Rana* – Maden, 1981; *Ambystoma* – Maden & Goodwin, 1980, Thoms & Fallon, 1980,

Muneoka & Bryant, 1982; *Notophthalmus* – Thoms & Fallon, 1980; chick – Saunders, Gasseling & Gfeller, 1958) and blastemas (*Ambystoma* – Tank, 1978; *Notophthalmus* – Iten & Bryant, 1975; *Pleurodeles* – Lheureux, 1978) tested. Similarly, 180° rotations of buds (*Xenopus* – Cameron & Fallon, 1977, Maden 1981; *Rana* – Maden, 1981; *Ambystoma* – Maden & Goodwin, 1980, Thoms & Fallon, 1980; chick – Saunders *et al.*, 1958) and blastemas (*Ambystoma* – Maden & Turner, 1978; *Notophthalmus* – Bryant & Iten, 1976) result in the production of supernumerary limbs, although their location and handedness can vary from system to system (see below). Removal of the apical ectodermal thickening from the limb bud (chick – Saunders, 1948; *Xenopus* – Tschumi, 1957) or blastema (Thornton, 1957; Stocum & Dearlove, 1972) causes the cessation of further limb outgrowth and distally deficient limbs.

Despite these numerous similarities, there are a few differences between developing and regenerating limbs which have led some to conclude that different mechanisms must be at work. Unlike regenerating limbs, developing limbs do not require the presence of nerve fibers for normal outgrowth (see Wallace, 1981, for review). Unlike developing limbs, regenerating limbs must go through a process of dedifferentiation prior to regeneration. During regeneration only partial limbs are formed after amputation, i.e. only the part which was removed, whereas during the development the complete limb structure is formed. Finally, in some cases, developing and regenerating limbs show different responses to experimental intervention. These differences have recently been reviewed by Tickle (1981). In almost all instances, the difference in response between the bud and the blastema involves the failure of the developing bud of some animals to show a response which is readily displayed by the regenerating limb. For example, grafting of a chick limb bud to the contralateral stump, so as to appose dorsal and ventral positions of graft and stump, fails to stimulate supernumerary outgrowths (Saunders *et al.*, 1958), whereas similar experiments performed on regenerating amphibian limbs lead to the formation of supernumerary limbs at dorsal and ventral positions (Bryant & Iten, 1976; Tank, 1978). Similarly, grafting to create a gap in the proximal-distal sequence of the developing *Xenopus* and chick limb bud (Summerbell *et al.*, 1973; Maden, 1981) is not followed by intercalary regeneration to form a normal and complete limb, unlike the situation in regenerating amphibian limbs (Iten & Bryant, 1975; Stocum, 1975). We will return later to consider the relevance of this type of data for conclusions about the universality of mechanisms.

MODELS FOR DEVELOPMENT AND REGENERATION

Over the years, much emphasis has been placed on the differences cited above between the behaviour of regenerating and developing limbs under similar circumstances, the result of which is that until recently the two fields of study have remained relatively isolated from one another. One consequence of this has been the formulation of pattern formation models for developing limbs which are very different from those proposed for regenerating limbs. This is unfortunate since, as we will show, there is now strong evidence that the bud and blastema utilize the same underlying mechanism for the establishment of the limb pattern. The most widely accepted view of how the developing bud is patterned is based on a combination of two models: the progress zone model (Summerbell et al., 1973) and the polarizing zone model (Tickle, Summerbell & Wolpert, 1975). The most prevalent model for the regenerating limb is the polar co-ordinate model (French, Bryant & Bryant, 1976; Bryant, French & Bryant, 1981); however, it must be mentioned that in recent years at least one laboratory (see Iten, 1982) has been testing the applicability of the polar co-ordinate model to chick limb development. A brief description of these models follows.

Progress Zone/Polarizing Zone Model

A combination of these two models has been proposed to explain how the chick limb bud establishes its pattern along the proximal-distal (progress zone model) and the anterior-posterior (polarizing zone model) axes during outgrowth. No formal model has been put forward to account for patterning of the dorsal-ventral axis. Both the progress zone model and the polarizing zone model assume that initially limb bud cells do not possess any information about their position along the proximal-distal and anterior-posterior axes prior to limb outgrowth. The progress zone model suggests that there is a specialized region at the distal tip of the limb bud (the progress zone) where cells are actively dividing. The size of the progress zone is thought to be controlled by the overlying apical ectodermal ridge. All cells within the progress zone at any given time are assumed to possess the same proximal-distal positional information. With the passage of time, the positional values of the cells within the progress zone become progressively more distal. It has been proposed that cells could measure time spent in the progress zone by counting the number of cell divisions they pass through (Summerbell & Lewis, 1975). As a cell leaves the progress zone at its proximal boundary, its proximal-distal positional value becomes fixed. Thus, cells which leave the progress zone after only a few cell divisions form more proximal limb structures than those which leave after many divisions.

The polarizing zone model suggests that limb cells acquire positional information for the anterior-posterior axis by reading the concentration of a diffusible morphogen which is produced at the posterior margin of the bud. The localized production of this morphogen establishes a gradient with the high end posteriorly and the low end anteriorly. Cells close to the posterior margin of the bud would respond to high morphogen concentrations by becoming posterior limb structures. Similarly, anterior cells would respond to low morphogen concentrations by becoming anterior limb structures. The assignment of position values for the anterior-posterior axis occurs as cells leave the progress zone (Summerbell, 1974).

Polar Co-ordinate Model

This model has been proposed to explain the pattern of regeneration in amphibian limbs, cockroach limbs and imaginal discs of *Drosophila*. It is assumed that blastema cells possess information about their position along the proximal-distal limb axis and the circumference of the limb. Central to this model is the assumption that tissues have the general property of intercalary regeneration, i.e. localized growth in response to positional disparities. The positional value of a cell is viewed as stable and new positional values are generated by cell division.

The polar co-ordinate model proposes two rules governing the behaviour of cells in blastemas. (i) The shortest intercalation rule states that when normally non-adjacent cells are juxtaposed, growth occurs at the junction until all intermediate positional values have been intercalated. In the case of the circumferential values where the sequence is continuous and there are two possible sets of intermediate values, the route taken is the shortest of the two. (ii) The distalization rule states that during intercalary regeneration of circumferential values, if a newly intercalated cell is assigned a circumferential value which is identical to that of a pre-existing, adjacent cell, then the new cell adopts a value which is more distal in the proximal-distal axis than that of the pre-existing cell. To account for limb growth, it is not necessary for cells at the wound surface to come together in an organized way. All that is necessary is that during healing, and subsequent distalization, cells from different positions around the circumference interact across short arcs (Bryant & Baca, 1978), thus intercalating values which are already present at that proximal-distal level. These newly intercalated cells take on positional values of a more distal level. Circumferential intercalation and further distalization is repeated until the entire limb structure has been generated.

THE EXPERIMENTAL TEST

It is clear from the descriptions in the previous section that the divergent models proposed to account for limb regeneration and limb development differ not only in the types of mechanisms thought to be at work during outgrowth but also in the initial assumptions from which these views have evolved. We have performed a simple experimental test of whether we can validly assume that development and regeneration are using different patterning mechanisms (Muneoka & Bryant, 1982).

We have made use of the underlying similarity in the patterning mechanisms of the hind- and forelimb (Rollman-Dinsmore & Bryant, 1982) to investigate whether developing limb buds and regenerating blastemas also share a common patterning mechanism. In the axolotl, *Ambystoma mexicanum*, hind-limb development is retarded compared to that of the forelimb thus making it possible to have regenerating forelimbs and developing hind-limb buds on animals of similar age. Grafts between the regenerating forelimb and the developing hind-limb can readily be accomplished.

The experimental design for the work reported here is shown in Fig. 1. In this experiment we exchanged forelimb blastemas and hind-limb buds either ipsilaterally (control series) or contralaterally (experimental series). Ipsilateral grafts were made in such a way that no positional disparity existed between the graft and the stump. Contralateral grafts were made so as to appose anterior and posterior limb cells while keeping dorsal and ventral positions aligned. This type of graft was chosen because it is well documented that blastema–blastema exchanges (Iten & Bryant, 1975; Tank, 1978) and limb bud–limb bud exchanges in urodeles (Maden & Goodwin, 1980; Thoms & Fallon, 1980; Muneoka & Bryant, 1982) lead to a similar result – the formation of supernumerary limbs. Under both of these experimental conditions supernumerary limbs formed at a high frequency, at regions of maximum positional disparity (anterior and posterior) and were consistently of stump handedness. If the developing limb bud and the regeneration blastema are using similar patterning mechanisms then the interaction between the two should result in the formation of supernumerary limbs.

The results of our experiment are summarized in Table I. In ipsilateral control grafts where no positional disparity existed after blastema-limb bud exchanges, development or regeneration continued to form a single

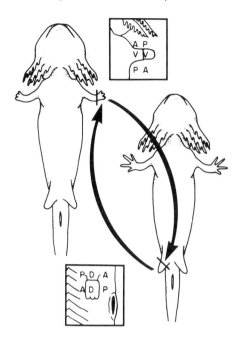

FIG. 1. Diagram of the experimental grafting procedure. Axolotls are viewed from the ventral aspect. Blastemas at the stage of early digits on the forelimbs were exchanged with the contralateral developing hind-limb bud of a sibling axolotl to appose anterior and posterior limb cells. The top inset shows a closer ventral view of the resulting limb bud to blastema-stump graft. The lower inset shows a closer lateral view of the resulting blastema to limb bud stump graft. In control grafts (not shown) blastemas and limb buds were exchanged ipsilaterally aligning all limb axes. A, anterior; D, dorsal; P, posterior; V, ventral.

TABLE I
Grafts between developing and regenerating limbs

	Total	Supernumerary limbs
Control series:		
Hind-limb bud to forelimb blastema stump	17	0 (0%)
Forelimb blastema to hind-limb bud stump	12	0 (0%)
Experimental series:		
Hind-limb bud to forelimb blastema stump	20	14 (70%)
Forelimb bud to hind-limb bud stump	13	8 (62%)

normal limb of graft origin (Figs 2 & 3). However, when contralateral grafts were made so as to create regions of positional disparity at the anterior and posterior graft junction, supernumerary limbs were produced (Figs 4 & 5). These supernumerary limbs formed at a high frequency and with the same orientation and handedness as supernumerary limbs which result from exchanges between either regenerat-

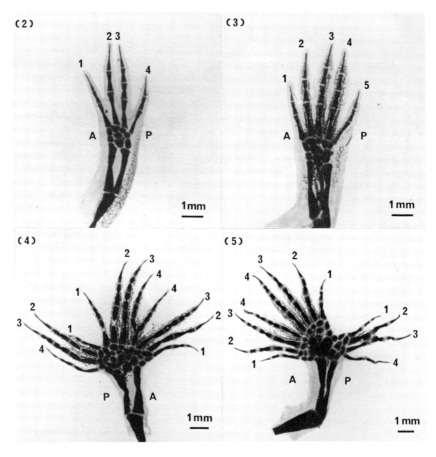

FIGS 2–5. (2) Skeletal preparation of a limb resulting from a control graft of a right forelimb blastema onto a right hind-limb bud stump. The graft developed autonomously to produce a normal four-digit right forelimb on a right hind-limb stump. A, anterior; P, posterior. (3) Skeletal preparation of a limb resulting from a control graft of a right hind-limb bud onto a right forelimb blastema stump. The graft developed autonomously into a normal five-digit right hind-limb on a right forelimb stump. A, anterior; P, posterior. (4) Skeletal preparation of a limb resulting from an experimental graft of a right forelimb blastema to a left hind-limb bud stump. The resulting limb consists of supernumerary limbs both anterior and posterior to the graft. The digital sequence (from left to right or posterior to anterior) is 4, 3, 2, 1 (posterior supernumerary), 1, 2, 3, 4 (grafted limb), 4, 3, 2, 1 (anterior supernumerary). A, anterior; P, posterior. (5) Skeletal preparation of a limb resulting from an experimental graft on a left hind-limb bud grafted to a right forelimb blastema stump. The resulting limb produced supernumerary limbs anterior and posterior to the graft. The digital sequence from anterior to posterior (left to right) was 1, 2, 3, 4 (anterior supernumerary limb), 4, 3, 2, 1 (grafted limb), 1, 2, 3, 4 (posterior supernumerary limb). A, anterior; P, posterior.

ing blastemas or developing limb buds (Table II). These results indicate that developing and regenerating limb cells can co-operate in forming an organized supernumerary limb pattern and suggest strongly that the regenerating blastema and the developing limb bud in the axolotl are using common mechanisms for pattern regulation during limb outgrowth.

TABLE II
Contralateral grafts in urodeles: anterior-posterior apposed

Type	Animal	Limb	Percentage supernumerary	Reference
Blastema-blastema	Axolotl	Forelimb/forelimb	57%	Tank, 1978
	Newt	Forelimb/forelimb	66%	Iten & Bryant, 1975
	Newt	Forelimb/hind-limb	82%	Rollman-Dinsmore & Bryant, 1982
Limb bud-limb bud	Axolotl	Forelimb-forelimb	70%	Maden & Goodwin, 1980
	Axolotl	Forelimb/forelimb	92%	Thoms & Fallon, 1980
	Axolotl	Hind-limb/hind-limb	72%	Muneoka & Bryant, 1982
	Newt	Forelimb/forelimb	100%	Thoms & Fallon, 1980
Limb bud-blastema	Axolotl	Hind-limb/forelimb	70%	Muneoka & Bryant, 1982
Blastema-limb bud	Axolotl	Forelimb/hind-limb	62%	Muneoka & Bryant, 1982

In a subsequent study we have utilized the triploid cell marker in the axolotl to analyse the cellular contribution to supernumerary limbs formed from the interaction between developing and regenerating limb cells (Muneoka & Bryant, in preparation). Triploid animals were made by hydrostatic pressure following the protocol of Gillespie & Armstrong (1979). Animals were screened for the number of nucleoli per cell either by analysis of tail tip tissue squashed and viewed by phase microscopy or by fixing and staining limb tissue following the protocol of Muneoka, Fox & Bryant (1983). Triploid cells possess three nucleoli per cell while diploid cells possess two nucleoli per cell. Cellular contribution was analysed in bismuth stained (Locke & Huie, 1977; Muneoka, Fox *et al.*, 1983; Figs 7 & 8) whole mount preparations of isolated dermis (Muneoka, Wise, Fox & Bryant, in preparation; Fig. 6) and the limb pattern was analysed following Victoria blue B staining (Bryant & Iten, 1974). The extent of the cellular contribution to supernumerary limbs was determined by scoring the percentage of triploid cells in the dorsal and ventral dermis of each digit of the supernumerary limb. Figures 9 & 10 show an example of a limb analysed in this manner. Twenty supernumerary limbs (10 limb bud to blastema stump; 10 blastema to limb

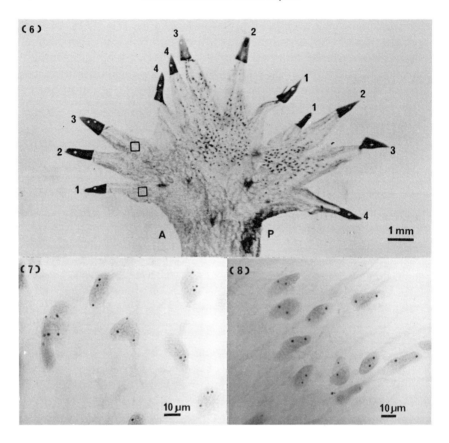

FIGS 6–8. (6) Bismuth stained dermal preparation of the experimental limb shown in Fig. 4. The dermis was isolated manually from the epidermis and underlying limb tissues and analysed as a whole mount preparation to ascertain the cellular contribution to the supernumerary limbs. The ventral dermis is shown here. Epidermis was left attached at the tips of the digits to prevent curling up of the preparation. This limb formed supernumerary limbs anterior (left four digits) and posterior (right four digits) to the graft (central four digits). A, anterior; P, posterior; a, b, insets shown in Figs 7 and 8. (7) Higher magnification of inset (a) of the dermal preparation shown in Fig. 6. Triploid cells each with three nucleoli are present in the dermis of digit 1 of the anterior supernumerary limb. (8) Higher magnification of inset (b) of the dermal preparation shown in Fig. 6. Diploid cells each with two nucleoli are present in the dermis of digit 3 of the anterior supernumerary limb.

bud stump) were analysed in this manner and all limbs revealed a substantial contribution from both the developing and the regenerating components of the interaction (Fig. 11). These results show definitively that developing and regenerating cells can interact and participate in an organized manner to form supernumerary limbs. We feel these studies show that the developing limb bud and regenerating blastema utilize the same mechanism for pattern regulation and are essentially

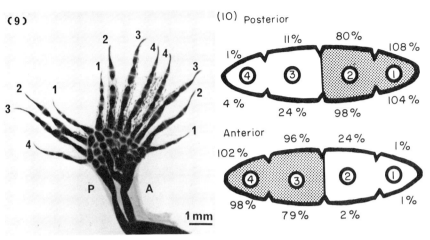

FIGS 9 & 10. (9) Skeletal preparation of an experimental limb where a triploid right hind limb bud was grafted to a diploid left forelimb blastema stump. The resulting limb produced supernumerary limbs anterior and posterior to the graft. The posterior to anterior (left to right) digital sequence was 4, 3, 2, 1 (posterior supernumerary limb), 1, 2, 3, 4 (grafted limb), 4, 3, 2, 1 (anterior supernumerary limb). A, anterior; P, posterior. (10) The cellular contribution to the anterior (lower) and posterior (upper) supernumerary limbs shown in Fig. 9 was analysed in dorsal and ventral dermal preparations. The frequency of triploid cells per digit was plotted across the anterior-posterior axis of the supernumerary limb and the boundary between 3N and 2N cells was arbitrarily determined as the point where cells matched 50% of the control frequency (50% boundary). The contribution line shown is the line connecting the 50% boundary points of the dorsal and ventral dermis. Percentage of control frequencies of triploid cells for each digit are shown for the dorsal and ventral dermis.

equivalent to each other in the way the limb pattern is established during limb outgrowth.

DISCUSSION

As we pointed out earlier, the two fields, pattern formation during limb development and pattern formation during limb regeneration, have remained reasonably separate. It now seems clear from a variety of lines of evidence but particularly from the results outlined above, that we must re-evaluate the situation and attempt to integrate the two fields. Earlier in this chapter we presented evidence in favour of Wolpert's universality concept as it applies to the vertebrate limb. The conclusion we are led to is that all developing limb buds and regenerating blastemas share common patterning mechanisms during limb outgrowth. The essential difference between the patterning models discussed above is the basic assumption that cells of the developing field do not possess any information concerning their position when limb bud

(a) Blastema / Limb bud   (b) Limb bud / Blastema

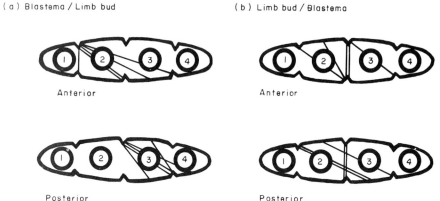

(c) Total

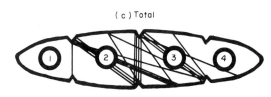

FIG. 11. Diagram showing the cellular contributions from developing and regenerating cells to the 20 supernumerary limbs analysed in this study. The contribution lines connect the dorsal and ventral 50% boundary points (see Fig. 10). (a) Cellular contribution to 10 supernumerary limbs (five anterior and five posterior) resulting from experimental grafts of blastema to limb bud stumps. (b) Cellular contribution to 10 supernumerary limbs (five anterior and five posterior) resulting from experimental grafts of limb bud to blastema stump. (c) Cumulative diagram showing all 20 supernumerary limbs described in (a) and (b). Note that in all cases studied there were substantial cellular contributions from both developing and regenerating cells. Fifteen of the 20 supernumerary limbs possessed four digits, two possessed five digits, two possessed three digits, and one possessed two digits.

formation is initiated, whereas cells in the regenerating blastema do possess positional information when blastema formation is initiated. In the regenerating limb this idea is easily testable and the results of such tests indicate that limb cells do carry with them information about their previous position in the mature limb as they form the blastema (see Tank & Holder, 1981, for a review). This idea is not as readily tested in the developing limb bud; however, there is good evidence that the cells of the limb field possess information about the anterior-posterior and dorsal-ventral limb axes long before there is any indication of an actual limb bud (urodeles – Swett, 1937; chick – Chaube, 1959). Furthermore, Iten and co-workers (Iten & Murphy, 1980; Javois & Iten, 1981, 1982; Javois, Iten & Murphy, 1981; Iten, 1982; Iten, Murphy & Muneoka, 1983) have shown that chick limb bud cells do retain information about their previous positions along the anterior-posterior and dorsal-ventral axes after grafting to different regions of the limb bud. Carlson (1983) has recently shown that chick limb bud cells retain positional information after *in vitro* culture. In this experiment chick limb bud cells which had been placed in culture, re-aggregated, and then grafted back into a limb bud, responded in a position-dependent manner by forming supernumerary limb structures.

We conclude that cells of both limb bud and blastema possess positional information about anterior-posterior and dorsal-ventral axes prior to the morphological appearance of the bud or blastema and that patterning mechanisms at work during limb outgrowth involve the distal elaboration of these positional values. We propose that the behaviour of limb cells (blastema or limb bud) during normal outgrowth and after experimental manipulation can be understood using the assumption and rules of the polar co-ordinate model (French *et al.*, 1976; Bryant *et al.*, 1981).

We must now focus on the differences mentioned on pp. 178 ff. in experimental results obtained with regenerating versus developing limbs. The first point to make is that among urodele amphibians the results of identical experiments on the limb bud (Maden & Goodwin, 1980; Maden, 1981) mimic exactly the experimental results with blastemas. So the central question is: why are there differences in the results from similar experiments performed on the limb buds of a variety of organisms, when the evidence points to the existence of similar underlying mechanisms for all vertebrate limbs? The key differences, pointed out by Tickle (1981) and discussed on pp. 178–181, between the behaviour of blastemas (and in fact urodele limb buds as well) and the behaviour of higher vertebrate limb buds such as those of the chick, revolve around the lack of a patterning response to some experimental alterations in the chick limb bud. We would argue, however, that the failure of a limb bud to respond to experimental spatial reorganization does not yield very useful information about the types of mechanisms involved in limb outgrowth. For example, grafting a blastema contralaterally to appose dorsal and ventral limb cells results in the formation of supernumerary limbs at dorsal and ventral positions. Similar experiments on the chick limb bud do not result in supernumerary limb formation, a fact which is used to support the conclusion that the two systems are utilizing different mechanisms. However, Javois & Iten (1982) have shown that indeed, when dorsal and ventral positions of the chick limb bud are apposed, a response is elicited, but it is not in the form of an externally visible outgrowth. They have attributed the more limited, internal response to the absence of an apical ectodermal ridge (essential for outgrowth) in the vicinity of the positional disparity. In another example, shifting blastemas (or urodele limb buds) from distal to proximal levels results in intercalary regeneration and the production of a normal limb. Similar grafting operations in the chick limb bud (Summerbell et al., 1973) result in a limb with a deletion in the proximal-distal pattern predicted by the progress zone model. However, if the grafting operation is performed at an earlier stage, intercalary regeneration occurs (Kieny & Pautou, 1977; Summerbell, 1977). What is clear is that the extent to which limb buds and limbs can regulate their pattern varies from system to system. We interpret this variation as *differences in ability of limb cells of different ages to recall their patterning mechanisms into use*. In animals which can regenerate, this ability is maintained throughout the life of the animal, but in higher vertebrates this ability is lost early in limb development.

If experiments which show a lack of patterning response are excluded, there is little reason to argue a difference between regenerating and developing limbs. The one striking difference, which has recently been described, deals with the dorsal-ventral muscle patterns of supernumerary limbs resulting from 180° rotation of the urodele blastema (Maden, 1980, 1982; Tank, 1981; Maden & Mustafa, 1982) versus similarly rotated chick limb buds (Javois & Iten, in preparation). The experimental manipulation in urodeles results in supernumerary limbs in which the muscle patterns are confused in the dorsal-ventral axis. Maden & Mustafa (1982) have hypothesized that these patterns result from the fusion of multiple supernumerary outgrowths arising around the circumference of the graft junction. Similar experiments in the chick (Javois & Iten, in preparation) yield supernumerary limbs with normal muscle patterns, similar to those observed by Maden (1980) in urodeles after single axis rotation. This puzzling difference can possibly be explained by assuming that the tissue interaction, and thus the number of possible supernumerary limbs formed after 180° rotation in the chick, is limited by the location of the apical ectodermal ridge anterior and posterior to the graft. Thus the 180° rotation of the chick limb bud would be functionally equivalent to the anterior-posterior contralateral blastema grafts in the urodele where the number of regions of positional disparity is restricted by the nature of the graft.

We are left, therefore, with the conclusion that developing and regenerating limbs probably use similar patterning mechanisms. This should make us cautiously optimistic about the possibility of limb regeneration in animals that do not normally regenerate (e.g. humans), since all animals with limbs clearly possess the information necessary to co-ordinate limb development. What they may lack is a way to recall this information into action beyond the period of embryonic development. We feel that studying the conditions necessary for pattern regulation during the embryonic stages of limb development, coupled with more detailed information about how this ability is lost, could lead to constructive approaches to prolonging the period during which limb cells are able to respond to positional cues. From this we might eventually be able to uncover ways to enable the cells of mature limb tissues to gain access to the information we believe they have at their disposal to build a new limb.

ACKNOWLEDGEMENTS

I (Susan Bryant) would like to take this opportunity to publicly acknowledge my former thesis adviser, Angus Bellairs, for his many and varied contributions to my growth and development as a biologist and as a person. Angus introduced me to research on regeneration, natural history, sherry, antiques, military history, the zoo, novels, movies, academic politics, poetry, his family, friends and colleagues, and his many original views about all sorts of subjects. In these days of ultraspecialization I am forever grateful for having had as my mentor a scholar with such broad-ranging interests and knowledge about biology as a whole. I wish him a productive and happy post-academic career.

Both authors would like to acknowledge the valuable comments and help in preparation of the manuscript by our colleagues Christine Rollman-Dinsmore, David Gardiner, Lorette Javois, Warren Fox, and Arlene Lum.

Research supported by NIH grant HD 06082. K. M. supported by NIH training grant HD 07029.

REFERENCES

Balcuns, A., Gasseling, M. T. & Saunders, J. W., Jr (1970). Spatio-temporal distribution of a zone that controls antero-posterior polarity in the limb bud of the chick and other bird embryos. *Am. Zool.* **10**: 323.

Bryant, S. V. & Baca, B. A. (1978). Regenerative ability of double-half and half upper arms in the newt, *Notophthalmus viridescens*. *J. exp. Zool.* **204**: 307–324.

Bryant, S. V., French, V. & Bryant, P. J. (1981). Distal regeneration and symmetry. *Science, Wash.* **212**: 993–1002.

Bryant, S. V. & Iten, L. E. (1974). The regulative ability of the limb regeneration blastema of *Notophthalmus viridescens*: Experiments *in situ*. *Wilhelm Roux' Archiv* **174**: 90–101.

Bryant, S. V. & Iten, L. E. (1976). Supernumerary limbs in amphibians: Experimental production in *Notophthalmus viridescens* and a new interpretation of their formation. *Devl Biol.* **50**: 212–234.

Cameron, J. & Fallon, J. F. (1977). Evidence for polarizing zone in the limb buds of *Xenopus laevis*. *Devl Biol.* **55**: 320–330.

Carlson, B. M. (1983). Positional memory in vertebrate limb development and regeneration. In *Limb development and regeneration, Part A*: 433–444. Fallon, J. F. & Caplan, A. I. (Eds). New York: Alan R. Liss, Inc.

Chaube, S. (1959). On axiation and symmetry in transplanted wing of the chick. *J. exp. Zool.* **140**: 29–77.

Chevallier, A., Kieny, M. & Mauger, A. (1977). Limb-somite relationship: origin of the limb musculature. *J. Embryol. exp. Morph.* **41**: 245–258.

Christ, B., Jacob, H. J. & Jacob, M. (1977). Experimental analysis of the origin of the wing musculature in avian embryos. *Anat. Embryol.* **150**: 171–186.

Fallon, J. F. & Crosby, G. M. (1977). Polarising zone activity in limb buds of amniotes. In *Vertebrate limb and somite morphogenesis*: 55–69. Ede, D. A., Hinchliffe, J. R. & Balls, M. (Eds). Cambridge: Cambridge University Press.

French, V., Bryant, P. J. & Bryant, S. V. (1976). Pattern regulation in epimorphic fields. *Science, Wash.* **193**: 969–981.

Gillespie, L. L. & Armstrong, J. B. (1979). Induction of triploid and gynogenetic diploid axolotls (*Ambystoma mexicanum*) by hydrostatic pressure. *J. exp. Zool.* **210**: 117–121.

Hinchliffe, J. R. & Johnson, D. R. (1980). *The development of the vertebrate limb*. Oxford: Clarendon Press.

Iten, L. E. (1982). Pattern specification and pattern regulation in the embryonic chick limb bud. *Am. Zool.* **22**: 117–129.

Iten, L. E. & Bryant, S. V. (1975). The interaction between the blastema and stump in the establishment of the anterior-posterior and proximal-distal organization of the limb regenerate. *Devl Biol.* **44**: 119–147.

Iten, L. E. & Murphy, D. J. (1980). Pattern regulation in the embryonic chick limb: Supernumerary limb formation with anterior (non-ZPA) limb bud tissue. *Devl Biol.* **75**: 373–385.

Iten, L. E., Murphy, D. J. & Muneoka, K. (1983). Do chick limb bud cells have positional information? In *Limb development and regeneration, Part A*: 77–88. Fallon, J. F. & Caplan, A. I. (Eds). New York: Alan R. Liss, Inc.

Javois, L. C. & Iten, L. E. (1981). Position of origin of donor posterior chick wing bud tissue transplanted to an anterior host site determines the extra structures formed. *Devl Biol.* **82**: 329–342.

Javois, L. C. & Iten, L. E. (1982). Supplementary limb structures after juxtaposing dorsal and ventral chick wing bud cells. *Devl Biol.* **90**: 127–143.

Javois, L. C. & Iten, L. E. (In prep.). *Supernumerary limbs after 180° rotation in the chick wingbud*.

Javois, L. C., Iten, L. E. & Murphy, D. J. (1981). Formation of supernumerary structures by the embryonic chick wing depends on the position and orientation of a graft in a host limb bud. *Devl Biol.* **82**: 343–349.

Kieny, M. & Pautou, M.-P. (1977). Proximo-distal pattern regulation in deficient avian limb buds. *Wilhelm Roux' Archiv* **183**: 177–191.

Lheureux, E. (1978). *Recherches expérimentales sur la morphogénèse régénératrice du membre de pleurodèle (Amphibien urodèle)*. Ph.D. Thesis: University of Lille, Lille, France.

Lheureux, E. (1983). The origin of tissues in the X-irradiated regenerating limb of the newt, *Pleurodeles waltlii*. In *Limb development and regeneration, Part A*: 455–466. Fallon, J. F. & Caplan, A. I. (Eds). New York: Alan R. Liss, Inc.

Locke, M. & Huie, P. (1977). Bismuth staining for light and electron microscopy. *Tiss. Cell* **9**: 347–371.

MacCabe, J. A. & Parker, B. W. (1976). Polarizing activity in the developing limb of the Syrian hamster. *J. exp. Zool.* **195**: 311–317.

Maden, M. (1980). Structure of supernumerary limbs. *Nature, Lond.* **286**: 803–805.
Maden, M. (1981). Experiments on anuran limb buds and their significance for principles of vertebrate limb development. *J. Embryol. exp. Morph.* **63**: 243–265.
Maden, M. (1982). Supernumerary limbs in amphibians. *Am. Zool.* **22**: 131–142.
Maden, M. & Goodwin, B. C. (1980). Experiments on developing limb buds of the axolotl, *Ambystoma mexicanum. J. Embryol. exp. Morph.* **57**: 177–187.
Maden, M. & Mustafa, K. (1982). The structure of 180° supernumerary limbs and a hypothesis of their formation. *Devl Biol.* **93**: 257–265.
Maden, M. & Turner, R. N. (1978). Supernumerary limbs in the axolotl. *Nature, Lond.* **273**: 232–235.
Muneoka, K. & Bryant, S. V. (1982). Evidence that patterning mechanisms in developing and regenerating limbs are the same. *Nature, Lond.* **298**: 369–371.
Muneoka, K. & Bryant, S. V. (In prep.). *Cellular contribution to supernumerary limbs in the axolotl*, Ambystoma mexicanum: *II. Blastema – limb bud interactions*.
Muneoka, K., Fox, W. F. & Bryant, S. V. (1983). Modified bismuth staining procedure for axolotl tissue. *Axolotl Newsl.* **12**: 2–7.
Muneoka, K., Wise, L. D., Fox, W. F. & Bryant, S. V. (In prep.). *Improved techniques for use of the triploid cell marker in the axolotl*, Ambystoma mexicanum.
Namenwirth, M. (1974). The inheritance of cell differentiation during limb regeneration in the axolotl. *Devl Biol.* **41**: 42–56.
Neufeld, D. A. (1980). Partial blastema formation after amputation in adult mice. *J. exp. Zool.* **212**: 31–36.
Pescitelli, M. J., Jr & Stocum, D. L. (1980). The origin of skeletal structures during intercalary regeneration of larval *Ambystoma* limbs. *Devl Biol.* **79**: 255–275.
Rollman-Dinsmore, C. & Bryant, S. V. (1982). Pattern regulation between hind- and forelimbs after blastema exchanges and skin grafts in *Notophthalmus viridescens. J. exp. Zool.* **223**: 51–56.
Saunders, J. W., Jr (1948). The proximo-distal sequence of origin of the parts of the chick wing and the role of the ectoderm. *J. exp. Zool.* **108**: 363–403.
Saunders, J. W., Jr, Gasseling, M. T. & Gfeller, M. D., Sr (1958). Interactions of ectoderm and mesoderm in the origin of axial relationships in the wing of the fowl. *J. exp. Zool.* **137**: 39–74.
Smith, S. D. (1981). The role of electrode position in the electrical induction of limb regeneration in subadult rats. *Bioelectrochem. Bioenerget.* **8**: 661–670.
Stocum, D. L. (1975). Regulation after proximal or distal transposition of limb regeneration blastemas and determination of the proximal boundary of the regenerate. *Devl Biol.* **45**: 112–136.
Stocum, D. L. (1980a). Autonomous development of reciprocally exchanged regeneration blastemas of normal forelimbs and symmetrical hindlimbs. *J. exp. Zool.* **212**: 361–371.
Stocum, D. L. (1980b). Intercalary regeneration of symmetrical thighs in the axolotl, *Ambystoma mexicanum. Devl Biol.* **79**: 276–295.
Stocum, D. L. (1982). Determination of axial polarity in the urodele limb regeneration blastema. *J. Embryol. exp. Morph.* **71**: 193–214.
Stocum,. D. L. & Dearlove, G. E. (1972). Epidermal-mesodermal interaction during morphogenesis of the limb regeneration blastema in larval salamanders. *J. exp. Zool.* **181**: 49–62.
Summerbell, D. (1974). Interaction between the proximo-distal and antero-posterior co-ordinates of positional value during the specification of positional information in the early development of the chick limb-bud. *J. Embryol. exp. Morph.* **32**: 227–237.

Summerbell, D. (1977). Regulation of deficiencies along the proximal distal axis of the chick wing-bud: A quantitative analysis. *J. Embryol. exp. Morph.* **41**: 137–159.
Summerbell, D. & Lewis, J. H. (1975). Time, place and positional value in the chick limb-bud. *J. Embryol. exp. Morph.* **33**: 621–643.
Summerbell, D., Lewis, J. H. & Wolpert, L. (1973). Positional information in chick limb morphogenesis. *Nature, Lond.* **244**: 492–496.
Swett, F. H. (1937). Determination of limb-axes. *Q. Rev. Biol.* **12**: 322–339.
Tank, P. W. (1978). The occurrence of supernumerary limbs following blastemal transplantation in the regenerating forelimb of the axolotl, *Ambystoma mexicanum*. *Devl Biol.* **62**: 143–161.
Tank, P. W. (1981). Pattern formation following 180° rotation of regeneration blastemas in the axolotl, *Ambystoma mexicanum. J. exp. Zool.* **217**: 377–387.
Tank, P. W. & Holder, N. (1981). Pattern regulation in the regenerating limbs of urodele amphibians. *Q. Rev. Biol.* **56**: 113–142.
Thoms, S. D. & Fallon, J. F. (1980). Pattern regulation and the origin of extra parts following axial misalignments in the urodele limb bud. *J. Embryol. exp. Morph.* **60**: 33–55.
Thornton, C. S. (1957). The effect of apical cap removal on limb regeneration in *Ambystoma* larvae. *J. exp. Zool.* **134**: 357–381.
Tickle, C. (1981). Limb regeneration. *Am. Scient.* **69**: 639–646.
Tickle, C., Shellswell, G., Crawley, A. & Wolpert, L. (1976). Positional signalling by mouse limb polarising region in the chick wing bud. *Nature, Lond.* **259**: 396–397.
Tickle, C., Summerbell, D. & Wolpert, L. (1975). Positional signalling and specification of digits in chick limb morphogenesis. *Nature, Lond.* **254**: 199–202.
Tschumi, P. A. (1957). The growth of the hindlimb bud of *Xenopus laevis* and its dependence upon the epidermis. *J. Anat.* **91**: 149–173.
Wallace, H. (1981). *Vertebrate limb regeneration*. Chichester: John Wiley & Sons.
Wolpert, L. (1969). Positional information and the spatial pattern of cellular differentiation. *J. theoret. Biol.* **25**: 1–47.
Wolpert, L. (1971). Positional information and pattern formation. *Curr. Topics Devl Biol.* **6**: 183–224.

Pattern Formation during Development of the Amniote Limb

LAWRENCE S. HONIG

Laboratory for Developmental Biology,
University of Southern California, Los Angeles,
California 90089-0191, USA*

SYNOPSIS

Vertebrate limb development has many unifying features. The patterned outgrowths of the limbs of amniotes – reptiles, birds, and mammals – possess strong similarities, but also significant differences from lower vertebrates such as amphibians. Amphibians possess major intercalative and regenerative abilities lacking in higher vertebrates; and the region responsible for positional information along the anteroposterior limb axis is located in the flank rather than in the limb bud. Recent studies by several investigators have shown that retinoid compounds of the vitamin A family have strong, but different, effects on pattern formation in amphibian and avian limb systems. It seems improbable that fundamentally different cellular/molecular mechanisms operate during limb development in amphibians and amniotes, but the distinction is useful, particularly when attempting to understand the biochemical basis for the specification of positional values. All amniote embryos studied possess a region at the posterior margin of the limb bud which exhibits a positional signalling activity. This activity may be demonstrated in embryonic chicks using tissue from various other birds, mammals, and as is shown here, reptiles. The embryonic limbs of both alligators and turtles have a posterior signalling region, and this region, analogous to the polarizing region of chicks, has unique morphogenetic capabilities within the limb bud.

VERTEBRATE LIMB DEVELOPMENT

The vertebrate limb first appears as a bulge from the flank of the embryo and consists of lateral plate mesoderm covered by ectoderm (Fig. 1). In amniotes, the epithelium always has a specialized thickened apical ridge; likewise some amphibians possess a specialized apical structure. The limb bud protrudes from the flank at an angle which depends upon the species and developmental stage. At early stage (bud length ≃ width), the bud extends laterally in alligators, ventrally in chicks, and intermediately in mice and turtles. The bud varies from a

---

*Present address: Department of Anatomy and Cell Biology (R-124), University of Miami Medical School, P.O. Box 016960, Miami, Florida 33101, USA.

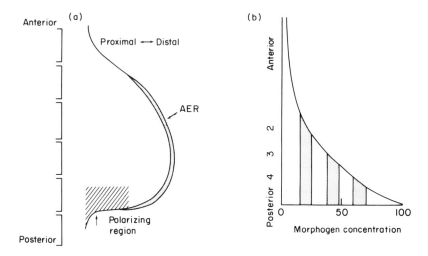

FIG. 1. The amniote limb bud. The limbs of birds, reptiles, and mammals all appear similar at the developmental stage in which bud length ≃ width. There are slight differences of contour shape, dorsoventral thickness, overall size, extent of apical ectodermal ridge (AER), angle of protrusion from the body and number of adjacent somites. The drawing in panel (A) shows a dorsal view of a right chick wing bud which lies opposite somites 15–20 at stage 21 (Hamburger & Hamilton, 1951). The anatomical axes, AER and polarizing region (striped) are shown. Panel (B) shows the morphogen gradient hypothesized to emanate from the polarizing region, and the concentration thresholds which would specify chick wing digits **2**, **3**, and **4**.

rather flat paddle shape in birds, to a form with a thicker cross-section in turtles and alligators, and an even more ellipsoid conical or cylindrical shape in amphibians. In those species possessing definitive limbs, the buds grow distally, and differentiate into a three-dimensional structure containing a two-dimensional set of skeletal elements, initially formed in the plane of the anteroposterior and proximodistal axes. Attaching to, and surrounding, the skeleton are muscles (which are of migratory somitic mesoderm lineage), tendons, and connective tissue as well as invading blood vessels and nerves. Despite the diverse functional modifications of the vertebrate limb for swimming, paddling, flying, walking or grasping and the plethora of epithelial coverings (including such individualized integumentary structures as the scales of avian feet), the basic plan of the vertebrate limb is remarkably constant (Fig. 2). The limb is conveniently divided into three "segments", called stylopod, zeugopod and autopod. The cartilaginous skeleton, which subsequently ossifies, consists of a single stylopodial upper limb element, two lower limb bones of the zeugopod, and a tiered array of up to 13 wrist/ankle elements and up to five digits which all together form the autopod. Each digit typically consists of a longer metacarpal/metatarsal and one to five shorter phalanges (Fig. 2).

Pattern Formation in the Amniote Limb

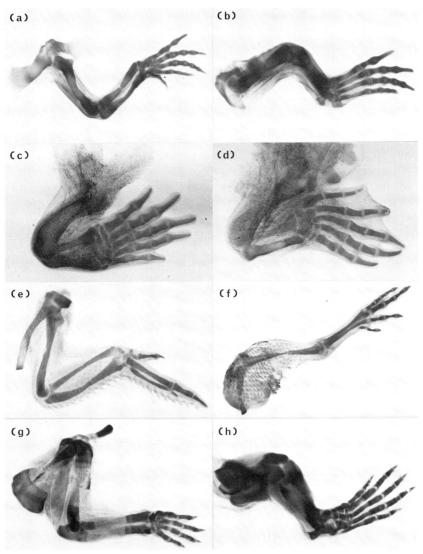

FIG. 2. Comparison of the limbs of selected amniotes. Alcian green stained and cleared wholemounts of late embryonic fore- and hind-limbs of alligator (A) and (B), turtle (C) and (D), chick (E) and (F), and skinned mouse (G) and (H).

While the skeletal pattern initially forms in a plane, the three-dimensional nature of the limb is important. Dorsal and ventral muscles, nerve patterns and integument are reproducibly organized, and physiologically of great importance. Some recent models have considered the limb as a two-dimensional field consisting of proximo-distal and circumferential co-ordinates (French, Bryant & Bryant,

1976; Bryant, French & Bryant, 1981). In part, this was prompted by analysis of *Drosophila* development where, unlike the situation in vertebrates, the appendages do indeed originate from the approximately two-dimensional imaginal discs. However, the product of imaginal disc evagination – the definitive insect limb – differs from the vertebrate limb inasmuch as it is an essentially hollow epidermal structure, centrally perfused. Muscles attach within the cylindrical appendage, but the pattern of these muscles depends on that of the ectodermal covering (Lawrence, 1982) which contains the (two-dimensional) information. In vertebrates, it is the three-dimensional mesodermal tissues that contain the positional information and signalling capabilities, as well as providing the bulk of the limb. With perhaps one exception regarding the dorsoventral axis, the epithelial covering provides no positional information to other limb tissues; it is usually the recipient of mesenchymal instructions.

In amphibians, limb buds form late in development, first appearing in the free-swimming, self-feeding larval tadpole which already has well-formed organ systems. In amniotes, i.e. reptiles, birds and mammals, limb outgrowth starts early: following heart and amnion formation and during the latter part of somite segmentation. Thus amniote limbs develop during the major phase of embryonic organogenesis. An abbreviated comparison of the temporal sequence of limb development in several amniote species is shown in Table I. The period of time that elapses from first bulging of the limb to completion of the cartilage model of the skeleton ranges widely. The process takes less than a week for the mouse or chick, about four weeks for humans and as long as ten weeks for the turtle developing at 20°C (but only three weeks during incubation at 30°C). The amniote embryo entering the "fetal growth period" has limbs which show the full complement and definitive pattern of differentiated tissues.

THE AVIAN EMBRYONIC LIMB

Amniote limb bud development has mostly been studied in chicks. Embryological manipulations of limb bud tissues show that the mechanisms of determination along the three anatomical axes are different. Through surgical transplantations of small pieces of embryonic tissue one can provoke the formation of duplicate limbs along either the anteroposterior axis or the dorsoventral axis, and can equally produce limbs with serial proximodistal repetitions. In each of these cases a single limb axis can be affected, while the arrangements of the tissue are harmonious in such a way that patterns viewed across the other axes are

TABLE 1
Comparative forelimb develpment in different amniote species

	Amphibians	Reptiles		Birds	Mammals	
	Xenopus (23°C)	Alligator (31°C)	Chelydra (20°C)	Chicken (38°C)	Mouse (37°C)	Human (37°C)
Limb description	Days (NF[a])	Days	Days (Y[b])	Days (HH[c])	Days (Th[d])	Days (Ca[e])
Mesenchymal condensation	7½ (48)	6	16 (9)	2¼ (16)	9 (14)	25
Bud protrusion	12 (49)	8	25 (11)	2½ (18)	9½ (15)	26 (12)
Width = length	15 (49)	9	30 (12)	3 (20)	10 (16)	27
Central condensation	17 (51)	10	32	3½ (22)	10½ (17)	28 (13)
Proximal differentiation	21 (52)	12	35 (13)	4 (24)	11 (18)	30 (14)
Elbow formation	22	14	42 (14)	4½ (25)	11½ (19)	33 (15)
Wrist	24 (53)	16	49 (15)	5 (27)	12 (20)	36 (16)
Metacarpals	26 (54)	18	56 (16)	6 (29)	13 (21)	41 (17)
Proximal phalanges	32 (55)	22	70 (18)	7 (31)	14 (22)	45 (18)
Distal phalanges	38 (56)	27	84 (20)	8 (34)	15 (23)	52 (21)

[a]Nieuwkoop & Faber (1967) stages. [b]Yntema (1968) stages. [c]Hamburger & Hamilton (1951) stages. [d]Theiler (1972) stages. [e]Chronology of Moore (1982), Carnegie stages. *Note*: For amphibians and reptiles, indicated developmental ages are days after egg-laying. For mammals, the approximate number of days post-fertilization is shown.

normal. For each limb axis, a different region of the limb bud (Fig. 1) seems to have prime importance: for the proximodistal axis, the apical ectodermal ridge (AER); for the anteroposterior axis, the posterior limb margin (ZPA; polarizing region); and for the dorsoventral axis, the ectoderm. Different mechanisms appear responsible for positional specification along the three axes. However, stable positional values for the three co-ordinate axes seem to be fixed concomitantly for any given region. Proximal tissues are so determined first and, as a rule, differentiate earlier.

The Proximodistal and Dorsoventral Axes

Early experiments of Saunders (1948) showed that removal of the apical ectodermal ridge (AER), at different stages of development, caused truncations of the resulting limbs. The later the removal of the ridge, the more distal the level of truncations. Ridge removal operations and combinations of proximal and distal limb slices at different stages of development (Wolpert, Lewis & Summerbell, 1975) have shown that such segments behave relatively autonomously. A young distal limb tip on an old stump gives serial duplications, whilst an older tip on a young stump yields limbs with deficiencies. Presumptive chick limb elements emerge from the growing tip of the bud in proximal to distal order: the humerus is specified first, then the radius and ulna, followed by the wrist elements, and lastly the digits. Truncated limbs have proper anteroposterior and dorsoventral polarity; proximodistal specification is thus independent of the information along the other axes. The "progress zone" model (Summerbell, Lewis & Wolpert, 1973) has been proposed to explain the specification of proximodistal positional values. It supposes that the distalmost histologically undifferentiated cells of the limb, which proliferate more actively, become of more distal positional value as they divide. As cells leave the "progress zone", they retain their most recent value. The time-dependent nature of proximodistal specification can also be accounted for by models postulating a morphogen-diffusion gradient such as the reaction-diffusion model of Meinhardt (1982) which involves a feedback-provoked increase in morphogen level at the growing tip.

Evidence suggests that very early in limb development (before the limb bud forms), the dorsoventral polarity of the ectoderm is defined and that subsequently the ectoderm controls dorsoventral polarity in the developing limb bud (e.g. J. A. MacCabe, Errick & Saunders, 1974), with stable specification occurring at about the stages at which proximodistal values are fixed.

The Anteroposterior Axis

The polarizing region

Rotation of the distal tip of the chick limb prompts formation of a mirror-image limb duplicated along the anteroposterior axis. Responsibility for this dramatic twinning can be assigned to tissue at the posterior margin of the limb bud, near its junction with the flank. If a small piece of this tissue, called the polarizing region (A. B. MacCabe, Gasseling & Saunders, 1973) is grafted to the anterior margin of the limb, a duplication along the anteroposterior axis results such that digits **4 3 2 2 3 4** form instead of the normal pattern **2 3 4** [Fig. 3(A), (B)]. The extra digits consist substantially of host tissues; the graft contributes few cells. The type of duplication formed depends on the position of the grafted extra polarizing region. Tissue closest to the polarizing region forms digit **4**, and that farthest, forms digit **2**. Tickle, Summerbell & Wolpert (1975) proposed a diffusible morphogen model to explain such behaviour [Fig. 1(B)]. The polarizing region acts as a source of morphogen, and is thus the high point of a concentration gradient. Adjacent to the polarizing region, high morphogen concentrations are at the threshold for production of posteriormost structures such as ulna and chick wing digit **4**, while at distances farther from the source, morphogen levels are lower, within threshold limits for anterior structures.

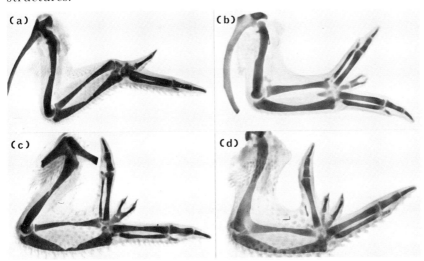

FIG. 3. Duplicated limbs resulting from polarizing region grafts. Cubes of chick anterior tissue (A) or posterior tissue from chick (B), turtle (C), or alligator (D) were grafted anteriorly opposite somite 16 of host chick wings. The posterior tissue from reptiles can produce mirror-image chick duplications (C), (D), of the same type as that caused by chick (B) grafts.

Long-distance signalling action

When a polarizing region is grafted to the anterior margin of the chick limb, several local and distant effects on the host limb can be noted. The portion of apical ridge immediately overlying the graft flattens, while ridge adjacent to the graft thickens. There are local increases in cell division in the vicinity of the graft, but also increased mitoses in areas substantially distant to the graft: over 20–30 cell diameters removed (Cooke & Summerbell, 1980). Limb outgrowth becomes more symmetrical and eventually a fully reduplicated limb results.

It can be shown that the pattern-signalling activity of the grafted polarizing region, like the proliferation-stimulating activity, occurs at a distance and is not merely the effect of immediate neighbour-neighbour interactions. Double grafts can be performed, with polarizing region placed on the anterior margin and identifiable "marked" responding limb tissue posteriorly adjacent to it, separating the signalling region from host responding tissue (Honig, 1981). Tissue from the Japanese quail (*Coturnix coturnix japonica*) is so distinguishable, possessing the distinctive heterochromatic nucleolar marker, as also is avian leg tissue, which differentiates into toes which can be discriminated by their form (as well as integumentary scales) from wing digits. When such double grafts are performed, the influence of the polarizing region extends through the marked tissue "barriers" into the host responding tissue. For example, using a leg barrier, limbs of digit pattern **4 III II 2 2 3 4** (anterior to posterior) may form. Digit **4** is a duplicate wing digit derived from the polarizing region graft. Digits **III** and **II** are extra toe digits formed by the barrier tissue (which was anterior tissue originally not fated to produce digits). The duplicate digit **2** is produced by host tissue posterior to the barrier, but influenced through the tissue. The maximum range of the signal from the polarizing region, as shown by using barriers of different widths, is about 300 μm. This distance is consistent with a number of estimates (reviewed by Summerbell & Honig, 1982) of the width of the limb digit field: by fate map, by impermeable barrier experiments interpretable as fate maps, through performance of multiple polarizing region grafts, and by similar tissue barrier experiments using quail-marked tissue.

The long-range nature of the polarizing region signal could derive from a morphogen freely diffusible in the extracellular space or diffusing/transported intracellularly; an ionic, electrochemical, or electrical gradient; or even a cell surface-to-surface propagated physiological phenomenon. Lightly irradiated tissue barriers ($<$ 100 Gy) transmit the signal equally as well as unirradiated tissue, while tissue barriers subjected to heavy doses of γ-irradiation ($>$ 250 Gy), sufficient to radically reduce polarizing activity, show no signal transmission (Honig, in preparation a). One interpretation is that signal propagation might need to be conducted via intracellular, rather than extracellular, space. An alternative explanation is that irradiation causes cells to engage in active destruction of morphogen.

The long-distance pattern action of the polarizing region (Honig, 1981) and the long-distance influence of cell division (Cooke & Summerbell, 1980) may be totally separate effects. Firstly, the largest increases in cell division are local, not at a long distance (Cooke & Summerbell, 1980, 1981; Smith & Honig, 1983). Secondly, when short-term polarizing region grafts were performed, in which the grafted region was allowed to reside in the host for periods of time (6–12 h) short enough that no duplication was produced, the increased distant cell division was nonetheless observed. This rapid effect may be mediated by the apical ectodermal ridge, unlike the pattern signal, which seems to pass through barrier tissue over which there is no ridge (Honig, in preparation a).

Cellular basis of signalling

Of foremost interest is the nature of the polarizing region signal at a cellular and molecular level. The observation that high, supralethal doses of γ-radiation (e.g. 160 Gy (= 16 000 rad) for quail) do not significantly inhibit polarizing region activity shows that polarizing region cell proliferation is not essential for signalling (Smith, Tickle & Wolpert, 1978). In a fashion similar to γ-radiation, ultraviolet radiation inhibits signalling: an e-fold reduction in signalling occurs after doses of $18\,\mathrm{jm}^{-2}$ (Honig, 1982). While γ-radiation causes reduction in polarizing activity only at doses well above those affecting cell survival, ultraviolet radiation inhibited activity at doses comparable to those reducing prolonged viability. The commonality between the two dose response curves is that the two agencies cause inhibition at comparable effective *nucleic acid* dosages.

Experiments with biochemical inhibitors have been used to examine whether any specific cell functions are particularly vital for polarizing activity (Honig, Smith, Hornbruch & Wolpert, 1981; Honig & Hornbruch, 1982). A wide range of inhibitor treatments have been used, and in an effort to gauge non-specific drug effects, tissue protein, RNA and DNA synthesis were assayed in parallel with examination of polarizing activity. Investigation as to whether the inhibition of any particular cell process necessarily was accompanied by disappearance of polarizing activity was to some extent hampered by the possibility that unwanted, unmeasured toxic side effects were responsible. So of greatest interest were cases in which: (i) a cell process was maintained at normal levels under conditions in which polarizing activity was heavily reduced; i.e. the cell function was not sufficient to ensure polarizing activity; and (ii) the abolition of a cell function did not result in major effects on polarizing activity; i.e. the process definitely was not required for polarizing activity. By these criteria, oxidative phosphorylation did not appear to be required for polarizing activity, nor did DNA synthesis. Protein synthesis, which was probed by the inhibitors puromycin, abrin, emetine, and diphtheria toxin, appeared to be necessary but not sufficient for signalling. Inhibitors of RNA synthesis yielded the most striking results since inhibition of polarizing activity by both α-amanitin and actinomycin D, which nominally affect different classes of RNA synthesis, nevertheless eliminated polarizing activity at doses where bulk RNA synthesis was comparatively unaffected. Viewed in the company of the irradiation results, which suggested that DNA integrity (although not DNA synthesis) was required for signalling, the possibility arises that production of some small subclass of RNA might be critically involved in signalling.

Polarizing region activity is unaffected when the tissue is totally disaggregated to a single cell suspension and reaggregated by centrifugation (Saunders, 1972). Supracellular architecture is not required. Nor is intracellular architecture. Agents which disrupt the cytoskeleton such as cytochalasin B, colchicine or vinblastine do not inhibit polarizing activity provided they are sufficiently washed out of the cells after treatment, and provided that non-toxic doses are used (Honig & Hornbruch, 1982). If grafted to the host limb margin without breaking the apical ectodermal ridge, surprisingly few polarizing region cells ($\sim$ 100) are required to signal (Tickle, 1981). While such small grafts do undergo an undetermined amount of cell proliferation, this result nonetheless lends support to the notion of an autocatalytic component to the signalling entity, as proposed by the models of Meinhardt (1982).

Both the nuclear and cytoplasmic cell compartments appear to be important in signalling. Polarizing region cells have been enucleated using cytochalasin B and the resultant karyoplast and cytoplast preparations grafted into recipient host limbs (Honig, 1983a). Neither of these membrane-enclosed, dye-excluding, subcellular fractions had signalling activity, which suggests that both cytoplasmic and nuclear information are required. No other subcellular preparations embedded in agar, or infiltrated into various chromotography beads, have been capable of producing limb duplications. Nor did a wide variety of biochemical inhibitors (L. S. Honig, unpublished observations) or hormones (Tickle, Alberts, Wolpert & Lee, 1982) have signalling activity. Thus it proved a great surprise when Tickle, Alberts *et al.*

polarizing region. Perhaps it acts as an analog of a Gierer-Meinhardt activator, or causes the destruction of inhibitor to levels so low as to provoke autocatalytic creation of a new activator peak. Such "resetting" action might be a result of the profound growth-inhibitory effects of the molecule. In chicks, there is little of the early limb widening that normally accompanies a polarizing region graft (and indeed at high doses there is loss of structures). In amphibians, it is only *subsequent* to the removal from vitamin A that regeneration (with duplication) proceeds. The situation is probably analogous in chicks, since owing to proximodistal outgrowth the drug-infiltrated graft falls behind, removed from a position where it may influence the limb tip. In addition, the efficacy of retinoic acid as compared to retinol (Summerbell & Harvey, 1983) may be due to the rapid metabolic removal of the former, in comparison to the long lifetime of retinol. (Interestingly, the fat-soluble retinol enters cells more freely than retinoic acid. It also seems that retinol is metabolizable to retinoic acid but the reverse process does not occur. Not enough is known about the relationship of these molecules.) "Resetting" might be at the cell surface, cytoplasmic, glycosylation or gene level.

THE REPTILIAN EMBRYONIC LIMB

Bellairs has pioneered work on embryonic reptiles. But, limb development in reptiles, which morphologically proceeds similarly to chicks, has been understudied, mostly owing to difficulties in obtaining a uniform and adequate experimental supply (New, 1966). Evolutionarily, reptiles rank between amphibians and birds, so knowledge about reptile development is clearly of interest. When it was first shown that posterior limb margin material from other species, such as mouse (Tickle, Shellswell, Crawley & Wolpert, 1976) and Syrian hamster (J. A. MacCabe & Parker, 1976) could signal in the chick limb, Fallon & Crosby (1977) noted that grafts from two species of turtle showed polarizing activity in chick hosts. Some studies of polarizing activity in alligators and map turtles are described below.

Presence of the Polarizing Region

Alligator embryos at stages (Ferguson, 1982) corresponding to post-laying incubations of eight to ten days at 31°C were used to prepare cube-shaped grafts of limb tissue from the region at the junction of the posterior margin of the alligator fore- or hind-limb bud with the embryonic flank. The presence of polarizing activity was tested for by performing operations consisting of replacement grafts to the anterior margin of the chick limb. These resulted in duplications of the chick autopod in 80% of the 20 cases. Zeugopodial duplications also occurred in 50% of the cases while stylopodial duplications were rare (5%). The duplicated digit patterns obtained included infrequent complete 4-duplications (10%) [Fig. 3(C)], and more frequent 3-duplications (25%) and 2-duplications (45%) yielding percentage polarizing activities of 44% (forelimbs) and 38% (hindlimbs). This moderate level of polarizing activity may be compared with levels obtained from other heterospecific grafts (see Table II).

Grafts of map turtle (*Graptemys*) posterior tissue from Yntema stages 12 and 13 embryos (Yntema, 1968) were also performed, both as replacement cubes, and as wedge grafts into slits in host chick embryos in the fashion of Iten & Murphy (1980). The 22 surviving operations gave results similar to those with the alligator regions (Fig. 3). Fore- and hind-limb regions showed 43% and 50% polarizing activity respectively, with 4-duplications (23% of cases), 3-duplications (14%) and 2-duplications (45%). Duplications of autopod, zeugopod, and stylopod were present in 82%, 45% and 23% of cases respectively.

While grafts of neither turtle tissue nor alligator tissue resulted in as high a frequency of duplications or as high polarizing activity in the

TABLE II
Compilation of results of grafts of the polarizing regions of diverse amniotes into chick host wing buds

Class	Species (limb)	Polarizing activity (number of grafts)	Reference
Aves	Chick (FL)	100 (8)	Honig (in preparation b)
	Chick (HL)	100 (9)	Honig (in preparation b)
	Quail (FL)	100 (8)	Honig (in preparation b)
	Quail (HL)	100 (8)	Honig (in preparation b)
Mammalia	Mouse (FL/HL)	33 (20)	Tickle, Shellswell *et al.* (1976)
	Mouse (FL)	30–43 (66)	Fallon & Crosby (1977)
	Mouse (HL)	28–38 (13)	Fallon & Crosby (1977)
	Ferret (FL/HL)	42–58 (15)	Fallon & Crosby (1977)
	Human (FL)	22–33 (3)	Fallon & Crosby (1977)
	Human (HL)	26–37 (9)	Fallon & Crosby (1977)
	Pig (FL)	26–32 (21)	Fallon & Crosby (1977)
	Syrian Hamster	42 (37)	J. A. MacCabe & Parker (1976)
Reptilia	Turtles (FL/HL)	"good" (NA)	Fallon & Crosby (1977)
	Alligators (FL)	44 (12)	Honig (in preparation b)
	Alligators (HL)	38 (8)	Honig (in preparation b)
	Turtle (FL)	43 (10)	Honig (in preparation b)
	Turtle (HL)	50 (12)	Honig (in preparation b)

FL – forelimb, HL – hindlimb, NA – not available. Range of values listed for data from Fallon & Crosby (1977) are activities assuming that all "good" duplications either contain extra digits **4** or only contain extra digits **3**.

Pattern Formation in the Amniote Limb 213

chick as is obtained with avian polarizing regions, the activity matched that of mammalian grafts into chicks. The less than maximum activity is a quantitative, not qualitative, failure as can be seen from the number of full reduplications obtained. There are several possible explanations for the reduced reptilian activity.

(i) The performance of the reptile tissue suffers at the unnaturally high incubation temperature (37°C). In fact, neither alligator (Ferguson & Joanen, 1982) nor turtle embryos develop to hatching at this temperature. However, Ferguson, Honig, Bringas & Slavkin (1982) have shown that reptile tissue in organ cultures will survive and differentiate at this temperature. Furthermore, while grafts of whole turtle or alligator limb rudiments to the chick chorioallantoic membrane show poor development, some differentiation of stylopod and zeugopod can occur. On the anterior margin of the chick wingbud, which is a more favourable site, considerable turtle limb development ensues at 37°C, including partial digit formation during a nine-day incubation period (Fig. 5).

(ii) Owing to slower growth, reptile tissue is left behind the progress zone and thus may have inadequate time to signal. That the described grafts included reptilian AER is also relevant. Operations placing reptilian mesodermal tissue under the chick AER have not yet been performed.

(iii) It is possible that the strength of the signal source in reptiles simply is lower than that of birds.

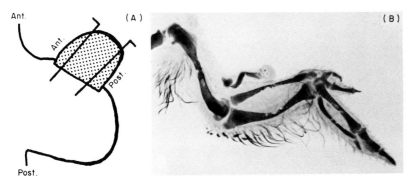

FIG. 5. Turtle forelimb buds can develop at 37°C in a chick egg. Panel (A) shows the grafting operation in which an Yntema (1968) stage 12½ turtle left hindlimb bud (stippled) was grafted to a site, prepared by minimal tissue removal, on the anterior margin of a stage 24 right chick wing, and secured by two platinum pins. Panel (B) contains an example of the results of such a graft. Note that because turtle posterior tissue was adjacent to host anterior tissue, the turtle polarizing region has provoked a minor duplication, 2 2 3 4, in the chick (despite the latter's late stage of development). During the nine-day incubation, the turtle hindlimb has differentiated shoulder, stylopod, zeugopod, wrist and nodules for metatarsals.

Ideally, we would like to use reptile embryos as hosts which would add greatly to our understanding. It is feasible to test whether avian or mammalian polarizing region cells are capable of signalling at reptilian incubation temperatures of 20–30°C. This would provide important information on the cellular mechanism of signalling. But after the performance of over 50 embryonic operations on alligator eggs only one embryo survived more than three days. The result of the graft of an alligator polarizing region to the anterior margin of an alligator forelimb bud, was a retarded, incompletely differentiated limb which, however, did show evidence of duplication (Fig. 6). The major difficulty with

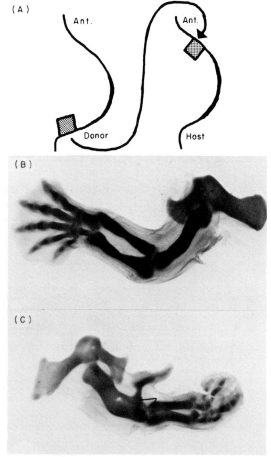

FIG. 6. Alligator polarizing region has activity in the alligator limb bud. Posterior tissue from the hindlimb of a nine-day alligator embryo was grafted to the anterior margin of a nine-day alligator forelimb bud as diagrammed in panel (A). The control unoperated left limb (B) and operated right limb (C) of a surviving embryo are shown. The operated limb is retarded but shows a duplicated humerus, zeugopodial element and extra digit.

operations on alligator eggs seems to stem from natural membrane-shell adhesions. About the time of egg-laying, the membranes of the early embryo (alligator eggs are laid at an advanced stage having 18 somite-pairs) stick firmly to the eggshell. With practice, windows in the eggshell can be made without damage, but partly because of postoperative volume changes in the egg, the embryo which remains attached at the periphery of the windows eventually suffers membrane rupture and death. Increased incubator humidity has not solved this problem. Presumably the use of shell-less or semi-shell-less culture (Ferguson, 1982) could alleviate such difficulties but because membrane-to-shell adhesion rapidly follows egg-laying, such experiments must be performed at a laboratory in close proximity to the breeding site. Numerous operations on turtle eggs have resulted in the occurrence of similar problems. The flexible eggshells lack any dimensional stability and in these eggs keeping the proper level of hydration, neither too much nor too little, is extremely difficult. An alternative to shell-less culture would be use of chick limb anterior margin or chorioallantoic membrane grafts (discussed above and disadvantageous owing to the problem of elevated incubation temperatures) or the use of organ culture with which experimentation is in progress.

We also still hope to surmount dehydration and membrane-adhesion *in ovo*. Demonstrations that tissue in homologous positions of reptilian and mammalian limb buds shows polarizing activity in chicks is persuasive indication, but not proof, of an identical role for the polarizing region in these species. Of great value would be the performance of operations on non-avian hosts. At present this appears more feasible in reptiles than in mammals.

The polarizing regions of alligator and turtle embryos grafted into chick embryos were able to cause full chick duplications despite their reduced growth rate. In a fashion, the reptilian polarizing regions resembled irradiated avian regions (Smith *et al.*, 1978), which tissues are unable to proliferate or contribute to the outgrowth, but signal nonetheless. This similarity was exploited to examine the uniqueness of the polarizing region.

Uniqueness of the Polarizing Region

In the developing chick limb, the only limb tissue that will cause duplications when grafted anteriorly is tissue from the polarizing region, at the posterior margin of the limb. However, extra limb structures will result if *anterior* tissue is grafted posteriorly (Iten & Murphy, 1980) into the polarizing region. Unlike the situation following polarizing region grafts, the extra structures are substantially donor-derived (Honig,

1983b). Furthermore, use of irradiated anterior tissue shows that even the lowest doses that stop cell multiplication (12 Gy = 1200 rad) but do not inhibit polarizing activity are sufficient to prevent the formation of anterior tissue-provoked supernumerary structures. However, to avoid the use of irradiation, a comparison was made between the signalling behaviour of turtle anterior and posterior tissue in the chick limb (Fig. 7).

Wedge grafts of turtle posterior tissue were grafted opposite somites

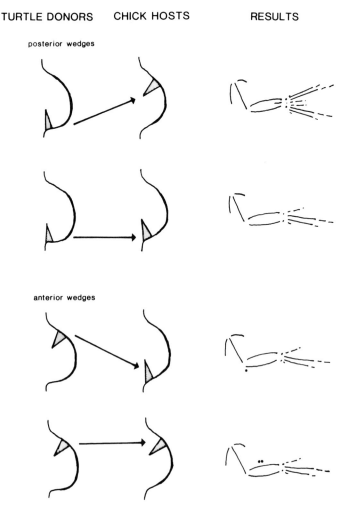

FIG. 7. Turtle posterior limb bud tissue (polarizing region) has unique signalling capabilities in the chick limb bud. Anterior tissue does not show any morphogenetic influence on the chick wing, but does frequently result in the formation of one or two extra cartilage nodules as shown.

16/17 in the chick resulting in a majority of limb duplications, as mentioned above. Posterior (polarizing activity) was 48% and the average number of extra digits was 1.3. By contrast, anterior turtle tissue grafted to a posterior chick host position opposite somite 19 gave rise to no extra digits and 0% anterior activity. Thus turtle anterior responding tissue, unlike chick, does not cause supernumerary elements. Clearly only the posterior tissue signalled, and presumably the anterior reptilian tissue was relatively unable to respond (probably because of suboptimal growth conditions) to the avian signalling region in the living chick embryo. In about half the cases, a single small nodule or spur of cartilage, or a slight, soft tissue protrusion did occur when anterior tissue grafts were used. However, when anterior tissue grafts were grafted orthotopically to anterior chick positions, the same results were obtained. By contrast, morphogenetic activity was nil when turtle posterior wedges were grafted orthotopically (to posterior chick positions). These results show that without resort to irradiation, using normal tissues of the embryonic turtle limb, posterior tissue has unique signalling properties, not possessed by other limb tissue, when assayed in the chick limb.

CONCLUSIONS

Vertebrate limbs are all constructed from the same basic pattern. Embryonic limb outgrowth is very similar between amniotes, and occurs at similar developmental stages even if the real time taken to traverse the determinative stages varies from six to 60 days. Early limb development in amphibians differs temporally (occurring at a later development stage) and structurally (in bud shape and ectodermal specializations). Furthermore, the region responsible for anteroposterior positional signalling in the amniote limb, whether in birds, mammals, or reptiles, is located within the growing limb bud. In amphibians, it appears to be located in the flank adjacent to the primordial limb rudiment (Slack, 1980) and apparently signals prior to limb outgrowth, having no role during the subsequent elongation and determination (Stocum & Fallon, 1982). This behaviour implies the possession of fixed anteroposterior (and likely dorsoventral) positional values during amphibian limb outgrowth, unlike the situation in higher vertebrates where the polarizing region plays its role during proximodistal elongation and specification. Amphibian limbs during development (and also during adulthood in urodeles) have dramatic powers of intercalary regeneration. All positions (but not epidermis) of the limb are capable of interacting during regeneration: no special signalling regions are apparent. Retinoic acid, which, following transient treatment in the chick limb seems to reset anteroposterior positional values to "highest" value, may behave similarly in certain anuran limbs. However, retinoid compounds consistently reset the proximodistal axis of the amphibian limb, an effect not observed in the higher vertebrate. This phenomenon may be related to the proximodistal intercalative abilities of amphibians which are not possessed by chicks. Universality, or mechanistic similarities between species, can be anticipated, and is both theoretically gratifying and experimentally useful. But we must be watchful over necessarily always demanding identical mechanisms from disparate species.

ACKNOWLEDGEMENTS

Support was provided by US NIH grants DE-02848, DE-07006, and HL-28325. I thank Dr J. J. Bull (University of Texas) for sending turtle eggs collected in Wisconsin, and thank Dr M. W. J. Ferguson (Queen's University of Belfast) and Mr T. Joanen of the Rockefeller Wildlife Refuge in Louisiana, through whose efforts freshly laid alligator eggs were obtained.

REFERENCES

Bryant, S. V., French, V. & Bryant, P. J. (1981). Distal regeneration and symmetry. *Science, N.Y.* **212**: 993–1002.

Cooke, J. & Summerbell, D. (1980). Cell cycle and experimental pattern duplication in the chick wing during embryonic development. *Nature, Lond.* **287**: 697–701.

Cooke, J. & Summerbell, D. (1981). Control of growth related to pattern specification in chick wing bud mesenchyme. *J. Embryol. exp. Morph.* **65** (Suppl.): 169–185.

DeLuca, L. M. & Shapiro, S. S. (1981). Modulation of cellular interactions by vitamin A and derivatives (retinoids). *Ann. N.Y. Acad. Sci.* **359**: 1–430.

Dhouailly, D., Hardy, M. H. & Sengel, P. (1980). Formation of feathers on chick foot scales: a stage-dependent morphogenetic response to retinoic acid. *J. Embryol. exp. Morph.* **58**: 63–78.

Elias, P. M. & Friend, D. S. (1976). Vitamin A-induced mucous metaplasia: an *in vitro* system for modulating tight and gap junction differentiation. *J. Cell Biol.* **68**: 173–188.

Elias, P. M., Grayson, S., Caldwell, T. M. & McNutt, N. S. (1980). Gap junction proliferation in retinoic acid-treated human basal cell carcinoma. *Lab. Invest.* **42**: 469–474.

Fallon, J. F. & Crosby, G. M. (1977). Polarizing zone activity in limb buds of amniotes. In *Vertebrate limb and somite morphogenesis*: 55–69. Ede, D. A., Hinchliffe, J. R. & Ball, M. (Eds). Cambridge: Cambridge University Press.

Ferguson, M. W. J. (1982). *The structure and development of the palate in* Alligator mississippiensis. Ph.D. Thesis: Queen's University of Belfast, Northern Ireland.

Ferguson, M. W. J., Honig, L. S., Bringas, P., Jr & Slavkin, H. C. (1982). *In vivo* and *in vitro* development of first branchial arch derivatives in *Alligator mississippiensis*. In *Factors and mechanisms influencing bone growth*: 275–286. Dixon, A. D. & Sarnat, B. G. (Eds). New York: A. R. Liss, Inc.

Ferguson, M. W. J. & Joanen, T. (1982). Temperature of egg incubation determines sex in *Alligator mississippiensis*. *Nature, Lond.* **296**: 850–853.

Frederick, J. M. & Fallon, J. F. (1982). The proportion and distribution of polarizing zone cells causing morphogenetic inhibition when coaggregated with anterior half wing mesoderm in recombinant limbs. *J. Embryol. exp. Morph.* **67**: 13–25.

French, V., Bryant, P. J. & Bryant, S. V. (1976). Pattern regulation in epimorphic fields. *Science, N.Y.* **193**: 969–981.

Ganguly, J., Rao, M. R. S., Murthy, S. K. & Sarada, K. (1980). Systemic mode of action of vitamin A. *Vitamins Horm.* **38**: 1–54.

Geelen, J. A. G. (1979). Hypervitaminosis A induced teratogenesis. *CRC Crit. Rev. Toxicol.* **6**: 351–375.

Hamburger, V. & Hamilton, H. L. (1951). A series of normal stages in the development of the chick embryo. *J. Morph.* **88**: 49–92.

Honig, L. S. (1981). Positional signal transmission in the developing chick limb. *Nature, Lond.* **291**: 72–73.

Honig, L. S. (1982). Effects of ultraviolet light on the activity of an avian limb positional signalling region. *J. Embryol. exp. Morph.* **71**: 223–232.

Honig, L. S. (1983a). Polarizing activity of the avian limb examined on a cellular basis. In *Limb development and regeneration, part A*: 99–108. Fallon, J. F. & Caplan, A. I. (Eds). New York: A. R. Liss, Inc.

Honig, L. S. (1983b). Does anterior (non-polarizing region) tissue signal in the developing chick limb? *Devl Biol.* **97**: 424–432.

Honig, L. S. (In prep. a). *Transmission of the positional signal of the limb polarizing region through irradiated responding tissue*.

Honig, L. S. (In prep. b). *A unique positional signalling region in the reptilian embryonic limb bud can be demonstrated by interspecies grafts in the chick*.

Honig, L. S. & Hornbruch, A. (1982). Biochemical inhibition of positional signalling in the avian limb bud. *Differentiation* **21**: 50–55.

Honig, L. S., Smith, J. C., Hornbruch, A. & Wolpert, L. (1981). Effects of biochemical inhibitors on positional signalling in the chick limb bud. *J. Embryol. exp. Morph.* **62**: 203–216.

Hurmerinta, K., Thesleff, I. & Saxen, L. (1980). *In vitro* inhibition of mouse odontoblast differentiation by vitamin A. *Arch. Oral Biol.* **25**: 385–393.

Iten, L. E. & Murphy, D. J. (1980). Pattern regulation in the embryonic chick limb: supernumerary limb formation with anterior (non-ZPA) limb bud tissue. *Devl Biol.* **75**: 373–385.

Lawrence, P. A. (1982). Cell lineage of the thoracic muscles of *Drosophila*. *Cell* **29**: 493–503.

Lotan, R. (1980). Effects of vitamin A and its analogs (retinoids) on normal and neoplastic cells. *Biochem. Biophys. Acta* **605**: 33–91.

MacCabe, A. B., Gasseling, M. T. & Saunders, J. W., Jr (1973). Spatiotemporal distribution of mechanisms that control outgrowth and anteroposterior polarization of the limb bud in the chick embryo. *Mech. Ageing Develop.* **2**: 1–12.

MacCabe, J. A., Errick, J. & Saunders, J. W. (1974). Ectodermal control of the dorsoventral axis in the leg bud of the chick embryo. *Devl Biol.* **39**: 69–82.

MacCabe, J. A. & Parker, B. W. (1976). Polarizing activity in the developing limb of the Syrian hamster. *J. exp. Zool.* **195**: 311–317.

Maden, M. (1982). Vitamin A and pattern formation in the regenerating limb. *Nature, Lond.* **295**: 672–675.
Maden, M. (1983). Vitamin A and the control of pattern in regenerating limbs. In *Limb development and regeneration, part A*: 445–454. Fallon, J. F. & Caplan, A. I. (Eds). New York: A. R. Liss, Inc.
Marin-Padilla, M. & Ferm, V. H. (1965). Somite necrosis and developmental malformations induced by vitamin A in the Golden hamster. *J. Embryol. exp. Morph.* **13**: 1–8.
Meinhardt, H. (1982). *Models of biological pattern formation.* London: Academic Press.
Moore, K. L. (1982). *The developing human – clinically oriented embryology*, 3rd Edn. Philadelphia: W. B. Saunders Co.
New, D. A. T. (1966). *The culture of vertebrate embryos.* London: Logos Press.
Niazi, I. A. & Saxena, S. (1978). Abnormal hind limb regeneration in tadpoles of the toad, *Bufo andersoni* exposed to excess vitamin A. *Folia biol., Kraków* **26**: 3–8.
Nieuwkoop, P. D. & Faber, J. (1967). *Normal tables of* Xenopus laevis *(Daudin).* Amsterdam: North Holland Publishing Co.
Pennypacker, J. P., Lewis, C. A. & Hassell, J. R. (1978). Altered protoglycan metabolism in mouse limb mesenchyme culture treated with vitamin A. *Arch. Biochem. Biophys.* **186**: 351–358.
Pitts, J. D., Burk, R. R. & Murphy, J. P. (1981). Retinoic acid blocks junctional communication between animal cells. *Cell Biol. Int. Rep.* **5** (Suppl. A): 45.
Prutkin, L. (1975). Mucous metaplasia and gap junction in the vitamin A acid-treated skin tumor, keratoacanthoma. *Cancer Res.* **35**: 364–369.
Saunders, J. W., Jr (1948). The proximo-distal sequence of origin of the parts of the chick wing and the role of the ectoderm. *J. exp. Zool.* **108**: 363–408.
Saunders, J. W., Jr (1972). Developmental control of three-dimensional polarity in the avian limb. *Ann. N.Y. Acad. Sci.* **193**: 29–42.
Slack, J. M. W. (1980). Regulation and potency in the forelimb rudiment of the axolotl embryo. *J. Embryol. exp. Morph.* **57**: 203–217.
Smith, J. C. & Honig, L. S. (1983). Growth and the origin of additional structures in reduplicated chick wings. In *Limb development and regeneration, part A*: 57–65. Fallon, J. F. & Caplan, A. I. (Eds). New York: A. R. Liss, Inc.
Smith, J. C., Tickle, C. & Wolpert, L. (1978). Attenuation of positional signalling in the chick limb by high doses of γ-radiation. *Nature, Lond.* **272**: 612–613.
Stocum, D. L. & Fallon, J. F. (1982). Control of pattern formation in urodele limb ontogeny: a review and a hypothesis. *J. Embryol. exp. Morph.* **69**: 7–36.
Summerbell, D. & Harvey, F. (1983). Vitamin A and the control of pattern in developing limb. In *Limb development and regeneration, part A*: 109–118. Fallon, J. F. & Caplan, A. I. (Eds). New York: A. R. Liss Inc.
Summerbell, D. & Honig, L. S. (1982). The control of pattern across the anteroposterior axis of the chick limb bud by a unique signalling region. *Am. Zool.* **22**: 105–116.
Summerbell, D., Lewis, J. H. & Wolpert, L. (1973). Positional information in chick limb morphogenesis. *Nature, Lond.* **244**: 492–496.
Theiler, K. (1972). *The house mouse.* Berlin: Springer-Verlag.
Tickle, C. (1981). The number of polarizing region cells required to specify additional digits in the developing chick wing. *Nature, Lond.* **289**: 295–298.
Tickle, C., Alberts, B., Wolpert, L. & Lee, J. (1982). Local application of retinoic acid to the limb bud mimics the action of the polarizing region. *Nature, Lond.* **296**: 564–566.
Tickle, C., Shellswell, G., Crawley, A. & Wolpert, L. (1976). Positional signalling by mouse limb polarizing region in the chick wing bud. *Nature, Lond.* **259**: 396–397.

Tickle, C., Summerbell, D. & Wolpert, L. (1975). Positional signalling and specification of digits in chick limb morphogenesis. *Nature. Lond.* **254**: 199–202.

Wolpert, L., Lewis, J. H. & Summerbell, D. (1975). Morphogenesis of the vertebrate limb. In *Cell patterning* (Ciba Foundation Symp. 29, new series): 95–130. Amsterdam: Elsevier.

Yntema, C. L. (1968). A series of stages in the embryonic development of *Chelydra serpentina*. *J. Morph.* **125**: 219–252.

Symp. zool. Soc. Lond. (1984) No. 52, 223-273

Craniofacial Development in *Alligator mississippiensis*

MARK W. J. FERGUSON

*Anatomy Department, The Queen's University of Belfast,
Medical Biology Centre, 97 Lisburn Road, Belfast BT9 7BL,
Northern Ireland*

SYNOPSIS

Overt facial morphogenesis in the alligator embryo commences at day 7 when the optic, otic and nasal placodes appear and three branchial arches are present. The optic, otic and nasal placodes progressively deepen to form the respective pits, and by day 9 five branchial arches are present. These branchial arches are initially separated by patent branchial clefts from the pharynx to the exterior. The opposing epithelial cells of these clefts develop cilia which join together and so facilitate cleft closure. At first, the medial nasal, lateral nasal and maxillary processes are separated by a nasal pit slit but the latter is closed by merging of these facial processes. There are differences in the relative sizes of these facial processes in different crocodilian species: e.g. in *Crocodylus johnstoni* the medial nasal process is larger and the lateral nasal process smaller than in *Alligator mississippiensis*. The primary palate is developed by day 12 and secondary palatal shelves bud off the maxillary processes around day 18. These shelves grow horizontally above the dorsum of the tongue and approximate each other in the mid-line behind the primary palate by about day 20. Palatal closure commences behind the primary palate and spreads progressively posteriorly, the entire palate being closed by day 25. This closure involves contact and adherence of the medial edge epithelial cells (MEE) of each shelf, limited MEE death, posteronasal migration of MEE out of the closure zone and mesenchymal continuity across the palate. Alligator MEE show a characteristic phenotype of cobblestoned, villous, migrating cells. This sequence of palatal closure is in contrast to that observed in mammals, where the MEE die, and in birds, where the MEE keratinize resulting in cleft palate. MEE differentiation and the mechanisms of palatal closure (or non-closure in the case of chickens) are identical under defined *in vitro* conditions to those observed *in vivo*. Classical epithelial mesenchymal recombination experiments between alligator, chicken and mouse embryonic palatal shelves reveal that the differentiation of MEE is regulated by the underlying mesenchyme. Spontaneous malformations are related to maternal age (as determined by cyclical osseous growth lines in the femur mid-diaphysis) and diet. The frequency of malformations is highest amongst the offspring of young or old alligators whereas middle-aged females produce eggs with a rate of malformation less than 0.5%. Using such eggs, cleft lip and palate, cyclopia, absent lower jaw and tongue and a wide variety of other craniofacial malformations have been reliably induced in alligator embryos by the application of 5-fluoro-2-desoxyuridine. A combination of semi-shell-less culture techniques and the teratogenic induction of absent lower jaw and tongue in alligator embryos has facilitated the first longitudinal study of palatogenesis in any animal.

INTRODUCTION

Congenital craniofacial birth defects represent a significant health care problem, such that in man approximately one baby in every 500 live births has a clefting malformation. These impersonal statistics translate into enormous human suffering and significant health care costs, especially when it is realized that the incidence of cleft palate has doubled in the past 25 years (Fogh Andersen, 1968). It is clear that basic scientific information on normal developmental mechanisms in palate formation, and their disruption in cleft palate, is necessary for a proper understanding and rational prevention of cleft palate in man. Thus development of the mammalian secondary palate has been the object of intense research interest, with many investigations emphasizing the study of teratogens that induce cleft palate (see reviews by Burdi, Feingold, Larsson, Leck, Zimmerman & Fraser, 1972; Greene & Kochhar, 1975; Greene & Pratt, 1976; Shah, 1979; Ferguson, 1978a, b, 1981a, b, c; Salomon & Pratt, 1979). More recently the model system of palatogenesis has been used as a tool to investigate a wide variety of basic developmental phenomena, e.g. morphogenetic movements (Wee, Wolfson & Zimmerman, 1976; Ferguson, 1978a, b; Wee & Zimmerman, 1980; Kuhn, Babiarz, Lessard & Zimmerman, 1980); extracellular matrix production (Pratt & Hassell, 1975; Hassell & Orkin, 1976; Wilk, King & Pratt, 1978; Ferguson, 1978a, b); epithelial mesenchymal interactions (Tyler & Koch, 1975, 1977a, b; Tyler & Pratt, 1980; Ferguson & Honig, 1984); cell adhesion (Greene & Pratt, 1976; Morgan, 1976; Morgan & Pratt, 1977; Shah, 1979); and cell death (Mato, Smiley & Dixon, 1972; Pratt & Martin, 1975; Pratt & Greene, 1976; Greene & Pratt, 1978; Tassin & Weill, 1980). Clearly the two approaches are complementary.

A number of recent reviews summarize much of the existing scientific data on normal and abnormal palatal development (Burdi *et al.*, 1972; Greene & Kochhar, 1975; Greene & Pratt, 1976; Ferguson, 1978a, b, 1981a, b, c; Shah, 1979; Salomon & Pratt, 1979; Pratt & Christiansen, 1980; Pratt, 1980; Pratt & Salomon, 1981). Succinctly, in mammals, the sequence of events is as follows: the ectomesenchymal cells of the future palate are derived from the mesencephalic neural crest (Johnston & Hazelton, 1972; Johnston, Bhakdinavonk & Reid, 1974; Le Lièvre, 1974, 1978; Le Lièvre & Le Douarin, 1975; Johnston, Morriss, Kushner & Bingle, 1977). These cells migrate from the neural tube to the oral cavity on a matrix rich in glycosaminoglycans, collagen and fibronectin (Weston, 1970; Pratt, Larsen & Johnston, 1975; Greenberg & Pratt, 1977; Weston, Derby & Pintar, 1978; Pintar, 1978; Derby, 1978;

Bolender, Seliger & Markwald, 1980; Löfberg, Ahlfors & Fällström, 1980; Newgreen & Thiery, 1980). There in association with pharyngeal ectoderm they form the bilateral palatal shelves. At first these shelves grow vertically down the lateral sides of the tongue, but at a precise stage in development they elevate to a horizontal position above the dorsum of the tongue (Ferguson, 1978a, b; Diewert, 1980; Zimmerman, Wee, Clark & Venkatasubramanian, 1980; Brinkley, 1980). The principal forces causing this elevation are thought to be generated by the hydration of glycosaminoglycans (Ferguson, 1978a, b; Wilk *et al.*, 1978; Brinkley, 1980) in the palatal mesenchyme and the contraction of the mesenchymal cells themselves (Wee *et al.*, 1976; Wee & Zimmerman, 1980; Kuhn *et al.*, 1980; Zimmerman *et al.*, 1980). The epithelial cells of the tips of the elevated palatal shelves contact and adhere to each other by means of a carbohydrate-rich surface coat (Greene & Kochhar, 1974; Pratt & Hassell, 1975; Souchon, 1975; Meller & Barton, 1978) and desmosomes (Morgan, 1976; Morgan & Pratt, 1977). The fused cells die and are removed by macrophages, so that mesenchymal continuity is established between the two shelves (Mato, Aikawa & Katahiva, 1966; Farbman, 1968, 1969; Hayward, 1969; Mato, Smiley *et al.*, 1972; Chaudhry & Shah, 1973; Greene & Pratt, 1976; Shah, 1979). The shelf mesenchyme differentiates into bony, fibrous and muscular elements (Ferguson, 1978a) whilst the cells on the nasal surface of the fused palate differentiate into pseudostratified ciliated columnar epithelia and those on the oral surface into keratinized stratified squamous epithelia (Tyler & Koch, 1975). Clearly palatal shelf epithelium consists of three distinct regions, nasal, medial and oral, with different developmental fates: to become pseudostratified ciliated, to die, and to keratinize respectively (Tyler & Koch, 1975). This regional differentiation is probably mediated by an epithelial mesenchymal interaction (Tyler & Koch, 1975, 1977a, b; Ferguson & Honig, 1984).

Although mammalian palatogenesis has been extensively studied there are still a number of outstanding problem areas (Ferguson, 1981a, b, c). Succinctly, these problems all accrue from the inaccessibility of the mammalian embryo for precise surgical and teratological experiments, and for longitudinal studies *in vivo*, as well as the severe difficulties of culturing intact mammalian embryos, late in gestation (during the period of palatogenesis). All these problems can be overcome by studying palatogenesis in an animal which develops in an external calcified egg, yet which possesses a mammal-like secondary palate. Crocodilians are the only animals which exhibit this unique combination (Ferguson, 1981a, b, c; Fig. 16) and so are of great interest to experimental craniofacial embryologists. Crocodilian facial development is also of general zoological interest in view of the evolutionary longevity of the species and their postulated affinity with birds. Recent studies have concentrated on elucidating some features of alligator facial development (Ferguson, 1979, 1981a, b, c, 1982c, in press; Ferguson, Honig, Bringas & Slavkin, 1981, 1982, 1983; Ferguson & Honig, 1984; Ferguson, Honig & Slavkin, in press) as well as related investigations into other aspects of their embryology (Ferguson, 1981d, 1982a, b, c, in press; Ferguson & Joanen, 1982, 1983) which can be exploited in experimental craniofacial studies, e.g. sexual differences in teratogenic responses and semi-shell-less culture. This paper will describe: some normal features of alligator craniofacial development (*in vivo* and *in vitro*); variations in facial development between some crocodilian species; the relationship between maternal age and spontaneous malformations; and the pathogenesis of teratogenically induced facial malformations.

NORMAL DEVELOPMENT

Branchial Arches

Parker (1883), Clarke (1891), Voeltzkow (1899) and Reese (1908, 1912, 1915) described the macroscopic appearances of some embryos of *Alligator mississippiensis* and *Crocodylus niloticus*. The development of five branchial arches and clefts was noted, and it was generally believed that three, and possibly four, of the branchial clefts opened to the exterior (Clarke, 1891; Reese, 1908, 1912, 1915). Clarke (1891) described a Y-shaped bifurcation of the first branchial cleft in *A. mississippiensis*: the stem of the Y and the anterior arm forming the true cleft, whilst the posterior arm was merely an external groove. However, Reese (1908, 1915) found no evidence of this bifurcation in his study of alligator embryos. Parker (1883) described the fates of the branchial arches and clefts in several crocodilian species and compared there with those of other reptiles, mammals and fish.

Crocodilian embryos are laid at an advanced developmental stage (c. 17–19 somites, primitive gut tube open between somites 8 and 18, posterior neuropore patent, simple S-shaped heart tube) and the first branchial arch is present at the time of egg-laying. This stage at egg-laying is virtually identical in embryos of *A. mississippiensis*, *C. johnstoni*, *C. porosus* and *C. niloticus*. The other branchial arches develop in a progressive craniocaudal sequence (Fig. 1). Two branchial arches are evident macroscopically at the 20 somite stage, i.e. day 2 (day 0 equals the day of egg-laying and all subsequent days are based on a

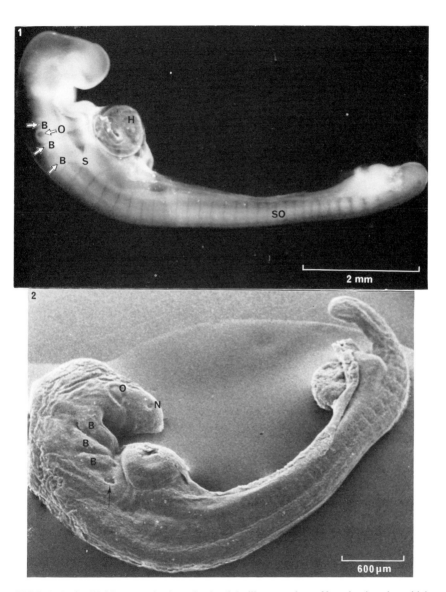

FIGS 1 & 2. (1) Macroscopic view of a day 3.5 alligator embryo. Note the three branchial arches (B) and clefts, the branchial sinus (S), the ganglia and neurites of the trigeminal, facial and glossopharyngeal nerves (arrowed), the otic pit (O), heart tube (H) and somites (SO). (2) SEM of a day 7 alligator embryo illustrating the nasal placode (N), optic vesicle (O), three branchial arches and clefts (B) and the branchial sinus (arrowed).

standard series incubated at 30°C and 100% humidity – Ferguson, 1982c). At this stage three pairs of cranial somitomeres are visible macroscopically on both sides of the neural tube, either when embryos are viewed with oblique incident illumination or in whole mount preparations of embryos stained with alcian green or azocarmine. These somitomeres are located cranial to the otocyst. Likewise in late stage 8 (Hamburger & Hamilton, 1951) chicken embryos there are seven somitomeres stacked in tandem along each side of the neural tube, the compacted eighth somitomere forming the first somite (Meier, 1979, 1981; Anderson & Meier, 1981). However, owing to the contiguity of the first seven cranial somitomeres in the chick, they cannot be resolved by light microscopy and were discovered by the use of scanning electron microscopy (Meier, 1979, 1981; Anderson & Meier, 1981). In *A. mississippiensis* (and also in *C. johnstoni* and *C. porosus*) the three somitomeres visible macroscopically may well be subdivided, but such divisions are beyond the resolution of the light microscope. The precise number and arrangement of cranial somitomeres in crocodilian embryos are unknown and await investigation by microdissection and scanning electron microscopy.

Three branchial arches are evident at the 25 somite stage (day 7), four branchial arches at the 30 somite stage (day 9) and five branchial arches at the 35 somite stage (day 11). The arches decrease in size as one moves caudally so that the first is the largest and the fifth is but a very small structure (Figs 1, 2, 4 & 9). At day 7 in *A. mississippiensis*, three branchial arches and clefts, the branchial sinus and optic placodes are evident, whilst the nasal placodes have just appeared and are beginning to invaginate (Figs 1 & 2). When viewed with transilluminated or oblique incident lighting, the ganglia and emerging nerve fibres of the cranial nerves which innervate each arch are evident (mandibular division of trigeminal for the first arch, facial for the second arch, glossopharyngeal for the third arch – Fig. 1). The ectodermal epithelial cells covering the embryonic head and branchial arches are smooth, have a single central cilium and occasional microvilli at their cell junctions. In day 7 embryos, the three branchial arches are V-shaped with the apex of the 'V' pointing ventrally (Fig. 2). The three arches are separated by two continuous grooves and the branchial sinus is located at the caudal extremity of the third arch (Fig. 2). The invaginating epithelial cells of the branchial sinus are pierced, separated by numerous circular pits and have only occasional cilia. The specialized appearance of the invaginating branchial sinus epithelium is identical to that seen in other areas of epithelial invagination (at different embryonic stages), e.g. nasal placodes and pits, and it is tempting to speculate that this is a general phenotype of invaginating alligator facial epithelium.

In the dorsal third of the first branchial groove, there is a patent branchial cleft connecting the embryonic pharynx to the surface (Figs 2 & 4). This cleft is lined by smooth surfaced epithelial cells, each with a central cilium, similar to those found on the surfaces of the branchial arches.

By day 9 four branchial arches are visible in surface view (Fig. 4). Five arches are present in histological section (Fig. 9) but the small fifth arch only becomes visible macroscopically on day 11. Each branchial arch consists of a core of mesenchyme and contains an aortic arch artery, a vein, branches of the appropriate cranial nerve, and later cartilaginous, osseous and muscular blastemae (Fig. 9). They are covered on their outer surfaces by ectoderm and on their inner pharyngeal surfaces by thicker endoderm (Fig. 9). The junction between ectoderm and endoderm is at the external cranial margin of the first branchial arch (Fig. 9).

The arches are separated from each other by surface and pharyngeal grooves (Figs 1–4 & 9), which are covered respectively by ectoderm and endoderm and contain a variable amount of mesenchyme (Fig. 9). However, in the dorsal half of the first and third branchial grooves there are true clefts which connect the surface of the embryo with the pharynx (Figs 4 & 9). The first cleft migrates progressively dorsally until it lies above the otocyst. There it forms the external meatus of the ear. Auricular hillocks develop along its dorsal margin which becomes the superior flap of the external ear (Ferguson, in press). The third cleft is transitory, eventually closing and disappearing completely. During this closure process the cilia of the opposing epithelial cells join together across the branchial cleft prior to epithelial contact and migration out of the cleft (Fig. 5). It is tempting to speculate that these cilia may be involved in pulling the epithelial cells together, so facilitating cleft closure. Cell processes similar to these have been described previously between the closing medial and lateral nasal processes in chick embryos (Yee, 1976; Yee & Abbott, 1978; Millicovsky & Johnston, 1981) but not between the branchial arches (Tamarin & Boyde, 1977; Yander & Searls, 1980). Doubtless their role in epithelial cell adherence and migration during closure of various embryonic processes and elevations is similar in many regions of many vertebrates (Yee, 1976; Yee & Abbott, 1978; Millicovsky & Johnston, 1981). In fish all the branchial clefts are patent (for gill development), whereas in mammals none are (even the first cleft has a thin limiting membrane which eventually forms the tympanic membrane). The remodelling of the first cleft in alligator embryos to form elements of the ear appears logical but it would be interesting to investigate how the tympanum develops. Why the third cleft should be patent for a transient period and then close

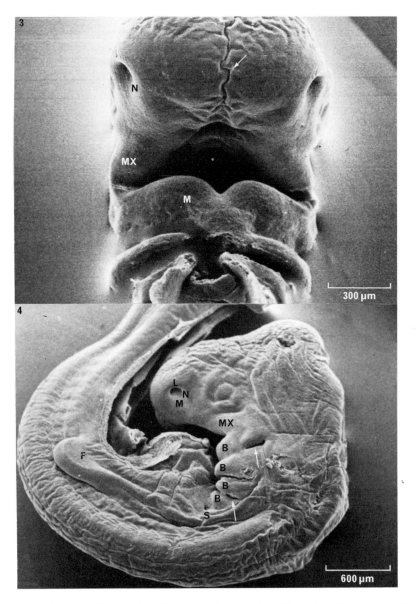

FIGS 3 & 4. (3) Face-on view of a day 8 alligator embryo illustrating the nasal pits (N), mid-line fissure (arrowed), maxillary processes (MX) and lobulated mandibular arch (M). (4) SEM of a day 9 alligator embryo. Note the nasal pit (N), medial nasal (M), lateral nasal (L) and maxillary (MX) processes, developing eye, four branchial arches (B), first and third branchial clefts (arrowed), branchial sinus (S) and forelimb bud (F). The cleft between the second and third arches is a shrinkage artifact.

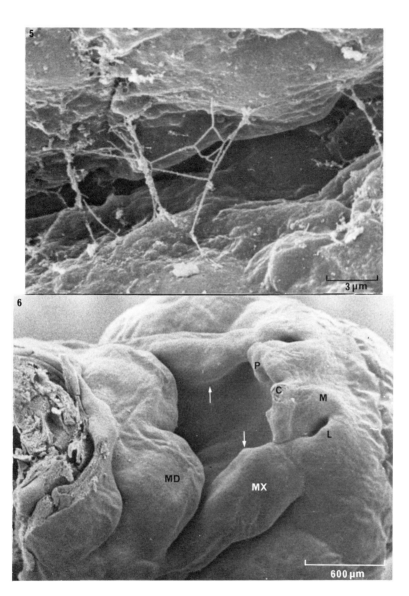

FIGS 5 & 6. (5) View of the two epithelial covered walls of the third branchial cleft in a day 9 alligator embryo. Note that the cilia of opposing epithelial cells have joined so that cell processes now extend across the closing cleft. (6) View of a day 15 alligator embryo illustrating closure of the nasal pit slits by the medial nasal (M), lateral nasal (L) and maxillary (MX) processes. Note the external caudal elevations of the medial nasal processes (C), the anterior intra-oral bulges of the primary palate (P), the intra-oral bulging of the maxillary processes (arrowed) and the developing lower jaw (MD).

remains a mystery. Presumably it is indicative of an intermediate state and/or function between the completely patent fish clefts and the completely closed mammalian clefts.

Bilateral maxillary processes bud off the first branchial arch (mandibular process) about day 8 (Figs 2 & 4). They rapidly enlarge and extend progressively forwards from the angle of the mouth (Fig. 2) to beneath the eye (Fig. 4) eventually appearing as club-shaped processes which form the upper lateral boundaries of the developing oral cavity (Fig. 6). Their later development will be considered on pp. 236–245.

The branchial arches continue to enlarge, partly as a result of mesenchymal cell migration from the neural crest and partly as a result of intrinsic mesenchymal cell division. As the branchial arches enlarge, so the grooves between them tend to temporarily deepen before the tissue joining them (Fig. 9) is progressively filled out by mesenchyme. During this process the epithelial cells at the bases of the branchial grooves "migrate" towards the surface as the underlying mesenchyme expands. These epithelial cells are cobblestoned and have numerous microvilli: appearances which are characteristic of migrating epithelial cells anywhere in the alligator embryo, as will be evident later (see pp. 240–241).

The branchial arches become less evident as development progresses and the definitive neck and lower jaw are formed. The first and second branchial arches are approximately four times larger than arches 4 and 5 (Figs 4 & 9). This size difference is accentuated during the merging and ultimate disappearance of the surface topography of the branchial arches, so that the first arch tends to overgrow the second, which in turn tends to overgrow the third. This latter is important as the surface overgrowth of the second arch extends caudally over arch 3 to the junction between arches 3 and 4. This overgrowth is pronounced at the 50–60 somite stages (*c.* day 11) and results in the submergence of arch 3. Arches 4 and 5 are small: they merge together and begin to disappear macroscopically at the 50–60 somite stages.

The merged conglomerate of arches 1, 2, and 3 forms the base of the lower jaw (Fig. 6). Viewed from the facial aspect, the lower jaw has a lobulated horse-shoe shaped anterior margin which represents the swellings of the first branchial arch on each side. Viewed from the side, the lower jaw tapers from its anterior first arch margin to a wedge-shaped base formed by the merged conglomerate of arches 1–3. The lower jaw commences a fairly rapid forward growth spurt around day 15 (Fig. 6). Prior to this, e.g. at day 10, the embryo is tightly coiled with the facial elevations approximating the pericardium. As development progresses the embryo uncoils and the neck develops, so allowing the initially retrognathic lower jaw to grow forwards beneath the upper jaw. At day 14 the lower jaw is approximately one-fifth the way beneath the upper, at day 16 one-third, at day 18 three-quarters, at day 19 it is behind the premaxillary bulge and at day 22 it is beneath the premaxillary bulge. Meckel's cartilages appear around day 17.

Nasal Placodes, Pits and Processes

At day 7 the nasal placodes are evident as thickened areas of epithelium on the surface of the forebrain bulge, which at this stage forms the rudimentary face (Fig. 2). In surface view these placodal epithelial cells have a different appearance to the surrounding ectodermal cells, but an identical appearance to the invaginating epithelial cells of the branchial sinus (see p. 228, Fig. 2). The nasal placodes burrow inwards to form the nasal pits (Fig. 3). At first (*c.* day 8) these are blind-ending and shallow but they rapidly invaginate towards the roof of the primitive stomatodeum (Fig. 3). During this process, two rims of tissue are embossed out around the surface openings of the nasal pits to form the medial nasal and lateral nasal processes respectively (Fig. 3). The area between the two nasal pits (including the medial nasal processes of each side) is known as the frontonasal process and is divided by a mid-line surface groove (Fig. 3). The maxillary processes, which arise from the first branchial arch at the angle of the mouth, also show rapid mesenchymal cell division and migration, so that their surfaces likewise become embossed outwards thus forming a groove between themselves and the developing eye above (Figs 4, 6 & 7).

During days 9 and 10, the nasal pits deepen towards the roof of the mouth and simultaneously the medial nasal, lateral nasal and maxillary processes become more pronounced (Figs 3, 4 & 6). At this and subsequent ages, differential growth of the head results in the nasal pits and eyes migrating from their early lateral locations towards each other in the mid-line. By day 11 the nasal pits have invaginated to reach and fuse with the epithelium covering the roof of the primitive mouth, so forming the primitive nasal cavities with anterior and posterior nasal choanae. The latter open into the anterior roof of the mouth – a few millimetres behind the anterior choanae. By a combination of extremely rapid outward proliferation of the medial and lateral nasal processes and a caudal extension of nasal pit invagination, true spaces develop between the ipsilateral medial and lateral nasal processes. These spaces are bounded posteriorly by the rapidly growing club-shaped maxillary processes, are first evident at day 12, and are called the nasal pit slits. The rapid outward embossing of the bilateral medial nasal processes also causes a V-shaped groove to develop between them (Fig. 6). The club-shaped maxillary processes rapidly grow forwards from their broad bases beneath the eye; their anterior margins lie

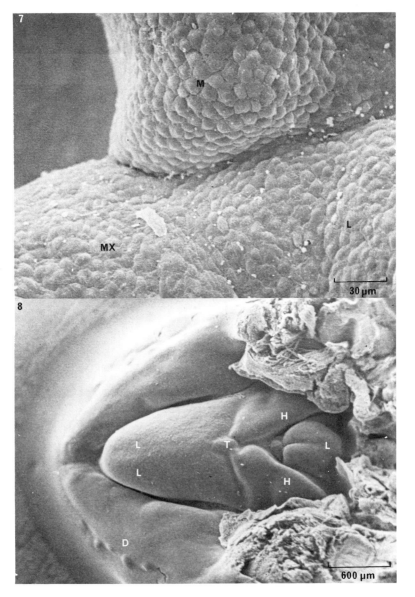

FIGS 7 & 8. (7) View of the junction between the closing medial nasal (M), lateral nasal (L) and maxillary (MX) processes in a day 15 alligator embryo. There is limited epithelial cell death and the closing epithelial cells appear cobblestoned and villous. (8) The developing tongue and lower jaw of a day 20 alligator embryo illustrating the paired lingual swellings (L), tuberculum impar (T), hypobranchial eminences (H), denticles (D), developing larynx and epiglottis (L).

behind the nasal pit slits at day 12, but beneath the nasal pit slits and lateral nasal processes at day 14 (Fig. 6). The maxillary processes merge with the lateral nasal processes at day 14 and the latter merge with the medial nasal processes at day 15 (Figs 6 & 7). There is only a small area of initial contact between the medial nasal and maxillary process (Figs 6 & 7). In the region of closure of the nasal pit slits, the epithelial cells of the medial nasal, lateral nasal and maxillary processes show a surface morphology characteristic of migrating cells (Figs 6 & 7). They become cobblestoned, have surface microvilli and actively migrate out of the epithelial seam (Figs 6 & 7). Epithelial cell death is very limited and located mostly in the regions of initial contact between the medial nasal, lateral nasal and maxillary processes (Figs 6 & 7). Once mesenchymal continuity has been established, even over a very small area, the remainder of the intervening epithelium migrates out of the closure zone (Figs 6 & 7) – a process known as merging. Presumably this epithelium is utilized to cover the ever growing face. The fate of the basement membranes during this process is unknown but currently under investigation. In this way the nasal pit slits are closed and the definitive upper boundary of the primitive mouth formed (Fig. 6).

A mid-line furrow separates the medial nasal processes of each side (Fig. 6). The caudal edges of the medial nasal processes are markedly embossed outwards so creating the effect of two elevated ridges at the lower borders of each of the medial nasal processes (Fig. 6). These caudal elevations enlarge so that by day 17 they merge with each other and the adjacent maxillary processes to form the upper boundary of the mouth. This has the effect of shifting the anterior nasal choanae from the tip to the dorsal aspect of the snout which is, of course, their adult position.

Yee (1976) and Yee & Abbott (1978) have described the early development of the chick face using scanning electron microscopy. It is similar to the sequence described here for alligator, e.g. chicks also show an outward embossing of the medial and lateral nasal processes with the formation of nasal pit slits (Yee, 1976; Yee & Abbott, 1978), although these are absent in mammals (Tamarin & Boyde, 1977). Minkoff & Kuntz (1977, 1978) and Minkoff (1980a, b), have demonstrated that the outward embossing of the chick medial and lateral nasal processes is due to the rates of mesenchymal cell division in these regions declining at a slower rate than those in adjacent regions of the face. A similar mechanism probably operates in alligators.

Primary Palate

At day 15, two bulges arise from the intra-oral aspects of the medial nasal processes. These grow posteromedially, approximate each other in the mid-line and merge at day 17, so forming the primary palate (Fig. 10). The development of the primary palate displaces the posterior nasal choanae posteriorly (Fig. 10). Olfactory epithelium and nerves differentiate from the superior nasal epithelium and olfactory bulbs respectively (Fig. 16). Superiorly, the bilateral primary palate bulges each merge with a downgrowth of the frontonasal process – the primary nasal septum. In so doing they form a paired nasopharyngeal duct, which is connected inferomedially with the primary nasal cavities. These paired nasopharyngeal ducts run posteriorly and from days 17 to 20 open into the mouth behind the primary palate which is thus the new location of the posterior nasal choanae. Subsequently, cartilage differentiates in the primary nasal septum. The primary nasal septum is in continuity posteriorly with downgrowths of the maxillary tectoseptal processes (Figs 10 & 19), which in anterior regions form the secondary nasal septum. The posterior nasal choanae remain located behind the primary palate until closure of the secondary palatal shelves displaces them posteriorly (Figs 10, 14, 15, 16 & 19).

Secondary Palate and Tectoseptal Processes

At day 15 the maxillary processes are large, club-shaped and have intra-oral and extra-oral projections (Fig. 6). One intra-oral front of maxillary mesenchymal cells has migrated upwards between the floor of the brain and the roof of the oral cavity to form the tectoseptal processes (Figs 10, 12–14 & 19). The tectoseptal processes of each side meet and merge in the mid-line, closure taking place progressively from days 15 to 24 in an anteroposterior gradient (Figs 10, 12–14 & 19).

Bilateral secondary palatal shelves arise from the maxillary processes about 17 days after egg-laying as thick, blunt-ended structures (Figs 10 & 12). Unlike mammalian palatal shelves, those of alligators normally grow horizontally from their first appearance except in the posterior one-fifth of the palate where they are more vertically orientated (Figs 10 & 12). With later development these posterior shelves gradually flow over the tongue to become truly horizontal (Figs 10–14). This reorientation process involves the migration of both mesenchymal and epithelial cells. The alligator palatal shelves have a loose mesenchymal matrix containing a large quantity of glycosaminoglycans (Figs 12–14). Furthermore, the amount of hydrated hyaluronic acid in the palatal mesenchyme increases during the period of palatogenesis, as evidenced by progressively heavier staining with alcian blue at critical electrolyte concentrations, and a corresponding increase in the size of the extracellular spaces between the mesenchymal cells; this resembles the

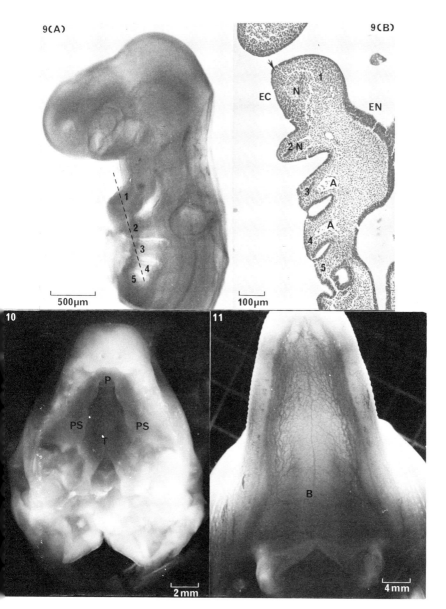

FIGS 9–11. (9) Whole mount (A), stained with alcian green, of a day 10 alligator embryo and a histological section (B) at the indicated level. Note the five branchial arches (labelled 1–5), the aortic arch arteries (A), branchial nerves (N) and junction (arrowed) between ectoderm (EC) and endoderm (EN). (10) Macroscopic view of the primary palate (P), secondary palatal shelves (PS) and tectoseptal processes (T) in a day 18 alligator embryo. (11) Macroscopic view of a day 40 embryonic alligator palate, injected with India ink to highlight the profuse plexus of palatal blood vessels. Note also the basihyal valve (B).

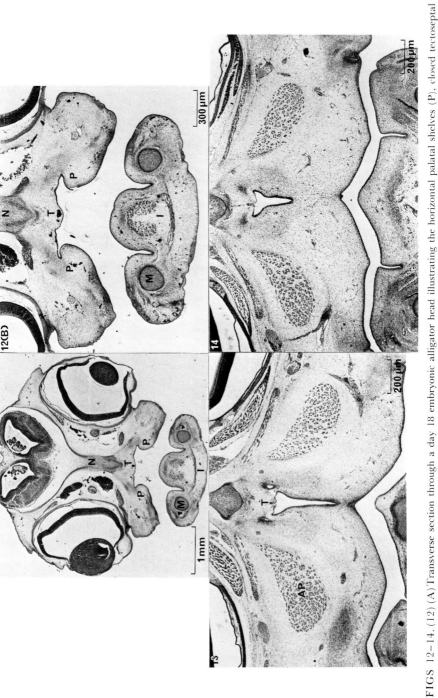

FIGS 12–14. (12) (A) Transverse section through a day 18 embryonic alligator head illustrating the horizontal palatal shelves (P), closed tectoseptal processes (T), inter-orbitonasal septum (N), tongue, Meckel's cartilages (M) and intermandibularis muscle (I). (B) Greater detail of the section in (A). (13) Transverse section through the posterior region of the closing secondary palatal shelves of a day 22 embryonic alligator. Mesenchymal continuity has been

sequence of events observed in mammals (Ferguson, 1977, 1978a, b). Indeed, part of the increase in shelf length and breadth during the period of palatal closure can be accounted for by swelling due to the hydration of palatal glycosaminoglycans. These glycosaminoglycans are probably involved in the re-orientation and merging of the two shelves, particularly in providing a milieu favourable for cell migration and later for osteogenesis (Fig. 17).

At day 18, the horizontal palatal shelves have not contacted each other (Figs 10 & 12). However, by day 19 the shelves have approximated each other anteriorly behind the primary palate and from there palatal closure spreads progressively backwards (Figs 10–14 & 19). The palate is macroscopically half closed by day 20, three-quarters closed by day 22 and fully closed by day 24: the superior flap of the basihyal valve is complete by day 29 (Figs 10–14 & 19). Concomitant with palatal closure, the tectoseptal processes of each side are merging with each other in the midline, so that between days 18 and 24 two gradients of anteroposterior closure can be seen in the alligator oronasal cavity: the more advanced superior one is the closure of the tectoseptal processes and the inferior one the closure of the palatal shelves (Figs 10–14 & 19). Anteriorly the palatal shelves merge with the primary palate, a process which is complete by day 20.

As palatal closure spreads posteriorly there is always a V-shaped gap where the posterior margins of the opposing shelves are approximating each other (Figs 10–14 & 19).

The process of palatal closure involves contact and adherence of the epithelial cells of the two shelves (Figs 10–14 & 19). This usually occurs first on the oral aspect of the shelf margins and then spreads nasally (Figs 13 & 14). The epithelial cells of the two shelves migrate, in both nasal and posterior directions, out of the region of closure, so that there is never an extensive epithelial seam (Figs 13 & 14). There is little evidence of epithelial cell death and epithelial remnants are never seen histologically following palatal closure (Fig. 14). Large numbers of small blood vessels were present in the shelf mesenchyme adjacent to the area of closure (Figs 11–15), and this increased vasculature, plus fluid exudate, in addition to the hydrophilic glycosaminoglycans, is likely to be involved in pushing the shelves together and the epithelium out of the area of closure in a posteronasal direction. The presence of these numerous blood vessels is not surprising for they represent the earliest development of the palatal plexus of blood vessels which is evident in late embryos (Figs 11 & 15) and adult alligators (Fig. 18). Mesenchymal continuity is usually established on the oral edges of the shelves before the nasal edges have even contacted each other (Figs 13 & 14). Anteriorly the shelves merge with themselves and with the

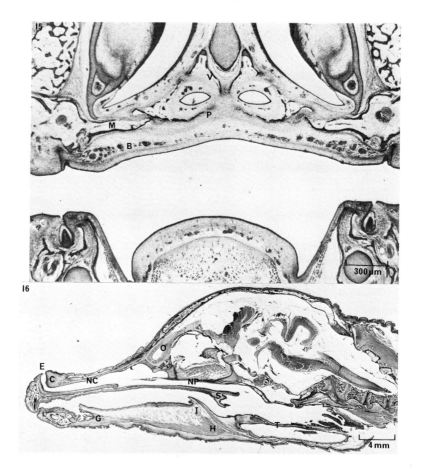

FIGS 15 & 16. (15) Tranverse section through the anterior region of the palate of a day 27 alligator embryo illustrating the palatal plexus of blood vessels (B) and osteogenic blastemata for the maxillary (M), palatal (P), and vomerine (V) bones. (16) Parasagittal section through a six-month old alligator head. Note the external nares (E), constrictor nares muscle (C), nasal cavities (NC), nasopharyngeal duct (NP), olfactory bulbs (O), superior (S) and inferior (I) flaps of the basihyal valve, trachea (T), hyoglossus (H) and genioglossus (G) muscles.

bulging downgrowth of the secondary nasal septum, so forming a continuation of the partitioned nasopharyngeal duct. Posteriorly the palatal shelves only merge with each other to produce an initially unpartitioned duct. This is subsequently partitioned by the fusion of internal mid-line ductal bulges which appear at day 25.

When viewed with the scanning electron microscope, the epithelial cells of the medial edges of the closing alligator palatal shelves display interesting surface topographies (Figs 19–24). When the palatal shelves first appear at day 17, their medial edge epithelial cells (MEE) are flat with a rather featureless surface (Fig. 22). By day 18 the MEE are more bulbous, have developed surface microvilli and their cell boundaries are distinct as everted tight junctions (Fig. 23). From days 19 to 22 the MEE are markedly cobblestoned, have numerous microvilli and can be seen actively migrating out of the region of palatal closure (Figs 19–21, 23 & 24). It will be remembered that cobblestoned, villous epithelial cells are characteristic of migration in the branchial arches and nasal pit slits (Fig. 7). In the region of closure of the palatal shelves, the cobblestoned villous MEE actively migrate out of the approximation zone and very little epithelial cell death is observed (Figs 19–21, 24). These epithelial cells appear to be recruited onto the nasal and oral surfaces of the closed palate, as well as onto the lateral margins of the approximating shelves (Figs 19–21). Since the palate is expanding in both length and breadth during the period of closure, it is assumed that the migrating epithelial cells are utilized to cover the ever enlarging palatal area. The fate of the basement membranes during this process is unknown but under investigation. No epithelial seam is present, neither is there extensive epithelial cell death (Figs 13, 14, 19–21) which is in marked contrast to the sequence of events observed during palatal fusion in mammals. Instead a mid-line oral ridge marks the site of palatal closure (Fig. 19) but this is covered by flatter, less villous, epithelial cells than those seen in the region of palatal closure.

It is interesting to compare the developmental events (particularly MEE differentiation) in alligator palatal closure with those in birds and mammals. In mammals MEE show characteristic morphological changes before contact and during fusion. These include: the appearance of a carbohydrate rich glycoprotein surface coat (Greene & Kochhar, 1974; Pratt & Hassell, 1975; Souchon, 1975; De Paola, Drummond, Lorente, Zarbo & Miller, 1975; Meller & Barton, 1978; Pratt, Figueroa, Greene, Wilk & Salomon, 1979; Bordet, 1981); a loss of distinct epithelial cell boundaries, flattening and necrosis of the superficial epithelial cells, the appearance of intercellular gaps and the appearance of long slender cell protrusions (Waterman, Ross & Meller, 1973; Waterman & Meller, 1974; Tassin & Weill, 1977, 1980; Meller, De Paola, Barton & Mandella, 1980; Meller, 1980; Bordet, 1981); a decrease in the amount of glycogen (Meller & Barton, 1979); a cessation of MEE DNA synthesis (Hudson & Shapiro, 1973; Pratt & Martin, 1975); an increase in the quantity of lysosomal bodies and the glycosylation of intracellular lysosomal enzymes (Mato, Aikawa et al., 1966; Farbman, 1968; Hayward, 1969; Smiley, 1970; Mato, Smiley et al., 1972; Chaudhry & Shah, 1973; Pratt & Greene, 1976; Greene & Pratt, 1978).

Contrariwise in birds, the closest living phylogenetic relatives of crocodilians, the MEE of each palatal shelf contact each other but neither adhere, fuse, migrate nor die, but rather keratinize, so that all birds have a physiological cleft palate with a mid-line choana joining the oral and nasal cavities (Shah & Crawford, 1980; Koch & Smiley, 1981). Avian MEE do not become thinner, do not show glycogen deposits, intracytoplasmic tonofilaments, lysosomes or intracellular autophagosomes, do not become cobblestoned, do not die, but rather differentiate into keratinized stratified squamae with a marked infolding of the epithelial cell membranes (Shah & Crawford, 1980; Koch & Smiley, 1981).

The different MEE phenotypes in these three vertebrates result in differing patterns of palatal morphology; shelf fusion in mammals, shelf merging in alligators and cleft palate in birds. Moreover differentiation of these species-specific MEE phenotypes occurs *in vitro* (Ferguson, Honig & Slavkin, in press) under defined culture conditions (Ferguson, Honig, Bringas & Slavkin, 1982, 1983). Experiments involving a myriad of heterotypic, heterochronic, homotypic, isochronic epithelial-mesenchymal recombinations within and between chick, alligator and mouse embryos have revealed that palatal MEE differentiation is regulated by an instructive epithelial-mesenchymal interaction (Ferguson, Honig, Bringas & Slavkin, 1981; Ferguson & Honig, 1984, in preparation.

The numerous blood vessels, present at the time of palatal closure (Fig. 11), develop rapidly after this event to form the extensive palatal plexus of blood vessels (Figs 15 & 18). There are two large lateral palatine arteries which enter the posterolateral aspects of the palatal submucosa and give off longitudinal and transverse branches which anastomose with each other and with an unpaired mid-line vessel (Figs 11 & 18). In this way the morphology of the adult palatal vascular plexus is established by day 40 of embryonic life (Figs 11, 15 & 18). This plexus may be involved in thermoregulation. Osteogenesis advances rapidly from the blastemata of the maxillary, palatine and pterygoid bones (Fig. 17). The basihyal valve arises between days 24 and 29 by a postero-inferior extension of palatal shelf closure (Fig. 11). It does not ossify, rather the mesenchyme differentiates into fibrous tissue. Denticles are present along the lateral margins of the upper and lower jaws (Fig. 19). These are epithelial covered spicules of disorganized dentine. With later development they sink beneath the surface epithelium and are resorbed. The dental laminae for the developing alligator dentition arise directly beside the denticles.

There are a few scanty and incomplete accounts of crocodilian palatal development in the literature. Thus, some macroscopic drawings of the palates of crocodilian embryos can be found in the publications of

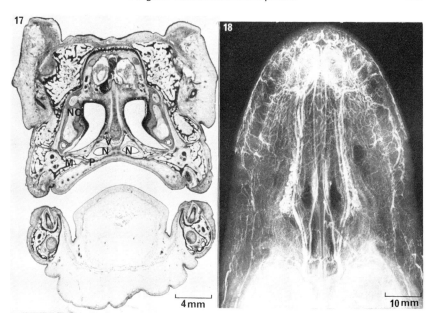

FIGS 17 & 18. (17) Transverse section through the head of a day 40 embryonic alligator. Note the maxillary (M), palatine (P) and vomerine (V) osteogenic blastemata, the nasal capsular cartilages (NC) and the paired nasopharyngeal duct (N). (18) Radiograph of a demineralised one-year old alligator skull which was injected with vital radio-opaque dye (micro-opaque) to highlight the blood vessels. Note the extensive palatal plexus and the cascades of vessels supplying the teeth.

His (1892), Voeltzkow (1899), Göppert (1903), Sippel (1907), and Fleischmann (1910), but these provide no details of the process of palatogenesis, and no histological studies were performed. Fuchs (1908) and Müller (1965, 1967) describe the development of the nasopharyngeal duct and secondary palate, but these papers are both inaccurate and idiosyncratic. Müller (1967) proposes a new terminology of "primary", "secondary" and "original" palate and choanae. The nasopharyngeal duct is described as arising from two separate "presumptive spaces", one rostral (just posterior to the original choanae) and one caudal (just anterior to the pharynx). These spaces arise by a dorsal invagination of the "original roof of the mouth" (cranial floor) in such a way that the definitive palate lies in the same plane as the "original roof of the mouth" (Müller, 1967). Müller (1967) alleges that this process is fundamentally different from that found in mammals where, it is asserted, the palate divides the oronasal cavity at a more ventral level than the "original roof of the mouth". Although this argument is central to Müller's philosophy of the embryogenesis of the crocodilian nasopharyngeal duct, no evidence (in the form of cephalometric tracings) to

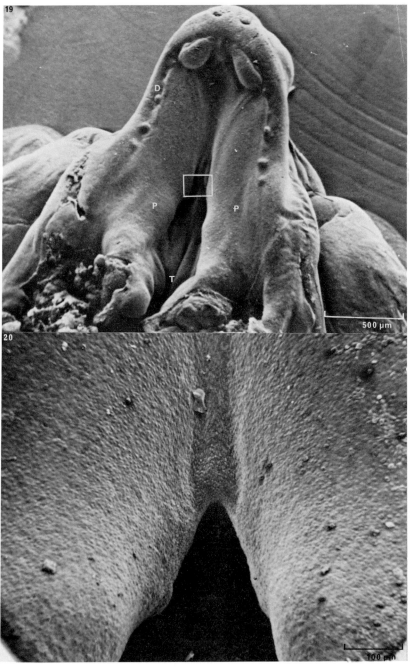

FIGS 19 & 20. (19) SEM of the closing palatal shelves (P) of a day 20 embryonic alligator. D = dentricle, T = closing tectoseptal process. Area shown in higher power in Fig. 20. (20) Higher power view of the region of palatal closure illustrated in Fig. 19. Note the altered cellular topography in this region, but absence of marked cell death.

support it is presented, and both this and her speculations about mammals are erroneous. It would appear that Müller (1967) has confused the near simultaneous closure of the tectoseptal processes and secondary palatal shelves in crocodilians, instead interpreting these as a primary (superior) palate and a secondary (inferior) palate. Perhaps the reason for this state of affairs is that Müller (1967) never studied complete dissected embryonic heads (either macroscopically or with the SEM) whilst her histological sections were embryos of unknown age from different species of *Crocodylus*. Moreover, Müller (1967) never studied any complete adult specimens, only sectioned embryos and skulls; thus she makes no comment on the development of the basihyal valve and her interpretations of the adult anatomy of this region are wrong (Ferguson, 1981a).

Tongue

The development of the crocodilian tongue is very poorly described. Rathke (1866), Voeltzkow (1899), Göppert (1903), Taguchi (1920), Sewertzoff (1929) and Wettstein (1954) make some reference to tongue embryology, but this is in late development when the muscle fibres have formed.

In *A. mississippiensis* the tongue arises in the classical vertebrate fashion from swellings on the pharyngeal aspects of the first three branchial arches (Fig. 8). These are the paired lingual swellings of the first branchial arch anteriorly, the mid-line tuberculum impar of the second arch and the paired hypobranchial eminences of the third arch posteriorly (Fig. 8). Behind the latter lie the developing epiglottis and larynx (Fig. 8). In the region of merging of these lingual elevations, the migrating epithelial cells show the classic cobblestoned, villous appearances described previously for the closing palate (Figs 19–24), nasal pit slits (Fig. 7) and branchial arches. During palatogenesis the tongue is small and lies low in the oronasal cavity (Figs 12–14), a feature in keeping with its adult morphology (Ferguson, 1981a). The lingual mesenchyme differentiates into anlage for the genioglossus, hyoglossus and intermandibularis muscles as well as fibrous tissue and fat deposits, intrinsic lingual musculature being completely absent even in the adult (Ferguson, 1981a).

Longitudinal Studies

In many embryological investigations (e.g. surgical and teratological experiments) it is very useful to be able to perform longitudinal studies,

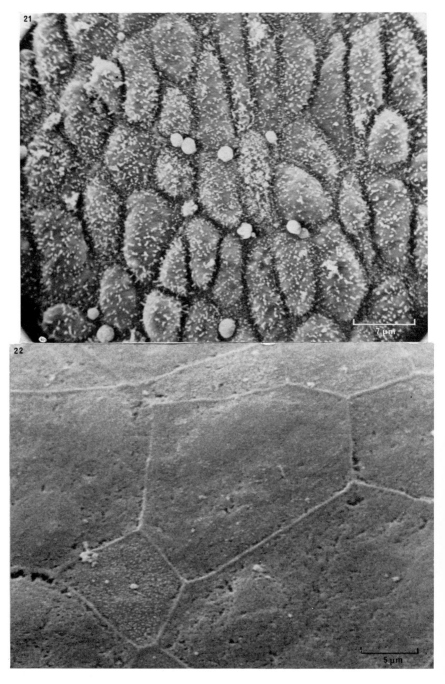

FIGS 21 & 22. (21) Higher power view of the closure zone area outlined in Fig. 20. Note the limited epithelial cell death and the cobblestoned, villous appearance of the migrating epithelial cells. (22) SEM of the medial edge epithelial cells (MEE) on the margins of day 17 embryonic alligator palatal shelves. They are flat and rather featureless.

i.e. to follow the development of the same embryo through incubation. This depends upon being able to visualize and manipulate the embryo through the eggshell. Numerous investigators (e.g. Auerbach, Kubai, Knighton & Folkman, 1974; Dunn, 1974; Dunn & Boone, 1976, 1977, 1978; Narbaitz & Jande, 1978; Dunn & Fitzharris, 1979; Slavkin, Slavkin & Bringas, 1980; Tuan, 1980, 1983; Dunn, Fitzharris & Barnett, 1981; Dunn, Graves & Fitzharris, 1981; Jaskoll & Melnick, 1982) have cultured chicken embryos with associated yolk and albumin outside of the eggshell and shell membranes – a technique known as shell-less culture. In contrast to windowing techniques, shell-less culture has a number of advantages: e.g. opportunities for continuous observation and access to developing embryos and extra embryonic membranes; ability to perform injections and operations at precise stages and to monitor resultant physiological (e.g. heart beat), behavioural or structural changes continuously without having to kill the embryo; ability to perform multiple injections or operations on any embryo without the complications of membrane adherence to the eggshell window; opportunities to monitor continuously the differentiation, morphogenesis, growth, vascularization, etc. of grafts placed on the chorio-allantoic membranes (CAM) of embryos in shell-less culture without having to sacrifice either the host or the graft tissue. Variations of the chick shell-less culture technique have been reported (Auerbach *et al.*, 1974; Dunn, 1974; Dunn & Boone, 1976; Dunn, Fitzharris & Barnett, 1981) and in general the embryos appear to develop fairly normally until 10 days of incubation. However, few embryos hatch, while some have gross abnormalities (Dunn, Fitzharris & Barnett, 1981), and others more subtle ones (Jaskoll & Melnick, 1982). Some of these malformations result from altered embryonic tensile forces as a result of lateral compression and embryonic constraint (Jaskoll & Melnick, 1982), whilst others may be related to: mineral deficiencies (Dunn & Boone, 1977; Narbaitz & Jande, 1978; Tuan, 1980, 1983, Dunn, Graves & Fitzharris, 1981); inadequate albumin uptake (Dunn & Boone, 1978); excessive humidity (Dunn, Fitzharris & Barnett, 1981); or abnormal gas exchange (Dunn, Fitzharris & Barnett, 1981). Fragments of eggshell and eggshell membrane placed on the CAM of embryos in shell-less culture can apparently restore calcium transport and are utilized for embryonic mineralization (Tuan, 1983).

However, the developing alligator embryo should be particularly suitable for such shell-less techniques, since it only obtains about 1.7–2.4 times as much calcium from the shell as from the egg contents, whereas turtles obtain four times as much and birds about five times as much (Jenkins, 1975). Furthermore, a study of the structure, chemical composition and calcium kinetics of the alligator eggshell and embryonic membranes (Ferguson, 1982b) has resulted in the development of a reliable technique for the shell-less and semi-shell-less culture of embryos throughout the incubation period.

For the shell-less culture technique, alligator eggs are collected within 12 h of egg laying – before the embryo has attached to the shell (Ferguson, 1981b, 1982b, in press). An incision is made around the longitudinal axis of the shell and the top one-third of the shell removed, followed by a separate, sterile excision of the underlying eggshell membrane. These cuts are made carefully using stork-bill microbiological scissors (Arnold Horwell Ltd., London) with the fixed blade innermost, so that the egg contents are not damaged. The egg contents are then carefully explanted into a 120 mm sterile, vented petri dish under a laminar flow hood. This dish is covered and placed inside a larger sterile petri dish which contains a sheet of sterile filter paper, saturated with sterile water. Once the lid is placed over this outer dish, high humidity surrounds the shell-less egg contents. The sterile dishes are artificially incubated at 30°C and 100% humidity. Development proceeds normally (as assayed by macroscopy) in these shell-less embryos up to approximately day 25 of the 65-day developmental period, after which they begin to exhibit malformations, particularly in the snout, cranium and limbs. These malformations are probably the result of altered embryonic tensile forces due to the drastically changed geometry of the egg contents in shell-less culture. Since the yolk and albumen of alligator eggs are much more viscous than those of avian eggs, the present author evolved a technique of semi-shell-less culture of alligator embryos.

In this technique sterile incisions are made around the longitudinal axis of the shell (as described earlier) and the top one-third of the shell and eggshell membrane removed (for figures see Ferguson, 1981b, in press). The yolk and albumen are viscous enough to remain intact inside the lower two-thirds of the shell so that the natural geometry of the egg contents is maintained. Incubation of these semi-shell-less culture embryos in sterile incubators at 30°C and 100% humidity results in normal development (montage illustrated in Ferguson, 1982c, in press), which can be filmed and experimentally altered. The modest extrinsic calcium needs of the embryo seem to be met by the intact lower two-thirds of the shell – similar to the eggshell supplementation results of Tuan (1983) for the chick embryo. The technique must be performed within 24 h of egg-laying – before the embryo has attached to the top of the eggshell membrane or developed an extensive chorio-allantois: such embryos will develop normally and hatch approximately 65 days later. There is no statistically significant difference between the weights and crown rump length of these embryos as compared to those recovered from normal incubating eggs of the same age.

It is possible to study the longitudinal development of the alligator palate *in situ* by combining the semi-shell-less culture technique with a precise teratogenic experiment. As will be evident (pp. 256–259) the alligator embryo is very sensitive to teratogens whose application can be precisely timed. Thus by applying 0.01 mg of 5 fluoro-2-desoxyuridine (dissolved in 0.1 ml of sterile injectable water) to the CAM overlying the head region of a day 10 embryo in semi-shell-less culture, it is possible to produce an alligator embryo with virtually no lower jaw or tongue but a normal palate (Figs 25–27 and see also Ferguson, 1981b). Initially the treated embryos are developmentally retarded, but they exhibit rapid compensatory growth and apart from an extremely small lower jaw, they exhibit no other macroscopically detectable abnormalities. The treated embryos remain viable and "hatch" at 65 days but die about two weeks thereafter owing to an inability to feed. The developmental events involved in palate ontogenesis can be continuously observed and filmed in these semi-shell-less, lower jaw-less embryos (Fig. 24). The morphogenetic details of palatal closure described on pp. 235–240 have been confirmed by such longitudinal studies – the first of palatal closure in any animal.

Species Variations

Craniofacial development has also been studied macroscopically (and to a lesser extent histologically) in an extensive series of *C. johnstoni* and *C. porosus* embryos (collections of Dr G. Webb and Dr M. W. J. Ferguson) and in a few specimens of *C. niloticus*. Remarkably, the first half of development (period of organogenesis) is extremely similar in all these species and at many of the earlier stages embryos of different species appear identical. Naturally, there are differences in absolute size, e.g. *C. porosus* embryos are considerably larger at all stages than any of the others, but since it is known that embryonic size is closely related to egg size (Ferguson, in press) and since *C. porosus* eggs are also larger than those of *A. mississippiensis*, *C. johnstoni* or *C. niloticus*, this is not surprising. A detailed account of crocodilian embryonic staging, embryo morphometrics and species differences is in preparation (Ferguson & Webb, in preparation) but a few of the more obvious craniofacial differences will be reported here. In general, species differences are minor and the foregoing account of alligator development may be considered to be representative of "the crocodilian schema".

In *C. porosus* the first brancial arch is both relatively and absolutely

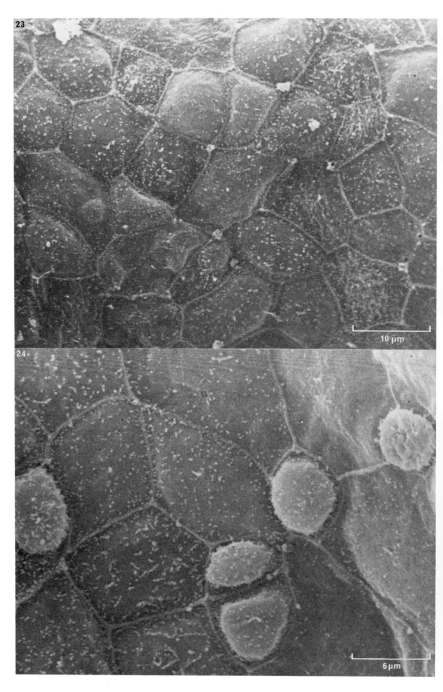

FIGS 23 & 24. (23) Appearance of MEE on day 18 embryonic palatal shelves. The cells are more bulbous, have developed microvilli and distinct everted cell boundaries. (24) Appearance of MEE on day 19 embryonic alligator palatal shelves. They have a cobblestoned appearance, microvilli and everted cell boundaries, and migrating cells are prevalent.

larger and extends further ventrally at all ages than in either *C. johnstoni* or *A. mississippiensis*. In *C. johnstoni* the lateral nasal processes are relatively smaller and the medial nasal processes relatively larger than in either *A. mississipiensis* or *C. porosus*. Adult *C. johnstoni* have long tapering snouts compared to those of *A. mississippiensis* or *C. porosus*, which are much broader. Evidently this pattern is present at the onset of snout development, the large medial and small lateral nasal processes contributing to the development of the elongated central snout. It would be interesting to compare the susceptibility of *C. johnstoni* embryos to teratogenically induced cleft lip and palate with that of either *A. mississippiensis* or *C. porosus* embryos. In mice it is known that face shape and the relative sizes of the nasal processes affect the susceptibility of different strains to cleft lip and palate (Trasler, 1968; Juriloff & Trasler, 1976; Trasler & Fraser, 1977; Trasler, Rearden & Hajchgot, 1978) and a similar association is suspected for man (Millard, 1976).

In late embryos (after day 30) of *C. johnstoni* and *C. porosus* the nostrils and nasal discs are elevated higher dorsally than they are in *A. mississippiensis* and the upper jaw margins are notched to receive the large fourth dentary teeth. In *C. johnstoni* embryos the lower jaw is much more tapering and remains lobulated into right and left segments for a much longer time than in either *A. mississippiensis* or *C. porosus*. All these embryonic differences are of course reflections of the varying morphologies of the adult species. Such variations extend outside the craniofacial region, e.g. *C. johnstoni* embryos have much larger hind limbs and *C. porosus* embryos much longer tails.

SPONTANEOUS MALFORMATIONS

Introduction

In an embryological or teratological investigation it is imperative to know the spontaneous malformation rate in the eggs under study. Such data are available for avian eggs (Romanoff & Romanoff, 1972), but not for those of any crocodilian species. Morover, since the spontaneous malformation rates of many human birth defects, e.g. spina bifida, anencephaly, cleft lip and palate, are related to the age of the mother (Ingalls, Pugh & MacMahon, 1954; Carter & Evans, 1973; Janerich, 1974), it would be important to determine if such an effect occurs in alligators. This in turn requires the ability to accurately age adult alligators.

Techniques for Aging Adult Alligators

Since 1959, Louisiana alligators of known age (usually hatchlings) have been permanently marked and released into the wild. Subsequent recaptures and measurements of these animals have generated data on growth rates (Chabreck & Joanen, 1979). A crude relationship exists between total length and age, although the accuracy of this decreases as the animal gets older and growth rates decline (Chabreck & Joanen, 1979). However, extensive surveys of ectotherms (including *C. niloticus* and *C. siamensis*) have shown that their bone is deposited in a cyclical fashion, with zones of well vascularized, poorly organized bone (deposited rapidly during summer growth) alternating with annuli of well organized, lamellar bone (deposited during winter hibernation or food restriction) (Peabody, 1961; Warren, 1963; Castanet, 1974, 1978; De Ricqlès, 1976; Castanet & Cheylan, 1979; Hemelaar & Van Gelder, 1980; De Buffrenil, 1980a, b; De Buffrenil & Buffetaut, 1981). It has been suggested that the dense annuli could be considered as growth rings and their number used to age the animals (Castanet, 1978; Castanet & Cheylan, 1979; Hemelaar & Van Gelder, 1980; De Buffrenil, 1980a, b). This possibility was investigated in *A. mississippiensis* by Ferguson (1982c). Using animals of known age from the mark/recapture studies, most parts of the skeleton were sectioned in order to determine which parts contained the maximum number of growth lines, and how closely this number reflected the age of the animal. It transpired that serial sections from a 40-mm long cylinder of the femur mid-diaphysis were most useful. Sections through the dorsal neck scutes (which could be removed without killing the animal) were disappointing. Annuli were detected until about six to eight years of age, after which the scutes underwent massive remodelling with the formation of numerous vascular sinuses and channels and the formation of much vascular non-lamellar bone. Such changes were more marked in females than in males and it may be that the scutes are demineralized as a calcium source during egg formation.

Most skeletal specimens showed some annuli but their number was variable (owing to remodelling) and bore little relation to the age of the alligator, except for the sections of the femur mid-diaphysis (Fig. 32, Table I). Here the correlation was close (Table I) and the numbers of annuli lost by medullary resorption could be calculated accurately (Ferguson, 1982c). It is most important to serially section the femur mid-diaphysis, to count annuli in all sections and to use the maximum number, i.e. the section with the least medullary resorption and the thickest bone (Fig. 32, Table I). Implicit in the idea of using such annuli

TABLE I
The known age of alligators recovered from mark/capture studies, correlated with the number of growth lines (annuli) visible in sections of femur mid-diaphysis

Number of animals	Age of alligator (years)	Average number of growth lines (annuli) in sections of femur mid-diaphysis
5	0 (hatchling)	0 ± 0
10	1	1 ± 0
1	5	4
1	8	6
5	16	14 ± 1
2	18	16 ± 0
1	25	22

as growth rings is the concept that one zone is deposited during the period of rapid summer growth and one annulus during the period of slow winter growth and hibernation (Fig. 32). This has been confirmed using timed injections of the fluorescent skeletal markers Tetracycline and D.Caf. (Ferguson, 1982c; Ferguson, Honig, Bringas & Slavkin, 1982). Therefore, using this aging technique, data could be generated on the relationship between maternal age and spontaneous malformation rates.

Relationship between Maternal Age and Spontaneous Malformation Rate

Table II illustrates the statistically significant difference in either the numbers of eggs per clutch, or their maximum dimensions, or the malformation or infertility rates between young, middle-aged and old female alligators. In general, young females lay small clutches of small eggs, middle-aged females lay large clutches of large eggs and old females lay small clutches of even larger eggs (Table II). Thus even if the laying female is unseen and her age therefore unknown, it is possible to group her into one of the three reproductive age classes (Table II) upon the basis of the number and size of the eggs in her clutch.

Moreover, of the eggs laid by young females approximately 40% are infertile, 10% contain malformed embryos (e.g. spina bifida, anencephaly, cleft lip and palate) which frequently die during incubation, and only 50% are viable and hatch (Table II). Contrariwise, the eggs of middle-aged females nearly all (98%) contain normal viable embryos with very low numbers of malformed embryos (0.5%) or infertile eggs (2%). The eggs of old females contain a higher number of malformed embryos (5%) but a similar number of infertile eggs (1.5%) to those of middle-aged alligators (Table II).

TABLE II
Differences in alligator clutch size, egg size and egg quality as a function of the age of the female. These figures are based on data (ecological and embryological) collected from 31 nests in the marshes of Louisiana and from measurements on the 1182 eggs which the nests contained. All the original data (egg measurements, etc.) are available for consulation in the Department of Anatomy, The Queen's University of Belfast

Age of female alligator	Average number of eggs per clutch ± 1 S.D.[a]	Average maximum length of the eggs (mm) ± 1 S.D.	Average maximum width of the eggs (mm) ± 1 S.D.	Average percentage of eggs containing normal embryos per clutch ± 1 S.D.	Average percentage of infertile eggs per clutch ± 1 S.D.	Average percentage of eggs containing malformed embryos per clutch ± 1 S.D.
Young (up to 15 years)	26 ± 6	61 ± 2	37 ± 3	50 ± 8	40 ± 10	10 ± 1.5
Middle (15–20 years) – peak reproductive performance	42 ± 5.5	70 ± 3.5	42 ± 1.6	98 ± 1.5	2 ± 0.5	0.5 ± 0.2
Old (older than 30 years)	20 ± 4	78 ± 3.4	44 ± 2.5	95 ± 2	1.5 ± 0.7	5 ± 1

[a] S.D. = Standard deviation.

These relationships correlate closely with the reproductive and behavioural biology of young, middle-aged and old alligators (Ferguson, 1982c, in press). For the present paper two points are important. First, the incidence of spontaneous birth defects in alligator embryos is related to maternal age, as it is in many human malformations. Second, the large clutches of large eggs laid by middle-aged females show a very low basal level of spontaneous malformation (less than one evident deformity in every 200 eggs) and are optimum for teratological experiments.

TERATOLOGY

Introduction

Experimental teratological investigations in reptiles are rare. Bellairs (1981) has summarized much of the literature (mostly case reports) on many reptilian developmental abnormalities. The cranial morphology of mature cleft lip and palate embryos has been described for lizards (*Lacerta vivipara* and *L. lepida*) and snakes (*Eunectes murinus*, *Natrix natrix* and *Vipera berus*) (Bellairs, 1965; Bellairs & Boyd, 1957; Bellairs & Gamble, 1960). Case reports exist of cleft lip in garter snakes, *Thamnophis elegans* (Fox, Gordon & Fox, 1961), bilateral cleft lip and palate in prairie rattlesnakes, *Crotalus viridis viridis* (Dean, Glenn & Straight, 1980) and in some unspecified turtle species (Ewert, 1979) but not in any crocodilian species. Conversely there is a voluminous literature on experimental mammalian and avian teratology.

In general all phases of palatal development in mammals are subject to perturbation by various substances. Thus the pathogenesis of mammalian cleft palate may involve: (a) abnormalities of mesencephalic crest cell numbers, and/or division, and/or migration, and/or determination, and/or differentiation; (b) reduced palatal shelf growth; (c) delay or failure of palatal shelf elevation; (d) excessive head width; (e) failure of palatal shelf fusion; (f) defective epithelial/mesenchymal interactions and (g) post-closure opening. Frequently cleft lip and palate are associated with other congenital malformations and a number of syndromes have facial clefting as one of their components. Unilateral clefts of the lip (± palate) are twice as frequent on the left side as on the right in both man and laboratory animals. Some authors have suggested that these axial differences are related to vascularity, the right side receiving the shortest and most direct blood supply from the heart, so that the left side is more susceptible to ischaemic necrosis (Braithwaite & Watson, 1949; Sanvenero Rosselli, 1953; McKenzie, 1968; Stark,

1968; Brescia, 1971; Poswillo, 1975). However, investigations of the vascular patterns of normal and cleft human foetuses have revealed conflicting results (Andersen & Matthiessen, 1967; Frederiks, 1972, 1973) so that the issue is unresolved.

The present author conducted teratological experiments on alligator embryos with the objective of discovering regimes which would reliably induce specific types of craniofacial malformations and then using such regimes to recover a series of affected embryos in which the pathogenesis of facial clefting could be elucidated (Ferguson, 1982c). Two drugs, hydrocortisone sodium succinate (cortisone) and 5-fluoro-2-desoxyuridine (FUDR) were used independently as potential craniofacial teratogens in alligator embryos.

Extensive trials (Ferguson, 1982c) revealed the critical doses, times of administration and routes of administration necessary to produce specific craniofacial malformations. These are summarized in Table III for FUDR.

Reduced Lower Jaw

These embryos exhibit a markedly reduced (virtually absent) lower jaw, but have a normal palate and face (Figs 25–27). The value of this regime for longitudinal studies of palatal development in embryos grown by semi-shell-less culture techniques was discussed earlier (pp. 245–249). The pathogenesis of the deformity is as follows. The first waves of neural crest cells which migrate from the mesencephalic region populate the mandibular arches and maxillary processes. The crest cells migrating into the maxillary processes initially divide more rapidly than those migrating into the mandibular arches. When the teratogen is administered it inhibits the migration of later waves of crest cells (from the caudal part of the mesencephalon and cranial part of the rhombencephalon) into the mandibular arches and reduces the rate of cell division in the existing mandibular mesenchymal cells. Since the maxillary processes have already received their major wave of crest cells and reached their peak of cell division, they are much less affected than the later developing mandibular arches. Thus development of the lower jaw is specifically arrested (Figs 25–27). The palate exhibits a normal closure sequence except that the palatal shelves are smaller than normal and palatal closure extends to day 30 (normally closed by day 24) (Fig. 25). Both the delay in palatal closure and the small size of the shelves are due to the effect of FUDR on mesenchymal cell migration and division in the maxillary processes. The small posterior lower jaw develops a solid bar of Meckel's cartilage. The posterior ends of Meckel's cartilage develop normal articulations with the posterior ends

TABLE III

Summary table indicating the doses of 5-fluoro-2-desoxyuridine (FUDR), times of administration and routes of administration to alligator embryos in order to reliably produce certain craniofacial malformations. The malformations produced by injecting into the albumen are highly variable, those by injecting into the yolk less variable, and those by application to the chorio-allantoic membrane of eggs in semi-shell-less culture very consistent and homogeneous.

Malformation required	Dose of FUDR (mg) and time (days after egg-laying) for injection into albumen to reliably produce the malformations	Dose of FUDR (mg) and time (days after egg-laying) for injection into yolk to reliably produce the malformations	Dose of FUDR (mg) and time of administration (days after egg-laying) to embryos in semi-shell-less culture to reliably produce the malformations
Cyclopia Monorrhinous Micro-ophthalmia	Unknown	0.06 mg; day 6	Unknown
Very small (virtually absent) lower jaw	Unknown	0.06 mg; day 8	0.01 mg; day 10
Cleft lip (± cleft palate) (a) On right side (b) Bilateral (c) On left side	1–0.1 mg; day 8 Highly variable clefting types	0.06 mg; day 10 Highly variable clefting types	0.01 mg; day 12 early; stage 12.2 0.01 mg; day 12 middle; stage 12.5 0.01 mg; day 12 late; stage 12.8
Cleft palate	1–0.1 mg; day 12	0.06 mg; days 14–15	0.01 mg; days 16–17
Temporary cleft palate (slowed time of palatal closure)	1–0.1 mg; day 14	0.06 mg; day 17	0.01 mg; day 19

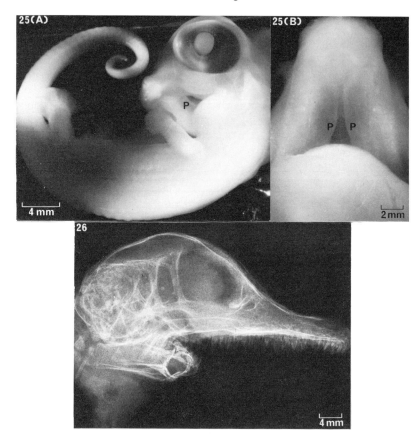

FIGS 25 & 26. (25) (A) Macroscopic view of the closing palate of a day 19 alligator embryo grown in semi-shell-less culture and treated on day 10 with 0.01mg of 5-fluoro-2-desoxyuridine (FUDR) to inhibit lower jaw development. These techniques enable longitudinal studies of palatal development (P). (B) Section shown in (A) in greater detail. (26) Lateral radiograph of a hatchling alligator head with FUDR-induced reduction of the lower jaw. Note the normal palate, lower jaw joint, and presence of teeth on the anterior margin of the lower jaw.

of the palatoquadrate cartilages (Figs 26 & 27). All these cartilages subsequently ossify so that a normal jaw joint is formed between the articular bone of the lower jaw and the quadrate bone of the skull (Figs 26 & 27). The large anterior pterygoid muscles attach to the posterior ends of Meckel's cartilages in the reduced lower jaw (Fig. 27) but, because of the small size and altered geometry of the latter, it is incapable of much movement and therefore functionless. It is remarkable that a normal jaw joint develops, in view of the widespread belief that early embryonic function is required for the normal development of the mammalian temporomandibular joint. The tongue, the inferior basihyal valve flap and the floor of the mouth are all absent (Figs

25–27). With subsequent ossifications, the bony pattern of the lower jaw is grossly distorted (Fig. 26). Three to six teeth may develop along the anterior margin of the reduced lower jaw (Fig. 26).

Facial Clefting

Utilizing the regime of administering 0.01 mg FUDR directly to the vascular cranial chorio-allantois of precisely staged embryos in semi-shell-less culture, one can induce very specific abnormalities, e.g. unilateral clefts of the lip and alveolus (with or without cleft palate) on either the right or left sides (Table III). The explanation for such precision is probably as follows. As with many embryological events, there is a slight difference in the timing and sequence of development on the right and left sides of the face. By incubating alligator eggs at 26°C between days 10 and 15 it is possible to slow development sufficiently that the time difference between closure of the nasal pit slit on the right and left sides is exaggerated with the right side being more advanced in most cases. Therefore, if one administers the teratogen early on day 12 it affects both right and left sides, but since the left closes later than the right, it recovers from the mild teratogenic insult and closes normally, whereas the earlier closing right nasal pit slit is affected by the teratogen resulting in unilateral right-sided clefts. Contrariwise, administration of the teratogen late on day 12 has little effect on the closure of the right nasal pit slit, which by this time has undergone most of its critical steps (e.g. cell proliferation), whereas the later closing left side is affected resulting in unilateral left-sided clefts. Administration in the middle of day 12 affects critical phases of both right and left sides resulting in bilateral cleft lip and palate. All cases of bilateral cleft lip and most cases of unilateral cleft lip are associated with cleft palate. The alligator is the first animal model in which it is possible to control the timing of teratogen administration with enough precision to permit the reliable production of either right-sided or left-sided clefts.

The pathogenesis of cleft lip (Figs 28–31, 33 & 34) involves a persistence of the nasal pit slits. Usually the lateral nasal process merges with the maxillary process as normal (Fig. 6) but these two processes do not merge with the medial nasal process, so that the nasal pit slit persists as a cleft lip (Figs 30 & 34). The term cleft lip is somewhat of a misnomer since alligators have no fleshy lips; nevertheless, the term is used to indicate a cleft of primary palate, alveolus and external bony margin extending into one of the external nares. In early embryos the anlage of the mammalian lip and the alligator primary palate, alveolus and bony margins are homologous, so that the term cleft lip seems appropriate. In the case of unilateral cleft lip,

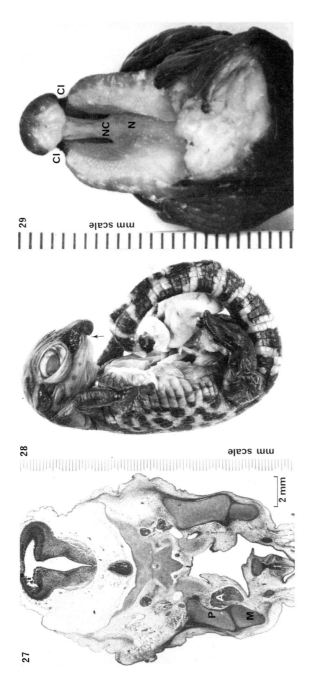

FIGS 27–29. (27) Transverse section through a day 24 alligator embryo with FUDR-induced reduction of the lower jaw. Note the embryonic articulation between the palatoquadrate (P) and Meckel's cartilages (M) and the anterior pterygoid muscle (A). (28) Lateral view of a hatchling alligator with FUDR-induced bilateral cleft lip and palate (arrowed). Note the lower jaw trapped behind the cleft premaxillary segment and the exophthalmia. (29) Intra-oral view of the hatchling illustrated in Fig. 28. Note the bilateral cleft lip (CL) and wide palatal cleft extending into the nasal cavities (NC) anteriorly and the nasopharyngeal duct (N) posteriorly.

without associated cleft palate, it appears that the medial nasal process is too small to meet the lateral nasal and maxillary processes and so a cleft arises. Contrariwise, in the case of unilateral or bilateral cleft lip and cleft palate the lateral nasal and maxillary processes appear smaller than normal. In this way not only does the nasal pit slit fail to close, but the hypoplastic palatal shelves arising from the defective maxillary processes fail to contact, so causing cleft palate. The inhibition of facial process size is relative, all processes being affected to varying degrees by the teratogen.

These data suggest that the distribution and frequency of unilateral cleft lip may be related principally to the timing of teratogenic insults, left-sided clefts being more frequent because normally the left nasal pit slit closes after the right. This means both that it is closer to a postulated threshold of head width beyond which the facial processes cannot fuse so resulting in cleft lip, and that it is susceptible to teratogenic insults for a greater period of time than the earlier closing right side.

By the same token, it is possible reliably to produce permanent isolated cleft palate (i.e. without cleft lip) by administering the FUDR on day 17 and temporary isolated cleft palate by administering it on day 19 (Table III). In the latter group, the palate is open for up to 10 days later than day 28 (when it is normally completely closed), but unlike the permanent cleft group (treated at day 17) the palates of embryos treated at day 19 eventually close, so that the clefting is due to a temporary developmental delay which is self-repairing. Facial clefting may be induced either in isolation or in association with other abnormalities such as polydactyly, syndactyly, tail anomalies and cardiac anomalies.

On the cleft side(s) there is continuity between the olfactory and ciliated columnar epithelia of the nose and the stratified squamous epithelia of the mouth and external jaw margins (Figs 30 & 31). In hatchlings, the junction between the ciliated columnar nasal epithelia and the stratified squamous oral epithelia is extremely sharp (Fig. 31). In bilateral cleft lip and palate, the mobilized premaxillary segment protrudes downwards and forwards with abnormal anteroposterior enlargement of the nasal septum (Figs 28 & 29). The lower jaw and tongue become trapped behind this cleft premaxillary segment resulting in inhibition of mandibular growth and compression of the tongue and floor of the mouth against the lower edge of the nasal septum anteriorly and the roof of the nasopharyngeal duct posteriorly (Fig. 31). This severely restricts and retards the development of the palatal shelves (Figs 31 & 33). Anteriorly, the shelves are hypoplastic (Fig. 31): in the mid-palate the shelves fail to become horizontal above the dorsum of the tongue but are rather compressed into a variety of shapes with one

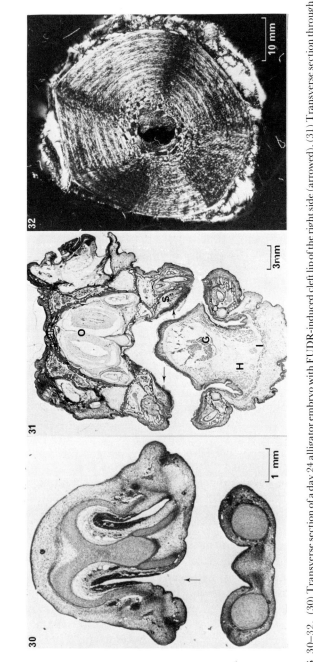

FIGS 30–32. (30) Transverse section of a day 24 alligator embryo with FUDR-induced cleft lip of the right side (arrowed). (31) Transverse section through a hatchling alligator with FUDR-induced bilateral cleft lip and palate. Note the wide palatal cleft, the wedge shaped distorted tongue, the sharp junction (arrowed) between the nasal ciliated columnar and oral stratified squamous epithelia, the hyoglossus (H), genioglossus (G) and intermandibularis (I) muscles, the palatal submucosa (S) and olfactory bulbs (O). (32) Undecalcified section from the femur mid-diaphysis of a 16-year-old alligator viewed in polarizing light (note the crossed nicol pattern). Fourteen annuli (white rings) are present.

shelf often having both vertically and horizontally directed projections (Fig. 33); posteriorly, the normally vertical palatal shelves do not flow over the dorsum of the tongue and the upper flap of the basihyal valve is cleft. Even in hatchlings, the tongue is firmly wedged between the margins of the cleft palate (Fig. 31). It is a narrow, highly arched wedge which exhibits distorted morphology (Fig. 31). The genioglossus, hyoglossus and intermandibularis muscles are compressed between the narrow lower jaw bones and expand as a ventral bulge of the floor of the mouth (Fig. 31). Dental abnormalities (e.g. rotated, misplaced or malaligned teeth) are prevalent. Interestingly, Trasler & Fraser (1963) suggested that trapping of the mammalian lower jaw and tongue behind the cleft premaxilla was the pathogenic factor in cleft palate associated with bilateral cleft lip.

The pathogenesis of permanent cleft palate without cleft lip (Table III) is severe inhibition of mesenchymal cell division, migration and extracellular matrix production within the developing maxillary processes and palatal shelves. As a result, palatal shelves are almost completely absent or reduced to small slender projections. In the cases of temporary cleft palate the palatal shelves are only slightly smaller than normal, but they exhibit a decrease in mesenchymal cell numbers, few mitotic figures, poor staining for glycosaminoglycans and a decreased vascular supply when compared with normal embryos of the same age. However, they rapidly recover from the teratogenic insult and exhibit catch-up development so that palatal closure is merely delayed by approximately 10 days, during which a temporary cleft palate exists. Thereafter, the structure of the palate appears normal.

ACKNOWLEDGEMENTS

I thank Mr Ted Joanen and staff of the Rockefeller Wildlife Refuge, Louisiana, for their outstanding help in the collection and transportation of alligator eggs. Dr G. Webb and the Conservation Commission of the Northern Territory, Australia, greatly facilitated my study of *C. porosus* and *C. johnstoni* embryos described on pp. 249–251. This work is supported by grant 8113610CB from the Medical Research Council of Great Britain, grant EP 109/74/75 from the Northern Ireland Eastern Health and Social Services Boards, a Wellcome Trust (London) Research Travelling Scholarship (1981) and grants DE-02848 and DE-03569 from the National Institutes of Health, USA. The collection and export of alligator and crocodile specimens were carried out with the appropriate permits. Misses A. Richardson and J. Smith kindly typed the manuscript.

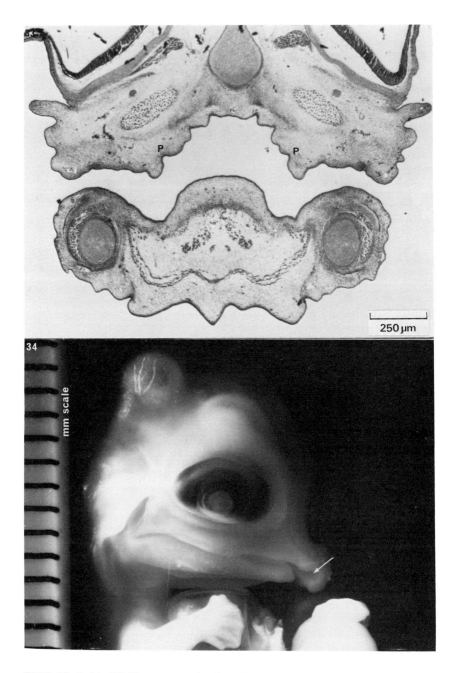

FIGS 33 & 34. (33) Transverse section through the mid-posterior snout of a day 20 alligator embryo with FUDR-induced bilateral cleft lip and palate. Note the distorted vertical palatal shelves (P). (34) Lateral view of a day 24 alligator embryo with FUDR-induced cleft lip on the right side (arrowed) and an extra digit on the right manus.

REFERENCES

Andersen, M. & Matthiessen, M. (1967). Histochemistry of the early development of the human central face and nasal cavity with specific reference to the movements and fusion of the palatine processes. *Acta. Anat. (Basel)* **68**: 473–508.
Anderson, C. B. & Meier, S. (1981). The influence of the metameric pattern in the mesoderm on migration of cranial neural crest cells in the chick embryo. *Devl Biol.* **85**: 385–402.
Auerbach, R., Kubai, L., Knighton, D. & Folkman, J. (1974). A simple procedure for the long term cultivation of chicken embryos. *Devl Biol.* **41**: 391–394.
Bellairs, A. d'A. (1965). Cleft palate, microphthalmia and other malformations in embryos of lizards and snakes. *Proc. zool. Soc. Lond.* **144**: 239–251.
Bellairs, A. d'A. (1981). Congenital and developmental diseases. In *Diseases of the Reptilia*, **2**: 469–485. Cooper, J. E. & Jackson, D. F. (Eds). London: Academic Press.
Bellairs, A. d'A. & Boyd, J. D. (1957). Anomalous cleft palate in snake embryos. *Proc. zool. Soc. Lond.* **129**: 525–539.
Bellairs, A. d'A. & Gamble, H. J. (1960). Cleft palate, microphthalmia and other anomalies in an embryo lizard. *Br. J. Herp.* **2**: 171–176.
Bolender, D. L., Seliger, W. G. & Markwald, R. R. (1980). A histochemical analysis of polyanionic compounds found in the extracellular matrix encountered by migrating cephalic neural crest cells. *Anat. Rec.* **196**: 401–412.
Bordet, G. (1981). Analyse ultrastructurale de la croissance supra-linguale des processus palatins chez *Oryctolagus cuniculus* (Fauve de Bourgogne). *J. Biol. Bucc.* **9**: 253–270.
Braithwaite, F. & Watson, J. (1949). A report of three unusual cleft lips. *Br. J. Plast. Surg.* **2**: 38–49.
Brescia, N. J. (1971). Anatomy of the lip and palate. In *Cleft lip and palate: surgical, dental and speech aspects*: 3–20. Grabb, W. C., Rosenstein, S. W. & Bzoch, K. R. (Eds). Boston: Little Brown Co.
Brinkley, L. L. (1980). *In vitro* studies of palatal shelf elevation. In *Current research trends in prenatal craniofacial development*: 203–220. Pratt, R. M. & Christiansen, R. L. (Eds). New York: Elsevier/North Holland.
Burdi, A., Feingold, M., Larsson, K. S., Leck, I., Zimmerman, E. F. & Fraser, F. C. (1972). Etiology and pathogenesis of congenital cleft lip and cleft palate, an NIDR state of the Art Report. *Teratology* **6**: 255–270.
Carter, C. O. & Evans, K. (1973). Spina bifida and anencephalus in Greater London. *J. med. Genet.* **10**: 209–234.
Castanet, J. (1974). Etude histologique des marques squelettiques de croissance chez *Vipera aspis* (L.) (Ophidia, Viperidae). *Zoologica Scr.* **3**: 137–151.
Castenet, J. (1978). Les marques de croissance osseuse comme indicateurs de l'âge chez les lézards. *Acta zool., Stockh.* **59**: 35–48.
Castanet J. & Cheylan, M. (1979). Les marques de croissance des os et des écailles comme indicateur de l'âge chez *Testudo hermanni* et *Testudo graeca* (Reptilia, Chelonia, Testudinidae). *Can. J. Zool.* **57**: 1649–1665.
Chabreck, R. H. & Joanen, T. (1979). Growth rates of American alligators in Louisiana. *Herpetologica* **35**: 51–57.
Chaudhry, A. P. & Shah, R. M. (1973). Palatogenesis in hamster. II Ultrastructural observations on the closure of palate. *J. Morph.* **139**: 329–350.
Clarke, S. F. (1891). The habits and embryology of the American alligator. *J. Morph.* **5**: 181–205.

Dean, J. N., Glenn, J. L. & Straight, R. C. (1980). Bilateral cleft labial and palate in the progeny of a *Crotalus viridis viridis* Rafinesque. *Herp. Rev.* **11**: 91–92.

De Buffrenil, V. (1980a). Données préliminaires sur la structure des marques de croissance squelettiques chez les crocodiliens actuels et fossiles. *Bull. Soc. zool. Fr.* **105**: 355–361.

De Buffrenil, V. (1980b). Mise en evidence de l'incidence des conditions de milieu sur la croissance de *Crocodylus siamensis* (Schneider, 1801) et valeur des marques de croissance squelettiques pour l'évaluation de l'âge individuel. *Archs Zool. exp. gén.* **121**: 63–76.

De Buffrenil, V. & Buffetaut, E. (1981). Skeletal growth lines in an eocene crocodilian skull from Wyoming as an indicator of ontogenic age and paleoclimatic conditions. *J. vert. Paleont.* **1**: 57–66.

De Paola, D. P., Drummond, J. F., Lorente, C., Zarbo, R. & Miller, S. A. (1975). Glycoprotein biosynthesis at the time of palate fusion by rabbit palate and maxilla cultured *in vitro*. *J. dent. Res.* **54**: 1049–1055.

Derby, M. A. (1978). Analysis of glycosaminoglycans within the extracellular environments encountered by migrating neural crest cells. *Devl Biol.* **66**: 321–336.

De Ricqlès, A. (1976). On bone histology of fossil and living reptiles with comments on its functional and evolutionary significance. In *Morphology and biology of the reptiles*: 123–150. Bellairs, A. d'A. & Cox, B. (Eds). (Linnean Society, Symposium **3**.) London: Academic Press.

Diewert, V. M. (1980). The role of craniofacial growth in palatal shelf elevation. In *Current research trends in prenatal craniofacial development*; 165–186. Pratt, R. M. & Christiansen, R. L. (Eds). New York: Elsevier/North Holland.

Dunn, B. E. (1974). Technique for shell-less culture of the avian embryo. *Poultry Sci.* **53**: 409–412.

Dunn, B. E. & Boone, M. A. (1976). Growth of the chick embryo *in vitro*. *Poultry Sci.* **55**: 1067–1071.

Dunn, B. E. & Boone, M. A. (1977). Growth and mineral content of cultured chick embryos. *Poultry Sci.* **56**: 662–672.

Dunn, B. E. & Boone, M. A. (1978). Photographic study of chick embryonic development *in vitro*. *Poultry Sci.* **57**: 370–377.

Dunn, B. E. & Fitzharris, T. P. (1979). Differentiation of the chorionic epithelium of chick embryos maintained in shell-less culture. *Devl Biol.* **71**: 216–227.

Dunn, B. E., Fitzharris, T. P. & Barnett, B. D. (1981). Effects of varying chamber construction and embryo pre-incubation age on survival and growth of chick embryos in shell-less culture. *Anat. Rec.* **199**: 33–43.

Dunn, B. E., Graves, J. & Fitzharris, T. P. (1981). Active calcium transport in the chick chorio-allantoic membrane requires interactions with the shell membrane and/or shell calcium. *Devl Biol.* **88**: 259–268.

Ewert, M. A. (1979). The embryo and its egg: development and natural history. In *Turtles, perspectives and research*: 333–413. Harless, M. & Morlock, H. (Eds). New York: J. Wiley & Sons.

Farbman, A. I. (1968). Electron microscope study of palate fusion in mouse embryos. *Devl Biol.* **18**: 93–116.

Farbman, A. I. (1969). The epithelium-connective tissue interface during closure of the secondary palate in rodent embryos. *J. dent. Res.* **48**: 617–624.

Ferguson, M. W. J. (1977). The mechanism of palatal shelf elevation and the pathogenesis of cleft palate. *Virchows Arch. path. Anat. Histol.* **375**: 97–113.

Ferguson, M. W. J. (1978a). Palatal shelf elevation in the Wistar rat fetus. *J. Anat.* **125**: 555–577.

Ferguson, M. W. J. (1978b). The teratogenic effects of 5 Fluoro-2-desoxyuridine (FUDR) on the Wistar rat fetus, with particular reference to cleft palate. *J. Anat.* **126**: 37–49.

Ferguson, M. W. J. (1979). The American alligator (*Alligator mississippiensis*): a new model for investigating developmental mechanisms in normal and abnormal palate formation. *Med. Hypoth.* **5**: 1079–1090.

Ferguson, M. W. J. (1981a). The structure and development of the palate in *Alligator mississippiensis*. *Archs oral Biol.* **26**: 427–443.

Ferguson, M. W. J. (1981b). The value of the American alligator (*Alligator mississippiensis*) as a model for research in craniofacial development. *J. craniofac. Genet. Devl Biol.* **1**: 123–144.

Ferguson, M. W. J. (1981c). Developmental mechanisms in normal and abnormal palate formation with particular reference to the aetiology, pathogenesis and prevention of cleft palate. *Br. J. Orthodont.* **8**: 115–137.

Ferguson, M. W. J. (1981d). Extrinsic microbial degradation of the alligator eggshell. *Science, Wash.* **214**: 1135–1137.

Ferguson, M. W. J. (1982a). Crocodilian embryology: an overview. In *The reproductive biology and conservation of crocodilians*: 1–27. Tryon, B. W. & Lang, J. W. (Eds). (SSAR Symposium, Milwaukie.) (Mimeo.)

Ferguson, M. W. J. (1982b). The structure and composition of the eggshell and embryonic membranes of *Alligator mississippiensis*. *Trans. zool. Soc. Lond.* **36**: 99–152.

Ferguson, M. W. J. (1982c). *The structure and development of the palate in* Alligator mississippiensis. Ph.D. Thesis: The Queen's University of Belfast, Northern Ireland.

Ferguson, M. W. J. (In press). The reproductive biology and embryology of crocodilians. In *Biology of the Reptilia*, **14** *Development*. Gans, C., Billett, F. S. & Maderson, P. (Eds). New York: J. Wiley and Sons.

Ferguson, M. W. J. & Honig, L. S. (1984). Epithelial-mesenchymal interactions during vertebrate palatogenesis. In *Current topics in developmental biology*, **19**. *Palate development: normal and abnormal, cellular and molecular aspects*: 137–164. Zimmerman, E.F. (Ed.). New York, Academic Press.

Ferguson, M. W. J., Honig, L. S., Bringas, P. & Slavkin, H. C. (1981). Epithelial-mesenchymal interactions in vertebrate secondary palate development. *Am. Zool.* **21**: 952.

Ferguson, M. W. J., Honig, L. S., Bringas, P. & Slavkin, H. C. (1982). *In vivo* and *in vitro* development of first branchial arch derivatives in *Alligator mississippiensis*. In *Factors and mechanisms influencing bone growth*: 275–286. Dixon, A. D. & Sarnat, B. G. (Eds). New York: Alan R. Liss.

Ferguson, M. W. J., Honig, L. S., Bringas, P. & Slavkin, H. C. (1983). Alligator mandibular development during long-term organ culture. *In Vitro* **19**: 385–393.

Ferguson, M. W. J., Honig, L. S. & Slavkin, H. C. (In press). Differentiation of cultured palatal shelves from alligator, chick and mouse embryos. *Anat. Rec.*

Ferguson, M. W. J. & Joanen, T. (1982). Temperature of egg incubation determines sex in *Alligator mississippiensis*. *Nature, Lond.* **296**: 850–853.

Ferguson, M. W. J. & Joanen, T. (1983). Temperature dependent sex determination in *Alligator mississippiensis*. *J. Zool., Lond.* **200**: 143–177.

Fleischmann, A. (1910). Uber den Begriff Gaumen. *Morph. Jb.* **41**: 681–707.

Fogh Andersen, P. (1968). Increasing incidence of facial clefts. Genetically or non-genetically determined. In *Craniofacial anomalies, pathogenesis and repair*: 27–29. Longacre, J. J. (Ed.). Philadelphia: J. B. Lippincott Co.

Fox, W., Gordon, C. & Fox, M. H. (1961). Morphological effects of low temperature during the embryonic development of the garter snake *Thamnophis elegans*. *Zoologica, N.Y.* **46**: 57–71.

Frederiks, E. (1972). Vascular patterns in normal and cleft primary and secondary palate in human embryos. *Br. J. Plast. Surg.* **25**: 207–223.
Frederiks, E. (1973). Vascular pattern in embryos with clefts of primary and secondary palate. *Ergebn. Anat. Entwgesch.* **46**: 1–50.
Fuchs, H. (1908). Untersuchungen uber Ontogenie und Phylogenie der Gaumenbildung bei den Wirbeltieren Zweite Mitteilung. Uber das Munddach der Rhynchocephalen, Saurier, Schlangen, Krokodile und Saüger und den Zusammenhang zwischen Mund und Näsenhohle bei diesen Tieren. *Z. Morph. Anthrop.* **11**: 153–248.
Göppert, E. (1903). Die Bedeutung der Zunge für den sekundären Gaumen und den Ductus Nasopharyngeus. *Morphol. Jb.* **31**: 311–359.
Greenberg, J. H. & Pratt, R. M. (1977). Glycosaminoglycan and glycoprotein syntheses by cranial neural crest cells *in vitro*. *Cell Diff.* **6**: 119–132.
Greene, R. M. & Kochhar, D. M. (1974). Surface coat on the epithelium of developing palatine shelves in the mouse as revealed by electron microscopy. *J. Embryol. exp. Morph.* **31**: 683–692.
Greene, R. M. & Kochhar, D. M. (1975). Some aspects of corticosteroid-induced cleft-palate: A review. *Teratology* **11**: 47–56.
Greene, R. M. & Pratt, R. M. (1976). Developmental aspects of secondary palate formation. *J. Embryol. exp. Morph.* **36**: 225–245.
Greene, R. M. & Pratt, R. M. (1978). Inhibition of epithelial cell death in the secondary palate *in vitro* by alteration of lysosome function. *J. Histochem. Cytochem.* **26**: 1109–1114.
Hamburger, V. & Hamilton, H. L. (1951). A series of normal stages in the development of the chick embryo. *J. Morph.* **88**: 49–92.
Hassell, J. R. & Orkin, R. W. (1976). Synthesis and distribution of collagen in the rat palate during shelf elevation. *Devl Biol.* **49**: 80–88.
Hayward, A. F. (1969). Ultrastructural changes in the epithelium during fusion of the palatal processes in rats. *Archs oral Biol.* **14**: 661–678.
Hemelaar, A. S. M. & Van Gelder, J. J. (1980). Annual growth rings in phalanges of *Bufo bufo* (Anura, Amphibia) from the Netherlands and their use for age determination. *Neth. J. Zool.* **30**: 129–136.
His, W. (1892). Die Entwickelung der Menschlichen und thierischer Physiognomien. *Arch. Anat. Physiol. (Anat.)* **1892**: 384–424.
Hudson, C. D. & Shapiro, B. L. (1973). An autoradiographic study of deoxyribonucleic acid synthesis in embryonic rat palatal shelf epithelium with reference to the concept of programmed cell death. *Archs oral Biol.* **18**: 77–84.
Ingalls, T. H., Pugh, T. F. & MacMahon, B. (1954). Incidence of anencephalus, spina bifida and hydrocephalus related to birth rank and maternal age. *Br. J. prev. soc. Med.* **8**: 17–23.
Janerich, D. T. (1974). Maternal factors in CNS malformations: The cohort approach. In *Congenital defects. New directions in research*: 73–93. Janerich, D. T., Skalko, R. G. & Porter, I. H. (Eds). New York: Academic Press.
Jaskoll, T. & Melnick, M. (1982). The effects of long-term fetal constraint *in vitro* on the cranial base and other skeletal components. *Am. J. Med. Genet.* **12**: 289–300.
Jenkins, N. K. (1975). Chemical composition of the eggs of the crocodile (*Crocodylus novaeguineae*). *Comp. Biochem. Phys.* **51**(A): 891–895.
Johnston, M. C., Bhakdinavonk, A. & Reid, Y. C. (1974). An expanded role for the neural crest in oral and pharyngeal development. In *Symposium on oral sensation and perception*, **4**: 37–52. Bosma, J. F. (Ed.). Washington D.C.: US Government Printing Office.
Johnston, M. C. & Hazelton, R. D. (1972). Embryonic origins of facial structures

related to oral sensory and motor functions. In *Symposium on oral sensation and perception. The mouth of the infant*, **3**: 76–97. Bosma, J. F. (Ed.). Springfield: C. C. Thomas.
Johnston, M. C., Morriss, G. M., Kushner, D. C. & Bingle, G. J. (1977). Abnormal organogenesis of facial structures. In *Handbook of teratology* **2**. *Mechanisms and pathogenesis*: 421–451. Wilson, J. G. & Fraser, F. C. (Eds). New York: Plenum Press.
Juriloff, D. M. & Trasler, D. G. (1976). Test of the hypothesis that embryonic face shape is a causal factor in genetic predisposition to cleft lip in mice. *Teratology* **14**: 35–42.
Koch, W. E. & Smiley, G. R. (1981). *In vivo* and *in vitro* studies of the development of the avian secondary palate. *Archs oral Biol.* **26**: 181–187.
Kuhn, E. M., Babiarz, B. S., Lessard, J. L. & Zimmerman, E. F. (1980). Palate morphogenesis 1. Immunological and ultrastructural analyses of mouse palate. *Teratology* **21**: 209–223.
Le Lièvre, C. (1974). Rôle des cellules mésectodermiques issues des crêtes neurales céphaliques dans la formation des arcs branchiaux et du squelette viscéral. *J. Embryol. exp. Morph.* **31**: 453–477.
Le Lièvre, C. S. (1978). Participation of neural crest derived cells in the genesis of the skull in birds. *J. Embryol. Exp. Morph.* **47**: 17–37.
Le Lièvre, C. S. & Le Douarin, N. M .(1975). Mesenchymal derivatives of the neural crest: Analysis of chimaeric quail and chick embryos. *J. Embryol. exp. Morph.* **34**: 125–154.
Löfberg, J., Ahlfors, K. & Fällström, C. (1980). Neural crest cell migration in relation to extracellular matrix organisation in the embryonic axolotl trunck. *Devl Biol.* **75**: 148–167.
Mato, M., Aikawa, E. & Katahiva, M. (1966). Appearance of various types of lysosomes in the epithelium covering lateral palatine shelves during a secondary palate formation. *Gunma J. med. Sci.* **15**: 46–56.
Mato, M., Smiley, G. R. & Dixon, A. D. (1972). Epithelial changes in the presumptive regions of fusion during secondary palate formation. *J. dent. Res.* **51**: 1451–1456.
McKenzie, J. (1968). The role of vascular anomalies. Failure of standard circulation in production of craniofacial anomalies. In *Craniofacial anomalies, pathogenesis and repair*: 61–63. Longacre, J. J. (Ed.). Philadelphia and Toronto: J. B. Lippincott Co.
Meier, S. (1979). Development of the chick embryo mesoblast. Formation of the embryonic axis and establishment of the metameric pattern. *Devl Biol.* **73**: 25–45.
Meier, S. (1981). Development of the chick embryo mesoblast. Morphogenesis of the prechordal plate and cranial segments. *Devl Biol.* **83**: 49–61.
Meller, S. M. (1980). Morphological alterations in the prefusion human palatal epithelium. In *Current research trends in prenatal craniofacial development*: 221–234. Pratt, R. M. & Christiansen, R. L. (Eds). New York: Elsevier/North Holland.
Meller, S. M. & Barton, L. H. (1978). Extracellular coat in developing human palatal processes: Electron microscopy and ruthenium red binding. *Anat. Rec.* **190**: 223–232.
Meller, S. M. & Barton, L. H. (1979). Distribution of glycogen in prefusion human palatal epithelium. *Anat. Rec.* **193**: 831–856.
Meller, S. M., De Paola, D. P., Barton, L. H. & Mandella, R. D. (1980). Secondary palatal development in the New Zealand white rabbit: A scanning electron microscopic study. *Anat. Rec.* **198**: 229–244.
Millard, D. R. (1976). *Cleft craft. The evolution of its surgery*, **1–3**. Boston: Little Brown & Co.

Millicovsky, G. & Johnston, M. C. (1981). Active role of embryonic facial epithelium: New evidence of cellular events in morphogenesis. *J. Embryol. exp. Morph.* **63**: 53–66.

Minkoff, R. (1980a). Regional variation of cell proliferation within the facial processes of the chick embryo: A study of the role of merging during development. *J. Embryol. exp. Morph.* **57**: 37–49.

Minkoff, R. (1980b). Cell proliferation and migration during primary palate development. In *Current research trends in prenatal craniofacial development*: 119–136. Pratt, R. M. & Christiansen, R. L. (Eds). Amsterdam: Elsevier/North Holland.

Minkoff, R. & Kuntz, A. J. (1977). Cell proliferation during morphogenetic change, analysis of frontonasal morphogenesis in the chick embryo employing DNA labelling indices. *J. Embryol. exp. Morph.* **40**: 101–113.

Minkoff, R. & Kuntz, A. J. (1978). Cell proliferation and cell density of mesenchyme in the maxillary process and adjacent regions during facial development in the chick embryo. *J. Embryol. exp. Morph.* **46**: 65–74.

Morgan, P. R. (1976). The fate of the expected fusion zone in rat fetuses with experimentally-induced cleft palate – an ultrastructural study. *Devl Biol.* **51**: 225–240.

Morgan, P. R. & Pratt, R. M. (1977). Ultrastructure of the expected fusion zone in rat fetuses with diazo-oxo-norleucine (DON) induced cleft palate. *Teratology* **15**: 281–290.

Müller, F. (1965). Zur Morphogenese des Ductus nasopharyngeus und des sekundären Gaumendaches bei den Crocodilia. *Revue suisse Zool.* **72**: 647–652.

Müller, F. (1967). Zur embryonalen Kopfentwicklung von *Crocodylus cataphracus*, C.U.V. *Revue suisse Zool.* **74**: 189–294.

Narbaitz, R. & Jande, S. S. (1978). Ultrastructural observations on the chorionic epithelium, parathyroid glands and bones from chick embryos developed in shell-less culture. *J. Embryol. exp. Morph.* **45**: 1–12.

Newgreen, D. & Thiery, J. P. (1980). Fibronectin in early avian embryos: synthesis and distribution along the migration pathways of neural crest cells. *Cell Tissue Res.* **211**: 269–291.

Parker, W. K. (1883). On the structure and development of the skull in the Crocodilia. *Trans. zool. Soc. Lond.* **11**: 263–311.

Peabody, F. E. (1961). Annual growth zones in living and fossil vertebrates. *J. Morph.* **108**: 11–62.

Pintar, J. E. (1978). Distribution and synthesis of glycosaminoglycans during quail neural crest morphogenesis. *Devl Biol.* **67**: 444–464.

Poswillo, D. (1975). Causal mechanisms of craniofacial deformity. *Br. Med. Bull.* **31**: 101–106.

Pratt, R. M. (1980). Involvement of hormones and growth factors in the development of the secondary palate. In *Development in mammals* **4**: 203–231. Johnson, M. H. (Ed.). Amsterdam: Elsevier/North Holland.

Pratt, R. M. & Christiansen, R. L. (Eds) (1980). *Current research trends in prenatal craniofacial development*. New York: Elsevier/North Holland.

Pratt, R. M., Figueroa, A. A., Greene, R. M., Wilk, A. & Salomon, D. S. (1979). Alterations in macro-molecular synthesis related to abnormal palatal development. In *Advances in the study of birth defects* **3**. *Abnormal embryogenesis cellular and molecular aspects*: 161–176. Persaud, T. V. N. (Ed.). Lancaster, MTP Press Ltd.

Pratt, R. M. & Greene, R. M. (1976). Inhibition of palatal epithelial cell death by altered protein synthesis. *Devl Biol.* **54**: 135–145.

Pratt, R. M. & Hassell, J. R. (1975). Appearance and distribution of carbohydrate rich

macromolecules on the epithelial surface of the rat palatal shelf. *Devl Biol.* **45**: 192–198.
Pratt, R. M., Larsen, M. A. & Johnston, M. C. (1975). Migration of cranial neural crest cells in a cell free hyaluronate rich matrix. *Devl Biol.* **44**: 298–305.
Pratt, R. M. & Martin, G. R. (1975). Epithelial cell death and cyclic AMP increase during palatal development. *Proc. natn. Acad. Sci. USA* **72**: 874–877.
Pratt, R. M. & Salomon, D. S. (1981). Biochemical basis for the teratogenic effects of glucocorticoids. In *The biochemical basis of chemical teratogenesis*: 179–200. Juchau, M. R. (Ed.). New York: Elsevier/North Holland.
Rathke, H. (1866). *Untersuchungen über die Entwickelung und den Korperbau der Krokodile.* Braunschweig: F. Vieweg & Sohn.
Reese, A. M. (1908). The development of the American alligator (*Alligator mississippiensis*). *Smithson. misc. Collns* **51** (1791): 1–66.
Reese, A. M. (1912). The embryology of the Florida alligator (*Alligator mississippiensis*). *Int Congr. Zool.* **7**: 535–537.
Reese, A. M. (1915). *The alligator and its allies*. New York and London: G. P. Putnam's Sons.
Romanoff, A. L. & Romanoff, A. J. (1972). *Pathogenesis of the avian embryo.* New York: Wiley Interscience.
Salomon, D. S. & Pratt, R. M. (1979). Involvement of glucocorticoids in the development of the secondary palate. *Differentiation* **13**: 141–154.
Sanvenero Rosselli, G. (1953). Developmental pathology of the face and dysraphic syndrome. *Plast. Reconst. Surg.* **11**: 36–38.
Sewertzoff, S. A. (1929). Zur Entwicklungsgeschichte der Zunge bei den Reptilien. *Acta zool., Stockh.* **10**: 231–341.
Shah, R. M. (1979). Current concepts on the mechanisms of normal and abnormal secondary palate formation. In *Advances in the study of birth defects*, **1**. *Teratogenic mechanisms*: 69–84. Persaud, T. V. N. (Ed.). Lancaster: M.T.P. Press Ltd.
Shah, R. M. & Crawford, B. J. (1980). Development of the secondary palate in chick embryo: A light and electron microscopic and histochemical study. *Invest. Cell Path.* **3**: 319–328.
Sippel, W. (1907). Das Munddach der Vögel und Saurier. *Morph. Jb.* **47**: 490–524.
Slavkin, H. C., Slavkin, M. D. & Bringas, P. (1980). Mineralization during long term cultivation of chick embryos *in vitro*. *Proc. Soc. exp. Biol. Med.* **163**: 249–257.
Smiley, G. R. (1970). Fine structure of mouse embryonic palatal epithelium prior to and after midline fusion. *Archs oral Biol.* **15**: 287–296.
Souchon, R. (1975). Surface coat of the palatal shelf epithelium during palatogenesis in mouse embryos. *Anatomy Embryol.* **147**: 133–142.
Stark, R. B. (Ed.) (1968). *Cleft palate. A multi-discipline approach.* New York, Evanston and London: Hoeber & Row Publishers.
Taguchi, H. (1920). Beitrage zur Kenntnis uber die feinere Struktur der Eingewadeorgane der Krokodile. *Mitt. med. Fak. K. Jap. Univ.* **25**: 119–188.
Tamarin, A. & Boyde, A. (1977). Facial and visceral arch development in the mouse embryo: A study by scanning electron microscopy. *J. Anat.* **124**: 563–580.
Tassin, M. T. & Weill, R. (1977). Changements de l'épithélium médian des bourgeons palatins de Souris au stade de préfusion. *W. Roux Arch. Dev. Biol.* **181**: 357–365.
Tassin, M. T. & Weill, R. (1980). Scanning electron microscopy study of the medio palatal epithelium: Simultaneous modifications characterizing fusion and degenerescence processes. *W. Roux Arch. Devl Biol.* **188**: 13–21.
Trasler, D. G. (1968). Pathogenesis of cleft lip and its relation to embryonic face shape in A/Jax and C57BL mice. *Teratology* **1**: 33–50.

Trasler, D. G. & Fraser, F. C. (1963). Role of the tongue in producing cleft palate in mice with spontaneous cleft lip. *Devl Biol.* **6**: 45–60.
Trasler, D. G. & Fraser, F. C. (1977). Time-position relationships with particular reference to cleft lip and cleft palate. In *Handbook of teratology*, **2**: 271–292. Wilson, J. G. & Fraser, F. C. (Eds). New York: Plenum Press.
Trasler, D. G., Rearden, C. A. & Hajchgot, H. (1978). A selection experiment for distinct types of 6-amino-nicotinamide induced cleft lip in mice. *Teratology* **18**: 49–54.
Tuan, R. (1980). Calcium transport and related functions in the chorio-allantoic membrane of cultured shell-less chick embryos. *Devl Biol.* **74**: 196–204.
Tuan, R. S. (1983). Supplemented eggshell restores calcium transport in chorio-allantoic membrane of cultured shell-less chick embryos. *J. Embryol. exp. Morph.* **74**: 119–131.
Tyler, M. S. & Koch, W. E. (1975). *In vitro* development of palatal tissue from embryonic mice. 1. Differentiation of the secondary palate from 12 day mouse embryos. *Anat. Rec.* **182**: 297–304.
Tyler, M. S. & Koch, W. E. (1977a). *In vitro* development of palatal tissues from embryonic mice. 11. Tissue isolation and recombination studies. *J. Embryol. exp. Morph.* **38**: 19–36.
Tyler, M. S. & Koch, W. E. (1977b). *In vitro* development of palatal tissues from embryonic mice. 111. Interactions between palatal epithelium and heterotypic oral mesenchyme. *J. Embryol. exp. Morph.* **38**: 37–48.
Tyler, M. S. & Pratt, R. M. (1980). Effect of epidermal growth factor on secondary palatal epithelium *in vitro*: Tissue isolation and recombination studies. *J. Embryol. exp. Morph.* **58**: 93–106.
Voeltzkow, A. (1899). Beiträge zur Entwicklungsgeschichte der Reptilien. I. Biologie und Entwicklung der äusseren Körperform von *Crocodilus madagascariensis* Grand. *Abh. senckenb. naturforsch. Ges.* **26**: 1–150.
Warren, J. W. (1963). *Growth zones in the skeleton of recent and fossil vertebrates*. Ph.D. Thesis: University of California, Los Angeles, USA.
Waterman, R. E. & Meller, S. M. (1974). Alterations in the epithelial surface of human palatal shelves prior to and during fusion: a scanning electron microscope study. *Anat. Rec.* **180**: 111–136.
Waterman, R. E., Ross, L. M. & Meller, S. M. (1973). Alterations in the epithelial surface of A-Jax mouse palatal shelves prior to and during palatal fusion: A scanning electron microsope study. *Anat. Rec.* **176**: 361–376.
Wee, E. L., Wolfson, L. G. & Zimmerman, E. F. (1976). Palate shelf movement in mouse embryo culture: evidence for skeletal and smooth muscle contractility. *Devl Biol.* **48**: 91–103.
Wee, E. L. & Zimmerman, E. F. (1980). Palate morphogenesis. II Contraction of cytoplasmic processes in ATP-induced palate rotation in glycerinated mouse heads. *Teratology* **21**: 15–27.
Weston, J. A. (1970). The migration and differentiation of neural crest cells. *Adv. Morph.* **8**: 41–114.
Weston, J. A., Derby, M. A. & Pintar, J. E. (1978). Changes in the extracellular environment of neural crest cells during their early migration. *Zoon* **6**: 103–113.
Wettstein, O. V. (1954). Crocodilia (2). In *Handbuch der Zoologie* **7**(1): 321–424. Kükenthal, W., Krumbach, T., Helmcke, J. G. & Langerken, H. V. (Eds). Berlin: De Gruyter.
Wilk, A. L., King, C. T. G. & Pratt, R. M. (1978). Chlorcyclizine induction of cleft palate in the rat: Degradation of palatal glycosaminoglycans. *Teratology* **18**: 199–210.

Yander, G. & Searls, R. L. (1980). A scanning electron microscopic study of the development of the shoulder, visceral arches and the region ventral to the cervical somites of the chick embryo. *Am. J. Anat.* **157**: 27–39.

Yee, G. W. (1976). *Facial development in mutant and normal chick embryos.* Ph.D. Thesis: University of California, Davis, USA. Published by University Microfilms Ltd.

Yee, G. W. & Abbott, U. K. (1978). Facial development in normal and mutant chick embryos. 1. Scanning electron microscopy of primary palate formation. *J. exp. Zool.* **206**: 307–322.

Zimmerman, E. F., Wee, E. L., Clark, R. L. & Venkatasubramanian, K. (1980). Neurotransmitter and teratogen involvement in cell mediated palatal elevation. In *Current research trends in prenatal craniofacial development*: 187–202. Pratt, R. M. & Christiansen, R. L. (Eds). New York: Elsevier/North Holland.

Amelogenesis in Reptilia: Evolutionary Aspects of Enamel Gene Products

HAROLD C. SLAVKIN, MARGARITA ZEICHNER-DAVID, MALCOLM L. SNEAD, EDWARD E. GRAHAM, NELSON SAMUEL and MARK W. J. FERGUSON†

Laboratory for Developmental Biology, Graduate Program in Craniofacial Biology, Graduate School, and Department of Basic Sciences, School of Dentistry, University of Southern California, Los Angeles, California 90089-0191, USA; and † Department of Anatomy, The Queen's University of Belfast, Medical Biology Centre, 97 Lisburn Road, Belfast BT9 7BL, Northern Ireland

SYNOPSIS

Evidence has accumulated which suggests that the developmental processes associated with epithelial differentiation and subsequent enamel production appear to be homologous among various vertebrate classes. This new evidence follows nearly 100 years of controversy regarding the histogenesis and evolutionary implications of the outermost extracellular matrix layer covering lower and higher vertebrate teeth. Enamel production appears to be homologous among the classes Chondrichthyes (e.g. shark), Teleostei (e.g. ballan wrasse, common eel, pike), and Mammalia (e.g. mouse, rat, rabbit, hamster, pig, cow, monkey and man) (Hertwig, 1874; Poole, 1967; Moss, 1977; Piesco, 1979; Slavkin, Zeichner-David & Siddiqui, 1981). Recently, Herold and his colleagues prepared antisera in rabbit produced against fetal bovine enamel proteins. They demonstrated the localization of cross-reactive antigenic determinants within the developing enamel extracellular matrices of Chondrichthyes, Teleostei, Reptilia and Mammalia (Herold, Graver & Christner, 1980). Our laboratory has confirmed and extended these studies by using (i) specific polyclonal antibodies directed against purified enamelins and amelogenins; (ii) biochemical characterizations of enamel proteins isolated from selected vertebrate species (including hagfish, spiny dogfish, alligator, mouse, rabbit and man); and (iii) methods from recombinant DNA technology which characterize features of structural gene homology and/or divergence among selected vertebrate enamel-bearing creatures. On the basis of these investigations, we suggest that the developmental processes associated with enamel gene product formation (i.e. amelogenesis) throughout the subphylum Vertebrata are homologous. The enamel proteins are highly conserved during vertebrate evolution, over the period of almost 500 My of history, as represented by the extant groups studied. At least two major families of structural gene products have been identified – amelogenins and enamelins. From the present evidence, it would appear that the enamelins are predominant in the aquatic vertebrates (e.g. hagfish, sharks, teleosts), whereas in terrestrial vertebrates (e.g. American alligator, mouse, hamster, pig, cow and man), the enamelins are detected in only low proportions relative to amelogenins. The amelogenins predominate during enamel extracellular matrix formation in higher vertebrates. We propose that amelogenins became the dominant protein class associated with enamel formation during the evolution of Reptilia.

INTRODUCTION

The mechanism by which eucaryotes co-ordinately express unique sets of structural genes is not as yet known. To understand the fundamental processes of differential structural gene expression during epidermal organ development has become of paramount importance in cellular, molecular and developmental biology. Several examples of epidermal organogenesis in which a number of structural genes are co-ordinately expressed both in time and within the same cell type are currently under investigation. Of particular interest in our laboratory are the questions as to how extracellular instructions are received and perceived by a potentially responsive cell type, and how these epigenetic instructions actually initiate *de novo* transcription of unique sets of structural genes during epithelial-mesenchymal interactions in the developing vertebrate tooth organ.

The enamel extracellular matrix genes within vertebrate tooth organs represent a system within which a relatively small group of structural genes are expressed in the enamel organ epithelia during craniofacial development. Two major classes of enamel proteins have been identified in a number of different mammalian species: (i) *enamelins*, which are acidic glycoproteins ranging in molecular weight from 70 to 80 Kd (kilodaltons), appear to contain high levels of serine, glycine, aspartic acid and arginine, and appear to be associated with the mineral phase of enamel; and (ii) *amelogenins*, which are hydrophobic glycoproteins ranging in molecular weight from 20 to 50 Kd, contain characteristically high levels of glutamic acid, proline, leucine and histidine, and are readily soluble in either acetic acid or guanidine hydrochloride solutions (Eastoe, 1963, 1965; Fukae & Shimizu, 1974; Termine, Torchia & Conn, 1979; Termine, Belcourt, Christner, Conn & Nylen, 1980; Fincham, 1980; Slavkin, Zeichner-David & Siddiqui, 1981). Available evidence indicates that both enamelins and amelogenins are synthesized and secreted from ameloblasts at the same time during odontogenic development (Lyaruu, Belcourt, Fincham & Termine, 1982; Zeichner-David, Slavkin, Lyaruu & Termine, 1983).

Enamel proteins are involved with the organization and function of all vertebrate enamel extracellular matrices. The enamel covering of vertebrate tooth organs and the physical and functional properties of this enamel, depend directly upon the chemical and organizational characteristics of the enamel proteins. It seems plausible that functionally distinct structures may require different proteins or different patterns of protein organization. In addition to providing phylogenetic information on enamel protein evolution, our studies attempt to establish molecular events associated with critical aspects of enamel morphological evolution – the development of new or different structures in different taxonomic groups, and the production of multiple enamel proteins. Comparisons of enamel proteins present in various vertebrate enamel matrices may demonstrate the extent to which the molecular evolution of enamel gene products is related to the morphological differentiation of the enamel extracellular matrix. This provides a model for the relationship between molecular and morphological adaptations in epidermal organ systems where extracellular matrix proteins are directly involved in the construction and evolution of functionally unique morphological structures.

SURVEY OF AMELOGENESIS IN SELECTED VERTEBRATE SPECIES

Developmental and Morphological Features

A number of excellent reviews regarding vertebrate odontogenesis are recommended (Hertwig, 1874; Gaunt, 1955; Gaunt & Miles, 1967; Reith, 1967, 1970; Peyer, 1968; Boyde, 1971; Koch, 1972; Kollar, 1972, 1978; Osborn, 1973; Mörnstad, 1974; Slavkin, 1974; Ruch & Karcher-Djuricic, 1975; Shellis, 1975, 1978; Moss, 1977; Slavkin, Trump, Schonfeld, Brownell, Sorgente & Lee-Own, 1977; Piesco, 1979; Thesleff & Hurmerinta, 1981). Several salient features are required to establish a general orientation for the process of tooth formation during vertebrate embryogenesis. In general, the initiation of tooth development begins with heterologous cell interactions between cranial neural crest-derived ectomesenchyme cells (within the forming maxillary and mandibular processes) and adjacent ectodermally-derived oral epithelia. These initial epithelial-mesenchymal interactions take place during the period of embryogenesis, and are precise with respect to developmental time, as well as to positional information within the forming maxilla and mandible. Of the many intercellular processes associated with these early stages of morphogenesis, oral epithelial cells in discrete regions appear to receive putative "instructions" from adjacent ectomesenchyme which result in significant epithelial cell proliferation and concomitant epithelial ingrowth into the adjacent ectomesenchyme region. Oral epithelial ingrowth results in the formation of a dental lamina which provides the progenitor epithelial component for subsequent odontogenesis. The timing and positional information features of this series of processes adheres to general embryological principles of "anteroposterior" gradients; anterior dental anlage form before posterior dental anlage (Gaunt & Miles, 1967; Kollar & Baird, 1969, 1970; Koch, 1972; Kollar, 1972, 1978; Osborn, 1973; Slavkin, 1974; Thesleff & Hurmerinta, 1981).

Following the dental lamina stage of tooth formation, a series of development stages are identified from bud, cap, bell through crown stages of odontogenesis (see Table I). Since detailed discussions of these developmental stages of odontogenesis are available in the literature (see earlier review citations), only a few critical features are now described. First, regional specification for the particular type or form of tooth organ (e.g. conical, incisor, canine, molar, etc.) appears to be determined by the dental papilla ectomesenchyme during the cap stage of odontogenesis (Kollar & Baird, 1969, 1970). All available literature indicates an inductive role of the dental ectomesenchyme in determining tooth organ form (see critical discussions by Huggins, McCarroll & Dahlberg, 1934; Koch, 1967, 1972; Kollar, 1972, 1978; Slavkin, 1974; Ruch & Karcher-Djuricic, 1975; Thesleff & Hurmerinta, 1981; Ruch, Lesot, Karcher-Djuricic, Meyer & Olive, 1982).

Recently, the inductive potential of the dental papilla ectomesenchyme was reported by Kollar & Fischer (1980) using heterotypic tissue recombinations between two different vertebrate species. Kollar & Fisher (1980) state that cap stage mouse dental papilla ectomesenchyme, when recombined with chick pharyngeal epithelium (modern birds do not express tooth formation), resulted in tooth organ formation which expressed both dentinogenesis (i.e. the differentation of odontoblast cells from the murine ectomesenchyme with the production of dentine extracellular matrix) and amelogenesis (i.e. the differentiations of ameloblast cells from the avian epithelium with the production of enamel extracellular matrix). Their intriguing evidence was interpreted to indicate that quiescent structural genes within the avian genome were activated and expressed as directed by the dental papilla ectomesenchyme. Comparable evidence has been reported for a number of different heterotypic epithelial-mesenchymal interactions between vertebrate species (e.g. reptile versus avian versus mammalian integument as discussed by Saxen, Karkinen-Jaaskelainen, Lehtoneu, Nordling & Wartiovaara, 1976; and Sengel & Dhouailly, 1977). It is

TABLE I
Comparative developmental patterns of tooth morphogenesis in selected vertebrates

	Developmental stages				Age (days)			
Developmental characteristics	Mouse (Theiler)[a]	Hamster	Rabbit	Man (Horizon)[b]	Mouse	Hamster	Rabbit	Man
Implantation	6	—	—	IV	4.5	4.5	4	7
Neural plate	11	—	—	IX	7.5	—	—	19
Forelimb bud	15	—	—	XII	9.5	8.5	9.5	26
Hind-limb bud	16	—	—	XIII	10	—	—	28
Dental lamina (molar)	16	—	—	XVII	10.5	10.5	12	45
Enamel organ bud (molar)		—	—	XX	13	12.5	15	53
Cap stage (molar)	24	—	—	XXIII	16	13.5	20	7 weeks
Bell stage (initial dentinogenesis)	25	—	—	—	19	Birth	23	13 weeks
Amelogenesis	—	—	—	—	Birth	2	24	18 weeks
Birth	—	—	—		19–21	16	32	36 weeks

[a]Theiler (1972).
[b]Slavkin (1979).

established that regional dermal or ectomesenchymal specificity determines "form" when combined with various sources of ectodermally-derived epithelia. However, the molecular expression by the epithelia reflects the genetic limitations of the specific epithelial genome, and is not acquired as a consequence of the heterotypic tissue interactions (Sengel & Dhouailly, 1977). Therefore, in the case described by Kollar & Fischer (1980) between chick epithelia and mouse dental mesenchyme, avian genomic DNA allegedly contains the structural genes for enamel proteins and expresses these enamel genes under the direction of dental ectomesenchyme-derived "instructions". Additional aspects of this intriguing feature of tooth development will be discussed on p. 295.

Following the cap stage of tooth formation, ectomesenchymal cells along the perimeter of the dental papilla ectomesenchyme become polarized, and differentiate into merocrine-type, secretory odontoblasts which produce the dentine extracellular matrix. The process of dentinogenesis precedes the differentiation of inner enamel epithelial cells into highly polarized, non-dividing, secretory ameloblasts which produce the enamel extracellular matrix (see recent review by Slavkin, Zeichner-David & Siddiqui, 1981). Ultrastructural bodies in selected higher vertebrates, from cap through bell and crown stages of odontogenesis, demonstrate that ectomesenchyme cells extend long cellular processes towards the undersurface (i.e. lamina diffusa) of the epithelial basal lamina. Evidence has been reported which indicates basal lamina removal, allegedly by enzymatic-mediated processes, and the formation of contacts between pre-odontoblast cellular processes and the outer surfaces of inner enamel epithelial plasma membranes, *prior to enamel matrix production* (Reith, 1967, 1970; Kallenbach, 1971, 1976; Slavkin & Bringas, 1976; Brownell, Bessem & Slavkin, 1981; Cummings, Bringas, Grodin & Slavkin, 1981). In contrast, however, this process of basal lamina degradation and direct epithelial-mesenchymal cell contact was not observed during selachian odontogenesis (Garant, 1970; Fosse, Risnes & Holmbakken, 1974; Mörnstad, 1974; Kerebel, Daculsi & Renaudin, 1977; Piesco, 1979; Nanci, Bringas, Samuel & Slavkin, 1983; Samuel, Nancy, Bringas, Santos & Slavkin, 1983).

The process of epithelial differentiation into functional ameloblasts with the production of enamel matrix is termed "amelogenesis". In most vertebrates studied, ameloblast differentiation follows the paradigm of merocrine-type, secretory cell differentiation, e.g. the nucleus is polarized towards the basal region of the cytoplasm. Ameloblast cells are highly polarized and appear as tall columnar epithelia, often having lengths of $40-60\mu$m. The secretion of enamel matrix constituents from ameloblasts appears to follow the outline of the perimeter of ameloblast Tomes processes (Boyde, 1971, 1978). Further, each mammalian species has a species-specific, cross-sectional pattern for secretory ameloblast cells. This genetically controlled pattern (e.g. polygon-shaped, hexagon-shaped, etc.) is also evident in the forming enamel prisms during mineralization of the enamel matrix (Boyde, 1971, 1978; Greenberg, Bringas & Slavkin, 1983). In general, the process of amelogenesis reflects a supramolecular phylogeny which can be identified during the late bell stages of odontogenesis in all vertebrates expressing prismatic enamel formation.

Biochemical Features of Enamel Matrix Proteins

Studies of the biosynthesis of enamel matrix proteins, using high resolution light microscopy autoradiography as well as transmission electron microscopy with autoradiography, indicate that enamel matrix proteins are glycosylated polypeptides which are synthesized and secreted from ameloblasts following the principles of merocrine-type, secretory processes (see Greulich & Slavkin, 1965; Weinstock, 1972; Slavkin, 1974; Shellis, 1975, 1978; Slavkin, Mino & Bringas, 1976). Biochemical studies of enamel protein biosynthesis also indicate merocrine-type, secretory processes (see Guenther, Croissant, Schonfeld & Slavkin, 1977; Slavkin, Trump et al., 1977; Slavkin, Weliky, Stellar, Slavkin, Zeichner Gancz, Bringas, Hyatt-Fischer, Shimizu & Fukae, 1979; Fukae, Tanabe, Ijiri & Masharu, 1980; Slavkin, Zeichner-David, MacDougall, Bessem, Bringas, Honig, Lussky & Vides, 1982; Slavkin, Zeichner-David, MacDougall, Bringas, Bessem & Honig, 1982; Slavkin, Zeichner-David & Siddiqui, 1982; Lyaruu et al., 1982; Zeichner-David, Slavkin et al., 1983). The evidence suggests that enamel protein polypeptides are initially synthesized on polysomes associated with the rough endoplasmic reticulum, are glycosylated in the Golgi apparatus via glycosyltransferases, packaged into secretory vesicles, phosphorylated via phosphokinases, secreted, and then degraded by extracellular enamel peptidases (e.g. serine esterases). This suggested sequence results in a large number of polypeptide fragments which can be isolated from various stages of enamel matrix formation and maturation. During the last two decades, a significant body of knowledge has been obtained related to the chemical and physical properties of enamel matrix proteins isolated from various stages of enamel matrix maturation (see Glimcher, 1961, 1979; Eastoe, 1963, 1965; Levine, Glimcher, Seyer, Huddleston & Hein, 1966; Poole, 1967, 1971; Eggert, Allen & Burgess, 1973; Elwood & Apostolopoulos, 1975).

Recently, Termine and his colleagues developed a novel approach for the study of enamel matrix protein chemistry, and also provided a classification with which to categorize enamel matrix proteins (see

Termine, Torchia et al., 1979; Termine, Belcourt et al., 1980). The approach included the use of a sequential dissociative extraction scheme, first using guanidine hydrochloride with protease inhibitors, and then re-extracting the remaining residue with guanidine hydrochloride plus EDTA. This sequential scheme resulted in two different classes of enamel matrix proteins based upon their relative solubility under these experimental conditions. The enamel proteins which were made solute by demineralization were the "enamelins" (Termine, Torchia et al., 1979). The amelogenins extracted by this scheme were identical to the class of enamel proteins first identified as amelogenins by Eastoe (1963, 1965). Therefore, in a number of species two classes of enamel proteins have been identified as being either enamelins: relatively high molecular weight acidic glycoproteins of approximately 72 Kd; or amelogenins: hydrophobic and proline-rich glycoproteins of approximately 30 Kd (Termine, Belcourt et al., 1980; Belcourt, Fincham & Termine, 1982; Fincham, Belcourt, Lyaruu & Termine, 1982; Fincham, Belcourt & Termine, 1982; Lyaruu et al., 1982). During the intitial process of enamel extracellular matrix formation, approximately 90% of total matrix proteins are represented by the amelogenins (Termine, Belcourt et al., 1980; Slavkin, Zeichner-David & Siddiqui, 1981; Fincham, Belcourt, Lyaruu et al., 1982). Enamel proteins have been extracted using either acetic acid solutions or sequential dissociation extraction using guanidine hydrochloride. Amino acid compositions of enameloid or enamel matrix proteins from selected vertebrate species are shown in Table II. In addition, the partial N-terminal amino acid sequences of amelogenins have been reported for a number of mammalian species (Glimcher, 1979; Fukae et al., 1980; Zalut, Henzel & Harris, 1980; Fincham, Belcourt, Termine, Butler & Cothran, 1981) and are summarized in Table III.

AMELOGENESIS IN THE AMERICAN ALLIGATOR

Developmental and Morphological Features

One of the most fascinating orders within the Reptilian class are the crocodilians. The crocodilians belong to the Archosauria subclass and are assumed to have evolved during the Triassic era, approximately 230 My ago. One very interesting aspect, ironically not receiving a significant amount of interest in the past, is the reproductive biology and embryology of the crocodilians. Selected highlights have recently been presented by Ferguson (1981, in press).

Of the three subfamilies of crocodilians, we have become interested in the American alligator (*Alligator mississippiensis*). According to Ferguson

TABLE II
Amino acid composition of enameloid or enamel matrix proteins from selected vertebrate species

Amino acid[a]	Hagfish[b]	Shark[c]	Alligator[d]	Rabbit[e]	Mouse[f]	Hamster[g]	Sheep[h]	Cow[i]	Pig[j]	Monkey[k]	Man[l]
Aspartic acid	89	148	55	85	93	52	58	38	28	28	34
Threonine	53	44	35	43	48	39	34	32	32	42	31
Serine	105	197	67	75	150	67	74	40	48	50	55
Glutamic acid	100	161	153	122	137	160	175	202	196	156	161
Proline	8	56	208	72	135	159	262	248	242	264	247
Glycine	110	174	78	127	84	77	70	50	51	55	66
Alanine	64	54	32	79	62	60	51	32	26	21	15
Valine	53	23	34	36	53	45	50	38	37	42	36
Methionine	23	4	13	15	30	35	30	41	46	51	37
Isoleucine	47	19	34	40	32	35	37	33	37	25	32
Leucine	98	33	77	61	81	85	100	92	94	92	92
Tyrosine	30	11	55	17	22	31	27	26	32	52	71
Phenylalanine	20	9	31	26	24	26	26	26	23	16	22
Histidine	14	32	41	15	35	58	33	16	11	66	59
Lysine	122	21	61	52	27	40	48	63	87	17	19
Arginine	62	19	27	27	13	27	25	18	10	16	23

[a] Results are expressed as residues 1000^{-1}.
[b] *Eptatretus stoutii*. Total acetic acid extraction.
[c] Hammerhead (Levine, Glimcher et al., 1966).
[d] *Alligator mississippiensis*. Total acetic acid extraction.
[e] *Oryctolagus cuniculus*. Total acetic acid extraction.
[f] *Mus musculus*. Total acetic acid extraction.
[g] *Cricetus auratus*.
[h] *Ovis aries*.
[i] *Bos taurus*.
[j] *Sus scrofa* (Fincham, Belcourt, Lyaruu et al., 1982).
[k] *Macaca mulatta* (Levine, Seyer, Huddleston & Glimcher, 1967).
[l] *Homo sapiens* (Fincham, Belcourt, Lyaruu et al., 1982).

TABLE III
Amino acid sequence of the amino terminal region of porcine and bovine amelogenins

Amelogenin	Amino acid sequence																			
	1	2	3	4	5	6	7	8	9	10	11	12	13	14	15	16	17	18	19	20
Porcine 2a[1]	MET-	PRO-	LEU-	PRO-	PRO-	HIS-	PRO-	GLY-	HIS-	PRO-	GLY-	TYR-	ILE-	ASP-	PHE-	SER-	TYP-	GLU-	VAL-	LEU-
Porcine 2b[1]	MET-	PRO-	LEU-	PRO-	PRO-	HIS-	PRO-	GLY-	HIS-	PRO-	GLY-	TYR-	ILE-	ASP-	PHE-	SER-	TYR-	GLU-	VAL-	LEU-
Bovine TRAP[2]	MET-	PRO-	LEU-	PRO-	PRO-	HIS-	PRO-	GLY-	HIS-	PRO-	GLY-	TYR-	ILE-	ASN-	PHE-	SER-	TYR-	GLU-	VAL-	LEU-
Bovine LRAP[2]	MET-	PRO-	LEU-	PRO-	PRO-	HIS-	PRO-	GLY-	HIS-	PRO-	GLY-	TYR-	ILE-	ASN-	PHE-	SER-	TYR-	GLU-	VAL-	LEU-

	21	22	23	24	25	26	27	28	29	30	31	32	33	34	35	36	37	38	39	40
Porcine 2a[1]	THR-	PRO-	LEU-	LYS-	X-	TYR-	GLU-	ASP-	MET-	ILE-	X-	HIS-	PRO-	TYR-	THR-	SER-	TYR-	GLY-	THR-	GLU-
Porcine 2b[1]	THR-	PRO-	LEU-	X-	X-	TYR-	GLU-	ASP-	MET-	ILE-	X-	HIS-	PRO-	TYR-	THR-	SER-	TYR-	GLY-	THR-	GLU-
Bovine TRAP[2]	THR-	PRO-	LEU-	LYS-	TRP-	TYR-	GLU-	ASP-	MET-	ILE-	ARC-	HIS-	PRO-	TYR-	SER-	PRO-	TYR-	GLY-	TYR-	GLU-
Bovine LRAP[2]	THR-	PRO-	LEU-	LYS-	TRP-	TYR-	GLU-	ASP-	MET-	ILE-	ARG-	HIS-	PRO-	PRO-	LEU-	pro-	PRO-	MET-	LEU-	PRO-

	41	42	43	44	45	46	47	48	49	50	51	52	53	54
Porcine 2a[1]	PRO-	MET-	GLY-	GLY-	X-	LEU-	HIS-	HIS-	GLU-	ILE-	ILE-	PRO-	VAL-	VAL
Porcine 2b[1]	PRO-	MET-	GLY-	GLY-	X-	LEU-	HIS-	HIS-	GLU-	ILE-	ILE-	PRO-	VAL-	VAL
Bovine TRAP[2]	PRO-	MET-	GLY-	GLY-	TRP-	ALA								
Bovine LRAP[2]	ASP-	LEU-	PRO-	LEU-	glu-									

[1]From Fukae et al. (1980). Fragments 2a and 2b are peptides obtained by cleavage with clostridiopeptidase B of amelogenins 26 Kd and 21 Kd respectively from porcine immature enamel.
[2]Fincham, Belcourt, Termine et al. (1981). Fetal bovine amelogenins rich in tyrosine (TRAP) or leucine (LRAP).

(1981, in press) the Crocodilia have undergone very few major changes in morphological features during their 230 My evolution and their embryogenesis, development and structure should reflect on that of the ancestral thecodonts. The thecodonts appear to have given rise to the now extinct ornithischian and saurischian dinosaurs, pterosaurs (flying reptiles), and modern birds (see discussion in Walker, 1972; Ferguson, 1981).

Alligator mississippiensis possesses maxillary and mandibular tooth organs arranged in a pseudoheterodont pattern (Ferguson, 1981). The teeth are continuously replaced throughout the life span of the animal. Alligator teeth exhibit a thecodont gomphosis mode of attachment. A major feature of the development of tooth organs in the American alligator is the striking similarity to mammalian odontogenesis [see Figs 1(A)–(D) and 2(A)–(C)].

The developmental stages of odontogenesis in mammals were observed in the American alligator. The sequence of developmental processes is homologous to those described in mammalian species. Figure 1(A)–(D) illustrates early and late cap stages, as well as early and late bell stages of odontogenesis in the developing American alligator. In Fig. 2(A)–(C) a series of transmission electron photomicrographs illustrates the apical region of the inner enamel epithelia during the process of ameloblast differentiation and enamel extracellular matrix production. Prior to overt differentiation, inner enamel epithelia possess a continuous basal lamina [Fig. 2(A)]. During the transition phases in this process of epithelial differentiation, pre-odontoblast cellular processes extend through the forming extracellular matrix (i.e. progenitor predentine) and make direct contacts with the undersurface of the differentiating inner enamel epithelium [see Fig. 2(B)]. At this stage the basal lamina is discontinuous and "pores" or spaces were observed [arrow, Fig. 2(B)]. Thereafter, inner enamel epithelial cells become tall columnar, non-dividing, secretory ameloblasts and produce enamel matrix constituents [see Fig. 2(C)]. Discrete apatite crystal formation is evident [Fig. 2(C), arrows indicate apatite crystal formation]. The sequence of inner enamel epithelial differentiation into ameloblasts with the production of enamel matrix is homologous with that observed in selected mammalian species (see Gaunt & Miles, 1967; Kallenbach, 1971, 1976; Weinstock, 1972; Moss, 1977; Fearnhead, 1979; Piesco, 1979; Slavkin, Zeichner-David & Siddiqui, 1981).

Preliminary Biochemical Features

Preliminary biochemical analysis of enamel extracellular matrices during alligator tooth development, indicates that both enamelins and

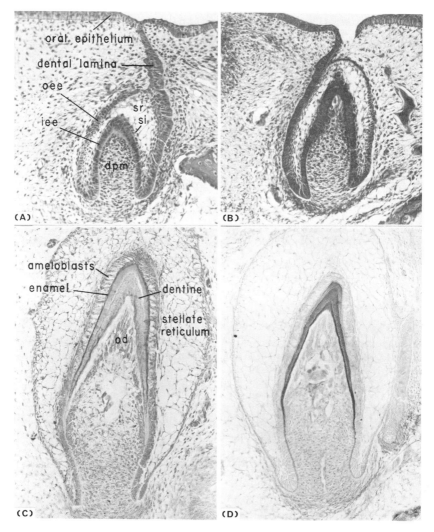

FIG. 1. Developmental and morphological features of American alligator odontogenesis. (A) Early cap stage. (B) Late cap stage. (C) Bell stage. (D) Late bell stage. Abbreviations: oee, outer enamel epithelia; iee, inner enamel epithelia; dpm, dental papilla mesenchyme; si, stratum intermedium; sr, stratum reticulum; od, odontoblasts. × 680.

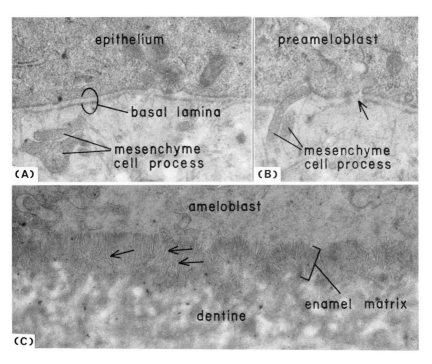

FIG. 2. Ultrastructural features of inner enamel epithelial differentiation into secretory ameloblasts during American alligator odontogenesis. (A). The basal lamina is continuous along the undersurface of the inner enamel epithelia during the "predifferentiative" phase of ameloblast differentiations. (B) Direct mesenchyme cell process/pre-ameloblast cell contacts form in discrete regions devoid of basal lamina, suggesting an enzyme-mediated removal of the basal lamina with resulting "pores" (arrow). (C) Secretory ameloblast apical region illustrates the secretion of enamel matrix constituents and the formation and growth of apatite crystals (arrows). Enamel matrix formation in the alligator tooth organ appears homologous with mammalian amelogenesis. × 10 400.

amelogenins are found in this species of reptile (see Fig. 3). The relative distribution of enamelin to amelogenin polypeptides in the alligator was similar to that observed for mammalian species including mouse, hamster and man (Fig. 3). The amelogenins in alligator enamel appear to consist of three polypeptides ranging in molecular weight from 36 Kd (kilodaltons) to 31 Kd. The amelogenins identified in mouse, hamster and man appear to have molecular weights below 30 Kd. In contrast, aquatic lower vertebrate species, such as hagfish and shark, do not contain amelogenins. The enameloid in these lower vertebrates appears to contain enamelins.

In order to study the biosynthesis of enamel proteins during alligator tooth organ development, we employed short-term radiolabeling studies

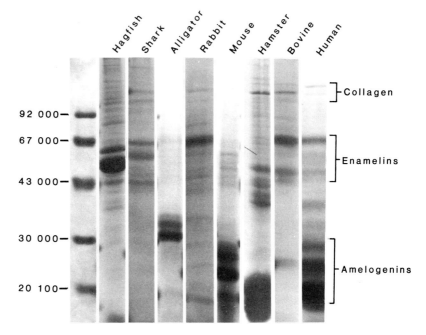

FIG. 3. Comparison of enameloid and enamel matrix proteins representing a number of aquatic as well as terrestrial vertebrate species. Equivalent concentrations of odontogenic extracellular matrix proteins were fractionated using sodium dodecylsulfate/urea 10% polyacrylamide gel electrophoresis (see Ornstein, 1964; Zeichner-David, Weliky & Slavkin, 1980). Amelogenins were not evident in either cyclostomes or elasmobranchs, whereas enamelins and amelogenins were detected in Reptilia and Mammalia. Note that lagomorphs (i.e. New Zealand white rabbits) contain enamelins but do not appear to contain amelogenins. Molecular weight markers are indicated in the range of 92–20 kilodaltons.

using [^{35}S]-methionine as an isotopic precursor during *in vitro* experiments. In Fig. 4 we report a comparison of the incorporation of [^{35}S]-methionine into the enamel organ epithelial-specific proteins from alligator, rabbit, mouse and hamster. Whereas alligator, mouse and hamster synthesize and secrete enamelins and amelogenins, rabbit tooth organs *in vitro* appear to synthesize and secrete only enamelins ranging in molecular weight from 67 to 43 Kd.

SURVEY OF IMMUNOLOGICAL DETERMINANTS OF ENAMEL PROTEINS DURING EVOLUTION

The first evidence which indicated the antigenicity of enamel proteins was reported by Nikiforuk & Gruca (1971). They showed that fetal bovine amelogenin was antigenic in rabbit; purified IgG fractions of the

COMPARISON OF ^{35}S-METHIONINE PROTEINS
FROM DIFFERENT VERTEBRATES

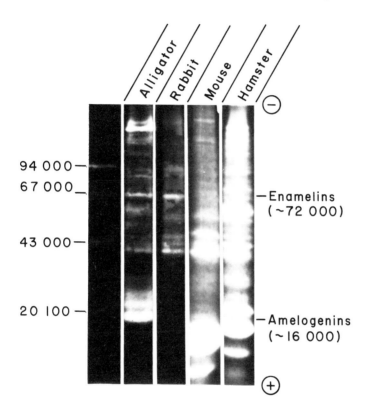

FIG. 4. Comparison of [^{35}S] methionine metabolically labeled enamel organ epithelial-specific proteins from selected vertebrates. Alligator, mouse and hamster tooth organs synthesize and secrete enamelins and amelogenins both *in vivo* and *in vitro*. Only enamelins appeared to be sythnesized in the lagomorph example. Following fractionation of radiolabeled proteins using electrophoresis, polyacrylamide gels were processed for fluorography for the detection of radioactivity. The units on the figure are molecular weight.

rabbit antiserum when reacted with amelogenin in a double-diffusion agar plate formed a single precipitin line. Elwood & Apostolopoulos (1975) reported on the antigenic properties of rat amelogenins. In the late 1970s, a number of studies were published indicating several different and interesting features of enamel antigenicity (Guenther *et al.*, 1977; Graver, Herold, Chung, Christner, Pappas & Rosenbloom, 1978; Herold, Christner, Graver & Rosenbloom, 1979; Hyatt-Fischer, Chrispens, O'Keefe & Slavkin, 1979; Schonfeld, 1979). Based upon these studies, two very significant results were published in 1980. First, polyclonal antibodies made in rabbits against purified fetal bovine amelogenins were found to be antigenically cross-reactive with purified fetal bovine enamelins (Termine, Belcourt et al., 1980). Second, and very pertinent to the exposition of this discussion, polyclonal antibodies for purified fetal bovine amelogenins were antigenically cross-reactive with the enameloid of teeth and dermal denticles of Chondrichthyes, the enameloid of Teleostei and Amphibia, the enamel of Reptilia and a number of species within Mammalia (Herold, Graver et al., 1980).

Mouse Amelogenins

During mouse tooth development the secretory ameloblasts differentiate and produce enamelin acidic glycoproteins and hydrophobic amelogenin glycoproteins which function in the formation of the extracellular enamel and organic matrix. Amelogenin-enriched fractions were isolated from molar and incisor tooth organs during early post-natal enamel matrix production using an extraction procedure with acetic acid. Major polypeptides were separated by sodium dodecylsulfate/polyacrylamide gel electrophoresis, amelogenin polypeptide bands (c. 16–20 Kd molecular weight) were excised, amelogenins were eluted from the gel bands, and then these purified proteins were used as immunogens for immunization of rabbits (Slavkin, Zeichner-David, MacDougall, Bringas et al., 1982). Rabbit antibodies raised against amelogenins showed extensive cross-reactivity with lower molecular weight amelogenin polypeptides of 16–20 Kd. These polyclonal antibodies (IgG fraction) were used with indirect immunofluorescence microscopy to show the specific localization of amelogenins during the synthesis and secretion of the enamel matrix in situ as well as in vitro (Slavkin, Zeichner-David, MacDougall, Bringas et al., 1982). No cross-reactivity was detected in keratinized or non-keratinized epidermal tissues, dental papilla mesenchyme, odontoblasts, dentine matrix, cartilage, bone, heart, brain, lung or salivary glands (Slavkin, Zeichner-David, Ferguson, Termine, Graham, MacDougall, Bringas, Bessem & Grodin, 1982; Slavkin, Zeichner-David, MacDougall, Bessem et al., 1982; Slavkin, Zeichner-David, MacDougall, Bringas et al., 1982).

During studies designed to investigate the biosynthesis of enamel proteins, we observed that polyclonal antibodies against purified mouse amelogenins were also cross-reactive with antigenic determinants of enamelins (Slavkin, Zeichner-David, MacDougall, Bessem et al., 1982; Slavkin, Zeichner-David, MacDougall, Bringas et al., 1982). During mouse tooth development, secretory ameloblasts synthesize and secrete both enamelins and amelogenins, and both enamel polypeptides appear to share antigenic determinants which cross-react with anti-amelogenin antibodies. This immunologic relatedness between enamelin and amelogenin antigenic sites has also been reported for fetal bovine enamel matrix proteins (Termine, Belcourt et al., 1980).

Rabbit Enamelins

The major fetal rabbit enamel protein has a molecular weight of 70 Kd (Guenther et al., 1977; Slavkin, Weliky et al., 1979; Zeichner-David, Weliky & Slavkin, 1980). The amino acid composition of this 70 Kd glycoprotein appears homologous with that reported for fetal bovine enamelin (Termine, Belcourt et al., 1980; Slavkin, Zeichner-David & Siddiqui, 1982). Recently we isolated fetal rabbit enamelin using sodium dodecylsulfate/urea 10% polyacrylamide gel electrophoresis, cut bands containing enamelin from the gels, and used this purified protein as an immunogen for subsequent antibody production (Slavkin, Zeichner-David & Siddiqui, 1982). Polyclonal antibodies directed against fetal rabbit enamelins were found to be cross-reactive with both enamelins as well as with amelogenins.

The immunogenicity of enamel proteins has recently been reviewed (Schonfeld, 1979). Of particular interest is the observation that allo-antibodies against fetal enamel proteins can be produced by injecting young adult female New Zealand white rabbits with fetal rabbit enamel immunogens (see discussions in Guenther et al., 1977; Hyatt-Fischer et al., 1979; Schonfeld, 1979; Slavkin, Zeichner-David & Siddiqui, 1981, 1982). Rabbit anti-fetal rabbit enamel protein antisera have been produced using the procedures published by Gomer & Lazarides (1981). The specificity of this polyclonal antibody against fetal enamel proteins (i.e. enamelins and amelogenins) has been determined using indirect immunofluorescent microscopy, immunoprecipitation assays, immuno-electrophoresis, and micro-ELISA assays. The polyclonal antibody is cross-reactive with enamel matrix proteins. No cross-reactivity has been detected using a number of heterologous fetal and adult tissues, and a number of different antigens including albumin, keratins, collagens and fibronectins.

Indirect Immunohistochemical Localization of Enamel Protein Antigens in Selected Vertebrate Species

The sensitivity of immunohistochemical methods is described as the lowest detectable concentration of one or more tissue antigens. Therefore, under laboratory conditions, sensitivity is represented by the lowest staining intensity that can be distinguished from background.

The confidence with which such a distinction can be made depends upon the immunohistochemical method being employed. We have used both fixed/frozen as well as fixed/paraffin-embedded tissue sections in our studies. Each technique has resulted in comparable data. In general, the lower limit of detection appears to be approximately 10^{-16} g of protein antigen.

The specificity using indirect immunohistochemical staining methods is difficult to define. In general, one assumes that specificity reflects the ability of antibodies (i.e. IgG fractions) to detect one antigen determinant to the exclusion of all others present in the histologic tissue sections. We use two sets of specificity criteria: (i) method specificity, and (ii) antibody specificity. Method specificity indicates the lack of staining due to interactions of tissue components with staining reagents other than the antibodies directed against the tissue antigen to be localized. For example, calcium hydroxyapatite can non-specifically bind immunoglobulins which could result in artifact. A number of classical procedures have evolved which are designed to screen for potential artifacts and which enable some degree of standardization of immunohistochemistry. A much more difficult problem is antibody specificity. It must be emphasized that antibodies recognize very small sites or regions on the surface of large antigenic macromolecules. Therefore, polyclonal antibodies, such as those used in our studies for either amelogenins or enamelins, are inherently incapable of identifying the enamelin or amelogenin molecules themselves. It must be clearly understood that immunohistochemistry specificity assays reflect only site specificity and not molecular specificity. If a recognized site, or a site very similar to it, occurs in several different molecules, then it is not reasonable to predict that immunohistochemistry with antibodies directed to that site will be able to distinguish between such molecules.

These comments regarding the sensitivity and specificity of immunohistochemical methods should be kept in mind when considering the following results. The indirect method of immunofluorescence was used to detect and localize amelogenin antigenic determinants within the same enameloid teeth and dermal denticles of Chondrichthyes, in the enameloid of Teleostei and Amphibia, in the enamel of Reptilia, and in the enamel of a number of mammalian species (Herold, Graver et al., 1980). We confirmed these results using polyclonal antibodies directed against mouse enamel organ (epithelial-derived) antigens in the enameloid of teeth of elasmobranchs (e.g. spiny dogfish and blue nurse shark), in the enamel of Reptilia (e.g. *Alligator mississippiensis*), and in the enamel of Mammalia (e.g. rabbit, hamster and mouse) (Slavkin, Zeichner-David, Ferguson et al., 1982).

More recently, we extended these studies using purified mouse amelogenins as immunogens to produce polyclonal antibodies (Gomer & Lazarides, 1981; Slavkin, Zeichner-David, MacDougall, Bessem et al., 1982; Slavkin, Zeichner-David, MacDougall, Bringas et al., 1982). Evidence was obtained which indicated that enameloid in the spiny dogfish (*Squalus acanthias*) contained antigenic determinants which cross-reacted with polyclonal antibodies for mouse amelogenins (Slavkin, Samuel, Bringas, Nanci & Santos, 1983). We have also demonstrated the localization of amelogenin antigenic sites in the outer covering of the lingual teeth of Pacific hagfish (*Eptatretus stoutii*) (Slavkin, Graham, Zeichner-David & Hildemann, 1983). The Pacific hagfish is one of the cyclostomes (the other being lampreys) and represents one of the only living jawless (agnathic) aquatic vertebrates. Their nearest relatives are the extinct agnathan Paleozoic fishes known as the ostracoderms.

Whereas a number of different mammalian species possess enamelin and amelogenin polypeptides (e.g. marine, bovine, porcine, man) (Fincham, 1980; Fukae et al., 1980; Termine, Belcourt et al., 1980; Fincham, Belcourt, Lyaruu et al., 1982; Fincham, Belcourt & Termine, 1982; Lyaruu et al., 1982; Snead, Zeichner-David, Chandra, Robson, Woo & Slavkin, 1983), lower aquatic vertebrates contain enamelin-like molecules but do not appear to have amelogenins (Slavkin, Zeichner-David & Siddiqui, 1981; Slavkin, Zeichner-David, Ferguson et al., 1982; Slavkin, Graham et al., 1983). Therefore, one possible explanation for the biochemical and immunological data obtained from lower and higher vertebrates is that both enamelins and amelogenins share antigenic determinants. These antigenic determinants appear to be highly conserved during evolution.

Micro-ELISA Assay To Compare Enamel Protein Antigens in Lower and Higher Vertebrates

Staphylococcus aureus protein A reacts with the $F'c$ portion of IgG molecules from all mammalian species (Engvall, 1978). Therefore, protein A is an extremely sensitive probe in that it can function as a purified anti-antibody of restricted specificity without any species specificity. This remarkable property has been used to prepare enzyme-labeled protein A for the detection of antibodies (Engvall, 1978). In the last few years this technology has advanced to provide reliable qualitative as well as quantitative determinations of remarkably small amounts of protein antigen. We have used the enzyme-linked immunosorbent assay (ELISA) microplates (micro-ELISA) to investigate the relative antigenicity of enamel matrix proteins isolated from lower and higher vertebrate species (Fig. 5). Shark, alligator, mouse, hamster and human enamel proteins each contain related antigenic determinants within seemingly diverse enamel molecules. In a comparison of relative ELISA Units (the relative optical density of the reaction product measured in a spectrophotometer at 449 nm) with a serial dilution of polyclonal antibodies directed against purified mouse amelogenins and their avidity for various vertebrate enamel antigens, we observed that shark enameloid-derived antigens were "weak" as compared with other vertebrate enamel antigens. The American alligator enamel protein antigens were significantly greater than the selachian antigens. No statistically significant differences were observed between mouse, hamster and human enamel antigens.

MOLECULAR INVESTIGATIONS OF ENAMEL GENE EXPRESSION

A crucial problem in developmental biology is to determine the mechanisms which control the activation and expression of structural genes. A multiphasic model has been suggested as the paradigm for

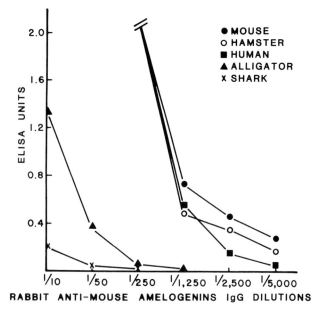

FIG. 5. Micro-ELISA assays were used to identify and compare the antigenicity of enamel matrix proteins from selected vertebrate species. Polyclonal antibodies directed against mouse purified amelogenins were antigenically cross-reactive with determinants of elasmobranch, Reptilia and Mammalia enamel proteins.

epithelial-mesenchymal interactions resulting in differential epithelial gene expression (Rutter, Clark, Kemp, Bradshaw, Sanders & Ball, 1968; Rutter, Pictet, Harding, Chirgwin, MacDonald & Przybyla, 1978). During epidermal organogenesis (e.g. pancreas, salivary gland, mammary gland, thyroid gland, tooth organ), epithelial cells are termed "undifferentiated" when their respective biochemical phenotype cannot be detected (e.g. insulin, amylase, casein, thyroxin, enamelins or amelogenins). At some time and position during epithelial-mesenchymal interactions, epithelial cells are not determined to express a unique set of structural genes and, therefore, no unique mRNAs or polypeptides are detected. The *de novo* "induction" of the epithelial cell genome to differentially activate and transcribe a unique set of structural genes appears to be regulated by adjacent mesenchymal-derived "instructions". The first event of the multiphasic model is the induction and initial determination of the epithelial phenotype at the molecular level. Such an induced epithelial cell is in the "protodifferentiated" state according to Rutter and his colleagues (Rutter, Clark *et al.*, 1968; Rutter, Pictet *et al.*, 1978). Low levels of unique mRNAs are processed in the nucleus during post-transcriptional phases in the protodifferentiated epithelial cells. Subsequently, the "differentiative" phase of this multiphasic model reflects the amplification or optimization of differential gene expression resulting in an optimal level of concentration of gene product. The differentiative phase may be regulated by short-range interactions between heterotypic tissue types vis-à-vis the extracellular matrix and humoral co-factors (Ruch & Karcher-Djuric, 1975; Saxen *et al.*, 1976; Sengel & Dhouailly, 1977; Rutter, Pictet *et al.*, 1978; Brownell *et al.*, 1981; Cummings *et al.*, 1981; Piatigorsky, 1981; Thesleff & Hurmerinta, 1981; Ruch *et al.*, 1982; Slavkin, 1982). The final or "differentiated" phase represents that state requiring maintenance of an already established differentiated phenotype. The developing vertebrate tooth organ, in particular the process of amelogenesis, is an exceptionally suitable epithelial-mesenchymal interacting system for qualitative and quantitative studies of determination, differentiation and morphogenesis (see reviews by Koch, 1972; Kollar, 1972; Slavkin, 1974; Ruch & Karcher-Djuricic, 1975; Thesleff & Hurmerinta, 1981).

Experimental Strategy

Structural genes are defined as a unit of DNA coding for a specific protein that is flanked by nucleotide sequences which specify regulatory signals for transcription and translation (Dawid, Britten, Davidson, Dover, Gallwitz, Garcia-Bellido, Kafatos, Kauffman, Moritz, Ohno, Schmidtke & Schutz, 1982; Flavell, 1982; Inana, Shinohara, Maizel &

Piatigorsky, 1982; James, Bond, Maack, Applebaum & Tata, 1982; Jeffreys, 1982; Takahashi, Ueda, Obata, Nikaido, Nakai & Honjo, 1982). In all eucaryotic cells but not in procaryotic cells, many genes are interrupted by non-coding intervening nucleotide sequences which are termed *introns* (see extensive discussions by Dawid *et al.*, 1982; Hunkapiller, Huang, Hood & Campbell, 1982). The sequences of nucleotides in DNA which code for mRNAs are termed *exons*. A significant number of structural genes are composed of exon and intron regions. For example, all active vertebrate globin genes contain two intervening sequences (i.e. introns) interrupting the protein coding sequence (Jeffreys, 1982; Flavell, 1982). In every case so far examined (e.g. Amphibia, Aves and Mammalia), the intervening sequences occur at precisely homologous positions in the genes (Jeffreys, 1982). The discontinuous structural gene organization appears to have been in existence at least 500 My ago, before the alpha and beta-globin gene duplication arose.

The functional mRNA found in the cytoplasm of differentiating cell types is formed in the nucleus from a larger transcript that often includes both the exon as well as the intron region sequences. During post-transcriptional processing, endo- and exonucleases or restriction nucleases remove intron sequences and splice together exon sequences to produce the functional mRNA. In most functional mRNAs which code for unique structural proteins (e.g. insulin, casein, amylase, imunoglobulins, globins), the 3'-end of the mRNA as well as the primary transcript appears to be determined by polyadenylation (i.e. poly[A]-containing RNAs) (Dawid *et al.*, 1982).

Multigene families consist of genetic information units in DNA which are (i) homologous in nucleic acid structure, (ii) overlap in function, and (iii) appear to be generally linked in tandem (Hunkapiller *et al.*, 1982). Multigene families range in size from the few gene copies of the hemoglobin families, to thousands of copies of ribosomal genes. Examples of multigene families include immunoglobulin genes, silkmoth chorion (i.e. eggshell) genes, vitellogenin (i.e. egg-yolk) genes, actin genes and keratin genes (see Dawid *et al.*, 1982; Hunkapiller *et al.*, 1982; Jeffreys, 1982; Kafatos, 1982; Takahashi *et al.*, 1982). Knowing that genes are composed of multiple discrete exons that encode distinct functional or structural peptide domains possibly indicates that new combinations of exons and/or introns can generate novel combinations of nucleotide sequences. The evolutionary implications of this information are enormous.

The availability of methods for hybridization of complementary sequences of DNA (cDNA) with total genomic DNA representing a specific vertebrate species further facilitated by the advent of recombinant DNA technology, now makes it possible to evaluate the extent of evolutionary divergence of coding sequences (i.e. exons) directly at the level of the structural gene. Recently, we have approached the question of possible evolutionary conservation of coding sequences in enamel structural genes by performing homologous and heterologous hybridizations between selected vertebrate species' genomic DNA and enzymatically synthesized cDNA to amelogenin mRNAs from mouse (Snead et al., 1983). Using the cloned mouse amelogenin cDNA probes, we have looked for evolutionary conservation of amelogenin genes by Southern blot analysis of restriction nuclease-digested genomic DNA from several vertebrate species.

Preliminary Results of Enamel Gene Expression

Evolutionary sequence divergence is known to be high for genes coding for proteins with a low degree of constraint, either structurally as in fibrinopeptides or on regional amino acid sequences within a single protein as with the C fragment of insulin (see discussions by Dawid et al., 1982; James et al., 1982; Plapp, 1982). Further, there is evidence for the high degree of conservation of sequences in the genes coding for proteins under stringent structural and functional constraints such as histones, globins, crystallins, collagens, vitellogenins and the constant regions of the immunoglobulins. In our strategy we plan to determine if the mouse genomic DNA contains coding sequences for both enamelins as well as amelogenins. Further, we plan to determine if enamelins and amelogenins share coding regions since previous evidence has demonstrated immunologic relatedness between these two classes of enamel proteins. Finally, we plan to ask if other vertebrates, such as alligators and humans, also contain amelogenin coding sequences in their genomic DNA.

First, we have isolated the mRNAs for fetal rabbit enamelins (Zeichner-David, Weliky et al., 1980). Studies are now in progress to complete the identification of cDNA clones authentic for fetal rabbit enamelins. Second, the mRNAs for mouse amelogenins, representing approximately 90% of the total enamel proteins, have been isolated and partially characterized by specific immunoprecipitation (Slavkin, Zeichner-David, MacDougall, Bessem et al., 1982; Snead et al., 1983). Mouse enamelins were also detected. The poly(A)-containing RNAs were used for the synthesis and eventual cloning of the mouse amelogenin cDNA (Snead et al., 1983). Recombinant colonies containing amelogenin DNA sequences were identified by differential hybridization, hybrid-selected translation, and blot hybridization analyses. The mouse amelogenin cDNA was capable of identifying the expression of amelogenins during tooth development.

The mouse cDNA sequence hybridized to mouse as well as human genomic DNAs (Snead *et al.*, 1983). Preliminary evidence suggests that mouse amelogenin cDNA sequences also hybridize with American alligator genomic DNA.

SUMMARY

The enamel proteins are highly conserved during vertebrate evolution over the period of almost 500 My, as represented by the extant groups studied. Two major classes of structural gene products have been identified in a large number of higher vertebrate species – amelogenins and enamelins. The enamelins appear to be the predominant enamel matrix constituent in the aquatic vertebrates (e.g. cyclostomes, elasmobranchs and teleosts), whereas in terrestrial vertebrates (e.g. Reptilia and Mammalia) both enamelins and amelogenins are detected. The major enamel proteins identified in Reptilia and many species of Mammalia are the amelogenins, often representing 90% of the total enamel extracellular matrix protein during amelogenesis. Both enamelins and amelogenins appear to share related antigenic determinants. Polyclonal antibodies directed against purified mouse amelogenins are cross-reactive with all vertebrate species so far examined in these studies. Preliminary evidence from hybridization studies, facilitated by recombinant DNA technology, suggests a high degree of conservation of sequences in the genes coding for amelogenins in Reptilia and Mammalia. On the basis of these investigations, we suggest that the developmental processes associated with enamel gene product formation throughout the subphylum Vertebrata are homologous. We propose that an enamel multigene family is a primary organizational unit for control and regulation of vertebrate enamel formation. We postulate that this multigene family consists of multiple structural genes for enamelins as well as amelogenins in higher vertebrates including Reptilia and Mammalia. Cyclostomes and elasmobranchs appear to contain only enamelin structural genes. During the evolution of the Reptilia, genomic variations associated with enamel structural genes may represent an ancient duplication (approximately 230 My ago) which gave rise to the ancestors of enamelin and amelogenin genes.

ACKNOWLEDGEMENTS

We wish to thank Dr Richard J. Krejsa for his many insights into the phylogeny of enamel and his enthusiastic tutorials regarding dental problems in ichthyology. We very much appreciate the continued technical assistance of Pablo Bringas, Jr, Conny Bessem, Valentino Santos and Julia Vides. This research was supported in part by research grant DE-02848 (H.C.S.) and training grant DE-07006 (H.C.S.) from the NIDR, NIH, USPHS.

REFERENCES

Belcourt, A. B., Fincham, A. G. & Termine, J. D. (1982). Acid-soluble bovine fetal enamelins. *J. dent. Res.* **61**: 1031–1032.
Boyde, A. (1971). Comparative histology of mammalian teeth. In *Dental Morphology and evolution*: 81–94. Dahlberg, A. (Ed.). Chicago: The University of Chicago Press.
Boyde, A. (1978). Development of the structure of the enamel of the incisor teeth in the three classical subordinal groups of the Rodentia. In *Development, function and evolution of teeth*: 43–58. Butler, P. M. & Joysey, K. A. (Eds). New York: Academic Press.
Brownell, A., Bessem, C. & Slavkin, H. C. (1981). Possible functions for mesenchyme cell-derived fibronectin during basal lamina formation. *Proc. natn. Acad. Sci. USA* **78**: 3711–3715.
Cummings, E. G., Bringas, P., Grodin, M. S. & Slavkin, H. C. (1981). Epithelial-directed mesenchyme differentiation *in vitro*. Model of murine odontoblast differentiation mediated by quail epithelia. *Differentiation* **20**: 1–9.
Dawid, I., Britten, R. J., Davidson, E. H., Dover, G. A., Gallwitz, D. F., Garcia-Bellido, A., Kafatos, F. C., Kauffman, S. A., Moritz, K., Ohno, S., Schmidtke, J. & Schutz, G. (1982). Genomic change and morphological evolution. In *Evolution and development*: 19–39. Bonner, J. T. (Ed.). New York: Springer-Verlag.
Eastoe, J. E. (1963). The amino acid composition of proteins from the oral tissues: II. The matrix proteins in dentine and enamel from developing human deciduous teeth. *Arch. oral Biol.* **8**: 449–458.
Eastoe, J. E. (1965). The chemical composition of bone and tooth. *Adv. Fluoride Res. Dental Caries Prev.* **3**: 5–16.
Eggert, F. M., Allen, G. A. & Burgess, R. C. (1973). Amelogenins – purification and partial characterization of proteins from developing bovine dental enamel. *Biochem. J.* **131**: 471–484.
Elwood, W. K. & Apostolopoulos, A. X. (1975). Analysis of developing enamel of rat. III. Carbohydrate, DEAE-Sephadex, and immunological studies. *Calcif. Tiss. Res.* **17**: 337–347.
Engvall, E. (1978). Preparation of enzyme-labeled Staphlococcal protein A and its use for detection of antibodies. *Scan. J. Immunol.* **8** (suppl. 7): 25–31.
Fearnhead, R. W. (1979). Matrix-mineral relationships in enamel tissues. *J. dent. Res.* **58**(B): 909–921.
Ferguson, M. W. J. (1981). The value of the American alligator *(Alligator mississippiensis)* as a model for research in craniofacial development. *J. Craniofac. Genet. Devl Biol.* **1**: 123–144.
Ferguson, M. W. J. (In press). The reproductive biology and embryology of crocodilians. In *Biology of the Reptilia*, **14** *Development*. Billett, F. S. & Gans, C. (Eds). New York and London: J. Wiley and Sons.
Fincham, A. G. (1980). Changing amino acid profiles of developing dental enamel in

individual human teeth and the comparison of developing human and bovine enamel. *Arch. oral Biol.* **25**: 669–674.

Fincham, A. G., Belcourt, A. B., Lyaruu, I. M. & Termine, J. D. (1982). Comparative protein biochemistry of developing dental enamel matrix from five mammalian species. *Calcif. Tissue Int.* **34**: 182–189.

Fincham, A. G., Belcourt, A. B. & Termine, J. D. (1982). The composition of bovine fetal enamel matrix. In *Chemistry and biology of mineralized connective tissues*: 523–529. Veis, A. (Ed.). New York: Elsevier/North-Holland.

Fincham, A. G., Belcourt, A. B., Termine, J. D., Butler, W. G. & Cothran, W. C. (1981). Dental enamel matrix: sequences of two amelogenin polypetides. *Biosci. Rep.* **1**: 771–778.

Flavell, R. B. (1982). Sequence amplification, deletion and re-arrangement: major sources of variation during species divergence. In *Genome evolution*: 164–189. Dover, G. A. & Flavell, R. B. (Eds). New York: Academic Press.

Fosse, B., Risnes, S. & Holmbakken, N. (1974). Mineral distribution and mineralization pattern in the enameloid of certain elasmobranchs. *Arch. oral Biol.* **19**: 771–780.

Fukae, M. & Shimizu, M. (1974). Studies on the proteins of developing bovine enamel. *Arch. oral Biol.* **19**: 381–386.

Fukae, M., Tanabe, T., Ijiri, H. & Masharu, S. (1980). Studies on porcine enamel proteins. *Dent. J.* **6**: 88–94.

Garant, P. (1970). An electron microscopic study of the crystal-matrix relationship in the teeth of the dogfish *Squalus acanthias*. *J. Ultrastruct. Res.* **30**: 441–449.

Gaunt, W. A. (1955). The development of the molar pattern of the mouse (*Mus musculus*). *Acta Anat.* (Basel) **24**: 249–268.

Gaunt, W. A. & Miles, A. E. W. (1967). Fundamental aspects of tooth morphogenesis. In *Structural and chemical organization of teeth*, **1**: 151–198. Miles, A. E. W. (Ed.). New York: Academic Press.

Glimcher, M. J. (1961). The molecular structure of the protein matrix of bovine dental enamel. *J. molec. Biol.* **3**: 541–546.

Glimcher, M. J. (1979). Phosphopeptides of enamel matrix. *J. dent. Res.* **58**: 790–806.

Gomer, R. H. & Lazarides, E. (1981). The synthesis and deployment of filamin in chicken skeletal muscle. *Cell* **23**: 524–532.

Graver, H., Herold, R., Chung, T., Christner, P., Pappas, C. & Rosenbloom, J. (1978). Immunofluorescent localization of amelogenins in developing bovine teeth. *Devl Biol.* **63**: 390–401.

Greenberg, G., Bringas Jr, P. & Slavkin, H. C. (1983). The epithelial genotype controls the pattern of extracellular enamel prism formation. *Differentiation* **25**: 32–43.

Greulich, R. C. & Slavkin, H. C. (1965). Amino acid utilization in the synthesis of enamel and dentin matrices as visualized by autoradiography. In *The use of radioautography in investigating protein synthesis*: 199–214. Leblond, C. P. & Warren, K. B. (Eds). New York: Academic Press.

Guenther, H. L., Croissant, R. D., Schonfeld, S. E. & Slavkin, H. C. (1977). Identification of four extracellular matrix enamel proteins during embryonic-rabbit tooth-organ development. *Biochem. J.* **163**: 591–603.

Herold, R., Christner, P., Graver, H. & Rosenbloom, J. (1979). Immunofluorescent evidence for the similarity of amelogenins in calf, mouse, and pig teeth. *J. dent. Res.* **58**(B): 992–993.

Herold, R., Graver, H. & Christner, P. (1980). Immunohistochemical localization of amelogenins in enameloid of lower vertebrate teeth. *Science, Wash.* **207**: 1357–1358.

Hertwig, O. (1874). Ueber Bau und Entwickelung der Placoidschuppen und der Zahn der Selachier. *Jena Z. Naturw.* **8**: 331–404.

Huggins, C. B., McCarroll, H. R. & Dahlberg, A. (1934). Transplantation of tooth germ elements and the experimental heterotypic formation of dentin and enamel. *J. exp. Med.* **60**: 199–210.

Hunkapiller, T., Huang, H., Hood, L. & Campbell, J. H. (1982). The impact of modern genetics on evolutionary theory. In *Perspectives on evolution*: 164–189. Milkman, R. (Ed.). Sunderland, Massachusetts: Sinauer Associates, Inc.

Hyatt-Fischer, H., Chrispens, J., O'Keefe, D. & Slavkin, H. C. (1979). An antisera for the fluorescent labeling of mouse amelogenesis. *J. dent. Res.* **59**(B): 1008–1009.

Inana, G., Shinohara, T., Maizel Jr, J. V. & Piatigorsky, J. (1982). Evolution and diversity of the crystallins: nucleotide sequence of a b-crystallin mRNA from the mouse lens. *J. biol. Chem.* **257**: 9064–9071.

James, T. C., Bond, U. M., Maack, C. A., Applebaum, S. W. & Tata, J. R. (1982). Evolutionary conservation of vitellogenin genes. *DNA* **1**: 345–353.

Jeffreys, A. J. (1982). Evolution of globin genes. In *Genome evolution*: 157–176. Dover, G. A. & Flavell, R. B. (Eds). New York: Academic Press.

Kafatos, F. C. (1982). The developmentally regulated chorion gene families of insects. In *Embryonic development Part A: Genetic aspects*. Burger, M. M. & Weber, R. (Eds). New York: Alan R. Liss, Inc.

Kallenbach, E. (1971). Electron microscopy of the differentiating rat incisor ameloblast. *J. Ultrastruct. Res.* **35**: 508–531.

Kallenbach, E. (1976). Fine structure of differentiating ameloblasts in the kitten. *Am. J. Anat.* **145**: 283–317.

Kerebel, B., Daculsi, G. & Renaudin, S. (1977). Ameloblast ultrastructure during enameloid formation in selachians. *Biol. Cell* **28**: 125–130.

Koch, W. E. (1967). *In vitro* differentiation of tooth rudiments of embryonic mice. I Transfilter interaction of embryonic incisor tissues. *J. exp. Zool.* **165**: 155–170.

Koch, W. E. (1972). Tissue interaction during *in vitro* odontogenesis. In *Developmental aspects of oral biology*: 151–164. Slavkin, H. C. & Bavetta, L. A. (Eds). New York: Academic Press.

Kollar, E. J. (1972). Histogenetic aspects of dermal-epidermal interactions. In *Developmental aspects of oral biology*: 125–149. Slavkin, H. C. & Bavetta, L. A. (Eds). New York: Academic Press.

Kollar, E. J. (1978). The role of collagen during tooth morphogenesis: some genetic implications. In *Development, function and evolution of teeth*: 1–12. Butler, P. M. & Joysey, K. A. (Eds). London: Academic Press.

Kollar, E. J. & Baird, G. R. (1969). The influence of the dental papilla on the development of tooth shape in embryonic mouse dental papilla. *J. Embryol. exp. Morph.* **21**: 131–148.

Kollar, E. J. & Baird, G. R. (1970). Tissue interactions in embryonic mouse tooth germs. II. The inductive role of the dental papilla. *J. Embryol. exp. Morph.* **24**: 173–186.

Kollar, E. J. & Fischer, C. (1980). Tooth induction in chick epithelium: Expression of quiescent genes for enamel synthesis. *Science, Wash.* **207**: 993–995.

Levine, P. T., Glimcher, M. J., Seyer, J. M., Huddleston, J. I. & Hein, J. W. (1966). Noncollagenous nature of the proteins of shark enamel. *Science, Wash.* **154**: 1192–1194.

Levine, P. T., Seyer, J., Huddleston, J. & Glimcher, M. J. (1967). The comparative biochemistry of the organic matrix proteins of developing enamel – I. Amino acid composition. *Archs oral Biol.* **12**: 407–410.

Lyaruu, D. M., Belcourt, A., Fincham, A. G. & Termine, J. D. (1982). Neonatal hamster molar tooth development: extraction and characterization of amelogenins, enamelins, and soluble dentin proteins. *Calcif. Tissue Int.* **34**: 86–96.

Mörnstad, H. (1974). On the histogenesis of shark enamel. *Odot. Revy* **25**: 317–325.

Moss, M. L. (1977). Skeletal tissues in sharks. *Am. Zool.* **17**: 335–342.
Nanci, A., Bringas Jr, P., Samuel, N. & Slavkin, H. C. (1983). Selachian tooth development III. Ultrastructural features of secretory amelogenesis in *Squalus acanthias*. *J. Craniofac. Genet. Devl Biol.* **3**: 53–73.
Nikiforuk, G. & Gruca, M. (1971). Immunological and gel-filtration characteristics of bovine enamel protein. In *Tooth enamel II: Composition, properties and fundamental structure*: 95–98. Fearnhead, R. W. & Stack, W. V. (Eds). Baltimore: Williams & Wilkins.
Ornstein, L. (1964). Disc electrophoresis – I. Background and theory. *Ann. N.Y. Acad. Sci.* **121**: 321–348.
Osborn, J. W. (1973). The evolution of dentitions. *Am. Sci.* **61**: 548–559.
Peyer, B. (1968). Histology and developmental history of the teeth in Actinopterygii. In *Comparative odontology*: 89–99. Peyer, B. (Ed.). Chicago: The University of Chicago Press.
Piatigorsky, J. (1981). Lens differentiation in vertebrates. A review of cellular and molecular features. *Differentiation* **19**: 134–153.
Piesco, N. P. (1979). *A comparative fine-structural study of cytodifferentiation and dental hard tissue formation in selected vertebrates with a note on the evolution of enamel.* Ph.D. Thesis: University of Florida, Gainsville, USA.
Plapp, B. V. (1982). Origins of protein structure and function. In *Perspectives on evolution*: 129–147. Milkman, R. (Ed.). Sunderland Massachusetts: Sinauer Associates, Inc.
Poole, D. F. G. (1967). Phylogeny of tooth issues: enameloid and enamel in recent vertebrates, with a note on the history of cementum. In *Structural and chemical organization of teeth*: 111–150. Miles, A. E. W. (Ed.). London: Academic Press.
Poole, D. F. G. (1971). An introduction to the phylogeny of calcified tissues. In *Dental morphology and evolution*: 65–79. Dahlberg, A. A. (Ed.). Chicago: The University of Chicago Press.
Reith, E. J. (1967). The early stages of amelogenesis as observed in molar teeth of young rats. *J. Ultrastruct. Res.* **17**: 503–526.
Reith, E. J. (1970). The stages of amelogenesis as observed in molar teeth of young rats. *J. Ultrastruct. Res.* **30**: 111–151.
Ruch, J. V. & Karcher-Djuricic, V. (1975). On odontogenic tissue interactions. In *Extracellular matrix gene influences on gene expression*: 549–554. Slavkin, H. C. & Greulich, R. C. (Eds). New York: Academic Press.
Ruch, J. V., Lesot, H., Karcher-Djuricic, V., Meyer, J. M. & Olive, M. (1982). Facts and hypotheses concerning the control of odontoblast differentiation. *Differentiation* **21**: 7–12.
Rutter, W. J., Clark, W. R., Kemp, J. D., Bradshaw, W. S., Sanders, T. C. & Ball, W. D. (1968). Multiphasic regulation in cytodifferentiation. In *Epithelial-mesenchymal interactions*: 114–131. Baltimore: The Williams & Wilkins Co.
Rutter, W. J., Pictet, R. L., Harding, J. D., Chirgwin, J. M., MacDonald, R. J. & Przybyla, A. E. (1978). An analysis of pancreatic development: role of mesenchymal factor and other extracellular factors. In *Molecular control of proliferations and differentiation*: 205–227. Papaconstantinou, J. & Rutter, W. J. (Eds). New York: Academic Press.
Samuel, N., Nanci, A., Bringas Jr, P., Santos, V. & Slavkin, H. C. (1983). Selachian tooth development I. Histogenesis, morphogenesis and anatomical features in *Squalus acanthias*. *J. Craniofacial Genet. Devl Biol.* **3**: 29–41.
Saxen, L., Karkinen-Jaaskelainen, M., Lehtoneu, E., Nordling, S. & Wartiovaara, J. (1976). The cell surface in animal embryogenesis and development. In *Inductive tissue interactions*: 331–407. Poste, G. & Nicolson, G. L. (Eds). New York: North-Holland Publishing Co.

Schonfeld, S. E. (1979). Immunogenicity of enamel protein. *J. dent. Res.* **58**(B): 810–816.
Sengel, P. & Dhouailly, D. (1977). Tissue interactions in amniote skin development. In *Cell interactions in differentiation*: 153–169. Karkinen-Jaaskelainen, M., Saxen, I. & Weiss, L. (Eds). New York: Academic Press.
Shellis, R. P. (1975). A histological and histochemical study of the matrices of enameloid and dentine in teleost fishes. *Arch. oral Biol.* **20**: 183–187.
Shellis, R. P. (1978). The role of the inner dental epithelium in the formation of the teeth in fish. In *Development, function and evolution of teeth*: 31–42. Butler, P. M. & Joysey, K. A. (Eds). New York: Academic Press.
Slavkin, H. C. (1974). Embryonic tooth formation: a tool in developmental biology. *Oral Sci. Rev.* **4**: 7–136.
Slavkin, H. C. (1979). *Developmental craniofacial biology*. Philadelphia: Lea & Febiger.
Slavkin, H. C. (1982). Combinatorial process for extracellular matrix influences on gene expression: a hypothesis. *J. Craniofacial Genet Devl Biol.* **2**: 179–189.
Slavkin, H. C. & Bringas Jr, P. (1976). Epithelial-mesenchymal interactions during odontogenesis: IV. Morphological evidence for direct heterotypic cell–cell contacts. *Devl Biol.* **50**: 428–442.
Slavkin, H. C., Graham, E., Zeichner-David, M. & Hildemann, W. (1983). Enamel-like antigens in hagfish: possible evolutionary significance. *Evolution, Lawrence, Kans.* **37**: 404–412.
Slavkin, H. C., Mino, W. & Bringas Jr, P. (1976). The biosynthesis and secretion of precursor enamel protein by ameloblasts as visualized by autoradiography after tryptophan administration. *Anat. Rec.* **185**: 289–312.
Slavkin, H. C., Samuel, N., Bringas, P., Nanci, A. & Santos, V. (1983). Selachian tooth development. II. Immunolocalization of amelogenin polypeptides in epithelium during secretory amelogenesis in *Squalus acanthias*. *J. Craniofacial Genet. Devl Biol.* **3**: 43–52.
Slavkin, H. C., Trump, N., Schonfeld, S., Brownell, A., Sorgente, N. & Lee-Own, V. (1977). Epigenetic regulation of enamel protein synthesis during epithelial-mesenchymal interactions. In *Cell interactions in differentiation*: 209–226. Saxen, L. & Weiss, L. (Eds). New York: Academic Press.
Slavkin, H. C., Weliky, B., Stellar, W., Slavkin, M. D., Zeichner-Gancz, M., Bringas Jr, P., Hyatt-Fischer, H., Shimizu, M., & Fukae, M. (1979). Ameloblast differentiation: protein synthesis and secretion in fetal New Zealand White rabbit molar tooth organs and isolated epithelia *in vitro*. *J. biol. Buccale* **6**: 309–326.
Slavkin, H. C., Zeichner-David, M., Ferguson, M. W. J., Termine, J. D., Graham, E., MacDougall, M., Bringas Jr, P., Bessem C. & Grodin, M. (1982). Phylogenetic and immunogenetic aspects of enamel proteins. In *Oral immunogenetics and tissue transplantation*: 241–251. Hildemann, W. & Riviere, G. (Eds). New York: Elsevier/North-Holland.
Slavkin, H. C., Zeichner-David, M., MacDougall, M., Bessem, C., Bringas, P., Honig, L. S., Lussky, J. & Vides, J. (1982). Enamel gene products during murine amelogenesis *in vivo* and *in vitro*. *J. dent. Res.* **61**: 1467–1471.
Slavkin, H. C., Zeichner-David, M., MacDougall, M., Bringas, P., Bessem, C. & Honig, L. S. (1982). Antibodies to murine amelogenins: localization of enamel proteins during tooth organ development *in vitro*. *Differentiation* **23**: 73–82.
Slavkin, H. C., Zeichner-David, M. & Siddiqui, M. A. Q. (1981). Molecular aspects of tooth morphogenesis. *Molec. Asp. Med.* **4**: 125–188.
Slavkin, H. C., Zeichner-David, M. & Siddiqui, M. A. Q. (1982). Enamel extracellular matrix: differentiation specific gene products and the control of their synthesis and

accumulation during development. In *Current advances in skeletogenesis*: 24–33. Silbermann, M. & Slavkin, H. C. (Eds). Amsterdam: Elsevier Science Publishing Co.

Snead, M. L., Zeichner-David, M., Chandra, T., Robson, K. J. H., Woo, S. L. C. & Slavkin, H. C. (1983). Construction and identification of mouse amelogenin cDNA clones. *Proc. natn. Acad. Sci. USA.* **80**: 7254–7258.

Takahashi, N., Ueda, S., Obata, M., Nikaido, T., Nakai, S. & Honjo, T. (1982). Structure of human immunoglobin gamma genes: implications for evolution of a gene family. *Cell* **29**: 671–679.

Termine, J. D., Belcourt, A. B., Christner, P. J., Conn, K. M. & Nylen, M. U. (1980). Properties of dissociatively extracted fetal tooth matrix proteins. I. Principal molecular species in developing bovine enamel. *J. biol. Chem.* **255**: 9760–9768.

Termine, J. D., Torchia, D. A. & Conn, D. M. (1979). Enamel matrix: structural proteins. *J. dent. Res.* **58**(B): 773–778.

Theiler, K. (1972). *The house mouse*. New York: Springer-Verlag.

Thesleff, I. & Hurmerinta, K. (1981). Tissue interactions in tooth development. *Differentiation* **18**: 75–88.

Walker, A. D. (1972). New light on the origin of birds and crocodiles. *Nature, Lond.* **237**: 257–263.

Weinstock, A. (1972). Matrix development in mineralizing tissues as shown by radioautography: formation of enamel and dentin. In *Developmental aspects of oral biology*: 202–238. Slavkin, H. C. & Bavetta, L. (Eds). New York: Academic Press.

Zalut, C., Henzel, W. J. & Harris, H. W. (1980). Microquantitative Edman manual sequencing. *J. Biochem. Biophys. Meth.* **3**: 11–30.

Zeichner-David, M., Weliky, B. G. & Slavkin, H. C. (1980). Isolation and preliminary characterization of epithelial-specific messenger ribonucleic acids and products during embryonic tooth development. *Biochem. J.* **185**: 489–496.

Zeichner-David, M., Slavkin, H. C., Lyaruu, D. M. & Termine, J. D. (1983). Biosynthesis and secretion of enamel proteins during hamster tooth development. *Calcif. Tissue Int.* **35**: 366–371.

The Evolution of Sex Chromosomes and Chromosomal Inactivation in Reptiles and Mammals

K. W. JONES

Institute of Animal Genetics, University of Edinburgh, King's Buildings, West Mains Road, Edinburgh EH9 3JN, Scotland

SYNOPSIS

The evolution of genetic systems encompasses striking examples of the functional specialization of an entire chromosome to control a single developmental pathway, namely primary sex determination. In some primitive vertebrate species, for example alligators, sex is determined by gene-environment interaction but in others, for example snakes, it is under genetic control. The snakes exemplify important aspects of the evolutionary elaboration of specialized sex-determining chromosomes, since these chromosomes have evolved in some, but not all, species. Thus the relatively primitive species of the family Boidae exhibit no specialized sex-determining chromosomes whereas the more advanced families of the Crotalidae, Elapidae and Viperidae possess such chromosomes in varying degrees of structural elaboration ranging from morphologically identical to extremely heteromorphic Z and W chromosomes. It therefore appears that the evolution of these important chromosomal mechanisms, which we have been studying at the DNA level, has occurred within the Order Serpentes. How, or even why, chromosomal specialization has arisen, why this should have happened in some but not other species within the same Order and whether or not is conferred an advantage are unanswered questions. One possible explanation which will be proposed is that such specialization was a potentially catastrophic side effect of the process whereby sex-determining genes evolved to assert genetic control of sexual development.

INTRODUCTION

Systems of Sex Determination

As a model system for unravelling the mysteries of development and evolution primary sex determination has a unique combination of advantages. For example, it concerns developmental pathways which, in different species of organisms, are variously responsive to environmental, genetic or chromosomal factors. In many lowly organisms sex ratio is determined by environmental factors. This is known to be the case in many species of reptiles in which temperature during the early stages of embryogenesis has a determinative effect on the sex ratio, for example in alligators (Ferguson & Joanen, 1982). In the Amphibia, however, whilst it is possible experimentally to interfere with the determinative process by, for example, hormone administration or grafting of gonadal rudiments, the sex ratio is normally genetically determined.

These different systems embody the evolutionary transition from an ancestral condition in which environmental factors prevailed to those in which genetic mechanisms of sex determination are the rule. The advantages of genetic sex determination are suggested to arise both from the stable sex ratio which it ensures and from the fact that species so equipped are liberated to explore a wider range of habitats. Thus, the evolution of a mechanism of primary sex determination under purely genetic control enabled the spread of species to new environments.

The genetic mechanisms of sex determination which characteristically are found in Amphibia, with rare exceptions (M. Schmidt, pers. comm.) do not involve obvious specialization of the chromosomes carrying the sex determinants. In fact, the grafting of gonadal primordia between female and male embryos in Amphibia (Humphrey, 1942) has allowed successful mating between genetic females to be demonstrated. In such cases, if we designate the female as ZW and the male as ZZ, it has been shown that offspring of the constitution WW are fully viable and fertile. Similar experiments, involving sex inversion by hormone treatment, have given comparable results in the fish *Oryzias latipes* (Aida, 1921). These examples show that the difference between the Z and W (or X and Y) chromosomes in some groups has remained limited to the sex determinants carried by these chromosomes. However, in other groups, genetic control of primary sex ratio has involved the specialization of the entire sex chromosome in respect of this one function, and has led to the virtual loss of all its other genetic functions. The mechanism whereby this occurred, its possible advantages over a purely genetic control and the fact that it is limited to some species have so far received no plausible explanation.

The reptiles in particular offer unique opportunities to elucidate the mechanisms involved in the evolution and functions of these chromosomes. This arises from the fact that in the various families of snakes, this evolutionary process is exemplified in different degrees of elaboration of sex chromosomes. These range from species which totally lack them through species exhibiting intermediate stages to species in which they are as well differentiated as in mammals.

Sex Determination in Snakes

The chromosomes of snakes exhibit a narrow range of variation in their karyotypes. There is a preponderance of species with 36 chromosomes which occur typically as eight pairs of macrochromosomes and 10 pairs of microchromosomes (Beçak, Beçak, Nazareth & Ohno, 1964; Singh, 1972). In snakes female heterogamety is the rule. Thus we refer to the genetic constitution of the females as ZW and of the males as ZZ as far as sex determination is concerned and to the specialized sex-determining chromosome, where it occurs, as the W chromosome. In many of the evolutionarily more advanced species chromosomal differences are visible when males and females are compared, signifying the presence in the female of a differentiated W sex-determining chromosome pairing with the Z chromosome. The W chromosome in these species is usually reduced in size relative to the Z chromosome. In all cases the W chromosome is out of synchrony in DNA synthesis with respect to the rest of the karyotype (Ray-Chaudhuri & Singh, 1972) and is entirely heterochromatic (Ray-Chaudhuri, Singh & Sharma, 1971). Even in species in which the Z and W are morphologically identical (homomorphic), it is important to note that the other differences with respect to the W remain. The Z chromosome has remained evolutionarily stable at about 10% of the total chromosome complement on a length basis. These observations indicate that it is the W chromosome which is evolving to specialization and that the Z chromosome has remained relatively unchanged over the same time. However, snakes of the family Boidae (constrictors) show no such sex differences and apparently have not evolved specialized chromosomal sex determination (CSD). It therefore appears that the evolution of these chromosomes is of independent, and of relatively recent, origin in this Order. The likelihood that such evolutionary developments have occurred independently within several major groups is substantiated by the fact that in birds, whilst most species show a well differentiated W chromosome in the female, there are also species of the family Ratitae which appear to lack such chromosomes (Tagaki, Itoh & Sasaki, 1972). Therefore, in both Orders, sex chromosomes have arisen independently since the differentiation of extant families.

The entire W chromosome in snakes remains condensed and heterochromatic for much of the life-cycle whereas the Z chromosomes in the male snake remain euchromatic like the autosomes. However, it has been shown that the W chromosome decondenses totally during oogenesis in those oocytes which have received it following meiosis (Singh, Purdom & Jones, 1979). Thus, it seems that the W is reactivated only during early oogenesis and it is especially important to note that the inactive and active phases involve the chromosome as a whole, rather than just parts of it. In this sense, the inactive W chromosome strongly resembles the inactive X chromosome in mammals since, in both, control appears to be at the chromosomal level. In both the W and the mammalian X chromosome a heterochromatic change and out-of-phase DNA synthesis accompany the inactivation phenomenon. Moreover, like the W chromosome in snakes, the mammalian X chromosome is active in the oocyte (Gartler, Liskay, Campbell, Sparkes & Grant, 1972).

The characteristics of being heterochromatic and of synthesizing DNA out-of-phase, as mentioned above, are already fully established in the homomorphic ZW bivalent. In this descriptive sense, the homomorphic ZW bivalent behaves exactly analogously to the XX bivalent in the mammalian female except that it is always the W which is inactive, in contrast to the random inactivation exhibited by one or other of the X chromosomes. However, the analogy holds even in this detail if we restrict the comparison to the metatherian XX female in which there is non-random inactivation of the paternal X (Sharman, 1971). The existence of homomorphic ZW bivalents makes it virtually certain that the inactivation of the W chromosome occurred at a very early stage in its evolution and that its morphological divergence from the Z chromosome occurred subsequently.

The present author's laboratory has adopted a direct approach to understanding the evolution of chromosomal inactivation based on analysing the DNA of the inactivated W chromosome.

W CHROMOSOME HETEROCHROMATIN AND REPEATED DNA

One of the molecular features of some heterochromatic regions of chromosomes which stain relatively more darkly with Giemsa stain, is the presence within them of highly repeated DNA. This was originally shown in the case of the mouse (Jones, 1970; Pardue & Gall, 1970) in which the paracentromeric regions of the autosomes and the X chromosome, which stain in this manner, contain the highly repetitious satellite DNA. This led us to examine snake species for the presence of repeated DNA in the intensely Giemsa-staining W chromosome (Singh & Ray-Chaudhuri, 1975). It was found that female snakes with CSD have a satellite DNA fraction which is missing from males of the same species (Singh, Purdom & Jones, 1976). This was then shown by *in situ* hybridization to be derived from the W chromosome, and to be located in a interspersed manner throughout its length. Moreover, the sequences concerned, which we refer to as Bkm DNA (Banded krait minor satellite), have been conserved throughout all snakes (Singh, Purdom & Jones, 1980). When examined in detail, it was found that these conserved sequences are in fact present in both sexes but have become especially concentrated upon, and interspersed throughout, the W chromosome in all species examined. Snakes lacking CSD also possess Bkm-related sequences but both sexes appear to have fewer and there is no quantitative sex difference. The conclusion drawn from these findings is that the evolution of CSD has been accompanied by the magnification of certain repeated mobile DNA sequences especially focused on the W chromosome. Similar experiments in which a probe of Bkm DNA was used to look for related sequences in birds remarkably also showed their presence, which was quantitatively correlated with the presence of a W chromosome in the female (Jones & Singh, 1981a). As mentioned, both snakes and birds include primitive species which lack CSD, whereas the majority of these species possess a well defined chromosomal sex-determining system. It therefore seems that two major groups exhibiting female heterogamety, in which CSD has apparently evolved independently, have magnified related repeated DNA in connection with the evolution of this chromosomal mechanism. It seems possible on these grounds that this DNA has some special significance for its evolution and/or that some kinds of conserved DNA sequences prefer to reside on sex chromosomes.

Thus it is apparent that there may well have been very significant changes in DNA composition in advance of the structural evolution of the W chromosome. This favours the view that heterochromatinization preceded structural evolution of the W chromosome, contrary to the classical view. It is extremely unlikely that such a radical alteration in the DNA composition of the W chromosome could have come about whilst there was normal crossing over between it and the Z, or in the absence of radical alterations in its genetic functions. A change in the W chromosome which abolished normal genetic interaction with the Z chromosome must therefore have occurred as an early event in its evolution. Presumably, since the female is the heterogametic sex in all species of snakes, birds and Lepidoptera, these changes arose from the dominant sex genes of the W chromosome.

In summary, it seems probable, from the existence of homomorphic sex bivalents including a euchromatic Z and heterochromatic W, that bivalents including a euchromatic Z and heterochromatic W, that major structural divergence between these chromosomes was the last step in a process which included the inactivation and functional specialization of the W chromosome; withdrawal of the W from genetic exchange with the Z and the interspersion of novel repeated DNA throughout the W chromosome. It seems probable that the W and Y chromosomes, both of which control the heterogametic sex, have evolved by the same mechanism. The fact that Bkm-related sequences are also to be found on the mouse Y chromosome (Jones & Singh, 1981b), specifically in the sex-determining locus (Singh & Jones, 1982), is obviously very interesting in this regard. To account for the whole range of intermediate forms of the W chromosomes within the snakes, it would seem that their morphological evolution must have been rapid, contrary to the usual supposition of a slow evolutionary process of sex chromosome specialization.

A rapid process of chromosome inactivation postulated in the evolution of the W chromosome requires no unusual assumptions, since X chromosome inactivation in mammalian females is a cyclical occurrence at each generation. The present author has suggested (Jones, 1983) that X chromosome inactivation may be informatively regarded not only to be the formal paradigm of this evolutionary phenomenon but to involve similar mechanisms.

X CHROMOSOME INACTIVATION

As is well known, in the development of every mammalian female one X chromosome becomes substantially inactivated (reviewed in Gartler & Andina, 1976). This has led to the suggestion (Lyon, 1961) that the essential significance of this phenomenon is to produce a balance of X-linked gene expression in males and females (Lyon, 1961). This has come to be referred to as "dosage compensation", a term first used to describe the fact that the males and females of *Drosophila* exhibit similar levels of expression of X-linked genes (Muller, League & Offerman, 1931; Muller, 1950). X chromosome inactivation in the female is viewed as a device to balance the effects of evolving a largely genetically inert Y chromosome. Since autosomal monosomy is generally lethal, it has been supposed that the evolution of the Y chromosome and of the corresponding X inactivation must have been a very complicated, difficult and protracted process. It is postulated that it involved step by step adaptation, or "compensation", for the loss of Y genes by the development of a mechanism of inactivation of the corresponding genes on the X chromosome (e.g. Charlesworth, 1978) in response to selection. To account for steady progress towards this ultimate goal the net advantage in evolving chromosomal sex determination, which is currently obscure, must be assumed to have been overwhelming. However, if the observations of snake sex chromosomes indicating early inactivation and rapid divergence of the W chromosome are correctly interpreted, and can be generalized, such a model seems doubtful. In species with female heterogamety, despite the inertness and specialization of the W chromosome, apparently there is no compensatory inactivation of the Z chromosome in the homogametic sex. Studies have so far shown that males express twice as many Z-linked genes as females (Ohno, 1967; Johnson & Turner, 1979; Baverstock, Adams, Polkinghorn & Gelder, 1982). Whilst not explaining the mechanism, this lack of a requirement for Z gene balance obviously would have facilitated rapid inactivation of the W chromosome and must cast doubt upon the necessity of X inactivation as a concomitant of the evolution of the Y chromosome. It seems more likely that there is something particular to the mechanism of inactivation of the sex chromosomes which overcomes the lethality normally inherent in monosomy. Presumably this reflects the special function of these chromosomes in sex determination.

CHROMOSOMAL HIJACKING

The mechanism proposed to explain the evolution of CSD by Jones 1983) and a more formal model incorporating it (Jones, in preparation), suggest the following sequence of events. Firstly, evolution from the environmentally determined sex of "primitive vertebrates", such as the Crocodilia, must have involved a mutation which placed the sex determinants under genetic control. This gave rise to the system of sex determination commonly met with in the Amphibia. There are several mechanisms whereby this may have occurred. For example, it may have resulted from a DNA rearrangement perhaps involving the insertion of a mobile DNA element such as a retroviral sequence or retroposon (Rogers, in preparation), or maybe some sequences included in the Bkm satellite, which affected the controlling sequence of the sex alleles. The insertion of mobile sequences which are hotspots for rearrangements by general recombination (Calabretta, Robberson, Barrera-Saldana, Lambrou & Saunders, 1982) could also have facilitated inversions, which could have juxtaposed the sex alleles with genes from which they were previously separated by some distance. The consequences of this would critically have depended on the genetic composition of these newly covalent regions in the individuals concerned and the precise nature of the rearrangement.

To account for the association of sex chromosomes with the phenomenon of chromosomal inactivation it is assumed (Jones, 1983, in preparation) that one type of neighbouring gene which became affected is normally involved in controlling chromosome condensation. Evidence for the existence of this type of gene, or centre, has been found from studies of the mouse X chromosome where it is referred to as Xce (Cattanach & Isaacson, 1967). It may be envisaged that, perhaps owing to the placement of the breakpoint of an inversion, the Xce came under the promotor control of the dominant female sex determiner so that a "readthrough" became established from the sex allele to the Xce. Thus, the entire chromosome became condensed and decondensed in response to the functional state of the sex determiner. Accidentally, but effectively, the sex allele thereby assumed control of chromosomal conformation. Since mitotically condensed chromosomes are metabolically inert, this chromosomal "hijacking" event would ensure that all genes on the W chromosome could subsequently only express at a time and place appropriate to sex determination, if at all. Inability to influence the phenotype would have led to the maximum rate of genetic degeneration of the W-linked genes and would have permitted the rapid reduction in size of this chromosome, as is seen in many species with CSD. The fact that the hijacking occurred via the Xce, responsible for the control of a mitotic function, could obviously account for the fact that all inactivated chromosomes, including the W chromosome, show an altered DNA synthesis pattern. The complete condensation of the W chromosome except in oocytes, in which it is completely decondensed, is also consistent with an Xce functioning under the control of one, or a few, genes such as the sex determinants.

The accumulation of novel Bkm DNA throughout the W chromosome might have been a consequence of the unhindered spread of sequences originally involved in controlling the expression of the dominant sex determiner and/or to the opportunistic colonization of the chromosome by a mobile type of "selfish" DNA sequence (Orgel & Crick, 1980; Doolittle & Sapienza, 1980). However, until the functions of Bkm DNA have been ascertained such possible explanations cannot be evaluated.

REGULATORY GENE FUNCTION BY ALLELIC EXCLUSION

A critically important question is why such an abrupt metabolic withdrawal of the W chromosome and the similar phenomenon of X chromosome inactivation does not lead to the usually lethal consequences of autosomal monosomy? Two essential factors have been hypothesized (Jones, in preparation). First, that the surviving functional chromosome must have been relatively free from seriously deleterious recessive genes. Second, that the alleles of major regulatory genes, including the sex determinants, operate by somatically irreversible allelic exclusion in which only one of a pair of alleles is expressed in a given cell. Such genes obviously will be "dosage sensitive". At present the best model for this is the immunoglobulin gene family (see Coleclough, 1983) only one allele of each of which becomes expressed in cells of the immune system. However, it is not necessary to suppose that all regulatory genes operate in precisely the same way as the immunoglobulin genes which are obviously specialized for particular functions. Simple aneuploidy or other chromosomal imbalance is therefore lethal because of upsets in the feedback mechanism involved in exclusion and resulting failure of regulatory gene function. However, it is envisaged that chromosomal inactivation is not lethal because it constitutes a collective allelic exclusion and thus provides the essential positive feedback which ensures the normal functioning of regulatory alleles on the active homologue. Thus chromosomal inactivation under sex gene control will not result in functional disturbances. However, in such circumstances, the sex alleles would be constrained to operate either before any other regulatory locus on the same chromosome, or at a time when earlier linked regulatory genes have already completed their function and shut down. Otherwise, since X inactivation is random, the elimination of previously established ongoing vital functions could occur. The fact that chromosomal inactivation is a developmentally early event is consistent with this aspect. In contrast to the assumed mode of function of regulatory alleles, the alleles of non-regulatory ("housekeeping") genes normally both operate and, if codominant, are expressed simultaneously. Inactivation of one allele would therefore not be expected to exert a dramatic effect on the fitness of the individual, unless the allele on the homologous chromosome was a deleterious allele for an essential pathway, which might be expected to be at a low frequency in a wild population. Although there may be some physiological consequences of dosage with respect to "housekeeping" alleles, these might not be incompatible with viability and, with few exceptions, would not be cell-lethal. For example, many tissue culture cell lines are aneuploid as are many tumours. Also some genes evidently escape inactivation in X chromosomes of some mammals. For example in humans the STS (steroid sulphatase) gene located at the distal end of Xp is not inactivated but this same X-linked gene appears to be inactivated in the female mouse (Crocker & Craig, 1983); thus, it seems not to be critical whether there are one or two doses of this particular gene. The more dramatic example, however, has already been mentioned, and is to be seen in all species with female heterogamety in which the males have two active Z chromosomes whilst the females have only one. Thus Z-linked genes apparently show dosage effects (Johnson & Turner, 1979; Baverstock et al., 1982) without serious consequence.

ACCOUNTING FOR THE DIFFERENCES IN CHROMOSOMAL INACTIVATION IN REPTILES, MAMMALS AND INSECTS

To summarize and extend the model to other systems involving chromosomal inactivation: in the instance of snakes such as the Boidae it is assumed that in evolving genetic control of sex alleles the changes involved occurred without reference to the chromosomal Xce. Thus there was no further evolution of a specialized W chromosome. In the case of the reptilian ZW chromosome pair, the model assumes that the sex alleles, in evolving genetic control, interfered with the Xce on the W chromosome but not on the Z, possibly because of the character of the inversions involved in their divergence. Thus, the W inactivated in the heterogametic sex but there is no Z inactivation in the homogametic sex. In mammals, however, the model assumes that the evolutionary changes in the XY bivalent involved such intereference in both. Thus, the inactivation of one X chromosome in the female signifies the operation by allelic exclusion of its sex allele which also controls the mitotic Xce.

The postulate that Xce, which are known to exist on the mammalian X chromosome (Cattanach & Isaacson, 1967), may be potentially functionally autonomous on any given chromosome implies that there is a higher centre which normally co-ordinates their functions during mitosis. If this master regulator or co-ordinator resides on a chromosome bearing a dominant sex gene, it could itself be hijacked by a similar process to that postulated for the Xce. In this case, the entire genome would inactivate at the behest of the sex gene. In certain insects, such as mealy bugs (reviewed in Brown & Chandra, 1977), the entire haploid paternal genome does become condensed and permanently inactivated, consistent with the predictions of the model. Thus the model, which owes its origin to studies of reptilian chromosomes, has the merit of being able to provide a unifying explanation of phenomena involving sex chromosome specialization and inactivation, including the vertebrate systems of male and female heterogamety, and insect systems. Such a common conceptual framework suggests new initiatives for elucidating the underlying molecular causation and evolutionary mechanism of a wide range of chromosomal phenomena of interest to the fields of genetics and evolution.

REFERENCES

Aida, T. (1921). On the inheritance of color in a fresh-water fish *Aplocheilus latipes* Temminck and Schlegel, with special reference to sex-linked inheritance. *Genetics* **6**: 554–573.

Baverstock, P. R., Adams, M., Polkinghorn, R. W. & Gelder, M. (1982). A sex-linked enzyme in birds: Z chromosome conservation but no dosage compensation. *Nature, Lond.* **296**: 763–766.

Beçak, W., Beçak, M. L., Nazareth, H. R. S. & Ohno, S. (1964). Close karyological kinship between the reptilian suborder Serpentes and the class Aves. *Chromosoma* **15**: 606–617.

Brown, S. W. & Chandra, H. S. (1977). Chromosome imprinting and the differential regulation of homologous chromosomes. In *Cell biology: A comprehensive treatise*, **1**: 109–198. Goldstein, L. & Prescott, D. (Eds). New York: Academic Press.

Calabretta, R., Robberson, D. L., Barrera-Saldana, H. A., Lambrou, T. P. & Saunders, G. F. (1982). Genome instability in a region of human DNA enriched in Alu repeat sequences. *Nature, Lond.* **296**: 219–225.

Cattanach, B. M. & Isaacson, J. M. (1967). Controlling elements in the mouse X chromosome. *Genetics* **57**: 331–346.

Charlesworth, B. (1978). A model for evolution of Y chromosomes and dosage compensation. *Proc. natn. Acad. Sci. U.S.A.* **75**: 5618–5622.

Coleclough, C. (1983). Chance, necessity and antibody gene dynamics. *Nature, Lond.* **303**: 23–26.

Crocker, M. & Craig, I. (1983). Variation in regulation of steroid sulphatase locus in mammals. *Nature, Lond.* **303**: 721–722.

Doolittle, F. W. & Sapienza, C. (1980). Selfish genes, the phenotype paradigm and genome evolution. *Nature, Lond.* **284**: 601–603.

Ferguson, M. W. J. & Joanen, T. (1982). Temperature of egg incubation determines sex in *Alligator mississippiensis*. *Nature, Lond.* **296**: 850–853.

Gartler, S. M. & Andina, R. J. (1976). Mammalian chromosome inactivation. *Adv. hum. Genet.* **7**: 99–140.

Gartler, S. M., Liskay, R. M., Campbell, B. K., Sparkes, R. & Grant, N. (1972). Evidence for two functional X chromosomes in human oocytes. *Cell diff.* **1**: 215–218.

Humphrey, R. R. (1942). Sex inversion in the Amphibia. *Biol. Symp.* **9**: 81–104.

Johnson, M. S. & Turner, R. G. (1979). Absence of dosage compensation for a sex-linked enzyme in butterflies (*Heliconius*). *Heredity* **43**: 71–77.

Jones, K. W. (1970). Chromosomal and nuclear localisation of mouse satellite DNA in individual cells. *Nature, Lond.* **225**:912–915.

Jones, K. W. (1983). Evolution of sex chromosomes. In *Development in mammals*, **5**: 297–320. Johnson, M. H. (Ed.). North Holland: Elsevier Science Publishers.

Jones, K. W. (In prep.). *Chromosomal hijacking: A model to explain the evolution of chromosomal sex determination and inactivation.*

Jones, K. W. & Singh, L. (1981a). Conserved sex-associated DNA in vertebrates. In *Genome evolution*: 135–154. Dover, G. & Flavell, R. (Eds). London: Academic Press.

Jones, K. W. & Singh, L. (1981b). Conserved repeated DNA sequences in vertebrate sex chromosomes. *Hum. Genet.* **58**: 46–53.

Lyon, M. F. (1961). Evolution of X chromosome inactivation in mammals. *Nature, Lond.* **250**: 651–653.

Muller, H. J. (1950). Evidence of the precision of genetic adaptation. *Am. J. Human Genet.* **2**: 111–176.
Muller, H. J., League, B. B. & Offerman, C. A. (1931). Effects of dosage changes of sex-linked genes, and the compensatory effects of other gene differences between male and female (abstract). *Anat. Rec.* (Suppl.) **51**: 11D.
Ohno, S. (1967). *Sex chromosomes and sex-linked genes*. Berlin-Heidelberg-New York: Springer.
Orgel, L. E. & Crick, F. H. C. (1980). Selfish DNA: The ultimate parasite. *Nature, Lond.* **284**: 604–607.
Pardue, M. L. & Gall, J. G. (1970). Chromosomal localization of mouse satellite DNA. *Science, Wash.* **168**: 1356–1358.
Ray-Chaudhuri, S. P. & Singh, L. (1972). DNA replication pattern in sex chromosomes of snakes. *Nucleus (Calcutta)* **15**: 200–210.
Ray-Chaudhuri, S. P., Singh, L. & Sharma, T. (1971). Evolution of sex chromosomes and formation of W chromatin in snakes. *Chromosoma* **33**: 239–251.
Rogers, J. (In prep.). *The structure and evolution of retroposons*.
Sharman, G. B. (1971). Late DNA replication in the paternally derived X chromosome of female kangaroos. *Nature, New Biol.* **230**: 231–232.
Singh, L. (1972). Evolution of karyotypes in snakes. *Chromosoma* **38**: 185–236.
Singh, L. & Jones, K. W. (1982). Sex reversal in the mouse (*Mus musculus*) is caused by a recurrent non-reciprocal crossover involving the X and an aberrant Y chromosome. *Cell* **28**: 205–216.
Singh, L., Purdom, I. F. & Jones, K. W. (1976). Satellite DNA and evolution of sex chromosomes. *Chromosoma* **59**: 43–62.
Singh, L., Purdom, I. F. & Jones, K. W. (1979). Behaviour of sex chromosome-associated satellite DNAs in somatic and germ cells in snakes. *Chromosoma* **71**: 167–181.
Singh, L., Purdom, I. F. & Jones, K. W. (1980). Sex chromosome associated satellite DNA: Evolution and conservation. *Chromosoma* **79**: 137–157.
Singh, L. & Ray-Chaudhuri, S. P. (1975). Localisation of C-band in the W sex chromosome of common Indian krait, *Bungarus caeruleus* Schneider. *Nucleus (Calcutta)* **18**: 166–171.
Tagaki, N., Itoh, M. & Sasaki, M. (1972). Chromosome studies in four species of *Ratitae* (Aves). *Chromosoma* **36**: 281–291.

Physiological ecology

Sex Ratio and Survivorship in the Australian Freshwater Crocodile *Crocodylus johnstoni*

GRAHAME J. W. WEBB and
ANTHONY M. A. SMITH*

*School of Zoology, University of New South Wales,
P.O. Box 1, Kensington, New South Wales 2033, Australia,
and Conservation Commission of the Northern Territory,
P.O. Box 38496, Winnellie, Northern Territory 5789, Australia*

SYNOPSIS

Crocodylus johnstoni embryos have their sex determined by incubation temperature. Females result from the highest and lowest temperatures at which embryo survivorship is possible. Consequently, embryos with the highest probability of dying from inadequate temperatures become females. Low temperature incubation prolongs development, increasing the probability of death (especially of females) through drowning in wet season floods. Sex ratio selection may be constrained by survivorship considerations. Both time of hatching and the sex ratio of hatchlings vary between river systems and within the one river system over time. A possible advantage of temperature-dependent sex determination is speculated upon, but not demonstrated.

INTRODUCTION

The endemic Australian freshwater crocodile, *Crocodylus johnstoni*, is widely distributed in the lagoons, rivers and streams which drain the north coast of the Australian continent (Worrell, 1952, 1964; Cogger, 1979). It is one of the more abundant extant crocodilians (Groombridge, 1982), and is neither endangered nor threatened (Jenkins, 1979). The species is particularly amenable to demographic study because individuals congregate during annual dry seasons and nest within a contracted two- to three-week period (Webb, 1982). In addition, it is small in size (maximum sizes are 2–3 m total length and 30–60 kg body weight), and compared to some crocodilians, is easy to handle.

---

*Present address: Department of Population Biology, Research School of Biological Sciences, The Australian National University, P.O. Box 475, Canberra City, Australian Capital Territory 2601, Australia.

Of particular relevance to the present study, *C. johnstoni* has temperature (environmental?) dependent sex determination (TSD) (Webb, Buckworth & Manolis, 1983d). This complicates demography, because it means that the sex of offspring in any one year or area is likely to be the interactive result of: nest sites available in a particular area or year; nest sites selected; time of nesting relative to environmental temperature cycles (and perhaps cycles of other parameters); the bias of selected nests (and perhaps eggs; Bull, Vogt & Bulmer, 1982) to one sex or the other; and possible long-term selection influences on the sex ratio (Bull & Vogt, 1979, 1981; Bull, 1980, 1981a, b; Mrosovsky, 1980; Mrosovsky & Yntema, 1980; Miller & Limpus, 1981; Bull, Vogt & Bulmer, 1982; Bull, Vogt & McCoy, 1982; Bulmer & Bull, 1982; Ferguson & Joanen, 1982, 1983; Vogt & Bull, 1982; Vogt, Bull, McCoy & Houseal, 1982). Many of the same factors affect embryo survivorship (Webb, Buckworth & Manolis, 1983d; Webb, Sack, Buckworth & Manolis, 1983), and it would be surprising if survivorship, sex determination and the sex ratio of annual recruits were independent of each other.

In the present study we investigate the extent to which these factors are interrelated in *C. johnstoni*, and specifically examine:

(i) Whether or not TSD, as determined by constant temperature incubation in the laboratory, reflects accurately the way in which sex determination operates in the field.

(ii) Whether or not as a result of TSD there is differential survivorship of eggs destined to become males or females.

(iii) The extent to which differential survivorship interacts with TSD to vary sex ratio at hatching (recruit sex ratio) within the one population over time, and between populations.

(iv) The extent to which survivorship constraints could be expected to limit sex ratio selection.

The three lines of evidence pursued are: an examination of field nests from laying to hatching; an examination of geographical variation in the time of hatching and sex ratio of recruits; and an analysis of variation in the population age structure. On the basis of these and other results we hypothesize that: the relationship between temperature and sex may be an indirect one mediated through the profound effect temperature has on embryonic metabolic and development rates; sex ratio selection is constrained by selection for maximizing embryo survivorship; sex-specific survivorship may be a fitness character consistent with the model of Charnov & Bull (1977), but a selective advantage in directing low probabilities of surviving at females would need to be demonstrated; an advantage of TSD over genotypic sex determination (GSD) would exist if TSD was itself an unavoidable consequence of a mechanism related primarily to enhancing embryo survivorship.

THE McKINLAY RIVER *C. JOHNSTONI* POPULATION: A SELECTED REVIEW

Study Area and History of Protection

The McKinlay River drains an extensive black soil plain of grassland, savannah and light eucalypt forest (Webb, Manolis & Buckworth, 1983). The substrate of the mainstream bed is typically sand, and sandbanks are often associated with flood plain billabongs (lagoons). During the dry season (April/May to November) water is restricted to isolated pools in the mainstream bed and flood plain creek lines, where *C. johnstoni* congregate, and ultimately nest (August/September). The wet season is usually heralded by isolated storms in September/October, but 92% of the 1400 mm annual average rainfall is between November and April, and this is associated with widespread flooding.

Crocodylus johnstoni was legally hunted throughout the Northern Territory in 1960–1963, when it was protected. Density has been steadily increasing since that time.

Population Size and Age Structure

In 1979 the McKinlay River population was estimated as 1000 animals (Webb, Manolis & Buckworth, 1983), and the approximate age structure (Fig. 1) was derived from a sample of 240 recaught individuals (Webb, Buckworth & Manolis, 1983a). The method of predicting age was based on the relationship between age and size, but with two corrections. First, the difference between the observed growth rate of an individual of given size, and the predicted mean growth rate of that sized individual in the population, was used to compute an approximate age-size relationship for the individual (a major correction). Second, predicted ages were rounded to the nearest November hatching period. Capture methods (almost exclusively fine nets; Webb & Messel, 1977) resulted in random capture of most size groups. In the younger animals (< six years), ages were considered accurate to the month.

The Population Sex Ratio

Of 697 *C. johnstoni* caught and sexed, the sex ratio (expressed as the proportion of males) has been 0.33 (0.31 among recaught individuals).

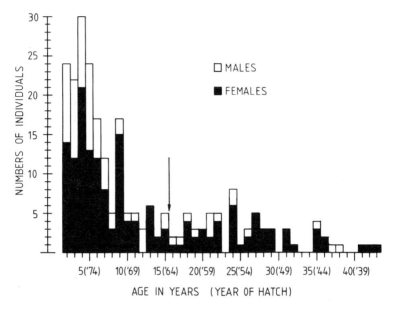

FIG. 1. The estimated 1979 age structure of 240 McKinlay River *C. johnstoni*. The arrow indicates when hunting ceased (protection).

The sex ratio of immatures (0.36) is significantly higher than that of matures (0.17) (Webb, Buckworth & Manolis, 1983a, d).

Post-hatching Mortality and Growth

Mortality between hatching and one year of age occurs mainly during the first wet season. It was estimated as 98% in one experiment (Webb, Buckworth & Manolis, 1983d), although 90–95% may be a more typical range (unpublished data).

Post-hatching male and female survivors grow at the same rate until about five years of age after which males grow faster than females (Webb, Buckworth & Manolis, 1983a). Recapture results demonstrate no significant sex-specific mortality in immature animals (Webb, Buckworth & Manolis, 1983a).

Movement and Dispersal

A high percentage (96.4%) of recaught animals are within 6 km of where they had been marked a year before. Long-distance movements (for example between the upstream and downstream parts of the study area) were rare (< 1%). The proportions of males and females which move are not significantly different, and both sexes have a well-developed homing ability (Webb, Buckworth & Manolis, 1983b).

Reproduction and Nesting

Females mature at 74–78 cm snout-vent length (SVL) and 11–14 years of age, and males at about 87 cm SVL and 16–17 years of age (Webb, Buckworth & Manolis, 1983a, d). Reproductive senescence is about 40–45 years of age. Each year 84.4% of females > 11 years of age and 8% of females ≤ 11 years of age lay eggs. Mean clutch size is 13.2 ± 3.2 eggs. The egg-laying period is contracted (August 21 ± 6 days (SD)), and appears constant (within two weeks) from year to year. Nest sites are chosen after a series of "test" excavations in sand or other friable substrates close to water. The availability and location of friable substrates is dependent on water levels, which themselves depend on the extent of the previous wet season. Nesting banks used in one year may be inundated and unavailable in another.

Eggs are typically deposited within a distinct moisture band (~ 5% moisture in oven dried samples), and depth to the top egg (19.6 ± 7.5 cm) is partly dependent on substrate friability. There is a significant tendency for large clutches of large eggs (from older or larger females) to be laid early in the egg-laying period.

Clutch temperatures show considerable daily variation (3.5°C range in one nest) and baseline levels depend on the degree of exposure, depth, moisture content and thermal characteristics of the substrate around and beneath the clutch. Some of the hottest nests at the time of egg-laying have been those in which sand overlies a solid clay bank. Incubation takes 9–14 weeks. Females rarely attend nests during this period, but excavate nests from which young are calling at the end of incubation.

Egg mortality is high (60–70% in one year). It results from varanid predation, inundation and/or washing out of nests with subsequent exposure of eggs, females excavating the nests of other females, trampling by buffalo, wild pig predation, overheating and/or possibly desiccation. Predation is significantly greater at colonial nesting sites than at sites where only one or two females nest.

Temperature-dependent Sex Determination

Sex varies with incubation temperature (Fig. 2) (Webb, Buckworth & Manolis, 1983d; additional data), but differs from *Alligator mississippiensis* (Ferguson & Joanen, 1982, 1983) and *C. porosus* (unpublished data) in the following features:

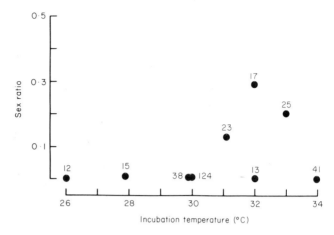

FIG. 2. The general trend in the relationship between sex ratio (proportion of males) and incubation temperature (< ± 1°C) as indicated by laboratory trials. Numbers are sample sizes.

(i) Females have been produced at all temperatures so far tested.

(ii) High (34°C) and low (26–30°C) temperatures produce exclusively females.

(iii) There is a narrow range of temperatures in which males are produced (31–33°C), and thus there are likely to be two temperature thresholds (where 0.5 sex ratios result) for sex determination.

At 31–32°C sex is determined at approximately 20–30 days of age (unpublished data), although the times and embryonic stages at which sex is determined are probably temperature-dependent (Yntema, 1979; Bull, 1980; Bull & Vogt, 1981; Ferguson & Joanen, 1983; unpublished data on *C. porosus*).

At the time of hatching, the gonads of most males [Fig. 3(A), (B)] and females [Fig. 3(C), (D)] are well differentiated. Unlike *A. mississippiensis* (Ferguson & Joanen, 1983), the cliteropenis is also differentiated, and most *C. johnstoni* hatchlings can be externally sexed (Webb, Manolis & Sack, in press).

The gonads and clitoris of females produced at the highest incubation temperatures (34°C) are macroscopically similar to those of 30°C females. Histologically [Fig. 3(E)] these females have a thin cortex and a greatly vacuolated and distorted medulla. This is consistent with gonadal development being severely retarded (relative to embryo stage) and the relatively young (small) gonad being distorted by enhanced development of the mesonephros, metanephros and adrenal to which it is attached.

In two of some 40 females produced at an extremely (~ 34°C) high

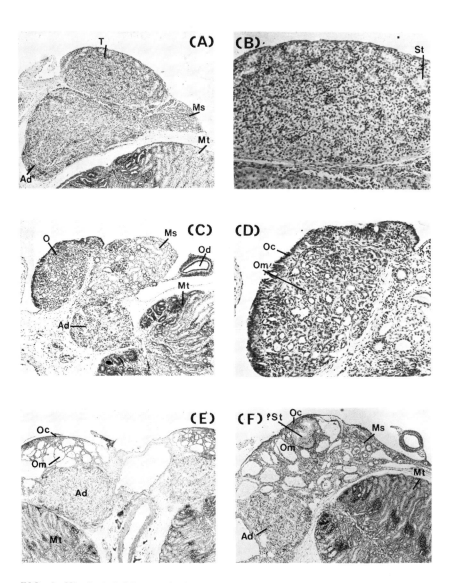

FIG. 3. Histological differences in the gonad region of hatchling *C. johnstoni*. (A), (B) A wild male, (C), (D) A wild female, but with similar gonads to a 30°C incubated female, (E) A 34°C incubated female with a thin cortex and greatly vacuolated medulla, (F) A 34°C female which contains an apparent seminiferous tubule. Ad, adrenal; Ms, mesonephros; Mt, metanephros; O, ovary; Oc, cortex; Om, medulla, Od, Oviduct; St, seminiferous tubule; T, testis.

incubation temperature that were examined histologically, an apparent seminiferous tubule was present [Fig. 3(F)], suggesting some may be hermaphroditic. That others retain their female sex was indicated by five 34°C females incorporated into a raising experiment (Webb, Buckworth & Manolis, 1983c). All retained female characteristics. Two had ovaries, oviducts and a distinct clitoris when examined after four months, and three had a distinct clitoris when released after seven months.

Temperature-dependent Survivorship

At 30°C, 63–100% of fertile eggs develop successfully. Incubation time is 84–85 days, and few hatchlings are abnormal (< 1%). In contrast, at 34°C only 21% hatch, and 74% of hatchlings have obvious abnormalities; incubation takes 68–69 days. At 26°C all embryos die before hatching, and most advanced dead embryos are abnormal. Based on development rates at 26°C (2.1 times slower than at 30°C), total incubation time would be 170–180 days if any animals survived (Webb, Buckworth & Manolis, 1983d).

Annual Variation in the Number and Sex Ratio of Hatchlings

On 10 November 1980, 1981 and 1982, all hatchlings sighted in the same 7 km section of the Mary River (adjacent to the McKinlay River) were collected. The number and sex ratio varied from year to year (Webb, Buckworth & Manolis, 1983d):

	Males	Females	Sex ratio	Total
1980	57	54	0.51	111
1981	22	63	0.26	85
1982	49	40	0.55	89

METHODS

Field Nests

In August–September 1982, 28 nests were located in the McKinlay River area soon after laying. Clutch temperatures (one to two eggs deep) were measured, and the top egg was removed for aging (Webb, Buckworth & Manolis, 1983d) unless a clutch was known to have been laid the previous night (prior visits; abundant mucus on the eggs). Most clutches were 1–6 days old ($N = 15$), six clutches were 7–13 days old and five were 14–24 days old. Two clutches could not be aged because the eggs removed were infertile. Nests were re-covered and 1m² pieces of wire mesh pegged over them to restrict predators.

At or near hatching, each nest was revisited. Predators had burrowed beneath the mesh at eight nests and removed all eggs. At the others, a second clutch temperature was measured, then each egg was removed, numbered and its position mapped in relation to other eggs in the clutch (Ferguson & Joanen, 1982, 1983). Eggs which were not hatching were returned to the laboratory and incubated at 31–32°C (13 clutches); laboratory incubation took < 6 days for four clutches, 7–13 days for two clutches and 14–21 days for seven clutches. On the basis of the second clutch temperature measured, and information on development rates at different temperatures (Webb, Buckworth & Manolis, 1983d; Webb, Buckworth, Manolis & Sack, 1983), minor corrections were made to the laboratory incubation times so that total field incubation times could be estimated. Two new nests were located at the time the others were revisited.

Hatchlings were consecutively numbered (independent of their clutch of origin), measured, weighed and sexed by examination of the cliteropenis. They were maintained in a heated raising chamber (Webb, Buckworth & Manolis, 1983c) for up to three weeks before being sexed again. If any doubts or inconsistencies existed in the sexing, individuals were killed and their gonads examined macroscopically and histologically (haematoxylin and eosin). Individuals which either hatched prematurely, or were abnormal, or were dead in eggs at advanced embryonic stages, were also examined histologically. All 22 animals thus examined were of the sex allocated by cliteropenis examination.

Geographic Variation in Hatchling Sex Ratios and Time of Hatching

In November–December 1982, 4569 hatchlings were collected in repetitive trips to a number of different Northern Territory rivers (Fig. 4). All were weighed and had snout-vent length (SVL) and the width of the umbilical scar measured, and 4556 were sexed by examination of the cliteropenis (in 13 animals sex was unclear). On the basis of information from hatchlings released and recaught, yolk scar width was used as an indicator of week of hatch: ≥ 5 mm = day of hatch; > 1.8 mm = one week; 0.3–1.7 mm = one to two weeks; 0.0–0.2 mm = two to three weeks plus. In 12.5% of animals the yolk scar was healed at the time of sexing. The majority of such hatchlings were in the same size class as animals with a 0.1 mm yolk scar from the same batch: they were assigned an age of two to three weeks, but in a few cases their size, and

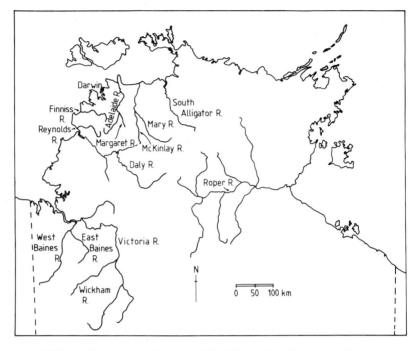

FIG. 4. The location of rivers in which *C. johnstoni* hatchlings were collected.

information from prior visits, were used to age them at three to four or four to five weeks.

Age Structure Analysis

General

Implicit within the population age structure (Fig. 1) is variation in sex ratio and survivorship (in the one area) over a number of years. In order to examine that variation we made the following assumptions:

(a) The aging method was sufficiently accurate for year class sex ratios up to nine years (on Fig. 1) to be accepted as realistic estimates of cohort sex ratio, if cohorts were represented by > 10 individuals. This assumption is supported, because the aging method was accurate to the month up to six years of age and contained few errors up to nine years. There is a high proportion (24%) of each population year class represented in the age structure. Using > 10 individuals effectively rendered one cohort sex ratio unusable (eight years).

It should be recognized that Fig. 1 is the 1979 age structure, whereas that originally presented (Webb, Buckworth & Manolis, 1983a) was the 1978 age structure.

(b) Recaptured animals were a random sample of the population. This assumption is supported by: the capture method (fine nets) being random for most size groups; the high recapture rates achieved (61.5% of animals marked in 1978 were recaught once in two subsequent annual recapture efforts); the higher proportion (24%) of the population represented in the age structure; and the sex ratio of initial captures (0.33) in comparison to that of recaptures (0.31).

(c) That age-specific survivorship is an important variable affecting the size of cohorts on Fig. 1 (Caughley, 1976).

(d) That for the purposes of estimating age-specific survivorship, and examining environmental variables which may have had a significant influence on survivorship, the size of year classes up to 16 years could be considered correct. This assumption is supported by the age-specific survivorship curve being primarily dependent on the younger year classes, and by the normal distribution of aging errors up to 16 years. Very high or very low survivorship should be reflected, even in the less accurately aged year classes (10–16 years).

(e) That up to 16 years, sex-specific trends in mortality, emigration and immigration would be negligible compared to the extent of annual variation in the number and sex ratio of recruits. This assumption is supported by the recapture data, which indicated no significant sex-specific mortality or dispersal predisposition, and by the extent of annual variation in the number and sex ratio of hatchlings in one area (see p. 326).

Estimating the number and sex ratio of recruits in past years

On Fig. 1, each year class effectively represents the animals that have survived from the number of eggs laid in a particular year to two years of age, three years, and so on. If the population had been at equilibrium in the past, and the number of eggs laid each year could be assumed to have been more or less constant, those survivors could be plotted as the proportion of the number of eggs laid which survived to different years of age. The difference in survival proportion (SP) between, for example years 3 and 4, contains an estimate of age-specific mortality from three to four years of age.

To use this approach with the McKinlay River population, which has been recovering since protection, the number of eggs laid in past years (the number of mature females present) cannot be assumed to have remained constant, and needs to be estimated (more females have been maturing than dying as the population was recovering and expanding). Animals > 11 years were assumed to be mature, and 1.35 females were assumed to have died each year for the previous 15 years (the mean number of animals per year class between 27 and 43 years on Fig. 1; animals mature at the time of protection). By recapitulating the age structure one year at a time and making an addition for dead animals, the number of mature females (> 11 years) in past years was estimated (Table I). It was assumed that 92.4% of these nested annually (the estimate was increased from the recorded 84.4% to account for younger females nesting), with a mean clutch size of 13.2 eggs. Annual egg production for each year was calculated (Table I), and survivorship from eggs to two, three, and up to 16 years of age (as discussed above) was calculated (Table I; Fig. 5). The relationship between SP from eggs to consecutive years (Y) was fitted to an exponential curve:

$$SP = 0.0505 \, e^{-0.1632Y} \; (r^2 = 0.75; P < 0.001).$$

Seventy-five per cent of the variation in year class size between two and 16 years (on Fig. 1) could be accounted for by age-specific mortality. In the results, the extent to which the remaining variation can be attributed to environmental conditions during the relevant incubation periods is examined.

On the basis of Fig. 5, the SP to one year of age can be approximated for each year (Table II), by assuming the curve could be extrapolated to year 1 (two years was the youngest year class). However, it could not be extrapolated to zero, because mortality from zero (hatching) to one year is known to be much greater than that estimated by the curve, and may be equally variable. Accordingly, SP_1 reflects the combined egg to hatching and hatching to one year survivorship, and SP_0 (survival proportion to hatching) cannot be estimated.

RESULTS

Field Nests

Sex ratios from individual clutches (Fig. 6) paralleled constant temperature incubation results (Fig. 2), with a female bias in the fast and slow developing nests. The peak of the male bias (sex ratios near 1.0) was in nests taking 72–74 days to incubate. At constant temperature, this incubation time would be achieved at 32–33°C (Webb, Buckworth & Manolis, 1983d), which is similar to the male peak in Fig. 2. The

TABLE 1
Estimated egg production and survival proportions to increasing years of age

Year of hatch	No. mature females still alive in 1979	Addition for dead females (1.35 per year)	Total no. of mature females per year	Estimated number of eggs produced	No. of offspring alive in 1979	Survival proportion to (N) years
1977	54	2.70	56.7	691.6	23	0.0347 (2)
1976	52	4.05	56.05	683.6	22	0.0322 (3)
1975	49	5.40	54.40	664.7	30	0.0451 (4)
1974	48	6.75	54.75	667.8	24	0.0359 (5)
1973	47	8.10	55.10	672.0	17	0.0253 (6)
1972	43	9.45	52.45	639.7	12	0.0188 (7)
1971	41	10.80	51.80	631.8	5	0.0079 (8)
1970	38	12.15	50.15	611.7	17	0.0278 (9)
1969	36	13.50	49.50	603.7	5	0.0083 (10)
1968	32	14.85	46.85	571.4	5	0.0087 (11)
1967	32	16.20	48.20	587.9	3	0.0051 (12)
1966	26	17.55	43.55	531.2	6	0.0113 (13)
1965	25	18.90	43.90	535.4	2	0.0037 (14)
1964	23	20.25	43.25	527.5	5	0.0095 (15)
1963	18	21.60	39.60	483.0	2	0.0041 (16)

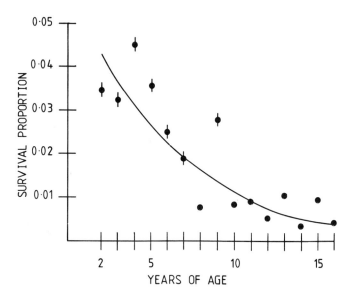

FIG. 5. The estimated proportion of eggs laid in previous years which are represented by survivors in subsequent years. Bars indicate the estimates for which there is also a sex ratio estimate.

distribution of clutches with regard to sex ratio (Fig. 7) was skewed towards females, and was not bimodal.

Females produced from the fastest developing clutches (65 days) were histologically similar to those from 34°C constant temperature incubation, although vacuolation of the medulla was not as pronounced. No animals with hermaphroditic tendencies were detected.

Within nests that produced both males and females, nest maps (Fig. 8) indicated a tendency for eggs producing either sex to be clumped, though not necessarily at the top or bottom of a nest. This is consistent with thermal or other gradients not necessarily being in the vertical plane.

With clutches lumped by mean incubation time (Fig. 9), survivorship of fertile eggs was reduced in the fast and slow developing nests, and within one group (76–79 days) in the middle of the range. The frequency of advanced dead embryos was distributed similarly.

Final clutch temperatures (T_f; 34.3 ± 1.8°C (SD); $N = 17$) were significantly higher than initial clutch temperatures (T_i; 30.2 ± 1.6°C; $N = 17$), and this difference was unchanged when both T_i and T_f were corrected (± 1°C) according to time of day at which they had been measured. T_i and T_f were not significantly correlated with each other ($r^2 = 0.003$) but the change in temperature ($\Delta T = T_f - T_i$) was predictable from T_i: $\Delta T = 34.2 - 1.0 T_i \pm 1.5$°C ($r^2 = 0.54; 0.01 > P > 0.001$).

TABLE II
Water height (WH) and rainfall characteristics of previous years, the estimated survival proportion from eggs to one year of age and sex ratio. Numbers in brackets are based on samples of less than 10 individuals.

Year of hatch	Estimated survivorship to one year of age	Sex ratio	Maximum Aug. WH (m)	Maximum Sept. WH (m)	Maximum Oct. WH (m)	WH Oct.–Aug. (m)	Total October rainfall	Total October raindays
1977	0.041	0.42	0.74	0.69	0.67	−0.07	12	3
1976	0.045	0.45	0.80	0.76	0.74	−0.06	18	2
1975	0.074	0.30	0.71	0.68	1.43	0.72	137	7
1974	0.069	0.46	0.93	1.00	1.63	0.70	57	4
1973	0.053	0.29	0.57	0.27	0.60	0.03	27	6
1972	0.050	0.33	0.47	0.49	0.29	−0.18	31	4
1971	0.025	(0.40)	0.03	0.01	1.69	1.66	145	12
1970	0.103	0.12	0.20	0.02	1.42	1.22	102	7
1969	0.036	(0.20)	0.72	0.61	2.66	1.94	74	10
1968	0.045	(0.20)	–	0.88	0.69	–	54	7
1967	0.031	(1.00)	–	–	–	–	14	3
1966	0.080	(0.00)	0.70	–	–	–	65	6
1965	0.031	(0.00)	0.95	0.91	0.91	−0.04	27	2
1964	0.093	(0.40)	0.93	0.91	1.70	0.77	143	7
1963	0.047	(0.50)	0.95	0.91	0.91	−0.04	64	6

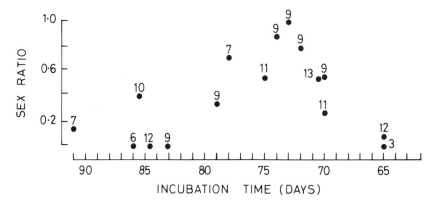

FIG. 6. Sex ratios of wild *C. johnstoni* clutches as a function of their mean incubation time. Numbers are the number of individuals in each clutch which could be sexed.

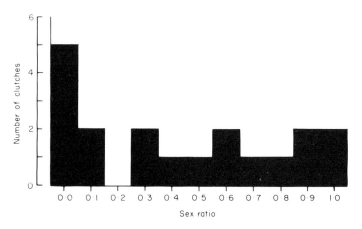

FIG. 7. Frequency distribution of clutch sex ratios for 19 *C. johnstoni* nests.

This indicates that independent of the range of T_i's, temperatures would tend to standardize at the same T_f's.

When nests were lumped according to T_i and T_j (Fig. 10), some trends in sex ratio and survivorship were apparent:

(i) Sex ratio was normally distributed with T_j, which is consistent with T_j being measured closer to the time that sex was actually determined.

(ii) The most commonly selected T_j range (87 viable eggs; 47%) was 30.0–30.9°C, which corresponded with a sex ratio of 0.49.

(iii) High and low T_j's gave a pronounced female bias, and only one T_i range (31.0–31.9°C) gave a male bias (0.60; 17% of viable eggs).

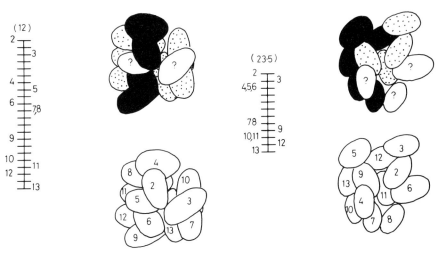

FIG. 8. Maps showing the distribution (when seen from above) of eggs with male (black) and female (dots) embryos. The number in brackets is the distance between the substrate surface and the top mid-point of the highest egg (egg 1 was removed for aging), and the vertical scale (5 mm divisions) shows the vertical distribution of eggs relative to each other. The nests took 85–86 days (left) and 70 days (right) to incubate. '?' indicates where embryos had died at a young age, and/or eggs were infertile.

(iv) The association between survivorship and T_i was the opposite to that between survivorship and T_f, if the five animals with T_i's in the 29.0–29.9°C category are considered anomalous.

(v) Minimum survivorship associated with T_i and T_f both corresponded with sex ratios strongly skewed towards females.

These results indicate that more animals selected nests in the T_i range 30.0–30.9°C than any other, and gained high survivorship and a nett sex ratio near 0.5. The normal distribution of temperatures selected around that mean resulted in most T_i's giving high survivorship, the exception being 34.0–34.9°C (female biased); in these, mortality probably reflects the effect of high temperature on young, rapidly developing embryos. Because T_i is independent of T_f, a range of T_i's could theoretically be utilized without compromising survivorship associated with T_f, but to shift by +1°C the normal distribution of T_i's selected would place even more animals in the critical range.

Few, if any, of the nests selected exceeded T_f values which were lethal to the advanced embryos: nests with the highest T_f's had maximal survivorship. In contrast, low survivorship tended to be associated with low T_f (mainly females). This trend is consistent with advanced embryos needing high temperatures to optimize growth and yolk utilization, both of which are impaired at low temperatures (unpublished data). It

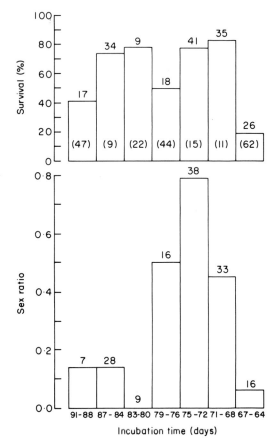

FIG. 9. The relationship between incubation time, sex ratio, and the percentage of viable eggs which survived in wild *C. johnstoni* nests. Upper: numbers are the numbers of viable eggs, and the percentage of viable eggs which contained advanced dead embryos (in parentheses). Lower: numbers are the numbers of individuals sexed.

is also consistent with nests being so selected that independent of T_i, they have a high T_f.

Temperature also affects survivorship of embryos through its influence on development rate, and thus time of hatching relative to the probability of there being a flood. From the nests studied 59 males and 67 females hatched successfully and were considered potential survivors (sex ratio = 0.47). As in *Graptemys* spp. (Vogt & Bull, 1982) there was a tendency for the fastest developing nests to hatch earliest, with males and high temperature females tending to hatch before lower temperature females. The greatest disparity in the hatching of the total complement of each sex was during 8–22 November (Fig. 11), which

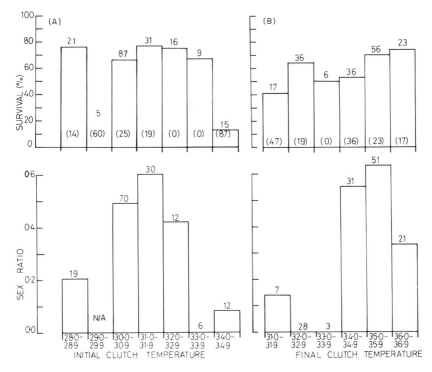

FIG. 10. The relationship between individual (A) and final (B) clutch temperatures, and the sex ratio and percentage survivorship of viable eggs. Upper: numbers are the number of viable eggs and the percentages of viable eggs which contained advanced dead embryos (in parentheses). Lower: numbers are the numbers of individuals sexed.

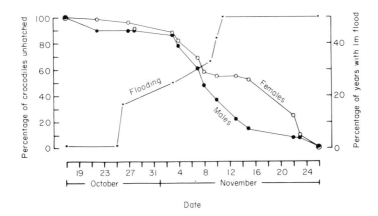

FIG. 11. The percentage of the total complement of male and female *C. johnstoni* from the field nests being monitored, which had hatched at different times, related to the increasing probability of there being a 1 m flood to inundate nests.

corresponded with an increased probability of flooding (based on 12 years' data).

The absolute height of clutches above water level at the time of laying in one part of the McKinlay River was 56 ± 18 cm ($N = 24$). This level varies within the system, as does absolute water height. Thus, when a 1 m water rise is recorded in one part of the study area, the actual rise in specific ponds varies around this recording. Nevertheless, such a water rise does inundate many nests, thereby drowning a greater proportion of females than males (no flood occurred during the study year).

Summarizing, the field results demonstrate:

(i) Nest temperatures affect *C. johnstoni* sex determination in the field in the same way that constant temperatures influence embryonic sex determination in laboratory incubators.

(ii) Nest temperatures affect embryo survivorship. There is a direct effect, which probably reflects the influence of temperature on developing embryos of different ages, and an indirect effect, mediated through development rates, the time of hatching and the probability of flooding.

(iii) The effects of nest temperature on sex determination and survivorship are not independent. Embryos most likely to die are females, and consequently, increased survivorship is likely to be associated with decreased sex ratios.

(iv) Within the range of temperatures where survivorship is not compromised, intitial and final nest temperatures are independent. However, the temperatures selected are normally distributed, and shifting the mean (for any reason) would shift that distribution. Thus, unless the thermal environment changed, a greater proportion of eggs would exceed lethal limits if the mean nest temperature was increased or decreased.

Geographic Variation

The time of hatching (Table III), the rate at which the total complement of hatchlings appeared (Table IV), and the sex ratio of hatchlings (Table V), varied between rivers. The following trends existed:

(i) Hatching in the south was later than in the north or central areas (Table III); nesting is also later (unpublished data).

(ii) The proportion of the total complement of hatchlings which appeared in the first week was higher in the south (40.4%; Table IV) than in the north (29.1%) or central (9.1%) areas.

(iii) The sex ratio of hatchlings was higher in the south (0.40) than in either the north (0.31) or central (0.31) areas.

In all rivers, sex ratio varied as a function of week of hatch (Table V).

TABLE III
The percentages of the total number of hatchlings from each river which hatched in different weeks

Area (no. of hatchlings)	River	No. hatchlings per river	Oct. 23–29	Oct. 30 –Nov. 5	Nov. 6–12	Nov. 13–19	Nov. 20–26	Nov. 27 –Dec. 3	Dec. 4–10	Dec. 11–17
North (795)	South Alligator	34	–	–	–	–	97.0	–	3.0	–
	Mary	629	–	8.9	19.4	39.9	5.4	23.1	3.3	–
	McKinlay	63	6.3	34.9	41.3	9.5	7.9	–	–	–
	Margaret	36	–	2.8	50.0	47.2	–	–	–	–
	Adelaide	33	–	–	–	30.3	66.6	3.0	–	–
Central (1858)	Finniss	337	–	–	22.9	43.0	22.3	8.3	3.6	–
	Reynolds	222	–	–	1.7	12.9	41.0	28.2	14.0	2.2
	Daly	1299	0.9	17.0	28.1	41.2	7.9	3.9	1.1	–
South (1916)	Roper	163	–	–	–	–	59.5	30.7	9.8	–
	Victoria	1134	–	–	1.7	12.9	41.0	28.2	14.0	2.2
	Wickham	235	–	–	2.1	2.1	13.6	44.3	26.4	11.5
	Wickham/Victoria	83	–	–	–	15.7	39.8	14.5	28.9	1.2
	East Baines	50	–	–	–	–	74.0	18.0	8.0	–
	West Baines	211	–	–	–	–	77.3	21.8	1.0	–
	E/W Baines	40	–	–	–	–	52.5	40.0	7.5	–
North	5 rivers	–	1.3	9.3	22.1	25.4	35.4	5.2	1.3	–
Central	3 rivers	–	0.3	5.6	18.2	35.4	18.3	20.4	1.6	0.2
South	7 rivers	–	–	–	0.5	4.4	51.1	28.2	13.7	2.1
North and Central	8 rivers	–	0.9	8.0	20.7	29.2	29.0	10.9	1.4	0.1
All areas (4569)	15 rivers	–	0.5	4.2	11.3	17.6	39.3	19.0	7.1	1.0

TABLE IV
The percentages of the total number of hatchlings from each river system which hatched in consecutive weeks relative to the first week of hatch in each area

Area	River	No. of hatchlings	Week of hatch relative to first hatchlings appearing						
			1	2	3	4	5	6	7
North	South Alligator	34	97.0	–	3.0	–	–	–	–
	Mary	629	8.9	19.4	39.9	5.4	23.1	3.3	–
	McKinlay	63	6.3	34.9	41.3	9.5	7.9	–	–
	Margaret	36	2.8	50.0	47.2	–	–	–	–
	Adelaide	33	30.3	66.6	3.0	–	–	–	–
Central	Finniss	337	22.9	43.0	22.3	8.3	3.6	–	–
	Reynolds	222	3.6	22.1	24.8	49.1	–	2.2	–
	Daly	1299	0.9	17.0	28.1	41.2	7.9	3.9	1.1
South	Roper	163	59.5	30.7	9.8	–	–	–	–
	Victoria	1134	1.7	12.9	41.0	28.2	14.0	2.2	–
	Wickham	235	2.1	2.1	13.6	44.3	26.4	11.5	–
	Wickham/Victoria	83	15.7	39.8	14.5	28.9	1.2	–	–
	East Baines	50	74.0	18.0	8.0	–	–	–	–
	West Baines	211	77.3	21.8	1.0	–	–	–	–
	E/W Baines	40	52.5	40.0	7.5	–	–	–	–
North	–	795	29.1	34.2	26.9	3.0	6.2	0.7	–
Central	–	1858	9.1	27.4	25.1	32.9	3.8	1.5	0.4
South	–	1916	40.4	23.6	13.6	14.5	5.9	2.0	–
North and Central	–	2653	21.6	31.6	26.2	14.2	5.3	1.0	0.1
All areas	–	4569	30.4	27.9	20.3	14.3	5.6	1.4	0.1

The sex ratio of hatchlings from different areas of the Northern Territory as a function of the time that first hatchlings appeared

Area	River	No. hatchlings sexed	Total sex ratio	Week of hatch relative to first hatchlings appearing						
				1	2	3	4	5	6	7
North	South Alligator	34	0.12	0.09	–	1.00*	–	–	–	–
	Mary	629	0.31	0.57	0.30	0.42	0.26	0.09	0.0	–
	McKinlay	60	0.25	0.25*	0.37	0.19	0.00*	–	–	–
	Margaret	36	0.44	1.00*	0.44	0.41	–	–	–	–
	Adelaide	33	0.45	1.00	0.23	0.00*	–	–	–	–
Central	Finniss	337	0.29	0.45	0.23	0.28	0.21	0.17	–	–
	Reynolds	220	0.23	0.38*	0.23	0.20	0.24	–	0.00*	–
	Daly	1292	0.40	0.55	0.48	0.53	0.34	0.21	0.18	0.00
South	Roper	163	0.20	0.10	0.46	0.06	–	–	–	–
	Victoria	1134	0.34	0.32	0.44	0.41	0.26	0.27	0.00	–
	Wickham	235	0.36	1.00*	0.80	0.78	0.43	0.10	0.00	–
	Wickham/Victoria	83	0.25	0.38	0.36	0.25	0.04	0.00*	–	–
	East Baines	50	0.58	0.68	0.44*	0.00*	–	–	–	–
	West Baines	210	0.61	0.63	0.53	1.00*	–	–	–	–
	E/W Baines	40	0.48	0.80	0.13	0.00*	–	–	–	–
All North mean	<5 rivers	–	0.31	0.55 (3)	0.34 (4)	0.34 (3)	0.26 (1)	0.09 (1)	0.00 (1)	–
All Central mean	<3 rivers	–	0.31	0.50 (2)	0.31 (3)	0.34 (3)	0.26 (3)	0.19 (2)	0.18 (1)	0.00 (1)
All South mean	<7 rivers	–	0.40	0.49 (6)	0.38 (5)	0.38 (4)	0.24 (3)	0.19 (2)	0.00 (2)	–
North and Central mean	<8 rivers	–	0.31	0.53 (5)	0.33 (7)	0.34 (6)	0.26 (4)	0.16 (3)	0.09 (2)	0.00 (1)
Total mean	<15 rivers	–	0.35	0.51 (11)	0.35 (12)	0.35 (10)	0.25 (7)	0.17 (5)	0.05 (4)	0.00 (1)

*Sample size < 10, in which case not included in totals.
Numbers in parentheses are the number of rivers included in each computation.

In most rivers, highest sex ratios occurred early in the hatching period, and as hatching progressed, sex ratios became more and more female-biased.

The South Alligator and Roper Rivers were notable exceptions. Sex ratios in the first week of hatch (0.09 and 0.10 respectively) were strongly female biased, even though 97% and 59.5% respectively of the total hatchling complement appeared in the first week (Table IV). The small number of animals collected reflects a paucity of hatchlings in these areas, and not a lack of catch effort.

As in the field nests (Fig. 11), variation in sex ratio and the time and rate of hatching was reflected in a disparity in the rate at which the total complement of males and females hatched (Fig. 12). The same pattern characterized most rivers (it was the opposite in the South Alligator and Roper), and indicates that as the probability of flooding increases (as it does throughout the period), the probability of female survivorship decreases at a greater rate than does that of males.

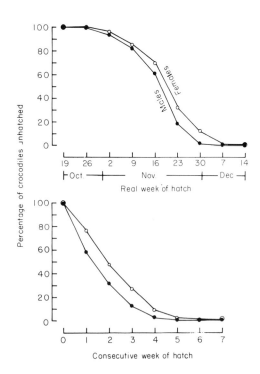

FIG. 12. The percentage of the total complement (4556) of male and female *C. johnstoni* examined in the geographic variation study, which hatched at different times (upper), and when related to the first week in which hatchlings appeared in each area (lower).

Age Structure Analysis

Sex ratio

Variation in cohort sex ratio (SR) could be examined using only seven year classes (Table II; Fig. 5). There was a highly significant trend for sex ratio to decrease (more females) with decreasing water levels at the times of nesting (August maximum water height: WH_{Aug}) (Fig. 13; $SR = 0.05 + 0.45\ WH_{Aug}$; $r^2 = 0.85$; $P = 0.003$), indicating that when water levels were low the nest sites selected (available?) had been more female-biased.

There was also a significant trend for sex ratio to decrease with increasing survival proportion from egg-laying to one year of age (SP_1) (Fig. 14; $SR = 0.60 - 4.14\ SP_1$; $r^2 = 0.56$; $P = 0.05$). Proportionally more females were associated with increased survivorship.

Multiple regression analysis predicting sex ratio from both WH_{Aug} and SP_1 explained a high proportion of the variation in sex ratio ($r^2 = 0.93$; $P = 0.005$), with both variables acting in the directions indicated on Figs 13 and 14 ($SR = 0.23 + 0.36\ WH_{Aug} - 1.92\ SP_1$; r^2 addition due to $SP_1 = 0.08$; $P = 0.09$). However, the relationship between SR and WH_{Aug} rendered the relationship between SR and SP_1 statistically insignificant; the relationship may be an artefact.

Although tested, no significant relationship between sex ratio and

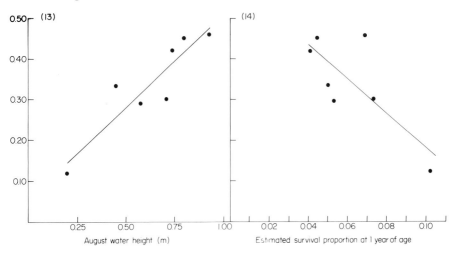

FIGS 13 & 14. (13) The relationship between recruit sex ratio and water height at the time of nesting, over seven years in the McKinlay river area. (14) The marginally significant trend ($P = 0.053$) between sex ratio and the number of one-year-olds (expressed as the estimated proportion of eggs laid which produced the one-year-old survivors), over seven years in the McKinlay River area.

monthly mean temperatures, nor the time of + 1 m floods, could be demonstrated. September maximum water height was highly correlated with WH_{Aug} and was thus similarly correlated with sex ratio, but no significant relationship between October or November water height (hatching) and either sex ratio or WH_{Aug} was apparent.

Survivorship

Using up to 15 years' data (Table II), an attempt was made to determine whether or not the variation in SP not already accounted for by age-specific mortality was attributable to environmental conditions during the incubation periods of each cohort (Fig. 5; age-specific mortality accounted for 75% of the variation in SP to consecutive years). The data on Fig. 5 were linearized by logarithmic transformation of SP (lnSP), and multiple regression analysis used to predict lnSP from the major variable, age (or years; Y), plus combinations of the temperature, water height and rainfall data from the incubation periods of each year class.

The number of days in which it rained in October (RD) and the change in water level between nesting and hatching (October–August water heights; ΔWH) were highly correlated with each other, and were the only variables which accounted for a significant proportion of the unexplained variation in lnSP. (October rain days: ln$SP = -3.59 - 0.17Y + 0.33RD - 0.03RD^2$; $r^2 = 0.93$, $P < 0.001$; r^2 addition attributable to $RD = 0.17$; $P = 0.005$) (change in water height; ln$SP = -3.59 - 0.17Y + 0.96 \Delta WH - 0.43 \Delta WH^2$; $r^2 = 0.90$; $P < 0.001$; r^2 addition attributable to $\Delta WH = 0.14$; $0.05 > P > 0.02$.)

In both cases, the relationship between RD, ΔWH and SP was parabolic. To demonstrate the trend graphically, the influence of age-specific mortality was removed (standardized) by plotting the relationship of SP_1 and ΔWH (Fig. 15; $SP_1 = 0.050 + 0.080 \Delta WH - 0.048 \Delta WH^2$; $r^2 = 0.65$; $0.05 > P > 0.01$).

This trend suggests that rain (water rises?) at or near hatching had enhanced survivorship (from egg-laying to one year of age), but that in the years where there was heavy rain (flooding?) survivorship had been reduced. It should be recognized that the seven years for which sex ratio and survivorship could be examined (Fig. 14) did not include the years with intense flooding on Fig. 15.

Decreased survivorship associated with heavy rains (Fig. 15) is probably a reflection of catastrophic nest inundation, and reduced survivorship of eggs. However, increased survivorship with rain and water level rises up to 1 m appears in direct contradiction to the known situation in the field, i.e. a + 1 m rise would inundate many eggs (the mean height of eggs above water level in one area was 56 ± 18 cm).

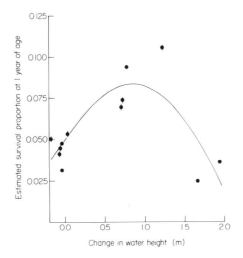

FIG. 15. The relationship between the number of one-year-old survivors (expressed as the proportion of eggs laid which produced them) and the change in water height between laying and hatching. The same parabolic trend exists with the number of rain days at or near hatching.

Two explanations, not mutually exclusive, suggest themselves, and both are dependent on the extent to which SP_1 reflects survivorship from eggs to hatching, or from hatching to one year of age (insufficient data are available to separate these two periods).

(i) Rains at the time of hatching may enhance survival of embryos destined to either die in eggs or hatch abnormally (as a result of desiccation and/or overheating), and this overcompensates for concurrent losses due to flooding.

(ii) Independent of survivorship from eggs to hatching, hatchling survivorship is greatly enhanced by early rains, the subsequent abundance of food, and availability of wet habitats. This survivorship overcompensates for embryo losses due to flooding.

Given the relationship between survivorship and high clutch temperatures at the time of hatching (Fig. 10), the former interpretation seems less likely than the latter.

Summarizing, the results of the age structure analysis demonstrate:

(i) A strong trend for sex ratio to decrease with water height at the time of egg-laying.

(ii) A weak trend for increased survivorship from egg-laying to one year old to be associated with increased survival of females, probably in the egg stage [(i) could render this insignificant].

(iii) Reduced survivorship, probably of eggs, with heavy rains at hatching.

(iv) Increased survivorship, probably of hatchlings rather than eggs, with some rain (flooding < 1 m) at the time of hatching.

DISCUSSION

Temperature-dependent Sex Determination

The mechanism by which incubation temperature determines sex in reptiles is unknown (Bull, 1980; Ferguson & Joanen, 1982, 1983), and it remains unclear whether temperature *per se* is an ultimate or proximate sex determining factor. The distinction is important, because results presented to date are consistent with sex being determined as a consequence of embryo development rates, the influence of temperature being pronounced because of its profound influence on development rate (at 26°C, *C. johnstoni* embryos develop 2.1 times more slowly than at 30°C; Webb, Buckworth & Manolis, 1983d; Webb, Buckworth, Manolis & Sack, 1983). If development rate and embryonic metabolic levels are indeed more proximally related to sex determination than is temperature, then factors such as the gaseous (Ackerman, 1980) and moisture (Packard, Packard, Boardman & Ashen, 1981; Morris *et al.*, 1983) environment of nests (which influence development rates) could also be expected to influence sex determination (Gutzke & Paukstis, 1983). Although perhaps minor variables compared to temperature, their existence could explain some experimental results.

When incubated at a constant temperature at or near the threshold temperature, there is a tendency for eggs of some clutches to be either male- or female-biased (Bull, Vogt & Bulmer, 1982; unpublished data on *C. johnstoni*). If temperature *per se* is the direct determiner of sex, then such results indicate a genetic predisposition to one sex or the other (Bull, Vogt & Bulmer, 1982). However, if development and metabolic rates are involved, such a finding could reflect minor inter-clutch variation in shell porosity, egg size or other factors likely to influence gas and moisture exchange; consequently, there are a number of pathways through which the variation could be environmentally rather than genetically determined.

A further implication is that shift experiments spanning the same temperature range (34–30°C; 30–26°C), or directed at the same base temperature (34°C and 32°–30°C), may not induce metabolic or development rate responses linearly related to temperature. Acclimation periods in which development rate stabilizes at a new temperature could be expected to be proportional to the extent of exposure to the initial temperature, the temperature gradient of the switch, and the absolute temperatures involved.

Nevertheless, temperature clearly does have a profound effect on both development rate and sex determination, and is the most important variable involved.

The relationship between incubation temperature and sex in *C. johnstoni* is unusual. The species has two threshold temperatures, and it is the 'high' temperature females that appear limited by temperature-related survivorship constraints. In the few turtles with two thresholds (*Chelydra serpentina*, *Macroclemys* sp., *Kinosternon* sp., *Sternotherus odoratus*; Yntema, 1976, 1979; Bull, 1980; Vogt *et al.*, 1982; Wilhoft, Hotaling & Franks, 1983) males are bounded by females, as in *C. johnstoni*, but it is the production of 'low' temperature females that is constrained by temperature-related survivorship.

It may not be coincidental that among species with single threshold temperatures, turtles produce males at low temperatures and females at high temperatures, whereas crocodilians and lizards do the reverse (Bull, 1980; Mrosovsky, 1980; Miller & Limpus, 1981; Ferguson & Joanen, 1982, 1983; unpublished data on *C. porosus*). If the one mechanism of sex determination characterizes reptiles with TSD, then the potential of a two-threshold pattern may exist in all reptiles with TSD. Survivorship constraints may tend to eliminate low temperature female turtles and high temperature female crocodilians and lizards.

Differential development of the gonad in relation to other organs may be involved in the mechanism of TSD. The embryonic stage at which sex is determined itself depends on temperature (or development rate), and with the exception of *Chelydra serpentina* (Yntema, 1979), the relationship operates in the opposite direction to what would perhaps be expected. At higher incubation temperatures (when general embryonic development is enhanced), sex is determined in later embryonic stages (Bull & Vogt, 1981; Ferguson & Joanen, 1983). In *C. porosus*, for example (unpublished data), sex is determined at 20–25 days at 30°C (0.8 g embryo in an average egg), and 40–45 days (30°C equivalent ages; 7 g embryo) at 32°C.

In *A. mississippiensis* (M. W. J. Ferguson, pers. comm.) and *C. johnstoni* (this study), as incubation temperature is increased, so gonadal development appears more and more retarded in relation to the external morphology of the embryo. *Crocodylus johnstoni* females incubated at high temperatures appear to have, at hatching, an ovary severely retarded in relation to that of females incubated at 30°C (Fig. 3). Asynchrony in gonadal development opens a number of pathways that could lead to a two-threshold pattern, especially if "female" is something of a sex-by-default and synchrony of development between the gonad and some other organ system is needed to induce "maleness". It also opens a number of pathways through which constraints on survival could operate. In *C. johnstoni*, the conditions resulting in "high" temperature females also lead to development so rapid that survival may be threatened. Selection against such extremes could result in the single-threshold pattern found in other crocodilians (low temperature females – high temperature males).

That selection can act on threshold temperatures is indicated by a comparison of turtle populations from the northern and southern United States (Bull, Vogt & McCoy, 1982). Threshold temperatures were slightly lower in the southern populations (about 1°C in *Chrysemys picta*), although ambient temperatures were significantly higher. This result would be expected if the higher ambient temperatures decreased the probability of surviving, so that sites which were cooler (at the time of sex determination) were needed in order to avoid lethal temperatures, even though these may occur at or near hatching. Without selection on the threshold temperature, a male bias (relative to northern populations) would be expected, and sex ratio selection would act to reduce the threshold, as was observed.

Sex Determination and Survivorship in the Field

The laboratory and field results with TSD complement each other quite closely, although under constant temperature incubation, few males were produced at any temperature. In this regard, *C. johnstoni* differs from *C. porosus* (unpublished data from the same incubators), *A. mississippiensis* (Ferguson & Joanen, 1982, 1983), and most turtles which have been examined (Bull & Vogt, 1979; Bull, 1980; Vogt & Bull, 1982) (all commonly give one sex or the other). The difference could reflect a stable versus a fluctuating incubation temperature (Wilhoft *et al.*, 1983), or perhaps the precision with which we could control incubation temperature. Additional studies are needed to clarify the issue. *Crocodylus johnstoni* clutch sex ratios in the field (Fig. 7) were also atypical when compared to other species; as a consequence of having two thresholds, rather than one, for sex determination, there was no pronounced bimodality (Bull & Vogt, 1979; Bull, 1980; Ferguson & Joanen, 1982, 1983).

The temperatures of nests selected by females influenced both survivorship and sex ratio of the offspring. If it is accepted that surviving involves stronger selection than producing an optimal sex ratio, then it is quite possible that the range of nest temperatures selected by *C. johnstoni* is based primarily on survivorship considerations. This hypothesis is testable, because it predicts that there would be no significant difference in the range of nest temperatures selected by two similar oviparous reptiles, one with genotypic and one with environmental sex determination.

When *C. johnstoni* nest temperatures are considered purely from the viewpoint of temperature-dependent survivorship, the species can be considered as having a preferred initial nest temperature (PT_i; 30.2 ± 1.6°C), which is 4–5°C below an initial lethal maximum temperature. An initial lethal minimum temperature exists, but is probably below the minimum temperatures available at the time of nesting in our particular study area. More important may be survival constraints associated with minimal temperatures delaying the time of hatching. If the distribution of temperatures around the PT_i reflects the variance or precision with which *C. johnstoni* of different sizes can or do select an optimal temperature, the consequences of raising or lowering PT_i are that a higher proportion of animals will surpass the inital lethal maximum, or die in floods.

That the PT_i corresponds with a sex ratio near 0.5 could be coincidental, but if not, it may indicate that sex ratio selection is constrained by survivorship considerations to the extent that synchronization represents the best compromise. If so, changes in PT_i associated with survivorship may be a prerequisite to changes in the threshold temperature/s.

Crocodylus johnstoni sex is determined up to a month after a nest (and thus nest temperature) has been selected, and accordingly the relationship between PT_i and the threshold temperatures must be predictive. Changes in time of nesting and the availability of nest sites could accordingly influence sex ratio, even though the same PT_i was chosen. For example, in a year with average water height, nests with a mean PT_i of 30.4°C may increase to 32.0°C, and a 0.5 sex ratio may result. However, if water levels were low at the time of nesting (Fig. 13), and the same PT_i only increased to 31.0°C (perhaps because the nest moisture levels, which affect temperature, remained more constant) a female bias would result, even though adults had passively nested in their normal manner.

If the relationships described above are correct, geographical variation in ambient temperatures and/or the rainfall/flood height patterns, should be reflected in significant sex ratio variation. Given that threshold temperatures tend to be conservative (Bull, Vogt & McCoy, 1982), PT_i may well vary substantially. The variation in sex ratio and time of hatching we found (Tables III, IV & V) were predictable on the basis of ambient temperatures.

Mean monthly 09.00 hours temperatures from the northern, central and southern areas respectively are: June, 22.8°C, 20.8°C, 20.6°C; July, 21.5°C, 20.3°C, 19.7°C; August, 24.3°C, 22.6°C, 22.5°C; September, 26.9°C, 26.0°C, 26.4°C; October, 28.5°C, 28.2°C, 29.5°C; and November, 29.1°C, 28.9°C, 30.5°C. Temperatures are cooler in the south during the period when ovulation occurs in the north (June, July), and this could be expected to retard nesting (Joanen & McNease, 1979). Sex in southern *C. johnstoni* would be determined in October rather than September, when temperatures in any one area are higher. In addition, October and November temperatures in the south are higher than in the north and central areas, and thus it is not surprising that more males were produced.

There is generally less rainfall in the south (860 mm versus 1400 mm annually), and the wet season occurs up to two weeks later (although data with which flood régimes in different rivers could be compared realistically were unavailable).

It was concluded before data analysis (i.e. during the period when hatchlings were being collected), that the South Alligator and Roper Rivers had been subjected to intense illegal poaching. Subsequently, poachers were apprehended, although exactly where they had been operating was not ascertained. In both rivers, the first week of hatch was associated with a pronounced female bias (Table V), irrespective of high proportions of the total complement of hatchlings appearing (Tables III & IV). The total hatchling sex ratios in these rivers (0.12 and 0.20 respectively) were the lowest recorded.

In the McKinlay River area, a similar reduction in density occurred during the hunting period (1960–1963), and within the limits of the age structure accuracy, we tried to determine whether or not there was any indication of a sex ratio bias at that time. August water heights in those years were used to predict sex ratios, and even when the animals on either side of that period were lumped, more females were present than were predicted. In addition, sex ratios in the most recent years are well above the population mean (Tables I & II), although no pronounced sex-specific mortality is known. If sex ratio in the McKinlay River has changed since protection, it has been steadily increasing from a female bias at the time of protection.

When it is considered that the time of nesting results partly from social ordering of animals (large or old females nest first; Webb, Buckworth & Manolis, 1983d), reduced densities could allow survivors to nest earlier, when it was cooler, regardless of their age or size. Alternatively, survivors from an intense culling may be wary animals, perhaps located and nesting in more vegetated and secluded sites.

Theoretical Considerations

TSD and its consequences in terms of variable primary sex ratios raise a number of theoretical problems (Charnov & Bull, 1977; Bull, 1980, 1981a, b; Bulmer & Bull, 1982; Ferguson & Joanen, 1982, 1983). That a diploid species has TSD rather than GSD is not in itself inconsistent with sex ratio/sex allocation models of Fisher (1930) and later workers (see Charnov, 1982). But as TSD often results in offspring sex ratios that are consistently and markedly skewed from 0.5 at the end of parental investment, a new theoretical basis may be required. Furthermore, in *C. johnstoni*, *A. mississippiensis* (Ferguson & Joanen, 1982, 1983) and a number of freshwater turtles (Vogt & Bull, 1982), skewed secondary sex ratios in wild populations appear to result mainly from skewed primary sex ratios. Such relationships may be explicable under a number of classical models (reviewed by Charnov, 1982), but to pursue such possibilities in the absence of an acceptable hypothesis explaining TSD at the primary sex ratio level may be premature.

Charnov & Bull (1977) proposed that environmental sex determination *per se* would be more advantageous than GSD if sex-specific fitness was a function of the environment in which eggs were laid. If a particular nest environment was likely to produce a fit male and an unfit female, the mechanism would allow males to be selectively produced. This idea has been expanded in a number of theoretical treatments (Bull, 1981a, b; Bulmer & Bull, 1982), which also indicate that TSD may not be more primitive than GSD, but could evolve from it, under certain circumstances.

In turtles and crocodilians, sex-specific fitness consistent with the model of Charnov & Bull (1977) has proved difficult to demonstrate. With *A. mississippiensis* (Ferguson & Joanen, 1982, 1983) the rate of yolk utilization at different temperatures has been proposed as a possible advantage, because more yolk is associated with faster post-hatching growth rates, and it is more advantageous for females to grow and attain maturity rapidly than it is for males.

This relationship does not apply directly to *C. johnstoni*, because females result from both the highest and lowest incubation temperatures, male and female hatchlings grow at the same rates in the field (Webb, Buckworth & Manolis, 1983a), and the amount of yolk embryos contain at hatching is positively correlated with incubation temperature and thus negatively correlated with total development time (unpublished data). Fast developing *C. johnstoni* embryos (females and males) hatch with more yolk than slow developing ones (the same trend occurs in *C. porosus* where there are no "high" temperature females; unpublished data), which is the opposite to what occurs in *A. mississippiensis*. This interspecific difference could have its roots in the very rapid development of *A. mississippiensis* relative to *C. johnstoni* and *C. porosus* (64 rather than 85–95 days at 30°C), or it could reflect a bias associated with *A. mississippiensis* yolk masses being measured just prior to hatching. Regardless, the selective advantage to *A. mississippiensis* does not appear to exist in *C. johnstoni* or *C. porosus*.

Survivorship in *C. johnstoni* embryos is clearly sex-specific, because mortality is biased towards females. In areas prone to overheating, "low" temperature females could be expected to be more fit than either "high" temperature females or possibly males, whereas in flood-prone situations the reverse would be true. But these relationships represent consequences of having TSD, and do not constitute an evolutionary advantage for the mechanism.

A survivorship advantage consistent with the model of Charnov & Bull (1977) would exist if:

(i) Females were more likely to survive at high and low temperature extremes than males.

(ii) Females were less likely than males to have their reproductive potential compromised – perhaps through structural abnormalities.

We have no evidence indicating that either condition occurs and accordingly can present neither yolk mass nor survivorship advantages for *C. johnstoni* that are definitively consistent with Charnov & Bull's (1977) model. From a sex ratio point of view *C. johnstoni* would seem equally if not better adapted with GSD. However, if embryo survivorship at the physiological rather than ecological level is examined, a possible explanation for TSD in reptiles can be attained, which may be broadly consistent with the above model. Evidence comes from two sources, the reproductive strategies of reptiles with TSD, and the timing of sexual differentiation in the embryos of those species.

(1) TSD is common in two vertebrate groups (chelonians and crocodilians), both of which are highly aquatic. Eggs are typically laid in an environment (on land) which is separated from that in which adults spend most of their time (water), and the ability to select optimal nest sites may accordingly be constrained (relative to terrestrial animals nesting in a terrestrial environment). There is virtually no manipulation of eggs between laying and hatching (as occurs in most birds and some squamates), the embryonic stage at laying is young (compared to squamates but not birds) and development times in the field may be greatly prolonged (relative to squamates and birds). *Embryos* characterized by TSD are exposed to

vagaries of the environment, over a large range of embryonic stages, without parental assistance, and for longer periods of time than is generally the case for related groups in which GSD predominates (squamates and birds).

(2) In embryos characterized by TSD, the time when part of the nephric system will be allocated to a gonad, and the type of gonad and ducts that will develop, both appear to be determined by the rate at which the embryos are themselves developing. The direction of timing is such that during rapid development, when nephric activity is presumably (and perhaps critically) maximized, gonadal development appears severely retarded.

If the latter facultative arrangement was primarily an adaptation towards enhancing embryo survivorship under extreme incubation conditions, and it was less readily attainable or unattainable with GSD, then the existence of TSD in groups where embryos are likely to incur environmental extremes would be explicable. TSD could be a somewhat unavoidable consequence of a mechanism which has been selected for, and maintained, for reasons other than those associated directly with sex determination.

ACKNOWLEDGEMENTS

The results presented have been gathered with the assistance of many people, and we would particularly like to thank George Sack, Rik Buckworth, Charles Manolis, Tony Spring, John Barker, Maria Gilham, Mike Bugler, Michael Beal and Karen Dempsey. Bill Freeland and Mark Ferguson discussed most aspects of the paper, and while being especially grateful for their assistance, advice and help, we take full responsibility for errors in the more speculative aspects. Jim Bull, Mark Ferguson and Jeff Miller were all implicated in stimulating our interest in the problem, and have encouraged our efforts. Special thanks go to Eve Kerr for typing numerous drafts of the manuscript.

Financial support for the study has come primarily from the Conservation Commission of the Northern Territory, with additional support from the University of New South Wales.

REFERENCES

Ackerman, R. A. (1980). Physiological and ecological aspects of gas exchange by sea turtle eggs. *Am. Zool.* **20**: 575–583.
Bull, J. J. (1980). Sex determination in reptiles. *Q. Rev. Biol.* **55**: 3–21.
Bull, J. J. (1981a). Sex ratio evolution when fitness varies. *Heredity* **46**: 9–26.

Bull, J. J. (1981b). Evolution of environmental sex determination from genotypic sex determination. *Heredity* **47**: 173–184.
Bull, J. J. & Vogt, R. C. (1979). Temperature-dependent sex determination in turtles. *Science, Wash.* **206**: 1186–1188.
Bull, J. J. & Vogt, R. C. (1981). Temperature-sensitive periods of sex determination in emydid turtles. *J. exp. Zool.* **21**: 435–440.
Bull, J. J., Vogt, R. C. & Bulmer, M. G. (1982). Heritability of sex ratio in turtles with environmental sex determination. *Evolution, Lawrence, Kans.* **36**: 333–341.
Bull, J. J., Vogt, R. C. & McCoy, C. J. (1982). Sex determining temperatures in turtles: a geographic comparison. *Evolution, Lawrence, Kans.* **36**: 326–332.
Bulmer, M. G. & Bull, J. J. (1982). Models of polygenic sex determination and sex ratio control. *Evolution, Lawrence, Kans.* **36**: 13–26.
Caughley, G. (1976). *Analysis of vertebrate populations.* London: Wiley.
Charnov, E. L. (1982). *The theory of sex allocation.* Princeton: Princeton University Press.
Charnov, E. L. & Bull, J. J. (1977). When is sex environmentally determined? *Nature, Lond.* **266**: 828–830.
Cogger, H. G. (1979). *Reptiles and amphibians of Australia.* (Revised edition.) Sydney: A. H. & A. W. Reed.
Ferguson, M. W. J. & Joanen, T. (1982). Temperature of egg incubation determines sex in *Alligator mississippiensis*. *Nature, Lond.* **296**: 850–853.
Ferguson, M. W. J. & Joanen, T. (1983). Temperature-dependent sex determination in *Alligator mississippiensis*. *J. Zool., Lond.* **200**: 143–177.
Fisher, R. A. (1930). *The genetical theory of natural selection.* Oxford: Clarendon Press.
Groombridge, B. (1982). *I.U.C.N. Amphibia – Reptilia Red Data Book. Part 1. Testudines, Crocodylia, Rhynchocephalia.* Gland, Switzerland: I.U.C.N. Publications.
Gutzke, W. H. N. & Paukstis, G. L. (1983). Influence of the hydric environment on sexual differentiation of turtles. *J. exp. Zool.* **286**: 467–469.
Jenkins, R. W. G. (1979). The status of endangered Australian reptiles. In *The status of endangered Australian wildlife*: 169–176. Tyler, M. J. (Ed.). Adelaide: Royal Zoological Society of South Australia.
Joanen, T. & McNease, L. (1979). Time of egg deposition for the American alligator. *Proc. A. Conf. S.E. Assoc. Fish Wildl. Ag.* **33**: 15–19.
Miller, J. D. & Limpus, C. J. (1981). Incubation period and sexual differentiation in the green sea turtle *Chelonia mydas* L. In *Proceedings of the Melbourne Herpetology Symposium*: 66–73. Melbourne: Royal Melbourne Zoological Gardens.
Morris, K. A., Packard, G. C., Boardman, T. J., Paukstis, G. L. & Packard, M. J. (1983). Effects of the hydric environment on growth of embryonic snapping turtles (*Chelydra serpentina*). *Herpetologica* **39**: 272–285.
Mrosovsky, N. (1980). Thermal biology of sea turtles. *Am. Zool.* **20**: 531–547.
Mrosovsky, N. & Yntema, C. L. (1980). Temperature dependence of sexual differentiation in sea turtles: implications for conservation practices. *Biol. Conserv.* **18**: 271–280.
Packard, G. C., Packard, M. J., Boardman, T. J. & Ashen, M. D. (1981). Possible adaptive value of water exchanges in flexible-shelled eggs of turtles. *Science, Wash.* **213**: 471–473.
Vogt, R. C. & Bull, J. J. (1982). Temperature controlled sex determination in turtles: ecological and behavioral aspects. *Herpetologica* **38**: 156–164.
Vogt, R. C., Bull, J. J., McCoy, C. J. & Houseal, T. W. (1982). Incubation temperature influences sex determination in kinosternid turtles. *Copeia* **1982**: 480–482.
Webb, G. J. W. (1982). A look at the freshwater crocodile. *Aust. nat. Hist.* **20**: 299–303.
Webb. G. J. W., Buckworth, R. & Manolis, S. C. (1983a). *Crocodylus johnstoni* in the

McKinlay River area, N.T. III. Growth, movement and the population age structure. *Aust. Wildl. Res.* **10**: 381–399.

Webb. G. J. W., Buckworth, R. & Manolis, S. C. (1983b). *Crocodylus johnstoni* in the McKinlay River area, N.Y. IV. A demonstration of homing. *Aust. Wildl. Res.* **10**: 401–404.

Webb, G. J. W., Buckworth, R. & Manolis, S. C. (1983c). *Crocodylus johnstoni* in a controlled-environment chamber – a raising trial. *Aust. Wildl. Res.* **10**: 421–432.

Webb, G. J. W., Buckworth, R. & Manolis, S. C. (1983d). *Crocodylus johnstoni* in the McKinlay River area, N.T. VI. Nesting biology. *Aust. Wildl. Res.* **10**: 607–637.

Webb, G. J. W., Buckworth, R., Manolis, S. C. & Sack, G. C. (1983). An interim method for estimating the age of *Crocodylus porosus* embryos. *Aust. Wildl. Res.* **10**: 563–570.

Webb, G. J. W., Manolis, S. C. & Buckworth, R. (1983). *Crocodylus johnstoni* in the McKinlay River area, N.T. II. Dry-season habitat selection and an estimate of the total population size. *Aust. Wildl. Res.* **10**: 371–380.

Webb, G. J. W., Manolis, S. C. & Sack, G. C. (In press). Cloacal sexing of hatchling *Crocodylus porosus* and *C. johnstoni*. *Aust. Wildl. Res.*

Webb, G. J. W. & Messel, H. (1977). Crocodile capture techniques. *J. Wildl. Mgmt* **41**: 572–575.

Webb, G. J. W., Sack, G. E., Buckworth, R. & Manolis, S. C. (1983). An examination of *Crocodylus porosus* nests in two northern Australian freshwater swamps, with an analysis of embryo mortality. *Aust. Wildl. Res.* **10**: 571–605.

Wilhoft, D. C., Hotaling, E. & Franks, P. (1983). Effects of temperature on sex determination in embryos of the snapping turtle, *Chelydra serpentina*. *J. Herpetol.* **17**: 38–42.

Worrell, E. (1952). The Australian crocodiles. *J. Proc. R. Soc. N.S.W.* 1951–52: 18–23.

Worrell, E. (1964). *Reptiles of Australia*. Sydney: Angus & Robertson.

Yntema, C. L. (1976). Effects of incubation temperature on sexual differentiation in the turtle *Chelydra serpentina*. *J. Morph.* **150**: 453–461.

Yntema, C. L. (1979). Temperature levels and periods of sex determination during incubation of eggs of *Chelydra serpentina*. *J. Morph.* **159**: 17–28.

Endocrinology of Reproduction in Male Reptiles

VALENTINE LANCE

Department of Medicine, Tulane University School of Medicine, New Orleans, Louisiana 70112, USA

SYNOPSIS

The common feature of gametogenesis in male reptiles is elevated androgen concentration in the plasma during spermiogenesis. There is more variability during the early stages of spermatogenesis, but in general plasma androgen is low at this time. A hypothesis is presented in which the onset of spermatogenesis in seasonally breeding reptiles is compared to the onset of puberty in mammals. It is proposed that initiation of meiosis in the resting spermatogonia is dependent upon a follicle stimulating hormone (FSH) and does not require high levels of androgen. Maturation of the spermatocytes into spermatozoa, however, does require androgen, and this is supplied by the interstitial (Leydig) cells in response to luteinizing hormone (LH). There is evidence that this may be the case in the Chelonia and the Crocodilia from which two gonadotropins similar to mammalian FSH and LH have been isolated. In turtles and in *Alligator mississippiensis* plasma testosterone is elevated at spermiogenesis and low at the onset of meiosis. Binding of radioactive FSH to turtle testicular tissue is also higher during meiosis than at later stages in the cycle. This model may not hold for the Squamata as there is evidence that there may be only a single gonadotropin, with both FSH and LH activities, in this group. Despite considerable species variation in both testicular cycles and in plasma testosterone concentrations, the androgen cycle in squamates is basically similar to that of the Chelonia and Crocodilia in that a peak occurs at spermiogenesis.

The mammalian hypothalamic decapeptide, luteinizing hormone releasing hormone (LH-RH) has no effect on gonadotropin release in turtles but is extremely potent in alligators. Evidence is presented that alligator hypothalamic LH-RH is different from the mammalian peptide, and probably different from that of the Chelonia, but may be identical to the recently characterized avian molecule.

INTRODUCTION

The major evolutionary advance of reptiles over amphibians was in the development of the amniote egg, an adaptation that freed them from a dependence on water for reproduction. However, another notable difference between amniotes and anamniotes, though of unknown adaptive significance, is in the organization of the testis. In fishes and amphibians, germ cell development takes place within membrane bound structures known as spermatocysts, or germinal cysts, which are located within tubules or lobules. Within each cyst all of the germ cells form a single clone and all develop in synchrony. In reptiles, birds and mammals, germ cell development takes place in the seminiferous tubules within which large numbers of clones develop in the same compartment. In both types of testis the developing germ cells are intimately associated with a supporting, or Sertoli, cell.

Reptiles, in common with other amniote vertebrates, have a well developed hypothalamic portal system (Green, 1966) and gonadotropin(s) composed of two subunits, chemically similar to those of mammals (Licht, 1979); and show evidence that gonadotropin release is controlled by factors from the hypothalamus (Ball, 1981). The testis of reptiles is of basically similar structure to that of mammals, and the limited biochemical data available indicate that testosterone biosynthesis by the testis is also similar to mammals (Kime, 1980). The considerable advances made in the past few years in unravelling the complexity of the hormonal control of spermatogenesis in mammals (see Parvinen, 1982) has not, however, been matched by similar advances in our understanding of the endocrinology of reproduction in male reptiles. In mammals it has been demonstrated that a barrier exists between blood bathing the interstitium and the fluid within the seminiferous tubules. This blood-testis barrier results in two separate compartments, with two different ionic and hormonal environments bathing the cells of each. Although such a barrier has been demonstrated in birds and fishes, a blood-testis barrier has yet to be demonstrated in reptiles. Futhermore, in not a single reptile or bird have the stages of spermatogenesis been studied in any detail (Roosen-Runge, 1977), and in only one species of lizard has the length of time required for the development of a mature spermatozoon from a spermatogonium been estimated (Joly & Saint Girons, 1975).

A large number of reviews have been published on reproduction in the Reptilia (Miller, 1959; Dodd, 1960; Forbes, 1961; Lofts, 1969, 1972, 1977; Licht, 1972a, 1974, 1977, 1979; H. Fox, 1977; Licht, Papkoff, Farmer, Muller, Tsui & Crews, 1977; Crews, 1979; Lance & Callard, 1980; Callard & Ho, 1980; Callard, Callard, Lance, Bolaffi & Rosset, 1978) and the reader is directed to these for more detailed coverage and bibliographies.

This brief review will diverge from the traditional method of lumping all reptiles together and considering each system in turn, but instead will present data on one or two species from the three orders about which we have any endocrinological information. As nothing is known on the endocrinology of the Rhynchocephalia, with the exception of the histological study by Gabe & Saint Girons (1964), they will not be considered further. The focus is on a very narrow topic in a small number of species. Many other aspects of the endocrine physiology of reptiles such as the hormonal control of behaviour, hypothalamic and pituitary immunocytochemistry, steroid hormone biosynthesis and metabolism, steroid hormone receptors, and the role of the pineal, thyroid and adrenal glands in reproduction have, of necessity, been ignored.

CHELONIA

Of the 12 living families of turtles (Pritchard, 1979) the spermatogenic cycle is known in only a few species from the Chelydridae, the Kinosternidae, the Testudinidae, the Emydidae and the Trionychidae (see reviews by Angelini & Picariello, 1975; Moll, 1979). These five families are, however, taxonomically widely separated, yet the spermatogenic cycle, taking into account variation due to latitude, appears to be essentially identical in representative species from all five. It is possible that the limited information we do have may be representative of the Order as a whole.

In general, spermatogenesis in temperate zone turtles begins in late spring as temperatures increase, reaches a peak in late summer, and is completed by early fall when mature spermatozoa move into the epididymides. During the winter months the testes remain in a regressed condition, with seminiferous tubules containing only resting spermatogonia, Sertoli cells and remnants of spermatozoa. During this period the Leydig cells are enlarged and full of lipid droplets. Mating takes place in early spring at which time the seminiferous tubules contain only resting spermatogonia and Sertoli cells. Females are thus inseminated with spermatozoa that were produced and stored in the epididymis the previous year. It appears, therefore, that in all turtles examined to date the peak in sexual activity occurs when the seminiferous tubules are in a regressed or inactive state, and Leydig cells present histological and histochemical signs of maximum activity. In contrast, the peak in spermatogenic activity takes place when Sertoli cells appear, by histological criteria, most active, and Leydig cells inactive or atrophic. For detailed descriptions of testicular cycles of individual species see Moll (1979).

Lofts & Tsui (1977) studied the testicular cycle of *Trionyx sinensis* in South China and concluded, on the basis of lipid and enzyme histochemistry, that Sertoli cells secrete androgen during the spermatogenic phase, and that Leydig cells secrete androgen during the mating period. They hypothesized that during the spring Leydig cells secrete androgen in response to luteinizing hormone (LH), and that this androgen initiates mating behaviour. They suggest that in midsummer at the peak of spermatogenesis, because Leydig cells appear atrophic, they probably do not produce androgen. However, since spermatogenesis is an androgen-dependent process, Sertoli cells must be secreting androgen within the seminiferous tubules at this time, probably in response to follicle stimulating hormone (FSH). Callard, Callard, Lance & Eccles (1976) presented data on the annual hormonal cycle of *Chrysemys picta* in North America which appear to support this theory. In this species maximum testosterone levels are found in April when the animals emerge from hibernation and commence breeding activities. Plasma testosterone then rapidly declines as spermatogenesis begins and remains low for the duration of the summer, rising again in the fall when spermiogenesis takes place. The authors conclude that testosterone produced by Sertoli cells during the summer, when spermatogenesis is at a peak, remains sequestered within the seminiferous tubules since circulating hormone levels are low to non-detectable at this time. Licht (1982) argues that such an interpretation may be incorrect, because *Chrysemys* shows increased sensitivity to injections of gonadotropins in the summer when testis mass is greatest (Lance, Scanes & Callard, 1977); and hormonal data from other species of turtles, with similar spermatogenic cycles, do not agree with the hormonal cycle seen in *Chrysemys*. In *Sternotherus odoratus* for example, plasma testosterone is low in early spring when mating occurs, increases to a maximum in late summer when spermatogenesis is at a peak, and declines again in the fall (Fig. 1) when spermiogenesis and a second mating period occur (McPherson, Boots, MacGregor & Marion, 1982). Not only are these two hormonal cycles different, but the maximum and minimum levels of testosterone are different. In *Sternotherus* maximum levels of over 100 ng ml^{-1} were recorded in midsummer, whereas in *Chrysemys* the maximum in spring is only 5 ng ml^{-1}, which is similar to the minimum levels in *Sternotherus* at this time. A slightly different cycle in plasma testosterone was observed in *Testudo hermanni* (Kuchling, Skolek-Winnish & Bamberg, 1981). In this species peak plasma testosterone levels (25 ng ml^{-1}) occur in August at spermiogenesis, but when testis mass is greatest in June testosterone levels are low (5 ng ml^{-1}). A second peak (10 ng ml^{-1}) occurs in April when the animals emerge from hibernation and mate. Since only two samples per month were assayed it is possible that the second peak is an artefact. Although the testicular cycle of the sea turtle *Chelonia mydas* has not been described, Licht, Wood, Owens & Wood (1979) noted that testosterone levels in captive males of this species were lower during mating (12 ng ml^{-1}) than they were about two months prior to the onset of sexual activity (33 ng ml^{-1}). In only one other species of turtle (*Trionyx sinensis*) has a seasonal hormonal cycle

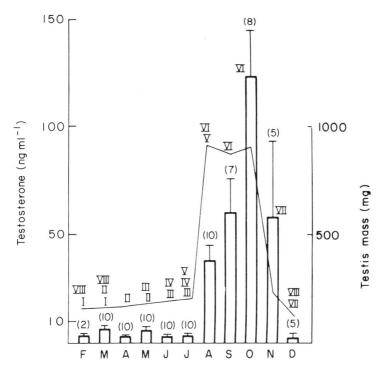

FIG. 1. Seasonal changes in testis mass and plasma testosterone levels in male stinkpot turtles, *Sternotherus odoratus*. Testis mass is adjusted for body size. Roman numerals indicate spermatogenic stage (see Table I) determined by histological examination. Vertical bars represent the mean concentration of testosterone and the vertical lines the standard error of the mean (SEM). Numbers in parentheses indicate sample size. (From McPherson *et al.*, 1982.)

been reported, and again, highest testosterone levels were seen during spermiogenesis (Licht, Tsui & Lam, unpublished, quoted by Licht, 1982). In the same species biosynthesis of testosterone from radioactively labeled pregnenolone by minced testicular tissue *in vitro* was also significantly greater during spermiogenesis than at other times of the year (Tsui, unpublished, quoted by Lofts, 1977). Thus, most hormonal data collected to date show a plasma androgen cycle in turtles markedly different from that published by Callard, Callard, Lance & Eccles (1976). The reason for these apparently contradictory results is no doubt due to the source of the animals studied. McPherson *et al.* (1982) used turtles collected in the field and sampled within 24 h of capture, whereas Callard, Callard, Lance & Eccles (1976) and Lofts & Tsui (1977) used turtles obtained from animal dealers. Not only was the locality from which the animal came unknown, but the length of time in captivity was also not known. As the stress of captivity has a rapid and extreme effect on circulating testosterone levels in both alligators and turtles (V. Lance, unpublished), we can assume that the published values for *Trionyx* (from less than 400 pg ml^{-1} to a maximum of less than 2 ng ml^{-1}) and *Chrysemys* (maximum about 5 ng ml^{-1}) are seriously in error.

The fact that highest testosterone levels occur when spermatogenesis is at a peak and at which time Leydig cells appear atrophic, does not, however, prove that Sertoli cells secrete androgen. It has been noted in a number of experimental conditions in rats (deKretser, 1982) and lizards (Pearson, Tsui & Licht, 1976) that the histological appearance of Leydig cells does not always correlate with testosterone secretion. The "atrophic" Leydig cells of turtles during spermatogenesis are probably capable of considerable hormone production. Mammalian Sertoli cells lack cholesterol side chain cleaving enzyme and the enzyme 3-beta-hydroxysteroid dehydrogenase (3B-HSD), and they are thus unable to synthesize testosterone from cholesterol; however, they are able to synthesize small amounts of testosterone and 5-alpha dihydrotestosterone (DHT) from progesterone (Tcholakian & Steinberger, 1978). These cells have been shown to be very active in converting testosterone to a number of C-19 metabolites, some of which appear to be essential for spermatogenesis (Weibe, 1977). Sertoli cells are also capable of converting androgens to estradiol (Dorrington, Fritz & Armstrong, 1978). Steroid metabolizing enzyme activity is stimulated by FSH which binds specifically to Sertoli cells (Dorrington *et al.*, 1978). The currently accepted theory is that testosterone production by Sertoli cells is negligible, but that these cells, in the presence of FSH, metabolize the testosterone that is secreted into the seminiferous tubules by Leydig cells in response to LH (Parvinen, 1982). The evidence for testosterone secretion by chelonian Sertoli cells is based solely on cytological criteria, and is unconvincing. There is no reason at present to assume that they are any different from those of mammals.

Since the majority of reptiles breed seasonally, and the beginning of seasonal breeding has been suggested as being analogous to the onset of puberty in mammals, in that meiosis is initiated in resting spermatogonia, an alternative interpretation of the turtle testicular cycle may be proposed, using as a model the onset of puberty in rats. Using pure cultures of Sertoli cells and Leydig cells, isolated from rat testis at different stages of sexual maturity, Weibe (1977, 1982) has recently shown that onset of the first meiotic division in resting spermatogonia is characterized by the appearance of a unique steroid produced by the Sertoli cells. Peak production of this steroid by Sertoli cells occurs well before significant production of testosterone by Leydig cells, and is associated with an increase in binding of FSH to Sertoli cells. Weibe

(1982) suggests that initiation of meiosis does not require testosterone, but does require FSH, and that testosterone secretion and LH binding to Leydig cells do not increase significantly until spermatogenesis is well advanced. Testosterone is necessary, however, for maturation of spermatids into mature spermatozoa. A number of observations suggest that a similar series of events may occur during the annual testicular cycle in turtles. Using the data of McPherson *et al.* (1982) for the seasonal cycle of *Sternotherus* (Table I, Figs 1 & 2) it is obvious that the highest testosterone levels are coincident with stage 6 of the cycle (spermiogenesis), and low testosterone levels with stages 2 and 3 of the cycle (onset of meiosis, or spermatocytogenesis). Dubois (1982) has shown that peak binding of FSH (Fig. 3) to *Chrysemys* testicular tissue occurs at stage 2 (equivalent to stages 2 and 3 of McPherson *et al.*, 1982), at which time meiosis is at a peak and testosterone levels low. Thus the pattern of circulating androgens and FSH binding in seasonally breeding turtles appears similar to that of rats at the onset of puberty.

Two gonadotropins, both of which, like mammalian LH and FSH, are glycoproteins composed of two subunits, have been isolated from pituitary glands of *Chelydra serpentina* and *Chelonia mydas* (Licht, Farmer & Papkoff, 1976; Papkoff, Farmer & Licht, 1976), and a radioimmunoassay (RIA) for chelonian LH has been developed (Licht, Wood *et al.*, 1979), but the seasonal variation in circulating levels in male turtles has not been studied. There is good evidence that gonadotropin release from the pituitary gland in turtles is controlled by a molecule similar to the decapeptide, luteinizing hormone releasing hormone (LH-RH) isolated from mammalian hypothalamus. However, synthetic LH-RH

TABLE I
Classification of spermatogenic stage [a]

Stage	Description of seminiferous tubules	Interstitial cells
1	Involuted with only spermatogonia; may have few to abundant spermatozoa in the lumens	Atrophied
2	Primary spermatocytes appearing; spermatogonia increasing, becoming abundant	Hypertrophied
3	Secondary spermatocytes and early spermatids abundant	Hypertrophied
4	Transforming spermatids with few spermatozoa	Hypertrophied
5	Spermatids and spermatozoa abundant	Atrophied
6	Spermatozoa abundant (maximum level of spermiogenesis)	Atrophied
7	Spermatozoa abundant but spermatids and spermatocytes greatly reduced	Atrophied
8	Few spermatozoa; few spermatids and spermatocytes, or spermatids and spermatocytes absent; may have abundant spermatozoa in the lumen	Atrophied

[a] From McPherson *et al.* (1982).

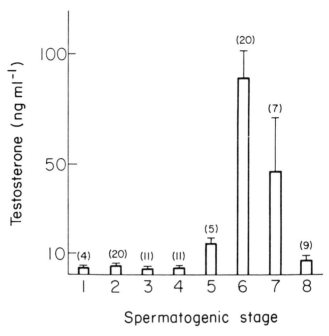

FIG. 2. Change in plasma concentration of testosterone associated with spermatogenic stage (Table I) in male *Sternotherus odoratus* (mean ± SEM). Numbers in parentheses indicate sample size. (From McPherson et al., 1982.)

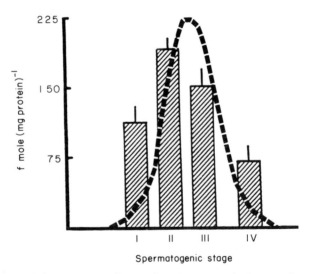

FIG. 3. Seasonal changes in testicular gonadotropin receptors in the painted turtle, *Chrysemys picta*. Specific binding of ^{125}I-rat FSH to crude membrane preparations of testicular tissue from stages I to IV of the spermatogenic cycle. Dotted line represents relative spermatogenic activity. (From Dubois, 1982.)

or synthetic agonists have no effect in sea turtles (Licht, 1982) or *Chrysemys picta* (V. Lance, unpublished). The LH-RH-like peptide in turtle hypothalamus is different from the known mammalian hormone (King & Millar, 1980).

In mammals a clear separation of function is evident between FSH and LH. LH binds specifically to Leydig cells and stimulates testosterone secretion; FSH binds specifically to Sertoli cells and stimulates production of androgen-binding protein and a number of other proteins, but does not stimulate testosterone secretion (Parvinen, 1982). When mammalian LH and FSH are injected into turtles, however, LH has no effect on testosterone secretion whereas mammalian FSH is highly stimulatory (see review by Lance & Callard, 1980). This biological activity of mammalian FSH is correlated with its ability to bind specifically to turtle testicular tissue, and the lack of biological activity of mammalian LH is likewise correlated with its lack of specific binding to turtle testicular tissue (Licht & Midgley, 1976; Dubois, 1982). In contrast, when chelonian FSH and LH are tested in turtles, both hormones have been shown to stimulate testosterone secretion, and both are able to displace specifically bound mammalian FSH from testicular tissue (Licht & Midgley, 1976; Licht, 1977, 1979). However, some experiments in which chelonian hormones are tested with chelonian tissues demonstrate a clear separation of LH and FSH activity. LH from *Chelonia mydas* is many times more potent than *Chelonia* FSH in stimulating testosterone secretion *in vitro* by testicular tissues from three species of turtle (Tsui & Licht, 1977). Radioiodinated FSH from *Chelonia* binds specifically to gonadal tissue from the same species and this bound FSH can be displaced by unlabeled mammalian FSH and unlabeled *Chelonia* FSH. Mammalian LH is ineffective in competing for the receptor and *Chelonia* LH is only 20–30% as effective as *Chelonia* FSH (Licht, Bona-Gallo & Daniels, 1977). These results suggest that, in *Chelonia* at least, the two hormones have separate receptors, and probably separate functions. In other experiments using turtle gonadotropins, however, no clear separation of biological or binding activity could be demonstrated (Licht, 1977).

It appears, therefore, that tissue receptors for gonadotropins in chelonian gonadal tissues are narrowly specific for FSH from a diverse array of mammalian species, but will respond equally to FSH and LH from birds and reptiles (Licht, 1977; Lance & Callard, 1980). What makes these observations difficult to interpret is the fact that mammalian FSH alone can apparently restore full gonadal function in hypophysectomized turtles (Licht, 1972b; Callard, Callard, Lance & Eccles, 1976), whereas restoration of gonadal function in hypophysectomized mammals requires both LH and FSH. It is possible that subtle distinctions in function between chelonian FSH and LH are being missed, and that more specific and sensitive bioassays are needed to assess their role in the Chelonia.

SQUAMATA

Lizards and snakes are believed to have evolved from a group of reptiles that branched off from the cotylosaurs more than 250 My ago (Bellairs, 1969; Carroll, 1970). They have evolved rapidly to fill a wide variety of niches and exhibit an extreme divergence from the main reptilian stock in both morphological and biochemical characteristics.

The male reproductive system shares most of the common amniote features, but it exhibits an unusual penial morphology and has a portion of the kidney that is androgen-sensitive. Del Conte (1972a), however, described a segment of the kidney in female *Cnemidophorus lemniscatus* which resembled the sex segment in males. He suggests that this unusual feature may be due to high plasma androgens in the female of this species. The male renal sex segment produces substances in response to testosterone which are added to semen and are believed to act as nutritive material for spermatozoa (Bishop, 1959; Prasad & Sanyal, 1969; Saint Girons, 1972). Other functions of these secretions have been demonstrated in snakes. In *Thamnophis* the semen forms a vaginal plug and prevents multiple matings (Devine, 1975); and in *Vipera berus*, it induces a hardening of the lower portion of the uterus (Nilson & Andren, 1982).

Histologically the squamate testis follows the basic amniote plan, although vasa efferentia connecting the testis and epididymis occur throughout the length of the structure. This feature is believed to be the reason why snakes of the families Leptotyphlopidae and Typhlopidae were able to evolve multilobed testes as an adaptation to a fossorial mode of life, yet still retain the basic squamate pattern of spermatogenesis (W. Fox, 1965; Werner & Drook, 1967). Another variation in histological structure of the testis is present in lizards of the genera *Cnemidophorus* and *Ameiva*. In these lizards a band of interstitial cells is located immediately beneath the testicular capsule (Goldberg & Lowe, 1966; Lowe & Goldberg, 1966; DeWolfe & Telford, 1966; Del Conte, 1972b). A similar layer of cells was described in *Agama tuberculata* (Duda & Kaul, 1976). These cells show seasonal morphological and histochemical changes which correlate with seasonal changes in secondary sex characteristics and are probably the major source of testicular androgens (Currie & Taylor, 1970; Neaves, 1976; Tsui, 1976).

Spermatogenic cycles have been described in many species of snakes and lizards (see Saint Girons, 1963, 1966, 1982; Fitch, 1970, 1982; Saint Girons & Pfeffer, 1971; Angelini & Picariello, 1975, for reviews). Although considerable variation exists amongst this group, spermatogenic cycles fall roughly into four basic types. Type I: Similar to that of turtles in that spermatogonial division starts in spring and spermiogenesis peaks in late summer and fall. Spermatozoa are stored in the vas deferens and epididymis until the following spring. Also, like the Chelonia, mating takes place when the seminiferous tubules are in a regressed state. Type II: Spermatogonial division begins in early summer, but spermatogenesis is interrupted by hibernation. Spermiogenesis and mating occur in spring shortly after emergence from winter torpor. Type III: The entire spermatogenic cycle occurs in spring or summer prior to mating. This type of cycle, which is common in desert lizards, has not been described in snakes. Type IV: Continuous reproduction. Not well documented, but probably occurs in some tropical species (Del Conte, 1972b; Gorman, Licht & McCollum, 1981; Fitch, 1982).

Serpentes

Testosterone levels have been reported in very few species of snakes. In general, it appears that plasma testosterone levels are highest at spermiogenesis, although in two species a second major peak occurs at mating. Weil & Aldridge (1981) studied the annual spermatogenic and hormonal cycle in *Nerodia fasciata*. They recorded a peak of testosterone (10–15 ng ml^{-1}) in April when testes are regressed and during which time mating occurs. This elevated testosterone and mating activity lasts for a period of four to six weeks, following which plasma testosterone levels decline during June and July as spermatocytogenesis proceeds. In late August a second peak in plasma testosterone coincides with maximum spermiogenesis [Fig. 4(A)]. A similar biphasic cycle was recently reported in *Opheodrys aestivus*, again with the major androgen peak in spring (J. J. Greenshaw, J. S. Jacob & M. V. Plummer, pers. comm.). In *Thamnophis sirtalis parietalis*, however, plasma testosterone was reported to be low in early spring when mating activity is observed and at which time testes are regressed, but showed a peak at spermiogenesis in late summer shortly before the snakes move into winter dens [Fig. 4(b)]. The actual plasma hormone concentrations were not presented (Hawley & Aleksiuk, 1975). In *Agkistrodon piscivorus* plasma testosterone levels are low (< 1 ng ml^{-1}) during mating in early spring when seminiferous tubules contain only resting spermatogonia and Sertoli cells, but rise to a peak of close to 3 ng ml^{-1} in late summer at spermiogenesis. Testosterone levels then decline again shortly before hibernation [Fig. 4(B)] (Johnson, Jacob & Torrance, 1982). The tropical marine snake *Acrochordus granulatus* in the Philippines shows a distinct seasonal breeding pattern in which mating and spermiogenesis coincide. A single peak of plasma testosterone occurs at this time (Gorman *et al.*, 1981). In this species, however, plasma testosterone ranges from a low of less than 1 ng ml^{-1} to a peak of close to 100 ng ml^{-1}, which is more than twice the highest testosterone concentration recorded in other snakes [Fig. 4(C)]. In *Cerberus rhynchops*, another tropical marine snake from the same habitat, the testes appear to undergo active spermatogenesis throughout the year. However, a single peak in testis mass and plasma testosterone occurs shortly before pregnant females appear in the population, suggesting a single mating period (Gorman *et al.*, 1981). Two other tropical species for which some hormonal data are available are of special interest. In *Thamnophis melanogaster*, a semi-aquatic garter snake, breeding occurs several times throughout the year. Male snakes are induced to mate when presented with pre-ovulatory females. The presence of vitellogenic or estrogen-injected females induces rapid testicular maturation and spermiation. Plasma testosterone increases from less than 0.5 to 2.5 ng ml^{-1} within six days (Garstka & Crews, 1982). It appears that spermatogenesis is arrested at some mid-point in the cycle until stimulated by the presence of a pre-ovulatory female. In the marine snake, *Laticauda colubrina*, testis weight does not vary during the year and gives an histological appearance of continuous spermatogenesis. Plasma testosterone, however, shows two distinct peaks (range 0.5–2.5 ng ml^{-1}). It is believed that the species is a year-round breeder (Gorman *et al.*, 1981). In the Asiatic cobra, *Naja naja*, in South China mating occurs in early spring when testis mass and spermiation are at a peak. Following mating testis mass rapidly declines, but spermatogenesis does not begin again until late summer, and is then arrested during the cooler winter months (Lofts, Phillips & Tam, 1966). Plasma testosterone shows a single narrow peak (range from less than 100 pg ml^{-1} to 2.5 ng ml^{-1}) coincident with peak spermiogenesis and mating (Bona-Gallo, Licht, MacKenzie & Lofts, 1980). Testicular tissue from the cobra also shows seasonal variation in testosterone synthesis *in vitro*. Conversion of radioactively labeled precursors to androgens is greater in testicular tissue from animals with testes at spermiogenesis than in testicular tissue taken from animals at other times in the year (Tam, Phillips & Lofts, 1969). The hormonal data on the cobra is taken from animals purchased from animal dealers and in all likelihood does not reflect the hormone levels in nature.

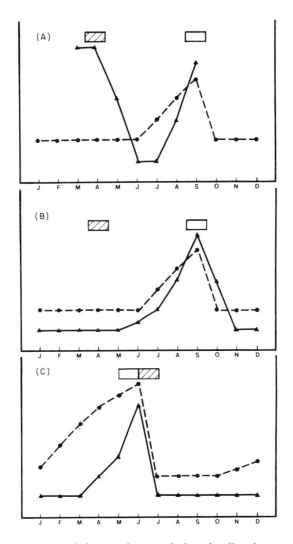

FIG. 4. Spermatogenetic and plasma androgen cycles in snakes. Open box at top of each graph represents approximate time of spermiogenesis and closed box approximate time of mating. The broken line represents relative testis mass and the closed lines the relative plasma testosterone. Months of the year are indicated at the bottom of each panel. Panel (A) represents the cycle in *Nerodia fasciata* and *Opheodrys aestivus*; panel (B), *Agkistrodon piscivorus* and *Thamnophis sirtalis*; and panel (C), *Acrochordus granulatus*. The months do not necessarily correspond to the cycle of the species indicated but have been normalized for simplicity. See text for details.

Lacertilia

Plasma testosterone has been measured in only four species of lizard throughout the annual cycle, and in all four, highest levels were recorded during spermiogenesis. In the Australian skink, *Tiliqua rugosa*, a single peak of 17β-hydroxysteroids (androgens) occurs at mating and is coincident with peak testis mass and spermiogenesis. The mean concentration in plasma reaches a maximum of 32 ng ml^{-1} at mating and remains below 10 ng ml^{-1} during the rest of the year (Bourne & Seamark, 1975). Testicular tissue from the same species shows a similar seasonal variation in steroid biosynthetic capacity. Testosterone production *in vitro* from radioactive precursors is higher in tissue collected during the mating period than at other times of the year (Bourne & Seamark, 1978). In another skink, *Liolaemus gravenhorsti* in Chile, plasma testosterone varies from only 200 to 600 pg ml^{-1}, though a clear seasonal spermatogenic cyle is evident. Although these values for testosterone are the lowest recorded in a seasonally breeding reptile, the peak (600 pg ml^{-1}) was recorded when testis mass was greatest (Leyton, Morales & Bustos-Obregon, 1977). Plasma testosterone in *Uromastix hardwicki* in Kashmir shows a single peak coincident with spermiogenesis and mating. Testosterone levels increase from approximately 5 ng ml^{-1} to 20 ng ml^{-1} (Arslan, Lobo, Zaidi, Jalali & Qazi, 1978). In the European viviparous lizard, *Lacerta vivipara*, the plasma androgen cycle is similar to other lizards in that peak levels occur during spermiogenesis and mating; however, the concentration of testosterone in the blood of this species is extraordinarily high. From a low of 2 ng ml^{-1}, levels rise to approximately 35 ng ml^{-1}, then increase dramatically to over 400 ng ml^{-1} at mating (Fig. 5). This level of androgen in *Lacerta* is the highest recorded for any vertebrate (Courty & Dufaure, 1979a). *Lacerta vivipara* has a high-capacity steroid binding protein in its plasma which may partially account for this extremely high hormone titer (Braux & Dufaure, 1982). Courty & Dufaure (1982) have carried out careful chromatographic separation of organic extracts of *Lacerta* plasma and testicular tissue at different stages of the reproductive cycle and have shown that several metabolites of testosterone are present in high concentrations in both tissues of this species (Courty & Dufaure, 1979b, 1980, 1982). Plasma androstenedione and DHT follow the pattern of testosterone in that elevations coincide with elevations in testosterone, but the concentrations of these steroids (approximately 30 ng ml^{-1}) are two orders of magnitude lower than testosterone concentration. When the concentration of androgen in testicular tissue is assayed, however, an unusual pattern emerges. Testicular testosterone and androstenedione are both high when plasma steroids are high during the mating period, but for a short period in June when spermatogenesis has ceased and the seminiferous tubules are regressing, a sudden increase in testicular testosterone and androstenedione is seen without a concomitant increase in plasma androgen (Fig. 5). DHT on the other hand is non-detectable in testicular tissue. During this period, of less than four weeks, the Leydig cells are enlarged and full of lipid droplets. The authors interpret the rise in testicular androgens, when peripheral levels are low, as being due to the steroids becoming trapped in lipid within Leydig cells (Courty & Dufaure, 1982). The lack of testicular DHT

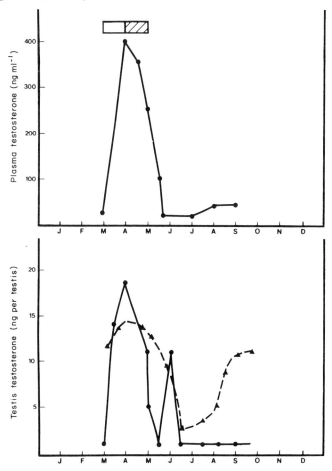

FIG. 5. Seasonal changes in plasma and testicular tissue testosterone in *Lacerta vivipara*. Open box represents approximate time of spermiogenesis and closed box time of mating. Broken line in lower panel represents testis mass and the closed line testicular testosterone concentration. Redrawn from Courty & Dufaure (1979a, 1980).

indicates peripheral conversion of testosterone to this metabolite and lack of testicular 5-alpha reductase. Hews & Kime (1978) reported the absence of 5-alpha reductase in testicular tissue of the closely related *Lacerta viridis*. Courty & Dufaure (1982) speculate that the onset of a new spermatogenic wave in July need not depend on an increase in gonadotropin secretion, but could be triggered by release of the androgens trapped within the testis. Arslan *et al.* (1978), however, noted a somewhat different series of events in *Uromastix*. In this species, testicular testosterone concentration was highest when plasma testosterone was elevated at spermiogenesis and mating. However, a second minor increase in testicular tissue testosterone occurred shortly before hibernation when plasma testosterone was low and spermatocytogenesis had already begun.

Bourne & Seamark (1978) showed an increase in production of epitestosterone, an inactive metabolite of androstenedione, by testicular tissue of *Tiliqua*, when testosterone production is low and testes are in a regressed state. They suggest that a regulatory mechanism for testosterone production may involve a seasonal switch in relative activity of the 17 alpha- and 17 beta-hydroxysteroid dehydrogenase enzymes. Both these enzymes use androstenedione as a substrate, the former converting it to biologically inactive epitestosterone, and the latter to testosterone. As the testes increase in size and androgen secretion increases, activity of 17 alpha-HSD decreases and activity of 17 beta-HSD increases. Whether this switch in enzyme activity is controlled by gonadotropin is not known.

Control of spermatogenesis in squamates is obviously under gonadotropin control, as it is in all vertebrates, but whether by an FSH, or an LH molecule, or by both has remained an issue of controversy for a number of years (Lofts, 1972, 1977; Licht, 1974; Callard & Ho, 1980). The initial attempts to purify gonadotropins from snake pituitaries suggested the presence of "FSH" and "LH" fractions, each of which induced testosterone production in squamate testis (Tsui & Licht, 1977). However, in a recent review of the chemical and biological properties of gonadotropins isolated from several species of snakes, Licht, Farmer, Bona Gallo & Papkoff (1979) conclude that there is but a single chemical entity, possessing full gonadotropic activity. Absence of LH-like activity in snake or lizard pituitaries is based on bioassays in which non-squamate animals are used to test the material; and since snake and lizard pituitary extracts are virtually inactive in all vertebrates tested with the exception of squamates (Licht, 1977; Lance & Callard, 1980), it is possible that a second hormone remains undetected. However, the bulk of the evidence suggests that there is probably only one (Licht, Farmer, Bona-Gallo *et al.*, 1979).

Although glycoprotein in nature, snake gonadotropin differs in amino acid composition from all other known tetrapod gonadotropins, and it is still not clear whether the molecule is composed of one or two subunits (Licht, Farmer, Bona-Gallo et al., 1979). This difference is further emphasized by the almost total lack of cross-reactivity of snake gonadotropin with antibodies to LH and FSH from a wide variety of vertebrates (Licht & Bona Gallo, 1978).

In contrast to turtles, lizards and snakes respond to both LH and FSH of mammalian origin and respond to pituitary extracts from a wide variety of vertebrates (Lance & Callard, 1980). Snake gonadotropin, however, appears to be biologically active only in squamates.

From the limited data available, it appears that the spermatogenic cycle in squamates is similar to that of turtles in that peak plasma testosterone occurs at spermiogenesis. However, the likelihood of a single gonadotropin molecule in this group makes comparisons difficult. The results of Bona Gallo et al. (1980) on plasma gonadotropin levels in the cobra throughout the spermatogenic cycle, although obtained from snakes purchased from animal dealers, show an interesting pattern which may be representative of the gonadotropin cycle in nature. A major peak (approximately 6 ng ml^{-1}) occurs in December shortly before the animals move into hibernation. This is the period during which there is no measurable testosterone in the plasma (Bona Gallo et al., 1980) and the resting spermatogonia in the seminiferous tubules are entering the first meiotic division (Lofts, Phillips & Tam, 1966). Following this peak, gonadotropin levels drop to about 4 ng ml^{-1} and remain at this level throughout spermatogenesis and mating. In August and October, four months after mating, and some months after testicular regression, gonadotropin concentration in the plasma drops to about 1 ng ml^{-1}. On the basis of these data it is difficult to ascribe a steroidogenic (LH-like) role to this hormone. However, the pattern of secretion is one which would be consistent with an FSH-like role, i.e. initiation of meiosis. It should be repeated, however, that these data are from snakes purchased from animal dealers and may be totally erroneous.

Several other pieces of data suggest gonadal tissue receptors for FSH and LH in the Squamata. Angelini, D'Uva, Picariello & Ciarcia (1978) injected mammalian gonadotropins into *Lacerta sicula* and noted stimulation of spermatogonial division in response to FSH with little effect on secondary sex characteristics. In contrast, they observed marked development of the secondary sex characteristics and stimulation of the later stages of spermatogenesis in response to LH. These results suggest that in this species LH stimulates testosterone secretion and FSH stimulates spermatocytogenesis. In the snake *Nerodia sipedon*, Weil

(1982) noted that response to mammalian LH and FSH depends on the time of year. In June, injection of LH results in elevated testosterone levels whereas FSH has no effect. In September, however, FSH is highly stimulatory and LH only moderately so. Whether these findings are indicative of the presence of two gonadotropins in squamates remains to be tested. An alternative explanation is that although Leydig cell and Sertoli cell gonadotropin receptors are different in this group, both receptors respond equally to a single squamate gonadotropin.

As in other reptiles, squamates possess a hypothalamic releasing factor immunologically related, but not identical, to mammalian LH-RH (King & Millar, 1980; Ball, 1981). As far as the present author is aware, synthetic LH-RH has not been tested for its gonadotropin-releasing effects in squamates, but it has been tested for behavioral effects in ovariectomized lizards. Alderete, Tokarz & Crews (1980) injected synthetic LH-RH, the tripeptide, thyrotropin releasing hormone (TRH), or saline into estrogen-primed, ovariectomized *Anolis carolinensis* and tested for sexual receptivity. Both LH-RH and TRH elicited sexual behavior in the experimental animals.

CROCODILIA

Apart from the unpublished study by Graham (1968) in which the testicular histology of *Crocodylus niloticus* is described, there are no published reports on the testicular cycle of the Crocodilia. Joanen & McNease (1980) collected male alligators throughout the year in southern Louisiana and noted that testis mass in mature animals shows a single peak in May at the height of the breeding season, and that following mating testis mass declines and does not vary during the rest of the year. Living spermatozoa are present in the penial groove during May and early June. These observations suggest that in this species the entire spermatogenic cycle occurs prior to mating. Lance (in press, and unpublished) studied the spermatogenic and hormonal cycle in alligators from the same locality and has shown that this is indeed the case (Fig. 6). The spermatogenic cycle begins in early spring, sometimes as early as February if temperature is higher then normal, and is completed by the first week in June. Following spermiation testicular regression is rapid; the testes remain in a regressed state for the remainder of the year. Plasma testosterone reaches a peak in April and May at the height of spermiogenesis and mating activity, and drops precipitously to non-detectable levels as the testes undergo regression in mid-June. An unusual feature of the hormonal cycle is the occurrence of a second peak of testosterone in September when the testes are still fully regressed.

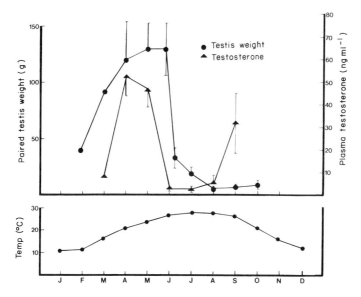

FIG. 6. Annual reproductive cycle of male *Alligator mississippiensis* in Louisiana. The lower graph represents the 10 year mean monthly temperature at Rockefeller Wildlife Refuge. Points on the upper graph represent mean monthly values ± SEM.

There is no evidence of spermatogonial division in the seminiferous tubules at this time and no evidence of sexual behavior. In turtles there is a peak of testosterone in the late summer, but this is when spermiogenesis occurs, and in some snakes a peak of testosterone occurs at mating when testes are regressed. The late summer peak of testosterone in the alligator does not coincide with mating behavior, spermiogenesis, or recrudescence of the seminiferous epithelium.

Testosterone levels in mature males are often in excess of 100 ng ml^{-1} at the peak of the breeding season. What was noted in this study was that the stress of captivity resulted in a rapid decline in plasma testosterone concentration. In some instances levels would decline from as high as 50 ng ml^{-1} to non-detectable within one day of captivity.

Two distinct glycoproteins similar in amino acid composition to mammalian LH and FSH have been isolated from alligator pituitary glands (Licht, Farmer & Papkoff, 1976). Tsui & Licht (1977) tested these purified hormones in an *in vitro* system in which testosterone production by minced alligator testicular tissue was used as a bioassay. Under these conditions, both gonadotropins stimulated testosterone production with alligator LH being about twice as effective as alligator FSH. When the same alligator testicular tissue was exposed to mammalian gonadotropins, ovine FSH was more than twice as potent as ovine LH (Tsui & Licht, 1977). These results, with tissue from a single immature alligator, are in contrast to experiments in which mammalian gonadotropins have been injected into mature alligators and plasma testosterone assayed as an end point. Under these conditions mammalian LH or human chorionic gonadotropin (hCG) have no effect on plasma testosterone, whereas mammalian FSH is extremely potent (Table II). The difference in results could be either due to cross-contamination of FSH in the LH preparations tested by Tsui & Licht (1977), or due to a difference in clearance rates of the two hormones *in vivo*.

TABLE II
Effect of mammalian gonadotropins on plasma testosterone concentration in male alligators [a]

Hormone injected	Plasma testosterone ng/ml^{-1}	
	Time 0	24 h
Saline	2.02 ± 1.34 (9)	0.24 ± 0.10 (9)
hCG	7.81 ± 6.74 (8)	0.19 ± 0.05 (8)
LH	1.56 ± 0.70 (8)	0.50 ± 0.22 (8)
FSH	3.88 ± 2.10 (10)	25.10 ± 3.77 (10)

[a] Blood samples were taken immediately before injecting 1 mg of hormone in 1 ml of 0.9% saline, and at 24 h post-injection and assayed for testosterone by RIA (V. Lance & K. Vliet, unpublished). Numbers in parentheses indicate sample size.

Unlike turtles, alligators show a marked response to the decapeptide, LH-RH. In a series of experiments in which immature and mature male alligators have been injected with LH-RH, and plasma testosterone measured by RIA, the synthetic peptide has been effective in every case (V. Lance & K. Vliet, unpublished). A single injection of 500 μg of LH-RH into male alligators ranging in size from 1.8 to 2.3 m in length resulted in a more than threefold increase in plasma testosterone at 24 h post-injection, whereas saline-injected controls remained unchanged (Fig. 7). However, the LH-RH-like molecule in alligator hypothalamus is not identical to the mammalian LH-RH. Using an antiserum which recognizes the entire decapeptide (Niswender R-42), and which recognizes analogues with substitutions in positions 2, 3, 5, 7 or 8, but with an absolute requirement for the aromatic ring in position 5, a C-terminal Pro^9, and Gly^{10}-NH$_2$ (Copeland, Aubert, Rivier & Sizonenko, 1979) a good parallel displacement curve is obtained with alligator hypothalamic extract. With this antibody, tissue levels of between 816 and 2780 pg mg^{-1} protein have been estimated in alligator hypothalamus (J. L. Bolaffi & V. Lance, unpublished). However, when alligator hypothalamic extract is tested against Jackson's antibody, I-J 29, no displacement of the radiolabeled LH-RH is observed. This antibody is

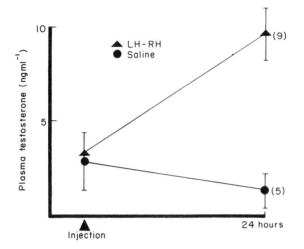

FIG. 7. Plasma testosterone response, after 24 hours, in male alligators to a single injection of 500 μg of synthetic LH-RH. Points represent the means ± SEM. Numbers in parentheses indicate sample size (V. Lance & K. Vliet, unpublished).

specific for the 7–10 fragment of the LH-RH. These results suggest that alligator LH-RH has some similarities to the mammalian decapeptide, but probably differs from it at least in the 8 position. The recent elucidation of the structure of avian LH-RH (King & Millar, 1982) shows that the decapeptide in birds differs from that of mammals in having a single amino acid substitution in the 8 position (Fig. 8). Given the close biochemical affinities of birds and crocodiles, and the immunological data cited above, it is quite possible that alligator LH-RH has a structure identical to that of birds.

SUMMARY AND CONCLUSIONS

In conclusion, it would appear that despite the wide biochemical and morphological divergence among the living orders of reptiles, the common feature of the male spermatogenic cycle is elevated testosterone production by the testes during the later stages of spermatogenesis. It is known from mammalian studies that high concentrations of testosterone within the testis are necessary for the successful development of spermatozoa from spermatids. This feature is probably of universal occurrence in amniote gametogenesis. What may also be true, though data in the Reptilia are scarce, is that initiation of the first meiotic division in resting spermatogonia requires the presence of an FSH-like hormone, and secretion of testosterone by Leydig cells requires an LH-like molecule.

1 2 3 4 5 6 7 8 9 10
pGlu-His-Trp-Ser-Tyr-Gly-Leu-Arg-Pro-Gly-NH$_2$

pGlu-His-Trp-Ser-Tyr-Gly-Leu-<u>Gln</u>-Pro-Gly-NH$_2$

FIG. 8. Amino acid sequence of mammalian LH-RH, upper figure, and avian LH—RH. The 8 position substitution is underlined.

In the Chelonia, the data on gonadotropin binding to membrane receptors, on circulating plasma steroids during the spermatogenic cycle and on the effects of chelonian LH on testosterone production suggest that regulation of spermatogenesis in this group is probably similar to that of mammals and birds. There are insufficient data on the Crocodilia, but given the close affinities of this group with birds, it is likely that the hormonal control of spermatogenesis is similar in the two groups. The situation in squamates is less clear. Spermiogenesis appears to be androgen-dependent as it is in all vertebrates, but the nature of pituitary control over steroid secretion and spermatogenesis is still unclear. It is possible that in lizards and snakes the entire process is regulated by a single gonadotropin as suggested by Licht (1977).

ACKNOWLEDGEMENTS

I would like to thank Dr Roger McPherson for generously supplying Figs 1 and 2, and Dr Wilfred Dubois for allowing me access to unpublished material and for supplying Fig. 3. I am especially grateful to Ted Joanen and Larry McNease, biologists at the Rockefeller Wildlife Refuge, Louisiana, for help in all aspects of the work on alligators. Part of the research on the alligators was financed by the Louisiana Department of Wildlife and Fisheries.

REFERENCES

Alderete, M. R., Tokarz, R. R. & Crews, D. (1980). Luteinizing hormone-releasing hormone and thyrotropin releasing hormone induction of female sexual receptivity in the lizard, *Anolis carolinensis*. *Neuroendocrinology* **30**: 200–205.

Angelini, F., D'Uva, V., Picariello, O. & Ciarcia, G. (1978). Effects of mammalian gonadotrophins and testosterone on the male sexual cycle of the lizard (*Lacerta s. sicula* Raf.) during the autumn spermatogenesis. *Monit. zool. ital.* **12**: 117–141.

Angelini, F. & Picariello, O. (1975). The course of spermatogenesis in Reptilia. 1. Classification of the types of spermatogenesis encountered in nature. *Mem. Acad. Sci. fis. mat. Soc. naz. Sci. Napoli* (3) **9**: 61–107.

Arslan, M., Lobo, J., Zaidi, A. A., Jalali, S. & Qazi, M. H. (1978). Annual androgen rhythm in the spiny-tailed lizard, *Uromastix hardwicki*. *Gen. comp. Endocr.* **36**: 16–22.
Ball, J. N. (1981). Hypothalamic control of the pars distalis in fishes, amphibians, and reptiles. *Gen. comp. Endocr.* **44**: 135–170.
Bellairs, A. (1969). *The life of reptiles.* London: Weidenfeld & Nicholson.
Bishop, J. E. (1959). A histological and histochemical study of the kidney tubule of the common garter snake, *Thamnophis sirtalis*, with special reference to the sexual segment in the male. *J. Morph.* **104**: 307–350.
Bona-Gallo, A., Licht, P., MacKenzie, D. S. & Lofts, B. (1980). Annual cycle in levels of pituitary and plasma gonadotropins, gonadal steroids, and thyroid activity in the Chinese cobra (*Naja naja*). *Gen. comp. Endocr.* **42**: 477–493.
Bourne, A. R. & Seamark, R. F. (1975). Seasonal changes in 17-hydroxysteroids in the plasma of a male lizard (*Tiliqua rugosa*). *Comp. Biochem. Physiol.* **50B**: 535–536.
Bourne, A. R. & Seamark, R. F. (1978). Seasonal variation in steroid biosynthesis by the testis of the lizard, *Tiliqua rugosa*. *Comp. Biochem. Physiol.* **59B**: 363–367.
Braux, J. P. & Dufaure, J. P. (1982). Liaison de la testosterone aux proteines plasmatiques chez le Lezard vivipare male. *C. r. Séanc. Soc. Biol.* **176**: 535–541.
Callard, I. P., Callard, G. V., Lance, V., Bolaffi, J. L. & Rosset, J. S. (1978). Testicular regulation in nonmammalian vertebrates. *Biol. Reprod.* **18**: 16–43.
Callard, I. P., Callard, G. V., Lance, V. & Eccles, S. (1976). Seasonal changes in testicular structure and function and the effects of gonadotropins in the freshwater turtle, *Chrysemys picta*. *Gen. comp. Endocr.* **30**: 347–356.
Callard, I. P. & Ho, S. M. (1980). Seasonal reproductive cycles in reptiles. *Progr. reprod. Biol.* **5**: 5–38.
Carroll, R. L. (1970). Problems on the origin of reptiles. *Biol. Rev.* **44**: 393–432.
Copeland, K. C., Aubert, M. L., Rivier, J. & Sizonenko, P. C. (1979). Luteinizing hormone-releasing hormone: sequential versus conformational specificity of antiluteinizing hormone-releasing hormone sera. *Endocrinology* **104**: 1504–1512.
Courty, Y. & Dufaure, J. P. (1979a). Levels of testosterone in the plasma and testis of the viviparous lizard (*Lacerta vivipara* Jacquin) during the annual cycle. *Gen. comp. Endocr.* **39**: 336–342.
Courty, Y. & Dufaure, J. P. (1979b). Androgènes testiculaires et cycles spermatogénétiques chez le Lezard vivipare. *C. r. Séanc. Soc. Biol.* **173**: 1083–1088.
Courty, Y. & Dufaure, J. P. (1980). Levels of testosterone, dihydrotestosterone, and androstenedione in the plasma and testis of a lizard (*Lacerta vivipara* Jacquin) during the annual cycle. *Gen. comp. Endocr.* **42**: 325–333.
Courty, Y. & Dufaure, J. P. (1982). Circannual testosterone, dihydrotestosterone and androstenediols in plasma and testis of *Lacerta vivipara*, a seasonally breeding viviparous lizard. *Steroids* **39**: 517–529.
Crews, D. (1979). Neuroendocrinology of lizard reproduction. *Biol. Reprod.* **20**: 51–73.
Currie, C. & Taylor, H. L. (1970). A histochemical study of circumtesticular Leydig cells of a teiid lizard, *Cnemidiphorus tigris*. *J. Morph.* **132**: 101–108.
deKretser, D. M. (1982). Sertoli cell-Leydig cell interaction in the regulation of testicular function. *Int. J. Androl.* (Suppl.) **5**: 11–17.
Del Conte, E. (1972a). Granular secretion in the kidney sexual segments of female lizards, *Cnemidophorus l. lemniscatus* (Sauria, Teiidae). *J. Morph.* **137**: 181–192.
Del Conte, E. (1972b). Variacion ciclica del tejido intersticial del testiculo y espermiogenesis continua en un largarto teido tropical: *Cnemidophorus l. lemniscatus* (L.). *Acta cient. venez.* **23**: 177–183.
Devine, M. C. (1975). Copulatory plugs in snakes: enforced chastity. *Science, Wash.* **187**: 844–845.

DeWolfe, B. B. & Telford, S. R. (1966). Lipid positive cells in the testis of the lizard, *Cnemidophorus tigris*. *Copeia* **1966**: 590–592.
Dodd, J. M. (1960). Gonadal and gonadotrophic hormones in lower vertebrates. In *Marshall's physiology of reproduction*: 417–582. Parkes, A. S. (Ed.). London: Longmans Green.
Dorrington, J. H., Fritz, I. B. & Armstrong, D. T. (1978). Steroidogenesis by granulosa and Sertoli cells. *Int. J. Androl.* (Suppl.) **2**: 53–63.
Dubois, W. (1982). *Testis structure and function in the fresh water turtle* Chrysemys picta. Ph.D. Dissertation, Boston University, USA.
Duda, P. L. & Kaul, O. (1976). Subtunic Leydig cells in the testes of *Agama tuberculata* (Gray). *Br. J. Herpet.* **5**: 539–541.
Fitch, H. S. (1970). Reproductive cycles in lizards and snakes. *Univ. Kans. Mus. nat. Hist. Misc. Publs* No. 52: 1–247.
Fitch, H. S. (1982). Reproductive cycles in tropical reptiles. *Occas. Pap. Mus. nat. Hist. Univ. Kans.* **96**: 1–53.
Forbes, T. R. (1961). Endocrinology of reproduction in cold-blooded vertebrates. In *Sex and internal secretions* **2**: 1035–1087. Young, W. C. (Ed.). Baltimore: Williams & Wilkins.
Fox, H. (1977). The urogenital system of reptiles. In *Biology of the reptilia*, **6**: 1–157. Gans, C. & Parsons, T. S. (Eds). New York: Academic Press.
Fox, W. (1965). A comparison of the male urogenital systems of blind snakes, Leptotyphlopidae and Typhlopidae. *Herpetologica* **21**: 241–256.
Gabe, M. & Saint Girons, H. (1964). Contribution à l'histologie de *Sphenodon punctatus* Gray. Paris: CNRS.
Garstka, W. R. & Crews, D. (1982). Female control of male reproductive function in a mexican snake. *Science, Wash.* **217**: 1159–1160.
Goldberg, S. R. & Lowe, C. H. (1966). The reproductive cycle of the western whiptail lizard (*Cnemidophorus tigris*) in southern Arizona. *J. Morph.* **118**: 543–548.
Gorman, G. C., Licht, P. & McCollum, F. (1981). Annual reproductive patterns in three species of marine snakes from the Phillippines. *J. Herpet.* **15**: 335–354.
Graham, A. (1968). *The Lake Rudolf crocodile* (Crocodylus niloticus *Laurenti*) *population*. M.Sc. Thesis: University College, Nairobi, Kenya.
Green, J. D. (1966). The comparative anatomy of the portal vascular system and of the innervation of the hypophysis. In *The pituitary gland*, **1**: 127–146. Harris, G. W. & Donovan, B. T. (Eds). London: Butterworths.
Hawley, A. W. L. & Aleksiuk, M. (1975). Thermal regulation of spring mating behavior in the red-sided garter snake (*Thamnophis sirtalis parietalis*). *Can. J. Zool.* **53**: 768–776.
Hews, E. A. & Kime, D. E. (1978). Testicular steroid biosynthesis by the green lizard *Lacerta viridis*. *Gen. comp. Endocr.* **35**: 432–435.
Joanen, T. & McNease, L. (1980). Reproductive biology of the American alligator in Southwest Louisiana. In *Reproductive biology and diseases of captive reptiles*: 153–159. Murphy, J. B. & Collins, J. T. (Eds). Lawrence, Kansas: Society for the Study of Amphibians and Reptiles.
Johnson, L. F., Jacob, J. S. & Torrance, P. (1982). Annual testicular and androgenic cycles of the cottonmouth (*Agkistrodon piscivorus*) in Alabama. *Herpetologica* **38**: 16–26.
Joly, J. & Saint Girons, H. (1975). Influence de la température sur la vitesse de la spermatogenèse, la durée de l'activité spermatogénétique et l'évolution des caractères sexuels secondaires du lézard des murailles, *Lacerta muralis* L. (Reptilia, Lacertidae). *Archs Anat. microsc. Morph. exp.* **64**: 317–336.

Kime, D. E. (1980). Comparative aspects of testicular androgen biosynthesis in non-mammalian vertebrates. In *Steroids and their mechanism of action in nonmammalian vertebrates*: 17–32. Delrio, G. & Brachet, J. (Eds). New York: Raven Press.

King, J. A. & Millar, R. P. (1980). Comparative aspects of luteinizing hormone-releasing hormone structure and function. *Endocrinology* **106**: 707–717.

King, J. A. & Millar, R. P. (1982). Structure of avian hypothalamic gonadotrophin-releasing hormone. *S. Afr. J. Sci.* **78**: 124–125.

Kuchling, G., Skolek-Winnish, R. & Bamberg, E. (1981). Histochemical and biochemical investigation on the annual cycle of testis, epididymis, and plasma testosterone of the tortoise, *Testudo hermanni hermanni* Gmelin. *Gen. comp. Endocr.* **44**: 194–201.

Lance, V. (In press). Reproduction in the alligator. *Int. Symp. comp. Endocr.* No. 9. Hong Kong: University Press.

Lance, V. & Callard, I. P. (1980). Phylogenetic trends in hormonal control of gonadal steroidogenesis. In *Evolution of vertebrate endocrine systems*: 167–232. Pang, P. K. T. & Epple, A. (Eds). Lubbock, Texas: Texas Tech Press.

Lance, V., Scanes, C. & Callard, I. P. (1977). Plasma testosterone levels in male turtles, *Chrysemys picta*, following single injections of mammalian, avian and teleostean gonadotropins. *Gen. comp. Endocr.* **31**: 435–441.

Leyton, C. V., Morales, C. B. & Bustos-Obregon, E. (1977). Seasonal changes in testicular function of the lizard *Liolaemus gravenhorsti*. *Archs Biol.* **88**: 393–405.

Licht, P. (1972a). Environmental physiology of reptilian breeding cycles: role of temperature. *Gen. comp. Endocr.* (Suppl.) **3**: 477–488.

Licht, P. (1972b). Actions of mammalian pituitary gonadotropins (FSH and LH) in reptiles II. Turtles. *Gen. comp. Endocr.* **19**: 282–289.

Licht, P. (1974). Reptilian endocrinology. The pituitary system. *Chem. Zool.* **9**: 399–448.

Licht, P. (1977). Evolution in the roles of gonadotropins in the regulation of the tetrapod testis. In *Reproduction and evolution*: 101–110. Calaby, J. H. & Tyndale-Biscoe, C. H. (Eds). Canberra: Australian Academy of Sciences.

Licht, P. (1979). Reproductive endocrinology of reptiles and amphibians: gonadotropins. *A. Rev. Physiol.* **41**: 337–351.

Licht, P. (1982). Endocrine patterns in the reproductive cycle of turtles. *Herpetologica* **38**: 51–61.

Licht, P. & Bona Gallo, A. (1978). Immunochemical relatedness among pituitary follicle-stimulating hormones of tetrapod vertebrates. *Gen. comp. Endocr.* **36**: 575–584.

Licht, P., Bona Gallo, A. & Daniels, E. L. (1977). *In vitro* binding of radioiodinated sea turtle (*Chelonia mydas*) follicle stimulating hormone to reptilian gonadal tissue. *Gen. comp. Endocr.* **33**: 226–230.

Licht, P., Farmer, S. W., Bona-Gallo, A. & Papkoff, H. (1979). Pituitary gonadotropins in snakes. *Gen. comp. Endocr.* **38**: 34–52.

Licht, P., Farmer, S. W. & Papkoff, H. (1976). Further studies on the chemical nature of reptilian pituitary gonadotropins: FSH and LH in the American alligator and green sea turtle. *Biol. Reprod.* **14**: 222–232.

Licht, P. & Midgley Jr, A. R. (1976). Competition for *in vitro* binding of radio-iodinated human follicle-stimulating hormone in reptilian, avian, and mammalian gonads by nonmammalian gonadotropins. *Gen. comp. Endocr.* **30**: 364–371.

Licht, P., Papkoff, H., Farmer, S. W., Muller, C. H., Tsui, H. W. & Crews, D. (1977). Evolution of gonadotropin structure and function. *Rec. Progr. Horm. Res.* **33**: 169–248.

Licht, P., Wood, J., Owens, D. W. & Wood, F. (1979). Serum gonadotropins and steroids associated with breeding activities in the green sea turtle *Chelonia mydas* 1. Captive animals. *Gen. comp. Endocr.* **39**: 274–289.

Lofts, B. (1969). Seasonal cycles in reptilian testes. *Gen. comp. Endocr.* (Suppl.) **2**: 147–155.

Lofts, B. (1972). The Sertoli cell. *Gen. comp. Endocr.* (Suppl.) **3**: 638–648.

Lofts, B. (1977). Patterns of spermatogenesis and steroidogenesis in male reptiles. In *Reproduction and evolution*: 127–136. Calaby, J. H. & Tyndale-Biscoe, C. H. (Eds). Canberra: Australian Academy of Science.

Lofts, B., Phillips, J. G. & Tam, W. H. (1966). Seasonal changes in the testes of the cobra, *Naja naja* (Linn). *Gen. comp. Endocr.* **6**: 466–475.

Lofts, B. & Tsui, H. W. (1977). Histological and biochemical changes in the gonads and epididymides of the soft-shelled turtle *Trionyx sinensis. J. Zool., Lond.* **181**: 57–68.

Lowe, C. H. & Goldberg, S. R. (1966). Variation in the circumtesticular Leydig cell tunic of teiid lizards (*Cnemidophorus* and *Ameiva*). *J. Morph.* **119**: 277–282.

McPherson, R. J., Boots, L. R., MacGregor, R. & Marion, K. R. (1982). Plasma steroids associated with seasonal reproductive changes in a multiclutched freshwater turtle, *Sternotherus odoratus. Gen. comp. Endocr.* **48**: 440–451.

Miller, M. B. (1959). The endocrine basis for reproduction in reptiles. In *Comparative endocrinology*: 499–516. Gorbman, A. (Ed.). New York: John Wiley & Sons Inc.

Moll, E. O. (1979). Reproductive cycles and adaptations. In *Turtles: perspectives and research*: 305–331. Harless, M. & Morlock, H. (Eds). New York: John Wiley.

Neaves, W. B. (1976). Structural characterization and rapid manual isolation of a reptilian testicular tunic rich in Leydig cells. *Anat. Rec.* **186**: 553–564.

Nilson, G. & Andren, C. (1982). Function of renal sex secretions and male hierarchy in the adder, *Vipera berus*, during reproduction. *Horm. Behav.* **16**: 404–413.

Papkoff, H., Farmer, S. W. & Licht, P. (1976). Isolation and characterization of follicle stimulating hormone and luteinizing hormone and its subunits from snapping turtle (*Chelydra serpentina*) pituitaries. *Endocrinology* **98**: 767–777.

Parvinen, M. (1982). Regulation of the seminiferous epithelium. *Endocr. Revs* **3**: 404–417.

Pearson, A. K., Tsui, H. W. & Licht, P. (1976). Effect of temperature on the production and action of androgens and on the ultrastructure of gonadotropic cells in the lizard *Anolis carolinensis. J. exp. Zool.* **195**: 291–303.

Prasad, M. R. N. & Sanyal, M. K. (1969). Effect of sex hormones on the sexual segment of kidneys and other accessory reproductive organs of the Indian house lizard, *Hemidactylus flaviviridis* Ruppell. *Gen. comp. Endocr.* **12**: 110–118.

Pritchard, P. C. H. (1979). Taxonomy, evolution and zoogeography. In *Turtles: Perspectives and research*: 1–41. Harless, M. & Morlock, H. (Eds). New York: John Wiley.

Roosen-Runge, E. C. (1977). *The process of spermatogenesis in animals*. London: Cambridge University Press.

Saint Girons, H. (1963). Spermatogenèse et evolution cyclique des caractères sexuels secondaires chez les squamata. *Annls Sci. Nat. (Zool. Biol. anim.)* **5**: 461–478.

Saint Girons, H. (1966). Le cycle sexuel des serpents venimeux. *Mems Inst. Butantan* **33**: 105–114.

Saint Girons, H. (1972). Morphologie comparée du segment sexuel de rein des Squamates. *Archs Anat. microsc. morph. exp.* **61**: 243–266.

Saint Girons, H. (1982). Reproductive cycles of male snakes and their relationships with climate and female reproductive cycles. *Herpetologica* **38**: 5–16.

Saint Girons, H. & Pfeffer, P. (1971). Le cycle sexuel des serpents du Cambodge. *Ann. Sci. Nat.* (Zool.) (12) **13**: 543–572.

Tam, W. H., Phillips, J. G. & Lofts, B. (1969). Seasonal changes in the *in vitro* production of testicular androgens by the cobra (*Naja naja* Linn). *Gen. comp. Endocr.* **13**: 117–125.

Tcholakian, R. K. & Steinberger, A. (1978). Progesterone metabolism by cultured Sertoli cells. *Endocrinology* **103**: 1335–1343.

Tsui, H. W. (1976). Stimulation of androgen production by the lizard testis: Site of action of ovine FSH and LH. *Gen. comp. Endocr.* **28**: 386–394.

Tsui, H. W. & Licht, P. (1977). Gonadotropin regulation of *in vitro* androgen production by reptilian testes. *Gen. comp. Endocr.* **31**: 422–434.

Weibe, J. P. (1977). Comparative gonadal steroid enzymology of vertebrates during sexual maturation. In *Reproduction and evolution*: 121–126. Calaby, J. H. & Tyndale-Biscoe, C. H. (Eds). Canberra: Australian Academy of Sciences.

Weibe, J. P. (1982). Identification of a unique Sertoli cell steroid as a 3a-hydroxy-4-pregnen-20-one (3a-dihydroprogesterone: 3a-DHP). *Steroids* **39**: 259–278.

Weil, M. R. (1982). Seasonal effects of mammalian gonadotropins (bFSH and bLH) on plasma androgen levels in male water snakes, *Nerodia sipedon. Comp. Biochem. Physiol.* **73A**: 73–76.

Weil, M. R. & Aldridge, R. D. (1981). Seasonal androgenesis in the male water snake, *Nerodia sipedon. Gen. comp. Endocr.* **44**: 44–53.

Werner, Y. L. & Drook, K. (1967). The multipartite testis of the snake, *Leptotyphlops phillipsi. Copeia* **1967**: 159–163.

Breeding the Gharial (*Gavialis gangeticus*): Captive Breeding A Key Conservation Strategy for Endangered Crocodilians

H. ROBERT BUSTARD

*Airlie Brae, Alyth, Perthshire PH11 8AX, Scotland**

SYNOPSIS

The need for captive breeding programmes as a part of the conservation strategy for endangered species of crocodilians is explained. Data are provided on key aspects of such captive breeding programmes including requisite safeguards. Details are set out of the design and construction of a complex which achieved the first captive breeding of the gharial, together with information on the breeding stock. The breeding results for the first four years of the project are reviewed. Some aspects of conditions and care important to breeding success are discussed. The chapter concludes with a brief account of gharial breeding in protected sanctuaries in the wild in India.

INTRODUCTION

This paper covers one aspect of crocodile conservation – captive breeding – particularly of the gharial, and the role that this can play in an overall conservation strategy. It is essential, therefore, to put the gharial captive breeding project in its context.

The gharial was considered to be one of the most endangered crocodilian species yet virtually nothing was known about its status in the wild (IUCN, 1971).

I was approached by the United Nations Development Programme, acting on behalf of the Government of India, to come to India to carry out a survey and to advise the Government what actions were required to save the gharial from extinction. I visited India in early 1974, when, as a result of excellent support from the Government of India, I was able to travel very extensively in the sub-continent and to observe both of India's crocodiles (*Crocodylus porosus* and *C. palustris*) as well as the

* Formerly: Chief Technical Adviser, Government of India Crocodile Breeding and Management Project, Hyderabad, India.

gharial. At the end of my survey I had very clear knowledge of the problems throughout India, and of what needed to be done. My report on behalf of the United Nations (Bustard/FAO, 1974) was accepted by the Government of India and I was invited to return to live in India and assist in its implementation. I was to spend almost seven years in India.

This large-scale project of the UNDP/FAO/Government of India has been extremely successful (Bustard/FAO, 1974, 1975; Bustard, 1981), and is continuing today, administered through the Government of India in association with the States of the Indian Union, with key personnel trained under the project, and currently employed in the Government Central Crocodile Breeding and Management Training Institute.

The key aim of the project was to protect India's remaining gharial (and other crocodiles) in especially gazetted sanctuaries and National Parks, and to provide crocodile inputs into existing Parks and sanctuaries as required – that is the conservation of India's crocodilians *in the wild* (Bustard/FAO, 1974, 1975; Bustard, 1981). Captive breeding was a minuscule part of the overall project. Indeed, since it is obviously much easier to protect crocodilians in captivity, a guideline was laid down that no wild crocodilians should be captured for captive breeding purposes, and that captive breeding attempts should be based only on animals held in captivity (mainly in zoos).

During my first extensive travels in India in 1974 I located two groups of adult or near-adult gharial and recommended that they be used for captive breeding attempts. One of these, said to comprise two females and a male, was located at Nandankanan Biological Park, in the Eastern State of Orissa. They were not well housed and were kept together with two saltwater crocodiles. The other gharial group belonged to the Maharaja of Baroda and was located in his private zoo at Baroda. H. H. Baroda, who is extremely interested in crocodilians, unfortunately did not have sufficient time to devote to the development of a breeding programme at Baroda, although his mugger (*C. palustris*) breed regularly. Therefore a gharial breeding complex for use in this project was established at Nandankanan.

The need for captive breeding programmes for all endangered crocodilians, not just the gharial, is now considered.

THE NEED FOR CROCODILIAN CAPTIVE BREEDING PROGRAMMES

There is an urgent need to build up captive breeding herds of all endangered crocodilians, because their continued survival in the wild in various parts of the world is seriously threatened. This is an important part of the conservation strategy for many diverse forms of wildlife. However, it is particularly suited to crocodilians because of their potentially great fecundity and longevity.

Anyone who has worked extensively in the tropics and subtropics will be aware of the many threats to sustained conservation programmes, even where these can be initiated. In many instances proper conservation strategies never get off the drawing board and for various reasons little is said about this by people in a position to exert influence. I would like to make it quite clear that none of these remarks refer to India, where the most extensive conservation programme for crocodilians has been undertaken, and where it is hoped the future is secure. However, there is a state of war, almost everywhere, between the forces for "development" (so often the modern euphemism for "alienation" of the land) and those (where any exist) for conservation. The problem, in this unequal contest, is that conservation is ultimately about land use which is extremely political. Governments everywhere have become the final arbitrators, and are, almost invariably, motivated by expediency. This is not to say that there have not been, and will not continue to be, some outstanding conservation success stories.

The crux of the problem can be summarized as follows: when conservation wins, it wins only one round in a protracted engagement. The decisions can always be set aside, later reviewed or modified in the light of changed Government policies. However, where the non-conservation forces (development) win and the bulldozers or loggers move in, there can be no second round. The war has been lost.

Captive breeding programmes are the policy of last resort. If adequate captive breeding programmes are in operation and the "war" is lost at least the species survives. There may also be the *hope* that, at some future time, sentiment may change and it will be possible to use the progeny of the captive breeding projects to reintroduce the species to the wild.

This need to build up captive breeding herds of endangered crocodilians led directly to the formation of the IUCN Crocodile Specialist Group, which, under different circumstances, might have been a strong force for effective crocodile conservation. Working in Australia, I felt that a Foundation for crocodiles located close to a major north–south highway in the Australian tropics – such as near Cairns or Townsville in North Queensland – would soon become self-supporting from entrance fees. The political stability, tropical climate, and affluence of Australia made it an ideal location. It was in correspondence with Sir Peter Scott in the mid-sixties about this proposal that the idea for the IUCN Crocodile Specialist Group was born. I mention this because Professor

Bellairs, to whom this volume pays tribute, became involved, and it was in Cambridge in 1970 that, in a meeting between Richard and Maisie Fitter of the Fauna Preservation Society, Dr Hugh B. Cott, Professor Bellairs and myself, the Group was born.

Crocodilians exhibit a number of important characteristics which make the creation of a "living gene bank" relatively straightforward, and also potentially ecologically effective (from the viewpoint of future release back into the wild). These advantages stand out if we compare them with another animal species also at the forefront of conservation programmes – the tiger. It is not a feasible proposition to breed tigers (or other large cats) in captivity for eventual introduction into the wild, because such animals tend to become familiar with man – lose their natural fear – and then attack humans after release. Furthermore, the breeding lifespan, and associated potential rate of increase, is very poor compared to crocodilians. It is the very adaptability of crocodilians, their longevity, their high potential rate of increase, and the fact – as the Indian projects have shown – that captive-hatched and reared individuals can be released back into the wild with a very high percentage survival rate, that make them ideal material for this technique. Finally, the ability to control the sex ratio of the progeny by manipulation of the incubation temperature (Ferguson & Joanen, 1982) offers tremendous scope. In rehabilitation work a sex ratio of three females to one male is recommended. Careful selection of the appropriate incubation temperature will prevent wastage through the production of excess males, as would occur in a one-to-one sex ratio.

If the above arguments in favour of captive breeding programmes for crocodilians are accepted, then it is necessary to ensure that such programmes are properly set up and operated.

INITIATION AND OPERATION OF CAPTIVE BREEDING PROGRAMMES AND REHABILITATION

It is essential at the outset to state what should be obvious. Proper records must be kept. Many zoos seem quickly to lose records – if they ever had them – on the locality in the wild from which their captive breeding stock came. Stud books should be maintained. Threats of inbreeding have been exaggerated but it would clearly be ill-advised to repopulate an entire river system or even an entire State using the progeny of only a single pair (as would be quite feasible on production considerations over a 10-year timespan). Such considerations affect the design of captive breeding programmes, and raise real problems in practice, since many zoos, which may contribute valuably to a captive-breeding programme, may only be able to house one pair of breeding adults. This will necessitate exchange of juveniles with another captive breeding project, a task at which most zoos (although there are a few world-famous exceptions) are extremely poor.

Captive breeding programmes should, wherever possible, be located within the natural geographical range of the species. There should be more than one programme for each species, or if this is not possible, then some of the progeny should be widely distributed to approved holding centres. It is perfectly reasonable that some well organized centres may breed species from other parts of the world but the main captive breeding thrust should be within the species' natural range. There are several good reasons for this. Not only is there then no need to worry about climatological considerations, but the breeding programme, assuming it is successful, can be used to assist the species' conservation in the wild. This was certainly an important facet, both biologically and politically, of the gharial conservation success story. When a species is on the verge of extinction, and extremely difficult to sight in the wild, it is very difficult to justify, to anyone other than the most ardent conservationist, the setting aside of large areas of land for the handful which survive, and without which recovery cannot take place. Natural recovery, where it occurs, is usually extremely slow. However, the fruits of a successful breeding programme, given good husbandry conditions, can very quickly become most impressive and they can rapidly outgrow the available facilities. This can be extremely useful in expediting the creation of special sanctuaries or reserves.

Captive breeding programmes should take great care to use only pure stock of the species, free from any risk that hybridization with another species has taken place in captivity. There are well known cases of extensive hybridization taking place in captivity, for instance *C. rhombifer* with *C. acutus* (Varona, 1966), and *C. siamensis* with *C. porosus* (Yang-prapakorn, 1971). Wherever possible the provenance of all the breeding stock should be known. Markedly different geographical races of certain crocodile species exist, and this is likely to prove true of all but the most geographically restricted species. Conservation programmes, wherever possible, should avoid mixing such geographical races during either breeding or reintroduction to the wild.

Some people may think that these apparently minor points are an extravagance, unnecessary, when the urgent problem is to save the species, but one small illustration will indicate why no soundly-based ecologist would agree with such a view. Gharial inhabit river systems which are either snow-fed (i.e. derive water from melting snow in the Himalayas) or not snow-fed. It has been shown that gharial inhabiting snow-fed river systems lay their eggs on average 1.1 m higher above the water level in the river at the time of nesting than do gharial from river systems not snow-fed (Chowdhury, 1981; Bustard, in preparation; Chowdhury, Bustard & Tandon, in preparation). Such information appears to be genetically coded in these races, and is vital for survival. The water level in snow-fed rivers rises during April, with the melting of the snow, while the eggs are incubating (gharial eggs are laid in late March or early April and hatch in June), whereas the water level in rivers which are not snow-fed continues to fall throughout the dry (incubation) season. If gharial from the genetic stock of the latter river system were used to rehabilitate a snow-fed river system, they would be unlikely to breed with great success because they would not lay their eggs sufficiently high above the water level at time of nesting to safeguard them from subsequent rises in water levels. This is a case which we know about, and which can be readily observed and quantified: many other effects may be less easy to detect. This is why the competent field ecologist is extremely cautious about doing anything which in any way interferes with the species' natural situation in the wild.

Of course, where the original stock is lost (locally extinct) this is not possible, and the choice becomes either to lose the species in that locality or to reintroduce the best available stock. In such circumstances the latter action is to be preferred.

DESIGN OF THE GHARIAL BREEDING ENCLOSURE

In this project, a great advantage was that the author was able to design a complete breeding complex *de novo* instead of having to make do with the modification of existing, often unsatisfactory, pools. This was a direct result of financial support under the project from the Government of India to the captive breeding scheme in Nandankanan Biological Park, Orissa. To date, captive breeding of this shy animal has not been successful anywhere in the world apart from Nandankanan, where six successful nests have now been laid.

An ideal enclosure for captive breeding will try to emulate what are considered to be key features of the natural habitat. Apart from physical features, temperature, climate and food may be important. Although certain individuals of a number of crocodilian species will breed under captive conditions which differ greatly from those in the wild, it must always be the aim to provide conditions as near ideal as possible and not just the bare minimum which *may* result in breeding. This approach may well reflect the difference between success and failure with difficult species, or species of which breeding is being attempted for the first time. It is always better to be ecologically cautious and to provide a safety margin.

Relevance of Gharial Biology

In order to simulate in captivity what are thought to be key aspects of the natural habitat, extensive knowledge of the species' ecological requirements in the wild is essential. For instance, in the case of the gharial, we are dealing with a riverine species occurring in the larger perennial rivers of northern and eastern India where it inhabits deep pools and basks on adjacent sandbanks, seldom moving more than its own length from the water's edge. High sandbanks are selected for nesting, and since these are in limited supply, the female may have to travel considerable distances from the preferred basking sites in order to nest.

Furthermore, the gharial is a very shy crocodilian. Although accustomed to hot weather during the summer (which it ameliorates by retreating to its favoured deep pools for much of the day), the gharial is also used to cold winters. In the wild in Orissa during the hot summer months gharial almost never emerge during daylight hours (they may do so after nightfall), and the closest they come to haul out, is to rest their snout, and sometimes one front foot, on the river bank. Deep water may be important in allowing them to maintain their preferred body temperature and to avoid overheating. This problem is easily aggravated in captivity where crocodile pools are usually very small and shallow. In small volumes of water, temperature changes are rapid, and the normal attribute of the water in nature – being cool in summer and warm in winter in relation to ambient air temperature – may be lost. Similarly, cold or cooler conditions during the winter may be important in bringing the species into breeding condition. In nature, gharial commence courtship in late January or at the start of February, after the two coldest months of the year, the eggs are laid at the end of March or early April, and incubate during the hottest months of the year.

It is important to stress that virtually nothing was known concerning any aspect of the natural history of the gharial, prior to the commencement of this project. Even the time of egg-laying and hatching had been wrongly given by Malcolm Smith (1931). When the captive breeding project commenced we did not know the age or size of a sexually mature gharial. Many assumptions or "informed guesses" had to be made. For example, in order to estimate the size of a first breeding animal extrapolations were made from other crocodilians of equally large 'final' adult size, such as *C. porosus*.

Ideal Breeding Conditions

Seclusion (privacy) is important in breeding many animals in captivity, and this might be expected to be very important in the case of an extremely shy animal such as the gharial. Safeguards to ensure this were incorporated in the design of the breeding complex.

From the above account of its natural history it may be concluded that ideal conditions would include a deep pool comprising a large body of water, which should preferably be flowing, and the provision of an extensive, high-rising sandbank. Water levels undergo extensive seasonal fluctuations in nature, the high nesting banks being completely inundated during the wet season. It was not known if seasonal fluctuations in water levels were important in bringing animals into breeding condition. However in the wild, gharial continue to breed in areas where, because of the presence of dams or barrages, the water levels do not fall markedly during the summer months, so no provision for this was included in the design.

Temperature conditions, although important, were not considered in the present design once the large, deep, pool had been allowed for, as the breeding enclosure is located only a few kilometres from the Mahanadi river, part of the gharial's natural habitat.

The Breeding Complex

Much of the potential for crocodilian captive breeding lies in zoos, but because of the size of the undertaking, (especially with the larger species) it is important to design pools which fulfil the biological needs of the inmates yet are aesthetically pleasing to visitors, and which unobtrusively control unruly behaviour by visitors towards the animals. Fortunately, these two apparently opposing goals need not be antagonistic, as the following design shows.

An undisturbed area of the Park had been set aside for the development of the crocodile project – which included captive breeding of all three species of Indian crocodilians. The gharial breeding complex was the first to be constructed. It consisted of a high (2 m) brick compound wall enclosing an area of 71×51 m. A restricted viewing area for the public 30 m wide (where the wall was 0.5 m high, in front of a dry moat to prevent the gharial's escape) was provided in the 220 m perimeter.

If this viewing area is located carefully in relation to the lay-out of the enclosure it will not seem restricted to the public who will feel they are "looking in" on a display, yet it will serve all the requirements of the species concerned. In the present display the road leading to the gharial breeding enclosure was aligned so that the first view the visitor saw was the huge sandbank (see below) as a backdrop to a wide expanse of water (the pool). The low front wall of the viewing area was disguised by low shrubby vegetation. There is considerable skill in designing enclosures to provide the aesthetic qualities so important for an understanding of an animal species, itself a key aspect of interesting the visitor in conservation; zoos are no longer adequate if they merely display animals in concrete enclosures behind bars.

To return to the gharial pool, the oval-shaped, concrete-lined pool is 59.4 m in length with a breadth varying from 5.5 to 29.6 m. The pool slopes from the narrow end towards its widest portion where it reaches a maximum depth of 9.2 m and the sides also slope. Sloping sides are more natural than vertical ones, but in the design of a pool, the relationship between the angle of slope and the strength of the final structure should be taken into account. By selecting the angle correctly, it is possible to greatly reduce the thickness of concrete required for the sides, and hence considerably reduce the cost. Pool capacity is 2.7 million litres.

The lay-out of the pool in relation to the rest of the enclosure, and especially the viewing area, is critical (see also Bustard, 1980a). Crocodilians suffer greatly from human disturbance whenever these is no physical barrier between the spectators and the animals. Because they typically spend long periods immobile on the basking grounds, people feel challenged to make them move by throwing things at them. Since basking is important (and in winter, critical) in maintaining body temperature, constant disturbance at the basking grounds, leading to frequent retreat to the water, interferes with this. However, this may be the tip of the iceberg. Continual disturbance cannot but destroy any feeling of security in the animal, which is often a key to bringing about successful breeding. The outstanding zoo keeper realizes that the animals' enclosure is their *sanctuary* and so arranges things that the animal is safe at all times therein.

Disturbance of the kind outlined above can be overcome by situating an area of land which the animals will seldom use between the viewing area and the pool (Fig. 1). Thus the animals are free from disturbance either when they are in the pool or when they are basking on the land to the rear of the pool. In the gharial breeding pool, a dry moat and then a land area separate the viewing area from the nearest portion of the pool. Clearly the above format depends on a high boundary wall which people cannot climb.

These essential features of the enclosure have been stressed because only too often such requirements are not appreciated. All too many crocodilian enclosures in zoos allow the public to walk round a small

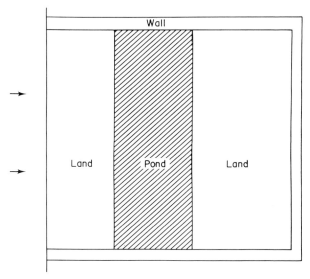

FIG. 1 A diagrammatic representation of a simple crocodile breeding enclosure suitable for outdoor construction in the tropics or subtropics. The design combines maximum privacy and minimum disturbance for the inmates. Viewing area arrowed.

enclosure which may consist of a small moat with a central land area. This provides none of the animal's basic requirements and there is nowhere in the enclosure where the animal cannot be 'stoned'.

Water Supply

The gharial pool design includes water circulation using two electric pumps. A 40 H.P. pump raises water from an existing lake within the Park beside which it is situated. The water enters the pool at the narrow end and provides a strong circulation across the pool. There is a broad but shallow (10 cm) overflow channel at the other end of the pool. This serves the important aesthetic function of removing any floating debris from the pool, and hence retaining its riverine appearance, when the 40 H.P. pump is in operation. This pump is also used to "top up" the pool to compensate for the substantial evaporation which occurs during the summer.

A second independent (10 H.P.) pump provides a means of removing excess water from the broad end of the pool. This water can then be used for other purposes within the Park, or be recirculated into the pool, entering it near the narrow end. This provides a second method for creating a water current across the pool. The second pump is located in a pump house just outside the perimeter wall; it is quiet in operation, the slight noise being muffled by the noise of water re-entering the pool.

When being recirculated the water removed from the pool is pumped into a header tank immediately outside the compound wall, and the water re-enters the enclosure under pressure through a grill in the compound wall. The rear of the enclosure, between the wall and the pool, was partially constructed as an artificial stream, concrete-lined, with rocks cemented in strategic positions so as to enhance the sound of running water. The overall effect is to give a pleasing sylvan setting to the complex – important because this is a public display in a State Zoo, as well as a captive breeding project.

One of its aims was to show crocodilians in an optimum setting and hopefully to generate conservation interest thereby. Moreover, the water recirculation stream is also appreciated by the gharial, which often come to rest in the shallow water in the pool facing the 'stream' when the recirculation pump is running.

A large sandbank, measuring 60 m in length by 7 m in breadth and rising to a height of 2.4 m, was constructed immediately behind the pool as the nesting sandbank. The natural vegetation of the area (scattered low trees) was carefully retained when the complex was constructed and the sandbank partly planted with coarse grasses to stabilize it. A diagrammatic representation of the gharial breeding pool is given in Fig. 2.

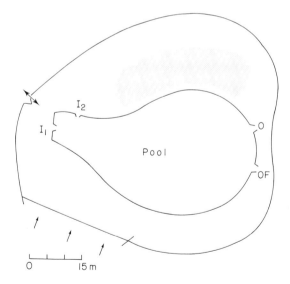

FIG. 2 The lay-out of the gharial breeding pool constructed at Nandankanan Biological Park, Orissa, in which gharial were bred for the first time in captivity. Stippled area is the sandbank of 2.4 m height. I_1, inlet for 40 H.P. pump; I_2, inlet for 10 H.P. pump; O, outlet for 10 H.P. pump; OF, overflow to clear surface debris. The viewing area (arrowed) is restricted to 30 m width in a 220 m perimeter wall. The only entrance in the 2 m high compound wall is marked thus ↔.

Food

Gharial are almost exclusively fish-eating, so the pool was stocked with naturally-occurring fish species of the area favoured by gharial including the genera *Anabas*, *Clarias* and *Ophiocephalus* and notoropterids. These fish are bottom-dwelling, so-called "scaleless" species as compared to large-scaled species such as carp.

A known weight of fresh live fish is added daily. No dietary supplements are given. The diet, therefore, consists entirely of fish.

Fish regularly breed in the pool and the fingerlings are taken by kingfishers and heron. Peacocks fly in to drink and troops of rhesus macaque (both species native to the area) also drink at the pool. These and other wild visitors greatly enhance the natural atmosphere created by judicious planting of additional trees, bamboo and tall grasses towards the rear of the enclosure. These serve the additional function of disguising the compound wall.

Summary and Conclusions

The enclosure design described above fulfilled the aim of providing:

(i) Seclusion – considered to be most important. Breeding displays for crocodilians should *not* be approachable from all sides (Bustard, 1980a) and should include a deep pool. A pool in which crocodilians are scarcely able to submerge completely and can never sink out of sight of spectators is totally unsatisfactory.

(ii) Key features of the natural habitat:
   (a) Deep pools.
   (b) Flowing water.
   (c) High nesting sandbanks.

While it is not suggested that the present water depth (9.2 m), or even the pool size, is the minimum required to breed gharial successfully, the large pool has marked advantages in an outdoor display in the extreme Oriyan climate. It provides a temperature retreat from the hottest weather of summer and ameliorates the cold Oriyan winter nights, as it does not undergo the large daily temperature fluctuations which characterize small, shallow pools in this climate.

The large pool also contributes to the aesthetics of the compound and so has a useful conservation input. Many people, even in India, never see a wild crocodile, so the conservation message must be put across in zoos. The purpose of captive breeding is ultimately conservation – initially to replace importation of more members of the species for display purposes, and ultimately to provide animals to put back into the wild where populations have become critically depleted or entirely wiped out.

Seeing several crocodilians lying motionless in an enclosure not much larger than themselves, with a small, perhaps stagnant, pool, completely fails to put across this message and is inferior to a good display of the animal in its "natural" surroundings. Thus, visitors to the display described here can see gharial "cruising" normally as they would in a natural river: this is impossible in the small, shallow pools of most captive displays. Good display and good breeding conditions need not be antagonistic: with proper planning both can go hand in hand.

THE GHARIAL CAPTIVE BREEDING PROGRAMME

The Breeding Stock

The original gharial complement held at Nandankanan Biological Park, Orissa, comprised one male and two females received from the Mahanadi river, Orissa, as juveniles and raised in the Park. The male was brought to the Park in March 1963 at a length of 1.35 m and the females are thought to have been received in November 1964 and November 1965 at lengths of 0.9 and 1.20 m respectively.

The sizes of the two females when they were received by the Park indicate that they were then 17 and 29 months old. The male was probably 33 months old when brought to the Park. These three individuals were transferred to the breeding pool on 13 February 1976. At this time, therefore, the male was about 16 years old and the two females probably in their twelfth and thirteenth years. At the time of transfer to the breeding pool the male measured 2.70 m and the females 2.65 and 2.50 m.

It should not be inferred from the above that gharial take this length of time to reach breeding size. From subsequent growth data (Bustard & Singh, 1980, in preparation) the above are seen to be very suboptimal growth rates. Crocodilians in captivity often show slower growth rates than in nature (Bustard, 1980a), although the best captive growth rates are likely to exceed those taking place in nature (Bustard, 1980a). Limited data suggest that female gharial commence breeding at a size of slightly over 3 m in total length (Bustard & Maharana, 1982).

Three subadults, thought to be females (juvenile gharial are very difficult to sex), were added to the breeding enclosure on 4 January 1979. These were all from the project's 1975 hatchling class and at the time of introduction measured 1.5–1.8 m. All were accepted by the three resident gharial paralleling the behaviour seen in the wild.

A further female subadult gharial measuring 2.30 m and weighing 64 kg, received from the Trivandrum Zoo, Kerala, for the breeding programme, was added to the breeding pool on 20 February 1979. This individual was accepted by the original three gharial, also without incident.

Courtship was observed in 1977 and 1978 but no eggs were laid in either year. It is not known if the females were sexually mature at this time. However, more problematic, the male suffered genital prolapse in January of both years. No courtship was observed in 1979. Because of the problem with the resident male, attempts were made to obtain a second male. Since it proved impossible to obtain a captive male from within India, and since it was against policy to attempt to capture one of the small number of wild-surviving males, negotiations were started to obtain one on a breeding loan from overseas.

The Frankfurt Zoological Society, West Germany, most kindly agreed to provide their superb 3.7 m male gharial on a breeding loan and Air India agreed to carry this large animal free of cost. This male, which had been the sole gharial at Frankfurt, was acclimatized to a large pool in India, initially by itself, and then with subadult animals, before being introduced to the breeding pool on 11 January 1980, in time for the 1980 breeding season.

The existing male was deliberately not removed from the enclosure prior to this introduction. It was uncertain how the new male, which had never seen an adult gharial before, would react to the mature females, so a close watch was kept for interactions. It was thought that competition between the males (if it occurred) might be important in bringing the Frankfurt male, (so long separated from members of its own species) into breeding condition. In both *C. palustris* and *C. porosus* conflict between two males is often immediately followed by courtship of any nearby female by the dominant male. This also proved to be the case in gharial and may have played a key role in the successful matings by the new male.

At the time of introduction the Frankfurt male measured 3.70 m as compared to the resident male's 2.90 m. However, the resident male had spent five years in the enclosure, and in any territorial conflict was fighting on home ground. The psychological effect of this in both captive and wild conditions is well documented (Bustard, 1965, 1968, 1970). It was somewhat surprising, therefore, when the Franfurt male quickly became dominant, took over the two females, and tried to keep the resident male out of the pool. At this stage the appropriate action would have been to remove the original male from the breeding complex; however, no action was taken which resulted in the Frankfurt male killing the original male (Bustard & Maharana, 1980).

The Frankfurt male also mated a number of times with both the original females, both of which were thought to be sexually mature.

The 1980 Breeding Season

One nest was laid in this first year of breeding. The eggs were laid in March in the artificial sandbank at a height of 1.6 m above ground level. The nest was closely guarded by the female, either from the base of the sandbank, or from the adjacent area of the pool approximately 6 m from the nest. The eggs were allowed to incubate *in situ* as the size of the sandbank curtailed excessive temperature fluctuations (particularly during the heat of the day). Since, unlike the natural riverine sandbanks, this sandbank rested on dry, red lateritic soil, instead of standing in the water, water was provided by spraying the sand surface in the vicinity of the nest on alternate days. At the warmest part of the summer, commencing in the second half of April, a coconut screen was erected over the area of the sandbank around the nest to provide shade during the heat of the day. Shortly before hatching was anticipated, the author carefully excavated the nest. In order to do this, the water level in the breeding pool was reduced to prevent the mother gharial emerging. The excavation was carried out in the early morning when ambient temperature approximated the temperature of the nest in order to avoid temperature shock to the eggs (it is most important to teach this practice in field work where, under somewhat primitive conditions, eggs may be exposed for longer than normal to ambient temperatures, which may be lethal). During excavation of the nest, the male as well as the female showed active interest, and both remained at the adjacent portion of the pool throughout the operation.

Egg handling

Not only are crocodile eggs vulnerable to temperature shock, they are also calcareous-shelled and thus easily damaged. They must therefore be handled carefully, especially during transport when shells of adjacent eggs should not touch but be separated by at least 2 cm of sand.

Furthermore, crocodilian eggs, like those of most reptiles, should not be rotated during incubation. If this happens the developing embryo will die, because it fixes itself to the inside of the shell during incubation and so cannot rotate if the eggs are rotated (Ferguson, 1982, in press). Recent work has shown this adfixure takes place very early, at about 12 hours following egg-laying (Ferguson, 1982, in press). Thus, the ideal time to lift eggs is immediately following laying. If they cannot be collected and transferred immediately then each egg must be carefully marked on the upper surface to avoid rotation when it is reburied in the transporter and in the incubation medium. However, even this degree of disturbance can reduce percentage hatch, and if eggs cannot be collected immediately following laying, it is better to leave them *in situ* until shortly before they are expected to hatch (provided, of course, they have been laid in a location where incubation can proceed normally, and they are subject to protection from human or non-human predators). When the embryo is large, much less care is needed in moving eggs as rotation is less injurious; however, normal temperature fluctuation precautions, and maintenance of "free water" in the incubation medium, should be maintained.

Hatchling and post-hatching care

The clutch laid in 1980 consisted of 25 eggs. Fifteen of these were removed and hatchery-incubated: 10 were left in the nest in order to observe female-hatchling interaction. The eggs hatched on 7 May 1980. All 15 hatchery eggs hatched, together with nine of the 10 eggs left in the natural nest. One of the latter hatchlings was apparently trampled by the mother.

The mother dug out the nest, remained in close attendance on the hatchlings and kept the other gharials, with the exception of the male, away from the area of the pool selected as the nursery. The hatchlings frequently rested on the body of the female [as observed by Bustard (1980b) in the wild], or on the body of the male.

Unfortunately, all the young left with the female died in the two months following hatching, some apparently as a result of attack by other members of the gharial group. All 15 hatchlings resulting from the eggs removed from the breeding compound, and hatchery-incubated, have survived. These young were reared in standard hatchling rearing pools (Bustard/FAO, 1975), which have been used by the project to raise several thousand gharial hatchlings. The young gharial resulting from captive breeding have shown outstanding growth – a mean of 92.3 cm at an age of 10½ months and 1.20 m at 15 months of age. This is considerably faster than the otherwise good captive growth rates recorded in gharial by Bustard & Singh (1980).

Further details of captive husbandry conditions for young gharial are given in Bustard (1980a).

The 1981 Breeding Season

The unfortunate results of the 1981 breeding season underline the fragility of such captive breeding programmes. In 1981 the same male courted the same two females, both of which are thought to have laid, although only one nest was located. The problem appears to have been disturbance. A film-maker, quite wrongly, was allowed access to the enclosure at the time of trial nesting. Before the author had him stopped, the episode had apparently interfered with the normal pre-nesting behaviour, by repeatedly frightening the two trial-nesting females back into the water.

The female which had laid, successfully guarded and hatched its clutch in 1980 (here called female 1) did not show normal nest-guarding behaviour in 1981 with the result that the nest could not be located (if eggs were indeed laid – it is possible that after repeated disturbances at the time of trial nesting the eggs were dropped in the pool). The second female, nesting for the first time, was disturbed during the actual egg-laying process and after laying four eggs returned to the pool. Shortly thereafter this female re-emerged to lay the balance of the egg clutch but was again disturbed, and dropping an egg on the sand surface, returned to the pool. Not unnaturally, such a disturbed female did not re-emerge to lay, and probably the balance of eggs may have been voided in the pool. This is the happier outcome of two possibilities. The other would be that the balance of the eggs were retained in the female's oviducts leading to death from peritonitis. An instance of this occurred in Nandankanan in 1974 (H. R. Bustard, unpublished). A *C. porosus* housed with no suitable area or materials in which to deposit its eggs laid several eggs on the red laterite, in which it is impossible to dig; the balance were retained in the female's oviducts, leading to death after some days.

None of the five eggs laid by female 2 hatched. It is unknown if they were fertile.

It is easy to locate the nest of an undisturbed female because she has strong nest guarding tendencies. When anyone enters the enclosure, the female at once leaves the pool and approaches the area of the nest. If the intruder approaches the section of the sandbank containing the nest, the female approaches until its elongated snout is pointing towards, and is close to, the actual nest. This behaviour was clearly demonstrated in 1980 with female 1 and in 1981 with female 2.

Clearly management of the programme in 1981 was inadequate. In a well-managed breeding programme staff would either witness the actual egg-laying or locate the nest by observing fresh moist diggings from a cursory daily examination of the poolside slope of the sandbank immediately after dawn.

The 1982 Breeding Season

Since the project was being handed over to local staff, the author was not present during the 1982 breeding season and the data set out below were obtained from Mr S. Maharana. Two nests were laid. The gharial are referred to as 1 and 2 following the terminology set out above. Gharial 1 laid 27 eggs on 10 March 1982. On 2 May the nest was opened and 11 eggs were removed to a hatchery, 16 being left in the nest. Of the 11 eggs brought to the hatchery four hatched but one hatchling died. Of the 16 eggs left in the natural nest seven hatched but two hatchlings died. This gives 11 hatchlings from 27 eggs compared to 24 hatchlings from 25 eggs for this same female two years previously. Furthermore, three of these 11 hatchlings died.

Gharial 2 laid 28 eggs on 20 March. The nest was opened on 1 May and 16 eggs were removed to a hatchery and 12 eggs left in the nest. Nine of the eggs removed from the nest hatched but one died. Of the 12 eggs left in the natural nest seven hatched but one died and one is stated as "missing". Hence 28 eggs gave 16 hatchlings, three of which were lost.

The results in 1982 were again confused by filming activities at the compound using floodlights, although attempts were made to accustom the gharial to this activity. Furthermore, approximately half of the eggs were left *in situ* because the same film-maker wanted to film the mother excavating the nest and demonstrating maternal behaviour. On conservation grounds all eggs should have been removed to the hatchery when the staff opened the nests. The gharial were again, quite wrongly, sacrificed to the interests of a commercial film-maker.

Full details of the hatching results are set out in Table I which allows comparison of the results between years. It is impossible to avoid loss of an occasional hatchling, for instance through trampling by the mother, but good post-natal care should minimize other losses (Bustard/FAO, 1975). Lack of care will result in loss of "weak" hatchlings or of individuals which, although strong, have hatched with a large amount of yolk still extruded. Such hatchlings should be individually housed in sterilized buckets of clean water. The water should be only 1 cm deep and the bucket, which should not be disturbed except for a 12-hourly inspection, should be covered by a clean towel. Such hatchlings should not be handled. Under this regime the extruded yolk should be absorbed after about 48–72 h; only in exceptional cases will it take longer. When the yolk is fully absorbed – but not before – such

TABLE I
Details of gharial clutch size, incubation success and hatchlings in the years 1980–1982 (inclusive). Percentages are bracketed where appropriate.

Year	Female	Clutch size	No. of eggs left in nest	No. of eggs removed to hatchery	No. of eggs hatched in nest	No. of eggs hatched in hatchery	Total no. of hatchlings	Total no. of surviving hatchlings
1980	1	25	10	15	9 (90)	15 (100)	24 (96)	23 (92)
1981	1	No nest located	–	–	–	–	0 (0)	0 (0)
	2	Only five eggs found (none hatched)	–	–	–	–	0 (0)	0 (0)
1982	1	27	16	11	7 (44)	4 (36)	11 (41)	8 (30)
	2	28	12	16	7 (58)	9 (56)	16 (57)	13 (46)

hatchlings can join their siblings in suitable hatchling pools (Bustard/ FAO, 1975).

In India the crocodile project has resulted in a much wider interest in all aspects of crocodilians and has led to a tremendous upsurge in breeding attempts. By late 1982, although the gharial had still only been bred at Nandankanan, two other centres in India were hoping to breed it, four were trying to breed the saltwater crocodile (unsuccessfully to date) and no less than 12 centres were involved in mugger breeding programmes. Mugger have been successfully bred at nine of these centres (Singh & Choudhury, 1982).

GHARIAL BREEDING IN SANCTUARIES

Sanctuaries, national parks, or wild life reserves are set up in order to *slow down* the rapid rate of habitat changes which have adversely affected so many animal and plant species, quite apart from direct hunting and collecting of the species. However, they cannot *stop* this change. Even within the best managed sanctuary there will be change. Even if intrusions within the sanctuary are kept to a minimum, changes outside it will still affect the ecological state of the sanctuary. It is important to realize that these effects may be the result of changes initiated at great distances outside the sanctuary boundaries. For instance, large-scale deforestation will greatly affect the characteristics (quality, rate of flow and seasonal volume) of the water supply passing through rivers within the sanctuary, and in the longer term may subtly alter the climate of the region. River "health" is clearly critical for a riverine species such as the gharial but here again damage may occur far upstream.

A major effect of the project has been the provision of better protection for the remaining wild gharial. This protection, combined with the banning of the use of set nets, and progressive banning of all commercial fisheries, has resulted in the survival of the remaining wild breeding stock and better survival of the immature year classes. This augurs well for breeding success in future years.

Wild breeding will also receive a massive boost as project-reared young released back into the wild commence breeding. This is likely to begin around 1985 (given a 10-year pre-breeding phase) and to accelerate over the next few years, owing to the increasing numbers of released animals. When it is realized that the total wild adult breeding population was estimated at 60–70 at the start of the project (Bustard/ FAO, 1974) and that by early 1982 a total of 855 gharial had been rehabilitated into protected areas, the scale of the potential increase becomes apparent. This is the positive side and is tremendously encouraging. It is hoped that large viable breeding populations will soon be re-established in a number of sanctuaries and that these, with the additional safeguard of captive-bred herds, will ensure the future survival of this fascinating crocodilian.

ACKNOWLEDGEMENTS

I wish to acknowledge many kindnesses from the Government of India during this work and from the many people involved would like to single out Mr N. D. Jayal, then Joint Secretary (Wildlife) in the Ministry of Agriculture & Irrigation, New Delhi, and the late Mr S. R. Choudhury, Conservator of Forests and Field Director Project Tiger, Orissa. Without their help the programme could not have developed. I also thank my students, my counterparts in the Indian States, and the FAO/UNDP for assistance. Special thanks are due to Dr R. Faust, Director, Frankfurt Zoological Society, West Germany, without whose help – the loan of their adult male gharial – the work described in this paper could not have been undertaken.

REFERENCES

Bustard, H. R. (1965). Observations on the life history and behaviour of *Chamaeleo hohnelii* (Steindachner). *Copeia* **1965:** 401–410.
Bustard, H. R. (1968). The ecology of the Australian gecko, *Gehyra variegata*, in northern New South Wales. *J. Zool., Lond.* **154**: 113–138.
Bustard, H. R. (1970). The role of behaviour in the natural regulation of numbers in the gekkonid lizard *Gehyra variegata. Ecology* **51**: 724–728.
Bustard, H. R. (1980a). Captive breeding of crocodiles. In *The care and breeding of captive reptiles*: 1–20. Townson, S., Millichamp, N. J., Lucas, D. G. D. & Millwood, A. F. (Eds). London: British Herpetological Society.
Bustard, H. R. (1980b). Maternal care in the gharial, *Gavialis gangeticus* (Gmelin). *Br. J. Herpet.* **6**: 63–64.
Bustard, H. R. (1981). Crocodile breeding project. In *Wildlife in India*: 147–164. Saharia, V. B. (Ed.). New Delhi: Government of India.
Bustard, H. R. (In prep.). *Nesting ecology of the gharial* (Gavialis gangeticus *(Gmelin)*) *in Narayani river, Nepal.*
Bustard, H. R./FAO (1974). *India: A preliminary survey of the prospects for crocodile farming.* Rome: F.A.O.
Bustard, H. R./FAO (1975). *Gharial and crocodile conservation management.* Rome: F.A.O.
Bustard, H. R. & Maharana, S. (1980). Fatal male-male conflict in the gharial (*Gavialis gangeticus* (Gmelin)) (Reptilia, Crocodilia). *J. Bombay nat. Hist. Soc.* **78**: 171–173.
Bustard, H. R. & Maharana, S. (1982). Size at first breeding in the gharial (*Gavialis gangeticus* (Gmelin)) (Reptilia, Crocodilia) in captivity. *J. Bombay nat. Hist. Soc.* **79**: 206–207.

Bustard, H. R. & Singh, L. A. K. (1980). Growth in the gharial. *Br. J. Herpet.* **6**: 107.
Bustard, H. R. & Singh, L. A. K. (In prep.). *Growth in the gharial* (Gavialis gangeticus).
Chowdhury, S. (1981). *Some studies on the biology and ecology of* Gavialis gangeticus, *the Indian Gharial (Crocodilia, Gavialidae)*. Ph.D. Thesis: University of Lucknow, India.
Chowdhury, S., Bustard, H. R. & Tandon, B. K. (In prep.). *The natural population and nesting ecology of* Gavialis gangeticus *(Crocodilia: Gavialidae) in Chambal River, North India*.
Ferguson, M. W. J. (1982). The structure and composition of the eggshell and embryonic membranes in *Alligator mississippiensis*. *Trans. zool. Soc. Lond.* **36**: 99–152.
Ferguson, M. W. J. (In press). The reproductive biology and embryology of crocodilians. In *Biology of the Reptilia*, **14**. *Development*. Billet, F. S. & Gans, C. (Eds). London: J. Wiley & Sons.
Ferguson, M. W. J. & Joanen, T. (1982). Temperature of egg incubation determines sex in *Alligator mississippiensis*. *Nature, Lond.* **296**: 850–853.
I.U.C.N. (Eds) (1971). Crocodiles. *IUCN Publs* N.S. Suppl. Pap. No. 32.
Singh, L. A. K. & Choudhury, B. C. (Eds) (1982). *Indian crocodiles – conservation and research*. Hyderabad: Forum of Crocodile Researchers.
Smith, M. (1931). *The fauna of British India. Reptilia and Amphibia*, **1**. London: Taylor & Francis.
Varona, L. S. (1966). Notas sobre los crocodílidos de Cuba y descripción de una nueva especie del Pleistoceno. *Poeyana* (A) No. 16: 1–34.
Yangprapakorn, U. (1971). Captive breeding of Crocodiles in Thailand. *IUCN Publs* N.S. Suppl. Pap. No. 32: 98–103.

Physiological Aspects of Lizard Growth: The Role of Thermoregulation

R. A. AVERY

Department of Zoology, The University, Bristol BS8 1UG, England

SYNOPSIS

Juvenile *Lacerta vivipara* were kept at different ambient temperatures, and allowed to thermoregulate for varying periods by altering the duration of exposure to incubator bulbs. Growth occurred in the absence of thermoregulation at higher ambient temperatures. It increased in thermoregulating lizards to an extent which was directly proportional to the period for which activity temperatures could be maintained, over the range $0-10$ h day^{-1}. The increase in growth was due entirely to increased voluntary food consumption: thermoregulation had no effect on net growth conversion efficiency, and gross conversion efficiency increased in lizards which were able to thermoregulate for long periods only to an extent which was determined by the balance between food intake and metabolic expenditure. Behavioural thermoregulation thus increases growth rates by increasing food intake, but has no qualitative effect on the fundamental mechanisms of growth physiology.

INTRODUCTION

The class Reptilia is categorized on morphological grounds, and the assemblage of about 4000 species which has survived to Modern times forms a distinct group of animals. They can be recognized by a combination of characters which is, by definition, unique. When one examines other biological aspects of reptiles, however, the uniqueness is less apparent. Nevertheless reptilian physiology, biochemistry, ecology and behaviour are adaptive, and it is therefore possible to investigate how they relate to the structures of the bodies within or through which they function. They also relate to the environment, and studies of the ways in which environmental factors have moulded the evolution of the dynamic aspects of function, subject to the constraints of morphology, have come to be called "physiological ecology". This essay concerns one aspect of reptilian physiological ecology – growth, and in particular, the environmental control of growth rates.

It has become increasingly realized over the past 40 years that just as some structural features have played a key role in reptilian evolution, for example scaly skin, circulatory system and the cleidoic egg, so too have some physiological and ecological factors. Foremost amongst these is *behavioural thermoregulation*. This is not unique to reptiles; it is seen in many other ectothermic animals, including butterflies, moths, beetles, dragonflies, spiders, fishes and amphibians. What is special about reptiles in this respect is that many species achieve a degree of control over body temperatures which is far more precise than in "lower" forms, from whose ancestors they have evolved. Small and medium-sized lizards in particular can often control their body temperatures at levels which are remarkable both for the magnitude of their excess over ambient levels and for their constancy.

The study of behavioural thermoregulation and its effects on the lives of reptiles (the mechanisms and costs of thermoregulating, and the effects of having done so) has become one of the central themes of herpetology (Avery, 1979; Gans & Pough, 1982). Amongst the aspects which have been explored in detail are phylogenetic and taxonomic considerations, i.e. which reptiles thermoregulate? (Avery, 1982); ethological considerations, i.e. how do reptiles thermoregulate? (Heatwole, 1976); physiological considerations, i.e. how are thermoregulation and its effects controlled? (Dawson, 1975; Bartholomew, 1982; Firth & Turner, 1982); and ecological considerations, i.e. what are the evolutionary consequences of thermoregulation, and how does it affect the daily lives of the animals? (Huey & Slatkin, 1976; Huey, 1982).

There are still, however, many facets of reptilian thermoregulation which remain to be investigated in detail. One of them is the way in which the physiology of the animals has become adapted, if at all, to the effects of operating in a biphasic temperature regime. Such a regime is illustrated diagrammatically in Fig. 1, which shows an idealized and simplified sequence of body temperature recordings from a lizard over a two-day period. Examples of such records may be seen in the following studies: *Podarcis muralis* in the laboratory (Licht, Hoyer & van Oordt, 1969) *Lacerta vivipara* in an outdoor enclosure (Avery, 1971); and *Sceloporus* sp. – an example of the kind of continuous recording which is made possible by radiotelemetry (McGinnis & Falkenstein, 1971). In reptiles which thermoregulate around relatively constant mean temperatures, the trajectory of body temperature over time approximates a square-wave cycle, provided that weather conditions remain favourable.

The purpose of the work reported here is to investigate the functioning of one specific process – growth – in relation to temperature cycles of this kind. The experimental animal is the lizard *Lacerta vivipara* Jacquin. This is a diurnal basker, maintaining activity temperatures around 33°C whenever sunshine is sufficiently intense to enable it to do

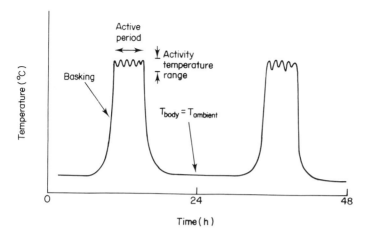

FIG. 1. Schematic and simplified representation of the body temperatures of a lizard which can bask over a two-day period, showing how the trajectory of body temperature readings approximates a square-wave.

so (Avery, 1971, 1976). Because the climates to which it is adapted, throughout its range from western Europe to eastern Siberia, are all characterized by cool air temperatures, the amplitude of the square wave is high. Note, however, that most of these climates are also rather cloudy, so there will be extended periods when the animals can only thermoregulate ineffectually, or not at all. Different periods of thermoregulation have been induced in the laboratory by exposing juvenile lizards to different daily periods of radiant heat from incubator lamps, and investigating the effects on food consumption and growth. The overall objective is to determine how much of overall growth occurs when a lizard is thermoregulating, and how much when it is not; also whether there is any qualitative or quantitative difference between the two.

MATERIALS AND METHODS

The lizards used in this study were derived from pregnant females captured in Somerset or Wiltshire and maintained in outdoor enclosures until after they had given birth. The hatchlings were separated to small containers, which were kept out of doors with exposure to sunshine. They were fed on a variety of live food, with crickets and mealworms forming a predominant part of the diet. Care was taken to see that each individual lizard obtained adequate food, and the largest and smallest individuals were maintained separately to prevent "bullying". All insects presented as food were sprinkled with multivitamin powder (Vionate).

The juveniles were transferred indoors during the middle of September; husbandry remained identical except that the animals were now exposed to illumination from "Trulite" fluorescent tubes for 13 h day^{-1} and to radiant heat and light from 275 W incubator lamps for 4 h day^{-1}; mealworms now formed only a very small proportion of the diet.

Experiments were carried out between mid-September and mid-February in constant temperature rooms (3, 9 and 15°C) or incubators (23, 28 and 33°C). Lizards were maintained individually in plastic boxes measuring 25 × 20 × 7.5 cm or 21 × 15 × 8.5 cm with lids of nylon mesh (gauge about 1.5 mm). Each box contained a "concertina" of cardboard to provide shade and shelter; a plastic petri dish containing foam rubber which was kept saturated with water was also provided. The boxes were illuminated by Trulites for 13 h day^{-1}; heat was provided with 275 W incubator lamps positioned about 0.5 m above the surface. Details of the experimental groups, together with the numbers of lizards in each, are shown in Table I.

Animals in each group were acclimated to the particular set of experimental conditions (ambient temperature, daily duration of radiant heat) for at least seven days. During this period, each animal was fed a constant weight of food each day. After the acclimation period

TABLE I
Lizard growth: details of experimental groups

Ambient temperature (°C)	Incubator bulb (h day^{-1})	Number of lizards
3	4	4
3	7	4
3	10	4
9	0	9
9	1	4
9	4	12
9	7	10
9	10	8
9	4*	4
15	0	6
15	1	23
15	4	23
15	7	8
15	10	8
23	0	4
28	0	7
33	0	4

* Bulb switched on during one day every four.

it was weighed, and the experiment continued with feeding at the same daily rates for 14 days (12 or 13 days in a small number of experiments), when it was reweighed. All weighings were carried out between 09.00 and 09.30 hours, before the lizards had been fed. All lizards weighed less than 700 mg at the beginning of an experiment.

Food in these experiments was small crickets, lightly sprinkled with Vionate. Crickets were anaesthetized with carbon dioxide to facilitate weighing, but were not presented to lizards until after they had recovered from anaesthesia. Feeding always took place during the period when the incubator bulbs were switched on. Uneaten crickets were removed 4–6 h later, anaesthetized, and weighed; the consumption of each lizard could hence be determined by difference. A separate series of experiments demonstrated that there was a small weight loss in uneaten crickets, and a large weight loss in those uneaten crickets which had died. Correction factors were applied for the crickets which died; the value of the factor varied between experimental treatments.

RESULTS

Growth under Constant Conditions

Four juvenile *L. vivipara* were maintained in an incubator at 23°C with illumination from 8 W fluorescent tubes for 13 h day^{-1}. They were fed *ad libitum*. All four lizards initially grew at rapid rates, but at times ranging from 21 to 42 days after the beginning of the experiment growth rates began to decline, the animals looked subjectively less healthy, and eventually they began to lose weight. This was probably a consequence of the artificial nature of the experimental environment (see Discussion). Figure 2 shows the growth, measured in terms of weekly changes in body mass, of the two individuals which continued to grow for the longest periods, here designated lizards A and B.

Lizard growth has been described by a logistic model of the general form:

$$W = a_w / (1 + be^{-rt}) \qquad (1)$$

where W is body mass, a_w the asymptotic mass, t is time, b and r are constants – the latter is sometimes called a "growth constant" (Schoener & Schoener, 1978; Andrews, 1982).

This can be written as a difference equation:

$$W_2 = \frac{a_w W_1}{W_1 + (a_w - W_1)\,e^{-rt}} \qquad (2)$$

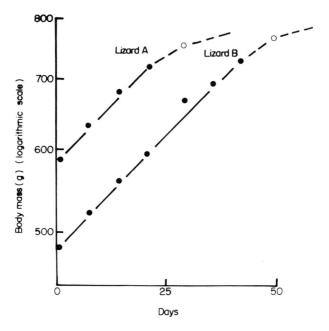

FIG. 2. Growth of two juvenile *L. vivipara* at 23°C. Continuous lines show curves fitted from equations (2) and (3) in the text.

which rearranges to:

$$r = \frac{1}{t} \ln \left[\frac{W_2(a_w - W_1)}{W_1(a_w - W_2)} \right] \quad (3)$$

Equation (3) was used to estimate values for r, taking W_2 as the body mass on the last occasion before growth rates began to decline (the right-hand solid symbols in Fig. 2) and $a_w = 3.5$ g. The values for r obtained by substitution are then:

lizard A: r = 0.011,
lizard B: r = 0.012.

Substituting these values into equation (2) gives growth curves which have been drawn as continuous lines in Fig. 2. Note that these curves are apparently linear when plotted on semilogarithmic co-ordinates. This suggests that there will be little loss in accuracy if growth *over short periods* is approximated as a simple exponential function:

$$W \approx be^{rt}. \quad (4)$$

Exponential growth has for convenience been assumed in the analysis which follows, and rates are expressed in units $\Delta W W^{-1}$ day^{-1}.

Growth Dynamics: the Role of Thermoregulation

Juvenile lizards were maintained at various ambient temperatures and exposed to different daily periods of heating from incubator lamps. Some individuals were fed *ad libitum*, others were permitted only restricted intakes. There was considerable day-to-day variation in the amount of food eaten by individual lizards, even within the same treatment group. Analysis is here confined to the individuals with the highest mean daily intake within each *ad libitum* group, since a preliminary inspection of the data showed that reduced intake results in reduced growth conversion efficiency (see later).

There were clear positive relationships between growth rates ($\Delta W W^{-1}$ day^{-1}) and the time for which lizards were able to thermoregulate (i.e. duration of radiant heat) from 0 to 10 h day^{-1} ($F > 14.13, P < 0.001$). Growth rates were consistently greatest at ambient temperatures of 15°C, lowest at 3°C (Fig. 3). Keeping the animals at their activity temperatures for 24 h day^{-1} (equivalent to continuous thermoregulation), however, resulted in a decrease in growth rate (solid square in Fig. 3). The open square in Fig. 3 shows the growth rate of a lizard at an ambient temperature of 9°C which was allowed to thermoregulate for 4 h on only one day in every four. Note that its growth rate is the same as that of the animal at the same ambient temperature which thermoregulated for the same aggregate period, but daily, i.e. for 1 h day^{-1}.

Feeding rates ($F W^{-1}$ day^{-1}) increased with the duration of thermoregulation in the same way as growth rates (Fig. 4; $F > 4.5, P < 0.001$). The lizard which was maintained at a constant 33°C showed the only discrepancy with the pattern of growth, having the highest recorded feeding rate (solid square in Fig. 4; cf. Fig. 3). Combining the data in Figs 3 and 4 to show the relationships between growth and food consumption reveals a more complex situation (Fig. 5). Most of the data points fall along a line which is equivalent to a gross growth conversion efficiency (i.e. growth/food, expressed in identical units) of 20%. The results from three lizards, all of which were able to thermoregulate for long periods (10 h at 9 and 15°C, 7 h at 15°C), deviate significantly from this line (*t*-tests, ($N + 1$) $P < 0.05$; Bliss, 1967). Gross growth conversion efficiency of the animal maintained at a constant 33°C also deviated significantly from this line, but the value was low (13.2%).

The increase in gross growth conversion efficiency resulting from long periods of thermoregulation could have a number of causes. One of the simplest hypotheses is that it results from differences in metabolic rate: a lizard can only devote to growth such energy as remains after

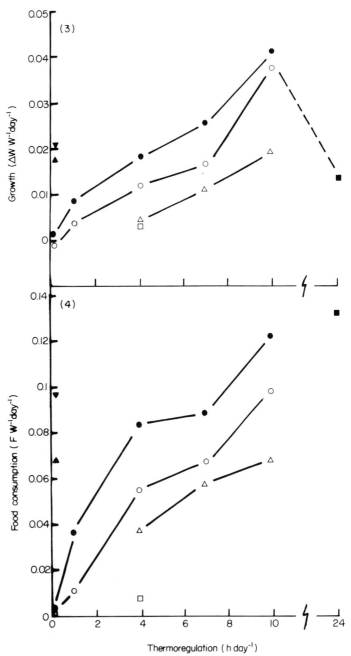

FIGS 3 & 4. (3) Growth rates of juvenile *L. vivipara* in relation to daily periods of thermoregulation. Symbols relate to ambient temperatures as follows: △, 3°C; O, 9°C; ●, 15°C; ▲, 23°C, ▼, 28°C; ■, 33°C. The open square shows the growth rate of a lizard at 9°C thermoregulating for 4 h every fourth day. (4) Food consumption of juvenile *L. vivipara* in relation to daily periods of thermoregulation. Symbols as in Fig. 3.

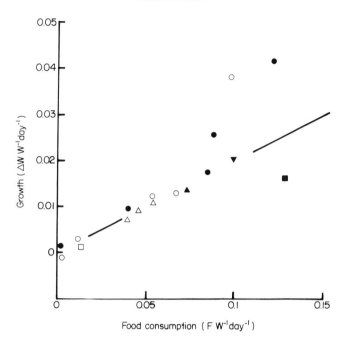

FIG. 5. Juvenile *L. vivipara*: growth in relation to food intake. Symbols as in Fig. 3. The line shows the relationship if gross growth conversion efficiency remained constant at 20%.

metabolic costs have been met. This was tested by calculating *net* growth conversion efficiencies in the following manner.

The oxygen consumption of lizards in relation to temperature was assumed to follow the curve given by Tromp & Avery (1977). The total oxygen consumption of an animal could hence be calculated by adding respiration carried out whilst it was at the ambient temperature to respiration during the period when it was thermoregulating; each lizard was assumed to have maintained a body temperature of 33°C throughout this period (Avery, 1982). Oxygen consumption was converted to energy units using an oxycaloric equivalent of 4.89 (Davis & Warren, 1968). The food required to sustain this level of energy expenditure was then calculated on the following further assumptions: energy value of food = 2.22 kJ g^{-1}; ratio dry/wet weight for food = 0.33; nitrogenous excreta corresponds to 7% of ingested food; assimilation efficiency = 89% (Avery, 1971, 1975). Simple thermodynamic considerations give:

> Energy available for growth = energy of food − [energy lost in faeces + energy lost as nitrogenous excreta + energy loss in respiration].

Figure 6 shows growth plotted against energy available for growth as calculated above. Units in both cases can be either g g^{-1} day^{-1} or J J^{-1} day^{-1} on the assumption that energy values of lizard tissue and crickets are identical. The relationship is linear, and is fitted by the regression:

$$y = 0.0004 + 0.368x. \quad (5)$$

The origin of the regression does not differ significantly from zero ($P < 0.001$, analysis of covariance) and hence the slope can be considered equivalent to the net growth conversion efficiency:

$$\frac{\text{observed growth}}{\text{energy available for growth (as defined above)}}.$$

One point is anomalous (t-test, $(N + 1)\ P < 0.05$) and has not been utilized in the calculation of equation (5). This is the net growth conversion efficiency at 33°C, which was 16.7% (Fig. 6).

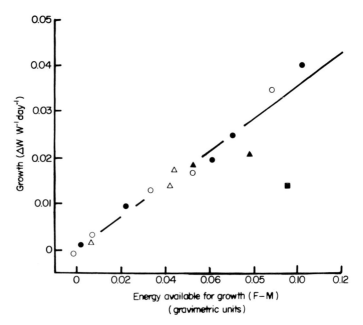

FIG. 6. Juvenile *L. vivipara*: growth in relation to available energy (as defined in the text). Symbols as in Fig. 3. The line shows the regression given by equation (5) in the text.

Growth Dynamics: Effects of Reduced Food Intake

The food intakes of many of the experimental animals were reduced, either as a result of their failure to find or capture the maximum possible amount, or as a result of decreased supply as part of the experimental design. Reduced intake resulted in a reduction of gross growth conversion efficiency in all experimental groups; Fig. 7 shows a representative set of data, in this case for the group of lizards kept at 9°C and given 4 h radiant heat per day. The dashed line in Fig. 7 shows the expected relationship if gross growth conversion efficiency remained constant at 20% for all levels of food intake.

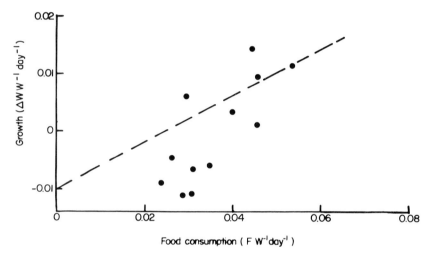

FIG. 7. Growth of juvenile *L. vivipara* kept at 9°C and allowed to thermoregulate for 4 h day^{-1}, in relation to food intake. Dashed line shows the expected relationship if gross growth conversion efficiency remained constant at 20%.

The data in Figs 3, 4 and 7, together with the remaining results relating to reduced food intakes, have been combined as a synoptic, three-dimensional model showing diagrammatically the effects of periods of thermoregulation on both food consumption and growth (Fig. 8). The response surface in Fig. 8 refers to a single ambient temperature; raising the ambient temperature would have the effect of raising the level of the surface, reducing it, the effect of lowering the level.

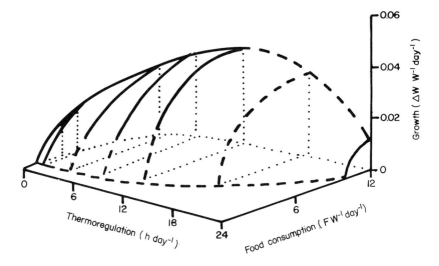

FIG. 8. Juvenile *L. vivipara*: three-dimensional model showing diagrammatically the relationships between thermoregulation, food consumption and growth at a single ambient temperature. Continuous heavy lines show relationships which were determined experimentally, dashed lines show interpolated or extrapolated relationships.

DISCUSSION

The Pattern of Growth in *L. vivipara*

The growth patterns of ectothermic vertebrates, including reptiles, have several features which distinguish them from those of endothermic birds and mammals. One is that growth tends to continue for very long periods, often well past the age of sexual maturity, although at ever-decreasing rates. Many reptiles continue growing throughout their lives; others do not. The difference appears to lie in the mechanisms of secondary ossification within the bones (Bellairs, 1969; Haines, 1969). Another characteristic feature is that growth is to a large extent indeterminate: although there must be underlying maximum rates which are genetically determined, the rates which are actually realized are usually lower, and are effectively set by environmental variables. There are a considerable number of factors which may affect reptilian growth rates; they include temperature, photoperiod, food availability (both in quantity and quality), water, competition, social stress and tail loss. Many of these factors have been studied, in varying degrees of detail (review by Andrews, 1982). There has apparently been no systematic study of the relationship between behavioural thermoregulation and growth.

Lacerta vivipara is a particularly suitable species in which to study the ecological and physiological correlates of thermoregulatory strategies, since it maintains relatively high activity temperatures (Avery, 1971, 1982) despite inhabiting climates in which ambient temperatures and solar radiation flux densities are lower than those experienced by the majority of reptiles (Avery, 1976). Basking is therefore an important part of the behavioural repertoire, and occupies a considerable fraction of an animal's total "active" time, i.e. the time during which it has emerged from its overnight retreat or hibernaculum. The use of incubator bulbs in the experiments described here simulates solar radiation with respect to heat exchange, but also has a number of shortcomings. Foremost amongst these is the fact that an individual animal may not necessarily thermoregulate for the whole of the period during which the bulb is switched on. Also, it may not always maintain body temperatures at the same mean level whilst it is thermoregulating. For the purposes of analysis, it was assumed that lizards always thermoregulated around 33°C when the bulbs were switched on. In order to determine the magnitude of the errors which these assumptions introduce it would be necessary to monitor body temperatures continuously. This was not attempted because the consequent disturbance to the animals would have distorted their behaviour. General observations on *L. vivipara* in captivity suggest that the errors are small, except in animals required to maintain activity temperatures continuously. The lizards in this group were consequently heated by controlling ambient air temperature rather than using radiant heat from incubator bulbs.

In the long run, growth in *L. vivipara* is a stepwise process. The juveniles are born in July–August and reach sexual maturity two or three years later. Growth is rapid during the first few months of life, ceases during the winter when the animals hibernate (whether this statement is strictly true, or is simply an approximation, has never been investigated), and continues during the spring and summer months at rates which decrease from year to year as the animals approach their maximum size (Smith, 1973). Whether an individual's pattern of growth, corrected for the lags produced by seasonal low temperatures, would fit one of the standard growth models (see Results) is not known. For the purposes of the present study this is unimportant, since experiments and analysis are confined to lizards less than 800 mg body mass, i.e. in their first four months of life.

Growth Parameters

Growth efficiencies in *L. vivipara* have not previously been measured directly, but there are two published studies from which values for the parameters can be derived. In juvenile lizards kept in the laboratory, fed on mealworms, and exposed to heat from incubator bulbs for 5 h day^{-1}, the relationship between food intake and growth (both in units g g^{-1} day^{-1}) was given by the regression:

$$G = 0.70F - 16.3 \qquad (6)$$

(Avery, 1971). Values of gross growth conversion efficiency derived from equation (6) for a lizard receiving maximum rations (0.07 g g^{-1} day^{-1} in this experiment), and also the value of net growth conversion efficiency calculated on the same assumptions as those used in deriving equation (5), are shown in Table II. Net and gross efficiencies have also been calculated from an energy budget of wild *L. vivipara* in central France (Pilorge, 1982), making corrections for the fact that data in this publication are expressed as fractions of assimilation, not intake. These too are shown in Table II. For comparison that Table also shows values for *Anolis carolinensis* feeding maximally on mealworms, from the very thorough laboratory study by Kitchell & Windell (1972). Gross and net growth conversion efficiencies obtained in the present study are in line with all of those from previous work quoted in Table II, and are within the ranges to be expected from the remaining, relatively sparse, relevant work on reptiles (Avery, in press), and from the much more extensive work on fishes (e.g. Elliott, 1982). Gross growth conversion efficiencies in *L. vivipara* fall with decreasing rations (Figs 7, 8) in much the same way as in various species of Salmonidae (Elliott, 1982).

TABLE II
Growth parameters for Lacerta vivipara *and* Anolis carolinensis

	Lacerta vivipara			*Anolis carolinensis*
	This study	Avery (1971)	Pilorge (1982)	Kitchell & Windell (1972)
Gross growth conversion efficiency (%)	19.0–34.0	32.4	33.4	30.5
Net growth conversion efficiency (%)	36.8	38.9	42.1	47.5

Growth Dynamics

The most important finding of this study is that periods of raised body temperature resulting from behavioural thermoregulation have no effect on the net growth conversion efficiency of *L. vivipara*. The effect on gross efficiency derives solely from the relationship between growth and respiration. There is thus no *qualitative* difference in growth physiology between thermoregulating and non-thermoregulating lizards; the former grow at higher rates simply because their voluntary food intake is greater. The group of lizards kept at 33°C were an exception to this generalization; their net growth conversion efficiency was reduced (Fig. 6). This was undoubtedly a consequence of stress; it is well known that many reptiles require a daily period at body temperatures lower than those normally maintained during activity (Regal, 1967) and that prolonged high temperatures can lead eventually to death, partly as a result of thyroid hyperactivity (Wilhoft, 1958). The temperature threshold at which the effects of such stress become apparent varies from species to species. *Podarcis muralis* (which is closely related to *L. vivipara*; the ranges of the two species overlap) does not survive for protracted periods at 33°C, but *P. sicula*, which has a geographical distribution extending further south, is able to do so (Licht *et al*., 1969). Constant temperature is also, in all probability, a major explanation for the failure of the lizards recorded in Fig. 2 to grow for long periods.

The mechanisms by which the body temperature regime determines voluntary food intake are not known. The effect is undoubtedly due in part to the influence of low temperature in reducing gut motility (Avery, 1973) and the rate of secretion of digestive enzymes (review: Skoczylas, 1978). Other factors might also be involved, including appetite. Control of appetite is a notoriously difficult field of study. In reptiles, the proximate factors involved include photoperiod (it was for this reason that all the experiments described here were performed at L : D 13 : 11) and temperature. Other factors are partly endocrinological, and include levels of growth hormone and prolactin (Licht & Hoyer, 1967). It may be that reductions in the levels of these hormones, resulting from negative feedback via the pituitary, were responsible for the reduced growth efficiencies observed in lizards with reduced food intakes (Figs 7 & 8).

Clearly there is much more work to be done in this area; knowledge of reptilian endocrinology is rudimentary. There are several assumptions which have been made in the present study which relate to a lack of understanding of endocrinological factors. Perhaps the most important is the assumption that although the experiments were carried out during the period September–February, when the lizards would in nature have been hibernating, growth rates and physiology remained similar to those during the previous season of activity. Both the temperature regime and the photoperiod to which the experimental lizards were exposed simulated conditions during the latter, but there is no direct evidence that the animals might not have been affected by intrinsic factors under the influence of a seasonal "clock". Indirect evidence that growth was not affected is provided by a comparison of the growth rates observed experimentally with those observed in the field. The only available data on growth in body mass in *L. vivipara* show a mean change of 325 mg by juvenile lizards over the period 17 August to 12 September (Avery, 1971). Assuming exponential growth (equation 4), this is equivalent to a rate of 0.027 g g^{-1} day^{-1}, which is almost identical with the rate observed experimentally under the regime which most closely approximated meteorological conditions in Britain in August–September, viz. ambient temperature 15°C, "sunshine" for 4 h day^{-1} (0.025 g g^{-1} day^{-1}: Fig. 3). Assumptions which remain untested are (a) that growth observed during the experiments was entirely somatic, i.e. the changes in mass were not due to the development of gonads or fat deposition (the latter is very small in juvenile *L. vivipara*, see Avery, 1974); (b) that there is no sex-linked difference in body mass up to 800 mg (the sex of lizards was not recorded in this study because there is no known non-destructive means of doing so in juvenile *L. vivipara* – the method described by Bauwens & Thoen (1982) is statistical and does not give unequivocal differentiation for individual animals); and (c) that the diet, together with illumination from "Trulite" fluorescent tubes, is nutritionally adequate.

General and Ecological Considerations

There has been much discussion about the evolutionary and ecological significance of growth rates (e.g. Case, 1978; Calow, 1982). The consensus is that under most circumstances, vertebrates are adapted and behave in ways that maximize growth, or which optimize growth in relation to realistically-attainable rates of gross energy intake (food). The results presented here demonstrate that this is indeed the case for *L. vivipara*. The animals invest a great deal of time in thermoregulating to attain high body temperatures (Avery, 1976), but despite the costs which this involves (Huey & Slatkin, 1976) the behaviour is adaptive because gross growth conversion efficiency is increased. Since food intake is also substantially increased, because gut throughput times are speeded (Avery, 1973) and hunting efficiency is greater (Avery, Bedford & Newcombe, 1982), the rates of growth which are achieved when thermoregulation is possible are higher than those in its absence (Fig. 3). This benefit only holds, however, so long as an individual can sustain high food intakes: the square-wave regime of daily body temperature shown in Fig. 1 results in a greater sensitivity to food shortage than would be experienced by a lizard in a temperate climate operating throughout at the ambient temperature (Fig. 8). Enhanced growth is not, of course, the only benefit of thermoregulation; there are many others, some of which are listed by Avery et al. (1982).

ACKNOWLEDGEMENTS

These experiments would not have been possible without the technical assistance of Jenny Smith and Doreen Bond, who bred the food insects and performed many of the weighings cheerfully and accurately. The 9°C : 0h and 3°C experiments are based partly on data from undergraduate project work carried out respectively by Julia Clarke and Kathryn Flower. Gary Duthie, Stephen Harris, Peter Miller and Stuart Reynolds commented on a draft of the manuscript. I am grateful to all of these people for their contributions.

REFERENCES

Andrews, R. M. (1982). Patterns of growth in reptiles. In *Biology of the Reptilia*, **13**: 273–220. Gans, C. & Pough, F. H. (Eds). London and New York: Academic Press.
Avery, R. A. (1971). Estimates of food consumption by the lizard *Lacerta vivipara* Jacquin. *J. Anim. Ecol.* **40**: 351–365.
Avery, R. A. (1973). Morphometric and functional studies on the stomach of the lizard *Lacerta vivipara*. *J. Zool., Lond.* **169**: 157–167.
Avery, R. A. (1974). Storage lipids in the lizard *Lacerta vivipara*: a quantitative study. *J. Zool., Lond.* **173**: 419–425.
Avery, R. A. (1975). Clutch size and reproductive effort in the lizard *Lacerta vivipara* Jacquin. *Oecologia, Berl.* **19**: 165–170.
Avery, R. A. (1976). Thermoregulation, metabolism and social behaviour in Lacertidae. *Linn. Soc. Symp. Series* No. 3: 245–259.
Avery, R. A. (1979). *Lizards – a study in thermoregulation. Studies in biology* No. 109. London: Edward Arnold.
Avery, R. A. (1982). Field studies of body temperatures and thermoregulation. In *Biology of the Reptilia*, **12**: 93–166. Gans, C. & Pough, F. H. (Eds). London and New York: Academic Press.
Avery, R. A. (In press). Reptiles. In *Animal energetics*. Vernberg, F. J. & Pandian, T. J. (Eds). London and New York: Academic Press.
Avery, R. A., Bedford, J. D. & Newcombe, C. P. (1982). The role of thermoregulation in lizard biology: predatory efficiency in a temperate diurnal basker. *Behav. Ecol. Sociobiol.* **11**: 261–267.
Bartholomew, G. A. (1982). Physiological control of body temperature. In *Biology of the Reptilia*, **12**: 167–211. Gans, C. & Pough, F. H. (Eds). London and New York: Academic Press.
Bauwens, D. & Thoen, C. (1982). On the determination of sex in juvenile *Lacerta vivipara* (Sauria, Lacertidae). *Amphibia-Reptilia* **2**: 381–384.
Bellairs, A. d'A. (1969). *The life of reptiles*, **2**. London: Weidenfeld and Nicolson.
Bliss, C. I. (1967). *Statistics in biology*, **1**. New York: McGraw-Hill.
Calow, P. (1982). Homeostasis and fitness. *Am. Nat.* **120**: 416–419.

Case, T. J. (1978). On the evolution and adaptive significance of postnatal growth rates in the terrestrial vertebrates. *Q. Rev. Biol.* **53**: 243–282.
Davis, G. E. & Warren, C. E. (1968). Estimation of food consumption rates. In *Methods for assessment of fish production in fresh waters*: 204–225. Ricker, W. E. (Ed.). Oxford: Blackwell Scientific Publications.
Dawson, W. R. (1975). On the physiological significance of the preferred body temperatures of reptiles. In *Perspectives of biophysical ecology*: 443–473. Gates, D. M. & Schmerl, R. B. (Eds). Berlin: Springer.
Elliott, J. M. (1982). The effects of temperature and ration size on the growth and energetics of salmonids in captivity. *Comp. Biochem. Physiol.* **73B**: 81–91.
Firth, B. T. & Turner, J. S. (1982). Sensory, neural and hormonal aspects of thermoregulation. In *Biology of the Reptilia*, **12**: 213–274. Gans, C. & Pough, F. H. (Eds.) London and New York: Academic Press.
Gans, C. & Pough, F. H. (1982). Physiological ecology: its debt to reptilian studies, its value to students of reptiles. In *Biology of the Reptilia*, **12**: 1–13. Gans, C. & Pough, F. H. (Eds). London and New York: Academic Press.
Haines, R. W. (1969). Epiphyses and the sesamoids. In *Biology of the Reptilia*, **1**: 81–116. Gans, C., Bellairs, A. d'A. & Parsons, T. S. (Eds). New York and London: Academic Press.
Heatwole, H. (1976). *Reptile ecology*. St. Lucia: University of Queensland Press.
Huey, R. B. (1982). Temperature, physiology, and the ecology of reptiles. In *Biology of the Reptilia*, **12**: 25–91. Gans, C. & Pough, F. H. (Eds). London and New York: Academic Press.
Huey, R. B. & Slatkin, M. (1976). Costs and benefits of lizard thermoregulation. *Q. Rev. Biol.* **51**: 363–384.
Kitchell, J. F. & Windell, J. T. (1972). Energy budget for the lizard, *Anolis carolinensis*. *Physiol. Zool.* **45**: 178–188.
Licht, P. & Hoyer, H. (1967). Somatotropic effects of exogenous prolactin and growth hormone in juvenile lizards (*Lacerta s. sicula*). *Gen. comp. Endocr.* **11**: 338–346.
Licht, P., Hoyer, H. E. & van Oordt, P. G. W. J. (1969). Influences of photoperiod and temperature on testicular recrudescence and body growth in the lizards, *Lacerta sicula* and *Lacerta muralis*. *J. Zool., Lond.* **157**: 469–501.
McGinnis, S. M. & Falkenstein, M. (1971). Thermoregulatory behaviour in three sympatric species of iguanid lizards. *Copeia* **1971**: 552–554.
Pilorge, T. (1982). Ration alimentaire et bilan énergétique individuel dans une population de montagne de *Lacerta vivipara*. *Can. J. Zool.* **60**: 1945–1950.
Regal, P. J. (1967). Voluntary hypothermia in reptiles. *Science, N.Y.* **155**: 1551–1553.
Schoener, T. W. & Schoener, A. (1978). Estimating and interpreting body-size growth in some *Anolis* lizards. *Copeia* **1978**: 390–405.
Skoczylas, R. (1978). Physiology of the digestive tract. In *Biology of the Reptilia*, **8**: 589–717. Gans, C. & Gans, K. A. (Eds). London and New York: Academic Press.
Smith, M. A. (1973). *The British amphibians and reptiles*, 5th edition revised by A. d'A. Bellairs & J. F. D. Frazer. London: Collins.
Tromp, W. I. & Avery, R. A. (1977). A temperature-dependent shift in the metabolism of the lizard *Lacerta vivipara*. *J. thermal Biol.* **2**: 53–54.
Wilhoft, D. C. (1958). The effects of temperature on thyroid histology and survival in the lizard, *Sceloporus occidentalis*. *Copeia* **1958**: 265–276.

How Metabolic Rate and Anaerobic Glycolysis Determine the Habits of Reptiles

R. A. COULSON

Department of Biochemistry, Louisiana State University Medical Center, New Orleans, Louisiana 70119, USA

SYNOPSIS

Individual cells of reptiles are quite capable of rapid chemical reactions as evidenced by the fact that the metabolic rates (MR) of small reptiles exceed those of large mammals. MR in all vertebrates seems to be determined primarily by the length of blood vessels, and secondarily by functional capacity of the lungs and cardiovascular system. The longer the blood vessels, the greater the resistance, the slower the capillary flow, and the slower the delivery of oxygen and substrates. Large reptiles produce too little aerobic energy for combat or rapid locomotion on land. However, since reptiles have as much muscle glycogen as mammals, and since they can also convert it to lactate as fast, for a brief time a reptile is as strong as a mammal of the same size. The problem for an exhausted reptile is the rate of restoration (aerobic) of muscle glycogen, which is 4000 times faster in a 2 g shrew than in a 700 kg alligator. As protein synthesis requires ATP in all animals, and as ATP production is a function of MR, protein synthesis is very slow in a large reptile. Low MR decreases the food and water requirements and increases survival time in anoxia. With respect to the evolution of homeothermy, development of respiratory and cardiovascular systems much more efficient than those found in modern reptiles would be the first stage. The difference between a warm-blooded animal and one that is cold-blooded is more one of chemical engineering than one of chemistry.

INTRODUCTION

In any reptile, metabolic rate will be of prime importance in determining the habits of the animal. It will affect directly the food and water requirements, the rate of locomotion over prolonged periods, the length of time aquatic species can remain submerged, etc. Unlike mammals, reptiles are always concerned with achieving a body temperature optimum for them. It is probable that if the means were available each reptile would maintain a temperature almost as constant as that of mammals and birds and that the temperature selected would differ among the species. A few might prefer temperatures as low as 20°C whilst others would select 35°C or even high. Generally speaking, those native to cool climates would probably thrive at lower temperatures than those from hot deserts.

FACTORS AFFECTING METABOLIC RATE

It would be difficult to exaggerate the number of misconceptions extant on metabolic rate in animals generally and in reptiles in particular. An understanding of metabolic rate requires clarification of the principles which govern it. Our own theories are presented here.

By way of chemical definition, metabolic rate is the average of the rates of all the oxidative reactions in the body of an animal. It is a variable, being lowest during sleep, or whatever condition approaches sleep, and highest during maximal physical exertion. In measuring metabolic rate, two general means are available. One could determine heat production, a method which is difficult, but not impossible, in a heterotherm, or one could measure the amount of oxygen consumed per unit time. The latter is a practical method but the duration of the oxygen consumption measurements needs to be fairly long to avoid confusion resulting from anaerobic glycolysis (reactions which will be explained later).

Size, Metabolic Rate and Blood Flow

The larger the animal, the lower the metabolic rate per unit body weight. If that well known fact did not mystify the first scientists that observed the phenomenon, it should have, for there is no obvious explanation for it. Most of the early work was done on mammals, and since they were warm, an almost immediate association was made between heat loss and surface area. The smaller the animal, the greater the surface area in relation to its weight and the higher the metabolic rate. By measuring surface areas and oxygen consumptions in scores of mammals the correlation was good, so good that it was believed that the problem was solved. However, not all animals are constantly warm and those that are not will lose heat to the air when the air is cooler than they are, and gain heat when the air temperature exceeds their own. Yet, in heterotherms, the larger the animal, the lower the metabolic rate, and their metabolic rates also correlate with their surface areas. Why should that be?

Since we were determining the rates of catabolism of various compounds in the alligator and other animals (Coulson & Hernandez, 1964), we needed to understand the factors controlling those rates. When, for example, glucose was given to an alligator, it took him longer to get rid of it than it would a rat, and the times required in each animal were inversely proportional to their metabolic rates. By the end of the last century, it was known that most reactions in life could only occur at a reasonable rate if they were catalysed by enzymes in the cells. Experiments *in vitro* showed that the rate of a particular reaction was proportional to three things; temperature, the concentration of the thing or things reacting (substrates), and the amount of functional enzyme present. Whether one was considering an alligator or a duck, these three factors were thought to be the only ones involved in determining the rate of any single reaction, or the average rate of all the oxidative reactions, which we call metabolic rate. In a long series of experiments and after considerable thought, we concluded that these factors alone could not explain the differences in metabolic rate among species (Coulson, Hernandez & Herbert, 1977). It was not that they were not important, for they were, but that enzyme and substrate concentrations (and even temperature) were so similar in all animals tested that some other variable was obviously of even greater importance.

Progress was made toward understanding by taking the reported oxygen consumption of the smallest mammal (a 2 g shrew) and computing how fast the blood must flow to deliver that amount of oxygen. It was immediately apparent that the very small mammals had high blood flows to each gram of tissue. Since the nature of the calculations is germane to the issue, let us look at the relationship between metabolic rate and blood flow in man.

A 70 kg sedentary man needs about 3000 kcal day^{-1}. For convenience, let us assume 2880. Each gram of glucose will provide about 4 kcal, therefore he could maintain his weight by eating 720 g which is 4 moles of glucose (molecular weight: 180) per day. To burn this amount he would need 24 moles of oxygen as is evident in the familiar equation:

$$1 \text{ glucose} + 6 \text{ O}_2 \rightarrow 6 \text{ CO}_2 + 6 \text{ H}_2\text{O}.$$

From the gas laws, 24 moles of oxygen have a volume of 537 600 ml. This is the amount he will use in a day, so the amount delivered by the blood cannot be less. Each 100 ml of blood (in man) leaves the lungs carrying 20 ml of gaseous oxygen, therefore each liter carries 200 ml. At the minimum, in one day 537 600/200 or 2688 litres of blood must be delivered if the cells use all the oxygen in the arterial blood and the venous blood returns to the heart with none. There are 1440 min in a day, so the heart must pump at least 1.867 litres of blood each minute. Venous blood actually contains about 62.5% as much oxygen as arterial blood in a sedentary man, therefore about 5 litres of blood are pumped per minute (1.867/0.375). A sedentary shrew uses 60 times as much oxygen per gram as a sedentary man (Pearson, 1948), and if it removes the same amount of oxygen from each liter of blood, its blood flow must also be 60 times that of a man. A 70 kg alligator at 38°C uses only 1/30 as much oxygen as a man, therefore its blood flow will be considerably less, for the magnitude of the difference between the oxygen content in venous blood and that in arterial blood is similar to that found in man.

Although metabolic rate is necessarily limited by the rate of blood flow, this does not constitute proof that blood flow is the only factor responsible for determining it. The relationship between the work performed by the vascular system and metabolic rate seems reasonable enough but when one thinks of the shrew pumping 60 times as much blood as man per kilogram, doubts arise. Why would the smaller blood vessels of the shrew not rupture? It took us several years to answer that to our own satisfaction, and even then we may still be wrong.

First and foremost is the issue of blood pressure. Do small animals, such as mice, have higher blood pressures than a man or an elephant? We found that, within fairly narrow limits, vertebrates have about the same aortal blood pressure, averaging close to 100 mmHg (Altman & Dittmer, 1968). Even an alligator's average blood pressure at 28°C was about 70 mmHg in the major arteries (Campos, 1964), and it is probable that it would have been close to 100 mmHg at 34°C. Difficult as it sounds, one animal can deliver blood at 60 times the rate (per gram) of another even though both have the same aortal blood pressure. The reason appears to lie in the difference in resistance in the blood vessels. The longer the vessel, the greater the resistance, for resistance is proportional to length. The average distance of body cells from the shrew heart is 2 cm: the average distance of body cells from the heart of the largest blue whale is several meters.

As the distance from the heart increases, the resistance also increases and the rate of flow through the arterioles decreases. In large animals, the slow rate of venous return means that the rate at which the right side of the heart fills will be slow and the heart rate will also be slow as it beats when the right side fills (Frank-Starling Law). In essence, according to our theory, metabolic rate is determined principally by the distance blood has to travel from the heart to the capillary, and the greater the distance, the greater the resistance and the slower the flow through the capillaries.

One of the more difficult aspects of the theory has been the need to explain why oxygen consumption by a tissue slice is usually so different from that of an equal weight of the same organ *in vivo*. In the last 50 years oxygen consumption has been determined on slices from various organs in scores of different animals. It can be said without exaggeration that more than 100 laboratories contributed data to the effect that it was primarily the organ that determined the rate of oxygen consumption and not the animal that contributed it (Altman & Dittmer, 1968). On the evidence obtained from oxygen consumption of slices, one would not suspect that the metabolic rate of a mouse differed from that of a cow. In a small live mammal, should the amount of oxygen delivered per minute be decreased to half, the animal would die within minutes. Yet isolated tissue slices from the same animal survive for hours in an oxygenated medium in a flask, and at only a small fraction of their oxygen requirement *in vivo*.

We have a theory to explain why oxygen demand is so much lower in a slice of an organ from a mouse than in an equivalent weight of the same organ in the mouse, and why organ slices of a cow consume more oxygen per unit weight than they would have in the cow. Simply put, rapid blood flow forces a flowing stream of interstitial fluid over and around tissue cells. The composition of the fluid outside the cells is very different from that inside and the two have a tendency to equilibrate rapidly. The greater the rate of flow the greater the requirement for energy to keep the composition of the two fluids dissimilar (Coulson, 1983). However, the greater the flow the greater the rate of delivery of the reactants producing energy, and therefore demand and supply remain in balance. The theory predicts that organ slices of a large reptile will actually consume more oxygen per gram than the intact organ. In a large reptile the rate of blood flow is low and therefore a slice in an oxygenated medium in a gently agitated flask behaves as though the flow had been increased.

With respect to metabolic rate in living reptiles, there are variables which must be considered. For example, alligator blood contains only about 40% as much oxygen as is carried in mammalian blood when it leaves the lungs (Coulson & Hernandez, 1964). Therefore at maximum exertion an alligator must depend primarily on an increased flow rate rather than on an increase in the amount of oxygen removed from each liter.

Table I gives our estimate of the metabolic rates of alligators weighing from 35 g to 700 kg, along with published estimates of metabolic rates of other species (Altman & Dittmer, 1968). Estimates for alligators above 7 kg were calculated from surface areas; below that size the figures are based on thousands of actual measurements (Hernandez & Coulson, 1952; Coulson & Hernandez, 1964, 1979, 1980, 1983; Coulson, Hernandez *et al.*, 1977; Coulson, Herbert & Hernandez, 1978). Unfortunately, the quoted values for some species may be unreliable, but it is probably safe to assume that most are within ± 50% of the correct value.

TABLE I
Power production per 70 kg body weight

Animal	Actual body weight (kg)	Temp (°C)	O$_2$ used[a] (l day^{-1})	Aerobic ATP production[b] (mol day^{-1})	Aerobic power[b] (W)	Max anaerobic ATP production[c] (mol)	Maximum anaerobic power[c] (W)	Anaerobic/Aerobic	Resting aerobic (W)	Maximum anaerobic (W)
Shrew	0.002	37	22 600	6054	2228	1.75	927	0.42	0.064	0.026
Mouse	0.020	37	5880	1575	580	1.75	927	1.60	0.166	0.265
Rat	0.300	37	1939	519	191	1.75	927	4.85	0.819	3.97
Skink	0.00016	28	700	188	69	1.75	927	13.4	0.00016	0.002
Dog	10.0	37	560	150	55	1.75	927	16.9	7.86	132
Man	70.0	37	377	101	37	1.75	927	25.1	37.0	927
Anolis	0.005	28	323	87	32	1.75	927	29.0	0.0023	0.066
Alligator	0.035	28	165	44	16	1.75	927	57.9	0.008	0.463
Alligator	0.050	28	145	39	14	1.75	927	66.2	0.010	0.662
Alligator	0.070	28	132	35	13	1.75	927	71.3	0.013	0.927
Alligator	0.090	28	124	33	12	1.75	927	77.3	0.015	1.19
Alligator	0.120	28	112	30	11	1.75	927	84.3	0.019	1.59
Alligator	0.160	28	107	29	10.5	1.75	927	88.3	0.024	2.12
Alligator	0.750	28	68	18	6.7	1.75	927	138.4	0.072	9.93
Alligator	1.0	28	60	16	5.9	1.75	927	157.1	0.084	13.24
Turtle	1.0	28	60	16	5.9	1.75	927	157.1	0.084	13.24
Blue whale	100 000	37	34	9.1	3.4	1.75	927	272.6	4857	1 324 300 (1775 H.P.)
Alligator	7.0	28	28	7.5	2.8	1.75	927	331.1	0.28	92.7 (12.4 H.P.)
Alligator	70.0	28	13	3.5	1.3	1.75	927	713.1	1.3	927
Alligator	700	28	6	1.6	0.6	1.75	927	1545	6.0	9270
Supersaur[d]	100 000[d]	28	1.2	0.32	0.12	1.75	927	7725	171.1	1 324 300 (1775 H.P.)
Alligator	700	5	0.8	0.21	0.08	1.75	927		0.80	

[a] Estimates of oxygen consumption are often unreliable. Values for alligators weighing from 35 g to 7 kg were obtained in our laboratory. Those for alligators above 7 kg were estimated by extrapolation from the curves.
[b] One liter of O$_2$ (44.64 mmol) will oxidize 7.44 mmol of glucose to give 7.44 × 36 or 267.8 mmol of ATP. Hydrolysis of 1 mol of ATP produces 7.6 kcal, and hydrolysis of 0.2678 mol will give 2.035 kcal. One kilocalorie is equivalent to 1.162 × 10^{-3} kWh; 1.162 Wh; 2.035 kcal are equivalent to 2.365 Wh, a power equal to that of a 0.0985 W bulb burning for one day.

Length of Fast

The longer an alligator has gone without food, the lower the metabolic rate (Coulson & Hernandez, 1979), but it is only by continuing the measurements for many days that the fall (which averages 2.5% day^{-1}) is detectable. Which value should one accept? In practice, rightly or wrongly, we used the results obtained the fifth day after feeding.

Reproducibility

Reproducible results on oxygen consumption may be obtained on any one reptile as soon as it becomes acclimated to the respiration chamber. Unfortunately, the values obtained will be characteristic for that animal alone, as the next one tested under apparently identical circumstances will differ by as much as 40%. Unlike man, individual variations are great. Two different laboratories each testing a reptile of the same size and species may well get different results, associated less with the method of testing than with individual variation.

Temperature

Years ago we measured oxygen consumption on small alligators and on one crocodile (*Crocodylus niloticus*) at temperatures from 5 to 30°C and plotted the relationship between oxygen consumption and temperature (Coulson & Hernandez, 1964). The slope was curious, indicating that increasing the temperature from 5 to 15°C caused little change in consumption. Above 15°C, oxygen consumption began to increase, and above 25°C it rose precipitously.

Feeding

The fact that a reptile is cold-blooded does not mean that it is free from the laws of thermodynamics, for, on a molar basis, each chemical reaction uses or produces the same amount of energy in all animals. The higher the metabolic rate, the faster energy is both produced and used. An animal at rest is in balance between energy income and expenditure. Should physical exertion or chemical reactions increase the demand for energy, none can be met from that produced at rest, all must come from increased production. The three principal causes of increased demand are muscle contraction, protein synthesis and gluconeogenesis from lactate. (In addition to these three, mammals and birds are forced to produce more heat in compensation for that lost when their environment gets cold.)

Force-feeding reptiles large amounts of fish muscle resulted in an immediate increase in metabolic rate, and both the amount of the increase and its duration were in rough proportion to the amount fed (Coulson & Hernandez, 1979). Purified proteins produced a response similar to that which followed feeding muscle in amounts equivalent in protein content, which made it appear that it was the protein in the food that was responsible. Neither fats nor carbohydrates were very effective.

Results indicated that chameleons, caimans, alligators, turtles (and presumably all vertebrates) converted most of the free amino acids released during digestion back into body protein as fast as they were absorbed. It was the energy cost of that protein synthesis that was responsible for the considerable metabolic rate increase. Each digested protein and absorbed amino acids at many times the rate they could be catabolized, and the immediate protein synthesis prevented toxic increases in plasma osmotic pressure. It is possible to show by calculation that failure to synthesize macromolecules immediately from absorbed amino acids would be lethal in any animal on a high-protein diet (Coulson & Hernandez, 1983).

To link each free amino acid to a developing polypeptide chain, energy equivalent to 4 ATP must be supplied. The actual rate of protein synthesis can be estimated by determining the increased energy production (and consumption) in an animal after feeding protein (Coulson, Herbert et al., 1978; Coulson & Hernandez, 1979), and the higher the metabolic rate, the more rapid the synthesis. The amount of extra ATP made available for peptide bonds seems to be a more or less fixed percentage of that produced by each animal at rest. For example, when chameleons and alligators were fed large amounts of protein, both increased oxygen consumption about three-fold even though the chameleon had four times the metabolic rate of the small alligator. When extra oxygen consumption was plotted as a function of time, the curve for the small lizard was tall and narrow in contrast to the low, wide curve for the alligator, but the total areas were the same for both animals (Coulson & Hernandez, 1983).

As the energy cost of protein synthesis will be the same in all animals, even though their metabolic rates may differ by a factor of hundreds, if the metabolic rate is known the time required to dispose of a given amount of dietary protein may be estimated. Table II illustrates the effect of metabolic rate on the rate of removal of amino acids absorbed from digesting protein. In compiling the figures, it was assumed that feeding 2 g protein kg^{-1} would double the metabolic rate, and that all the increase in ATP production was devoted to protein synthesis. From results of such experiments on several animals, these estimates would seem to be reasonable approximations. The shrew would produce enough extra ATP to convert all the absorbed free amino acids into body protein in less than 2 min compared with about two weeks in a 100-ton dinosaur.

ANAEROBIC GLYCOLYSIS

The amount of energy available from oxidative processes for any but the smallest alligators is very small, too small for them either to defend themselves or to catch food. As is also the case in very large mammals, significant power can only be developed by that remarkable series of reactions called glycolysis, reactions which proceed without the need for any oxygen and therefore without the need for blood flow. When the animal is working maximally, most of the glycogen in the muscle is converted to lactic acid and energy is produced at a rate up to hundreds of times as great as that which can be produced by oxidative processes. This burst of energy, common to all animals, lasts only about 2 min if they are at maximum exertion (Coulson, 1979), but work of lesser intensity may be performed at a slower rate, with a lower rate of drain on the muscle glycogen supply. The lactate, which is formed instantaneously, is either oxidized (20%) or converted by the liver back into glucose, and from there carried back to be reconverted into glycogen.

Cost of Resynthesis of Muscle Glycogen

It is quite easy to determine the cost of glycogen resynthesis (gluconeogenesis) in a reptile. The resting oxygen consumption is determined, the animal is forced to run until it is exhausted, and then oxygen consumption is determined again. The metabolic rate in a reptile will usually rise to two or three times what it was before work and then gradually fall, but remain elevated until all the lactate is gone from the blood. The cost of removal of the lactate can be determined by the amount of extra oxygen used. Gluconeogenesis is equally energy-demanding in a mouse and in a turtle for the reason that each produces the same amount of lactate from the same amount of glycogen during maximum exertion, and each must furnish the same amount of ATP to restore it. The energy cost to both is 8 mmol of ATP per millimole of glucose residue formed from 2 mmol of lactate.

Rate of Resynthesis of Muscle Glycogen

In theory, there need be no direct relationship between metabolic rate and the rate of gluconeogenesis. For example, an animal of low metabolic rate could increase energy production after work by several fold, while another of high metabolic rate might increase its energy production by only a few per cent with the result that both could restore glycogen at the same rate. However, gluconeogenesis seems to be a reaction of the highest priority. The more rapid the restoration, the better prepared an animal would be if another emergency arose. In the recovery phase after glycolysis, the 5 g chameleon and the 1 kg alligator both tripled their metabolic rates, for a short time in the lizard, and for a long time in the alligator. The difference between the two was in the length of time oxygen consumption remained elevated. As the lizard had four times the metabolic rate of the alligator, its metabolic rate was increased for only about a quarter as long (Coulson & Hernandez, 1980).

By again assuming that the metabolic rate would be doubled following work, as it was following feeding 2 g protein kg^{-1}, one can calculate how long it would take various animals to repay the oxygen debt. The range was from under 2 min in the shrew to over three weeks in the hypothetical dinosaur. We have observed the length of time needed for glycogen restoration in alligators weighing from 50 g to 200 kg and the rates were reasonably close to the estimates shown in Table II.

RELATIONSHIP BETWEEN METABOLIC RATE AND VARIOUS ACTIVITIES

Diving

Metabolic rate affects the length of time a reptile can remain submerged even though aerobic metabolism is secondary to anaerobic when the reptile is anoxic. Even if one had no advance knowledge of the diving abilities of animals, in general it would still be possible to make rough estimates of the diving time (maximum) if certain assumptions were allowed. Striated muscle is well supplied with glycogen and therefore capable of producing sufficient ATP over a prolonged period to enable the muscles to survive for a while. Unlike the maximum rate of glycolysis following extreme exertion, while the animal is submerged the rate of glycolysis would be slower and in proportion to energy need. In contrast to the muscle, the brain requires oxygen at all times as the

TABLE II
Energy cost

	Energy cost of protein digestion, amino acid absorption and body protein synthesis (per kg body wt). Each animal was fed (theoretically) 2 g protein kg^{-1} body wt					Energy cost of repayment of O$_2$ debt after maximum work		
Animal	Body weight (kg)	Resting ATP[a] (mmol day^{-1})	ATP needed[b] to dispose of 2 g of fed protein (mmol)	Cost of protein feeding in % of resting ATP production	At double the resting metabolic rate, minimum time needed to assimilate 2 g of protein	Total[c] ATP cost (mmol kg^{-1})	Cost of repayment in percentage of resting ATP production	At double the resting metabolic rate, minimum time needed to repay the O$_2$ debt
Shrew	0.002	86 500	69.6	0.08	1.16 min	98	0.11	1.6 min
Skink	0.00016	2685	69.6	2.60	37.3 min	98	3.64	53 min
Man	70	1440	69.6	4.84	69.6 min	98	6.81	98 min
Anolis	0.005	1240	69.6	5.60	80.8 min	98	7.90	114 min
Alligator	0.035	629	69.6	11.08	2.7 h	98	15.6	3.7 h
Alligator	1.0	229	69.6	30.4	7.3 h	98	42.8	10.3 h
Turtle	1.0	229	69.6	30.4	7.3 h	98	42.8	10.3 h
Blue whale	100 000	130	69.6	53.6	12.8 h	98	75.4	18.1 h
Alligator	70	50	69.6	13.9	1.39 days	98	196	1.96 days
Alligator	700	23	69.6	302	3.03 days	98	426	4.26 days
Supersaur	100 000	4.6	69.6	1523	15.1 days	98	2144	21.4 days

[a] For calculations, see Table I. Very little is needed for digestion and absorption.
[b] Two grams of protein contain 17.4 mmol of amino acids. Each amino acid added to a growing peptide chain requires energy equivalent to that provided by hydrolysing 4 ATP. Therefore, 69.6 mmol of ATP are needed to incorporate 17.4 mmol of amino acids into protein.
[c] On the basis of the breakdown of all the labile muscle glycogen (2.2 g glycogen kg^{-1} of total body weight). Resynthesis of each glucose residue (in muscle glycogen) from two lactates costs 8 mmol of ATP per millimole of glucose synthesized. 2.2 g (12.25 mmol) of glucose in glycogen yield 24.5 mmol of lactate in anaerobic glycolysis and 4 × 24.5 (98) mmol of ATP are needed for glycogen resynthesis from the lactate.

trace of glycogen present is not a practical source of energy. If we assume that oxygen in the blood at the time of the dive can be reserved for the use of the brain, and that the rest of the carcass will be supplied with anaerobic energy, we can then estimate how long blood oxygen will last. Giving an alligator a blood volume of 80 ml kg^{-1}, and a hematocrit half that of a mammal, blood at the beginning of the dive would carry about 8 ml of oxygen at standard temperature and pressure.

We have two variables – the metabolic rate, and the size of the brain. The following illustrates the remarkable decrease in the relative size of the brain in alligators as they grow: body weight 131 g, brain 0.67%; body weight 110 kg, brain 0.034% (Coulson & Hernandez, 1983). In proportion, the brain of the smallest was 20 times as big as that of the largest. If we assume that the alligator brain resembles those of mammals in its proportionately high oxygen consumption per gram, and that each gram of brain then uses 10 times as much oxygen as the average for the total body, even the smallest could remain submerged for over an hour. The 700 kg alligator not only has a much lower metabolic rate than the small one, but his brain is also disproportionately smaller, which would (theoretically) allow him to stay submerged for days.

A second estimate can be obtained by assuming that all organs use the oxygen in the blood in the same relative proportion as they did when the animal was on the surface. How long would the oxygen last if all of it could be extracted from the blood? The time would depend on the metabolic rate. From Table I, a 35 g alligator used 165 litres 70 kg^{-1} day^{-1}, or 2.36 litres kg^{-1} day^{-1} or 0.00164 litres kg^{-1} min^{-1} or 1.64 ml kg^{-1} min^{-1}. It would have sufficient oxygen for 8/1.64 or 4.88 min. On the other hand, a 700 kg alligator could have enough oxygen for 27.5 times that length of time (134 min), for the small alligator's metabolic rate is 27.5 times that of the larger. By this estimate, the small one could dive for only a few minutes while the larger could stay down over 2 h. These calculations were based on two incorrect assumptions: one, that no glycolysis occurred, and another, that all the oxygen could be extracted from the blood.

The true times are somewhere between the two estimates with large alligators capable of remaining below the surface for over a day and small ones poorly adapted for diving. Whereas one of our 200 kg animals routinely remains voluntarily on the bottom of the tank for 6 h (at 28°C) at a time, the smallest are reluctant to dive for more than a minute. At low water temperatures large alligators can stay down for a week or more.

Energy for Sustained Effort

With respect to reptiles generally, the smaller the reptile, the more capable it will be in terms of sustained work performance. Its capacity for work will equal or exceed that of man over a period of many hours, but only if the body temperature remains high. Should the temperature fall, the contribution of aerobic metabolism falls leaving it more dependent on anaerobic energy. It would not be an exaggeration to view large reptiles as being incapable of aerobic work of any consequence.

Basking

Solar radiation increases the capacity for oxidative work by raising the body temperature, which of course leads to increased respiratory rate, blood pressure and blood flow. As the temperature increases, appetite is increased, which although it increases the need for food, also gives the animal the strength to capture it. Should food be in short supply, continuous high ambient temperature leads to weight loss. This is limited, for prolonged inanition decreases metabolic rate, and by a sufficient amount to reduce the amount of food required for weight maintenance (see below).

Food Requirements

Since these are related to metabolic rate directly, the differences in demand among the Reptilia are enormous. The smallest lizards at 32°C require three times as much per kilogram as a man, and the largest crocodiles less than 2% as much. Some idea of how long a 700 kg crocodile could go without food may be estimated by assuming it could survive a weight loss of one-third the body weight (233 kg). This would be about 70 kg dry weight, and at 5 kcal g^{-1} it would supply about 350 000 kcal. Its oxygen consumption (at 28°C) is 60 litres day^{-1} (Table I), or 2.68 moles day^{-1}. This is sufficient to oxidize 0.67 moles of glucose (121 g) to provide 484 kcal day^{-1}. It would then have enough of a calorie store to last 723 days or about two years. Even this long time is an underestimate since fasting decreases metabolic rate and, in addition, it could probably survive a body weight loss as high as 50%. A more reasonable estimate of how long it could survive in starvation would be about four years. Using the same type of calculations, the 100-ton dinosaur could go without food for about 20 years at 28°C and much longer at a lower temperature.

Speculation on the Reason Reptiles are Cold-blooded

Recently interest in the evolution of homeothermy has been revived. Present reptiles are heterotherms and birds are homeotherms, therefore we must assume that some transitional reptile found the secret. Theories on the subject have been so controversial that one hesitates to participate in the discussion, but some recent research may have provided a key to the phenomenon. We know that a decrease in cellular ATP in any vertebrate signals the heart and lungs to increase the flow of oxygenated blood. In the reptile, increased demand for ATP will accompany physical exertion, or it may be provoked by either rapid protein synthesis or gluconeogenesis. These factors operate in mammals also, but mammals react to yet one more stimulus, and that is cold. As soon as the temperature over the body surface begins to fall, ATP is hydrolysed to provide heat, and the ATP deficit acts as a stimulus to promote an increase in the metabolic rate. In contrast, as a reptile is cooled, metabolic rate is allowed to fall for the reason that no deficit in ATP is perceived, and so no signals are sent to the heart and respiratory systems.

Where does the extra heat come from in a homeotherm exposed to cold? If the animal is not doing physical or chemical work of the sort mentioned above, the only reasonable source would be hydrolysis of ATP in the presence of some species of ATPase. This enzyme could not be present in quantity in an active state as it would tend to hydrolyse ATP ordinarily used for physical and chemical work. To continue the speculation, cooling the body surface sends a signal to the hypothalamus, which in turn sends a signal to the cells (by the way of the nerves), and ATPase is activated. ATP is hydrolysed, heat is produced, cellular ATP falls, and metabolic rate increases to restore the energy level.

Considering the lower metabolic rate at the preferred temperature of most reptiles, even if the proper conditions were obtained, could reptiles become homeothermic? Perhaps they could but the range of ambient temperatures that they could defend against would probably be rather narrow. Yet, size would work in their favor even as it does in a mammal. The largest ones have low metabolic rates but their correspondingly low surface areas would tend to decrease the rate of heat loss, and the converse is true for the smallest.

According to this hypothesis, an animal would be cold-blooded for one or more of the following reasons: failure of the hypothalamus to send a signal to the cells as the body temperature falls; failure of the signal (if sent) to activate an ATPase; or lack of the proper ATPase, active or inactive. Unfortunately, as is so often true, the fewer the known facts the more secure the theory appears. Further research may disprove it in part or *in toto*.

Energy Production in a Homeothermic 100-ton Dinosaur

Recent indirect evidence has been presented in support of a theory that at least some of the larger dinosaurs were warm-blooded. Restricting consideration to only the largest ones, they would have had such low metabolic rates that oxidative activity would still have been quite limited. Some blue whales have been observed weighing as much as 140 tons, a bulk considerably greater than the largest known dinosaur. The metabolic rate of the largest whales is only about 7% of that of man, and on a comparative basis even man is a large animal of limited oxidative capacity. Fortunately the whale is aquatic which relieves him of problems associated with gravity, and which also permits him to move aerobically by expending very little energy (Tucker, 1975). A terrestrial homeothermic giant dinosaur would be ill-equipped for locomotion on land if he depended on aerobic metabolism. The cost of maintaining an erect posture would be considerable, and that of moving 100 tons almost prohibitive. One problem that cannot be circumvented in any large animal, regardless of body temperature, is the resistance in the exceedingly long blood vessels. To achieve a high constant temperature, the animals would require more powerful hearts and lungs with a greater surface area, and, along with increased efficiency of circulation and respiration leading to a higher metabolic rate, a larger and more efficient kidney to dispose of the increased amount of solids needing excretion. The liver, pancreas, etc. would also be called upon to perform at a higher rate. In short, the animal would need to be redesigned, and even if it were, metabolic rate would still be low.

To illustrate, by our estimates a 100-ton dinosaur at 28°C had a metabolic rate only 3.6% that of a 100-ton blue whale. If we assume that the metabolic rate of the dinosaur would be doubled by increasing its body temperature to 37°C, its metabolic rate would still be only 7.2% of that of the whale at rest. To make its resting metabolic rate equal to that of the resting whale, it would be necessary to increase blood flow to 14 times the rate at 37°C, and to 28 times that at 28°C. If at maximum oxidative activity blood flow is necessarily further increased to five times that at rest, then the total increase in flow over that at rest at 28°C would be 140-fold. Perhaps very small dinosaurs could have developed homeothermy and their secret could have been passed on to larger and larger ones in the course of evolution. Then again, perhaps they were all cold-blooded.

Barring evidence to the contrary, cells of an immature, small animal are almost identical to those of a large one of the same species even though their metabolic rates may differ by 25 to 1. If, from a chemical standpoint, the work capacity of individual cells is similar in all, the actual work accomplished when those cells are incorporated in organs may vary considerably, depending on the rate at which the organ is perfused with blood. In both reptiles and mammals all systems are in balance. Those with low metabolic rates demand and produce less aerobic energy than their more energetic neighbours, but both react to an increase in energy demand in the same manner. Regardless of temperature, a large reptile cannot produce as much energy as a mammal of the same size for the reason that it can neither oxygenate the blood, nor pump it, fast enough to supply the required oxygen and substrates. The metabolic difference between a reptile and a mammal owes less to chemistry than to chemical engineering.

ACKNOWLEDGEMENTS

We are grateful to the Louisiana Department of Wildlife and Fisheries for financial support, and to T. Hernandez and J. D. Herbert for their contribution to the theories.

REFERENCES

Altman, P. L. & Dittmer, D. S. (1968). *Metabolism*. Bethesda: Federation of the American Society of Experimental Biology Press.
Campos, V. (1964). Efecto de los cambios de temperatura sobre las frecuencias cardiáca y resperatoria del lagarto. (*Alligator mississippiensis*). *Revta Biol. trop.* **12**: 49–57.
Coulson, R. A. (1979). Anaerobic glycolysis: the Smith and Wesson of the heterotherms. *Perspect. Biol. Med.* **22**: 465–479.
Coulson, R. A. (1983). Relationship between fluid flow and O_2 demand in tissues *in vivo* and *in vitro*. *Perspect. Biol. Med.* **27**: 121–126.
Coulson, R. A., Herbert, J. D. & Hernandez, T. (1978). Energy for amino acid absorption, transport and protein synthesis *in vivo*. *Comp. Biochem. Physiol.* **60A**: 13–20.
Coulson, R. A. & Hernandez, T. (1964). *Biochemistry of the alligator, a study of metabolism in slow motion*. Baton Rouge: Louisiana State University Press.
Coulson, R. A. & Hernandez, T. (1979). Increase in metabolic rate of the alligator fed proteins or amino acids. *J. Nutr.* **109**: 538–550.
Coulson, R. A. & Hernandez, T. (1980). Oxygen debt in two reptiles: relationship between the time required for repayment and metabolic rate. *Comp. Biochem. Physiol.* **65A**: 453–457.
Coulson, R. A. & Hernandez, T. (1983). Alligator metabolism, studies on chemical reactions *in vivo*. *Comp. Biochem. Physiol.* **74B**: 1–182.
Coulson, R. A., Hernandez, T. & Herbert, J. D. (1977). Metabolic rate, enzyme kinetics *in vivo*. *Comp. Biochem. Physiol.* **56A**: 251–262.

Hernandez, T. & Coulson, R. A. (1952). Hibernation in the alligator. *Proc. Soc. exp. Biol. Med.* **79**: 145–149.
Ostrom, J. H. (1978). A new look at dinosaurs. *Natn. geogr. Mag.* **154**: 152–185.
Pearson, O. P. (1948). Metabolism of small animals. *Science, N.Y.* **108**: 44–46.
Tucker, J. A. (1975). The energetic cost of moving about. *Am. Scient.* **63**: 413–419.

Thermoregulation in the Nile Crocodile, *Crocodylus niloticus*

J. P. LOVERIDGE

*Department of Zoology, University of Zimbabwe,
P.O. Box M.P. 167, Harare, Zimbabwe*

SYNOPSIS

Thermoregulation in wild and captive Nile crocodiles was studied in Zimbabwe. Whilst basking in the sun, juvenile crocodiles of 0.5–1.5 kg body weight had mean body temperatures (T_b) in the range 29.7–33.4°C. Biotelemetric studies suggest that basking temperatures of 32–34°C are also maintained by crocodiles up to 38 kg body weight. Crocodiles of body weight above 90 kg had mean T_bs of 25.1–28.4°C.

Juvenile crocodiles caught on land during the night had $T_b < T_w$ (water temperature) and $T_b > T_a$ (air temperature) during the cold months but $T_b < T_a$ during the hot months. Those caught in the water had $T_b = T_w$ at all seasons of the year.

In the absence of radiant heat, crocodiles in the weight range 75 g–4 kg did not heat significantly faster than they cooled. When heated by two 1000 W lamps the relationship between thermal time constant (τ) and body weight in the range of 0.22–314 kg was $\log_{10} \tau = 0.3259 + 0.4479 \log_{10}$ body weight (g). The aquatic weed *Salvinia molesta* affords a substantial insulation when draped on the backs of crocodiles. Windspeeds in the range 1.0–4.5 m s^{-1} significantly shortened the time constant for heating under two 1000 W lamps.

Although mouth gaping did not always occur in circumstances where there was need for cooling of the body, laboratory experiments showed that the oral mucosa (tongue and roof of the mouth) was a site of evaporative cooling. Tongue temperatures were about 5°C lower than back temperatures when a crocodile was heated from a T_b of 16°C to 34°C. Evaporative water losses in the range 2.8–21.2 mg cm^{-2}h^{-1} were measured by means of ventilated capsules on portions of the oral mucosa.

INTRODUCTION

Despite the obvious importance of thermoregulation in the lives of crocodilians, it has been studied in only a few species. The American alligator, *Alligator mississippiensis*, has received most attention (Colbert, Cowles & Bogert, 1946; Spotila, 1974; Smith, 1975; Lang, 1979a). The Australasian crocodiles, *Crocodylus novaeguineae* and *C. porosus* have been studied by Johnson (1974), Johnson, Webb & Tanner (1976), Lang (1981) and *C. johnstoni* by Grigg & Alchin (1976). Aspects of thermoregulation in *C. acutus* have received attention from Lang (1979a), and in *Caiman crocodilus* from Diefenbach (1975).

Thermal preferences and behavioural thermoregulation of crocodilians have been studied in a number of species (Spotila, 1974; Smith, 1975; Diefenbach, 1975; Johnson, Webb et al., 1976; Lang, 1981). Heating and cooling rates in *A. mississippiensis* (Smith, 1976a; Johnson, Voigt & Smith, 1978; Smith & Adams, 1978) and in *Crocodylus johnstoni* (Grigg & Alchin, 1976) have been measured, and mechanisms for differences in heating and cooling rates investigated (Smith, 1976b; Grigg & Alchin, 1976). Lang (1979a) has reported on the thermophilic behaviour after feeding in *A. mississippiensis* and *C. acutus* and Spotila, Terpin & Dodson (1977) have studied the effect of mouth gaping on heating rates in *A. mississippiensis*.

The Nile crocodile, *Crocodylus niloticus*, has not received the same attention as North American and Australasian species. Cott (1961) reported on behavioural studies of diurnal activity rhythm, thermoregulation and gaping of adult animals. Cloudsley-Thompson (1969) measured evaporative water loss of hatchling animals when heat-stressed and Diefenbach (1975) observed the effect of gaping in young *C. niloticus*. There have been no studies of the body temperatures of Nile crocodiles during thermoregulation or of heating and cooling rates in this species. Lang (1979b) has suggested that alligators, living in temperate climates, are thermoregulators and attempt to maintain high, stable body temperatures. By contrast, crocodiles living in tropical climates are thermoconformers, avoiding rapid heating during the day. In view of these possible differences it is important to extend our knowledge away from alligators towards the more tropical crocodiles.

In nearly all the work mentioned above, the size range of animals studied has been in the range of 60 g to 14 kg. As has been pointed out by Gans (1976) and by Smith (1979), very large animals should be included to confirm the validity of extrapolation of work done on hatchlings and juveniles to all sizes of crocodilians. This study attempts to extend the knowledge of crocodilan thermoregulation to another species, *C. niloticus*, and to include all sizes of animals from hatchlings to mature adults.

MATERIAL AND METHODS

Field observations of basking behaviour of crocodiles were made on the Zambezi river at Mana Pools, at Sinamwenda on Lake Kariba (wild populations) and at Kyle National Park and Kariba crocodile farm (captive populations). Environmental temperatures were recorded using a Grant 8-channel miniature temperature recorder while the crocodiles were being observed from a hide. Body temperatures of crocodiles captured in the wild were measured using a Yellow Springs Instrument telethermometer with a blunted type 418 soil thermistor probe inserted into the cloaca. Water and air temperatures were measured with type 418 and 405 thermistors respectively.

For telemetric studies of body temperatures, crocodiles were force-fed a calibrated temperature-sensitive transmitter (Mini-Mitter Co. type V). The signals were received up to 2 m away using an AM radio and timed using a stopwatch. Environmental temperatures were measured using a Yellow Springs telethermometer with appropriate probes and extension cables.

In laboratory experiments temperatures were recorded by a Grant 8-channel recorder. Cloacal probes were inserted to a depth of 2.5–13 cm depending on the size of the crocodile. Body surface temperatures were measured with a disc-shaped thermistor taped to the skin, tongue temperatures were measured with a needle probe inserted beneath the tongue epithelium and black bulb temperatures with a thermistor covered by a spherical steel bulb, 1.5 cm diameter, painted matt black. Crocodiles used in experiments came from a wide variety of sources, but most were maintained in the laboratory for a period of years. The larger animals were captured in the wild and the experiments were done before they were relocated on crocodile farms. All crocodiles above the weight of 30 kg used in laboratory experiments were immobilized using gallamine (Loveridge & Blake, 1972).

Crocodiles were cooled in a walk-in cold room with an air temperature (T_a) of 2°C until their cloacal temperature (T_b) was below 15°C. They were then rapidly transferred to a hot room ($T_a = 30°C$) and tied to a metal frame on a dry sand substrate. In one series of experiments substrates of pebbles, wet sand, dry and wet *Salvinia molesta* were also used. Air in the hot room was stirred by a ceiling fan rotating slowly (89 rpm). In experiments on the effect of wind speed on heating rate a wind tunnel was used to blow air over the crocodile, from head to tail, at speeds of 1.0, 2.5 or 4.5 m s^{-1}. In experiments requiring the use of a radiant heat source, two 1000 W lamps were suspended with their filaments 93 cm above the substrate and 30 cm apart. For all heating and cooling experiments, τ – the time take for T_b to change by 63% of the difference between T_b and T_a – was calculated (Smith, 1976a). The advantage of the use of τ (the thermal time constant) is that it is independent of the magnitude of the difference between T_b and T_a and remains constant with time during the exponential approach of T_b towards T_a (Grigg, Drane & Courtice, 1979).

Respiratory rates were observed and timed with a stopwatch. Heart rates were recorded by picking up the ECG with needle electrodes inserted subcutaneously into a forelimb and hind-limb. The signal was amplified and fed into a chart recorder. Rates of evaporative water loss from the tongue and roof of mouth were measured using a tight-fitting double capsule ventilated at either 150 or 270 ml min^{-1}. Upstream and downstream humidity was measured and water loss calculated as described by Loveridge & Crayé (1979).

RESULTS

Thermoregulatory Behaviour

The precise patterns of basking behaviour are dependent on the season of the year and the prevailing weather conditions. Generally speaking, crocodiles emerge from the water for basking once the sun is shining on the basking site, or the air temperature exceeds the water temperature. They may remain ashore, basking for the best part of the morning, retreating to the water for short periods of time when they may wholly or partly submerge (Fig. 1). This shuttling between sun and water is more pronounced in small crocodiles and in hot weather. During the hottest part of the day few crocodiles are out basking; rather they emerge from the water again around 15.00 hours and onwards. During the late afternoon, when black bulb and substrate temperatures are falling, crocodiles spend longer periods ashore and may remain there after sunset. The two main periods of basking, in early morning and middle to late afternoon, are reflected in a typically bimodal relationship between time and numbers of crocodiles basking in a captive population (Fig. 2). This bimodal pattern is evident in crocodiles of all ages.

Observations of adult crocodiles at Sinamwenda indicated that they frequently came ashore festooned with the floating aquatic weed, *Salvinia molesta* (Fig. 3). Basking crocodiles made no attempt to rid themselves of the weed, and the supposition that this insulating material influences the rate of heating gave rise to some of the experiments described later (p. 457).

A record was made of the incidence of mouth gaping during observations of basking behaviour. In both adult crocodiles at Sinamwenda and young crocodiles on the Zambezi, gaping frequently occurred on first emergence from the water as well as immediately before re-entry to the water (Fig. 1). In adults particularly, mouth gaping was used on crowded sandbanks and appeared to reflect a social interaction. Other observations suggest that mouth gaping may serve more than a thermoregulatory role. At Sinamwenda at 20.19 hours on 13 October 1971 during a night-capture exercise, a 1.6 m crocodile was seen gaping while

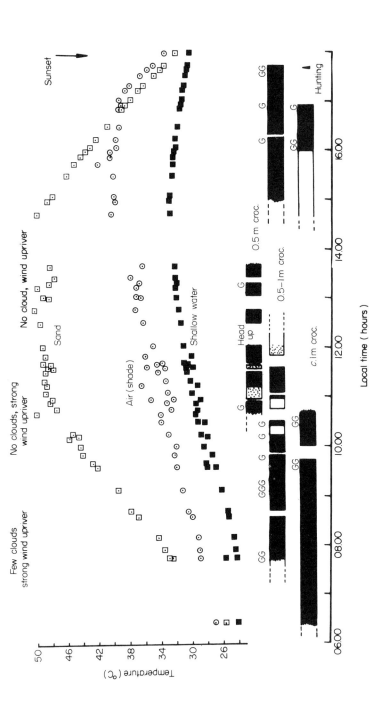

FIG. 1. Basking behaviour of young crocodiles and environmental temperatures, Zambezi River, 22–25 October 1968. Solid bars – crocodiles out of water; stippled – partly in water; G – gaping.

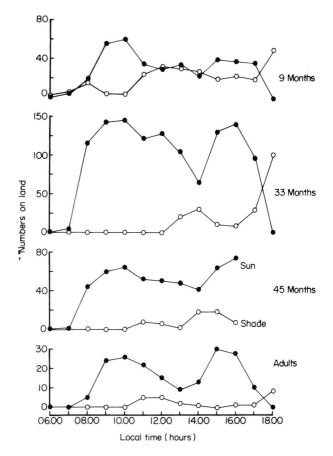

FIG. 2. Numbers of crocodiles of different ages basking in the sun (●) and in the shade (O) at Kariba crocodile farm, 24 August 1976.

out of water, on a rock. The substrate, dried *Salvinia*, was 23.9°C, T_a was 23.8°C and the water temperature was 25.2°C. At Kyle National Park at 08.25 hours on 3 June 1971 when T_a was 10.4°C and the water temperature was 13.4°C, a 2.8 m female crocodile lay gaping for 4 min in the water with the water wetting the tongue. This individual only emerged to start basking at 10.11 hours on that day.

Body Temperatures

Measurements of body temperatures of both captive crocodiles and wild juvenile crocodiles were made during mark and release studies.

FIG. 3. Adult crocodile partially covered with the aquatic weed, *Salvinia molesta*, basking on a sandbank, Mwenda River, 31 August 1968. The vertical wire to the left is the lead crossing the river to measure the substrate temperature.

Body temperatures during daytime, when on land

Most of the data available in this category are from captive crocodiles, and are summarized in Table I. The highest mean T_b of 33.4°C was recorded from a sample of 15 juvenile crocodiles on a very hot January day. The highest individual T_b in this group (and in the whole study) was 39.0°C. Smaller crocodiles, of body weight less than about 19 kg, appeared to have higher body temperatures than larger ones (Table I). In the two groups of crocodiles with mean body weights of 91 and 97 kg the highest T_b was 33.0°C and 29.5°C respectively.

There are some data on the body temperatures of wild juvenile crocodiles caught in a handnet while on land, at Ngezi National Park (J. Hutton, pers. comm.). The information is limited to the cold dry season, at the beginning of Hutton's study, when water temperatures were low (17.1 ± 0.7°C, range 16–18°C). After 09.30 hours the body temperature of the captured crocodiles started exceeding the water temperature by 3.1°C h^{-1} (Fig. 4) so that by 12.15 hours crocodiles had T_bs in the range 23–31°C.

Body temperatures during night-time, when on land

During studies at Sinamwenda, Lake Kariba, a proportion (14.7% in a sample of 197) of juvenile crocodiles was captured on land or in the

TABLE I
Body temperatures of basking captive crocodiles of different size (error estimates are ± 1 SD)

Locality	Month	T_b (°C)	Mean T_a (°C)	T_w (°C)	N	Mean body weight (kg)
Kariba Crocodile Farm	September	31.2 ± 1.9	28.6 ± 0.4	24.9 ± 0.8	18	0.5[a]
Binga Crocodile Farm	January	33.4 ± 3.5	32.1 ± 1.9	–	15	0.69 ± 0.51
University, Harare	Various	29.7 ± 2.6	26.5 ± 2.9	23.3 ± 2.5	26	1.51 ± 2.39
Victoria Falls	March	26.7 ± 3.6	23.3 ± 3.0	25.3 ± 1.7	45	19.54 ± 2.87
Victoria Falls	March, October	28.4 ± 2.5	28.6 ± 2.8	24.3 ± 1.2	17	91.4 ± 26.1
Various	Various	25.1 ± 3.8	25.4 ± 3.9	21.6 ± 3.1	19	97[a]

[a] Mean body weight estimated from mean body length of sample.

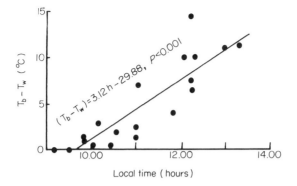

FIG. 4. The increase in body temperature (T_b) above water temperature (T_w) with time of juvenile crocodiles caught on land, Ngezi National Park during the cold season. (Data from J. Hutton.)

water after being on land. Nearly all these animals had a T_b below the water temperature, T_w (Fig. 5), both during the cold season (May–June) and the hot season (September–October). By contrast most of the crocodiles captured on land at night during the hot season had a T_b lower than T_a but during the cold season most had a T_b higher than T_a (Fig. 6). This no doubt arises from the fact that during the cold season $T_w > T_a$, which is reversed during the hot season.

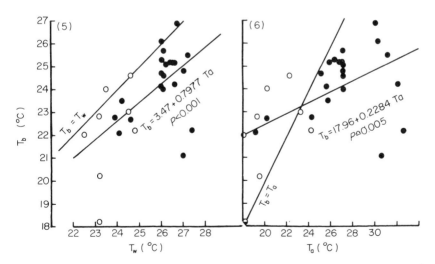

FIGS 5 & 6. (5) The relationship between water temperature and body temperature of juvenile crocodiles caught ashore at night, Sinamwenda. Solid symbols, September–October; open symbols, May–June. (6) The relationship between air temperature and body temperature of juvenile crocodiles caught ashore at night, Sinamwenda. Solid symbols, September–October; open symbols, May–June.

Body temperatures during night-time, when in water

Most crocodiles were caught in the water at night, and their T_b did not depart markedly from T_w, with the regression equation being $T_b = 0.1458 + 0.9851\ T_w$ ($r = 0.915$, $N = 168$). The relationship between T_b and T_a for crocodiles caught in the water is given in Fig. 7 which shows that below $T_a = 25.2°C$, $T_b > T_a$ and above $T_a = 25.2°C$, $T_b < T_a$.

Body Temperature and Thermoregulatory Behaviour

An attempt was made to unify the observations given in the preceding sections using telemetric techniques to monitor the body temperatures of captive crocodiles thermoregulating naturally in an outdoor enclosure. Sample results from different sized crocodiles and at different

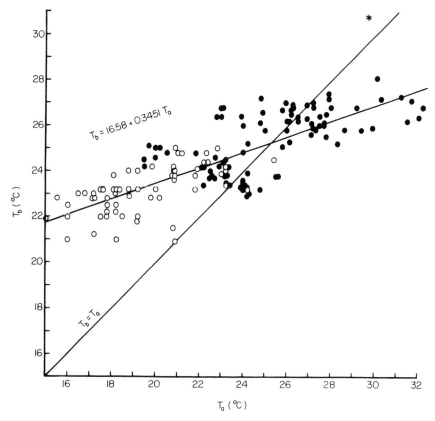

FIG. 7. The relationship between air temperature and body temperature of juvenile crocodiles caught in the water at night, Sinamwenda. Solid symbols, September–March; open symbols, May–June, asterisk, an individual caught during the day.

times of the year are given in Figs 8–10. On a cool, overcast day with intermittent drizzle [Fig. 8(A)], crocodiles came onto land for basking; but the T_b of the smaller (5.9 kg) only reached 28°C at 13.15 hours and the larger (38.2 kg) which started basking at 11.30 hours, reached a T_b of 26°C by 14.45 hours. Two days later when the weather was warm and partly overcast [Fig. 8(B)], the larger crocodile started basking earlier than the smaller and attained a $T_b > 34$°C during late afternoon. The smaller crocodile was the victim of aggression by other crocodiles in the enclosure and, despite numerous attempts, spent only two periods

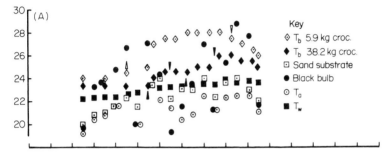

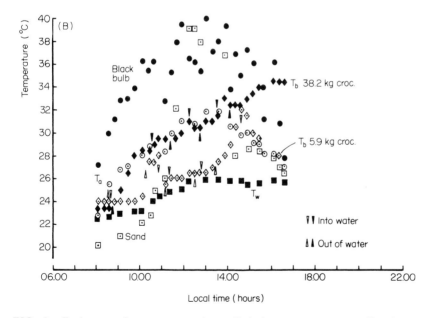

FIG. 8. Environmental temperatures and crocodile body temperatures measured by telemetry during February 1977, (A) on an overcast, intermittently drizzly day and (B) on a warm, partly overcast day.

basking, one in mid-morning, the other in early afternoon. In the latter period a T_b of 32°C was reached for a brief period.

On a day when there was much less cloud, but the weather was still warm, the body temperatures of both a large (37 kg) and small (5.6 kg) crocodile showed a distinctly bimodal pattern (Fig. 9). The T_b of both crocodiles rose rapidly to reach 32°C by mid-day, peaking at 34°C shortly thereafter in the case of the smaller animal. Both crocodiles spent about an hour in the water before emerging to bask again. The small crocodile after reaching a T_b of 34.7°C at 15.15 hours, entered the water once again, whereas the larger crocodile remained ashore in the shade until after 18.00 hours, when its T_b was still in excess of 31°C.

The bimodal pattern in T_b was not observed in the animal illustrated

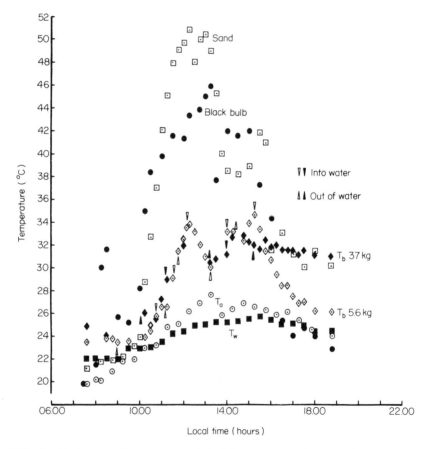

FIG. 9. Environmental temperatures and the body temperature of two crocodiles measured by telemetry during April 1973, on a warm, partly cloudy day. Note the bimodal pattern of T_b with time in both animals.

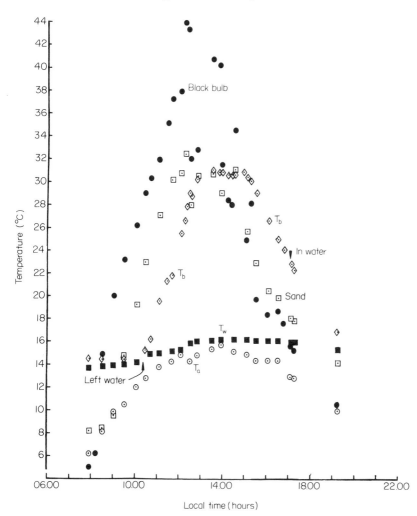

FIG. 10. Environmental temperature and the body temperatures of a 5.4 kg crocodile measured by telemetry on 29 June 1972, a warm day with intermittent cloud.

in Fig. 10. Basking commenced much later (10.00 hours), a T_b of 30°C was reached at 12.45 hours and remained above this level until 15.15 hours when it started dropping rapidly. The crocodile entered the water at 17.00 hours. On this day it is notable that at no time did T_a exceed T_w, but the clear sky allowed very high black bulb temperatures to be reached in early morning, contributing to the rapid heating once basking commenced.

All observations using telemetry indicated that crocodiles in water during the morning had T_bs which were 1–2°C in excess of T_w. This contradiction with the results from night-capture of crocodiles in water (p. 452) where T_b was the same as T_w, may perhaps be explained by the fact that these crocodiles were much smaller. The biotelemetry studies also show that crocodiles with a body weight of 37–38.2 kg may have body temperatures in excess of 32°C when basking [Figs 8(B) & 9].

Heating and Cooling Rates

Heating and cooling without radiation

For any particular size of crocodile in the range of body weight 75 g–4 kg, the time constant for heating was slightly greater than that for cooling (Fig. 11). The slopes and intercepts of the two regression lines are, however, not significantly different ($P > 0.5$).

Influence of radiation and windspeed on heating

In nature, crocodiles almost invariably heat up during basking under the influence of the sun's radiant energy. The effect of body size on the thermal time constant (τ) for heating was measured with two 1000 W lamps providing a radiant heat source. Wind was also introduced as a variable, "no wind" being slow stirring of air by the overhead fan and windspeeds of 1.0, 2.5 and 4.5 m s^{-1} being provided by a wind tunnel.

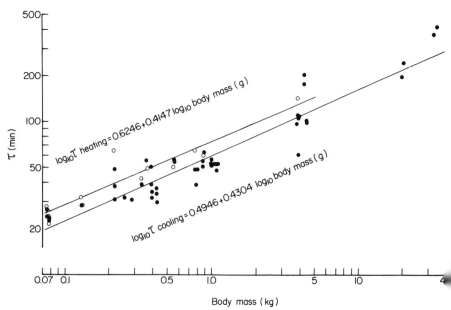

FIG. 11. The effect of body weight on the time constants for heating in the absence of a radiant source (open symbols) and cooling (solid symbols) of *C. niloticus*.

Results are given in the form of regression equations relating τ and body weight (Table II). The data were compared by analysis of covariance using an SPSS statistical package. The slopes of the regressions for windspeeds 1.0, 2.5 and 4.5 m s^{-1} were tested for homogeneity, and $F_{2,8}$ was 0.543 ($P > 0.25$). The intercepts for the three windspeeds were similarly tested, and $F_{2,8}$ was 0.948 ($P > 0.25$). Accordingly the data for all windspeeds between 1.0 and 4.5 m s^{-1} were combined to give the fifth regression equation in Table II. This was compared with the first regression equation; in a test for homogeneity of the slopes, $F_{1,19}$ was 0.071 ($P > 0.25$), but the intercepts were significantly different ($F_{1,19}$ = 34.647, $P > 0.005$). Thus, windspeeds between 1.0 and 4.5 m s^{-1} have no significant influence on the rate of heating of crocodiles under 4.0 kg body weight exposed to a radiant source. Crocodiles exposed to winds in the range 1.0–4.5 m s^{-1} have significantly lower time constants than those exposed to no wind; and wind or no wind does not significantly affect the slope of the relationship between the time constant and body weight.

TABLE II

The effect of body size and windspeed on the rate of heating of crocodiles < 4.0 kg body weight under two 1000 W lamps

Windspeed (m s^{-1})	N	Regression equation
"No wind"	9	$\log_{10} \tau$ (min) = 0.5796 + 0.3583 $\log_{10}$ weight (g)
1.0	5	$\log_{10} \tau$ (min) = 0.3897 + 0.3873 $\log_{10}$ weight (g)
2.5	4	$\log_{10} \tau$ (min) = 0.5153 + 0.3265 $\log_{10}$ weight (g)
4.5	5	$\log_{10} \tau$ (min) = 0.3096 + 0.4067 $\log_{10}$ weight (g)
1.0–4.5	14	$\log_{10} \tau$ (min) = 0.4061 + 0.3734 $\log_{10}$ weight (g)

The influence of substrate and insulation on heating

In these experiments a single crocodile was used over a short space of time, so keeping the body weight fairly constant. The thermal time constant for heating was similar when the substrates were dry sand or large pebbles, but τ increased by some 6 min when the sand was wet and by nearly 10 min when the substrate was wet *Salvinia* (Table III). Dry *Salvinia* substrate gave a similar thermal time constant to wet *Salvinia*. The most marked influence on τ was evident when wet *Salvinia* was piled on the back of the crocodiles, as is often seen in the wild (Fig. 3). In this case τ increased by 62–68% over the corresponding values for the same substrates without an insulation of *Salvinia*.

Heart and respiratory rates during heating

No consistent pattern of respiratory rate emerged during heating experiments. In most individuals the respiratory rate was high initially,

TABLE III
The effect of substrate type and insulation on the thermal time constants for a 1 kg crocodile exposed to two 1000 W bulbs and "no wind"

Body mass (kg)	τ (min)	Substrate	Insulation
1.038	44.8	Dry sand	–
1.091	43.0	Pebbles	–
0.990	51.0	Wet sand	–
0.987	54.7	Wet *Salvinia*	–
1.063	56.5	Dry *Salvinia*	–
1.091	72.5	Dry sand	*Salvinia* on back
0.993	86.0	Wet sand	*Salvinia* on back

becoming lower as T_b increased (Fig. 12). This pattern is attributed to handling stress at the start of the experiment. In other crocodiles the respiratory rate remained fairly constant or increased during heating (Fig. 13). In all the heating experiments in which heart rate was measured there was a steady increase in heart rate with the increase in T_b (Fig. 13).

Effect of body size on heating

Crocodiles of body weight in the range 220 g–314 kg were heated in air under two 1000 W lamps. On the graph (Fig. 14), crocodiles that were immobilized using gallamine are indicated. There does not seem to be any difference between immobilized and non-immobilized crocodiles in the thermal time constant, τ. One 4.25 kg animal had a τ of 75 min when immobilized and 81.5 min when not immobilized. The results of Grigg & Alchin (1976) for *Crocodylus johnstoni* heating in air are indicated on Fig. 14 (although not included in the regression analysis) as well as the regression line of Grigg *et al.* (1979) for lizard-shaped reptiles heating in air. Although the *C. niloticus* and *C. johnstoni* thermal time constants are similar, there is a large discrepancy between thermal time constants for crocodiles and those for other lizard-shaped reptiles (including *Alligator*), with the lizard-shaped reptiles heating much more rapidly than a crocodile of the same body weight.

The Significance of Gaping

It was found to be impossible to induce spontaneous gaping during laboratory experiments with crocodiles. Measurements were, however, made of the temperature below the tongue epithelium and of the rate of evaporative water loss from the epithelium of tongue and roof of mouth when the mouth was propped open. In three cases where the tongue

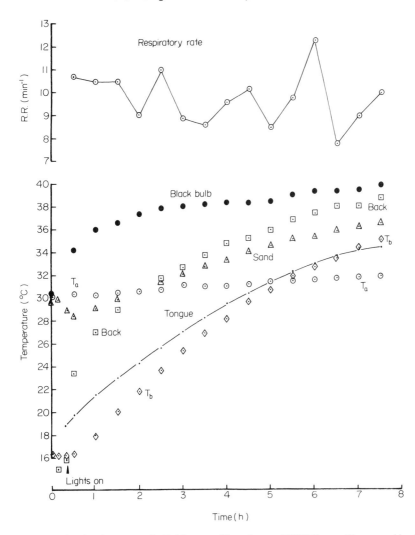

FIG. 12. The heating curve of a 19.9 kg crocodile under two 1000 W lamps. Tongue and back temperatures and the respiratory rate (R.R.) are also indicated.

temperature was measured during heating it was higher than T_b during the initial phase of heating (Fig. 12). As T_b increased, the difference between tongue temperature and T_b decreased until at a T_b of 34.3°C the two temperatures were equal and tongue temperature was lower than T_b thereafter (Fig. 12). In the other two experiments, where the crocodile was exposed to the sun's radiant heat, the points at which T_b was equal to tongue temperature were 35.8°C and 36.1°C. These data imply that when T_b increases to a certain level, evaporation of water

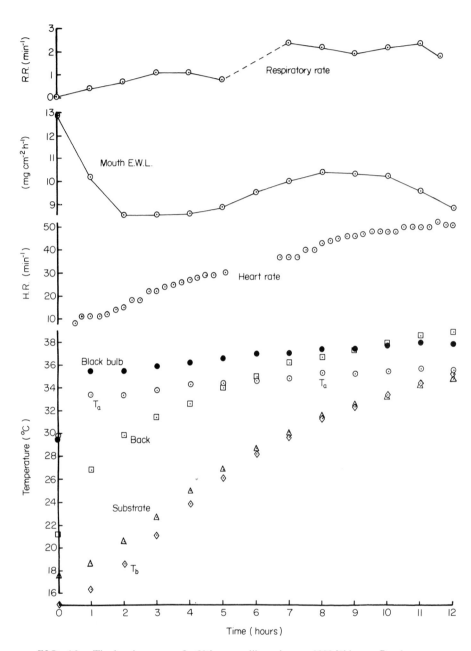

FIG. 13. The heating curve of a 60 kg crocodile under two 1000 W lamps. Respiratory rate (R.R.), evaporative water loss (E.W.L.) from the mouth and heart rate (H.R.) are also plotted as well as changes in environmental temperature variables.

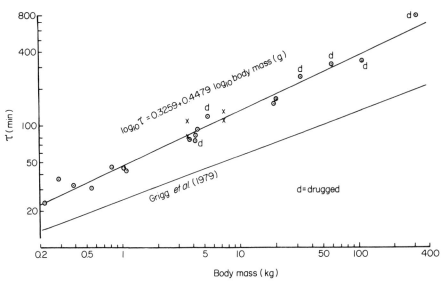

FIG. 14. The effect of crocodile bodyweight in the range 220 g to 314 kg on the time constant for heating under two 1000 W lamps. Time constants for heating *C. johnstoni* in air (Grigg & Alchin, 1976) are plotted (x) as well as the line for lizard-shaped reptiles (from Grigg *et al.*, 1979).

from the tongue is also increased, so cooling the lingual epithelium to a temperature below that of the body.

This hypothesis was not confirmed by data on the evaporative water losses from the oral epithelia. There was no trend for the water loss to increase as T_b rose during heating experiments (Fig. 13). Although water loss was usually fairly constant during an experiment, there was substantial variation between experiments. Water losses measured from tongue alone varied between 2.8 and 21.2 mg cm^{-2}h^{-1} whilst from tongue and roof of mouth variation was lower, in the range 1.4–9.6 mg cm^{-2}h^{-1} (Table IV). Tongue water loss certainly was much higher than tongue and roof of mouth combined in the 60 kg crocodile, the one case where comparable flow conditions were used.

DISCUSSION

Thermoregulatory Behaviour

Patterns of thermoregulatory behaviour described on pp. 446 and 452 serve to confirm the main conclusions of Cott (1961) in his study of adult *C. niloticus*. Cott (1961) did not, however, follow the behaviour of individual animals which is, for some purposes, more informative than

TABLE IV
Evaporative water losses from the oral epithelia of crocodiles measured with a ventilated capsule (error estimates are ± 1 SD)

Crocodile body weight (kg)	Evaporating surfaces	T_b (°C)	Flow rate (ml min^{-1})	Inflowing relative humidity (%)	Water loss (mg cm^{-2}h^{-1})
5.4	Tongue only	30.0	147 ± 0.4	5.8 ± 0.1	2.8 ± 0.1
5.4	Tongue only	16.6–36.4	148 ± 0.3	5.6 ± 0.1	4.1 ± 1.0
60	Tongue and roof of mouth	15.0–34.0	269 ± 1.1	3.6 ± 0.1	9.6 ± 1.2
60	Tongue only	30.0	269 ± 0.0	3.5 ± 0.0	21.2 ± 1.2
109	Tongue and roof of mouth	18.0–31.1	147 ± 2.9	7.0 ± 0.5	7.2 ± 1.1
109	Tongue only	32.1–34.6	269 ± 4.0	5.7 ± 0.3	18.9 ± 4.4
314	Tongue and roof of mouth	25.5–28.0	147 ± 0.8	5.0 ± 0.5	1.4 ± 0.1

the behaviour of large groups. On hot, clear days crocodiles emerge to bask once $T_a > T_w$ or the sunlight is on the basking site. By mid-day or early afternoon crocodiles frequently retreat to the water (Figs 2, 9) or to shade, if available (Fig. 2). This is followed by another period of basking in early to mid-afternoon (Fig. 9). This is the typical pattern of diurnal activity shown by Cott (1961) from counts of crocodiles on land, in the water or partly ashore. Basking behaviour is, however, modified by prevailing weather conditions so that on rainy days or during the coldest part of the year [Figs 8(A) & 10] emergence for basking may be delayed, with no return to the water in the middle of the day. Retreat to the water does not always occur by sunset (e.g. the larger crocodile in Fig. 9), and by no means all of the juveniles captured at Sinamwenda were found in the water, although nearly all adults were seen in the water during night capture exercises.

Body Temperatures

During the day, crocodiles emerge from the water to bask, thus behaving as typical heliotherms. This behaviour is anomalous in one respect, however, as crocodiles are not active during the day, but show peak activity during the early evening (Cloudsley-Thompson, 1964; Brown & Loveridge, 1981). Thus the preferred body temperature (PBT) as defined by Cowles (1962) would be that maintained during the active period of early evening, when the crocodiles are moving about and hunting. Body temperatures of basking crocodiles seem to show some relationship to body weight (Table I), with juveniles of 0.5–1.5 kg having mean body temperatures in the range 29.7–33.4°C, and larger crocodiles (above 19 kg) having lower body temperatures. The data for the larger crocodiles probably suffer from the fact that fewer individuals had achieved preferred (basking) body temperatures because of their slow heating rates. Biotelemetric studies [Figs 8(B) & 9] strongly suggest that crocodiles of 37–38.5 kg may achieve T_bs in the range 32–34°C during basking. Studies such as these should be undertaken on really large crocodiles to discover their body temperatures during basking. Cott (1961) gives some T_b data for adult *C. niloticus* shot in Uganda and Zambia. The mean T_b for those shot in the water was 24.7°C ($N = 4$) and on land was 26.8°C ($N = 3$). Studies on juvenile *Caiman crocodilus* (Diefenbach, 1975) indicated that the PBT in a thermal gradient was 29.9°C for 150–350 g animals, 32.8°C for 700–1050 g animals and 34.8°C for 2–6 kg animals. Lang (1981) showed similar ontogenetic changes in PBT in *Crocodylus novaeguineae* maintained in outdoor thermal gradients. Hatchlings up to two weeks old had T_bs of 33.4–33.9°C while those two to five weeks old had T_bs of

31.8–32.2°C. These levels declined so that two-year-old juveniles and adults had a T_b of about 30°C. Johnson, Webb et al. (1976) showed that the PBT of juvenile *Crocodylus johnstoni* was 31.3–32.5°C and of juvenile *C. porosus* was 32.2–33.1°C. Lang (1979a) demonstrated that fasted juvenile *Alligator mississippiensis* selected a T_b of 28.7°C (daytime) and 26.9°C (night-time); and fasted juvenile *Crocodylus acutus* selected a T_b of 27.9°C (daytime) and 28.0°C (night-time) in a thermal gradient. In both species there was a significant elevation in both daytime and night-time T_b due to thermophilic behaviour when the animals were fed.

In contrast to the large number of studies on crocodilian body temperatures during basking, there have been few studies of the T_b of crocodiles during the night, which is their active period. In this study, data were obtained for juvenile crocodiles in a wild population. Those caught on land had a T_b below that of the water at all seasons of the year (Fig. 5) whereas T_b generally exceeded T_a during the cold months but was lower than T_a during the hot months (Fig. 6). Crocodiles caught in the water, however, had a T_b equal to T_w at all seasons of the year. It appears, therefore, that in order to forage on the land at night, juvenile crocodiles must tolerate a lowering of T_b by about 2°C from T_w.

Preliminary indications from the biotelemetry studies are that in larger crocodiles, T_b may not be entirely influenced by T_w when the animals are in the water at night (Figs 8, 9 & 10). Even before basking started in the morning, $T_b > T_w$ by as much as 2°C. This may well be of great importance for larger crocodiles that are able, by physiological means, to prevent all the heat accumulated during the day from being lost to the water at night.

Heating and Cooling Rates

Work on *Alligator mississippiensis* (Smith, 1976a; Johnson, Voigt et al., 1978) has shown that juveniles heat very much more rapidly than they cool, but hatchlings (< 100 g) heat and cool at the same rate (Smith & Adams, 1978). Grigg & Alchin (1976) were unable to demonstrate a consistent difference in heating and cooling rates in juvenile *Crocodylus johnstoni*. In this study on *C. niloticus*, heating and cooling in air in the absence of radiant heat sources took place at the same rate (Fig. 11). Moreover, Nile crocodiles of 0.22–314 kg body weight took very much longer to heat than the equation of Grigg et al. (1979) predicts. The time constant for heating in a 300 kg *C. niloticus* is about 10 h (Fig. 14). This finding is supported by field observations, such as those of Cott (1961: 227) in which large crocodiles spend nearly all the daylight hours basking in one position without the need to return to the water.

The effect of different substrates and insulation on the time constant for heating can be significant (Table III). For example the time constant of 43 min for a 1 kg crocodile heating on pebbles is doubled when the crocodile is lying on wet sand with *Salvinia* on its back. Wind shortens the time constant for heating, although no difference is observed if the winds are increased from 1.0 to 4.5 m s^{-1} (Table II). All these environmental factors should be considered in the design of heating and cooling experiments or the interpretation of field observations.

The question of whether crocodiles and alligators have different heating and cooling strategies needs to be answered, and may provide a corollary to differences in thermal behaviour between the two groups suggested by Lang (1979b).

The Significance of Gaping

Cott (1961) was firmly of the opinion that gaping in *C. niloticus* exposed the moist oral mucosa to the air, so prompting evaporative cooling. He showed that as many as 60% of crocodiles ashore at 09.30 hours were gaping when the air temperature was as low as 26°C. Even at 06.30 hours when $T_a = 20°C$, 25% of crocodiles ashore were gaping and it hardly seems possible that these animals were heat-stressed. Observations during the present study indicate that gaping may occur immediately after crocodiles come ashore (Fig. 1), at night and while in the water, as well as during the heat of the day. Diefenbach (1975) showed that gaping and gular fluttering were insignificant cooling mechanisms in heat-stressed *Caiman crocodilus*, while Johnson, Voigt et al. (1978) showed that gaping had little effect on head temperatures in *Alligator mississippiensis*. Spotila et al. (1977), however, showed in *Alligator* that gaping was important in reducing the rate of heat gain in the head region and that the thermal time constants for heating in alligators with mouths open were much higher than those in which mouths were closed.

In experiments where tongue epithelium temperatures were measured during heating in crocodiles, tongue temperature exceeded T_b until T_bs of 34.3–36.1°C were reached. If, however, tongue temperature is compared with back temperature (Fig. 12), it is always about 5°C lower, indicating that substantial evaporative cooling of the tongue is occurring and that the rate of evaporation is roughly constant. This is supported by measurements of evaporative water losses from the oral epithelia which are in the range of 2.8–21.2 mg cm^{-2}h^{-1} and are much greater than the water loss from the general body surface of crocodiles of mean body weight 3.87 kg which was 0.114 mg cm^{-2}h^{-1} at 23°C, rising to 0.202 mg cm^{-2}h^{-1} at 35°C (Brown & Loveridge, 1981). It certainly seems that gaping allows evaporation of water from the oral mucosa. Whether this has any influence on head temperature or the rate of heating in *C. niloticus* is not known. It is possible that the site of the watery secretions from the tongue may be the salt glands recently described by Taplin & Grigg (1981) in *C. porosus* and known to occur in *C. niloticus* too. Gaping in crocodiles when there could be no advantage in evaporative cooling needs to be explained, perhaps in the social context.

ACKNOWLEDGEMENTS

The help of D. K. Blake, P. R. Dewhurst, T. Harris, S. L. Childs and D. Parry in the experiments and L. Madziwa in maintaining animals is gratefully acknowledged. I thank J. Hutton for permission to quote some of his unpublished results. Financial support came from the University of Zimbabwe Research Board.

REFERENCES

Brown, C. R. & Loveridge, J. P. (1981). The effect of temperature on oxygen consumption and evaporative water loss in *Crocodylus niloticus*. *Comp. Biochem. Physiol.* **69A**: 51–57.
Cloudsley-Thompson, J. L. (1964). Diurnal rhythm of activity in the Nile crocodile. *Anim. Behav.* **12**: 98–100.
Cloudsley-Thompson, J. L. (1969). Water relations of the young Nile crocodile. *Br. J. Herpet.* **4**: 107–112.
Colbert, E. H., Cowles, R. B. & Bogert, C. M. (1946). Temperature tolerances of the American alligator and their bearing on the habits, evolution and extinction of the dinosaurs. *Bull. Am. Mus. nat. Hist* **86**: 329–373.
Cott, H. B. (1961). Scientific results of an inquiry into the ecology and economic status of the Nile crocodile (*Crocodilus niloticus*) in Uganda and Northern Rhodesia. *Trans. zool. Soc. Lond.* **29**: 211–337.
Cowles, R. B. (1962). Semantics in biothermal studies. *Science, N.Y.* **135**: 670.
Diefenbach, C. O. da C. (1975). Thermal preferences and thermoregulation in *Caiman crocodilus*. *Copeia* **1975**: 530–540.
Gans, C. (1976). Questions in crocodilian physiology. *Zoologica Afr.* **11**: 241–248.
Grigg, G. C. & Alchin, J. (1976). The role of the cardiovascular system in thermoregulation of *Crocodylus johnstoni*. *Physiol. Zool.* **49**: 24–36.
Grigg, G. C., Drane, C. R. & Courtice, G. P. (1979). Time constants of heating and cooling in the eastern water dragon, *Physignathus lesueurii* and some generalizations about heating and cooling in reptiles. *J. therm. Biol.* **4**: 95–103.
Johnson, C. R. (1974). Thermoregulation in crocodilians – I. Head-body temperature

control in the Papuan-New Guinean crocodiles, *Crocodylus novaeguineae* and *Crocodylus porosus*. *Comp. Biochem. Physiol.* **49A**: 3–28.

Johnson, C. R., Voigt, W. G. & Smith, E. N. (1978). Thermoregulation in crocodilians – III. Thermal preferenda, voluntary maxima, and heating and cooling rates in the American alligator, *Alligator mississippiensis*. *Zool. J. Linn. Soc.* **62**: 179–188.

Johnson, C. R., Webb, G. J. W. & Tanner, C. (1976). Thermoregulation in crocodilians – II. A telemetric study of body temperature in the Australian crocodiles, *Crocodylus johnstoni* and *Crocodylus porosus*. *Comp. Biochem. Physiol.* **53A**: 143–146.

Lang, J. W. (1979a). Thermophilic response of the Amerian alligator and the American crocodile to feeding. *Copeia* **1979**: 48–59.

Lang, J. W. (1979b). Crocodilian thermal behaviors: alligators vs. crocodiles. *Am. Zool.* **19**: 975.

Lang, J. W. (1981). Thermal preferences of hatchling New Guinea crocodiles: effects of feeding and ontogeny. *J. therm. Biol.* **6**: 73–78.

Loveridge, J. P. & Blake, D. K. (1972). Techniques in the immobilisation and handling of the Nile crocodile, *Crocodylus niloticus*. *Arnoldia (Rhodesia)* **5**(40): 1–14.

Loveridge, J. P. & Crayé, G. (1979). Cocoon formation in two species of southern African frogs. *S. Afr. J. Sci.* **75**: 18–20.

Smith, E. N. (1975). Thermoregulation of the American alligator, *Alligator mississippiensis*. *Physiol. Zool.* **48**: 177–194.

Smith, E. N. (1976a). Heating and cooling rates of the American alligator, *Alligator mississippiensis*. *Physiol. Zool.* **49**: 37–48.

Smith, E. N. (1976b). Cutaneous heat flow during heating and cooling in *Alligator mississippiensis*. *Am. J. Phsyiol.* **230**: 1205–1210.

Smith, E. N. (1979). Behavioral and physiological thermoregulation of crocodilians. *Am. Zool.* **19**: 239–247.

Smith, E. N. & Adams, S. R. (1978). Thermoregulation of small American alligators. *Herpetologica* **34**: 406–408.

Spotila, J. R. (1974). Behavioural thermoregulation of the American alligator. In *Thermal ecology*: 322–334. Gibbons, J. W. & Sharitz, R. R. (Eds). Springfield, Virginia: U.S. Atomic Energy Commission.

Spotila, J. R., Terpin, K. M. & Dodson, P. (1977). Mouth gaping as an effective thermoregulatory device in alligators. *Nature, Lond.* **265**: 235–236.

Taplin, L. E. & Grigg, G. C. (1981). Salt glands in the tongue of the estuarine crocodile *Crocodylus porosus*. *Science, Wash.* **212**: 1045–1047.

Snake Venoms

FINDLAY E. RUSSELL

Department of Pharmacology and Toxicology, College of Pharmacy, University of Arizona, Tucson, Arizona, USA

SYNOPSIS

Snake venoms are complex mixtures, chiefly proteins, many of which have enzymatic properties. Some of the more lethal proteins of snake venoms are peptides, while other components are metalloproteins, glycoproteins, lipids, biogenic amines, free amino acids and inorganic substances, such as sodium, calcium, potassium, magnesium, zinc and other metals. The primary function of a snake venom is to immobilize and, in most cases, kill the prey. A second function relates to the digestion of the prey, and a third property implicates its use in a defensive posture against predators. The discharge of a snake venom as a 'marking' or 'tagging' agent for the location of food following envenomation seems questionable.

In studying venom it is common to use techniques that fractionate the poison into individual parts. Thus, a destructive process is used to study what evolved as a constructive process. The consequence is that the function of the whole venom is sometimes lost sight of in the preoccupation with the individual parts. This has led to labeling venoms (or even venom components) with such misleading terms as 'neurotoxins', 'cardiotoxins', 'hemotoxins', 'myotoxins' and the like, obscuring the evolution of the important synergism, metabolite and autopharmacological phenomena of the whole venom.

The function of a snake venom is closely interrelated to the evolution of the venom apparatus. It is difficult to consider these as separate phenomena. The present report treats of certain chemical and pharmacological properties of snake venoms and the relationships between these activities and the development of the venom apparatus.

INTRODUCTION

Snake venoms are considered to be the most complex poisons known to man. They are mixtures, chiefly proteins, many of which have enzymatic activities. Some of the more lethal proteins of snake venoms are peptides, while other components are metalloproteins, glycoproteins, lipids, biogenic amines, free amino acids and inorganic substances such as sodium, calcium, potassium, magnesium and zinc. The primary function of a snake venom is to immobilize and/or kill the prey. A second function is to digest the prey, or at least to make it more digestible. A further property rests in the venom's defensive posture, that is its use as a deterrent against predators, where it may function in a kill or underkill. There seems to be sufficient evidence that predators can 'learn' through experience or observation. Additional properties have been suggested for snake venoms, such as their possible role as 'tagging' agents, which would make it easier for the snake to locate its prey, or as 'marking' agents, which individualize the prey. However, these particular functions have not yet been adequately demonstrated. Suffice to say, snake venoms are highly successful compounds, if one measures success as how effectively a substance fulfils its specific functions. In studying the properties of these poisons over many years I never cease to be amazed at how well the snake venom components perform biological roles.

If success in chemical evolution is measured by the amount of energy conserved in survival, which obviously implies the getting of food, its assimilation and other factors, as well as defence, then snake venoms would appear to be the most successful adaptation among the reptiles. The venoms appear to have more survival value than adaptations in concealment, in speed, better eyesight, modifications of appetite or constriction, or other properties enjoyed by some non-venomous snakes. In the complexity of snake venoms rests an insight into their evolution. It is not possible to separate this evolution from that of the evolution of the venom apparatus. The relationship between the chemistry and pharmacology of the venom, on the one hand, and the structure of the venom apparatus – the venom glands, venom ducts and fangs – on the other, is a well-integrated process. It is difficult to consider one without the other. It is not likely that function could have preceded structure any more than structure could have preceded function. Whenever I thought I had discovered a deficit in the functional composition of snake venom, Angus Bellairs, Elazar Kochva, Sherman Minton, Charles Bogert, Carl Gans or the late Laurence Klauber reminded me that significant changes in food, environment, prey-predator relationships or immune mechanisms could well have been responsible for the modification. I agree.

DISCUSSION

In speaking of evolution I do not intend to treat it in its definitive biochemical, pharmacological, or structural terms. It certainly appears that certain toxins of snake venoms evolved from ancestral amino acids and that sequencing demonstrates nodes of the phylogenetic tree. Some biochemists hold that the most ancient form of non-enzymatic polypeptide is 'cardiotoxin' and that the two sizes of 'neurotoxin' evolved from it (Strydom, 1972, 1974). However, more recent work has shown that there are other chain-size polypeptides and these may make the initial scheme rather difficult. A further burden is classifying the various toxins as 'cardiotoxins' or 'neurotoxins', when definitive pharmacological studies indicate that they can have both neurotoxic and cardiotoxic activities, as well as other toxicological effects. Once again, this bears out the shortcoming of the present system of classifying venom activities and deters a more reasonable approach to their chemical evolution. There is no doubt, however, that on the simple basis of the differences in the sequence of amino acids in a protein, such as that presented by Fitch & Margoliash (1967), a comparison matrix (Needleman & Wunsch, 1970), and the diagram method proposed by Gibbs & McIntyre (1970), it is possible to program a system of venom fraction evolution from a common ancestral gene.

With respect to the venom glands of snakes, Gans (in Russell, Gans & Minton, 1978) suggests that they were probably derived from the buccal glands or enteric secretory tissues. The addition of enzymes to the secretions is known in Amphibia. He expresses the opinion that the first functions of such enzymes may have served as conditioning substances which could dissolve prey-derived materials, such as molluscan slime or the arthropod hemolymph that might adhere to the teeth or buccal surfaces of the snake. The type of venom often seems particularly suited both for the type of prey and for a particular age. Some snakes have venom that is specifically effective against snails, centipedes, other invertebrates and even lower vertebrates (Russell *et al.*, 1978). One might add that changes in food patterns which surely occurred over eons of time might have been associated with changes in the venom components, providing either a direct fraction-activity change or a synergism to modify the new or added function of the venom. One cannot help considering what role the development of immunity mechanisms in prey or predators might have had in modifying the venom.

Perhaps one of the most important evolutions in snake venoms has been in the functional synergisms between the various fractions. This property seems to have been almost entirely overlooked by chemists and pharmacologists alike. It is one of the unfortunate facts in the study of the biological activities of venoms that structure and design are most easily investigated by taking a venom apart. This has two shortcomings: it means that a destructive process must be substituted for a constructive, progressive, and integrative one; secondly, the essential quality of the whole venom may be destroyed before one has made a suitable acquaintance with it. Often the process of examination becomes so exacting that the end is lost sight of in the preoccupation with the means, so much so that in some cases the means becomes substituted for the end. The failure to recognize the synergisms between snake venom components has led to timely errors in experimental deductions, as well as therapeutic misunderstandings (Russell, 1980a).

There are yet other important biological properties of venoms frequently overlooked in the attempt to study the activities of individual components. One of the most important of these is the probability of the formation of intermediary metabolites, the rate at which they are formed, and the rate of their destruction and elimination. Unfortunately, we know little about such metabolites and even more discouraging is the fact that few investigators have pursued this subject. A further deficit in our knowledge concerns the mechanisms involved in autopharmacological reactions. Such substances as histamine, bradykinin, serotonin and perhaps others may play an important role in an organism's response to envenomation. On the other hand, overemphasis has sometimes been given to the data from these responses, which are usually based on specific single cell or single tissue preparations. The extrapolation of data from these highly sensitive preparations to a complex integrated system such as the intact mammal is fraught with danger; at least it should be seasoned with considerable forethought. As Rosenberg (1979) has pointed out: 'There is no way to extrapolate from *in vitro* assay systems to the *in vivo* biologic situation where ionic conditions, availability of substrate, etc., cannot be controlled'.

The immobilizing or lethal component of a snake venom has been the object of much research. It is not surprising to find that this activity has been attributed to a number of different fractions during the passing of time, the attribution more often than not dependent upon improvements in biochemical and pharmacological gadgeteering. Before 1930 the more toxic parts of snake venoms were little more than groups of proteins, lacking specificity, except for their exotic names. By the late 1940s, however, some semblance in the order of functions became apparent, as demonstrated by Zeller (1948). However at that time it was commonly held that the lethal and more deleterious properties of a snake venom could be attributed to one enzyme or another (Russell, 1980a). There were few studies which indicated that the more toxic properties of the snake venoms might be due to some non-enzymatic protein component. A notable exception, however, was the study of Slotta & Frankel-Conrat (1938), who isolated a crystalline protein from the venom of the tropical rattlesnake *Crotalus durissus terrificus*. This protein had a non-enzymatic protein portion but it was also found to contain hyaluronidase, phospholipase and possibly several other enzymes. This mixture, somewhat purer than previous isolations, was called 'crotoxin'. By removing the phospholipase A, a more toxic protein was found and named 'crotactin', while another component was named 'crotamine'. The latter was separated into several additional fractions (Gonçalves, 1956).

In 1965 the first amino acid composition of a snake venom was published (Yang, 1965) and the following year at the First Symposium of the International Society on Toxinology, Tamiya presented a paper on the chromatography, crystallization, electrophoresis, ultracentrifugation and amino acid composition of the venom of the sea snake *Laticauda semifasciata*. Almost all of the lethal activity was recovered as two toxins, erabutoxin *a* and *b*, using carboxymethylcellulose chromatography. Thirty per cent of the proteins of the crude venom were erabutoxins. By ultracentrifugation the molecular weights were found to be approximately 7430 for each toxin, with 61 amino acid residues (Tamiya, Arai & Sato, 1967). Additional polypeptides from other venoms were reported by Su and colleagues at the same meeting. These polypeptides were found to be seven times more lethal than the crude venom (Su, Chang & Lee, 1967). These and other studies on the polypeptides have been reviewed elsewhere (Elliott, 1978; Lee, 1979; Russell, 1980a).

In some snake venoms these polypeptides have a lethal index from two to ten times greater than the crude venom, while in other venoms their index is less, or unknown. It appears that the lethal property of a snake venom is in some fashion related to the nature and amount of polypeptide. This is not to say, however, that the venom enzymes do not contribute to the overall lethal property of the venom, nor that synergisms between the two groups do not exist. Further, some snake venom enzymes can be lethal (Rosenberg, 1979). Their lethal index varies considerably, and again it may be dependent upon synergisms or other factors. As gadgeteering advanced through the years one enzyme after another was implicated as a lethal unit. Earlier workers implicated the proteases, while currently the object of most indictments is phospholipase A. In Table I is a listing of the important enzymes of snake venoms. Reviews of the chemistry of snake venoms will be found in Zeller (1948, 1951), Russell (1967), Elliott (1978), Iwanga & Suzuki (1979), Russell (1980a), and Habermehl (1981).

The venoms of snakes contain at least 26 enzymes, although no single venom contains all of these. At least 10 enzymes are found in almost all snake venoms, while the remainder are scattered throughout the venoms of the five families of snakes. As would be expected, these latter enzymes tend to be characteristic to certain families or even genera. Certain esterases appear identical in both African and Asian cobras, the king cobra, two species of kraits, and the coral snake of North America, all of which, of course, belong to the family Elapidae. Elapid venoms are

TABLE I
Enzymes of snake venoms[a]

Proteolytic enzymes	Phosphomonoesterase
Arginine ester hydrolase	Phosphodiesterase
Thrombin-like enzyme	Acetylcholinesterase
Collagenase	RNase
Hyaluronidase	DNase
Phospholipase A_2 (A)	5'-Nucleotidase
Phospholipase B	NAD-Nucleotidase
Phospholipase C	L-Amino acid oxidase
Lactate dehydrogenase	

[a] See Russell (1980a) for discussion.

rich in acetylcholinesterase, while crotalid and viperid venoms lack this enzyme but are rich in endopeptidase.

Arginine ester hydrolase is found in many crotalid and viperid venoms, but is lacking in elapid and hydrophid venoms, with one or two exceptions. Crotalid and viperid venoms have significant amounts of thrombin-like enzyme, while elapid and sea snake venoms contain little or none. Collagenase is found in many crotalid and viperid venoms but is lacking in elapid venoms. Hyaluronidase and the phospholipases are found throughout many venoms, regardless of genus, while phosphomonoesterase is found in all families but the hydrophids and colubrids. Phosphodiesterase and 5'-nucleotidase have been identified in the secretions of all five families of snakes. RNase and DNase have a scattered distribution, while NAD Nucleotidase is found in *Agkistrodon* but not in *Crotalus*, and in *Bungarus* but not in *Naja*. L-Amino acid oxidase has been demonstrated in all snake venoms so far studied. It is responsible for the yellow colour of the venom.

Snake venoms probably have an effect on almost every cell, membrane and organ in the body and it is likely that individual fractions have several or even many tissue sites of action. The action of a venom fraction at a specific receptor site is dependent upon the manner in which the fraction affects specific tissue ions at that site, and while it is not yet generally conceded that these alterations in ion transfer are similar at various tissue sites – nerve, heart, blood, etc. – recent studies have shown that the changes produced on a muscle fiber may involve the same basic ion changes as produced on a nerve, or at a presynaptic junction or elsewhere. The various membranes of the body are remarkably similar in their chemical structure and spatial arrangement, regardless of their location. The so-called 'neurotoxins' have been shown to have hemotoxic properties and *vice versa*; and 'myotoxins' have been shown to have neurotoxic or cardiovascular actions, and *vice versa*. It is obvious that for the most part these misleading terms reflect little more than observations on the specific preparation on which they were studied.

Although the time has not yet come for discarding these words (only because they are used so extensively in the literature and their use forms a means of communication), their use in describing a venom or venom fraction remains questionably valid. More reasonable terminology might employ such words as neurotoxic, hemotoxic or cardiotoxic. These do not exclude identification with other tissue-site activities. Hopefully, the time will come when venom fractions will be labeled on the basis of their chemical composition, with their pharmacological and immunological characteristics to qualify. An unidentified protein might be labeled on the basis of the technique employed, such as: *Crotalus viridis helleri* Peptide Ic, 4490 (molecular weight), 43 (amino acid residues), etc. (Maeda, Tamiya, Pattabhiraman & Russell, 1978). Additional data could then be added as new data become available. This would abolish such meaningless terms as Mojave toxin, rattletoxin, Gesundheit toxin, and the like.

Another difficulty with these confusing labels is their effect on the clinician. Having treated over 750 cases of snake venom poisoning I am continually worried by the naive concept often put forth by physicians that such and such a venom is a 'neurotoxin' or a 'hemotoxin', etc. and that all therapeutic measures must be directed toward this calling. I recall being told that a certain cobra venom was a 'neurotoxin' and that alterations in circulating blood volume and electrolyte balance were not important, because the venom 'attacks the nervous system'. The neurological deficit was easily overcome with assisted ventilation and oxygen, and although the cardiovascular deficit was not so quickly remedied, it did respond to appropriate cardiac drugs, as did the patient's general condition. Some years ago I summarized this problem as follows:

"The clinician must never slight any symptom or sign his patient presents, or minimize any manifestation, on the naive assumption that the venom has to be either a 'neurotoxin,' 'cardiotoxin,' 'hemotoxin,' or 'myotoxin,' and its activity limited to one organ or system. While the patient may have respiratory distress from a 'neurotoxic venom' he can also have changes in cardiac dynamics or vascular permeability and these can become far more life-threatening situations, particularly if the physician centers his attention and therapy on the so-called neurotoxic activity of the venom (an effect that can often be adequately treated by simple positive pressure respiration). The physician must guard his knowledge and experience zealously and be aware of the limits of application of pharmacologic data based on animal experimentation. On the other hand, he must explore, carefully, the pharmacologic literature on venoms for those data that give him a greater knowledge of the mechanisms involved in venom poisoning and, hopefully, provide him with better methods of therapy" (Russell, 1963, 1980a).

Reid (1964), among others (see Russell, 1980a), also notes his concern. In studying the action of a venom or venom component, certain physiopharmacological factors must be considered. For instance, in determining the lethal median dose (LD_{50}) of a substance it is well known that the effective or lethal dose in one kind of animal will not be quantitatively equal to that in another kind of animal. For instance, the LD_{50} in the rat for the pesticide alphanaphylthiourea is 2.5 mg kg^{-1}, while in the guinea-pig it is 350 mg kg^{-1}, and in the chicken it is 2550 mg kg^{-1}. Some biologists have mistakenly assumed that by merely multiplying the mouse LD_{50} by X, one can arrive at the LD_{50} for the hippopotamus or the fruit fly. Such extrapolations attempt to prove that certain animals are 100 times more 'immune' to a certain snake venom than the mouse. Based on the mouse LD_{50}, and only the size of the animal, plus a dash of rueful conclusion-jumping, one can place the jack rabbit of eastern Arizona and the tarantula high among the animals 'immune' to *Crotalus scutulatus* venom, while the kangaroo rat and the wasp would be worse off than the laboratory mouse. Even field mice have a higher LD_{50} than laboratory mice for this venom. I trust that no one will be encouraged by these examples to look for an 'immune' elixir in tarantula hemolymph, or to cross the tarantula with the rabbit for the production of such an antidote (Russell, 1980b).

These variations in responses to snake venoms, as well as other substances, can be more probably explained by differences in bioavailability, membrane transport, site accumulation, absorption, metabolism and excretion of the substance in question. With respect to bioavailability, the physiochemical properties of the venom or its fraction, its pH, vehicle, particle size and concentration have different values in different kinds of animals. Also, it makes a considerable difference how or where the venom is injected. Even when given intravenously there is a difference in the bioavailability of a venom in various animals.

Once available, a substance may be transported across membranes by one or several ways, which differ from one animal to another. These ways include passive diffusion, facilitated diffusion, active transport, and pinocytosis. Studies to date indicate that the first two mechanisms are those usually involved in venom or venom component transport. When the toxin enters the circulation it is distributed to and partitioned in various tissues or organs. In the case of most venom fractions this distribution is rather unequal, being affected by protein binding, pH, membrane permeability and other factors. When the toxin reaches a particular tissue its entry is dependent upon the rate of blood flow into that tissue, the mass of tissue, and the partition relationships between the toxin and the tissue. On arrival at a tissue or receptor site the characteristic of the toxin, its amount and its rate of metabolism determine its pharmacological properties. A toxin produces its pharmacological effect when the quantity of the toxin attains a critical minimum level at a receptor, and, as previously noted, there may be several or many receptor sites. The differences in response between man and other animals do not appear to be due to an increased sensitivity on the part of man's target organs but are probably directly related to the rate of metabolism of the toxin.

A venom can be metabolized in several or many different tissues. In some cases, it may be more important in experimental work to evaluate to what extent each tissue contributed to the venom's metabolism, as demonstrated by the level of the toxin or its metabolites in the plasma or urine, than to rely solely on the dose given. The amount of a toxin which the tissues of various species of animals can metabolize without endangering the organism varies considerably. To demonstrate a specific tissue's ability to metabolize a venom, tissue slices or hemogenates and subcellular fractions of different tissues might be studied. However, in evaluating such data it must be remembered that organs usually consist of several different kinds of tissues, each of which may contain enzymes which catalyse different reactions. Enzymes which oxidize venoms by oxygenase mechanisms are, for the most part, localized in the parenchymal cells of the liver, while other enzymes may be found somewhat unevenly distributed in many tissues. Thus, species differences may play an important role in determining the metabolism of a toxin, on the simple basis that the relative population of a particular cell type in an organ or tissue varies from animal to animal.

The major organ for excretion of snake venoms is the kidney. The intestines play a minor role and the contribution of the lungs and biliary system has not yet been demonstrated. The excretion rate of some snake venoms is complicated by several factors affecting the kidneys, one of the most important of which is the direct effect of the venom on blood cells, and the resulting obstruction of the tubules. In addition, some venom may have a direct effect on the kidneys.

There are a number of loose ends relating to the problem of the functional evolution of snake venoms as they interlink structural evolution. What does one do with the 'venomous' colubrids, that is with respect to the composition of their saliva? The saliva of many colubrids has now been shown to be toxic (Minton, 1976). The primary morphological indicator is probably Duvernoy's gland and the secondary indicator is enlarged maxillary teeth, either grooved or solid. The saliva of *Rhabdophis* species, for instance, has a higher lethal index than the venom of many rattlesnakes and its bite may be sufficiently severe to cause serious bleeding defects, or even death (Mittleman & Goris,

1974). The toxicity of *Dispholidus typus* and *Thelotornis kirtlandi* are certainly well established but the status of toxicity for the secretions of most other colubrids is unknown. With respect to *Elaps* and *Atractaspis*, too little is known about their secretions from a biochemical and pharmacological viewpoint to be able to speculate.

Another area of concern is that of the relationship between structure and function, with respect to the question of how much venom a snake delivers during a biting act. Our studies in rattlesnakes would seem to indicate that in securing food this snake is rather consistent in the amount of venom it injects. It appears that the rattlesnake delivers a larger amount of venom when it preys upon a rat than upon a mouse. Preliminary studies indicate that the amount of venom ejected seems proportional to the surface area of the prey, at least as far as mice and rats are concerned. It is suggested that the dosage is determined by visual and pit receptor cues, and probably to a lesser extent by chemical and proprioceptive ones, and that these cues are interpreted and affect the amount of venom injected. However, Allon & Kochva (1974) suggest that in *Vipera palestinae* the amount of venom injected during a biting act or during successive bites is not necessarily influenced by the size of the prey.

It would be interesting to speculate on these differences. They may be more closely related to differences in techniques used by the different investigators than to any other factor, although differences between various families of snakes could be an important element (Morrison, Pearn & Coulter, 1982). It would seem that if the evolution of function and its correlation with structure were so highly developed, then to leave such an important factor as the amount of venom to be injected to chance would be inconsistent. Perhaps this impression has influenced the experiments. The convincing experimental model has not yet been developed but, hopefully, it is not far off.

In the case of a defensive strike or a strike in which the snake is not attempting to secure food there is no question that the amount of venom ejected may be considerable, perhaps nearly all of the venom gland contents. On the other hand, the massive clinical literature testifies to the fact that a venomous snake may inject very little venom, or even none at all. An individual snake may bite several persons and inject different amounts of venom into each and there does not seem to be any way in which one can predict the amount it injects in a defensive strike (Russell, 1978).

CONCLUSIONS

In this short review an attempt has been made to summarize some of the relationships between the venom glands of snakes and the functions of their secretions. The complexity of the venom indicates the diversity in both offensive and defensive statures and the probable evolution of various feeding patterns, and, perhaps, immunity responses. Amino acid sequencing should give insight into the evolution of the different toxic components from a common ancestral gene(s). The various fractions of snake venoms probably exert their biological effects on many if not most tissues of the body, the important consideration being how much of a specific component accumulates at an activity site and its rate of metabolism at that point. The important role of synergisms in these venoms is little understood, and an area of study badly neglected. Finally, the evolution of structure – the venom glands, ducts and fangs – is so carefully integrated with the evolution of function – the composition of the venom – that one could not have preceded the other.

REFERENCES

Allon, N. & Kochva, E. (1974). The quantities of venom injected into prey of different size by *Vipera palestinae* in a single bite. *J. exp. Zool.* **188**: 71–75.
Elliott, W. B. (1978). Chemistry and immunology of reptilian venoms. In *Biology of the Reptilia*, **8**: 163–436. Gans, C. (Ed.). London and New York: Academic Press.
Fitch, W. M. & Margoliash, E. (1967). Construction of phylogenetic trees. *Science, Wash.* **155**: 279–284.
Gibbs, A. J. & McIntyre, J. A. (1970). The diagram, a method comparing sequences. Its use with amino acid and nucleotide sequences. *Eur. J. Biochem.* **16**: 1–11.
Gonçalves, J. M. (1956). Purification and properties of crotamine. In *Venoms*: 261–274. Buckley, E. & Porges, N. (Eds). Washington: American Association for the Advancement of Science.
Habermehl, G. G. (1981). *Venomous animals and their toxins*. Berlin: Springer-Verlag.
Iwanga, S. & Suzuki, T. (1979). Enzymes in snake venoms. *Handb. exp. Pharmak.* **52**: 61–158.
Lee, C.-Y. (Ed.). (1979). Snake venoms. *Handb. exp. Pharmak.* **52**: 1–1130.
Maeda, N., Tamiya, N., Pattabhiraman, T. R. & Russell, F. E. (1978). Some chemical properties of the venom of the rattlesnake *Crotalus viridis helleri*. *Toxicon* **16**: 431–441.
Minton Jr., S. A. (1976). A list of colubrid envenomations. *Kentucky Herp.* **7**: 4.
Mittleman, M. B. & Goris, R. C. (1974). Envenomation from the bite of the Japanese colubrid snake *Rhabdophis tigrinis* (Boie). *Herpetologica* **30**: 113–119.
Morrison, J. J., Pearn, J. H. & Coulter, A. R. (1982). The mass of venom injected by two Elapidae: the taipan (*Oxyuranus scutellatus*) and the Australian tiger snake (*Notechis scutatus*). *Toxicon* **20**: 739–745.
Needleman, S. H. & Wunsch, C. D. (1970). A general method applicable to the search for similarities in the amino acid sequence of two proteins. *J. molec. Biol.* **48**: 443–453.

Reid, H. A. (1974). Cobra-bites. *Br. med. J.* **2**: 540–545.
Rosenberg, P. (1979). Pharmacology of phospholipase A_2 from snake venoms. *Handb. exp. Pharmak.* **52**: 403–447.
Russell, F. E. (1963). Venomous animals and their toxins. *Times Sci. Rev.* **49**: 10–11.
Russell, F. E. (1967). Pharmacology of animal venoms. *Clin. Pharmacol. Therap.* **8**: 849–873.
Russell, F. E. (1978). Consecutive bites on three persons by a single rattlesnake. *Toxicon* **16**: 79–80.
Russell, F. E. (1980a). *Snake venom poisoning*. Philadelphia: J. B. Lippincott. (Reprinted 1983 Great Neck, New York: Scholium International.)
Russell, F. E. (1980b). Pharmacology of venoms. In *Natural toxins*: 13–21. Eaker, D. & Wadström, T. (Eds). Oxford: Pergamon Press.
Russell, F. E., Gans, C. & Minton Jr, S. A. (1978). Poisonous snakes. *Clin. Med.* **85**: (1), 13–22; (2), 13–30.
Slotta, K. & Fraenkel-Conrat, H. (1938). Two active proteins from rattlesnake venom. *Nature, Lond.* **142**: 213.
Strydom, D. J. (1972). Phylogenic relationships of proteroglyphae toxins. *Toxicon* **10**: 39–45.
Strydom, D. J. (1974). Snake venom toxins. The evolution of some of the toxins fround in snake venoms. *Syst. Zool.* **22**: 596–608.
Su, C., Chang, C. & Lee, C.-Y. (1967). Pharmacological properties of the neurotoxin of cobra venom. In *Animal toxins*: 259–267. Russell, F. E. & Saunders, P. R. (Eds). Oxford: Pergamon Press.
Tamiya, N., Arai, H. & Sato, S. (1967). Studies on sea snake venoms: crystallization of "erabutoxins" a and b from *Laticauda semifasciata* venom, and of "laticotoxin" a from *Laticauda laticauda* venom. In *Animal toxins*: 249–258. Russell, F. E. & Saunders, P. R. (Eds). Oxford: Pergamon Press.
Yang, C. C. (1965). Crystallization and properties of cobrotoxin and their relationship to lethality. *Biochim. Biophys. Acta* **133**: 346–355.
Zeller, E. A. (1948). Enzymes of snake venoms and their biological significance. In *Advances in Enzymology*, **8**: 459–495. Nord, F. (Ed.). New York: Interscience.
Zeller, E. A. (1951). Enzymes as essential components of bacterial and animal toxins. In *The enzymes*, **1**: 986–1013. Sumner, J. F. & Myback, K. (Eds). New York: Academic Press.

Evolution

Scleral Ossicles of Lizards: An Exercise in Character Analysis

GARTH UNDERWOOD

Department of Biological Sciences, City of London Polytechnic, Old Castle Street, London E1 7NT, England

SYNOPSIS

The scleral ossicles of lizards are of three kinds: overlapping both adjacent ossicles (plus), overlapped by adjacent ossicles (minus) and imbricating. The commonest pattern has 14 ossicles with a ventral plus, two imbricating, a temporal minus, one imbricating, a dorsal plus-minus-plus, one imbricating, a nasal minus and four imbricating. With variations in the imbricating ossicles, this pattern is recognizable in birds, *Sphenodon* and 11 families of lizards. In three families of lizards are found patterns with only a single dorsal plus ossicle; some arbitrary decisions are required to make out correspondences with the standard pattern in the dorsal half of the ring. A system of scoring the ossicles is described and the scores are coded as 11 binary characters. Phylogenetic relationships between these patterns are worked out; two logical alternatives appear and are resolved by preferring the one which gives the better internal consistency with the whole data-set. Taking either birds or *Sphenodon* as the outgroup the *Varanus* pattern appears primitive to those of other lizards. The xantusiid pattern is directly derivable from the varanid, independently of other lizards. The agamid pattern is nearer to some skink patterns than to the iguanid patterns. Intraspecific variation in sceloporine iguanids spans as many as five steps of the dendrogram; however, the modal scores for the species concerned accord with generic assignments of *Callisaurus* and *Holbrookia* but divide *Uma*. Parallel evolution of six of the 11 characters is implied by the dendrogram. The limitations of an algorithmic approach to phylogenetic analysis are discussed. It is suggested that the most that we can hope to achieve is the best measure of internal consistency on the basis of a combination of numerical and biological considerations.

INTRODUCTION AND SOURCES OF DATA

In diurnal lizards the orbital portion of the eye is approximately hemispherical. The cornea on the other hand has a radius of curvature much less than that of the orbital hemisphere. There is, therefore, between the cornea and the orbital hemisphere an annular depression (sulcus) which is maintained against the intra-ocular fluid pressure [Fig. 1(B)]. The orbital hemisphere of the sclera contains cartilage which extends to the inflexion where it meets the sulcus. Embedded in the sclera between the orbital hemisphere and the cornea is a ring of small plates of bone, the scleral ossicles (Walls, 1942) [Fig. 1(A)]. Their function appears to be to maintain the annular depression against the intra-ocular pressure. The muscles of accommodation also originate in this area. The corneal–air interface is the principal refractive surface of the eye.

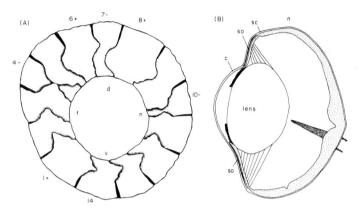

FIG. 1. (A) Scleral ring of right eye of a lizard (*Anolis lineatopus*), plus and minus ossicles indicated, numbered as in Gugg (1939). (B) Horizontal section of eye of a lizard. c, cornea; d, dorsal; n, nasal; sc, sclera, containing cartilage; so, scleral ossicle; t, temporal; v, ventral. (After Underwood, 1970.)

In a survey of the scleral ossicles of lepidosaurian reptiles (Plagiotremata) Gugg (1939) reported that most lizards have 14 ossicles. His data show that there are certain features of the pattern of overlap which recur in many different families. Gugg (1939) recognized three types of ossicle: those which overlap ossicles on either side – designated plus; those which are overlapped by the ossicles on either side – designated minus; and imbricating ossicles – like tiles on a roof. Gugg's survey was later extended and his system of scoring modified by Underwood (1970). Some ossicles show reciprocal overlaps; the overlap nearer the corneal margin is scored. Some ossicles extend across the adjacent ossicle to overlap the next but one. For scoring purposes only the relationships between immediately adjacent ossicles are counted since this gives greater consistency.

Presch (1969, 1970) studied the scleral ossicles of a number of iguanid lizards and reported differences within the family. DeQueiroz (1982) recently made a more extensive survey of the same family and gave some valuable information on intraspecific variation. Kluge (1967, 1976) has recorded scleral ossicle numbers for representatives of all genera of the families Gekkonidae and Pygopodidae. Moody (1980) has recently recorded similar information for representatives of all genera of the Agamidae. For this report three further species of *Varanus*, a *Lepidophyma* and a further *Xantusia* species have been checked.

Scleral ossicles represent attractive material for formal systematic and evolutionary analysis. In the majority of lizards the overlaps are well enough defined that every ossicle can unequivocally be classified as either imbricating, plus or minus. One pattern emerges which is clearly central, as it is found in representatives of the majority of lizard families. There are differences, nevertheless, between families and within families. There is a good general level of constancy but sufficient intraspecific variation to afford suggestions about transformation from one condition to another.

Gugg (1939) recognized that in all lizards with a regular pattern there is a plus ossicle on the ventral side of the ring, somewhat temporal of mid-ventral. It may be significant that the transversalis muscle has a temporoventral origin. This ossicle he called number one; from it he counted around the ring in a temporal-dorsal-nasal sequence (clockwise for a right eye). On the temporal and on the nasal side of the ring there is nearly always a minus ossicle. Mid-dorsally there is usually a plus-minus-plus sequence. The above pattern is recognizable in 11 families of lizards and in *Sphenodon*. The differences lie mainly in the numbers of imbricating ossicles between the plus and minus members of the ring but also in the sizes of some of the members. In three families are found forms which show similar regular patterns save mid-dorsally where there is only one plus ossicle. In the large families Gekkonidae and Scincidae the central pattern occurs but many forms show irregularity of number and pattern of ossicle.

This chapter attempts to assess the contribution which scleral ossicles can make to the phylogenetic analysis of lizards, drawing upon the above mentioned data. To this end an analysis is made free as far as possible of prior taxonomic and evolutionary assumptions. Initial reference to the patterns shown by various species of lizards is therefore only by numbers and consideration of the taxonomic status of these lizards is deferred until later. Other patterns are compared with the central pattern in terms of gains and losses without intending to imply evolutionary polarity, but simply to avoid cumbersome phraseology.

SCORING SYSTEM

Where the dorsal plus-minus-plus sequence is present the correspondences between different patterns are readily recognizable. Several such patterns are set out in Table I using the numbering system of Gugg (1939).

TABLE I
Some representative scleral ossicle patterns using the numbering system of Gugg (1939).

Pattern no.	Temporal					Dorsal				Nasal	
	+	i	−	i		+	−	+	i	−	i....
(1)	1	2,3,4	5	6,7		8,9,10		11	12		13,14,15,16
(2)	1	2,3,4	5	6		7,8,9		10	11		12,13,14,15
(3)	1	2,3	4	5		6,7,8		9	10		11,12,13,14
(8)	1	2,3	4	−		5,6,7		−	8		8,10,11,12
(14)	1	2,3	4	−		5,6,7		−	8		9,10,11
	A		X			B	Y	C		Z	

i = imbricating; − = no ossicle in that place. Proposed scoring by letters indicated by base of columns.

By reason of the variations in the total number of ossicles this system leads corresponding ossicles to bear different numbers; this is a problem encountered by deQueiroz (1982). A system of letters is therefore adopted: A, B and C for the plus ossicles and X, Y and Z for the minus ossicles. Thus pattern 1 would be represented as A, B and C for ossicles 1, 8 and 10 respectively and X, Y and Z for 5, 9 and 12. As the imbricating ossicles are not individually recognizable, only their numbers are indicated in each sector. Table III shows these five patterns, amongst others, scored according to the proposed system.

There are several regular patterns with a single dorsal plus ossicle where we might expect to see the plus-minus-plus sequence (Table II).

Plus ossicle no. 6 appears to correspond with the B plus ossicle of other lizards. Pattern 16 would be derivable from pattern 3 if the C plus ossicle (no. 8) passed under instead of over the Y minus ossicle (no. 7). This raises a question concerning the scoring of the imbricating ossicles between B and Z. In patterns 15 and 16 ossicles nos 7 and 8 appear to correspond to the 'lapsed' Y and C ossicles. Which ossicle is missing from pattern 17? Once the Y and C ossicles have lapsed into imbricating

TABLE II
Some scleral ossicle patterns with only a single dorsal ossicle

Pattern no.	Temporal				Dorsal		Nasal		
	+	i	−	i	+	i	−		i....
(15)	1	2,3	4	5	6	7,8,9,10		11	12,13,14
(16)	1	2,3	4	5	6	7,8,9		10	11,12,13,14
(17)	1	2,3	4	5	6	7,8		9	10,11,12,13
(18)	1	2,3	4	5	6	7,8		9	10,11,12
(19)	1	2,3	4	5	6	7		8	9,10,11
	A		X		B			Z	

The numbering system of Gugg (1939), proposed lettering at base of columns.

status they cannot be distinguished from adjacent imbricating ossicle no. 9. If we are to score patterns 17, 18 and 19 according to the system proposed above we have to make an arbitrary decision abour correspondences. Nos 7 and 8 are therefore designated as Yi and Ci respectively, indicating lapse to imbricating status. Following the numbering of pattern 16 this implies loss of 9 before 8. Table III shows the scoring adopted. It also shows the scoring for some further patterns with the plus-minus-plus dorsal sequence; many of these data are taken from deQueiroz (1982). Small sized ossicles are indicated by Bs, Cs and 1s.

TABLE III

Nineteen scleral ossicle patterns shown by lepidosaurian reptiles, scored according to the system here adopted.

Pattern no.	Temporal			Dorsal				Nasal		Total
(1)	A	3	X	2	B	Y	C	1	Z	4	16
(2)	A	3	X	1	B	Y	C	1	Z	4	15
(3)	A	2	X	1	B	Y	C	1	Z	4	14
(4)	A	2	X	1	B	Y	Cs	1	Z	4	14
(5)	A	2	X	1	Bs	Y	Cs	1	Z	4	14
(6)	A	2	X	1	Bs	Y	Cs	1s	Z	4	14
(7)	A	2	X	–	B	Y	Cs	1	Z	4	13
(8)	A	2	X	–	B	Y	C	–	Z	4	12
(9)	A	2	X	–	B	Y	Cs	–	Z	4	12
(10)	A	2	X	1	Bs	Y	Cs	–	Z	4	13
(11)	A	2	X	–	B	Y	Cs	1s	Z	4	13
(12)	A	2	X	1	B	Y	C	2	Z	3	14
(13)	A	3	X	1	B	Y	C	1	Z	3	14
(14)	A	2	X	–	B	Y	C	–	Z	3	11
(15)	A	2	X	1	B	Yi	Ci	2	Z	3	14
(16)	A	2	X	1	B	Yi	Ci	1	Z	4	14
(17)	A	2	X	1	B	Yi	Ci	–	Z	4	13
(18)	A	2	X	1	B	Yi	Ci	–	Z	3	12
(19)	A	2	X	1	B	Yi	–	–	Z	3	11

The numbers refer to imbricating ossicles between the plus and minus ossicles. Ci and Yi refer to C and Y ossicles lapsed to imbricating status; s = small ossicle; – = minus.

We have now to consider the relationships between these patterns. A–X and X–B series of imbricating ossicles appear simply to undergo gains or losses in number. The C–Z and Z–A series are less straightforward. If we compare patterns 3 and 12 in terms of gains and losses they are two steps apart. If however we allow that ossicle no. 11 (Table I) may develop as the nasal minus instead of no. 10 (transposition) then the patterns are only one step apart. Formally this is a more parsimonious interpretation and therefore to be preferred. However without knowledge of the regulation of development of the ossicles we have no way of knowing whether it is biologically plausible. Further, whilst it is easy to interpret pattern 12 as pattern 3 with a transposed Z ossicle, the author is unable to decide whether or not Z is transposed in patterns 13, 15, 18 and 19. The formally more parsimonious system is therefore abandoned because it would require unsupportable arbitrary interpretations.

OUTGROUP COMPARISON

It is clear that variations in different parts of the ring may occur separately. Each variable sector is therefore considered independently. As these sectors show discontinuous variation they lend themselves to additive binary coding (Farris, Kluge & Eckardt, 1970). As the aim is phylogenetic analysis we have to consider evolutionary polarity; this involves making an outgroup comparison.

Scleral ossicles are known, amongst recent groups, in teleost fish, chelonians, birds, *Sphenodon* and lizards. They are known in many fossil groups but in very few are they well enough preserved to admit of detailed comparison with recent forms. Amongst amniotes chelonians have rather fewer and less regular rings of ossicles, from 6 to 13, than do lizards (Underwood, 1970). Lemmrich (1932) made a survey of the scleral ossicles of birds. He reports a modal number of 15 with 14 a fairly close second; the range is from 10 to 17. A common pattern consists of two plus ossicles, one dorsal and one ventral, and two minus ossicles, one nasal and one temporal. There is a clear similarity to lizard ossicles. In some birds the resemblance goes further with three plus and three minus ossicles and a plus-minus-plus dorsal sequence interrupted by imbricating ossicles. For example with the notation here adopted the emu would be scored:

$$A\ 3\ X\ -\ B\ Y\ 1\ C\ 2\ Z\ 3 \qquad 15$$

Lemmrich (1932) did not himself observe a closed plus-minus-plus sequence.

Sphenodon is the only living survivor of the eosuchian grade (Evans, 1980); as such it is generally regarded as sister to the Squamata since they are all members of the Lepidosauria (Bellairs, 1969). Recent crocodilians have no scleral ossicles. Birds, as derivatives of the Archosauria, are diapsids, sister to the Lepidosauria. Chelonians as anapsids are sister to birds and other Recent reptiles. As no Recent mammals have scleral ossicles we do not have to take them into consideration.

In order to use *Sphenodon* for the outgroup comparison we have to make an *a priori* assumption, as above, about its relationships. The Squamata, including lizards and snakes, are a well characterized group (Bellairs, 1969). As snakes have no trace of scleral ossicles they need be considered no further. *Sphenodon* does not show several of the special features of the Squamata, most particularly the paired eversible hemipenes, which are known nowhere else. In these respects *Sphenodon* appears to be primitive to the Squamata.

Whilst several special features set the Archosauria apart from the Lepidosauria, relatively few features set the Lepidosauria apart; they show more primitive features. If the Lepidosauria were a paraphyletic group the birds could be sister to lizards and *Sphenodon* sister to both. As anapsids the chelonians should be more remote but, in this case, birds and *Sphenodon* admit of so much more detailed comparison with lizards that we would use them for the outgroup even were we to believe chelonians to be sister to lizards.

We have a much larger sample of the Archosauria than we have of the eosuchian grade. It is unlikely that we will ever have any eosuchian scleral ring, other than that of *Sphenodon*, well enough preserved to allow detailed comparison; a sample of a single species may well be unrepresentative. The *Sphenodon* pattern is taken as primitive to the lizard states because the uninterrupted plus-minus-plus dorsal sequence is accepted as a special resemblance, and also because to question this would require a full analysis of bird scleral ossicles.

CHARACTER ANALYSIS

In seven sectors of the scleral ring there is variation: the A–X imbricating, the X–B imbricating, the B plus, the Y minus, the C plus, the C–Z imbricating and the Z–A imbricating. For each of these the *Sphenodon* condition is coded as primitive, indicated by a zero (Table IV). The only character for which a question might arise concerning the relationship between the states is the C ossicle. It may be large plus, small plus, absent or imbricating. As the C ossicle necessarily imbricates when the Y ossicle imbricates there is no need to code separately for the imbrication of the C.

Decimals are used to number binary components of multistate characters (Underwood, 1982). We have a total of 11 binary characters. Table V shows the 19 patterns in binary coded form. Inspection shows that some of the binary characters, for example 6.3 and 7, are incompatible in that they show all four combinations of states: 0 0, 1 0, 1 1 and 0 1. This constitutes a failure of the LeQuesne test (1969). It is a

TABLE IV
Coding of the seven sectors of the scleral ring where there is variation

Character no. and description	Coding		
(1) A–Z imbricating ossicles	three 0		
	two 1		
	2.1	2.2	
(2) X–B imbricating ossicles	two 0	0	
	one 1	0	
	absent 1	1	
(3) B plus ossicle	large 0		
	small 1		
(4) Y ossicle	minus 0		
	imbricating 1		
	5.1	5.2	
(5) C ossicle	large 0	0	
	small 1	0	
	absent 1	1	
	6.1	6.2	6.3
(6) C–Z imbricating ossicles	two 1	0	0
	one large 0	0	0
	one small 0	1	0
	absent 0	1	1
(7) Z–A imbricating ossicles	four 0		
	three 1		

logically inescapable minimum conclusion that one of the characters has undergone parallel evolution. If, for example, we pass from the 0 0 condition to the 1 1 condition via 1 0 then the 0 1 condition can be derived from the 0 0 condition by a 0 to 1 transformation of character 7 in parallel to the 1 0 to 1 1 transition. This incompatibility is unconditional; it does not depend on scoring of evolutionary polarity.

Each binary character was tested against each other in turn, apart from binary components of the same multistate character. Characters 2.1 and 5.3 are singletons: only one pattern has a score differing from the majority score; they cannot therefore fail the test. All the rest fail at least once, character 7 more often than not (Table VI).

For each character the computer program counts the observed LeQuesne test failures. At each character comparison it computes the probability that the two characters would fail the test on the assumption that the states of the two characters are distributed at random, using a formula given by LeQuesne (1979). For each character these probabilities are summed. If a character shows many fewer observed failures than would be expected on the null hypothesis of random

TABLE V
The 19 patterns of Table III coded in additive binary form

Pattern nos	Character nos										
	1	2.1	2.2	3	4	5.1	5.2	6.1	6.2	6.3	7
(1)	0	0	0	0	0	0	0	0	0	0	0
(2)	0	1	0	0	0	0	0	0	0	0	0
(3)	1	1	0	0	0	0	0	0	0	0	0
(4)	1	1	0	0	0	1	0	0	0	0	0
(5)	1	1	0	1	0	1	0	0	0	0	0
(6)	1	1	0	1	0	1	0	0	1	0	0
(7)	1	1	1	0	0	1	0	0	0	0	0
(8)	1	1	1	0	0	0	0	0	1	1	0
(9)	1	1	1	0	0	1	0	0	1	1	0
(10)	1	1	0	1	0	1	0	0	1	1	0
(11)	1	1	1	0	0	1	0	0	1	0	0
(12)	1	1	0	0	0	0	0	1	0	0	1
(13)	0	1	0	0	0	0	0	0	0	0	1
(14)	1	1	1	0	0	0	0	0	1	1	1
(15)	1	1	0	0	1	0	0	1	0	0	1
(16)	1	1	0	0	1	0	0	0	0	0	0
(17)	1	1	0	0	1	0	0	0	1	1	0
(18)	1	1	0	0	1	0	0	0	1	1	1
(19)	1	1	0	0	1	1	1	0	1	1	1

0 = primitive, 1 = derived.

TABLE VI
LeQuesne test matrix of 11 binary characters compared pairwise

	Character nos									
	1	2.1	2.2	3	4	5.1	5.2	6.1	6.2	6.3
7	x	–	x	–	x	x	–	–	x	x
6.3	–	–	x	x	x	x	–			
6.2	–	–	x	x	x	x	–			
6.1	–	–	–	–	x	–	–			
5.2	–	–	–	–	–					
5.1	–	–	x	–	x					
4	–	–	–	–						
3	–	–	–							
2.2	–	–								
2.1	–									

x = failure, i.e. unconditional incompatibility.

distribution of states we have evidence of order of distribution of the states. If, on the other hand, a character shows about as many observed failures as would be expected on the null hypothesis we may suppose its states to be randomly distributed in relation to the states of the other characters considered. The ratio of observed to expected failures expressed as a percentage is LeQuesne's coefficient of character state randomness. Without beginning to speculate about phylogeny we have a distinction between those characters likely to be "good" and those likely to be "bad". We can see (Table VII) that three characters (6.2, 6.3, 7) have observed failures within a decimal fraction of those expected on the null hypothesis; they show negligible evidence of order. As part of a larger data set the scleral ossicle characters would be tested against all the other kinds of characters.

TABLE VII

Results of LeQuesne test. For each of the 11 binary characters the number of failures observed, the number of failures expected on null hypothesis of random distribution of states, and the ratio of observed to expected is given

Character no.	Failures observed	Failures expected	Ratio of observed to expected
1	1	4.91	0.2
2.1	0	–	–
2.2	4	6.21	0.64
3	2	4.91	0.41
4	5	6.21	0.81
5.1	5	6.94	0.72
5.2	0	–	–
6.1	1	2.35	0.43
6.2	5	5.5	0.91
6.3	5	5.29	0.95
7	6	6.54	0.92

PHYLOGENETIC ANALYSIS OF PATTERNS

Let us now consider the relationships between the patterns and the implications for the possessors of the patterns. This is on the basis that shared derived features are potential indicators of phylogenetic affinity. By comparing every pattern with every other pattern in turn and counting the number of characters in respect of which they have differing scores we can draw up a matrix of distances – city block or Manhattan distances (Table VIII). Each pattern has one, two or even three nearest neighbours. Some pairs of patterns are reciprocal nearest neighbours. We can use this information to construct a putative phylogeny of scleral ossicle patterns.

TABLE VIII

Pairwise comparison of 19 scleral ossicle patterns showing the number of character state differences, derived from Table V (Manhattan distances)

	1	2	3	4	5	6	7	8	9	10	11	12	13	14	15	16	17	18
19	8	7	6	5	6	5	6	5	4	4	5	6	6	4	5	5	3	2
18	6	5	4	5	6	5	6	3	4	4	5	4	4	2	3	3	1	
17	5	4	3	4	5	4	5	2	3	3	4	5	5	3	4	2		
16	3	2	1	2	3	4	3	4	6	5	4	3	3	5	2			
15	5	4	3	4	5	6	5	6	7	7	6	1	3	5				
14	6	5	4	5	6	5	4	1	2	4	3	4	4					
13	2	1	2	3	4	5	4	5	6	6	5	2						
12	4	3	2	3	4	5	4	5	6	6	5							
11	5	4	3	2	3	2	1	2	1	3								
10	6	5	4	3	2	1	4	3	2									
9	6	5	4	3	4	3	2	1										
8	5	4	3	3	5	4	3											
7	4	3	2	1	2	3												
6	5	4	3	2	1													
5	4	3	2	1														
4	3	2	1															
3	2	1																
2	1																	

Patterns 1, 2, 13, 3, 16 and 4 are unambiguously linked by a series of single steps (Fig. 2). From 4 arise two single step lineages: 5, 6, 10; and 7, 11, 9, 8, 14; again without ambiguity. Patterns 12 and 15 are reciprocal nearest neighbours. Pattern 12 is two steps removed from both 3 and 13; and 15 is two steps from 16. The 13–12 link involves parallel loss of an A–X ossicle, a character with a good randomness ratio, 0.2, here deemed implausible. The 16–15 link implies gain of a C–Z ossicle and loss of a Z–A. The 3–12 link implies loss of a C–Z and a Z–A ossicle. We have a closed loop of 3, 16, 12 and 15.

Lineages 3–12–15 and 16–15–12 both involve three steps with parallel evolution of character 4, imbrication of the Y ossicle; quite plausible with a poor randomness ratio of 0.81. The alternative 3–12 and 16–15 links involve four steps of characters 6.1 and 7; with virtually random ratios this is entirely plausible.

Pattern 17 is two steps from 8 and 16. Transition 8–17 involves gain of an X–B ossicle and imbrication of the Y. On the other hand 16–17 involves loss of a C–Z ossicle in parallel with other such losses and looks more plausible. Pattern 19 links with 18 at two steps removed.

The two lineages leading from 4 are marked by unique transformations of characters 3 and 2.2. The 9–8 transition implies reversal of character 5.1, enlargement of a small C ossicle. To avoid this reversal we would have to link 8 to 17 (two steps) or to 3 (3 steps).

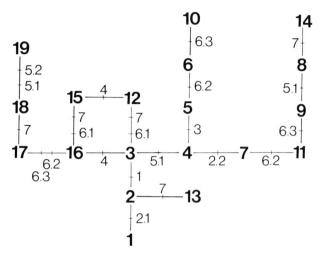

FIG. 2. Phylogenetic dendrogram of scleral ossicle patterns, showing character transformations. Heavy type – ossicle pattern nos. Light type – character nos.

TAXONOMIC IMPLICATIONS

We have a character complex tree some parts of which are unambiguous, some of which involve judgement between alternatives and one unresolved closed loop. Table IX shows the taxonomic distribution of the 19 patterns.

Variation can give us indications of the genetic relationships between the several states of a character (Underwood, 1982). DeQueiroz (1982) gives us information about variation involving two sides of the same individual (sharing the same genotype, barring mosaics) and members of the same species (sharing similar gene pools). If we recognize species by discontinuity of phenetic variation then we can use this evidence without begging any evolutionary questions.

From de Queiroz's data a summary of the alternative patterns which he found within a single species is extracted, indicating only the ossicles in respect of which they differ (Table X).

We see that *Holbrookia propinqua* and *Uma notata* embrace four steps with patterns 8 and 6 and 9 and 5, and *H. maculata* five steps with patterns 5 and 8. DeQueiroz (1982) further reports that one *Callisaurus* had a small 15th supernumerary ossicle positive to B and C; that one *Holbrookia propinqua* had the C ossicle extremely small in both eyes and that one *Petrosaurus* eye had a small 15th supernumerary ossicle positive to the Y and C of pattern 4. This evidence could be taken to suggest that we should amalgamate the 4–10 and 4–14 lineages. If, however, we

TABLE IX
Taxonomic distribution of the 19 patterns

Pattern	Taxonomic distribution
(1)	*Sphenodon*
(2)	*Varanus*, 6 spp.
(3)	Gekkonidae: *Dravidogekko, Geckonia, Gehyra, Ptychozoon, Tarentola*; Iguanidae: 32 genera; Teiidae: 3 genera; Scincidae: 5 genera; Lacertidae: 2 genera; Cordylidae: 3 genera; Anguidae, *Gerrhonotus*; Xenosauridae: *Shinisaurus, Xenosaurus*
(4)	Iguanidae: *Petrosaurus*, 2 spp.; *Sceloporus*, 5 spp.; *Urosaurus*, 4 spp.; *Uta*
(5)	Iguanidae: *Cophosaurus*, 1 sp.
(6)	Iguanidae: *Callisaurus*, 1 sp.; *Uma*, 2 spp.
(7)	Iguanidae: *Holbrookia*, 1 sp.
(8)	Iguanidae: *Holbrookia*, 2 spp.
(9)	Iguanidae: *Phrynosoma*, 7 spp.; *Uma*, 1 sp.
(10)	Iguanidae: *Callisaurus*, 1 sp.; *Uma*, 2 spp. (minority variant)
(11)	Iguanidae: *Holbrookia maculata, H. propinqua, Phrynosoma m'calli, Uma exsul* (minority variant)
(12)	Gekkonidae: *Hemidactylus*, 5 spp.; Lacertidae: *Acanthodactylus*, 1 sp.
(13)	Xantusiidae: *Xantusia*, 2 spp.; *Lepidophyma*, 1 sp.
(14)	Chamaeleonidae: *Chamaeleo*, 3 spp.
(15)	Gekkonidae: *Gehyra, Hemidactylus*
(16)	Scincidae: *Ablepharus, Scincus*
(17)	Scincidae: *Ablepharus*, 2 spp.; *Tiliqua*
(18)	Agamidae: all genera but one
(19)	Agamidae: *Phrynocephalus*

TABLE X
Summary of alternative patterns within a single species (from deQueiroz, 1982)

Pattern no.		Species	Showing patterns no.
(5):	1 Bs Y Cs 1	*Callisaurus draconoides* }	5.6
*(6):	1 BS Y Cs 1s	*Uma paraphygos* }	
(10);	1 Bs Y Cs –	*Uma exsul*	5,6,10
*(9):	– B Y Cs –	*Phrynosoma m'calli*	9,11
(11):	– B Y Cs 1s	*Uma notata*	9,11,5
(5):	1 Bs Y Cs 1		
*(8):	– B Y C –	*Holbrookia propinqua*	8,11,6
(11):	– B Y Cs 1s		
(6):	1 Bs Y Cs 1s		
(5):	1 Bs Y Cs 1	*Holbrookia maculata*	5,7,11,8
*(7):	B Y Cs 1		
(11):	– B Y Cs 1s		
(8):	– B Y C –		

* = modal condition.

consider only majority states we see *Callisaurus* and two *Uma* species on the 4–10 lineage and *Holbrookia, Phrynosoma* and one *Uma* on the 4–14 lineage, with more conservative sceloporines showing pattern 4.

DISCUSSION OF CHARACTER ANALYSIS

Taking the group Squamata as given and *Sphenodon* as a member of the sister group, an analysis has been attempted of the scleral ossicles as a character complex solely on the basis of the intrinsic evidence. To construct an analysis some choices had to be made.

In the case of the separation of the 4–10 and 4–14 lineages the choice was made on the basis of statistical evidence that parallel transformations of characters 6.1 and 6.2 were more plausible than of characters 2.2 and 3. If the scleral ossicle data were part of a larger data-set and other characters supported a *Cophosaurus, Callisaurus, Uma* lineage and a *Holbrookia, Uta, Phrynosoma* lineage then characters 2.2. and 3 should still get good randomness ratios. The degree of variation found by deQueiroz (1982) indicates the importance of a sample large enough to detect the variation first and then to recognize the modal condition for each species.

In the case of the 3, 12, 15, 16 closed ring, statistical considerations do not suggest a resolution. Reversals of characters 4, 6.1 and 7 were considered and none appeared to be biologically implausible. So "opening" of the ring must wait on the evidence of other characters for the taxa concerned. The transformation of 6.1 and 7 together, gain of a C–Z ossicle and loss of a Z–A ossicle, raises the question of transposition of the Z ossicle which was dismissed, not for biological reasons, but because of formal difficulties of applying this coding to the whole data-set. Had the data-set not included iguanids, agamids and chameleons this difficulty would not have been encountered, but to take that into consideration in the initial stages would prejudice the analysis.

For the rest of the dendrogram the guiding principle was parsimony; the aim was to produce a dendrogram with a minimum number of character transformations. There are no grounds for supposing that evolution is parsimonious and there are grounds for supposing that it is not. Parsimony is an analytical necessity; if we allow consideration of other than the most parsimonious analysis then we are confronted by an indefinite number of possibilities and an endless task.

Does parsimony defined in terms of character transformation steps have any biological meaning? Its meaning may well be limited. The number of steps is to some extent an artefact of the system of coding adopted, as with "transposition" of the Z ossicle, and to some extent an artefact of the information available; the C ossicle would have been coded as merely present or absent but for the observation of deQueiroz (1982) of some C ossicles which do not reach the corneal border. Sneath & Sokal (1973) suggest that if we have a large enough number of characters we have a representative sample of the phenetic expression of the genome. This prompts the thought that we might wish ideally to achieve parsimony in genetic terms. Were we to know the genetic basis of the phenotypic characters employed, rigorous parsimony would still elude us. At the level of DNA we encounter the same problem of equivalence of different transformations that we encounter at the phenetic level. One base substitution may be equivalent to another base substitution, but is deletion of 10 bases equivalent to one substitution, to 10 substitutions or to some other number (Fitch, 1977)? For the above reasons it is thought preferable to state the aim of analysis as the achievement of greatest internal consistency; this cannot be quantified because it is a partly formal and partly biological judgement.

It has been noted above that intraspecific variation can embrace as many as five character transformations, but that, although this calls in question the discreteness of the character states recognized, the modal states of each species do appear to show order. What about intraspecific variation in the rest of the dendrogram? There are few reports of it. From one point of view we may be grateful for the lack of recorded variation, for the species show clear-cut differences. However, unless we believe in saltational evolution we cannot avoid accepting that variation did occur in these lineages. It is ironic that the variation which is evidence of the phenomenon of evolution is an inconvenience when we seek to trace the course of that evolution.

DISCUSSION OF TAXONOMY AND EVOLUTION

The implications for the evolution of lizards may now be considered. There are no grounds for supposing *a priori* that the scleral ossicles are either more or less important indicators of phylogeny than other characters. We may expect to find that sometimes they coincide with indications of other characters and sometimes they conflict, sometimes they throw no light on outstanding problems and occasionally they afford the only evidence on a particular question.

The assumption that the *Sphenodon* (1) pattern with 16 ossicles is primitive leads to the conclusion that *Varanus* with 15 ossicles shows a pattern (2) primitive to all other lizards. Bellairs (1969) recognizes an infraorder Diploglossa (= Anguimorpha) with superfamilies Anguoidea and Varanoidea (= Platynota). This expresses views which have been generally supported for some years. The varanoids are more highly derived than the anguoids and within the Varanoidea the Varanidae are more derived than the other Recent families. It would be difficult to sustain the view that the *Varanus* pattern is primitive, in the face of evidence from other characters. Is the initial inference justified that the *Sphenodon* 16 ossicle pattern is primitive?

Birds have a modal number of 15 ossicles; this would suggest that 15 ossicles as in *Varanus* are primitive for lizards and that *Sphenodon* shows a secondary increase. One non-varanid diploglossan (*Gerrhonotus*) shows the central pattern 3 so there is little doubt that the evidence from all characters would persuade us that *Varanus* shows a secondary increase. Would we arrive at this conclusion if the Varanidae were the only living Anguimorpha?

An interesting finding is the unique pattern for the Xantusiidae (13), here confirmed for three species, only one step removed from the varanid pattern. Whilst placement of the Xantusiidae has been uncertain, they have never been associated with the Anguimorpha; Bellairs (1969) and Camp (1923) have them in the Scincomorpha, others have them near to the Gekkota (evidence cited in Underwood, 1971). If we do not accept the pattern as evidence of varanid affinities we may consider whether it bears upon gekkotan or scincomorph affinity. If pattern 2 is primitive then the xantusiid pattern might be thought to support gekkotan affinities because there are good grounds for regarding the Gekkota as archaic stock. Xantusiidae would, however, have to be primitive to the Gekkota, a view which could hardly be sustained in the face of other evidence (Camp, 1923). If we regard them as Scincomorpha, archaic by reason of the scleral ossicle pattern, then we imply that the central pattern 3 arose in the Scincomorpha in parallel to the Gekkota and Anguimorpha. It seems likely that in the end this unique pattern can be used to tell us no more than that the Xantusiidae are different from other lizards.

In the Gekkota there is a very wide range of variation of the scleral ossicles. Representatives of five genera show the central pattern. Within the Gekkonidae five species of *Hemidactylus* (12) show what looks like a transposition of the Z ossicle (here coded as two steps) and a *Hemidactylus* and *Gehyra* (15) show loss of imbrication of the Y ossicle. The evidence of the other features which characterize the Gekkonidae suggests that they go round the closed loop 3–12–15. Many other patterns have been reported for geckos, some implying increase in the numbers of plus and minus ossicles. They may well have value for the analysis of geckos but there is sufficient bilateral asymmetry and intraspecific variation that a detailed survey would be required. Underwood (1977) has argued that, in association with nocturnal habits, enlargement of the cornea and reduction of the annular sulcus, the geckos first show a trend to loss of regulation in the pattern of overlaps of the 14 ossicles and then a trend to loss of regulation in the number of ossicles.

The infraorder Iguania, with families Iguanidae, Agamidae and Chamaeleonidae, has achieved general recognition. Most of the Iguanidae show the central pattern (3). The chameleons (14) are linked to this pattern via five intermediate patterns all found in iguanids. The Agamidae (18) on the other hand are linked to this central pattern via two intermediate patterns (and four steps) represented by skinks! Agamids and chameleons have acrodont teeth, in which they differ from iguanids, and have been supposed on this account to have a common origin. The agamids show imbrication of ossicle Y and reduction in number. Chameleons show reduction in number of ossicles but retention of the plus-minus-plus sequence so they are associated with iguanids, 10 steps away from the agamids with some skink intermediates. If we had taken the Iguania as the given ingroup (instead of the Squamata) we would have arrived at a different conclusion with the same system of scoring. The agamid pattern (18) is, within the Iguania, closest to pattern 14 (two steps) shown by the chameleons and three steps away from pattern 14 shown by *Holbrookia*. We might well have invented a hypothetical agamid-chameleon ancestor with the score:

$$A\ 2\ X\ 1\ B\ Y\ C - Z\ 3,$$

two steps away from pattern B, one step from chameleons and one step from agamids. If the case for associating chameleons and agamids were to rest on only one or two characters it might be "outvoted" by the scleral ossicles considered in an all-lizard context.

In the Scincomorpha we find the central pattern 3 in many skinks, in teids, *Lacerta*, *Cordylus* and *Gerrhosaurus*. Within the lacertids one species of *Acanthodactylus* (pattern 12) looks as though it is derived from the central *Lacerta* pattern by transposition of the Z ossicle. The lacertids have not been extensively surveyed and the scleral ossicles may well provide further information. Within the Scincidae one species each of *Ablepharus* and *Scincus* show imbrication of the Y ossicle and one *Tiliqua* and two more *Ablepharus* species further show loss of an ossicle. Many skinks are modified in association with secretive or burrowing habits and reduction in the number of ossicles is reported for a few such forms. Less than 30 species of this large family have been reported and it might well repay further study.

In the Anguimorpha the central pattern 3 is known in one species each of *Gerrhonotus*, *Xenosaurus*, and *Shinisaurus*. The ossicles are reduced in *Anguis*, *Ophisaurus* and *Anniella*, in association with secretive or burrowing habits, and in *Heloderma* and *Lanthanotus* in association with nocturnal habits. . *Varanus* has already been discussed. The Anguidae are sufficiently diverse that they might repay further study.

GENERAL DISCUSSION

A fairly thorough analysis of this character complex and consideration of the results in relation to generally accepted views about lizards prompts some thoughts about phylogenetic analysis in general. In order to set up the analysis it was necessary to make some *a priori* assumptions about phylogenetic relationships between lizards and other amniotes. It is evident that the precise pattern of outgroup relationships can influence the interpretation of the primitive conditions within the ingroup. In order to score the data in a form in which they could be sorted according to a defined procedure it was necessary to make decisions of various degrees of arbitrariness. The small quantity of data considered here could have been sorted informally but were we to include them as part of a larger data-set we would hardly be able to escape the need for formal coding. Unless our procedures are defined, and thus made explicit, they are not accessible to critical scrutiny and cannot plausibly pass for scientific.

The scope of the ingroup chosen for analysis can influence the pattern of affinities suggested by the group of characters. Consideration of the distribution of states of a character in relation to a larger body of evidence (other characters) may persuade us that there has been parallel evolution. For example it seems likely that the Y ossicle has changed from minus to imbricating independently in geckos and skinks. To the idealistic taxonomist imbrication of the Y ossicle in the two groups is homoplasic, not the "same" condition. There is, however, no apparent way that intense scrutiny could at the outset reveal that imbrication in skinks is not descriptively the same as imbrication in geckos. At the outset our units of analysis are species; "geckos" and "skinks" would at this stage beg a question. The inference of parallelism in a single character is probabilistic, it is never inescapable. There are always alternatives which involve other characters. The small data-set considered here gives no comfort to the idealistic assumption that there is a pattern (in the singular) in nature. We are confronted by conflicting patterns.

The large gaps between some of the lizard family groups leave us uncertain of the significance of some of the evidence of the scleral ossicles. On the other hand, where we have a diversity of forms without large gaps, as in the Iguanidae, we have complications due to intraspecific variation. The particular pattern of survival of forms, which are thus available to provide data, can influence the indications afforded by out data. While the balance of the evidence would probably associate the Agamidae with the rest of the Iguania they would on this evidence alone remain only one step away from some skinks. Arnold (1980) has neatly demonstrated how accidents of extinction can influence the inference of phylogeny. One may suspect that phylogenies tend to be robust in proportion to the amount of evidence which is either ignored, or unavailable because the taxa which could furnish it are extinct and beyond recovery.

How then, if at all, are we to infer phylogeny? If we consider a large number of different characters we may hope that the above considered accidents will cancel one another out. Statistically however we must expect that these accidents will sometimes coincide. Parts of our overall analysis are likely to rest on single lines of evidence and again we have to accept that it is statistically likely that this evidence will sometimes be subject to accidental distortion.

There are grounds for believing that the processes of evolution are several and various. There are grounds for believing too that the accidents to which the interpretation of the course of evolution is liable are also several and various. It seems implausible that any computer algorithm could model all of these factors. We should sort our data, by all means with the assistance of computers, to achieve the best degree of internal consistency and should recognize the evolutionary interpretations placed upon the analysis as probabilistic speculations to be tested against the circumstantial evidence of the fossil record, geographical distribution and ecology.

REFERENCES

Arnold, E. N. (1980). Estimating phylogenies at low taxonomic levels. *Z. zool. Syst. EvolForsch.* **19**: 1–35.
Bellairs, A. d'A. (1969). *The life of reptiles*. London: Weidenfeld & Nicholson.
Camp, C. L. (1923). Classification of the lizards. *Bull. Am. Mus. nat. Hist.* **48**: 289–481.
Evans, S. E. (1980). The skull of a new eosuchian reptile from the Lower Jurassic of South Wales. *Zool. J. Linn. Soc.* **70**: 203–264.
Farris, J. S., Kluge, A. G. & Eckardt, M. J. (1970). A numerical approach to phylogenetic systematics. *Syst. Zool.* **19**: 172–189.
Fitch, W. M. (1977). The phyletic interpretation of macromolecular sequence information: simple methods. In *Major patterns in vertebrate evolution*: 169–204. Hecht, M. K., Goody, P. C. & Hecht, B. M. (Eds). N.A.T.O., Advanced Study Institute, Series A: Life Sciences **14**. New York and London: Plenum Press.
Gugg, W. (1939). Der Skleralring der plagiotremen Reptilien. *Zool. Jb.* (Anat.) **65**: 172–189.
Kluge, A. G. (1967). Higher taxonomic categories of gekkonid lizards and their evolution. *Bull. Am. Mus. nat. Hist.* **135**: 1–60.

Kluge, A. G. (1976). Phylogenetic relationships in the lizard family Pygopodidae: an evaluation of theory, methods and data. *Misc. Publs Mus. Zool. Univ. Mich.* No. 152: 1–72.

Lemmrich, W. (1932). Der Skleralring der Vogel. *Jena. Z. Naturwiss.* **65**: 513–584.

LeQuesne, W. J. (1969). A method of selection of characters in numerical taxonomy. *Syst. Zool.* **19**: 201–205.

LeQuesne, W. J. (1979). Compatibility analysis and the uniquely derived character concept. *Syst. Zool.* **28**: 92–94.

Moody, S. M. (1980). *Phylogenetic and historical biogeographical relationships of the genera in the family Agamidae (Reptilia, Lacertilia)*. Ph.D. dissertation: University of Michigan, USA.

Presch, W. (1969). Evolutionary osteology and relationships of the horned lizard genus *Phrynosoma* (family Iguanidae). *Copeia* **1969**: 250–275.

Presch, W. (1970). Scleral ossicles in the sceloporine lizards, family Iguanidae. *Herpetologica* **26**: 446–459.

deQueiroz, K. (1982). The scleral ossicles of sceloporine lizards: a reexamination with comments on their phylogenetic significance. *Herpetologica* **38**: 302–311.

Sneath, P. H. A. & Sokal, R. R. (1973). *Numerical taxonomy*. San Francisco: W. H. Freeman.

Underwood, G. L. (1970). The eye. In *Biology of the Reptilia*: 1–97. Gans, C. & Parsons, T. (Eds). London: Academic Press.

Underwood, G. L. (1971). *A modern appreciation of Camp's "Classification of lizards"*, in facsimile reprint of *C. L. Camp, Classification of the lizards*: vii–xvii. Lawrence, Kansas: Society for the Study of Amphibians and Reptiles.

Underwood, G. L. (1977). Simplification and degeneration in the course of evolution of squamate reptiles. *Colloq. int. Cent. natn. Rech. scient.* No. 266: 341–351.

Underwood, G. L. (1982). Parallel evolution in the context of character analysis. *Zool. J. Linn. Soc.* **74**: 245–266.

Walls, G. L. (1942). The vertebrate eye and its adaptive radiation. *Bull. Cranbrook Inst. Sci.* **19**: 1–785.

# Miniaturization of the Lizard Skull: Its Functional and Evolutionary Implications

OLIVIER RIEPPEL

Paläontologisches Institut und Museum der Universität, Künstlergasse 16, 8006 Zürich, Switzerland

SYNOPSIS

A concept of miniaturization is put forward, based on a comparative analysis of the head anatomy of fossorial lizards. Miniaturization involves the arrangement of the dermatocranium and neurocranium to a common level which allows for closure of the occiput and of the lateral braincase wall. Together with the reduction of the upper temporal arcade, this provides the possibility for the jaw adductor muscles to expand their area of origin in a posterodorsal direction on to the skull roof. Similar changes may have been involved in the origin of snakes from a small fossorial squamate ancestor. They may have also set the stage for an adaptive shift during the early evolution of snakes. This shift may have facilitated further adaptations for swallowing relatively large prey.

INTRODUCTION

In recent times the phenomenon of miniaturization, that is of phylogenetic size decrease, has received some attention as a modus of evolutionary change both in invertebrates (Seilacher, 1982) and in vertebrates (Hanken, 1980). Gould (1977: 332) discusses miniaturization by progenesis as an adaptive response to pressures for small size. Larval morphology may thereby result from the primary need for small size.

The aim of the present contribution is to describe the functional and evolutionary consequences of miniaturization of the lizard skull. It can be shown that quantitative change, namely phylogenetic size decrease, results in qualitative change, namely a modification of the *bauplan* of the lizard skull, which opens up new evolutionary possibilities. The acquisition of a new adaptational programme may thus lie at the heart of the origin of a new higher taxon (Wright, 1982: 442).

Formulated as such, the concept of miniaturization is another way of seeing Cope's Law (Gould, 1977: 285), and it is put in perspective with an adaptational programme. The adaptationist paradigm is, however, problematical (Lewontin, 1978; Gould & Lewontin, 1979; Brady, 1982). Essentially it must be assumed that evolution is adaptive, at least for those traits of the organism that are to be explained. The adaptive interpretation of an organism will always represent an *a posteriori* statement. These difficulties will be reflected in the present contribution. It is impossible to present one unifying model of miniaturization encompassing all lizards. There is more than one way for lizards to become small, and each of these different patterns will have to be explained in view of changing specific adaptations. Even among burrowing lizards, which present the most homogenous pattern of skull miniaturization, it is not possible to constantly correlate any one given feature with a similar size class. It is only possible to present an idea, a concept of miniaturization, which involves the restructuring of a number of structural and functional traits. It is possible to document each aspect of this model by reference to one or more lizard species, but it is not possible to find all aspects of miniaturization realized in any one single lizard species, except perhaps *Dibamus novaeguineae*. However, the concept is useful, for it explains the morphology of *Dibamus*, and it adds to our ideas on the origin of snakes.

A major drawback of the present study is the impossibility, as yet, to discuss the implications of this concept of miniaturization for lizard ontogeny. This is mainly due to the lack of information concerning the embryonic development of the lizard skull, viewed from the perspective of miniaturization. This deficit makes it extremely difficult, if not impossible, to evaluate at present the involvement of paedomorphosis in lizard miniaturization.

The institutional abbreviations used in the figure legends are the following: AMNH, American Museum of Natural History, New York; MBS, Natural History Museum, Basel; UMMZ, University of Michigan Museum of Zoology, Ann Arbor.

WHY BECOME SMALL?

It is intuitively obvious that a segregation into different size classes of sympatric, closely related species of the same habitat represents an ecological strategy to promote resource partitioning among these species. This strategy was empirically documented in a community of 12 species of Gekkonidae in the western Australian desert (Pianka & Pianka, 1976). A search through the ecological literature provides further examples (Pianka, 1974). For two sympatric species of limbless, fossorial lizards of the genus *Typhlosaurus* (*lineatus* and *gariepensis*) it was shown that the different head sizes of the two species correlate with a different size class of prey. This relationship represents a case of ecological character displacement (Huey, Pianka, Egan & Coons, 1974). However, ecological character displacement will work primarily between closely related species (sympatric congeners: Pianka, 1974), and does not adequately explain differences in size between higher taxa that amount to miniaturization. For instance, *Typhlosaurus*, as a genus, is represented by miniaturized forms, and ecological character displacement only accounts for shifts in size within that frame (Huey & Pianka, 1974).

Miniaturization must be viewed as an adaptive strategy on a larger scale characterizing the evolution of higher taxa, but obviously with different causes in different taxa. Thus, the miniaturized geckos of the subfamily Sphaerodactylinae occupy a different adaptive zone from the equally very small and fossorial pygopodid genera *Pletholax* or *Aprasia*. Selection pressure resulting in a miniaturized head is particularly obvious in limbless, fossorial lizards, since a reduction of the diameter of the head and body will reduce the amount of energy required to penetrate the soil (Gans, 1960: 181; Huey *et al.*, 1974: 305). The reduction of diameter results in a miniaturized skull, but it does not imply a reduction of the snout-vent length. Absolute skull length may decrease to values between 4 and 5 mm in some fossorial species such as *Aprasia repens* and *Dibamus novaeguineae*. Since that group lends itself particularly well to studies of miniaturization, the present analysis will be based mainly on limbless, fossorial lizards. However, this represents only one of many possible adaptations resulting in small size.

THE REDUCTION OF SKULL DIAMETER

The primitive (anapsid) reptile skull can be visualized as consisting of two cylinders, one smaller than the other, and the two pushed into one another. The smaller, inner cylinder represents the neurocranium, the outer cylinder is the dermatocranium. All the jaw adductor muscles have to be packed into the limited space between the two components of the skull. At the back of the skull, on the occiput, this space opens through the post-temporal fossae.

The situation did not change with the development of one or two temporal openings in early reptiles. By analogy with Modern reptiles these fenestrae must have been closed by fasciae, from the inner surface of which jaw adductor muscle fibres took their origin (cf. *Sphenodon*: Haas, 1973). The temporal fossae did not allow the jaw adductor muscles to escape out of the boundaries set by a primitively closed dermatocranium (Frazzetta, 1968).

Modern lizards have lost the lower temporal arcade, or rather replaced it by the quadrato maxillary ligament. Thus arose the possibility for the superficial fibres of the external adductor to expand on to the lateral surface of the lower jaw (Rieppel & Gronowski, 1981). However, the upper temporal arcade usually persists, and, together with the fascia of the upper temporal fossa, it still sets the primitive limits of the once closed (anapsid) dermatocranium. These limits prevent expansion of the origin of the jaw adductors, in a dorsal and posterodorsal direction, on to the skull roof.

However, as the absolute size of the skull decreases, the relative size of the inner cylinder, i.e. the neurocranium, increases. This relative size increase of the neurocranium may be correlated with brain size and also with the function of the semicircular canals. Jones & Spells (1963) have shown that the radii of curvature of the semicircular canals are remarkably constant in animals of very different body mass. They have also calculated that a reduction of the radius of curvature would raise the lower response threshold of a semicircular canal. This study would therefore predict a relative increase of the radii of curvature of the semicircular canals and therewith an increase of the diameter of the otic capsules of the neurocranium as the absolute size of the skull decreases. This prediction is easily corroborated by a comparison of related lizards at different absolute size (Fig. 1). It is also corroborated, at least as far as lower tetrapods are concerned, by a graph published by Carroll (1970: fig. 5). The present author has attempted to extend this work by measuring the internal transverse diameter of the otic capsules in a series of radiographed lizard skulls, and by relating these values to the total width of the respective skulls as measured across the quadrate suspensions (Fig. 2). Plotted against absolute skull length, a marked relative increase of the width of the neurocranium becomes evident as the absolute skull length decreases. As a consequence thereof, the inner cylinder, the neurocranium, approaches and reaches the same level as the dermatocranium. The space between these two skull components vanishes, the roof of the neurocranium (supraoccipital) comes to lie at the same level as that of the dermatocranium (parietal) and the post-temporal fossae are closed. It can be shown (Fig. 2) that at a skull length of 15 mm or less the post-temporal fossae are either very narrow and slit-like, or more frequently closed, irrespective of the phylogenetic relationships of the lizards concerned. However, the post-temporal fossae may also be closed at a larger skull size, which shows that historical (phylogenetic) factors must also be considered.

In fossorial lizards the small size of the head results in a closure not only of the occiput but also of the lateral braincase wall. This is effected by union of a descending process of the parietal with the alar process of

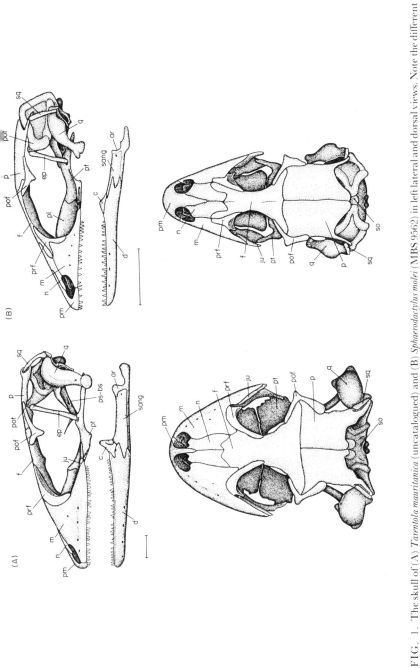

FIG. 1. The skull of (A) *Tarentola mauritanica* (uncatalogued) and (B) *Sphaerodactylus molei* (MBS 9562) in left lateral and dorsal views. Note the different relative size of the neurocranium and the prominent semicircular canals in *Sphaerodactylus molei*. Scale line equals 2 mm. (A key to abbreviations used in the figures is given on p. 520).

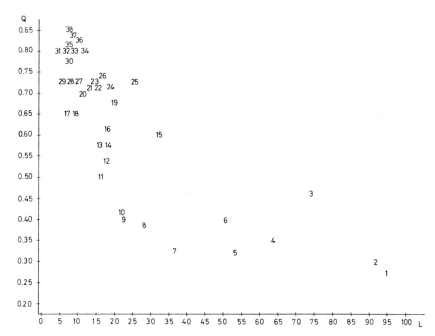

FIG. 2. The relative width of the otic capsule in relation to skull size. L, skull length in mm; Q, internal transverse diameter of the otic capsule divided by ½ total skull width as measured across the quadrate suspensions. The numbers refer to the following species: 1, *Tupinambis teguixin*; 2, *Varanus exanthematicus*; 3, *Iguana iguana*; 4, *Rhacodactylus trachyrhynchus*; 5, *Heloderma horridum*; 6, *Sphenodon punctatus*; 7, *Gekko gekko*; 8, *Ameiva ameiva*; 9, *Lanthanotus borneensis*; 10, *Tarentola mauritanica*; 11, *Bradypodion pumilus*; 12, *Lacerta lilfordi*; 13, *Chalcides ocellatus*; 14, *Lacerta agilis*; 15, *Acontias plumbeus*; 16, *Pygopus lepidopodus*; 17, *Sphaerodactylus molei*; 18, *Algyroides fitzingeri*; 19, *Acontias meleagris*; 20, *Acontias lineatus*; 21, *Acontias percivali*; 22, *Feylinia currori*; 23, *Anguis fragilis*; 24, *Ophiodes* sp.; 25, *Lialis jicari*; 26, *Acontias gracilicauda*; 27, *Dibamus novaeguineae*; 28, *Dibamus argenteus*; 29, *Aprasia striolata*; 30, *Feylinia elegans*; 31, *Aprasia repens*; 32, *Nessia layardi*; 33, *Typhlosaurus vermis*; 34, *Typhlosaurus auriantiacus*; 35, *Pletholax gracilis*; 36, *Typhlosaurus auriantiacus*; 37, *Typhlosaurus lineatus*; 38, *Anniella pulchra, Typhlosaurus braini*.

the prootic at the same level. The prootic may be expanded anteriorly by a crista alaris of variable size (Rieppel, 1981). In combination with the development of a closed lateral braincase wall the upper temporal arcade is usually lost, or at least it closely approaches the descending process of the parietal. If the upper temporal arcade is interrupted, the postorbitofrontal persists as a V-shaped element bracing the potentially movable (mesokinetic) frontoparietal suture laterally. Of the squamosal, only the posterior portion persists, if at all, in relation to the quadrate suspension.

However, not all small lizards develop a closed lateral braincase wall (*Sphaerodactylus*: Fig. 1; *Algyroides fitzingeri*: Fig. 3). This is rather typical of fossorial or burrowing forms of different phylogenetic relationships.

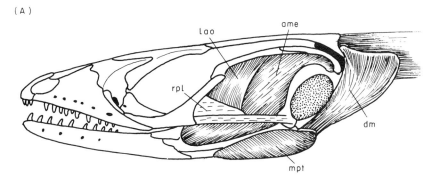

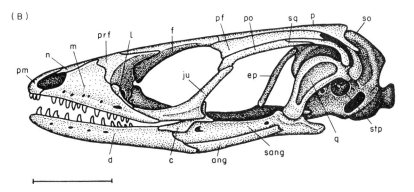

FIG. 3. The skull and superficial jaw adductor musculature of *Algyroides fitzingeri* (uncatalogued). Scale line equals 2 mm.

Thus, this trait cannot be viewed as only related to small size, but must also be considered in correlation with burrowing or fossorial habits (Bellairs & Underwood, 1951; Bellairs, 1972).

The development of a closed skull can be nicely documented by considering the Acontinae, a subfamily of the Scincidae from South Africa, which comprises three genera, *Acontias*, *Acontophiops* and *Typhlosaurus*. All species are limbless and fossorial. *Acontias plumbeus* is the only large representative of its genus, retaining an adult skull length of about 25 mm. The skull structure is primitive. Post-temporal fossae are present as well as a complete upper temporal arcade. Laterally descending processes of the parietals are present, but they are not as extensive as in other species of the genus, which are all distinctly smaller. These show an adult skull length of 15 mm or less and they document the step-wise closure of the post-temporal fossae (Broadley & Greer, 1969; Rieppel, 1981). The lateral wall of the braincase is more or less closed and the upper temporal arcade is incomplete.

All species of the genus *Typhlosaurus* have an adult skull length of less than 15 mm and hence show a closed occiput. A complete upper temporal arcade persists only in *Typhlosaurus lineatus* (Greer, 1970), but it closely approaches the descending process of the parietal, thus reducing the size of the upper temporal fossa. It is also overcrowded in its posterior part by the superficial fibres of the external jaw adductor (Rieppel, 1981). In the other species of *Typhlosaurus* the upper temporal arcade is incomplete and the squamosal closely applied to the laterally descending process of the parietal. Its step-wise reduction from front to back can be documented within the genus (Rieppel, 1981, 1982). The reduction of the upper temporal arcade might indeed represent a paedomorphic feature.

A similar pattern of change can be observed within the Anguidae. The skull of *Anguis fragilis* [Fig. 4(A)] with an adult length of around 15 mm retains a basically primitive structure. There are narrow post-temporal fossae and a complete upper temporal arcade, which closely approaches the parietal, however. The skull of the related *Anniella pulchra* [Fig. 4(B)] with an adult length of 7–8 mm shows the typical changes associated

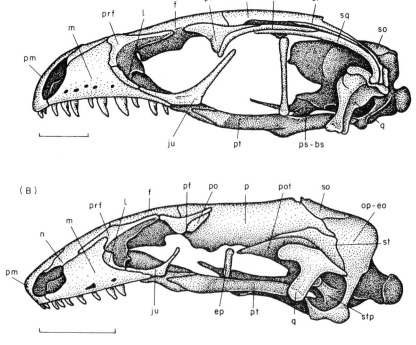

FIG. 4. The skulls of (A) *Anguis fragilis* (uncatalogued) and (B) *Anniella pulchra* (MBS 8331) in left lateral views. Scale line equals 2 mm.

with miniaturization. The occiput is completely closed, the parietal develops extensive lateral downgrowths, the upper temporal arcade is lost with the squamosal completely reduced, and the supratemporal is fully incorporated in the posterolateral braincase wall.

A third group of miniaturized fossorial lizards is found among the Pygopodidae. In the Gekkota the upper temporal arcade is always incomplete, irrespective of skull size. The reasons for its reduction are unknown. In the rather large *Pygopus lepidopodus* (Fig. 5), with an adult skull length of around 17 mm, narrow post-temporal fossae persist. There are descending processes of the parietals, but they are not very deep and they originate from the lower surface of the parietal as in a number of other lizards. Hence, the lateral edge of the parietal projects laterally beyond the descending process and thus prevents any expansion of the jaw adductor muscles in a posterodorsal direction. In the miniaturized skulls of the genera *Pletholax* (Fig. 6) or especially *Aprasia repens* (Fig. 7) with adult lengths of between 4 and 8 mm the neurocranium and dermatocranium come to lie at the same level. The post-temporal fossae are either reduced to a fissure between the parietal and supraoccipital (*Pletholax*) or are fully closed (*Aprasia*). The descending process of the parietal is somewhat enlarged, and lies in the same plane as the alar process of the prootic. Thus the descending process merges into the lateral edge, and not into the lower surface, of the parietal (Stephenson, 1962; Figs 6 & 7). If it persists, the squamosal is smoothly applied to the descending process of the parietal, and is partly overgrown by the origin of jaw adductor musculature. *Aprasia repens* (Fig. 7) is peculiar in that the posterior ramus of the V-shaped

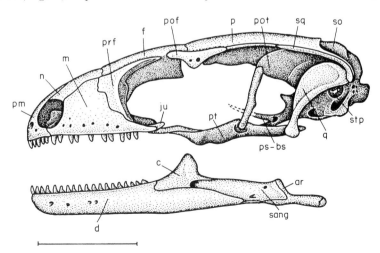

FIG. 5. The skull of *Pygopus lepidopodus* (MBS 19662) in left lateral view. Scale line equals 5 mm.

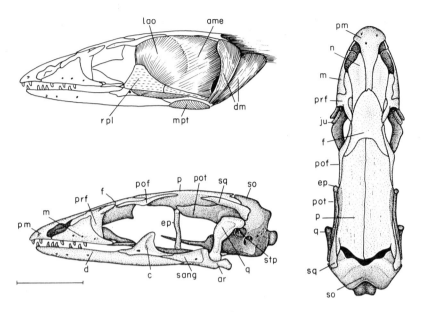

FIG. 6. The skull and superficial jaw adductor musculature of *Pletholax gracilis* (UMMZ 173966). Scale line equals 2 mm.

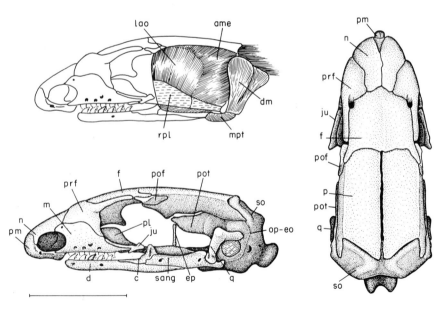

FIG. 7. The skull and superficial jaw adductor musculature of *Aprasia repens* (UMMZ 173865). Scale line equals 2 mm.

postorbitofrontal is twisted and smoothly applied to the parietal downgrowth. Its surface is invaded by the anterior fibres of the external jaw adductor.

THE JAW ADDUCTOR MUSCULATURE

Miniaturization may be correlated with changes in the arrangement of the jaw adductor musculature. Maintenance of an effective muscle fibre length as the lizard head becomes very small appears to be a problem. This may necessitate the expansion of the area of origin of the jaw adductors beyond the limits set by the primitively closed (anapsid) dermatocranium.

The degree to which a muscle fibre may be stretched, within a reasonable range of the length-tension curve, depends on its absolute length, i.e. on the number of sarcomeres of which it is composed. In a small skull the limited fibre length may restrict the excursion range of the jaw adductors and thus may restrict gape. However, the issue is rather complex and hence the problem appears to have been circumvented in a number of ways.

On the one hand, the relative length of the individual muscle fibre is determined not only by the size of the head but also by the internal structure of the jaw adductor complex. On the other hand, the degree of relative stretching of the muscle fibres upon jaw depression is correlated with the angle of insertion of these fibres and with the distance of the insertion from the lower jaw joint (Rieppel & Gronowski, 1981). Small geckos (Sphaerodactylinae) as well as the small lacertid *Algyroides fitzingeri* (Fig. 3) show no expansion of the jaw adductors in a posterodorsal direction. But, compared to other lizards, most Gekkonidae and Lacertidae show some reduction of the internal tendinous skeleton of the jaw adductors (bodenaponeurosis) which results in a less complex pinnation and hence in a lengthening of the individual fibres. In *Sphaerodactylus* [Fig. 1(B)] and *Algyroides* (Fig. 3) the coronoid process is very low. In *Algyroides*, the postorbital segment of the skull is relatively long, which results in an oblique orientation of the muscle fibres, thus reducing the relative degree to which they are stretched during depression of the lower jaw.

The situation is different in the small fossorial lizards of the subfamily Acontinae, in *Anniella* and *Dibamus*, and to some extent in the small pygopodids, *Aprasia* and *Pletholax*. These lizards retain an extensive and complex internal tendinous skeleton of the jaw adductors which results in an asymmetrical pinnate muscle architecture. This arrangement has the advantage of maximizing the number of muscle fibres and of broadening the range of the length-tension optimum (Gans & Bock, 1965; Alexander & Goldspink, 1977). The individual muscle fibres are shorter, however. This arrangement results in a greater adductive force, which may reflect dietary specializations. A possible reason for the multipinnate and extended jaw adductor muscles of all these lizards may be that small arthropods are relatively harder to crush than larger ones (H. F. Rowell, pers. comm.).

The shortness of the individual fibres, together with the increase of their number, may have necessitated the expansion of the external jaw adductor beyond the limits of a primitively closed (anapsid) dermatocranium. The expansion occurs in a posterodorsal direction, to what Säve-Söderbergh (1945: 9) has termed the "temporalis position". Such an expansion of the area of origin of the external adductor is made possible by the closed lateral wall of the braincase, the correlated reduction of the upper temporal arcade, and the closure of the occiput, in these miniaturized lizards. The expansion is least developed in the small pygopodids *Pletholax* (Fig. 6) and *Aprasia* (Fig. 7). There it involves the entire lateral downgrowth of the parietal, the lateral surface of the squamosal and of the quadrate, but the muscle fibres do not actually invade the broad skull roof. This may be correlated with the very low coronoid process.

In the Aycontinae and in *Anniella* the muscle fibres invade the entire vaulted lateral wall of the braincase, the entire lateral surface of the quadrate, squamosal and supratemporal as well as the posterolateral process of the parietal. The jaw adductors do not expand up to the dorsal mid-line of the skull, however, as they do in *Dibamus* (Fig. 8). In some cases, such as in advanced species of *Typhlosaurus* (Rieppel, 1981) and very extensively in *Dibamus* (Fig. 8), posterodorsal adductor fibres even invade the fascia covering the epaxial neck muscles, which themselves insert into the dorsal surface of the supraoccipital.

As a consequence of the posterodorsal expansion of the massive external jaw adductor in these attenuated skulls its main resultant force vector is rotated so as to form a very acute angle with the long axis of the lower jaw. On the one hand this reduces the relative stretching of the obliquely orientated individual fibres as the lower jaw is depressed, but on the other hand it implies a loss of mechanical advantage. This loss could be at least partly compensated by an elevation of the coronoid process above the level of the jaw articulation (DeMar & Barghusen, 1972). A higher coronoid process would, however, reduce the length of the anteriormost fibres of the external adductor which insert almost vertically into the apex of the coronoid process. Raising the coronoid process would shorten precisely those muscle fibres which are most extensively stretched upon jaw opening. In order to allow for a higher

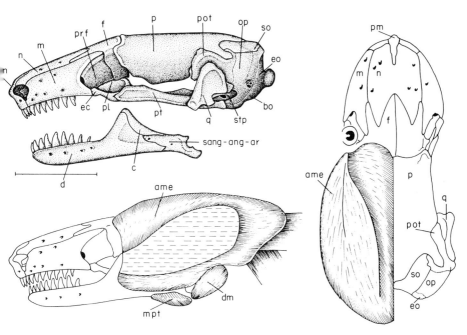

FIG. 8. The skull and superficial jaw adductor musculature of *Dibamus novaeguineae* (AMNH 86717). Scale line equals 2 mm.

coronoid process either these anterior fibres would have to be obliterated or they too would have to expand their area of origin up on to the skull roof. That is rendered impossible by the presence of the postorbitofrontals which laterally brace the potentially movable frontoparietal suture, an important mechanical trait in these fossorial lizards (Rieppel, 1981). Thus anteriorly the postorbitofrontal sets the old limits of the primitively closed dermatocranium. The anterior fibres of the external jaw adductor originate from its lower surface.

Dibamus (Fig. 8) is the only lizard which has lost the postorbitofrontal. This is correlated with the loss of mesokinesis at the complexly joined frontoparietal suture. It is interesting to note that, as compared to other lizards, the coronoid process of *Dibamus* is disproportionately high (Figs 8 & 9), and this is correlated with an origin of the anterior adductor fibres from the dorsal surface of the posterolateral part of the frontal and of the anterior part of the parietal. These features, which are unknown from any other lizard, provide the jaw adductor system of *Dibamus* with a slightly improved mechanical advantage (Fig. 9). With respect to the characteristics outlined above, *Dibamus* resembles the Amphisbaenia (Rieppel, 1979). They too show a very high coronoid

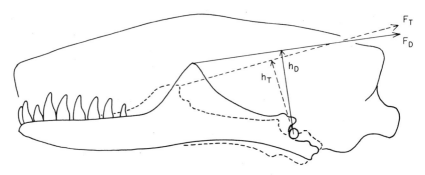

FIG. 9. The lower jaw of *Dibamus novaeguineae* (solid line) and of *Typhlosaurus cregoi* (broken line) in relation to idealized skull contours. F_T and F_D indicate the approximate resultant direction of the force exerted by the external jaw adductor in *Dibamus* (F_D) and in *Typhlosaurus* (F_T). The higher coronoid process in *Dibamus* results in an increased mechanical advantage (h_D) as compared to *Typhlosaurus* (h_T).

process, loss of the postorbitofrontal and an expansion of the anterior and posterior jaw adductor fibres up to the dorsal mid-line of the skull.

THE ORIGIN OF SNAKES

The concept of miniaturization can add to the formulation of an evolutionary scenario for the origin of snakes. Among squamate reptiles the Amphisbaenia as well as the Ophidia are characterized by a completely closed cranial box surrounding the brain. Again this is achieved by a combination of the dermatocranium and neurocranium. The parietal and frontal bones form extensive lateral downgrowths which actually meet the basicranium in a closed suture. The course of the trigeminal nerve indicates that the complete closure of the braincase has been achieved independently in the two groups. In the Amphisbaenia the ophthalmic branch passes anteriorly into the orbit, lateral to the parietal downgrowth, i.e. extracranially; in snakes the nerve runs intracranially and enters the orbit together with the optic nerve though the optic foramen (Bellairs & Kamal, 1981). It appears plausible to hypothesize a miniaturized, fossorial ancestral stage for both the Amphisbaenia and the Ophidia. As they regained a larger size, the burrowing habits were retained by the Amphisbaenia and the structural adaptations improved by certain traits peculiar to that group, such as the angulation of the facial versus the occipital region of the skull (Gans, 1960, 1974).

The origin of snakes from some fossorial squamate ancestor has been supported by a number of authors on account of numerous anatomical specializations, mainly of the optic system, but also of the skull (Walls,

1940; Bellairs & Underwood, 1951; Senn, 1966; Bellairs, 1972; Rieppel, 1978). A squamate ancestor, adapting to fossorial habits by miniaturization, would approach the ophidian *bauplan* of skull structure. The dermatocranium and neurocranium would come to lie at the same level, the occiput of the skull would be closed and the formation of a more or less closed lateral braincase wall, together with loss of the upper temporal arcade, would allow for expansion of the jaw adductor musculature to the temporalis position. Their origins would extend over the entire lateral wall of the parietal up to the dorsal mid-line of the skull.

All these modifications, which are to be considered a result of miniaturization, can be viewed as exaptations *sensu* Gould & Vrba (1982) for the future evolution of snakes. They provided the basis for an adaptive shift during the early evolution of snakes. The group retained the structural modifications gained by miniaturization as it returned to a supraterrestrial mode of life and grew to larger sizes once again. Snakes adapted to relatively large prey by the development of extreme mobility of the upper and lower jaws. The extensive cranial kinesis required an increased excursion range of the jaw adductors, and this was granted by the posterodorsally expanded muscles in these, now larger, snakes.

In the primitive fossorial Anilioidea (*Anilius* and *Cylindrophis*) cranial kinesis has not yet reached the degree of perfection characteristic of higher snakes. Prey diameter relative to the anilioidean snake's head diameter is relatively small, the "ingestion ratio" accordingly relatively low (Greene, in press). This is reflected in the structure of the jaw adductor muscles which retain a complex internal tendinous skeleton closely comparable to that of some fossorial miniaturized lizards (Rieppel, 1980a, in preparation). This complexly pinnate muscle architecture results in relatively shorter individual muscle fibres.

Higher snakes such as the Booidea and Colubroidea show a more perfected system of cranial kinesis (Gans, 1961; Frazzetta, 1966; Rieppel, 1980b) resulting in a higher ingestion ratio (Greene, in press). The increase of cranial mobility is correlated with the complete reduction of the internal tendinous skeleton of the jaw adductor muscles. This results in the very long, parallel-fibred muscle bundles characterizing the "three externi type" jaw adductors of higher snakes (Haas, 1973: fig. 145). With a further posterodorsal expansion of the jaw adductors, caused by a posterior expansion of the supratemporal (which now forms a free ending posterior process), the individual muscle fibres become long enough to provide the excursion range necessary for an extreme ingestion ratio.

ACKNOWLEDGEMENTS

I would like to express my sincere gratitude to PD Dr R. Otto, who made the radiography possible at the Röntgendiagnostisches Zentralinstitut, Zürich. Dr H. W. Greene kindly made available a preprint of his work on dietary correlates of the origin and radiation of snakes. The study is based on material generously made available by a number of persons and their respective institutions: Dr E. N. Arnold and Miss A. C. Grandison, British Museum (Natural History), London; Dr W. Böhme, Museum Alexander König, Bonn; Dr W. R. Branch, Port Elizabeth Museum, South Africa; Dr D. G. Broadley, Umtali Museum, Zimbabwe; Dr A. E. Greer, The Australian Museum, Sydney; Dr W. D. Haacke, Transvaal Museum, Pretoria; Dr A. G. Kluge, University of Michigan Museum of Zoology, Ann Arbor; Dr V. Mahnert, Muséum d'Histoire Naturelle, Geneva; Dr U. Rahm, Naturhistorisches Museum, Basel; Dr E. E. Williams and J. Rosado, Museum of Comparative Zoology, Cambridge (Mass.); Dr R. G. Zweifel, American Museum of Natural History, New York.

REFERENCES

Alexander, R. McN. & Goldspink, G. (1977). *Mechanics and energetics of animal locomotion.* London: Chapman & Hall.
Bellairs, A. d'A. (1972). Comments on the evolution and affinities of snakes. In *Studies in vertebrate evolution*: 157–172. Joysey, K. A. & Kemp, T. S. (Eds). Edinburgh: Oliver & Boyd.
Bellairs, A. d'A. & Kamal, A. M. (1981). The chondrocranium and the development of the skull in Recent reptiles. In *Biology of the Reptilia*, **11**: 1–263. Gans, C. & Parsons, T. S. (Eds). London and New York: Academic Press.
Bellairs, A. d'A. & Underwood, G. (1951). The origin of snakes. *Biol. Rev.* **26**: 193–237.
Brady, R. H. (1982). Dogma and doubt. *Biol. J. Linn. Soc.* **17**: 79–96.
Broadley, D. G. & Greer, A. E. (1969). A revision of the genus *Acontias* (Sauria: Scincidae). *Arnoldia* **4**: 1–29.
Carroll, R. L. (1970). Quantitative aspects of the amphibian-reptilian transition. *Forma functio* **3**: 165–178.
DeMar, R. & Barghusen, H. R. (1972). Mechanics and the evolution of the synapsid jaw. *Evolution, Lawrence, Kans.* **26**: 622–637.
Frazzetta, T. H. (1966). Studies on the morphology and function of the skull in the Boidae (Serpentes). Pt. II. Morphology and function of the jaw apparatus in *Python sebae* and *Python molurus. J. Morph.* **118**: 217–296.
Frazzetta, T. H. (1968). Adaptive problems and possibilities in the temporal fenestration of tetrapod skulls. *J. Morph.* **125**: 145–158.
Gans, C. (1960). Studies on amphisbaenids (Amphisbaenia, Reptilia). 1. A taxonomic revision of the Trogonophinae, and a functional interpretation of the amphisbaenid adaptive pattern. *Bull. Am. Mus. nat. Hist.* **119**: 133–204.

Gans, C. (1961). The feeding mechanism of snakes and its possible evolution. *Am. Zool.* **1**: 217–227.
Gans, C. (1974). *Biomechanics*. Philadelphia and Toronto: J. B. Lippincott Co.
Gans, C. & Bock, W. J. (1965). The functional significance of muscle architecture – a theoretical analysis. *Ergebn. Anat. EntwGesch.* **38**: 115–142.
Gould, S. J. (1977). *Ontogeny and phylogeny*. Cambridge, Mass.: Harvard University Press.
Gould, S. J. & Lewontin, R. C. (1979). The spandrels of San Marco and the panglossian paradigm: a critique of the adaptationist program. *Proc. R. Soc. Lond.* **205**: 581–598.
Gould, S. J. & Vrba, E. S. (1982). Exaptation – a missing term in the science of form. *Paleobiology* **8**: 4–15.
Greene, H. W. (In press). Dietary correlates of the origin and radiation of snakes. *Am. Zool.*
Greer, A. E. (1970). A subfamilial classification of scincid lizards. *Bull. Mus. comp. Zool. Harv.* **139**: 151–184.
Haas, G. (1973). Muscles of the jaws and associated structures in the Rhynchocephalia and Squamata. In *Biology of the Reptilia*, **4**: 285–490. Gans, C. & Parsons, T. S. (Eds). London and New York: Academic Press.
Hanken, J. (1980). Osteological variation at small size in salamanders of the genus *Thorius*. *Am. Zool.* **19**: 987 (abstract).
Huey, R. B. & Pianka, E. R. (1974). Ecological character displacement in a lizard. *Am. Zool.* **14**: 1127–1136.
Huey, R. B., Pianka, E. R., Egan, M. E. & Coons, L. W. (1974). Ecological shifts in sympatry: Kalahari fossorial lizards (*Typhlosaurus*). *Ecology* **55**: 304–316.
Jones, G. M. & Spells, K. E. (1963). A theoretical and comparative study of the functional dependence of the semicircular canal upon its physical dimensions. *Proc. R. Soc. Lond.* (B) **157**: 403–419.
Lewontin, R. C. (1978). Adaptation. *Scient. Am.* **239**: 156–169.
Pianka, E. R. (1974). *Evolutionary ecology*. New York and San Francisco: Harper & Row Publ.
Pianka, E. R. & Pianka, H. D. (1976). Comparative ecology of twelve species of nocturnal lizards (Gekkonidae) in the western Australian desert. *Copeia* **1976**: 125–142.
Rieppel, O. (1978). The evolution of the naso-frontal joint in snakes and its bearing on snake origins. *Z. zool. Syst. EvolForsch.* **16**: 14–27.
Rieppel, O. (1979). The external jaw adductor of amphisbaenids (Reptilia: Amphisbaenia). *Rev. suisse Zool.* **86**: 867–876.
Rieppel, O. (1980a). The trigeminal jaw adductors of primitive snakes and their homologies with the lacertilian jaw adductors. *J. Zool., Lond.* **190**: 447–471.
Rieppel, O. (1980b). The evolution of the ophidian feeding system. *Zool. Jb.* (Anat.) **103**: 551–564.
Rieppel, O. (1981). The skull and jaw adductor musculature in some burrowing scincomorph lizards of the genera *Acontias*, *Typhlosaurus* and *Feylinia*. *J. Zool., Lond.* **195**: 493–528.
Rieppel, O. (1982). The phylogenetic relationships of the genus *Acontophiops* Sternfeld (Sauria: Scincidae), with a note on mosaic evolution. *Ann. Transv. Mus.* **33**: 241–257.
Rieppel, O. & Gronowski, R. W. (1981). The loss of the lower temporal arcade in diapsid reptiles. *Zool. J. Linn. Soc.* **72**: 203–217.
Säve-Söderbergh, G. (1945). Notes on the trigeminal musculature in non-mammalian tetrapods. *Nova Acta R. Soc. Scient. upsal.* (4) **13**: 1–59.

Seilacher, A. (1982). Constraint and innovation in bivalve mollusc evolution. In *Evolution and development*: 302–305. Bonner, J. T. (Ed.). Berlin: Springer Verlag.
Senn, D. G. (1966). Ueber das optische System im Gehirn squamater Reptilien. *Acta Anat.* (Suppl.) **52**: 1–87.
Stephenson, N. G. (1962). The comparative morphology of the head skeleton, girdles and hind limbs in the Pygopodidae. *J. Linn. Soc.* (Zool.) **44**: 627–644.
Walls, G. L. (1940). Ophthalmological implications for the early history of snakes. *Copeia* **1940**: 1–8.
Wright, S. (1982). Character change, speciation and the higher taxa. *Evolution, Lawrence, Kans.* **37**: 427–442.

APPENDIX – KEY TO ABBREVIATIONS USED IN FIGURES

Skull

ang	angular	pf	post-frontal
ar	articular	pl	palatine
bo	basioccipital	pm	premaxilla
c	coronoid	po	postorbital
d	dentary	pof	postorbitofrontal
ec	ectopterygoid	pot	prootic
eo	exoccipital	prf	prefrontal
ep	epipterygoid	ps-bs	parasphenoid-basisphenoid
f	frontal	pt	pterygoid
ju	jugal	q	quadrate
l	lacrimal	sang	surangular
m	maxilla	so	supraoccipital
n	nasal	sq	squamosal
op	opisthotic	st	supratemporal
op-eo	opisthotic-exoccipital	stp	stapes
p	parietal		

Musculature

ame	m. adductor mandibulae externus	lao	levator anguli oris
		mpt	m. pterygoideus
dm	depressor mandibulae	rpl	rictal plate

On the Cranial Morphology and Evolution of Ornithopod Dinosaurs

DAVID B. NORMAN

Department of Zoology and University Museum,
South Parks Road, Oxford OX1 3PS, England

SYNOPSIS

It has been common practice to discuss the evolution of, and relationships between, ornithopod dinosaurs by reference to several distinct morphological 'grade-groups'. It is also generally accepted that these groups are merely convenient frames of reference (despite their taxonomic attribution to familial status) which serve to bring some, albeit artificial, order to the Ornithopoda but also reflect our incomplete knowledge of the group as a whole.

Despite the inconsistent phylogenetic nature of these 'grade-groups', there are nevertheless some benefits that can be derived from a broad consideration of the functional/biological trends represented in these groups. A brief outline of the comparative cranial morphology and function within various representative ornithopods demonstrates just such a potential benefit. As a result, it is tentatively proposed that a novel form of cranial kinesis (pleurokinesis) may be found in some ornithopods and its functional significance is discussed.

Finally, since our knowledge of ornithopods has greatly improved in recent years, a preliminary systematic review of the Ornithopoda is presented using a cladistic methodology.

INTRODUCTION

There is tacit approval of a very generalized evolutionary scheme for the ornithopod dinosaurs (Thulborn, 1971; Galton, 1972, 1974a, b). This relies mainly on the concept of grades of anatomical organization progressing from primitive to advanced ornithopods via four levels. The most primitive ornithopods, and indeed those implicated in the ancestry of the Ornithischia as a whole, are the 'fabrosauroids' (Fabrosauridae, Galton, 1972) – small, bipedal ornithischians which show no obvious cranial or post-cranial specializations. Derived from the former group are the 'hypsilophodontoids' (Hypsilophodontidae, Dollo, 1882) – a morphologically conservative group of small to medium sized ornithopods; these are in turn considered to be ancestral to several, supposedly iteratively derived, groups of larger and more anatomically specialized ornithopods called 'iguanodontoids' (Iguanodontidae *sensu* Galton, 1974b). Finally, the 'hadrosauroids' (Hadrosauridae, Cope, 1883), representing the most sophisticated ornithopods, are characterized by their dental and cranial specializations (Lull & Wright, 1942) and are derived from the 'iguanodontoids'.

The fifth group usually implicated in ornithopod phylogeny, the 'heterodontosauroids' (Heterodontosauridae, Romer, 1966) is excluded from the suborder Ornithopoda for reasons which are explained by Santa-Luca (1980) and with which the present author tentatively concurs.

From the point of view of attempting to understand the nature of biological trends within the Ornithopoda, it has proved useful to view these imperfectly known dinosaurs in terms of these levels of organization, artificial though they may be in many respects. Indeed this type of conceptual framework has been used to investigate some aspects of cranial function in this group (pp. 535 and 539). However, the concept of these grade groups as an evolutionary continuum ('fabrosauroid' → 'hypsilophodontoid' → 'iguanodontoid' → hadrosauroid') can be misleading from a phylogenetic/systematic viewpoint because the grade-group pattern can obscure phyletic patterns. Indeed it was partly conceded by Galton (1974b: 1065) that the Iguanodontidae, as currently defined, may be polyphyletic (see also Dodson, 1980). Since many ornithopods have been partly or wholly redescribed in recent years, it is considered timely to attempt to produce a preliminary model of the apparent systematic relationships between the various ornithopod genera. This, it is hoped, will serve as a starting point for the elucidation of the phyletic relationships of the Ornithopoda and, more generally, of the Ornithischia as a whole (Norman, in preparation).

GENERAL CRANIAL ARCHITECTURE

The following account forms a general comparative survey of skull form in the various ornithopod groups currently recognized. As previously mentioned these fall into distinct grades of organization which will be recognized by the informal titles used in the Introduction.

The 'Fabrosauroid'

There are several genera of 'fabrosauroid' currently recognized: *Fabrosaurus* Ginsburg, 1964; *Echinodon* Owen, 1861; *Nanosaurus* Marsh, 1877; *Alocodon* and *Trimucrodon* Thulborn, 1973; and *Scutellosaurus*

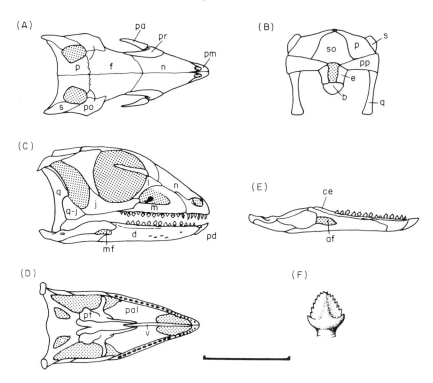

FIG. 1. *Lesothosaurus diagnosticus* Galton, 1978. Skull and lower jaw (after Thulborn, 1970 and BMNH.R.8501). (A) Dorsal view of skull. (B) Occipital view. (C) Lateral view. (D) Ventral view. (E) Lower jaw and teeth. (F) Representative tooth. Scale bar = 5 cm. For key to abbreviations, see p. 547.

Colbert, 1981. However, the only adequately preserved and described cranial material is that of *Lesothosaurus* Galton, 1978 [described as *Fabrosaurus* (Thulborn, 1970)]. *Lesothosaurus* will therefore be considered as representative of fabrosauroids (Fig. 1).

The skull table is essentially flat, and approximately rectangular in dorsal view [Fig. 1(A)]. The parietals form the posterior margin of the skull roof and, unusually for ornithopods, retain a clear parietal suture (Thulborn, 1970; this is not visible in BMNH.R.8501, a newly prepared skull of *Lesothosaurus*, and may therefore be an ontogenetic feature). The parietals are expanded laterally to form the posterior wall of the upper temporal fenestra, and meet the squamosals at an oblique suture at the rear corner of the skull roof. Each squamosal sends a slender process forward to meet a posterior extension of the postorbital, at an overlapping suture, forming the inter-temporal bar. The fused frontals (BMNH.R.8501) are sutured to the parietals posteriorly and the postorbitals posterolaterally; their lateral margins form part of the dorsal margin of the orbit, while anteriorly the frontals are overlapped by the extreme posterior tip of the prefrontal and medially by the nasals.

The occiput [Fig. 1(B)] is not very well known (Thulborn, 1970, and BMNH.R.8501). So far as can be seen, the foramen magnum was bordered laterally by pillar-like exoccipitals which extend dorsally and laterally from the occipital condyle forming the paroccipital wings; the extent to which the opisthotics are involved in paroccipital support is unknown, although Thulborn (1970) speculated on their posterior extension to support the anterolateral margin of the paroccipital process. Dorsally, the foramen magnum is bounded by the supraoccipital which is a broad, curved plate running obliquely forward from the exoccipitals beneath the parietals. The major part of the braincase and its relationship to the bones of the occiput and skull table were not revealed in the specimens described by Thulborn (1970) but are preserved in a newly prepared specimen (BMNH.R.8501) which is to be described by Charig & Crompton (in preparation).

The cheek region (the lateral wall of the skull behind the snout) is formed of an open frame of struts surrounding the lateral temporal fenestra and orbit [Fig. 1(C)]. The posterior margin of the cheek is formed by the slender, curved, pillar-like quadrate which fits into a notch in the ventral surface of the squamosal. A narrow splint of the squamosal descends along the anterior margin of the jugal wing of the quadrate; the lower half of this jugal wing is embayed and sutured to the robust, triangular quadratojugal which in turn forms an oblique suture with the posterior extremity of the jugal (BMNH.R.8501). The jugal is quite slender and curved; its anterior end tapers slightly, abuts the lachrymal and is involved in a long overlapping suture with the maxilla (BMNH.R.8501); dorsally, the jugal extends upward behind the orbit to form an overlapping suture with a ventral extension of the postorbital.

The snout (the facial region of the skull anterior to the orbit) tapers quite sharply anteriorly [Fig. 1(A), (C)]. The nasals are thin, arched, tapering bones roofing the nasal and buccal region, flanked posteriorly by the prefrontals (which incidentally support the base of the slender palpebral bone), meeting the maxillae and lachrymal laterally and the premaxillae rostrally. The premaxillae are apparently butt-jointed in the mid-line (BMNH.R.8501) and taper to a narrow, rounded tip. Posterodorsally, the premaxillae produce a slender curved prong which runs backward between the anterior tips of the nasal (BMNH.R.8501). Posterior and lateral to this median process there is another posterior extension of each premaxilla which lies obliquely over the maxilla, thus partially separating maxilla and nasal; between these two processes of the maxilla, there is an embayed area roofed by the curved nasal margin which marks the external naris.

The palate [Fig. 1(D)] is not very well preserved in the originally described material (Thulborn, 1970) but consists of thin mid-line vomers, broad moderately vaulted palatines and slender pterygoids. The latter are separated in the mid-line and connect the anterior palate to the suspensorium and the maxilla laterally; apart from the moderate vaulting of the palatines and separation of the pterygoids, the palate bones exhibit conventional reptilian relationships (the palate is well preserved in BMNH.R.8501).

The mandible [Fig. 1(E)] is long, slender, has a longitudinal sigmoid curve and is transversely compressed. The predentary is edentulous and arrow-head shaped, with a large ventral process which underpins the dentary symphysis. The adductor fossa is large and there is both a conventional archosaurian mandibular fenestra between the dentary, angular and surangular and also a surangular foramen (BMNH.R.8501). There is a low coronoid 'eminence' (Galton, 1978) posterior to the dental alveoli, and a well developed retroarticular process.

The dentition is heterodont. The premaxillary teeth, of which there are six, are relatively simple in structure. The most anterior have cylindrical roots and crowns which are laterally compressed, slightly recurved and sharply pointed. Later in the series the crowns become less tall, more triangular in outline, and bear tiny serrations on mesial and distal edges (Thulborn, 1970). These latter are transitional to the maxillary teeth which have roots which are ellipsoid in cross-section and slightly waisted beneath the crown. The crowns of maxillary and dentary teeth are broad and triangular in outline [Fig. 1(F)], transversely compressed and markedly serrated along mesial and distal edges. Bifacial wear facets on individual teeth have been described by Thulborn (1971). Contrary to the statements by Galton (1972, 1973, 1974b, 1978) the dentary teeth are inset from the jaw margin and there are dental foramina with an alveolar parapet (BMNH.R.8501).

The 'Hypsilophodontoid'

The past decade has seen a remarkable increase in knowledge of the 'hypsilophodontoids' largely through the work of Galton (1974a, 1975, 1977, 1980; Galton & Taquet, 1982). By far the best described of these dinosaurs is *Hypsilophodon* (Galton, 1974a) (Fig. 2) which is therefore taken as representative of this taxonomically extensive but morphologically conservative group.

The skull roof [Fig. 2(A)] is remarkably similar to that of *Lesothosaurus* apart from relatively minor differences of proportion.

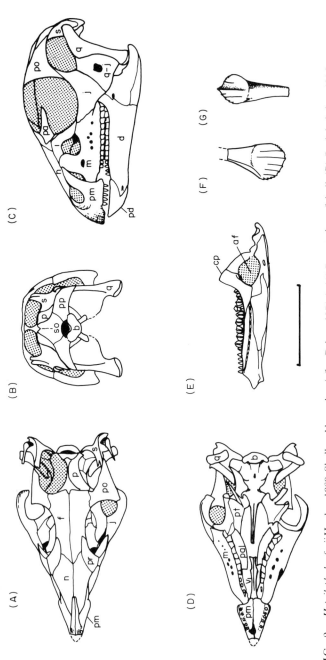

FIG. 2. *Hypsilophodon foxii* Huxley, 1869. Skull and lower jaw (after Galton, 1974a). (A) Dorsal view of skull. (B) Occipital view. (C) Lateral view. (D) Ventral view. (E) Lower jaw and teeth. (F) Maxillary tooth (labial). (G) Dentary tooth (lingual). Scale bar = 5 cm. For key to abbreviations, see p. 547.

The occiput [Fig. 2(B)] is by contrast to the 'fabrosauroid' reasonably well preserved. The supraoccipital spans the dorsal margin of the foramen magnum and runs anterodorsally to meet the ventral edges of the posterolateral wings of the parietal along a very weak suture; ventrolaterally, the supraoccipital rests against the base of the massive, transverse, paroccipital processes. These are probably formed of the fused exoccipitals and opisthotics (Norman, 1980) although Galton (1974a: fig. 9) described small wedge-like exoccipitals in the ventrolateral corner of the foramen magnum. The squamosals are quite extensively sutured to the parietals medially and the paroccipitals ventrally. The braincase (neurocranium) is well preserved in BMNH.R.2477 and comprises a rigidly fused box, the posterodorsal roof of which is formed by the supraoccipital, the lateral walls by proötic and opisthotic and the floor by basioccipital, basisphenoid and parasphenoid. The pleurosphenoid, as preserved, is fused to neither the anterior margins of the braincase nor the skull table (probably a juvenile feature). A feature of note is the fact that the anterolateral extremity of the pleurosphenoid has the form of a smooth, transverse, condyle which is received by a correspondingly smooth depression on the ventral surface of the frontal-postorbital suture. The more anterior elements of the braincase are absent and were presumably cartilaginous.

The cheek [Fig. 3(C)] region again closely resembles *Lesothosaurus*. There is a horizontal ridge on the external surface of the squamosal. The jugal is deeper than that of *Lesothosaurus* and there is a fenestra in the quadratojugal.

The snout [Fig. 2(C)] shows very similar proportions to that of *Lesothosaurus*. The principal differences in structure are found in the tooth-bearing elements. The premaxilla, while being edentulous rostrally, bears five teeth and its ventral border is displaced below that of the maxilla. The posterolateral process of the premaxilla resembles that of *Lesothosaurus* in that it does not completely separate maxilla and nasal. The maxilla is only very generally triangular in shape; its lateral surface overhangs the alveolar margin producing a pronounced cheek recess. The antorbital fenestra, although extensively developed within the body of the maxilla is restricted laterally by the development dorsally of the external surface of the maxilla – the latter is pierced by several large foramina that pass into the antorbital fenestra. Anteriorly, the maxilla develops a curious narrow, elongate rostral process, which fits into a recess in the posterior/medial surface of the premaxilla and is wedged in position by the rostral spine of the vomer. The maxilla has a short dorsal contact with the nasal and overlaps the lachrymal both externally and medially, where it forms the inner wall of the antorbital fenestra. The lachrymal, in turn, is sutured to the nasal anteriorly, fitting into a groove in the nasal, and posteriorly is grooved to receive the prefrontal.

The palate [Fig. 2(D)] is moderately vaulted, with the fused vomers anteriorly forming a narrow median plate of bone arched upward and deepening posteriorly. The palatines are sutured to the posteromedial edge of the maxilla flooring the orbital cavity and roofing the nasal and buccal cavities posterior to the vomer. The pterygoid is complex, connecting suspensorium to anterior palate, maxillae and braincase by means of alar processes and the splint-like ectopterygoid.

The mandible [Fig. 2(E)] is similar to that of *Lesothosaurus* in that it presents a sinuous outline in occlusal view, with the teeth lying along the medial edge. However, the body of the mandible is considerably more massive. The mandibular fenestra is absent. The coronoid process is moderately well developed and the articulation for the lower jaw is set below the occlusal plane of the teeth.

The dentition is heterodont, with five simple compressed, cylindrical premaxillary teeth (Galton, 1974a: fig. 13); broad, transversely compressed maxillary teeth (10–11), with crenelated edges and small, superficial, vertical ridges laterally [Fig. 2(F)]: while the dentary teeth (13–14) are again compressed and crenelated and the thickly enamelled lingual surface is marked by a large, median, vertical ridge and several smaller vertical striae [Fig. 2(G)]. The first three or four teeth in the dentary are apparently more simple, cone-like teeth (Galton, 1974a). One interesting aspect of the dentition is the fact that the mesial edge of each maxillary tooth bears vertical recesses which receive the distal edge of the crown immediately behind, forming a partially interlocking arrangement quite distinct from the simple dentition of *Lesothosaurus*. The 'cheek' teeth are also quite heavily worn, producing planar, oblique wear facets.

The 'Iguanodontoid'

As with the 'hypsilophodontoids', recent years have seen a great increase in knowledge of the 'iguanodontoid' dinosaurs – see for example papers by Rozhdestvenskii (1966: *Probactrosaurus*), Ostrom (1970: *Tenontosaurus*), Taquet (1976: *Ouranosaurus*), Norman (1980: *Iguanodon*), Galton & Powell (1980: *Camptosaurus*), and Bartholomai & Molnar (1981: *Muttaburrasaurus*).

The group is generally regarded as relatively conservative in postcranial skeleton: represented by medium to large-sized ornithopod dinosaurs which show 'graviportal' adaptations (*sensu* Galton) and lack the cranial specializations of the hadrosaurian dinosaurs. *Iguanodon* will serve as a representative of the group with comparative comments made where appropriate.

The skull table [Fig. 3(A)] is flat (slightly concave across the frontals) and rectangular with large ellipsoid upper temporal fenestrae flanked by the inter-temporal bar which is somewhat bowed medially. The arrangement of bones in the skull roof conforms very closely with that in previous examples. The postorbitals have a more elongate medial process, meeting the parietal, so that the frontal is not at all involved in bordering the upper temporal fenestrae. There is also a much reduced orbital exposure of the frontals.

The occiput [Fig. 3(B)] in *Iguanodon* is very robustly constructed when compared to that of *Hypsilophodon* (Fig. 2). The supraoccipital is lodged in a recess in the posterior dorsal wall of the braincase and is excluded from the foramen magnum by a transverse bar formed by the meeting of the exoccipitals; the latter form pillars, on either side of the foramen magnum, whose bases almost meet so as to exclude the basioccipital from the foramen magnum. The exoccipitals are greatly expanded laterally to form the massive, wing-like paroccipital processes, which partly support the squamosals; between squamosal, supraoccipital and exoccipital, the posterior end of the opisthotic is visible as a roughened convex boss, which forms the principal sutural attachment of the squamosal. The parietals are almost completely excluded from the occipital surface by the squamosals laterally and sit upon the rounded dorsal surface of the supraoccipital.

The conformation of the occiput shows variation in those 'iguanodontoids' for which there is adequate material. In *Camptosaurus* (Gilmore, 1909; Galton & Powell, 1980) the occiput is reminiscent of that in 'hypsilophondontoids', with a large, plate-like supraoccipital forming the dorsal border of the foramen magnum. This arrangement is also described in *Thescelosaurus* (Morris, 1976) and *Muttaburrasaurus* (Bartholomai & Molnar, 1981). However, *Tenontosaurus* (Ostrom, 1970) and *Ouranosaurus* (Taquet, 1976) have a similar arrangement to that in *Iguanodon*.

The cheek region [Fig. 3(C)] of *Iguanodon* differs little from the previously described pattern. The lateral temporal fenestra is inclined posteriorly and subelliptical in outline. There is a lateral shelf on the squamosal, and the latter forms a smooth socket for the head of the quadrate; the quadrate is elongate and curved with a quadrate foramen between quadrate and quadratojugal. The jugal forms the ventral margin of the lateral temporal fenestra and the orbit, and has a smooth overlapping suture with both the postorbital and the quadratojugal; the latter is in turn wrapped around the lateral jugal wing of the quadrate. The jugal has a complicated articulation with a finger-like process on

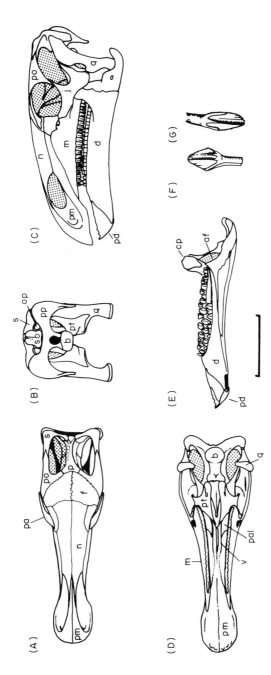

FIG. 3. *Iguanodon atherfieldensis* Hooley, 1925. Skull and lower jaw. (A) Dorsal view. (B) Occipital view. (C) Lateral view. (D) Ventral view. (E) Lower jaw and teeth. (F) Dentary tooth (lingual). (G) Maxillary tooth (labial). Scale bar = 20 cm. For key to abbreviations, see p. 547.

the maxilla: there is also a cup-like depression on its medial surface which received the head of the ectopterygoid (Norman, 1977). The anterior tip of the jugal contacts the lachrymal.

The snout [Fig. 3(C)] is notably elongated and is roofed by the elongate and arched nasals which rest on the dorsal edges of the posterolateral extensions of the premaxillae: they extend forward as long, curved, tapering spines which are sutured to the lateral edges of the median, dorsal premaxillary process. The prefrontal has a curved, overlapping suture with the nasal and rests upon the dorsal edge of the lachrymal, with which it has an apparently loose articulation. The palpebral, rather unusually, lies loosely against the ventrolateral surface of the prefrontal (Norman, 1980) and has no rôle in supporting the lachrymal-prefrontal suture. The maxilla is massive and has a well developed lateral recess above the teeth. The antorbital fenestra is reduced to a small foramen running in a channel to the dorsal surface of the maxilla; the channel is roofed by the lachrymal. The premaxilla is notably enlarged when compared to *Hypsilophodon*: it is transversely expanded to form a spoon-shaped 'beak' [Fig. 3(D)] while the external surface is excavated to produce an embayed area below the external naris. The posterolateral process of the premaxilla separates the maxilla and nasal completely, and overlaps the anterior lamina of the lachrymal.

The palate [Fig. 3(D)] is very deeply vaulted, continuing the trend seen in the 'fabrosauroid' and 'hypsilophodontoid' skulls. The vomers form a keel-like structure separating right and left nasal cavities. The palatine is sutured to the mediodorsal edge of the maxilla and separates nasal/orbital cavities. The pterygoids are complex bones; a thin ascending process supports the posterior margin of the palatine and has a flange on its medial edge which broadens posteriorly to separate the sphenoid region of the braincase from the buccal cavity. Posteriorly, the pterygoid develops a deep plate to meet the quadrate. The nature of the basal articulation has not been established.

The mandible [Fig. 3(E)] is very robust. The predentary is an edentulous, scoop-shaped bone with its occlusal margin modified by the presence of bony projections, which would have underlain the horny beak. The dentary is elongate and massive with the dentition arrayed along its medial edge; the alveoli extend posteriorly so that they lie medial to the very tall coronoid process. The glenoid for the jaw articulation is displaced below the occlusal plane of the teeth and, as in *Hypsilophodon*, the mandibular fenestra is absent although a small surangular foramen is retained.

The dentition of *Iguanodon* is more extensive than that of *Hypsilophodon*. There are many more vertical tooth series (20–29 maxillary series and

19–25 dentary series) and the individual teeth are more complex in their structure. Both maxillary and dentary teeth are grooved along their mesial and distal edges so that adjacent functional and replacement crowns partly interlock. The dentary teeth [Fig. 3(F)] are typically broad, transversely compressed and thickly enamelled lingually with a pattern of evenly spaced low vertical ridges running to the crenelated occlusal margin. The maxillary teeth are by contrast mesiodistally compressed and have a very prominent primary vertical ridge. [Fig. 3(G)].

The 'Hadrosauroid'

Hadrosaurs are the best known and best described of all ornithopods because of their abundant and well-preserved remains in the Upper Cretaceous deposits of North America (Lull & Wright, 1942; Ostrom, 1961). While they are quite conservative in their post-cranial anatomy, that of the skull is quite variable and only a relatively conservative form has been chosen as a representative of the group *Edmontosaurus* (Fig. 4) (Lambe, 1920).

The skull table [Fig. 4(A)] is massively constructed when compared to that of *Iguanodon* (Fig. 3). The upper temporal fenestrae are quite narrow, being enclosed by a transversely thickened inter-temporal bar composed equally of postorbital and squamosal. The squamosals are particularly massive and almost meet medially, being separated by a narrow finger of the parietal; the postorbitals are similarly massive with a prominent transverse swelling where they form an inflated posterior wall to the orbital cavity. The frontals, by contrast, are relatively small compared to those of *Iguanodon* with a small orbital exposure which is lost altogether in the lambeosaurine hadrosaurs. The frontals are rigidly fused together and also with the neighbouring nasals, prefrontals, postorbitals and parietals.

The occiput [Fig. 4(B)] is reminiscent of that in *Iguanodon* in which the supraoccipital is lodged in a recess between the exoccipital bar, the squamosals and the parietals. The exoccipitals are massive and form much of the paroccipital wing (Langston, 1960) and are overlain by the massive squamosals. The breadth of the occiput contrasts quite strikingly with the narrow, deep occiput of *Iguanodon*.

The cheek region [Fig. 4(C)] is dominated by the enlarged postorbital and jugal. The former is massive and triangular in lateral aspect, forming the posterodorsal orbital border, and making a weak overlapping suture with the jugal. Posteriorly, the postorbital overlaps the squamosal and occludes much of the upper half of the lateral temporal fenestra. The quadrate is essentially straight and its head fits into a

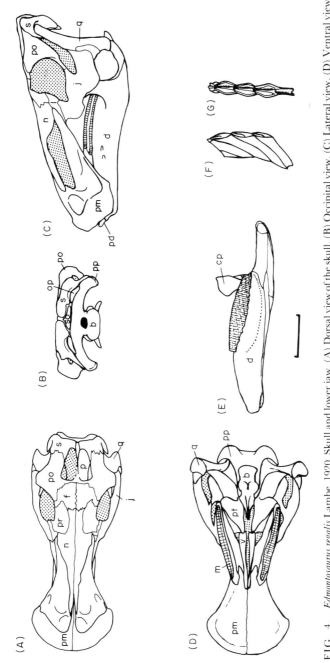

FIG. 4. *Edmontosaurus regalis* Lambe, 1920. Skull and lower jaw. (A) Dorsal view of the skull. (B) Occipital view. (C) Lateral view. (D) Ventral view. (E) Lower jaw and teeth. (F) Dentary tooth family (mesial) and (G) (lingual). Scale bar = 20 cm. For key to abbreviations, see p. 547.

smooth cotylus in the squamosal. The lower half of the temporal fenestra is somewhat wider than the upper half and is enclosed by the jugal ventrally and the quadrate posteriorly. The quadratojugal is a relatively small disc of bone separating the jugal and quadrate ventrally. There is no quadrate foramen. The jugal is extensively sutured to the lateral wall of the maxilla (Lambe, 1920).

The snout [Fig. 4(C), (D)] is about twice the length of the skull table and is dominated by the premaxillae which are enormously expanded transversely producing a 'muzzle' which is excavated externally, around the border of the nares. By contrast the lambeosaurine hadrosaurs have a relatively narrower muzzle, similar in shape but slightly broader than that of *Iguanodon*. The posterolateral process of the premaxilla again separates the maxilla from the nasal and, as in *Iguanodon*, this process overlaps the anterior lamina of the lachrymal. The lachrymal is overlapped by the prefrontal and premaxilla and underlies the jugal. The nasal has a long anterior spine which lies alongside the median premaxillary process and extends posteriorly to a point above the orbits. The maxilla is quite large, although it is not very extensively exposed, being overlapped considerably by premaxilla and jugal.

The anterodorsal surface of the maxilla forms a sutural surface for the premaxilla; this is a broad, smooth, shallow trough which receives the ventral surface of the premaxilla. Dorsally, the lachrymal and jugal lie against the external surface of the maxilla. The exposed external surface of the maxilla is recessed to form a cheek space above the alveolar margin.

The palate [Fig. 4(D)] is highly vaulted, as in *Iguanodon*, and the bones are similar in general shape and arrangement; they are also generally far better preserved (see Lambe, 1920; Heaton, 1972).

The mandible [Fig. 4(E)] is very robust even when compared to that of *Iguanodon* [Fig. 3(E)]. Nearly half of the occlusal margin of the dentary is edentulous, the remainder being occupied by the massive dental battery. The predentary is quite broad and more scoop-shaped than that of *Iguanodon*.

The dentition [Fig. 4 (E), (F), (G)] is remarkable because of the massive dentary and maxillary batteries which are composed of 50 or more vertical tooth series, each of which may have six lozenge-shaped functional and replacement teeth. The growing replacement teeth are open-rooted, but as growth continues the pulp cavity is progressively occluded with dentine (Horner, 1983) and the roots become cemented to those of adjacent teeth to form an integrated battery. Tooth wear produces a tessellated occlusal pavement which has a continuous curved surface. As in other ornithopods the dentary teeth are enamelled lingually. The maxillary battery is similar to that of the dentary but, when worn, forms an irregular rasp-like arrangement of enamel ridges across the occlusal surface of the battery (Ostrom, 1961). Observation of the few hadrosaur jaws and crania in the collections of the British Museum (Natural History) reveal a relatively constant pattern with the obliquely inclined shearing surface of the dentary battery concave transversely, while that of the maxillary battery is planar or convex transversely.

CRANIOLOGICAL OBSERVATIONS ON *IGUANODON*

The preceding general anatomical survey provides the basis for a review of functional design in the skull of ornithopods. The skull of *Iguanodon* has been chosen as the model with which comparison shall be made because it seems to demonstrate a combination of unusual features, and is in some respects mid-way in the 'morphological gradient' between the 'fabrosauroid' and 'hypsilophodontoid' with relatively short facial regions, and the hadrosaurs with their long and elaborate snouts.

Cranial Flexibility

There are two primary types of cranial flexibility: streptostylism, where the quadrate is able to move independently of the rest of the skull, and kinetism, in which transverse hinges (or regions of flexibility) are found in the skull roof. These two types of movement, although quite distinct, are frequently linked in recent squamates (Frazzetta, 1962).

Streptostyly

In *Iguanodon*, the quadrate appears to have had a *limited* streptostylic ability. The head of the quadrate is a smoothly rounded convex boss (Fig. 5 – based on BMNH.R.5764) which fits into a smoothly excavated cotylus in the squamosal. Anterior to the cotylus, there is a ventrally directed process which prevents anterior rotation of the quadrate. Posteriorly, the squamosal forms an oblique, posterolateral wing, where it lies against the paroccipital process. When the quadrate and squamosal are articulated (BMNH.R.5764) it is found that the quadrate is free to rotate posterolaterally. The medial surface of the head of the quadrate is scarred as if for the attachment of ligaments which presumably restricted the degree of rotation of the quadrate. In addition to its dorsal attachment, the quadrate has two other significant

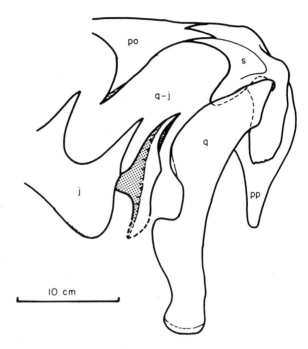

FIG. 5. *Iguanodon atherfieldensis* Hooley, 1925. Partly articulated jugal arch, quadrate and squamosal (from BMNH.R.5764, with details of quadratojugal from IRSNB.1536). Scale bar = 10 cm. Stipple indicates flexible sutures. For key to abbreviations, see p. 547.

sutural relationships: with the palatal complex anteromedially and the jugal arch anterolaterally.

The palatal articulation with the quadrate consists of an extensive overlap between the thin, deep, pterygoid wing of the quadrate and the quadrate ramus of the pterygoid. This suture is evidently firm, with the pterygoid fitting against a shallow recess on the internal surface of the quadrate. This suture is well adapted to resist both tensile and compressive stresses resulting from anteroposterior rotation of the quadrate. The jugal arch articulates with the quadrate in the following way (see Fig. 5). The quadratojugal is lodged between the jugal and quadrate and is grooved along its posterodorsal edge so as to fit against the jugal wing of the quadrate. Anteriorly, there is a long shallow rebate and a finger-like projection, which rest against the internal surface of the jugal. As with the palatal articulation, the jugal articulation is resistant to tensile and compressive stresses. However, none of these joints is adapted to resist transverse movements of the quadrate (abduction-adduction); while this range of movement is not normally within the scope of reptilian skulls (except for very specialized squamates) there is some evidence that such transverse rotation associated with a form of cranial kinesis may have occurred in *Iguanodon*.

Kinesis

Classical forms of cranial kinesis (Frazzetta, 1962) typically involve transverse hinges or flexible regions of bone across the skull roof which are linked with moveable joints between bones of the snout, suspensorium and braincase. These linkages operate a kinematic chain so that the snout can be elevated or depressed.

In *Iguanodon*, however, the median bones of the skull roof are rigidly sutured together so that flexure of the skull roof is impossible. Nevertheless, there does appear to be a rather novel system of cranial hinges which allows the cheek region of the skull to rotate transversely against the skull roof, an arrangement here termed pleurokinesis. Pleurokinesis is defined as a system of hinges that permit transverse rotation of the upper jaw. As can be seen (Fig. 6) the skull of *Iguanodon* can be divided into two principal functional units: the tooth-bearing maxillary unit (comprising maxilla, palatine, pterygoid and ectopterygoid, all of which are rigidly sutured together) and the median dermal skull roof (from premaxillae to braincase). There is also a subsidiary framework of struts comprising lachrymal, jugal, quadratojugal and quadrate.

Briefly, the nature of the various hinged areas is as follows (see Norman, 1977 and in preparation for details). The suture between premaxilla and maxilla is long and straight, and the transversely

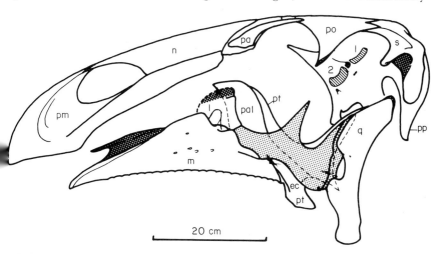

FIG. 6. *Iguanodon atherfieldensis* Hooley, 1925. Partly disarticulated skull showing the pleurokinetic hinge system. Scale bar = 20 cm. 1, M. levator pterygoideus; 2, M. protractor pterygoideus. Light stipple = jugal arch; dark stipple = pleurokinetic hinges; cross-hatching = muscle scars. For key to abbreviations, see p. 547.

rounded ventral surface of the premaxilla rests against a corresponding longitudinal groove in the dorsal edge of the maxilla. This suture is evidently not designed to resist rotational couples applied around its long axis. Indeed, it is not uncommon to find skulls which are otherwise fully articulated showing marked separation along this suture. Excessive rotation at this hinge was prevented by a complex mediodorsal process of the maxilla (Norman, 1977, in preparation). The relationship between lachrymal and maxilla is quite complex. The lachrymal is perched rather precariously atop the maxilla; it is held in position by two devices: the first is a peg-like projection from the dorsal surface of the maxilla, and the second is by the lachrymal being wedged between two thin laminae on the dorsal edge of the maxilla.

The maxilla and jugal articulate by means of a finger-like process of the maxilla which fits into a recess on the ventral surface of the jugal (BMNH.R.5764 – Norman, 1977, in preparation). There is also a ball-and-socket joint between jugal and ectopterygoid (the ball-like head of the ectopterygoid fitting into a smooth, concave depression in the medial surface of the jugal). The jugal arch forms a series of flexibly linked elements connecting the 'maxillary unit' to the posterior skull roof and quadrate, and permits torsion. Finally, the lachrymal forms a smooth, flexible linkage with the prefrontal. The jugal-postorbital suture is simple, smooth and overlapping.

Functional significance of the pleurokinesis

The existence of the pleurokinetic hinge system has a profound effect on considerations of cranial function and jaw action. The wear facets created on the teeth of upper and lower jaws indicate that the teeth sheared past each other vertically and obliquely during occlusion. The most important consequence of this type of jaw action (at least in relation to this discussion) is that during occlusion there is a significant transverse component of the forces acting on the teeth. This will inevitably result in reactions tending to cause simultaneous outward rotation of the maxilla and inward rotation of the mandible. Significant medial rotation of the mandible is resisted by the presence of the predentary, which is broad, slots between the mandibles and wraps around the symphysis (Norman, 1980), although mandibular adduction/abduction could not have been entirely prevented. By contrast, the skull is singularly poorly designed to resist transverse forces exerted at the edge of the maxilla. As is the case in most Ornithopoda, the skull (and snout in particular) is narrow and highly vaulted, forming a deep, inverted U-shape in cross-section. This can probably be correlated, in part, with improvements in the jaw musculature. However, since there is no transverse bracing between the upper jaws (such as is provided by the bony secondary palate of crocodiles and mammals) transverse forces at the level of the teeth must have exerted considerable stress on the median skull roof of *Iguanodon* by virtue of the long lever-arms of the deep 'cheeks'. It is proposed therefore that the development of pleurokinesis is correlated with the need to reduce the stresses acting across the skull roof during tooth occlusion, by allowing controlled passive, lateral displacement of the upper jaw.

Obviously, an unrestricted arrangement of this kind would not be conducive to effective mastication in *Iguanodon*, because the damping effect of the pleurokinetic hinge would reduce the shearing forces acting between opposing teeth. It is possible that a mechanism for maxillary adduction might be provided by the natural elasticity in the connective tissue binding the hinges. However, a muscular maxillary adduction system was probably also present. A large muscle scar on the lateral surface of the basisphenoid (Fig. 6), and another on the external surface of the proötic represent the origins of components of the constrictor dorsalis musculature (which are primarily involved in movements of the pterygoid in the kinetic skulls of squamates) – the M. protractor ptertygoideus and the M. levator pterygoideus respectively. Both these muscles were evidently very powerful (judged by the degree of scarring on the bone), pennate-fibred (the muscle-scars having a characteristically ridged appearance), and probably inserted on a broad medial shelf of the pterygoid, where they were well positioned to adduct the maxillae. The nature of the wear surfaces of the teeth can also best be explained in terms of maxillary rotation (Norman, in preparation).

Kinesis in other ornithopods

The various representative ornithopod groups were investigated for evidence of cranial kinesis. *Lesothosaurus*, the generalized 'fabrosauroid', has a lightly constructed, short-snouted skull which may possess considerable cranial flexibility. However, despite these suspicions, the sutural relationships of the skull bones await description by Charig & Crompton (in preparation). *Hypsilophodon* was described in detail by Galton (1974a) and restricted streptostylism as well as meso- and metakinetic hinges were identified. Norman (1977) re-analysed the skull of *Hypsilophodon* for evidence of pleurokinety. As was the case in *Iguanodon*, the skull, despite having an oblique-orthal jaw action, is not well constructed to resist transverse forces on the maxilla since the snout is moderately vaulted and lacks any inter-maxillary bracing. As in *Iguanodon*, a rigid 'maxillary unit' can be identified comprising: maxilla, lachrymal, jugal, palatine, pterygoid and ectopterygoid. This 'unit' is apparently flexibly sutured to the premaxilla, prefrontal and postorbital, while the squamosal cotylus permitted lateral rotation of the quadrate. It seems quite probable that *Hypsilophodon* had a pleurokinetic skull, with involvement of the constrictor dorsalis musculature, as in *Iguanodon*. Sues (1980) proposed that maxillary rotation may have occurred in the very closely related *Zephyrosaurus*. 'Hadrosauroids' are generally regarded as having akinetic skulls (Ostrom, 1961). Nevertheless, their skulls (particularly the snout) are relatively narrow and highly vaulted (Heaton, 1972) and jaw action, which was undoubtedly powerful and oblique/orthal (*contra* Ostrom, 1961 – who proposed a propalinal model of jaw action – see Norman, 1977; Hopson, 1980) would have produced considerable shear stresses in the skull roof. In the few hadrosaur crania available for study in the British Museum (Natural History) there is some evidence of lateral cranial flexibility, as originally suspected by Heaton (1972). As in *Iguanodon* there is a diagonal line of sutural weakness running from the anterior extremity of the maxilla along its premaxillary suture (which is broad and smooth, with little evidence of any powerful ligamentous attachment) to the lachrymal-prefrontal suture, across the orbit to the loose overlapping suture between postorbital and jugal, to the squamosal cotylus. A large muscle scar on the lateral surface of the basisphenoid which was interpreted as the area of origin of the M. levator bulbi (Ostrom, 1961) is consistent with the constrictor dorsalis musculature system of maxillary adduction in *Iguanodon*.

General conclusions

Within the anatomical groupings of the Ornithopoda that have been considered, it is possible to discern a clear progressive trend in the complexity of cranial organization. One of the primary influences on cranial form must be the distribution of stresses created by the combination of jaw musculature and the reactions produced within the jaws during occlusion. In this respect, the 'fabrosauroids' with their small, lightly built, short-snouted skulls, and slender jaws with simple triangular crowned teeth, represent the least specialized of the ornithopods. Plant food was presumably chipped from vegetation and processed relatively briefly in the mouth prior to being swallowed, much as in modern herbivorous lizards (cf. *Amblyrhynchus*).

By comparison, *Hypsilophodon* shows adaptations which are suggestive of a greater capability for food processing. The lower jaw is noticeably more massive than that of *Lesothosaurus* and has lost the mandibular fenestra; the coronoid process is slightly more elevated; the jaw articulation is depressed below the occlusional plane of the teeth; and the teeth show at least partial interlocking between adjacent crowns, so that continuous, if uneven, shearing blades are formed in each jaw. Despite these adaptations, however, *Hypsilophodon* possesses a remarkably flexible skull which (though possibly a juvenile feature) indicates that the stresses being transmitted via the skull bones were not so great as to threaten its structural integrity.

'Iguanodontoids', exemplified by *Iguanodon*, show considerable elaboration of the skull. The occiput is reinforced, the skull roof is rigidly constructed; the snout is lengthened and deepened, increasing the overall size of the tooth-bearing jaws, and the beak is considerably enlarged as a cropping device; the maxillary and dentary crowns interlock to a greater extent than in 'hypsilophodontoids'; and the lower jaw is notably massive, with a prominent, elevated coronoid process. Despite the reinforcement of the skull roof and occiput – an evident adaptation to withstand the increased stresses in the skull associated with more powerful mastication – the pleurokinetic system is present. This apparent paradox is explained as a potential adaptation which counteracted excessive stresses in the skull roof associated particularly with repetitive mastication.

The 'hadrosauroids', which culminate the trend in skull construction seen in *Iguanodon*, have developed sophisticated dental batteries and massive, reinforced skulls. There is some evidence that hadrosaur skulls were also pleurokinetic.

At a higher level of interpretation, the adaptive significance of pleurokinesis in ornithopods may have been considerable. The ability to chew fibrous plant tissues is dependent upon the combined shearing and crushing action of the dentition. If lateral rotation of the maxillae with its implicit bilateral occlusion occurred in reality, it would have resulted in a considerably enhanced shearing and crushing potential. It is quite probable that the ability to crush plant food more effectively may have been a major selective factor in the success of the ornithopod dinosaurs in the late Mesozoic, culminating in the numerically and taxonomically diverse hadrosaurs. Contrary to the opinions of Galton (1973) the development of cheek recesses in ornithopods may be of secondary importance in explaining the success of ornithopods compared to the specializations of the skull roof associated with jaw movements (Norman, in preparation).

SYSTEMATICS OF THE ORNITHOPODA

Various aspects of the classification and evolution of the ornithopod dinosaurs have been reviewed in several papers in recent years (Thulborn, 1971; Galton, 1972, 1973, 1974a, b, 1978; Taquet, 1975; Morris, 1976; Dodson, 1980; Santa Luca, 1980; Bartholomai & Molnar, 1981; Gow, 1981). The general consensus of much of the earlier work is that the ornithopods originated from Upper Triassic [or Lower Jurassic – Olsen & Galton (1977)] bipedal ornithischians – 'fabrosauroids' (Fabrosauridae). The 'hypsilophodontoids' (Hypsilophodontidae) evolved from unknown 'fabrosauroids' in the early Jurassic. The 'hypsilophodontoids' were a morphologically conservative lineage of ornithopods which persisted until the end of the Cretaceous period; they gave rise, iteratively, to several lineages of 'iguanodontoid' (Iguanodontidae). The 'hadrosauroids' (Hadrosauridae) in turn evolved monophyletically from an 'iguanodontoid' lineage, which included *Iguanodon*, in the later Lower Cretaceous (see Thulborn, 1971; Galton, 1974b, 1978).

Brief analysis of the various families defined by Galton (1974b, 1978) reveals that the family Fabrosauridae is characterized by a suite of plesiomorphic characters (see also Gow, 1981); the Hypsilophodontidae and Iguanodontidae (see Dodson, 1980) are similarly defined. The Hadrosauridae are defined by what seems to be a valid synapomorphy (the structure of the tooth batteries).

In an attempt to resolve the currently confused position of ornithopod systematics, a cladistic method of analysis of ornithopod genera has been used (see character cladogram – Fig. 7). The pattern of character-states is offered as a tentative pattern of relationships, but is deliberately neither reinforced by a classificatory scheme nor by a phylogenetic tree. The primary purpose of this cladogram is to serve as a basis for constructive analysis of the conventional phylogeny diagrams and systematic proposals of previous workers (Norman, in preparation).

General Observations and Comments

(i) Heterodontosaurids (*Heterodontosaurus, Abrictosaurus, Lycorhinus*) are not included in the clade Ornithopoda because they lack an obturator process on the ischium (Santa-Luca, 1980); for the same reason, ceratopians and pachycephalosaurs are similarly excluded (*contra* Thulborn, 1971; Galton, 1974b).

(ii) The 'fabrosauroids' (*Lesothosaurus, Echinodon, Nanosaurus, Fabrosaurus*) are confirmed as the primitive sister-group of the Ornithopoda. *Scutellosaurus* (Colbert, 1981) exhibits ornithischian traits, but lacks the character-states of ornithopods [as do *Alocodon* and *Trimucrodon* (Thulborn, 1973)]; all are provisionally *incertae sedis*.

(iii) The 'hypsilophodontoids' are rather restrictively characterized and include *Hypsilophodon*, '*Thescelosaurus*' (Galton, 1974b; Morris, 1976) and *Zephyrosaurus* (Sues, 1980) as well as the derived forms *Parksosaurus, Tenontosaurus* (Ostrom, 1970) and possibly *Mochlodon*.

(iv) The 'iguanodontoids' are characterized to include *Dryosaurus*/

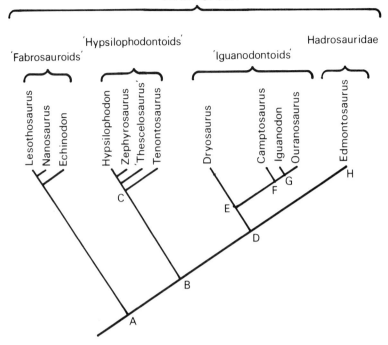

FIG. 7. A character cladogram of the Ornithopoda.
A. Ornithopoda
Obturator process on the ischium and proximally positioned; prepubic ramus long and narrow; digit V of pes reduced to a metatarsal splint; (?) jaw joint in approximately the same horizontal plane as the teeth; (?) long, splint-like lesser trochanter on femur.
B. 'Hypsilophodontoids' + 'iguanodontoids' + Hadrosauridae
Crowns of maxillary and dentary teeth transversely flattened, broad, and thickly enamelled on labial and lingual surfaces respectively; tooth crowns partly interlock; tooth wear produces oblique, continuous cutting surface along each jaw; median vertical ridge on lingual surface of dentary crown; crowns of teeth set at an angle to the root; mandibular fenestra occluded.
C. 'Hypsilophodontoids' = Hypsilophodontidae
Obturator process positioned on the middle of the shaft of ischium; lesser trochanter of femur small; (?) manus showing reduction of digits IV and V; sheath of ossified tendons surround the extreme end of the tail; spike-like anterior process of the maxilla which articulates with a socket in the posterior surface of the premaxilla; (?) transverse hinges in skull roof.
D. 'Iguanodontoids' + Hadrosauridae
Antorbital fenestra reduced; nasal and maxilla separated by premaxilla; deltopectoral crest of humerus reduced; median bones of skull roof rigidly fused (pleurokinetic hinge retained and more precisely controlled); deep inter-condylar groove on the dorsal surface of the distal end of the femur, becoming almost tubular in some.
E. 'Iguanodontoids'
Large median vertical ridge on crowns of maxillary teeth; distinct notch in the jugal wing of the quadrate; shaft of ischium with a rounded cross-section and bowed downward (decurved).
F. *Camptosaurus* + *Iguanodon* + *Ouranosaurus*
Manus robust, specialized by fusion of the carpals and fusion of metacarpal I with the radiale.
G. *Iguanodon* + *Ouranosaurus*
Pollex developed into a conical spike; 1st phalanx of digit I of manus thin and plate-like; (?) short postpubic ramus; digit I of pes reduced to splint-like metatarsal.
H. Hadrosauridae
Maxillary and dentary teeth arranged into batteries of lozenge-shaped teeth which are cemented together; more than one replacement crown in each vertical tooth series.

Dysalotosaurus (Galton, 1980) and *'Valdosaurus'* (Galton & Taquet, 1982) as well as *Camptosaurus, Iguanodon* and *Ouranosaurus. Muttaburrasaurus* and *Probactrosaurus* should probably also fall within the *Camptosaurus, Iguanodon, Ouranosaurus* grouping.

(v) Despite the many cranial and post-cranial similarities between *Iguanodon – Ouranosaurus* and hadrosaurs, they have not fallen as direct sister-groups as perhaps intuitively expected. Such incongruities require explanation: are the similarities parallel (perhaps size-related) adaptive traits, or do they require a re-assortment of the character-states in the cladogram? These and many other points will be dealt with at greater length elsewhere (Norman, in preparation).

CONCLUSIONS

Cranial form in ornithopod dinosaurs is reviewed and a general trend toward increasing complexity of teeth, jaws, jaw musculature and cranial strengthening is observed. Pleurokinesis is briefly described in *Iguanodon* and its functional significance is discussed. It is concluded that pleurokinesis represents an adaptation to withstand the stresses imposed on the skull roof of ornithopod dinosaurs which result from powerful and repetitive oblique-shearing jaw action. Maxillary rotation with median mandibular adduction of this type may also occur in 'hypsilophodontoids' and in hadrosaurid dinosaurs. Components of the constrictor dorsalis musculature (M. protractor pterygoideus and M. levator pterygoideus) may also have been involved in maxillary adduction, and if so, provide an explanation for very prominent muscle scars on the lateral wall of the braincase in these ornithopods. The adaptive significance of pleurokinesis in oral food processing was probably of far greater importance to the success of ornithopods than the presence of cheek recesses.

A preliminary cladistic analysis of the Ornithopoda has produced more restrictive character-state definitions of the 'fabrosauroid', 'hypsilophodontoid', 'iguanodontoid' groupings. The implied systematic regrouping of genera suggests that the phylogenetic trees of the ornithopods produced by Thulborn (1971) and Galton (1972, *et seq.*) need careful revision.

ACKNOWLEDGEMENTS

Some of the points raised in this article relate ultimately to work carried out whilst holding a postgraduate studentship at King's College,

London. The thesis which resulted from the studentship was ultimately examined (and approved!) by Professor Bellairs (among others) whom I thank for his kindness and support. I therefore hope that this article repays a debt of gratitude and also gains a measure of approval this time as well. This paper was typed by Mrs Sue Friend (Department of Zoology, Oxford).

REFERENCES

Bartholomai, A. & Molnar, R. E. (1981). *Muttaburrasaurus*, a new iguanodontid (Ornithischia: Ornithopoda) dinosaur from the Lower Cretaceous of Queensland, Australia. *Mem. Qd Mus.* **20**: 319–349.

Colbert, E. H. (1981). A primitive ornithischian dinosaur from the Kayenta Formation of Arizona. *Bull. Mus. nth. Ariz.* **53**: 1–61.

Cope, E. D. (1883). On the characters of the skull in the Hadrosauridae. *Proc. Acad. nat. Sci. Philad.* **35**: 97–107.

Dodson, P. (1980). Comparative osteology of the American ornithopods *Camptosaurus* and *Tenontosaurus*. *Mém. Soc. géol. Fr.* **139**: 81–85.

Dollo, L. (1882). Première note sur les dinosauriens de Bernissart. *Bull. Mus. r. Hist. nat. Belg.* **1**: 55–80.

Frazzetta, T. H. (1962). A functional consideration of cranial kinesis in lizards. *J. Morph.* **111**: 287–320.

Galton, P. M. (1972). Classification and evolution of the ornithopod dinosaurs. *Nature, Lond.* **239**: 464–466.

Galton, P. M. (1973). The cheeks of ornithischian dinosaurs. *Lethaia* **6**: 67–89.

Galton, P. M. (1974a). The ornithischian dinosaur *Hypsilophodon* from the Wealden of the Isle of Wight. *Bull. Br. Mus. nat. Hist. (Geol.)* **25**: 1–152.

Galton, P. M. (1974b). Notes on *Thescelosaurus*, a conservative ornithopod dinosaur from the Upper Cretaceous of North America, with comments on ornithopod classification. *J. Paleont.* **48**: 1048–1069.

Galton, P. M. (1975). English hypsilophodontid dinosaurs (Reptilia: Ornithischia). *Palaeontology* **18**: 741–751.

Galton, P. M. (1977). The ornithopod dinosaur *Dryosaurus* and a Laurasia-Gondwanaland connection in the Upper Jurassic. *Nature, Lond.* **268**: 230–232.

Galton, P. M. (1978). Fabrosauridae, the basal family of ornithischan dinosaurs (Reptilia: Ornithopoda). *Paläont. Z.* **52**: 138–159.

Galton, P. M. (1980). *Dryosaurus* and *Camptosaurus*, intercontinental genera of Upper Jurassic ornithopod dinosaurs. *Mém. Soc. géol. Fr.* **139**: 103–108.

Galton, P. M. & Powell, P. H. (1980). The ornithischian dinosaur *Camptosaurus prestwichii* from the Upper Jurassic of England. *Palaeontology* **23**: 411–443.

Galton, P. M. & Taquet, P. (1982). *Valdosaurus*, a hypsilophodontid dinosaur from the Lower Cretaceous of Europe and Africa. *Geobios* **2**: 147–159.

Gilmore, C. W. (1909). Osteology of the Jurassic reptile *Camptosaurus*, with revision of the genus and description of two new species. *Proc. U.S. natn. Mus.* **36**: 197–332.

Ginsburg, L. (1964). Découvert d'un Scélidosaurien (Dinosaure ornithischien) dans le Trias superieur du Basutoland. *C. r. hebd. Séanc. Acad. Sci. Paris* **258**: 2366–2368.

Gow, C. E. (1981). Taxonomy of the Fabrosauridae (Reptilia: Ornithischia) and the *Lesothosaurus* myth. *S. Afr. J. Sci.* **77**: 43.

Heaton, M. J. (1972). The palatal structure of some Canadian Hadrosauridae (Reptilia: Ornithischia). *Can. J. Earth Sci.* **9**: 185–205.
Hooley, R. W. (1925). On the skeleton of *Iguanodon atherfieldensis* sp. nov., from the Wealden of Atherfield (Isle of Wight). *Q. Jl geol. Soc. Lond.* **81**: 1–61.
Hopson, J. A. (1980). Tooth function and replacement in early Mesozoic ornithischian dinosaurs: implications for aestivation. *Lethaia* **13**: 93–105.
Horner, J. R. (1983). Cranial osteology and morphology of the type specimen of *Maiasaura peeblesorum* (Ornithischia: Hadrosauridae) with discussion of its systematic position. *J. Vert. Paleont.* **3**: 29–38.
Huxley, T. H. (1869). On *Hypsilophodon* a new genus of Dinosauria. *Abstr. Proc. geol. Soc. Lond.* **204**: 3–4.
Lambe, L. H. (1920). The hadrosaur *Edmontosaurus* form the Upper Cretaceous of Alberta. *Mem. Can. geol. Surv.* **120**: 1–79.
Langston, W. (1960). The vertebrate fauna of the Selma Formation of Alabama. VI, The dinosaurs. *Fieldiana (Geol.)* **3**: 314–360.
Lull, R. S. & Wright, N. E. (1942). Hadrosaurian dinosaurs of North America. *Spec. Pap. geol. Soc. N. Am.* **40**: 1–242.
Marsh, O. C. (1877). Notice of new dinosaurian reptiles of the Jurassic formation. *Am. J. Sci.* **14**: 514–516.
Morris, W. J. (1976). Hypsilophodontid dinosaurs: a new species and comments on their systematics. In *Athlon*: 93–113. Churcher, R. S. (Ed.). Toronto: University of Toronto Press.
Norman, D. B. (1977). *On the anatomy of the ornithischian dinosaur* Iguanodon. Ph.D. Thesis: University of London, UK.
Norman, D. B. (1980). On the ornithischian dinosaur *Iguanodon bernissartensis* from the Lower Cretaceous of Bernissart (Belgium). *Mém. Inst. r. Sci. nat. Belg.* **178**: 1–105.
Norman, D. B. (In prep.). *On the cranial morphology and relationships of the ornithischian dinosaur* Iguanodon.
Olsen, P. E. & Galton, P. M. (1977). Triassic-Jurassic tetrapod extinctions: are they real? *Science, Wash.* **197**: 983–984.
Ostrom, J. H. (1961). Cranial morphology of the hadrosaurian dinosaurs of North America. *Bull. Am. Mus. nat. Hist.* **122**: 33–186.
Ostrom, J. H. (1970). Stratigraphy and palaeontology of the Cloverly Formation (Lower Cretaceous) of the Big Horn Basin of Wyoming, Montana. *Bull. Peabody Mus. nat. Hist.* **35**: 1–234.
Owen, R. (1861). On the Reptilia of the Wealden and Purbeck formations. Part 5. *Palaeontogr. Soc. (Monogr.)* **12**: 31–39.
Romer, A. S. (1966). *Vertebrate paleontology*, 3rd edn. Chicago and London: University of Chicago Press.
Rozhdestvenskii, A. K. (1966). [New iguanodonts from Central Asia. Phylogenetic and taxonomic relationships between late Iguanodontidae and Hadrosauridae.] *Paleont. Zh.* **3**: 103–116 [in Russian].
Santa Luca, A. P. (1980). The postcranial skeleton of *Heterodontosaurus tucki* (Reptilia, Ornithischia) from the Stormberg of South Africa. *Ann. S. Afr. Mus.* **79**: 159–211.
Sues, H-D. (1980). Anatomy and relationships of a new hypsilophodontid dinosaur from the Lower Cretaceous of North America. *Palaeontographica (A)* **169**: 51–72.
Taquet, P. (1975). Remarques sur l'évolution des Iguanodontidés et l'origine des Hadrosauridés. *Colloque int. Cent. natn. Rech. scient.* **218**: 503–510.
Taquet, P. (1976). Géologie et paléontologie du gisement de Gadoufaoua (Aptien du Niger). *Cah. Pal. Cent. natn. Rech. scient.* **9**: 191.
Thulborn, R. A. (1970). The skull of *Fabrosaurus australis*, a Triassic ornithischian dinosaur. *Palaeontology* **13**: 414–432.

Thulborn, R. A. (1971). Origins and evolution of ornithischian dinosaurs. *Nature, Lond.* **234**: 75–78.
Thulborn, R. A. (1973). Contribuicao para conchecimento da fauna do Kimmeridgiano da Mina de Lignito Guimarota (Leiria, Portugal) III Parte VI. *Mems Servs geol. Port.* (N.S.) **22**: 89–134.

APPENDIX – KEY TO ABBREVIATIONS USED IN FIGURES

af	adductor fossa	p	parietal
b	basioccipital	pa	palpebral (supraorbital)
ce	coronoid eminence	pal	palatine
cp	coronoid process	pd	predentary
d	dentary	pm	premaxilla
e	exoccipital	po	postorbital
ec	ectopterygoid	pp	paroccipital process
f	frontal	pr	prefrontal
j	jugal	pt	pterygoid
l	lachrymal	q	quadrate
m	maxilla	q-j	quadratojugal
mf	mandibular fenestra	s	squamosal
n	nasal	so	supraoccipital
op	opisthotic	v	vomer

NOTES ADDED IN PROOF

Further analysis of the character cladogram (Fig. 7) has resulted in several changes:
(i) *Dryosaurus* is excluded from the 'Iguanodontoid' group and placed as the sister-group of the remaining 'Iguanodontoids' and Hadrosauridae.
(ii) *Node A* – small (1–2 m) bipedal ornithischians; triangular skull with large orbit; femur < tibia; anterior pubic spine; obturator process at midlength of ischium; (?) heterodonty.
(iii) *Node B* – hooked coracoid; thumb-print scar on femur; pleurokinetic skull; interlocking teeth; maxillae meet anteriorly; moderate coronoid eminence; median ridge on dentary teeth.
(iv) Dryosaurus + *'Iguanodontoids' and Hadrosauridae*. Premaxilla contacts prefontal and lachrymal; dentary forms elevated coronoid process; low postacetabular blade; well developed brevis shelf; reduced antorbital fenestra; median ridge on maxillary teeth; premaxillary teeth absent; alveoli medial to coronoid process; obturator process of ischium proximal; ischium curved and footed; quadrate notched.
(v) *'Iguanodontoids'* + *Hadrosauridae*. Metatarsal V lost; 4th trochanter low and crested; femur > tibia; thumb-print scar lost; anterior intercondylar groove developed; anterior pubic spine laterally flattened; elongate preorbital region.

From Reptile to Mammal: Evolutionary Considerations of the Dentition with Emphasis on Tooth Attachment

J. W. OSBORN

Department of Oral Biology, Faculty of Dentistry, University of Alberta, Edmonton, Alberta, Canada T6G 2N8

SYNOPSIS

The paper discusses the evolution of changes during the transition from reptilian to mammalian dentitions in terms of sequential changes in the cell populations which might be responsible for developing the tissue components. It adopts the hypothesis (Holtzer, Rubinstein, Fellini, Yeoh, Chi, Birnbaum & Okayama, 1975) that a particular cell has only two options when it differentiates. This finds its equivalent in the cladistic approach to evolution adopted by many palaeontologists whereby a taxon can be ancestral to only two different taxa. Thus, equivalent restrictions are applied to the differentiation of cell lineages during ontogeny and to the evolution of animal lineages during phylogeny.

Two different aspects of dental evolution are approached: the sequences and number of teeth which are initiated, and gradients in their shapes; and the attachment of teeth to the jaw. The first of these largely summarizes previous work. For the second, data relating to development in recent reptiles and mammals are combined with inferences drawn from descriptions of tooth attachment in a (fossil) mammal-like reptile. An evolutionary sequence of histological changes in the attachment apparatus is now constructed and matched to a hypothetical cladistic lineage of cell types which differentiate out of a parent clade. Each terminal clade produces a different tissue. Evolutionary changes are explained in terms of discrete changes in the lineage of cell types.

INTRODUCTION

Evolutionary changes between the dentitions of reptiles and mammals were reviewed by Kuhne in 1973. He identified 19 separate characteristics, listed their earliest appearance and briefly summarized what he considered to be the functional advantage of each. He was not concerned to trace the evolution of any particular feature, merely its first appearance and its possible advantage.

Although palaeontologists spend a considerable time on descriptive studies their usual objective is to establish the correct relationships between different fossil species. Judgements are based on morphology and the physiology which can be deduced from it. The present paper is mainly concerned with the developmental controls which can be deduced from changes in morphology. It first assumes that similar developmental controls have been operating on dentitions throughout their evolution. Second, it assumes that the development of recent reptiles and mammals is very like that of the ancestral reptiles and mammals involved in the transition between the two classes. The sequence of changes in morphology between the simplest reptile and the earliest mammal is then plotted. Combining the above assumptions and data with current embryological theories and data, it is possible to postulate a suite of developmental controls. The advantages of using data from fossils are first that they restrict the choice of controls to those which can account not just for the development of recent reptiles and mammals but for the intermediaries as well. Second, the intermediaries frequently suggest the types of control which may be required.

SEQUENCE, SHAPE AND NUMBER

Sequences of Tooth Initiation in Reptiles

Numerous data from recent animals (see Osborn, 1970) indicate that the sequences in which teeth erupt into the oral cavity usually match the sequences in which they have been initiated. Since in nearly all reptiles teeth continue to be replaced throughout life, fossil jaws can be used to infer the sequences in which teeth were being initiated. Edmund (1960) observed that most recent and fossil reptiles replace their teeth in waves which sweep through alternate tooth positions, usually from the back to the front of the jaw. To account for this phenomenon he proposed a control mechanism which has come to be identified as the Zahnreihe theory. He suggested that rows of teeth (*Zahnreihe*) are developed in sequence from the front to the back of the jaws (Fig. 1). New rows are initiated at regular intervals and, surprising though it may seem, if the intervals are adequately controlled the result is the wave replacement of alternate teeth from back to front. The Zahnreihe theory as visualized by Edmund (1960) requires a transmitter at the front of the jaw periodically emitting signals which travel slowly backwards through the jaw. Osborn (1971) pointed out that a further developmental control would be required to set up receivers at regularly spaced intervals along the jaw. When a receiver picks up a signal from the transmitter it initiates a new tooth.

The Zahnreihe theory was at one time accepted by most (e.g.

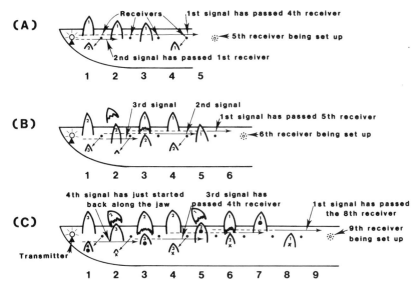

FIG. 1. Edmund's Zahnreihe Theory. A transmitter at the front of the jaw periodically emits a signal which travels back through the jaw. Regularly spaced receivers initiate a new tooth when they pick up the signal. Therefore, rows of teeth (*Zahnreihe* 1, 2, 3, 4, etc.) are initiated in sequence from the front to the back of the jaw. Provided the periodicity of the transmitter and the speed of signal transmission are within certain limits the result is the wave replacement of alternate teeth from the back to the front of the jaw [symbols ● and x in (C)]. Note that the teeth in a replacement wave are derived from different *Zahnreihe*. It might also be noted that the stimulus responsible for initiating a new tooth family at the back of the dentition (even in an adult) began its journey in the early embryo!

Cooper, Poole & Lawson, 1970; Hopson, 1971; Ziegler, 1971; De Mar, 1973) as the foundation for understanding the control of dental development and evolution. However, Osborn (1971) pointed out that no known vertebrate embryo initiates teeth in the sequence predicted by the Zahnreihe theory. The piranha (Berkowitz & Shellis, 1978) may prove to be an exception.

What is probably a typical sequence of tooth initiation in a reptile embryo has been described by Osborn (1971). The first tooth is initiated, for example, at position 5 (Fig. 2). Two new buds (3 and 1) are initiated in sequence towards the front of the jaw. Owing to interstitial jaw growth these spread apart. From this time forward, new buds are initiated in sequence in the spaces between earlier teeth starting with a new position at the back of the jaw.

Osborn (1971, 1973) suggested a biological control of tooth initiation and elaborated a clone model (Osborn, 1978), later renaming the units clades rather than clones (Osborn, 1979). The tissues involved in tooth development and their associated attachment apparatus (see later) are

J. W. Osborn

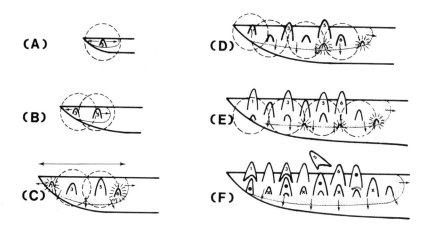

FIG. 2. Clade model (cf. the Zahnreihe Theory in Fig. 1). The dentition is derived from a clade of cells (outlined) which grows (arrows) by means of anterior, posterior and inferior progress zones (stippled margins). Each newly inititated tooth is surrounded by a zone (interrupted circle) which temporarily inhibits the initiation of a further tooth. A new tooth is initiated when a progress zone escapes from the inhibitory effect. The first teeth at positions 5, 3 and 1 are separated by interstitial growth of the jaws (A)–(D). Wave replacement of alternate teeth [e.g. symbols ● or x in (F)] is the result of the inhibitory zones.

isolated from the remainder of the jaw at an early stage. The ectomesenchymal cells associated with tooth development (which require to interact with the overlying jaw ectoderm) initially comprise a clade of identical cells. The clade grows by means of progress zones, a term used by Summerbell, Lewis & Wolpert (1973) for a model describing limb development, quite different from the model presented here.

For the dentition in Fig. 2 the clade of primitive dental (ecto)-mesenchyme differentiates at what will later be tooth position 5 and grows at different rates at anterior, posterior and inferior progress zones. A group of identical cells all having the same probability of growing and dividing in particular directions (Osborn, 1975) is visualized. The different growth rates of the whole clade in different directions are an expression of these initial probabilities. When the clade reaches a critical size a tooth bud is initiated near its centre. The newly initiated tooth inhibits the initiation of a new bud until the margins of the growing clade have grown beyond the inhibitory influence. In this way buds 3 and 1 are sequentially initiated in the anterior progress zone which now stops growing as it meets its equivalent from the other side of the jaw. These buds are spread apart by interstitial growth of the jaws before they have mineralized. Meanwhile bud 6 is appearing at the margin of the more slowly growing posterior progress zone. Provided each tooth bud is temporarily surrounded by a zone in which the initiation of a further bud is inhibited, new buds will automatically be initiated in sequences which pass from the back to the front of the jaw through alternate tooth positions. The rate at which new buds are initiated depends on the rates of growth at the inferior and posterior progress zones.

It should be noted that if adjacent tooth families (a family consists of all the teeth developed at a particular tooth position) become separated by bone, they no longer affect the rate of replacement at adjacent tooth positions. Under these circumstances the rate of tooth replacement is autonomously determined at each tooth position by the growth rate at the inferior progress zone of the tooth family.

Gradients of Shape

Osborn (1978) discussed the differences between two fundamentally different models which can be applied to morphogenetic problems: field models and clone (referred to here as clade) models. For field models, whose most eloquent proponent in terms of dentitions has been Butler (e.g. 1939, 1956, 1979), it is argued that all tooth primordia in a dentition are equivalent and contain an identical set of what will be called *cusp target* cells; each cusp target, if and when activated, produces a cusp whose equivalents can be recognized on any tooth in the dentition. The stimulus which activates cusp targets comes from an external source, a field transmitter. It will be noted that the Zahnreihe theory (see above) is another example of a field model.

A signal (or signals) from the field transmitter percolates through the growing jaw and, being slowly inactivated, develops a gradient of signal substance. Each (equivalent) tooth primordium is subjected to a different concentration of the signal substance or substances which may activate or suppress cusp targets, depending on their nature, to an extent directly related to this concentration. It can be visualized that the theory is capable of accounting for observed gradients in shape.

Osborn (1978) rejected field models and proposed that all tooth primordia in a dentition are different from the time they are initiated in the jaws. He observed that gradients in tooth shape seem always to match the sequences in which teeth are developed, rising to a peak of complexity and then falling away – but never falling and then rising. He likened this to the well known rise and fall of the embryonic competence of cells to react to an inducing stimulus and postulated a rise and fall in the *shape competence* of cells at the progress zone of a tooth clade. Only if they become isolated into a tooth bud are cells able to express their shape competence in the development of a tooth. Gradients in the shapes of teeth are an expression of the rise and fall in shape competence of the cells in a progress zone. Therefore, gradients of shape match the sequences in which teeth develop.

The size and complexity of the six or seven post-canines in the jaw quadrant of a generalized mammal usually peak somewhere amongst the permanent molars. Despite conforming to a gradient each post-canine has such a typical shape that even in isolation its position in the series can be predicted with near certainty. Intermediate shapes are very uncommon suggesting that differences in shape competence are quantal (abrupt) rather than continuous and hence that they are directly related to the genome although obviously being modifiable by polygenic effects. In contrast, with a field model it is the cusp targets, common to all tooth primordia, which would be directly related to the genome while the differences between teeth would be secondary and an expression of differences in the concentration(s) of field substance(s). Contrary to the data such a model seems to predict the existence of many intermediate shapes because of continuous rather than quantal differences in concentration(s). The *positional information* (Wolpert, 1969) variant of field models could lead to the observed quantal differences in shape but for other reasons has been rejected by Osborn (1978).

Tooth Number

The differences between the numbers of teeth developed in different dentitions seem not to have attracted the attention of those advocating field models. Osborn (1978) outlined a possibility based on the positional information variant of field models of Wolpert (1969), noting that such an explanation would imply that information for the position of every tooth, and hence tooth number, would be directly represented in the genome. The problem is the same as one inherent in the Zahnreihe theory; that of setting up a number of receivers at regularly spaced intervals along the jaw (see above).

For Osborn (1974), the increasing numbers of teeth along the growing jaw of a reptile and the replacement of teeth are both growth phenomena and explanations should be sought which relate to growth. Thus, the number of tooth positions sequentially added at the back of a growing jaw has no point-for-point relationship with the genome but is an expression of the amount of posterior growth of the tooth clade and the size of the inhibitory zones surrounding newly initiated teeth (Osborn, 1978).

Evolution of Mammalian from Reptilian Dentitions

It is first necessary to outline a sequence of dentitions, from the simplest reptile to the earliest mammal, and then to apply changes in the developmental controls suggested by the above model.

Osborn (1978: fig. 11) outlined the sequence of changes in the shapes of dentitions from the cotylosaurs through pelycosaurs and cynodonts to mammals. He added an important corollary to the rules described above. The teeth in a jaw quadrant may develop from more than one clade of cells. Boundaries between clades can be recognized in a homodont dentition by a constant disruption in the replacement pattern (e.g. at the premaxillary/maxillary suture in reptiles: Cooper, 1963) or in a heterodont dentition by a break in the gradient of shapes (e.g. the canine/premolar boundary in most mammals).

Using the above markers for recognizing the boundaries between clades, a sequence of dentitions containing known fossil and equivalent recent representatives can be constructed. The sequence is characterized by four features: (a) an increase in the number of clades from 1 to 3; (b) the evolution of heterodonty of size followed by heterodonty of shape; (c) a reduction from polyphyodonty to diphyodonty; and (d) the loss of wave replacement of alternate teeth. The following brief description explains the required changes at the clade level.

The development of a dentition similar to that of dentition (A) [Fig. 3(A)] has been described earlier (Fig. 2). In dentition (B) [Fig. 3(B)] a second clade, possibly initiated from the region of the egg tooth, has evolved anterior to the first clade. Around this time differences in the shape competence of cells at the progress zones began to appear and were expressed as heterodonty of size [Fig. 3(C)]. A third clade appeared in dentition (D), probably in the middle of the dentition. The homologues of the mammalian incisor, canine and molar series were now present. In dentition (E), well known from studies of *Thrinaxodon* (Crompton, 1963; Osborn & Crompton, 1973), the rate of tooth replacement has become greatly reduced, marked heterodonty of shape has appeared in the post-canines, and the canine clade contains a single tooth family. Finally, in mammals [dentition (F)], the dentition became diphyodont and alternation was lost because the developing deciduous teeth, now isolated in bony crypts, no longer influenced tooth initiation in adjacent tooth families. Note that the molar clade has anterior, posterior and inferior progress zones; the incisor clade has posterior and inferior progress zones; the canine clade has only an inferior progress zone (Osborn, 1978, 1979).

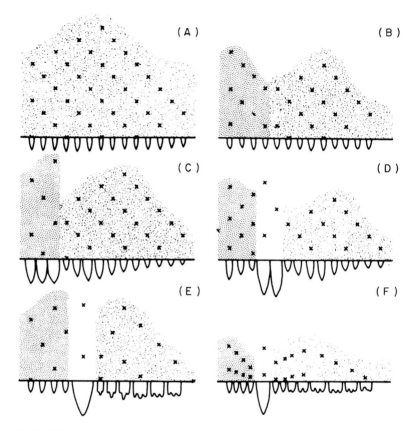

FIG. 3. The evolution of clades. (A) The whole homodont dentition develops from a single clade. (B) A new clade evolves at the front of the jaw. (C) Graded heterodonty of size evolves. (D) A third clade of caniniform teeth evolves between the two original clades. (E) Heterodonty of shape evolves in the posterior clade and the rate of tooth replacement is slowed down. (F) Diphyodonty evolves. In each diagram time runs vertically and x denotes the eruption time of a new tooth. Alternate teeth are replaced in waves (A)–(E). Different clades are differently stippled. The upper margin of a stippled region represents the time the clade appears.

TOOTH ATTACHMENT

Introduction

For the purpose of introducing terminology rather than presenting any well accepted sequence of evolution, different types of tooth attachment are shown in Fig. 4. Most teeth are either somewhat immovably attached to the jaw bone by mineralized tissue (*ankylosis*) or attached by unmineralized fibres embedded into the root surface and the adjacent

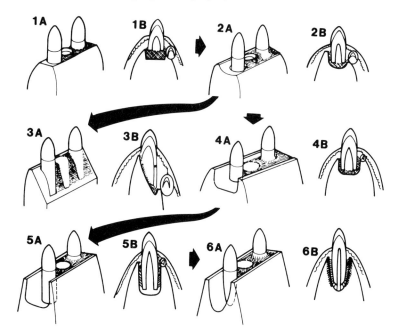

FIG. 4. Types of tooth attachment beginning with an acrodont ankylosis (1), two types of protothecodont ankylosis (2 and 4), pleurodont ankylosis (3) and ending with a gomphosis (6). A possible intermediary, based on evidence from a mammal-like reptile (*Thrinaxodon*) is represented in (5). In the (A) figures the tooth furthest from the viewer is surrounded by its bone of attachment (shaded). The bone of attachment has been removed from around the tooth nearest the viewer. In the mid-region a tooth has been removed. The jaw bone, as opposed to bone of attachment, is unshaded. The (B) figures show a section of the jaw in which bone of attachment is cross-hatched. The beginnings of a periodontal ligament are shown in 5 and completed in 6.

bone (*fibrous attachment*), thereby cushioning strains imparted to the teeth primarily during mastication and predation. In a mineralized attachment the tooth may be cemented to the somewhat horizontal surface of the bone (*acrodont ankylosis*) or to the lingual side of the bone (*pleurodont ankylosis*). In a *thecodont ankylosis* the tooth has a root embedded in the jaw. Intermediate conditions, where a labial wall of bone is built up on one side, or on both sides thereby developing a gutter in which teeth are ankylosed, are known as *protothecodont ankylosis*. The hard tissue which unites the tooth with the jaw bone is the *bone of attachment*. It is resorbed when a tooth is replaced and redeveloped around each newly erupted tooth. The bone of attachment often spreads anteriorly and posteriorly to unite adjacent teeth as well as the teeth to the jaw bone.

In a *gomphosis* the roots of teeth are embedded in a socket to whose walls they are attached by fibres. In the *thecodont gomphosis* of alligators and crocodiles, new teeth develop in, and erupt out of the socket of the preceding tooth. Berkowitz & Sloan (1979) suggest that some of the fibres attaching the predecessor to its socket wall may be re-utilized in the attachment of a new tooth. In the *gomphosis* of mammals the walls of the deciduous tooth socket are probably entirely replaced by new bone when the permanent successor erupts. The fibrous tissue suspending a tooth in its socket is the *periodontal ligament*.

In nearly all reptiles, including the ancestral stock, teeth are ankylosed to the jaw by a bone of attachment. At the reptile/mammal boundary the teeth were probably attached by a gomphosis with its associated periodontal ligament. The crucial intermediate forms existed only in the mammal-like reptiles of the Upper Permian and Lower Triassic. Knowing how tooth attachment develops in modern reptiles and mammals, it may be possible to deduce how the intermediate forms were developed by studying what remains of the attachment apparatus in fossilized jaws.

Acrodont Ankylosis

The simplest form of tooth attachment seen in some stem reptiles is an acrodont ankylosis [Figs 4 & 5(A)]. Shellis (1981) gives a clear description of the development of an ankylosed tooth in a recent fish *Labrus bergylta*. Odontoblasts differentiate under the inductive influence of the inner wall of *Hertwig's root sheath* (HRS), an open-ended cylinder of ectodermal cells surrounding the developing root of the tooth. A cylinder of root dentine is formed inside HRS. This mineralized cylinder is briefly continued beyond HRS by cells differentiated from the base of the dental papilla under the influence of some unknown inductive stimulus. This tissue will here be called *attachment dentine*. During this time the tip of the tooth is erupting into the oral cavity. A new bone-like tissue (called here *protocement*) now develops on the outer surface of the root base and a further bone-like tissue (bone of attachment) on the adjacent walls of bone surrounding the base. Protocement and bone of attachment each incorporate fibres contained in the soft tissue surrounding the developing root base. The mineralizing fronts meet, engulfing the soft tissue, with the result that the tooth is now ankylosed to the jaw.

The bone of attachment deposited on the surface of the jaw seems to be produced by cells derived from the soft tissue surrounding the developing tooth. The attachment dentine comes from pulpal cells. The cells producing protocement might come from either of the above sources.

The development of the tissue(s) cementing dentine to bone in the ankylosed teeth of reptiles appears not to have been described in detail. Cooper (1963) observes in Lacertilia that the attachment tissue is bone-like, may contain a few short dentinal tubules on its dentine side, is stronger than the tooth itself and takes about two days to develop. His histological description suggests at least a dual origin of the cementing tissue: from pulpal cells, some of which retain the capacity to develop processes and their surrounding tubules; and from bone cells derived from the soft tissues around the developing tooth. The tissue formed by the pulpal cells could be homologous with attachment dentine or protocement.

Gomphosis

The development of the tooth attachment in mammals is far better known [Fig. 5(D)]. A deciduous mammalian tooth germ lies in the middle of a fibrous dental follicle which separates it from the wall of its crypt in the developing jaw bone. The inner *investing layer* of the follicle is continuous with the cells of the dental papilla around the margin of HRS (and, prior to this, the cervical loop of the enamel organ from which HRS is derived). At the base of the tooth, both are apparently separated from the remainder of the follicle by a fibrous *pulp limiting membrane*. As in all tooth development, HRS outlines the presumptive root. Odontoblasts differentiate against its inner surface and lay down dentine matrix, most of which is mineralized, sometimes (or always ?) leaving a thin unmineralized remnant against HRS (Owens, 1976). As the tooth erupts away from HRS the latter breaks up along the sides of the tooth but, just prior to this, secretes a product onto the unmineralized dentine remnant (Slavkin & Boyde, 1974; Owens, 1978). The fragmented HRS cells move out towards the middle of the follicle. At the same time follicular cells from the investing layer move towards the root surface, differentiate into cementoblasts, and mineralize the dentine remnant with its included HRS secretion. Fibres derived from the follicle become included in further layers of (pure) *cement*, a bone-like tissue, produced by the cementoblasts.

While the tooth is erupting, cells from the investing layer of the follicle appear to migrate coronally along the ligament and out towards the surface of the bony crypt enclosing the tooth (Ten Cate & Mills, 1972; Perera & Tonge, 1981). The crown of the tooth begins to erupt past the old rim of its crypt. Some of the migrating cells differentiate into osteoblasts and lay down new bone on the rim of the crypt deepening it from above to accommodate the lengthening root. The new bone incorporates follicular fibres which are probably continuous with those

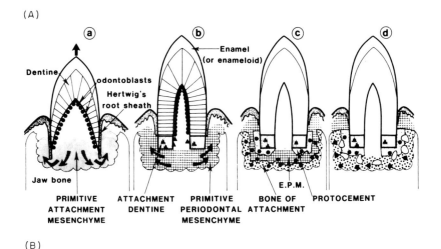

FIG. 5. Hypothetical sequence showing the evolution of a gomphosis from an acrodont ankylosis (cf. Fig. 6). (A) Four stages in the development of an acrodont ankylosis with attachment dentine (E.P.M. = early periodontal mesenchyme). (B) The development of a protothecodont ankylosis without attachment dentine.

embedded in the cement deposited on the neck of the tooth. The developing tooth erupts coronally out of its crypt faster than new bone is developed at the deepening rim with the result that its crown breaks out of the crypt and moves into the oral cavity, finally to meet its opponent. The root is now completed by growing back into the base of its resorbing crypt during the *intrusive phase* of root formation (Kovacs, 1971). This intrusive phase of root growth involves a marked narrowing of the *apical foramen* connecting the tooth pulp to the periodontal ligament, which contrasts strongly with the wide open apex seen in all ankylosed teeth.

With the exception of the hybrid tissue on the developing root surface

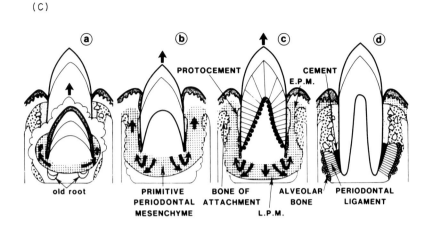

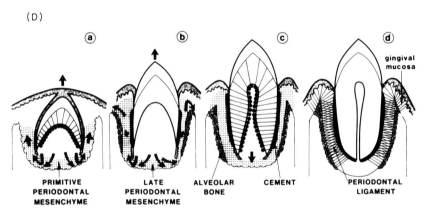

(C) The appearance of a periodontal ligament around the apical end of the root (L.P.M. = later periodontal mesenchyme. Pulp attachment mesenchyme produces attachment dentine). (D) The development of a gomphosis with periodontal ligament and attached gingiva.

(which does not concern us here) three tissues are responsible for attaching root dentine to the jaw bone in a gomphosis: cement derived from the investing layer of the follicle; fibres derived from both investing layer and the remainder of the follicle; and bone which grows up from the rim of the crypt and possibly lines much of the early crypt. The osteoblasts forming this latter bone, in common with cementoblasts, are derived from the investing layer of the follicle (Ten Cate & Mills, 1972; Yoshikawa & Kollar, 1981).

It is now possible to predict homologies between the attachment tissues in reptiles and mammals. Attachment dentine is not seen in mammals. Protocement could be homologous with cement and the bone of attachment with the bone formed by cells derived from the investing layer of the follicle (but see later). The (unmineralized) periodontal ligament is not seen in an ankylosis. This view contrasts with the alternative possibility that cement is homologous with the bone of attachment and alveolar bone is a mere upgrowth of jaw bone.

Mixed Ankylosis/Gomphosis

It is now possible to combine observations of tooth attachment in the fossilized jaws of mammal-like reptiles with the homologies predicted above in order to predict the sequence in which the mammalian gomphosis evolved.

In a study of tooth succession in the mammal-like reptile, *Thrinaxodon*, Hopson (1964) published drawings of the appearance of coronal sections of an upper and lower jaw. The following description is based on these drawings. In *Thrinaxodon's* oligophyodont dentition a replacement tooth developed lingually at the level of its predecessor's neck. By inducing resorption of material ahead of it the replacement moved labially as it grew and now occupied a hole directly beneath its predecessor. Further tooth growth was probably expressed mainly as apical movement of the root, filling the old socket, with some occlusal movement of the crown, again by resorbing material in the way. Occasionally the predecessor was incompletely resorbed because a few shards of old roots remained embedded in the jaws. When the old socket was filled by lengthening root, further growth became accommodated by movement of the crown into the oral cavity until it reached its definitive position. The apex of the root remained wide open to the pulp, the reptilian as opposed to mammalian condition. The attachment between the jaw bone and root is not clear. Crompton (1963), in a study of tooth succession and tooth morphology in *Thrinaxodon*, clearly describes a spongy bone which developed around all teeth and ankylosed their necks to the jaw bone. In the reconstructions of Hopson (1964) of whole jaws the roots of younger fully erupted teeth are often clearly separated from the jaw bone while those of older teeth touch the jaw bone. In a drawing of actual sections as opposed to reconstructions the roots of a canine and of a post-canine, both adjaccent to crypts occupied by their replacements, are widely separated from the jaw while those of two precanines appear to be united with the jaws. The data of Crompton (1963) and Hopson (1964) suggest the possibility that the necks of canines and post-canines become ankylosed to the jaw but an unmineralized region persisted around the rest of their roots.

If the above interpretation is correct, *Thrinaxodon's* teeth may have developed in the following way [Fig. 5(C)]. The *dental lamina*, the ectodermal contributor to tooth development, occupied a groove in the jaw bone just lingual to the necks of the post-canines (Crompton, 1963). New tooth buds were initiated at the free margin of the dental lamina. As a tooth germ moved labially beneath its predecessor it was surrounded by mesodermal cells, some being carried along from the site at which the tooth was initiated and some being derived from the resorbed tissues whose position it was usurping [Fig. 5(C), a]. As the tooth lengthened in its new site cells were being extruded from the widely open pulp apex into the region between HRS and the jaw bone. These cells travelled along the walls of the socket between the tooth and the jaw bone. The earliest extruded cells moved to the region between the neck of the tooth and the rim of the old socket. Later extruded cells lined the deeper walls of the socket. The earlier ones laid down protocement on the dentine and bone of attachment on the jaw, finally meeting in the middle of the soft tissue. The attachment tissue contained vascular spaces because the cells forming the two hard tissue layers, protocement and bone of attachment, needed to maintain a blood supply when they met back-to-back. Apically, the later extruded cells laid down hard tissue on the surface of the root dentine and on the socket wall, incorporating fibres from the soft tissue, but the two mineralizing surfaces did not meet and remained separated by a periodontal ligament. It is not clear whether a ligament was maintained because the hard tissue forming cells were different from those forming protocement and bone of attachment or because some inhibitor (such as maintained remnants of HRS) evolved. Incidentally this problem remains unanswered in mammals.

Evolution and (differentiation) Clades

The evolutionary sequence can now be represented by a diagram (Fig. 6) which combines some of the rules of cladistic taxonomists with observations and predictions relating to the differentiation of embryonic cell populations. Osborn (1979) suggested that during tooth morphogenesis cell populations pass through differentiation compartments similar to those described by Dienstman & Holtzer (1975), which can be likened to clades. All the cells within a particular clade are identical and differ from those in other clades.

It should be noted that the remains of a parent clade can co-exist with its two daughter clades. The point to be emphasized here is that a cell in a particular clade has only two options if it differentiates further. Just as cladistic taxonomists find that a similar restriction requires them to postulate the existence of as yet undiscovered or unrecognized fossil

564   J. W. Osborn

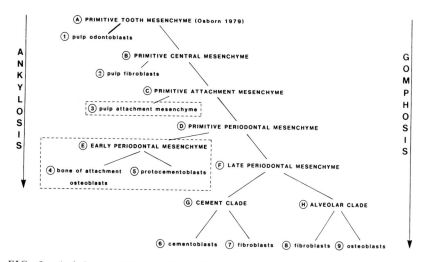

FIG. 6. A cladogram of the hypothetical 'family' of cells evolved to produce a tooth with its differing forms of attachment apparatus. The 'family' comprises two types of 'species', terminally differentiated cells (1–9) and intermediary cells (A–H). The lineages from which an ankylosis (left) and a gomphosis (right) develop can be traced back to the 'parent' primitive tooth mesenchyme whose own ancestry has been suggested by Osborn (1979). The 'species' surrounded by dotted lines have been deleted in the development of a gomphosis while the late periodontal mesenchyme (F) evolved in some mammal-like reptiles and provides the attachment apparatus in mammals.

taxa of animals or plants, so a cladistic interpretation of development requires postulating that groups of cells exist at precisely defined stages of differentiation which have not yet been recognized. Incidentally a cladistic interpretation must reject the widely held concept of a totipotent or pluripotent cell. Only sequential bipotency exists. For example, suppose A-type cells apparently give rise to three different cell types, B, C, and D, and that the B cells are the earliest to differentiate. Before the remaining A cells can produce the C and D types they must enter a new clade which is a precisely defined, as yet unrecognized, differentiation compartment, X. Only now can the C and D types be differentiated. This differentiation of A-type into X-type, or indeed any differentiation, does not necessarily require an external stimulus. It may merely depend on an already programmed cell division. The point here is that the critical cell division produces a different cell type. Dienstman & Holtzer (1975) have documented a sequence of such cell divisions in the differentiation of myoblasts and chondroblasts from primitive mesenchyme.

The development of an acrodont ankylosis in a reptile is shown in Figs 5(A), (B) and 6. For descriptive convenience it can be said that a clade of mesenchymal cells, *tooth family (ecto)mesenchyme*, is responsible for producing all the teeth developed at a particular tooth position by successively budding new tooth germs from its inferior progress zone. A more accurate description is that all teeth in a reptilian lower jaw develop out of a clade of mesenchyme, from whose long inferior progress zone (Fig. 2) new buds are initiated on an alternating pattern controlled by the inhibitory zones developed around each. A clade of *primitive tooth mesenchyme* (Fig. 6) buds out from the parent clade. A clade of *pulpal odontoblasts* is differentiated from this leaving a *primitive central mesenchyme* clade. This produces the *pulp fibroblast* clade and a *primitive attachment mesenchyme* clade. The latter now generates a *pulp attachment mesenchyme* clade which produces the attachment dentine [Fig. 5(A) a], or, for example, the osteodentine which fills the pulp of many ankylosed teeth such as the marginal teeth of the pike. The remainder of the primitive attachment mesenchyme clade becomes a *primitive periodontal mesenchyme* clade. This migrates to fill the space between the outside of the root and the bone of the jaw and from it, via an *early periodontal mesenchyme* clade, differentiate the *protocement* and *bone of attachment* clades which ankylose the tooth to the jaw bone. A *late* periodontal mesenchyme clade does not develop for an ankylosed tooth (see later).

During the next stage of evolution the role of pulp attachment mesenchyme (osteodentine formation, for example) in attaching the tooth to the jaw became progressively less important and finally terminated when this function was entirely undertaken by the protocementoblasts and osteoblasts of the periodontal mesenchyme clades. The effect of deleting pulp attachment mesenchyme (and lengthening the root of the tooth) is shown in Fig. 5(B).

If the cladistic approach to ontogeny is extended to the phylogeny of cell populations there is no way to delete the apparently redundant primitive attachment mesenchyme clade (C in Fig. 6) from animals which no longer produce pulp attachment mesenchyme (3 in Fig. 6). Evidence in support of the maintenance of this clade comes from the genetically determined condition, radicular dentinal dysplasia. In affected individuals the bulk of the dentine and what would have been the pulp consists of osteodentine like that filling the pulps of many ankylosed teeth (Perl & Farman, 1977). This tissue might be developed via the normally unexpressed potential of the transitory primitive attachment mesenchyme clade. The less dramatic but frequently observed calcifications in the pulps of human teeth could have the same origin.

The postulated evolution of a periodontal ligament around the apices of some mammal-like reptile teeth presents an awkward problem for a cladistic interpretation. The possible evolution of an inhibitor of calcification from remnants of Hertwig's root sheath around the lower part of the root has been rejected (but not abandoned) because it would not account for the evolution of the fibroblasts in the periodontal ligament: these would have to be new cells from new protocement or bone attachment clades. The evolution of a *late periodontal mesenchyme clade* (Fig. 6) was the chosen solution (this necessitated the assignment of an *early* periodontal mesenchyme clade which may, till now, have seemed awkward to the reader). The earlier cells migrating coronally from the primitive periodontal mesenchyme clade (D in Fig. 6) ultimately produce an ankylosis around the neck of the tooth; the later migrating cells have entered a new clade which produces an attachment via a gomphosis [Fig. 5(C)].

The early periodontal mesenchyme clade (E in Fig. 6) becomes progressively reduced and is absent in mammals (cf. pulp attachment mesenchyme: 3 in Fig. 6).

The above solution suggests the existence of at least three unrecognized stages in the differentiation of cementoblasts, for example: primitive attachment, primitive periodontal and late periodontal mesenchyme clades. It is suggested that none of these stages of differentiation is induced by an external stimulus but rather that each stage is entered and exited autonomously in response to a (genetically) predetermined number of cell mitoses.

All the cell "species" represented in the "cladogram" (Fig. 6) can be allotted to a tooth mesenchyme "family". As might be expected from such an analogy with zoological taxonomy, the cell "species" existing in the adult are closely related in terms of their metabolism. For example, all share the ability to produce type I collagen and similar protoglycans and glycosaminoglycans. One can visualize that cells entering the primitive tooth mesenchyme clade (A in Fig. 6) have potentially opened a pathway for transcribing the DNA associated with these products. Progression through the lineage represented in the cladogram, by autonomous or externally induced differentiation, modulates the resultant behaviour of terminal cell types within the limitations of a shared heredity. It seems possible that any repair or maintenance processes in the adult could be accomplished via the existence of residual cells occupying intermediate differentiation compartments (A, B, C, D, F, G and/or H in Fig. 6). By dividing and differentiating in response to some unknown stimulus such cells can repair and/or maintain the adult tissues.

Functional Advantages

Before a model of an evolutionary sequence can gain even the most limited acceptance it is important to identify a functional advantage for critical stages in the proposed sequence. The loss of attachment dentine may be unimportant for two reasons. First it seems a somewhat trivial change. Second it is quite possible that an ankylosis with attachment dentine or its equivalent evolved from an ankylosis without attachment dentine rather than the reverse and that such a stage need not be included in the transition from reptiles to mammals.

It would be an advantage to lengthen roots and to build up more bone of attachment in order to attach a tooth more firmly to the jaw. It would be a further advantage to build up the inner and outer flanges of jaw bone seen in the protothecodont condition to form a gutter in which teeth could develop and be replaced. The walls of the gutter would not need to be removed by osteoclasts during tooth replacement, thereby facilitating the succession of teeth. It was still necessary to fill up the gutter between adjacent teeth with bone of attachment. It will be argued later that the inter-dental bone separating mammalian teeth is alveolar bone, from late periodontal mesenchyme, and that much of the ancestral inner and outer walls of the gutter persists and is jaw bone rather than alveolar bone.

The postulated appearance of a periodontal ligament around the apices of teeth in mammal-like reptiles is the most critical stage in the evolution of a gomphosis. Species such as *Thrinaxodon*, the closest known cynodont to mammals, had developed marked heterodonty in their post-canines. They also possessed masseter muscles (Crompton & Parker, 1978) which might have been able to move the jaws laterally. The teeth may therefore have been primitively involved in processing food rather than merely grasping it as is the case for most carnivorous reptiles. If this is true the teeth would have been subjected to more lateral forces than those of earlier groups (Fig. 7). Laterally directed loads potentially cause a tooth to rotate in its socket around an axis whose position largely depends on the stress/strain properties of the tissues attaching it to the jaw bone. If the axis were somewhere near the middle of the tooth [Fig. 7(C)] it can be seen that on the buccal sides of the upper teeth and the lingual sides of the lower teeth the coronal attachment tissues would have been compressed and the apical attachment tissues tensioned. In these regions it would have been an advantage to capitalize on the physical properties of bone and collagen by maintaining (pressure resistant) bone coronally and evolving a (tension resistant) collagenous ligament apically.

The above argument can account for the evolution of a periodontal ligament apically on the buccal sides of upper teeth and the lingual sides of lower teeth. If, despite the absence of occlusion, the struggles of prey and the stresses produced by rudimentary masticatory movements led to both buccal and lingual tilting of teeth it would have been

568   J. W. Osborn

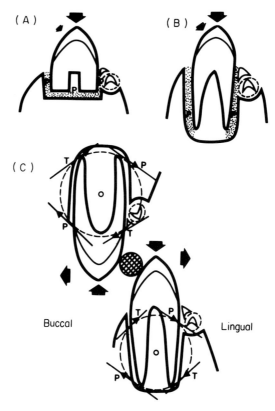

FIG. 7. The evolution of a periodontal ligament. (A) In most acrodont ankyloses the tooth is subjected largely to axial loads. (B) Protothecodont ankylosis may have evolved in response to increased axial loads. (C) It is argued that lateral forces were generated on the teeth of *Thrinaxodon*, with food (cross-hatched) wedging between the teeth. The lateral forces caused the teeth to rotate in their sockets with the result that areas of tension (T) and pressure (P) were produced. A periodontal ligament evolved apically in tension regions. If the teeth were more vigorously rocked buccally and lingually the ligament spread around the whole apical region [Fig. 5(C)]. Ankylosis remained around the necks of teeth in order that the teeth would be retained as long as possible during their resorption and replacement – or to resist compression forces until a more sturdy ligament had evolved.

advantageous to surround the whole apical half of the root with periodontal ligament.

The apical presence of a ligament raises the axis of tooth rotation towards the neck of the tooth and encourages further spread of the ligament. Perhaps only with the near complete evolution of mammalian muscles of mastication in an as yet undiscovered form at the reptile/mammal boundary did a full periodontal ligament finally evolve.

Cause and effect now become difficult to unravel, partly because there is no agreement on the details of how a (full) periodontal ligament works in mammals. There is no proof that the primary function of a ligament, as opposed to an ankylosis, is to act as a stress breaker protecting a functioning tooth from forces which might fracture it, although this hallowed argument seems reasonable. Nor is there any evidence to support the importance of a ligament converting potential pressure on the alveolar bone, which might initiate bone resorption, into tension which encourages bone deposition. A major advantage of a periodontal ligament seems to be that it permits an erupted, functioning tooth to move through the jaws, by removing bone in the direction of movement and depositing bone behind, in order to take up its most effective position. Clearly an ankylosed tooth cannot move without being shed and the only way advantageously to change the relationship between upper and lower teeth is to erupt replacement teeth in new positions. The (diphyodont) mammals have this chance once in a lifetime so the opportunity of tooth drift presented by a full periodontal ligament may be very important. But it is obviously impossible for the evolving periodontal ligament to begin to confer this advantage. Only the completed ligament can allow tooth movement.

Eruption

Many different views have been proposed to account for the force which causes mammalian teeth to erupt but none seems to have taken into account tooth eruption in non-mammals. There seems no reason to believe that during the gradual evolution of mammalian teeth there should have been any change in the mechanical forces which cause a tooth to move from the site it develops to a position under its successor and thence up into the oral cavity. It has been argued here that periodontal attachment mesenchyme (Figs 5 & 6) is extruded out of the bases of developing teeth. This suggests the tooth is pushed into the oral cavity by pulpal growth, a mechanism which is generally listed but dismissed in reviews of tooth eruption. Admittedly a pulpal growth force does not explain the continued eruption of a rodent incisor when its growing end has been removed but there are objections to every mechanism yet proposed. Osborn (1974) has argued that tooth replacement is equivalent to tooth growth and therefore the force which produces eruption is probably related to growth.

Root Morphology

The short, open apex, ankylosed roots of reptiles evolved into the long, tapering apex, gomphosed roots of mammalian molars. The lengthening of roots can be related to an increase in the security of attachment. A disadvantage would be the greater effort required to resorb a tooth in order to replace it with a successor.

When, and if, a ligament evolved around the apices of the teeth of *Thrinaxodon* the apical part of the root was stressed as tilting movements potentially pulled it away from the bone to which it was attached (see above). It would have been an advantage to thicken the apical root dentine, thereby narrowing the apex and tapering the root, in order to limit the possibility of root fracture.

Noble (1969) noted that the post-canine roots of a cynodont, *Cynognathus crateronotus*, are single but deeply grooved. This grooving could be an advantage on the lingual side of the tooth because it would increase the space available for the development of a successor (always on the lingual side) and postpone the resorption which progressively weakens the attachment of the incumbent tooth. If this is carried to an extreme, the roots of the predecessor become fully divided thereby providing a space which temporarily houses the developing successor. If true, this argument leads to the somewhat obvious conclusion that the homologues of the permanent molars of mammals, all of which have two or more roots, were at one time replaced by successors which developed within the root bifurcation. The increased attachment area presented by a bifurcated root provided a secondary, rather than primary, advantage. Root bifurcation is not so widespread in premolars as in deciduous molars. First, premolars are successional teeth and second, because they are subjected to less force the weakening of their roots by resorption (to accommodate a third tooth series) may not have been sufficient to induce the evolution of grooving and subsequent root bifurcation. However, if one seeks evolutionary change at the level of the cellular clade, it will be recalled that both premolars and molars develop from the same clade. A change in the properties of cells at the initial focus of the post-canine clade may not only be expressed for the advantage of dividing the roots of molars developed out of the posterior progress zone but also, without advantage, in the deciduous molars and premolars developed out of the anterior and inferior progress zones.

Alveolar Bone

It was argued above that alveolar bone evolved from late periodontal mesenchyme and that it develops from a clade of cells whose lineage is different from that of the jaw bone (Osborn, 1978, 1979). Observations of the development of the bone which surrounds teeth (e.g. Scott & Symons, 1974) and the evolutionary sequence proposed here suggest the alveolar bone clade produces osteoblasts whose product fills a gutter between buccal and lingual plates of jaw bone and spreads over the coronal crest of the jaw bone.

Mummery (1924) observed that the lower posterior molars of the manatee develop within a shell of (alveolar) bone which is separated by soft tissue from the ramus of the jaw. Possibly equivalent, but less striking, is the alveolar bulb (e.g. Noble, 1969) which extends behind the maxillae of man and the pig and temporarily accommodates the developing permanent molars. A further example of the separation between alveolar bone and jaw bone is provided by a young elephant skull in the author's possession, the tusks of which are about 2 inches long (Fig. 8).

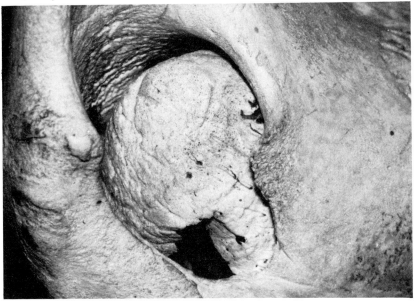

FIG. 8. An elephant's lower successional teeth develop within the ramus of the mandible behind the functioning dentition. A developing tooth is encased in (alveolar) bone which is quite separate from the bone of the jaw.

No one has shown a significant difference between the histology of alveolar bone and jaw bone. Nor are there any convincing data to prove they are physiologically different although orthodontists, commonly involved in moving teeth around the jaws by means of mechanical appliances, have long observed that teeth cannot be successfully moved beyond the margins of what they believe to be alveolar bone, which implies some special difference between it and jaw bone. The suggestion, therefore, is that alveolar bone is more plastic than jaw bone in that it can be more readily remodelled by osteoblasts and osteoclasts in order to allow the teeth to drift into their most functionally appropriate positions. For example, when a new elephant molar moves forwards into the oral cavity it passes into a gutter of jaw bone lined by alveolar bone which permits the tooth a freedom of movement, by bone remodelling, which cannot be matched by jaw bone. Also, when human molars drift forwards through the jaws the partitions of alveolar bone separating adjacent teeth appear to move with them. In fact, the partition is being resorbed on one side and built up on the other side.

The clade which produces alveolar bone leaves a rump of fibroblasts which form the outer fibres of the periodontal ligament. It seems possible that some of these fibroblasts migrate over the alveolar crest (formed of alveolar bone) and down as far as the mucogingival junction which separates alveolar from gingival mucosa [Fig. 5(D)]. This suggestion is made to account for the abrupt change between the gingival mucosa, which is keratinized without a submucosa, and the alveolar mucosa, which is non-keratinized and has a submucosa. Perhaps the difference in the histology of the mesoderm is the expression of a difference between the (soft tissue) cell residues of the clades responsible for producing alveolar bone and jaw bone.

REFERENCES

Berkowitz, B. K. B. & Shellis, R. P. (1978). A longitudinal study of tooth succession in piranhas (Pisces: Characidae) with an analysis of the tooth replacement cycle. *J. Zool., Lond.* **184**: 545–561.
Berkowitz, B. K. B. & Sloan, P. (1979). Attachment tissues of the teeth in *Caiman sclerops* (Crocodilia). *J. Zool., Lond.* **187**: 179–194.
Butler, P. M. (1939). Studies of the mammalian dentition. Differentiation of the postcanine dentition. *Proc. zool. Soc. Lond.* **109**: 1–36.
Butler, P. M. (1956). The ontogeny of molar pattern. *Biol. Dev.* **31**: 30–70.
Butler, P. M. (1979). The ontogeny of mammalian heterodonty. *J. Biol. Buccale* **6**: 217–227.
Cooper, J. S. (1963). *The dental anatomy of the genus* Lacerta. Ph.D. Thesis: University of Bristol, England.
Cooper, J. S., Poole, D. F. G. & Lawson, R. (1970). The dentition of agamid lizards with special reference to tooth replacement. *J. Zool., Lond.* **162**: 85–98.
Crompton, A. W. (1963). Tooth replacement in the cynodont *Thrinaxodon liorhinus* Seeley. *Ann. S. Afr. Mus. Cape Town* **46**: 479–521.
Crompton, A. W. & Parker, P. (1978). Evolution of the mammalian masticatory apparatus. *Am. Sci.* **66**: 192–201.
De Mar, R. E. (1973). The functional implications of the geometrical organization of dentitions. *J. Paleont.* **47**: 452–461.
Dienstman, S. R. & Holtzer, H. (1975). Myogenesis: a cell lineage interpretation. In *Results and problems in cell differentiation*: 1–25. Reinert, J. & Holtzer, H. (Eds). Berlin: Springer-Verlag.

Edmund, A. G. (1960). Tooth replacement phenomena in the Lower Vertebrates. *Contr. Life. Sci. Div. R. Ont. Mus.* **52**: 1–90.
Holtzer, H., Rubinstein, N., Fellini, S., Yeoh, G., Chi, J., Birnbaum, J. & Okayama, M. (1975). Lineages, quantal cell cycles, and the generation of cell diversity. *Q. Rev. Biophys.* **8**: 523–557.
Hopson, J. A. (1964). Tooth replacement in cynodont, dicynodont and therocephalian reptiles. *Proc. zool. Soc. Lond.* **142**: 625–654.
Hopson, J. A. (1971). Postcanine replacement in the gomphodont cynodont *Diademodon*. *Zool. J. Linn. Soc.* **50** (Suppl.): 1–21.
Kovacs, I. (1971). A systematic description of dental roots. In *Dental morphology and evolution*: 211–256. Dahlberg, A. A. (Ed.). Chicago: University of Chicago.
Kuhne, W. G. (1973). The evolution of a synorgan. Nineteen stages concerning teeth and dentition from the pelycosaur to the mammalian condition. *Bull. Grpmt. int. Rech. scient. Stomat.* **16**: 293–325.
Mummery, J. H. (1924). *The microscopic and general anatomy of the teeth, human and comparative*. Oxford University Press.
Noble, H. W. (1969). The evolution of the mammalian periodontium. In *Biology of the periodontium*: 1–26. Melcher, A. H. & Bowen, W. H. (Eds). London: Academic Press.
Osborn, J. W. (1970). New approach to Zahreihe. *Nature, Lond.* **225**: 343–346.
Osborn, J. W. (1971). The ontogeny of tooth succession in *Lacerta vivipara* Jacquin (1787). *Proc. R. Soc. Lond.* (B) **179**: 261–289.
Osborn, J. W. (1973). The evolution of dentitions. *Am. Scient.* **61**: 548–559.
Osborn, J. W. (1974). On the control of tooth replacement in reptiles and its relationship to growth. *J. theor. Biol.* **46**: 509–527.
Osborn, J. W. (1975). The control of tooth shape. *J. dent. Res. (I.A.D.R. Abst. L.)* **57**. (Special volume.)
Osborn, J. W. (1978). Morphogenetic gradients: fields versus clones. In *Development, function and evolution of teeth*: 171–201. Butler, P. M. & Joysey, K. A. (Eds). London: Academic Press.
Osborn, J. W. (1979). A cladistic interpretation of morphogenesis. *J. Biol. Buccale* **6**: 327–337.
Osborn, J. W. & Crompton, A. W. (1973). The evolution of mammalian from reptilian dentitions. *Breviora* No. 399: 1–18.
Owens, P. D. A. (1976). The root surface in human teeth: a microradiographic study. *J. Anat.* **122**: 389–401.
Owens, P. D. A. (1978). Ultrastructure of Hertwig's epithelial root sheath during early root development in premolar teeth in dogs. *Archs oral Biol.* **23**: 91–104.
Perera, K. A. S. & Tonge, C. H. (1981). Fibroblast cell population kinetics in the mouse molar periodontal ligament and tooth eruption. *J. Anat.* **133**: 281–300.
Perl, T. & Farman, A. G. (1977). Radicular (Type 1) dentin dysplasia. *Oral Surg.* **43**: 746–753.
Scott, J. H. & Symons, N. B. B. (1974). *Introduction to dental anatomy*. Edinburgh: Churchill Livingstone.
Shellis, R. P. (1981). Comparative anatomy of tooth attachment. In *The periodontal ligament in health and disease*: 3–24. Berkowitz, B. K. B., Moxham, B. J. & Newman, H. N. (Eds). Oxford: Pergamon Press.
Slavkin, H. C. & Boyde, A. (1974). Cementum: an epithelial secretory product? *J. dent. Res.* **53**. [Spec. Issue: Proc. 52nd Session Int. Assoc. Dent. Res.: 157 (Abstr.).]
Summerbell, D., Lewis, J. H. & Wolpert, L. (1973). Positional information in chick limb morphogenesis. *Nature, Lond.* **244**: 492–496.

Ten Cate, A. R. & Mills, C. (1972). The development of the periodontium: the origin of alveolar bone. *Anat. Rec.* **173**: 69–77.

Wolpert, L. (1969). Positional information and the spatial pattern of cellular differentiation. *J. Theor. Biol.* **25**: 1–47.

Yoshikawa, D. K. & Kollar, E. J. (1981). Recombination experiments on the odontogenic roles of mouse dental papilla and dental sac tissues in ocular grafts. *Arch. oral Biol.* **26**: 303–307.

Ziegler, A. C. (1971). A theory of the evolution of therian dental formulas and replacement patterns. *Q. Rev. Biol.* **46**: 226–249.

The Relationships and Early Evolution of the Diapsida

MICHAEL J. BENTON

University Museum, Parks Road, Oxford OX1 3PW, England

SYNOPSIS

Living reptiles fall into several natural groupings: turtles, crocodiles, lizards and snakes. The last three groups have, or their ancestors had in the past, two openings in the skull behind the eye – the diapsid condition – and fossil evidence strongly supports their association into the Subclass Diapsida. The diapsids arose 300 million years (My) ago and evolved as two major lineages during the Permo-Triassic (215–285 My ago): the Archosauromorpha [Pterosauria, Prolacertiformes, Archosauria, Rhynchosauria] and the Lepidosauromorpha [Younginiformes, *Sphenodon*, Squamata (early "lizards", lizards, amphisbaenians, snakes)]. A new classification of these groups is presented here on the basis of a cladistic analysis. There was a succession of diapsid radiations in the Permo-Triassic, including some bizarre gliding and swimming forms. Ecologically, the thecodontians and rhynchosaurs were important in the middle and late Triassic when they partly took over major carnivore and herbivore niches from mammal-like reptiles. All these groups died out in the late Triassic (220 My ago) and the dinosaurs subsequently radiated opportunistically world-wide into all major terrestrial niches. The assumption of competition between mammal-like reptiles and various diapsid groups during the Triassic is not supported here.

INTRODUCTION

Reptiles with two temporal arches – the diapsid condition – have had a confused taxonomic history. Osborn (1903) coined the name Diapsida to include lizards, snakes, *Sphenodon*, crocodiles, dinosaurs, thecodontians and pterosaurs, as well as pelycosaurs, procolophonids, ichthyosaurs and *Mesosaurus*. Williston (1925) removed the last four groups from the Subclass Diapsida, but also placed the lizards and snakes in his Subclass Parapsida with *Mesosaurus* and the ichthyosaurs. All of these forms have only an upper temporal opening, but Broom (1925) argued strongly that lizards and snakes were true diapsids that had lost the lower temporal bar. Since then, Romer (1933, 1956, 1966, 1971) has maintained the view that the diapsids really consist of two subclasses – the Lepidosauria (basal "eosuchians", lizards, snakes, *Sphenodon*, rhynchosaurs) and the Archosauria (thecodontians, crocodiles, pterosaurs, dinosaurs), each of which probably had a separate origin.

New studies on early diapsid reptiles have shown that they probably all derived from a single ancestral stock, and that there was a series of adaptive radiations of diapsids during the Permo-Triassic (215–285 My ago). The initial radiations were of small terrestrial, aquatic and gliding forms that made little impact on a world dominated by the mammal-like reptiles. However, diapsids achieved larger size and greater abundance during a multiphase replacement of the mammal-like reptiles in the Triassic which culminated in the successful radiation of the dinosaurs.

In this paper, new work on the classification and evolution of early diapsids is reviewed. The application of a cladistic methodology to the classification of all well known early diapsids has produced a scheme of relationships already hinted at by several authors, but rather different from the standard notion (e.g. Romer, 1966). A consideration of the composition and stratigraphic position of the major Permo-Triassic reptile faunas has suggested an interpretation of the diapsid take-over different from the usual competitive models.

CLASSIFICATION OF THE DIAPSIDA

All reptiles from the Permo-Triassic that have been called "diapsid" at one time or another were considered and an attempt was made to include all but the most scrappy specimens in a review of their relationships. At first, it was expected that Romer's "archosaurs" and "lepidosaurs" (Romer, 1966) would not appear as distinct groups, but that the Subclass Diapsida would contain several major lineages radiating in the Permian (Benton, 1982). However, there is strong evidence for two lines that diverged in the mid- to late Permian – the Prolacertiformes (*Protorosaurus*, *Prolacerta*, etc.) and the Younginiformes (*Youngina*, tangasaurids). The Prolacertiformes show close relationships with the rhynchosaurs of the Triassic and with the archosaurs and pterosaurs. This assemblage is the Archosauromorpha (von Huene, 1946). The Younginiformes show close relationships with the "early lizards" of the Permian and Triassic, as well as with subsequent Squamata and Sphenodontidae. This assemblage is the Lepidosauromorpha (Gauthier, in press). The Archosauromorpha and Lepidosauromorpha share numerous derived characters, some of which were absent in the late Permian *Claudiosaurus* and the gliders *Weigeltisaurus* and *Coelurosauravus*. Some archosauromorphs could not be confidently placed (*Noteosuchus*, *Malerisaurus*, *Trilophosaurus*); others could have been archosauromorphs or lepidosauromorphs (*Heleosaurus*, *Monjurosuchus*), and some "lepidosaurs" could not be clearly placed at all (Claraziidae, Champsosauridae, Pleurosauridae).

A cladistic methodology was used in this study. An attempt was made to determine shared derived characters between various genera or families as an indication of closest relationship. Closely related animals, of course, share numerous primitive characters, but these do not help to resolve taxonomic questions. This kind of analysis was possible since the lowest operational units (usually genera) are well defined because of the relative incompleteness of the record of fossil reptiles, and because many characters may be recorded.

Many problems typical of such taxonomic exercises were encountered: assessment of true homology of characters as opposed to parallelism or convergence; ancestral groups which display few derived characters; rapidly radiating stocks which contain many genera going in different directions and possessing few *shared* derived characters; highly modified groups that stand in isolation which may have lost numerous characters considered to be diagnostic of their large monophyletic group of relatives; and lack of key characters in poorly preserved fossil material. Several of the early diapsid taxa could not be placed confidently since they appeared to share derived characters with two or more separate groups – e.g. the Proterosuchidae – or because of the absence of critical characters in the fossils or descriptions – e.g. *Protorosaurus*, *Heleosaurus*, *Mesosuchus* and the "Paliguanidae".

The results of the study are presented in the form of a cladogram (Fig. 1; full details to be published elsewhere), which is a tentative statement of relationships that is readily open to testing and modification by future work. The classification given below is based closely on the cladogram, but several conventions are used in order to avoid: the mechanical problems of a proliferation of new taxon and category names for all dichotomies; the stability problems of introducing new taxa, and particularly fossil taxa, to an established classification (or of revised opinions regarding relationships); and the problem of confusing other biologists with constantly changing and unfamiliar classifications. These conventions (sequencing, indented lists, plesions) have been discussed by Patterson & Rosen (1977), Eldredge & Cracraft (1980) and Wiley (1981). Extinct taxa are indicated by daggers (†). No new taxonomic names are introduced here, although some less familiar ones are reinstated. The names "Eosuchia", "Protorosauria" and "Rhynchocephalia" have no agreed meaning and are applied to variable assortments of unrelated forms. They are not used here.

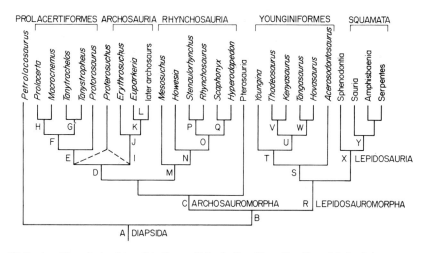

FIG. 1. The relationships of the Permo-Triassic diapsid reptiles. Shared derived characters at each dichotomy are summarized below (full details: Benton, in preparation).
A. Diapsida. Superior temporal fenestra; shape of bones in temporal region; suborbital fenestra; shape of palate bones around fenestra; true Jacobson's organ; olfactory bulbs of brain on stalks; one or more nasal conchae; Huxley's foramen at end of extracolumella.
B. Archosauromorpha + Lepidosauromorpha. Reduced lacrimal; ventromedial flanges on parietal; absence of "caniniform" maxillary teeth; reduced quadratojugal; quadrate exposed in lateral view; quadrate emarginated; stapes slender; reduction in pterygoid teeth; no parasphenoid teth; retroarticular process; ulna lacks well developed olecranon; acetabulum rounded; femur sigmoidal and slender; distal articular surfaces on femur level; femur longer than humerus.
C. Archosauromorpha. Premaxilla extends up behind naris; nares close to mid-line and elongate; quadratojugal mainly behind lower temporal fenestra; loss of tabulars; stapes rod-like and without foramen; vertebrae not notochordal; dorsal transverse processes project; cleithrum absent; no entepicondylar foramen; loss of perforating foramen in carpus; lateral tuber on calcaneum; complex concave-convex articulation between astragalus and calcaneum; fifth distal tarsal lost; fifth metatarsal hooked in one plane.
D. Prolacertiformes + Archosauria. Long snout and narrow skull; nasals longer than frontals; post-temporal fenestrae small or absent; recurved teeth; parasphenoid/basisphenoid participates in side wall of braincase; long thin tapering cervical ribs with anterior process.
E. Prolacertiformes. Lower temporal bar incomplete; 7–12 elongate cervical vertebrae; cervical vertebrae have long low neural spines; short ischium.
F. Prolacertidae + Tanystropheidae. Quadratojugal much reduced or absent; (?)partially streptostylic quadrate.
G. Tanystropheidae. Very long neck with 12 cervical vertebrae; post-cloacal bones; fifth metatarsal very short.
H. Prolacertidae. Squamosal has tetraradiate shape; choanae and bones of palate very long; mid-line gap in palate between pterygoids and long cultriform process of parasphenoid.
I. Archosauria. Antorbital fenestra; orbit triangular; teeth laterally compressed; fourth trochanter.
J. Archosauria (excluding *Proterosuchus*). Skull is high; antorbital fenestra close to naris; loss of supratemporal; lateral mandibular fenestra; coronoid reduced or absent; scapula very tall

and narrow; coracoid small and glenoid faces largely backwards; deltopectoral crest of humerus extends far down shaft; distal end of humerus reduced in width; hand is short; pubis has strong anterior tuber; iliac blade has small anterior process; ischium has large posteroventral process; tarsus contains only four elements.

K. *Euparkeria* and later archosaurs. Antorbital fenestra large and lies in a depression; parietal foramen absent; otic notch well developed; posterior border of lower temporal fenestra is kinked forward; dentition thecodont; pelvis markedly three-rayed; hind-limbs brought in under body; rotation between astragalus and calcaneum; dermal armour.

L. Later archosaurs. Postparietals absent; pterygoids meet medially; palatal teeth absent; presence of pleurosphenoid (?); presacral inter-centra absent.

M. Rhynchosauria. Downturned premaxilla bearing acrodont teeth or no teeth; single median naris; fused parietals; three proximal tarsals.

N. Rhynchosauroidea. Premaxilla beak-like and toothless; parietal foramen absent; teeth ankylothecodont; batteries of functional teeth on maxilla and dentary.

O. Rhynchosauridae. Loss of supratemporal; interlocking groove and blade jaw apparatus; centrale large and united with astragalus.

P. Rhynchosaurinae. Two grooves on maxilla; occipital condyle set well forward; single row of teeth on pterygoid.

Q. Hyperodapedontinae. Skull broader than long; jugal large and ridged; single longitudinal groove on maxilla; no teeth on lingual side of maxilla; no teeth on pterygoid; lower jaw deep; dentary has only one or two rows of teeth; coracoid has no posterior process; humerus about as long as femur.

R. Lepidosauromorpha. Post-frontal in border of upper temporal fossa; accessory inter-vertebral articulations; cervical centra short; dorsal ribs single-headed; co-ossification of paired sternal plates; specialized sternal rib connections.

S. Younginiformes. Distinctive sutures on parietal for frontal and post-frontal; reduced rod-like quadratojugal below temporal fenestra; dorsal neural spines high and rectangular; entepicondyle of humerus well developed; lateral centrale loses contact with third distal carpal.

T. Younginoidea. Short neck, 4–5 cervicals; (?) specialized intervertebral articulations; radius longer than shaft of ulna (Currie, 1982).

U. Tangasauridae. Humerus as long as, or longer than, femur; scapula low and mainly ventral; coracoid as large as scapula; fifth distal tarsal not a discrete element (Currie, 1982).

V. Kenyasaurinae. 19–28 pairs of caudal ribs and transverse processes present, all of which taper distally (Currie, 1982).

W. Tangasaurinae. Neural spines high; 9–12 pairs of caudal ribs; anterior caudal ribs expanded distally; haemal spines large and plate-like; presacral inter-centra do not ossify until animal is mature (Currie, 1982).

X. Lepidosauria. Determinant growth; specialized articulating surfaces of long bones (bony epiphyses); specialized joint between ulna and ulnare; lacrimal reduced or absent; postparietal and tabular absent; supraparachordal course of notochord; median hypocentral occipital condyle; thyroid fenestra in pelvis; fusion of astragalus and calcaneum; loss of centrale; loss of distal tarsals 1 and 5; hooking of 5th metatarsal in two planes.

Y. Squamata (living). Mid-line skull roof bones often fused; post-frontal and postorbital often fused, or one missing; pterygoids do not reach vomers; pterygoids do not meet in the mid-line; supratemporal situated deep betwen squamosal and parietal above quadrate; specialized articulation surface for dorsal wing of quadrate; squamosal reduced to slender bar or absent; no lower temporal bar; no quadratojugal; quadrate ramus of pterygoid reduced and no suture between quadrate and pterygoid; quadrate has tympanic conch; mesokinesis; fenestra rotunda; vidian canal; ossification of braincase anterior to otic capsule; pre-articular fused with articular; vertebrae usually procoelous; all ribs holocephalous; dorsal intercentra seldom developed; hypapophyses on cervical vertebrae; no true sacral ribs; loss of entepicondylar foramen in humerus; fenestration of anterior margin of scapulocoracoid (as well as some soft-part characters). Fossil 'squamates' show selections of these derived characters.

Classification of the Permo-Triassic Diapsids

Subdivision Diapsida Osborn 1903
  plesion †Petrolacosauridae Peabody 1952 *Petrolacosaurus*
  plesion †Galesphyridae Currie 1981 *Galesphyrus*
  plesion †Weigeltisauridae Romer 1933 *Weigeltisaurus, Coelurosauravus*
  plesion †Claudiosauridae Carroll 1981 *Claudiosaurus*
  Infradivision Neodiapsida nov.
    Neodiapsida, *incertae sedis*
      †Family Heleosauridae Haughton 1924 *Heleosaurus*
      †*Lacertulus*
      †Family Kuehneosauridae Romer 1966 *Kuehneosaurus, Kuehneosuchus, Icarosaurus*
      †Family Monjurosuchidae Endo 1940 *Monjurosuchus*
      †Family Thalattosauridae Merriam 1904 *Askeptosaurus, Thalattosaurus*
    Cohort Archosauromorpha Huene 1946
      Archosauromorpha, *incertae sedis* † *Noteosuchus*
      plesion †Pterosauria Owen 1840 (Kaup 1834)
      plesion †Trilophosauridae Gregory 1945 *Trilophosaurus*
      plesion †Rhynchosauria Osborn 1903 (Gervais 1859)
        Suborder Mesosuchidia Haughton 1924
          Family Mesosuchidae Haughton 1924 *Mesosuchus*
        Suborder Rhynchosauroidea Nopcsa 1928 (Gervais 1859)
          Family Howesiidae Watson 1917 *Howesia*
          Family Rhynchosauridae Huxley 1887 (Cope 1870)
            Subfamily Rhynchosaurinae Nopcsa 1923 *Stenaulorhynchus, Rhynchosaurus*
            Subfamily Hyperodapedontinae Chatterjee 1969 *Hyperodapedon, Scaphonyx*
      plesion †Prolacertiformes Camp 1945
        Prolacertiformes, *incertae sedis* ?*Cosesaurus, Malerisaurus*
          Family Protorosauridae Baur 1889 (Cope 1871) *Protorosaurus*
          Family Prolacertidae Parrington 1935 *Prolacerta, Macrocnemus,* ?*Boreopricea,* ?*Kadimakara*
          Family Tanystropheidae Romer 1945 (Gervais 1859) *Tanystropheus, Tanytrachelos*
    Incertae sedis (Prolacertiformes or Archosauria)
      †Family Proterosuchidae Huene 1908 *Chasmatosaurus, Proterosuchus, Chasmatosuchus,* etc.
    Superorder Archosauria Cope 1869
      plesion †Thecodontia Owen 1859
        Suborder Erythrosuchia Goodrich 1930
          Family Erythrosuchidae Watson 1917 *Erythrosuchus, Vjushkovia, Garjainia, Shansisuchus,* etc.
        Suborder Pseudosuchia Zittel 1887–1890
          Family Euparkeriidae Huene 1920 *Euparkeria*
        (Suborder Parasuchia Huxley 1875)
        (?Suborder Ornithosuchia Bonaparte 1971)
        (?Suborder Lagosuchia Chatterjee 1982)
      (plesion †Saurischia Seeley 1888)

(plesion †Ornithischia Seeley 1888)
(Order Crocodylia Gmelin 1788)
(Class Aves Linnaeus 1758)
Cohort Lepidosauromorpha Benton 1983
Lepidosauromorpha, *incertae sedix,* †*Palaeagama;* †*Paliguana,* †*Blomosaurus,* †*Kudnu,* †*Colubrifer*
plesion †Younginiformes Romer 1933
plesion †*Acerosodontosaurus*
Superfamily Younginoidea Currie 1982
Family Younginidae Broom 1914 *Youngina*
Family Tangasauridae Camp 1945 (Piveteau 1926)
Subfamily Kenyasaurinae Currie 1982 *Kenyasaurus, Thadeosaurus*
Subfamily Tangasaurinae Piveteau 1926 *Tangasaurus, Hovasaurus*
plesion †Saurosternidae Haughton 1924 *Saurosternon*
Superorder Lepidosauria Haeckel 1866 (Duméril & Bibron 1839)
Order Sphenodontia Williston 1925
Family Sphenodontidae Cope 1870 *Sphenodon, Brachyrhinodon, Clevosaurus, Homoeosaurus, Toxolophosaurus,* etc.
?Family Sapheosauridae Baur 1895 *Sapheosaurus*
plesion †Gephyrosauridae Evans 1980 *Gephyrosaurus*
Order Squamata Oppel 1811
(Suborder Sauria Macartney 1802)
(Suborder Amphisbaenia Gray 1844)
(Suborder Serpentes Linnaeus 1758)
Diapsida, *incertae sedis*
?†Family Claraziidae Peyer 1936
(†Family Champsosauridae Cope 1876)
(†Family Pleurosauridae Lydekker 1888)

An evolutionary tree of the early diapsids is also given (Fig. 2) – this incorporates the data from the cladistic analysis, as well as stratigraphic information.

PERMO-TRIASSIC DIAPSIDS AND FAUNAL EVOLUTION

Origin of the Diapsids

The earliest known diapsid is *Petrolacosaurus* from the late Carboniferous of Kansas [Fig. 3(A)]. *Petrolacosaurus* was 60–70 cm long and it had a long neck and was probably an agile terrestrial reptile that may have fed on large insects and other arthropods. It shows typical diapsid characters – two temporal fenestrae, suborbital fenestra in the palate, relatively small skull, long limbs, locked tibio-astragalar joint (Reisz, 1981). In the late Carboniferous and early Permian, several major lines of reptiles were diverging (Pelycosauria, Protorothyrididae, Captorhinidae, Diapsida), and the diapsids seem to show closest relationship to the protorothyridids.

Diapsids are not known from the early Permian (except for one

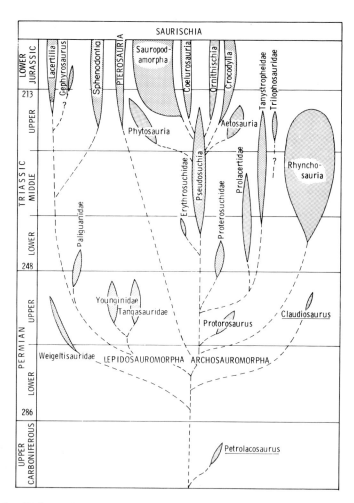

FIG. 2. Phylogenetic tree of the Permo-Triassic diapsid reptiles. Data on relationships are taken from the cladogram (Fig. 1). Stratigraphic information from Anderson & Cruickshank (1978) and Tucker & Benton (1982). Age dates from Harland *et al.* (1982). The spindles are drawn to indicate the known stratigraphic range of each group, and the width represents the relative abundance (a subjective measure of numbers of individuals and numbers of genera present in typical faunas at particular times) – data from various sources, reviewed in Benton (1983a). Dotted lines connecting groups indicate uncertainty about relationships.

possible fragment), but several forms have been found in later Permian faunas of Germany, England, Russia, Tanzania, South Africa and Malagasy. In most of these, the diapsids are represented by only one or two specimens, but they are abundant in the Malagasy fauna.

Late Permian: the Lower Sakamena Formation, Malagasy

The Lower Sakamena Formation (?late Permian: ?Tatarian) of the region between Mount Eliva and Ranohira, south-west Malagasy, has yielded hundreds of fossil reptile specimens. The reptiles consist of the procolophonid *Bavarisaurus*, isolated bones of mammal-like reptiles (dicynodont, theriodont), and the diapsids *Coelurosauravus*, *Claudiosaurus*, *Acerosodontosaurus*, *Thadeosaurus* and *Hovasaurus* [Fig. 3(B)–(E)]. Associated fossils include plants (equisetales, cycads and coniferopsids typical of the *Glossopteris* Flora), bivalves, crustaceans, fish (*Atherostonia*) and the amphibian *Rhinesuchus*. There were probably several environmentally controlled faunas (Currie, 1981). The overwhelming dominance of the faunas by diapsids is remarkable in view of their rarity in faunas of similar age in South Africa, where mammal-like reptiles (dicynodonts, gorgonopsians) were abundant.

Coelurosauravus (*Daedalosaurus*) was a small, 30 cm long animal with hugely expanded dorsal ribs. It has been suggested that these ribs were joined by a membrane, as in the living lizard *Draco* (Carroll, 1978; Evans, 1982). *Coelurosauravus*, represented by only three or four specimens, had a short skull, with pleurodont teeth and a large orbit. The trunk and tail were long, and the ribs were up to 16 cm long. A close relative is known from the late Permian of England and Germany (*Weigeltisaurus*).

Claudiosaurus, a 60 cm long animal with a small head, a long neck, and heavy paddle-like hands and feet, has been interpreted as a plesiosaur ancestor by Carroll (1981). The 20 or so specimens certainly show adaptations for swimming in the hands and feet, but the interpretation of *Claudiosaurus* as an early member of the nothosaur/plesiosaur group is less certain. Carroll (1981) notes several features in which *Claudiosaurus* could be seen as intermediate between a younginiform and a nothosaur, but many of these are general aquatic adaptations. *Claudiosaurus* is a diapsid, and it probably classifies as sister-group to the Archosauromorpha + Lepidosauromorpha.

The other three Malagasy diapsids are younginiforms. *Acerosodontosaurus* (Currie, 1980) was probably 60–70 cm long, but the single known specimen is incomplete. It had a generalized "younginid" skull with pointed teeth. *Thadeosaurus* (Carroll, 1981), based on about 10 specimens, was also a terrestrial animal with heavy limbs. The skull is poorly

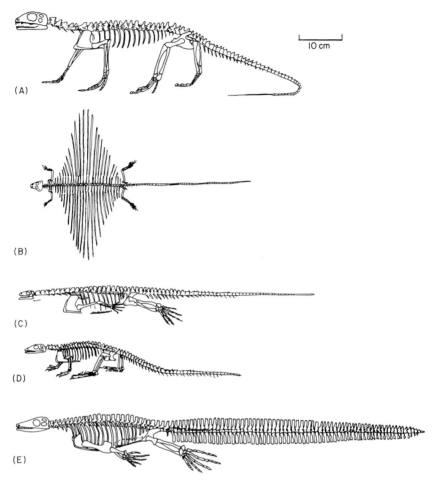

FIG. 3. Diapsid reptiles of the late Carboniferous and late Permian. (A) *Petrolacosaurus* from the late Carboniferous of Kansas. (B) *Coelurosauravus*. (C) *Claudiosaurus*. (D) *Thadeosaurus*. (E) *Hovasaurus* from the Lower Sakamena Formation of Malagasy. (A, after Reisz, 1981; B, after Carroll, 1978 and Evans, 1982; C, D, after Carroll, 1981; E, after Currie, 1981.)

known, but *Thadeosaurus* shows close similarities with *Kenyasaurus* and with the tangasaurid *Hovasaurus*. *Hovasaurus* is the most abundant reptile known from the Lower Sakamena Formation, being represented by more than 300 specimens (Currie, 1981). The skeleton shows clear aquatic adaptations: large paddle-like hands and feet; long and deep tail with high neural spines and long haemapophyses; and ballast pebbles in the body cavity. An adult was 30–35 cm long (snout–vent length) with a 50–60 cm tail, but a series of juveniles is known, the smallest of which had a total length of about 20 cm.

Early Triassic: the *Lystrosaurus* and *Cynognathus* Zones, South Africa

Diapsid reptiles are known from the late Permian of South Africa (e.g. the lepidosauromorphs *Youngina, Palaeagama, Saurosternon*). The first archosauromorph, the proterosuchid *Archosaurus*, is known from the late Permian (Tatarian) of European Russia, but the Archosauromorpha only became well known in the Triassic.

The *Lystrosaurus* Zone (= *Lystrosaurus* Assemblage Zone of Keyser & Smith, 1979; lowest Scythian) is heavily dominated by the dicynodont *Lystrosaurus* (over 90% of all specimens collected), but four diapsids are present: *Prolacerta, Proterosuchus, Paliguana* and *Noteosuchus* [Fig. 4(A)–(C)]. *Prolacerta* was a 60 cm long quadruped with a low skull, very long neck, long slender limbs and a deep tail (Gow, 1975). The teeth are recurved and pointed, and the lower temporal bar is broken – *Prolacerta* has been regarded as an early lizard (Robinson, 1967; Wild, 1980), but the archosauromorph features shared with thecodontians and rhynchosaurs are overwhelming. The broken lower temporal bar character, often regarded as a diagnostic squamate feature, occurs in many unrelated groups, and is a parallelism (*Coelurosauravus, Claudiosaurus*, Prolacertiformes, the sphenodontids *Clevosaurus* and *Planocephalosaurus*, (?) nothosaurs and plesiosaurs). *Proterosuchus* (*Chasmatosaurus*), a varanid-shaped carnivore with a 25 cm skull, has frequently been placed at the foot of the archosaur radiation on the basis of its antorbital fenestra. The snout is long and narrow, the teeth are sharp and recurved, and the orbit is high and archosaur-like (Cruickshank, 1972). However, there are numerous prolacertiform characters: the downturned snout tip, elongate snout and palatal bones, long inter-pterygoid vacuity, elongate cervical vertebrae, and broad and deep haemapophyses. Its taxonomic position is uncertain. *Paliguana*, a small reptile, is known only from its 2.5 cm long skull which shows some lizard-like features (squamosal reduced, no lower temporal bar, quadrate with conch-like tympanic notch), but it lacks key lepidosaur and squamate characters. *Noteosuchus*, redescribed as the earliest rhynchosaur (Carroll, 1976), could belong anywhere among primitive archosauromorphs. Elements of the *Lystrosaurus* Zone fauna have been found in Antarctica, Australia, China, Russia and India. In most of these, diapsids were minor elements – small insectivores and small to medium-sized carnivores – comprising only about 1% of all specimens known in each fauna.

The *Cynognathus* Zone (= *Kannemeyeria* Assemblage Zone of Keyser & Smith, 1979; late Scythian) has yielded some more significant diapsids: the thecodontians *Erythrosuchus* and *Euparkeria* [Fig. 4(D), (E)], and the

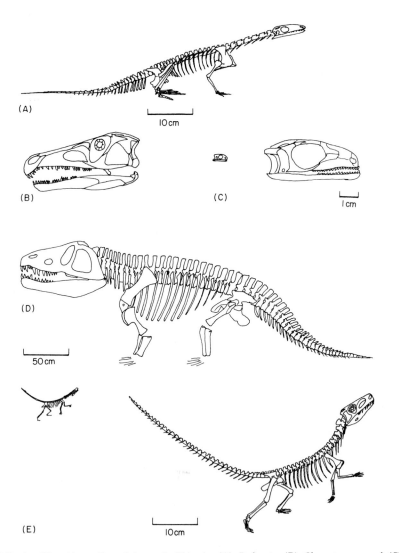

FIG. 4. Diapsid reptiles of the early Triassic. (A) *Prolacerta*, (B) *Chasmatosaurus* and (C) *Paliguana* from the *Lystrosaurus* Zone of South Africa. (D) *Erythrosuchus* and (E) *Euparkeria* from the *Cynognathus* Zone of South Africa. Note the scales: *Paliguana* and *Euparkeria* are shown at two sizes. (A, after Gow, 1975; B, after Cruickshank, 1972; C, after Carroll, 1975; D, after von Huene, 1956; E, after Ewer, 1965.)

rhynchosaurs *Mesosuchus* and *Howesia*. *Erythrosuchus*, a 5 m long heavily-built quadruped with a massive 1 m long skull, must have been a prodigious carnivore, preying on the contemporary herbivorous dicynodonts and smaller mammal-like reptiles and amphibians. *Erythrosuchus* is often classed with proterosuchids, but it shows numerous advanced characters shared with later thecodontians that are absent in *Proterosuchus* (Fig. 1: I). Close relatives of *Erythrosuchus* are known from Russia and China. *Euparkeria*, a small 65 cm long quadruped and facultative biped shows more advanced features in the skull and hind-limbs. The teeth are thecodont and relatively large. The early rhynchosaurs, *Mesosuchus* and *Howesia*, are inadequately known at present. Of these small animals, *Howesia* is more of a typical rhynchosaur, with multiple rows of deeply rooted teeth. Thecodontians were clearly important carnivores in the *Cynognathus* Zone fauna (14% of all specimens found).

Middle Triassic: the Grenzbitumenzone, Switzerland

During the Middle Triassic, diapsids radiated in several adaptive zones: the rhynchosaurs became dominant herbivores in the Manda Formation, Tanzania; pseudosuchian thecodontians of various kinds appeared (rauisuchids, proterochampsids, lagosuchids, etc.) and became the dominant carnivores world-wide, although the cynognathoid mammal-like reptiles continued to diversify at the same time.

One exceptional middle Triassic diapsid fauna is that of the Grenzbitumenzone (Anisian–Ladinian boundary) of Monte San Giorgio, Tessin, Switzerland. The largely aquatic fauna contains early ichthyosaurs, nothosaurs, placodonts and the diapsids *Macrocnemus*, *Tanystropheus*, *Ticinosuchus*, *Askeptosaurus*, *Clarazia* and *Hescheleria*, as well as fish and marine invertebrates (Kuhn-Schnyder, 1974) (Fig. 5).

Macrocnemus and *Tanystropheus* are classed as prolacertiforms. *Macrocnemus* was a small lizard-like animal, with a 7 cm skull, very like *Prolacerta* in certain features, and with long light limbs. *Tanystropheus*, represented by a series of juvenile to adult skeletons, is the most remarkable reptile present. Adults were up to 6 m long, of which half was made up of an elongate neck. The 12 cervical vertebrae were up to 30 cm long, and the neck was clearly not enormously flexible. Despite its bizarre appearance, *Tanystropheus* shows prolacertiform features in the skull and limbs. Wild (1973) has suggested that *Tanystropheus* fed on insects as a juvenile, when the neck was relatively much shorter, and on fish as an adult.

Ticinosuchus, a 2.5 m long rauisuchid thecodontian with large teeth, a long neck and long limbs, must have been a fearsome predator (Krebs,

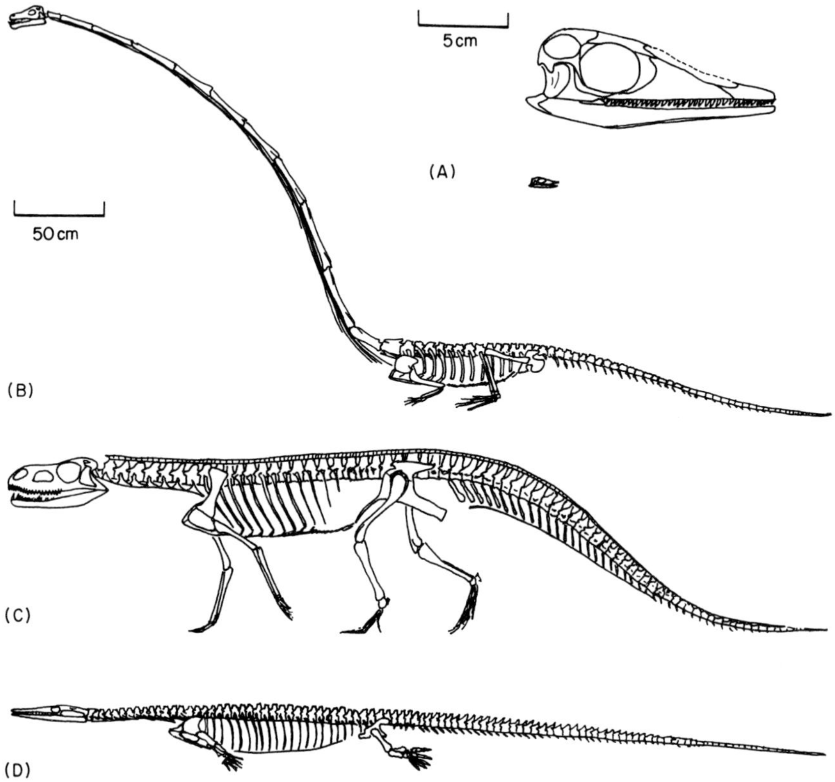

FIG. 5. Diapsid reptiles of the middle Triassic. (A) *Macrocnemus*, (B) *Tanystropheus*, (C) *Ticinosuchus* and (D) *Askeptosaurus* from the Grenzbitumenzone of Switzerland. Note the scales: *Macrocnemus* is shown at two sizes. (A, after Kuhn-Schnyder, 1974; B, after Wild, 1973; C, after Krebs, 1965; D, after Kuhn, 1952.)

1965). *Askeptosaurus*, a 2.5 m long reptile with a long narrow skull, very long neck, trunk and tail and tiny limbs, was probably a good swimmer (Kuhn, 1952). *Askeptosaurus* was clearly a diapsid, but it cannot be confidently placed in the Lepidosauria, as has often been done. Finally, *Clarazia* and *Hescheleria*, both poorly known, had broad skulls, heavy grinding teeth, long bodies and reduced limbs. These were aquatic forms, but they cannot confidently be identified as 'lepidosaurs' (Romer, 1966), or even as diapsids at present.

Late Triassic: the Lossiemouth Sandstone Formation, Scotland

By the middle of the late Triassic (Carnian – Norian), several faunas are known around the world in which rhynchosaurs and/or aetosaurs were dominant herbivores, and pseudosuchians and/or phytosaurs were dominant carnivores (e.g. Santa Maria Formation, Brazil; Ischigualasto Formation, Argentina; Maleri Formation, India; Lossiemouth Sandstone Formation, Scotland; Dockum Group, Texas; Chinle Formation, Arizona). In the southern-continent and American faunas of this age, mammal-like reptiles were still significant elements: dicynodonts and diademodontoids as herbivores, and cynognathoids as carnivores. However, these are absent from the Lossiemouth Sandstone Formation, which consists of the procolophonid *Leptopleuron*; a selection of thecodontians: the aetosaur *Stagonolepis*, the pseudosuchians *Ornithosuchus*, *Erpetosuchus* and *Scleromochlus*; the primitive coelurosaur dinosaur *Saltopus*; the rhynchosaur *Hyperodapedon*, and the sphenodontid *Brachyrhinodon* (Fig. 6).

Stagonolepis, a 2.7 m long quadruped with short limbs, a shovel-snouted skull and extensive dermal armour, was a relatively abundant herbivore that may have fed on tubers and roots (Walker, 1961). *Ornithosuchus*, represented by several individuals, ranging in size up to 3.5 m long, was a bipedal or quadrupedal carnivore with heavy jaws (Walker, 1964). It could probably have fed on *Stagonolepis* and *Hyperodapedon*. *Erpetosuchus*, a small carnivore with an 8 cm skull, is poorly known. *Scleromochlus* was also tiny (25 cm long) and is remarkable for its relatively large skull and very long limbs which may have been adapted for rapid running over sand. *Saltopus*, the small dinosaur, is known from only one incomplete skeleton without a skull. These last three genera probably fed on juvenile *Stagonolepis* and *Hyperodapedon*, as well as *Leptopleuron* and *Brachyrhinodon*.

The rhynchosaur *Hyperodapedon*, a 1.3 m long animal with strong digging claws on the foot, a beaked premaxilla, and strong slicing dentition, could have fed on a variety of tough vegetation (Benton, 1983b). The small sphenodontid *Brachyrhinodon* had a very short snout, and is the only lepidosauromorph known from Elgin.

Latest Triassic: the Knollenmergel, Germany

Some time between the early and the middle Norian (220–225 My ago: Harland, Cox, Llewellyn, Pickton, Smith & Walters, 1982) the majority of the early archosauromorphs and the remaining mammal-like reptiles disappeared: rhynchosaurs, thecodontians, dicynodonts, diademodontoids and cynognathoids. They were replaced world-wide by dinosaurs as medium to very large-sized herbivores (prosauropods, ornithischians) and carnivores (coelurosaurs), which dominated all terminal Triassic terrestrial faunas.

A typical early dinosaur fauna is that of the Knollenmergel (late

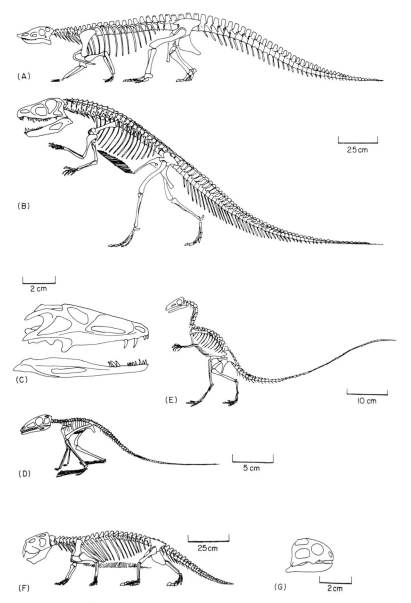

FIG. 6. Diapsid reptiles of the late Triassic. (A) *Stagonolepis*, (B) *Ornithosuchus*, (C) *Erpetosuchus*, (D) *Scleromochlus*, (E) *Saltopus*, (F) *Hyperodapedon* and (G) *Brachyrhinodon* from the Lossiemouth Sandstone Formation of Elgin, Scotland. Note the scales: (A), (B) and (F) are drawn to the same scale, as are (C) and (G). (A, after Walker, 1961; B, after Walker, 1964; C, D, E, after von Huene, 1956; F, original; G, after von Huene, 1956.)

Norian) of south-west Germany. A few turtles (*Proganochelys*) are present, as well as abundant specimens of the dinosaur *Plateosaurus* (Fig. 7). This large prosauropod (5–10 m long) had a relatively small skull with peg-like teeth, a long neck and tail and heavy limbs. It was probably quadrupedal and facultatively bipedal. The "Knollenmergel" has also yielded remains of the coelurosaurs *Halticosaurus* and *Pterospondylus*.

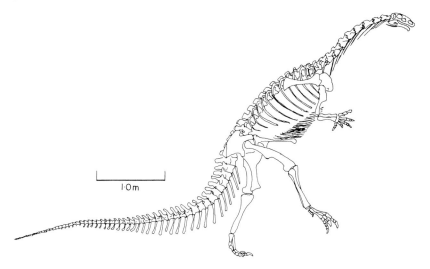

FIG. 7. Diapsid reptile of the latest Triassic. *Plateosaurus* from the Knollenmergel, south-west Germany (after von Huene, 1956).

It has been suggested elsewhere (Tucker & Benton, 1982; Benton, 1983a) that the rise of the dinosaurs need not have been the result of prolonged and successful competition with mammal-like reptiles and thecodontians, as has been assumed (e.g. Charig, 1979, 1980; Bonaparte, 1982). There is evidence that the elements of the immediately "pre-dinosaur" faunas of the late Triassic died out as a result of floral and/or climatic changes. The dinosaurs, already present as small to medium-sized, but rare, faunal elements (e.g. *Saltopus*) radiated rapidly to fill empty ecological space. It seems clear that the dinosaurs achieved their dominance and large size in 2–3 My or less. There is no evidence for sustained competition throughout the Triassic between "inferior" mammal-like reptiles and "superior" archosaurs, whatever the "superior" feature of the latter group is – improved locomotory capability (Ostrom, 1969; Bakker, 1971; Charig, 1972, 1979, 1980), endothermy (Bakker, 1971, 1975), or ectothermic inertial homeothermy (Spotila, Lommen, Bakker & Gates, 1973; Benton, 1979).

Late Triassic/Early Jurassic: the Bristol fissures, England

The Triassic-Jurassic fissures in the region of Bristol, south-west England, and south Wales have yielded a fascinating sample of small animals from typical early dinosaur faunas. The exact ages of the fissures are hard to determine, and there were probably several generations of infilling in the Norian, Rhaetian and early Jurassic. Great interest has focused on these fissures for the early mammal-like reptiles and mammals that they have yielded (*Oligokyphus, Haramiya, Thomasia, Eozostrodon, Morganucodon, Kuehneotherium*) and the small reptiles. The dominance of small reptiles is probably the result of preservational sorting – dependent on which animals fell into the fissures – rather than an indication of a specialized "upland" fauna (cf. Robinson, 1957; Tarlo, 1962).

The diapsids present (Fig. 8) include the phytosaur or aetosaur *Rileya*, the prosauropod dinosaur *Thecodontosaurus*, and a selection of other undescribed archosaurs (Marshall & Whiteside, 1980; D. I. Whiteside, pers. comm.; N. C. Fraser, pers. comm.). Several sphenodontids and lizards also await description. Two sphenodontids with broken lower temporal bars have been described; *Clevosaurus* (Swinton, 1939; Robinson, 1973) and *Planocephalosaurus* (Fraser, 1982), with tiny 2–3 cm long skulls. A remarkable gliding animal, *Kuehneosaurus* (Robinson, 1962), is also represented by several specimens – it has a high 3 cm long skull and expanded dorsal rib, giving a span of 25–30 cm. *Kuehneosaurus* has been called a lizard, but it lacks most squamate and lepidosaur characters. It is not even certain that it is a lepidosauromorph. *Gephyrosaurus* (Evans, 1980, 1981), a 25–30 cm long lizard-like animal, is probably the sister-group of the Squamata. Some jaw fragments with batteries of broad herbivorous teeth have also been described: *Tricuspisaurus* and *Variodens*, which may or may not have belonged to diapsids.

CONCLUSIONS

Recent work on Permo-Triassic diapsids has shed new light on their relationships and evolution. The taxonomic and evolutionary aspects that have been touched upon here, are discussed in more detail elsewhere (Benton, 1983b, in preparation). The new taxonomic outline presented here may be tested and modified by redescriptions of old material, and by the discovery of new specimens. The views expressed on faunal evolution in the Permo-Triassic and the opportunistic radiation of the

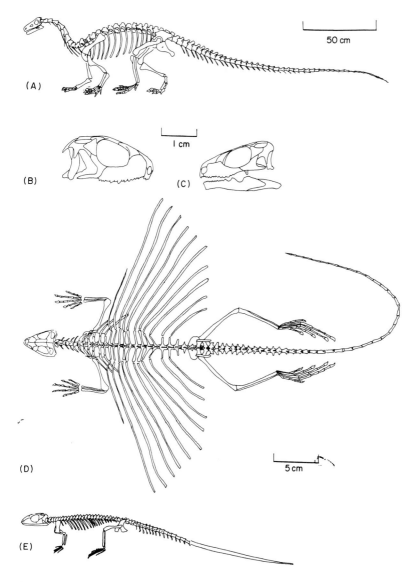

FIG. 8. Diapsid reptiles of the late Triassic/early Jurassic. (A) *Thecodontosaurus*, (B) *Clevosaurus*, (C) *Planocephalosaurus*, (D) *Kuehneosaurus* and (E) *Gephyrosaurus* from the fissures of the Bristol region, England, and south Wales. Note the different scales. (A, after von Huene, 1956; B, after Robinson, 1973; C, after Fraser, 1982; D, after Robinson, in Romer, 1966; E, after Evans, 1981.)

dinosaurs in the mid-late Norian are also cast in a testable form – the discovery of significant numbers of medium to large mammal-like reptiles or rhynchosaurs together with prosauropod dinosaurs in the same fauna would disprove the hypothesis and suggest that competition was involved.

ACKNOWLEDGEMENTS

I thank G. R. Chancellor, A. R. I. Cruickshank, S. E. Evans, T. S. Kemp, G. M. King, D. B. Norman, and R. Wild for helpful comments on the manuscript. Denise Blagden drafted Fig. 2. I thank the President and Fellows of Trinity College, Oxford, for the financial support of a Junior Research Fellowship.

REFERENCES

Anderson, J. M. & Cruickshank, A. R. I. (1978). The biostratigraphy of the Permian and Triassic. Part 5. A review of the classification and distribution of Permo-Triassic tetrapods. *Palaeontol. afr.* **21**: 15–44.
Bakker, R. T. (1971). Dinosaur physiology and the origin of mammals. *Evolution, Lawrence, Kans.* **25**: 636–658.
Bakker, R. T. (1975). Dinosaur renaissance. *Scient. Am.* **232** (4): 58–78.
Benton, (1979). Ectothermy and the success of dinosaurs. *Evolution, Lawrence, Kans.* **33**: 983–997.
Benton, M. J. (1982). The Diapsida: revolution in reptile relationships. *Nature, Lond.* **296**: 306–307.
Benton, M. J. (1983a). Dinosaur success in the Triassic: a noncompetitive ecological model. *Q. Rev. Biol.* **58**: 29–55.
Benton, M. J. (1983b). The Triassic reptile *Hyperodapedon* from Elgin: functional morphology and relationships. *Phil. Trans. R. Soc.* (B) **302**: 605–717.
Benton, M. J. (In prep.). *Phylogeny and classification of the early diapsid reptiles.*
Bonaparte, J. F. (1982). Faunal replacement in the Triassic of South America. *J. vertebr. Paleont.* **2**: 362–371.
Broom, R. (1925). On the origin of lizards. *Proc. zool. Soc. Lond.* **1925**: 1–16.
Carroll, R. L. (1975). Permo-Triassic "lizards" from the Karroo. *Palaeont. afr.* **18**: 71–87.
Carroll, R. L. (1976). *Noteosuchus* – the oldest known rhynchosaur. *Ann. S. Afr. Mus.* **72**: 37–57.
Carroll, R. L. (1978). Permo-Triassic 'lizards' from the Karoo system. Part 2. A gliding reptile from the Upper Permian of Madagascar. *Palaeont. afr.* **21**: 143–159.
Carroll, R. L. (1981). Plesiosaur ancestors from the Upper Permian of Madagascar. *Phil. Trans. R. Soc.* (B) **293**: 315–383.
Charig, A. J. (1972). The evolution of the archosaur pelvis and hind-limb: an explanation in functional terms. In *Studies in vertebrate evolution*: 121–155. Joysey, K. A. & Kemp, T. S. (Eds). Edinburgh: Oliver & Boyd.
Charig, A. J. (1979). *A new look at the dinosaurs*. London: Heinemann.

Charig, A. J. (1980). Differentiation of lineages among Mesozoic tetrapods. *Mém. Soc. géol. Fr.* **139**: 207–210.
Cruickshank, A. R. I. (1972). The proterosuchian thecodonts. In *Studies in vertebrate evolution*: 89–119. Joysey, K. A. & Kemp, T. S. (Eds). Edinburgh: Oliver & Boyd.
Currie, P. J. (1980) A new younginid (Reptilia: Eosuchia) from the Upper Permian of Madagascar. *Can. J. Earth Sci.* **17**: 500–511.
Currie, P. J. (1981). *Hovasaurus boulei*, an aquatic eosuchian from the Upper Permian of Madagascar. *Palaeont. afr.* **24**: 99–168.
Currie, P. J. (1982). The osteology and relationships of *Tangasaurus mennelli* Haughton (Reptilia, Eosuchia.) *Ann. S. Afr. Mus.* **86**: 247–265.
Eldredge, N. & Cracraft, J. (1980). *Phylogenetic patterns and the evolutionary process*. New York: Columbia University Press.
Evans, S. E. (1980). The skull of a new eosuchian reptile from the Lower Jurassic of South Wales. *Zool. J. Linn. Soc.* **70**: 203–264.
Evans, S. E. (1981). The postcranial skeleton of the Lower Jurassic eosuchian *Gephyrosaurus bridensis*. *Zool. J. Linn. Soc.* **73**: 81–116.
Evans, S. E. (1982). The gliding reptiles of the Upper Permian. *Zool. J. Linn. Soc.* **76**: 97–123.
Ewer, R. F. (1965). The anatomy of the thecodont reptile *Euparkeria capensis* Broom. *Phil. Trans. R. Soc.* (B) **248**: 379–435.
Fraser, N. C. (1982). A new rhynchocephalian from the British Upper Trias. *Palaeontology* **25**: 709–725.
Gauthier, J. A. (In press). Phylogenetic relationships of the Lepidosauromorpha and the origin of the lizards. *Am. Zool.*
Gow, C. E. (1975). The morphology and relationships of *Youngina capensis* Broom and *Prolacerta broomi* Parrington. *Palaeont. afr.* **18**: 89–131.
Harland, W. B., Cox, A. V., Llewellyn, P. G., Pickton, C. A. G., Smith, A. G. & Walters, R. (1982). *A geologic time scale*. Cambridge: University Press.
Huene, F. von (1946). Die grossen Stämme der Tetrapoden in den geologischen Zeiten. *Biol. Zbl.* **65**: 268–275.
Huene, F. von (1956). *Paläontologie und Phylogenie der niederen Tetrapoden*. Jena: Gustav Fischer Verlag.
Keyser, A. W. & Smith, R. M. H. (1979). Vertebrate biozonation of the Beaufort Group with special reference to the Western Karoo Basin. *Ann. geol. Surv. S. Afr.* **12**: 1–35.
Krebs, B. (1965). Die Triasfauna der Tessiner Kalkalpen. XIX. *Ticinosuchus ferox* nov. gen. nov. sp. *Schweiz. paläont. Abh.* **81**: 5–140.
Kuhn, E. (1952), Die Triasfauna der Tessiner Kalkalpen. XVII. *Askeptosaurus italicus* Nopcsa. *Schweiz. paläont. Abh.* **69**: 6–73.
Kuhn-Schnyder, E. (1974). Die Triasfauna der Tessiner Kalkalpen. *NeujBl. naturf. Ges. Zürich* **1974**: 1–119.
Marshall, J. E. A. & Whiteside, D. I. (1980). Marine influence in the Triassic 'uplands'. *Nature, Lond.* **287**: 627–628.
Osborn, H. F. (1903). The reptilian subclasses Diapsida and Synapsida and the early history of the Diaptosauria. *Mem. Am. Mus. nat. Hist.* **1**: 449–507.
Ostrom, J. H. (1969). Terrestrial vertebrates as indicators of Mesozoic climates. *Proc. N. Am. Paleont. Conv.* **D1969**: 347–376.
Patterson, C. & Rosen, D. E. (1977). Review of ichthyodectiform and other Mesozoic teleost fishes and the theory and practice of classifying fossils. *Bull. Am. Mus. nat. Hist.* **158**: 81–172.
Reisz, R. (1981). *Petrolacosaurus kansensis* Lane, the oldest known diapsid reptile. *Univ. Kans. Mus. nat. Hist. Spec. Publs* **7**: 1–74.

Robinson, P. L. (1957). The Mesozoic fissures of the Bristol Channel area and their vertebrate faunas. *J. Linn. Soc. (Zool.)* **43**: 260–282.
Robinson, P. L. (1962). Gliding lizards from the Upper Keuper of Great Britain. *Proc. geol. Soc. Lond.* **1061**: 137–146.
Robinson, P. L. (1967). The evolution of Lacertilia. *Colloq. int. Cent. natn. Rech. scient.* **163**: 395–407.
Robinson, P. L. (1973). A problematic reptile from the British Upper Trias. *J. geol. Soc. Lond.* **129**: 457–479.
Romer, A. S. (1933). *Vertebrate paleontology*. Chicago: University of Chicago.
Romer, A. S. (1956). *Osteology of the reptiles*. Chicago: University of Chicago.
Romer, A. S. (1966). *Vertebrate paleontology*, 3rd edn. Chicago: University of Chicago.
Romer, A. S. (1971). Unorthodoxies in reptilian phylogeny. *Evolution, Lawrence, Kans.* **25**: 103–112.
Spotila, J. R., Lommen, P. W., Bakken, G. S. & Gates, D. M. (1973). A mathematical model for body temperatures of large reptiles: implications for dinosaur ecology. *Am. Nat.* **107**: 391–404.
Swinton, W. E. (1939). A new Triassic rhynchocephalian from Gloucestershire. *Ann. Mag. nat. Hist.* (11) **4**: 591–594.
Tarlo, B. (1962). Ancient animals of the upland. *New Scient.* **15**: 32–34.
Tucker, M. E. & Benton, M. J. (1982). Triassic environments, climates and reptile evolution. *Palaeogeogr., Palaeoclimatol., Palaeoecol.* **40**: 361–379.
Walker, A. D. (1961). Triassic reptiles from the Elgin area: *Stagonolepis, Dasygnathus* and their allies. *Phil. Trans. R. Soc.* (B) **244**: 103–204.
Walker, A. D. (1964). Triassic reptiles from the Elgin area: *Ornithosuchus* and the origin of carnosaurs. *Phil. Trans. R. Soc.* (B) **248**: 53–134.
Wild, R. (1973). Die Triasfauna der Tessiner Kalkalpen. XXIII. *Tanystropheus longobardicus* (Bassani) (Neue Ergebnisse). *Schweiz. paläont. Abh.* **95**: 1–162.
Wild, R. (1980). Die Triasfauna der Tessiner Kalkalpen. XXIV. Neue Funde von *Tanystropheus* (Reptilia, Squamata). *Schweiz. paläont. Abh.* **102**: 1–43.
Wiley, E. O. (1981). *Phylogenetics*. New York: Wiley.
Williston, S. W. (1925). *The osteology of the reptiles*. Cambridge, Mass: Harvard University.

Competition Between Therapsids and Archosaurs During the Triassic Period: A Review and Synthesis of Current Theories

ALAN J. CHARIG

Department of Palaeontology, British Museum (Natural History), London SW7 5BD, England

SYNOPSIS

During the Triassic the archosaurs replaced the therapsids as the dominant group of large terrestrial tetrapods. The main features of that replacement, as seen in the fossil record, are briefly recapitulated. Narrative explanations of such phenomena are called "scenarios" and can be formulated scientifically in accordance with certain guiding principles, requiring in particular a sound factual basis and a minimum of conjecture.

Contrary to a recently expressed opinion, the Triassic faunal replacement was certainly competitive – at least in the sense that, of two sympatric lineages occupying the same broad adaptive zone and subjected to the same environmental pressures, one waxed to total dominance while the other correspondingly waned to near-extinction. Explanations of this replacement rely essentially upon some alleged difference of anatomy or physiology between the two lineages in question, a difference which, in the conditions then prevailing, would prove advantageous to the archosaurs and disadvantageous to the therapsids. "Anatomical" explanations are based exclusively on the striking improvements to the locomotor apparatus of Triassic archosaurs. "Physiological" explanations are variously based on the alleged advantages of ectothermy, endothermy, uricotelic nitrogen excretion and the ability to withstand high ambient temperatures; these are detailed and critically reviewed. There are also a number of combination theories, uniting locomotor improvements with one of the physiological factors.

The physiological explanations (and therefore the combination theories too) are all characterized by a deplorable lack of firmly established fact. They depend upon the conjectured effects of conjectured environmental conditions upon conjectured physiological features, concerning none of which do all authorities agree. Such speculative argument cannot be regarded as science but only as fantasy. Without good evidence of other factors of obvious significance it is reasonable to prefer the tangible testimony of the locomotor improvements and of the great increases in size which those improvements made possible.

INTRODUCTION

The fossil record, despite its manifest imperfection, shows very clearly that at the beginning of the Triassic period the large vertebrates on land were nearly all therapsids (advanced mammal-like reptiles) and that at the end of that period they were almost exclusively archosaurs (dinosaurs and their kin). This applied both to the carnivorous element of the fauna and to the herbivores, and is disputed by nobody. Likewise nearly everyone accepts that the therapsids and the archosaurs constitute two partly contemporaneous yet entirely separate groups in the phylogeny; only Reig (1970) and Gardiner (1982) have claimed any close relationship between them, and very few palaeontologists support either one of their heterodox beliefs.

It may therefore be said that during the Triassic the archosaurs *replaced* the therapsids as the dominant group of large terrestrial tetrapods. Various workers have put forward many different ideas as to the reasons for this replacement; their interest is doubtless enhanced by the fact that mammal-like reptiles (presumably already possessing many mammal-like characters, including a tendency towards "warm-bloodedness") were forced to yield their dominance to "reptile-like" reptiles (presumably lacking those characters), thus contradicting the apparent superiority of modern mammals over modern reptiles. (Since we too are mammals, might there not be a personal element of pique in the scientists' attitude?) Some of the explanations proposed will be briefly reviewed below.

REPLACEMENT OF THERAPSIDS BY ARCHOSAURS: THE MAIN CHARACTERISTICS

The most striking features of the therapsid-archosaur replacement as seen in the fossil record and as detailed previously (Charig, 1975, 1979, 1980; see Fig. 1) are as follows:

(1) The replacement occurred first among the carnivores, being almost completed before the herbivore replacement had even begun. The change-over point is typified by the fauna of the Ischigualasto Formation (Carnian) of Argentina, which consists essentially of archosaurian carnivores (i.e. replacement almost completed) co-existing with therapsid and rhynchosaurian herbivores (i.e. replacement scarcely begun); no other fauna like it is known. Every other Triassic deposit is clearly classifiable as either "pre-Ischigualasto", with more carnivorous therapsids but without herbivorous archosaurs, or "post-Ischigualasto", without carnivorous therapsids and with more herbivorous archosaurs.

Whether the last of the therapsid carnivores overlapped in time with the first of the archosaurian herbivores, or whether it just failed to do so,

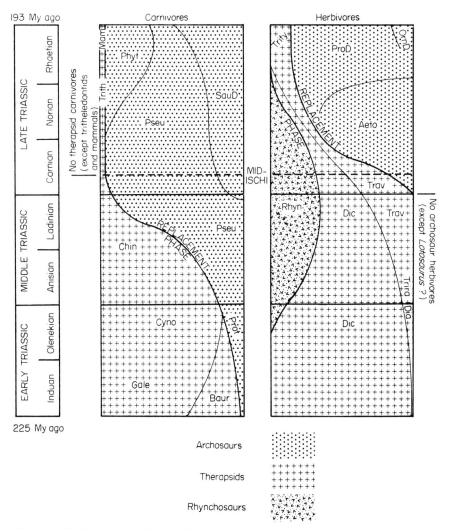

FIG. 1. Block diagrams to illustrate the replacement of the therapsids by the archosaurs during the Triassic. The rhynchosaurs too are shown. The carnivore fauna (left-hand block) and herbivore fauna (right-hand block) are shown separately. The vertical time-scale runs from the beginning of the Triassic (c. 225 My ago) at the base, to the end of the Triassic (c. 193 My ago) at the top.

No phylogenetic implications should be read into these diagrams, which together represent nothing more than the succession of reptiles occupying the adaptive zone for large terrestrial tetrapods.

Abbreviations:
Aeto	Aetosauria	ProD	prosauropod dinosaurs
Baur	Bauriamorpha	Pseu	Pseudosuchia
Chin	Chiniquodontidae	Rhyn	Rhynchosauria
Cyno	Cynognathidae	SauD	saurischian dinosaurs
Dia	Diademodontidae		(other than prosauropods)
Dic	Dicynodontia	Trav	Traversodontidae
Gale	Galesauridae	Trira	Trirachodontidae
Mam	Mammalia	Trith	Tritheledontidae
OrnD	ornithischian dinosaurs	Trity	Tritylodontidae
Phyt	Phytosauria	MID-ISCHI	middle of the Ischigualasto
Prot	Proterosuchia		Formation

is not quite clear. There must have been a lineage of carnivorous, or at least omnivorous, therapsids leading on through Late Triassic time to the tiny near-mammals and earliest true mammals of the earliest Jurassic; but they were so insignificant during the Late Triassic that they are virtually unknown in the several beds of that age. Those apart, the last carnivorous therapsid known is the cynodont *Chiniquodon* von Huene 1936 in the bottom third of the Ischigualasto Formation. The first possibly herbivorous archosaur known (the strange edentulous *Lotosaurus* Zhang 1975 from the Middle Triassic of China is ignored) could be *Aetosauroides* Casamiquela 1960, the first aetosaur, also in the bottom third of the Ischigualasto Formation; but *Aetosauroides* has recurved, laterally compressed, sharp-pointed teeth of carnivorous type, not at all like the straighter, more cylindrical, herbivorous-type teeth of all later aetosaurs (see Bonaparte, 1978: 302; Casamiquela, 1960: fig. 1). Alternatively the first herbivorous archosaur could be *Pisanosaurus* Casamiquela 1967 (see Bonaparte, 1976) from the middle third of the Ischigualasto Formation; the unresolved question of whether or not *Pisanosaurus* is an ornithischian dinosaur (Charig, 1982: 120) is irrelevant to the present discussion, the animal is certainly an archosaur and for the purposes of our discussion we are interested only in the nature of its diet.

(2) The carnivorous archosaurs concerned were characterized by progressive increases in their abundance, diversity and maximum size and by a general improvement in their locomotor apparatus (tending towards the "semi-improved" and then the "fully improved" stance and gait, followed in some cases by bipedality: see Charig, 1972). By contrast, the carnivorous therapsids became relatively fewer and less diverse with the passage of time, their maximum known size decreased, and their locomotor apparatus improved little if at all (see Jenkins, 1970, 1971a).

(3) The rate at which the carnivorous archosaurs were evolving as described was itself increasing exponentially, reaching a maximum by the end of Middle Triassic (or the beginning of Late Triassic) times.

(4) The subsequent replacement of herbivorous therapsids by archosaurs took place much more rapidly.

(5) There is good evidence that therapsids and archosaurs lived sympatrically. The author's own experience shows that, in the Manda Formation (Anisian) of Tanzania, the remains of carnivorous cynodonts and pseudosuchians are all found at the same sites, together with herbivorous cynodonts, dicynodonts and rhynchosaurs. An even better example of this sympatric occurrence in Middle Triassic beds is to be found in the Chañares Formation of Argentina (see Bonaparte, 1978).

THE NATURE OF "SCENARIOS"

Any attempt at a detailed explanation of the therapsid-archosaur replacement will be what is fashionably called a "scenario" – an unpleasing word, but one for which there is no precise substitute. A scenario is defined in its palaeobiological sense (Charig, in preparation, read to the Linnean Society of London on 9 December 1982) as a sequential narrative of hypothesized prehistorical events based on conjectural interpretations of palaeontological data. More specifically, it is concerned with the ecological interplay of various groups of organisms and of the changing relationships between those groups through time and space, as affected by changes in geography and climate and by geological (and perhaps extraterrestrial) events. Such a narrative rests only partly on speculation and is otherwise derived from the use of the data as circumstantial evidence; it therefore offers, at least in part, a possible explanation of some of those data.

However, if a single factor – or even a pair of factors acting either in conjunction or separately – be put forward as the sole cause of some major event such as the faunal replacement discussed here, it can hardly be described as a "scenario"; the term "scenario" is not synonymous with "explanation" or "theory". A scenario must be a *sequential* narrative; it must contain a time element, like the account of the Triassic explosion in tetrapod evolution by Charig (1975, 1979, 1980, in preparation), and it is therefore a more complex idea than a simple explanation. Any scenario explaining the therapsid-archosaur replacement should *include* some explanation of archosaur superiority, like the various suggestions discussed below, but there must be more to it than that.

Some workers (e.g. Patterson, 1980) claim that the explanation of nature by the production of unprovable narratives is empty rhetoric, a waste of time. Yet scenarios *can* be formulated scientifically, even though they cannot be proven, provided that certain guiding principles are carefully followed:

(1) The facts on which the scenario ultimately rests (mostly data from the fossil record) must themselves be soundly based.

(2) The interpretations offered as constituting the scenario should be the most parsimonious conceivable.

(3) It must be made crystal clear exactly what is fact and what is interpretation or scenario.

(4) It must be emphasized that no scenario should be regarded as definitive. Future additions to our basic information and changes in

attitudes and approach and to supporting hypotheses may lead us to modify or even reject the scenario, in whole or in part.

WAS THE REPLACEMENT COMPETITIVE?

It is first necessary to examine the recent and fundamental proposition (Tucker & Benton, 1982; Benton, 1983a, b, c) that most other workers are wrong in viewing the therapsid-archosaur replacement as "competitive". Those workers include not only the present author but also Benton himself (1979a, b) and Bonaparte (1982).

To obviate criticism, the examination must begin with an explanation of the use here of the words "competitive" and "competition". They are used in a broad, general sense, not in the narrow sense of the ecologists, who restrict the concept of competition to interactions between individuals or species and do not recognize its occurrence between higher groups. Further, the therapsid-archosaur replacement might not have been "competitive" in the same sense as would a present-day situation where, for example, a wildebeest drinking at a water-hole might find his foreparts seized by a crocodile at the very moment when a lion was pouncing on his rear; the two predators would be tugging the same hapless victim in opposite directions. This is what Bakker called "interference competition" (1980: 361). The Triassic replacement might not even have been due to what Bakker called "exploitative competition" (1980: 361), where two individuals, or two species, are exploiting the same limited food source without coming into direct confrontation. The very different types of teeth possessed by, say, carnivorous cynodonts and by pseudosuchians suggest (no more than that) that they might have fed upon *different* sorts of prey; likewise traversodontid cynodonts, dicynodonts, rhynchosaurs, aetosaurs and prosauropods, with their highly diverse dental equipment, might well have fed upon different sorts of plants. But the therapsid-archosaur replacement must nevertheless have been competitive in the sense that two lineages living in the same place at the same time, possibly sharing the same *broad* dietary habits, were subjected to the same environmental pressures; one responded by waxing until it achieved total dominance, the other waned *pari passu* almost to extinction. That seems hard to deny. Benton (1983b: 40) proposes the excellent term "differential survival" for this, defining it as the result of the [simultaneous] occupation, by a group of native animals and a group of invaders, of similar adaptive zones; he regards it as a synonym of "competition".

Incidentally, Bakker (1980: 361) did well to stress the potential importance of the intense competition between large predators in modern East African faunas, both interspecific and intraspecific, as an agent in natural selection.

It might therefore be thought that the difference between Charig (and various other authors) on the one hand, and Tucker & Benton and Benton *solo* on the other, was simply a semantic quibble and without real substance. That is not so. The main premise of Tucker & Benton (1982) and Benton (1983a, b, c) is that there was no competition between therapsids and rhynchosaurs on the one hand and dinosaurs on the other, and that the major causes of the Triassic extinctions, replacement and radiation were extrinsic factors (such as climatic or floral change) rather than intrinsic (relating to differences between the two groups, for example in their locomotor ability or their thermoregulatory physiology). The argument of Tucker & Benton (1982) is challenged here on four important points. First, several details in their revised correlation of the Triassic reptiliferous strata are disputed; there is, however, no space to criticize their correlation here and in any case it has little bearing on the subject under discussion, although it should be noted that they consider the Ischigualasto Formation to be of Norian age rather than earliest Carnian, which explains their extension of the range of the carnivorous therapsids by several million years. Secondly, the main point they make, that there was no competition between therapsids and dinosaurs, is not at issue; what the present author and others have said is that there was competition between therapsids and *archosaurs*, more specifically the thecodontians which must have included the dinosaurs' ancestors. Nor is it disputed that, with few exceptions, mammal-like reptiles and rhynchosaurs are not found together with dinosaurs, nor, in consequence, that dinosaurs radiated "opportunistically" only after the virtual extinction of the other groups mentioned. The present author's own works on the subject (Charig, 1975, 1979, 1980) all bear this out, and there is therefore no difference of opinion on this matter. But, thirdly, he cannot accept the additional point made by Benton very recently (1983a, b) that there was not even competition between therapsids and *thecodontians*.

Benton himself sets out very clearly (1983b: 42) four pairs of opposing characteristics by which competitive (differential survival) replacement may be distinguished from opportunistic replacement. The competitive hypothesis requires (Benton's no. 2) that the replacement be gradual rather than rapid, from which his no. 3 necessarily follows, that the two groups concerned should be found together, and likewise his no. 1, that one group must become more abundant with the passage of time while the other becomes less so. Those three requirements, which in any case are no more than different aspects of the same thing, are satisfied to perfection by our general record of the therapsid-archosaur replacement. The characteristics of the replacement may be established without details of the numbers of the various genera involved or of the numbers of individuals in each; apart from that, estimates of relative abundance based on scientific papers, on lists in reference books and on "nose counts" in museum collections are liable to be affected by so many unascertainable factors and subjective considerations that their apparent accuracy must be spurious and their scientific value diminished accordingly.

As to the fourth point at issue, it is not apparent how one could separate evolutionary changes due to "extrinsic" factors from those due to "intrinsic"; surely the whole neo-Darwinian position is that natural selection generally operates through the effect of extrinsic factors upon intrinsically variable populations, producing differential changes within those populations. With respect to any particular character of a sympatric population (morphological, physiological, behavioural, etc.) there have to be heritable differences between the various components of that population if the same extrinsic factors are to have a differential evolutionary effect upon them through the mechanisms of natural selection. All this operates at various taxonomic levels, not only at species level but infraspecifically at the level of the deme and supraspecifically at the level of the genus, the family or even higher. In the case under review the competing components are subclasses or infraclasses; this might truly be regarded as a study in macroevolution.

Further, even if "non-competitive" replacement caused by extrinsic factors were possible, the present author cannot envisage any such factors that would have led the carnivorous archosaurs to have replaced their therapsid counterparts almost completely before the corresponding replacement of the herbivores had even begun. (It is nevertheless possible to imagine the opposite, a set of extrinsic factors – climatic and floral – that would affect the herbivore fauna first; but in that case there would still be an intrinsic factor too, namely the difference between an innate requirement to eat vegetable matter and a requirement to eat flesh.) Indeed, the only scenarios that have attempted to explain this two-phase "carnivores first" phenomenon, that recognize explicitly that it happened, are those of Charig (1975, 1979, 1980) and Benton (1979b); it was also referred to by Bonaparte (1982).

EXPLANATIONS PROPOSED

Every explanation of the faunal replacement proposed hitherto relies essentially upon an alleged difference between Triassic therapsids and contemporary archosaurs in some feature of their anatomy or physiology – a difference which, in the conditions then prevailing, would

Therapsids versus Archosaurs in the Triassic

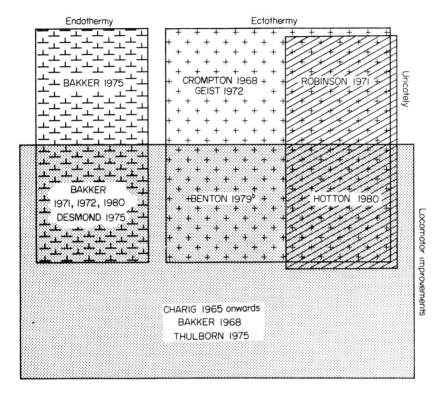

FIG. 2. The various published explanations of the "archosaur success" in the Triassic faunal replacement. "Anatomical" explanations are at the bottom, "physiological" explanations are at the top, and "combination theories" are in the middle where the rectangles overlap.

prove advantageous to the archosaurs and correspondingly disadvantageous to the therapsids.

Which authors have suggested which causes, either alone or in various combinations, is shown very simply in Fig. 2.

Anatomical Explanations: Charig's/Thulborn's

Of these, the only one worthy of serious consideration concerns the locomotor apparatus, in which the Triassic archosaurs soon showed improvement (Charig, 1965, 1972): first to the "semi-improved" stance and gait, then to the "fully improved" (the acquisition of which is arbitrarily regarded as the distinguishing character of a dinosaur) and finally, in some cases, to habitual bipedality. These improvements not only made the archosaurs faster and more agile than other contemporary tetrapods but also enabled them to grow much bigger than any earlier land animals without loss of speed or agility and thus to dominate the fauna. The suggestion that these developments were the main reason for the archosaurs' success appears in various forms in Charig (1965, 1966, 1972, 1975, 1979, 1980), Bakker (1968) and Thulborn (1975); Thulborn emphasized the particular importance of the enarthrodial hip-joint. A prerequisite of this, of course, is the lack of any comparable development in the contemporary therapsids. There seems to be a vague and intuitive idea that some Triassic therapsids, as the precursors of mammals, must already have been showing a parallel trend towards locomotor improvement; but Jenkins (1970, 1971a, b), while accepting that the therapsids were "incipiently mammalian" in their locomotor functions, opined that they were still a long way below the level of cursorial mammals. This means that they must have been far less developed in that respect than were the most advanced archosaurs of the time. It was held also by Bakker (1968: 22) that mammals did not achieve an equivalent improvement of their locomotor system until the Late Cretaceous, while Ostrom (1970: 366) likewise opined that the therapsids did not possess upright limb posture and gait.

As for the jaws and teeth, any suggestion that the dentition might have played some part in the faunal replacement can easily be discounted. The primary change-over took place among the carnivores; and the homodont dentition of archosaurian carnivores (effective only for stabbing and seizing the prey) is not nearly so specialized as the heterodont dentition of therapsid carnivores (some teeth of which, though not truly mammal-like, are nevertheless adapted for comminuting the flesh of their victims).

If the improvements to the stance and gait of the Middle to Late Triassic archosaurs were indeed the essential factor which enabled them to become spectacularly effective predators, then the following outline scenario (sequential interpretation of the fossil record) might be considered likely:

(1) The carnivorous archosaurs of the Middle Triassic were outstandingly successful in competing for food with the contemporary therapsids. Some of the omnivorous therapsids, being unable in those circumstances to obtain an adequate supply of meat, evolved an advanced form of herbivory; those were the traversodontid cynodonts.

(2) Before mid-Ischigualasto (mid-Carnian) times the carnivorous archosaurs had virtually eliminated those therapsids that had retained the carnivorous habit (apart from the tritheledontid and true mammal lineages). By then the locomotor apparatus of some of the archosaurs had already attained the "fully improved" condition.

(3) The carnivorous archosaurs preyed *too* successfully on the herbivorous therapsids and rhynchosaurs, causing a reduction in the numbers of those groups (through "overkill") and consequently their virtual extinction (apart from the tritylodontids) by Norian times.

(4) As a consequence of (3) there was less animal food available in the early part of the Late Triassic but much more plant food.

(5) As a consequence of the selection pressure due to (4) some of the "improved" archosaurs changed their diet, from carnivory to herbivory. The herbivorous archosaurs possessed much the same "improvements" as had the carnivorous archosaurs from which they had evolved; they were therefore far better equipped than were the herbivorous therapsids to withstand the predations of their carnivorous cousins and thus to maintain and increase their populations.

If this hypothesized sequence of events actually occurred, then we might expect to find that the replacement of carnivorous therapsids by archosaurs took place fairly slowly, with some overlap (i.e. as a result of competition). By contrast, we might also expect to find that the replacement of herbivorous therapsids by archosaurs took place later, much more quickly and without overlap (i.e. without competition; the archosaurs were filling an empty ecological niche). That, in fact, is what we do find. The Chañares and Manda faunas (see p. 600 above) includes carnivorous cynodonts and pseudosuchians existing sympatrically, with neither group showing dominance; no one, on the other hand, has ever described a deposit with herbivorous therapsids and archosaurs in a similar situation.

It must be emphasized that these ideas constitute nothing more than a "scenario", a plausible narrative derived from conjectured interpretations of such data as are presently available; those interpretations, however, are the most parsimonious conceivable by the present author. New, even more plausible ideas would be welcomed by everyone; more welcome still would be new relevant data.

Physiological Explanations

Explanations based on physiology are more numerous than are those based on anatomy. They concern thermoregulation, nitrogen excretion and susceptibility to potentially lethal high temperatures.

Thermoregulation: general remarks

These explanations rely essentially upon the supposition that one of the groups in question was ectothermic in Triassic times and the other endothermic, or that, if both were considered to have been endothermic, they differed significantly in the degree or nature of their endothermy. Unfortunately there is considerable disagreement as to the type of thermal physiology in the two groups. For example, Robinson (1971: 141) was almost certain that both therapsids and archosaurs were ectothermic in Triassic times, neither group achieving endothermy until the Cretaceous or even the early Tertiary; yet Bakker (1975: 68,

70) held even more strongly that both groups already included endotherms early in the Triassic. If both *were* ectothermic, then it is hard to see how thermoregulation could have played any part in the faunal replacement. Moreover, it appears that no one has ever made the combined suggestion that Triassic therapsids were ectothermic while Triassic archosaurs were endothermic. We are therefore left with only two possible categories of explanation based on thermoregulation.

The more important category requires:

(1) that Triassic therapsids were endothermic, possessing a significant (though not necessarily highly developed) degree of homoiothermy;
(2) that Triassic archosaurs were ectothermic;
(3) that Late Permian and Early Triassic climates were cool, so that endothermy was advantageous in those times; and
(4) that Triassic climates became increasingly hotter, so that endothermy became progressively less advantageous until it was positively *dis*advantageous.

The changes in (4) would favour the archosaurs at the expense of the therapsids.

The other category of explanation has to be invoked if it is believed that Triassic archosaurs, as well as the therapsids, were endothermic. In those circumstances therapsid endothermy must have differed significantly from archosaur endothermy in some particular so that, once again, the increasingly hot Triassic climate would have selected for the archosaurs and against the therapsids.

Thermoregulation: archosaur ectothermy

Crompton's explanation

This explanation (Crompton, 1968) is no longer fully supported by its author, but it is included here in order to clarify the development of our ideas on the subject. Crompton suggested that the therapsids might have developed a high body temperature but not the ability to cool themselves effectively; they would therefore have been less able to endure the hotter conditions of the later Triassic. Yet their sprawling posture required the expenditure of a great deal of energy during locomotion, with the concomitant production of body heat. A second factor (not mentioned by Crompton) is that, in order to secure the food that their high energy expenditure demanded, the animals would have been forced to even greater activity, even more energy expenditure and even more heat production; thus they would have been trapped in a "vicious circle". On the other hand, Crompton (1968) presumed that the therapsids possessed none of the mammals' special cooling mechanisms (sweating, licking the fur, controlling the blood flow to the skin, superventilation of the lungs, etc.) and could have avoided overheating only by a curtailment of their activity. The ectothermic archosaurs, by contrast, were under no such restriction; they would have been able to tolerate wide ranges of body temperature, which they could, in any case, adjust by behavioural adaptations (e.g. seeking shade or sun) and, up to a point, they would have become more active with an increase in ambient – and hence body – temperature. Crompton was careful to emphasize the highly speculative nature of his explanation.

He also believed that the therapsid line was perpetuated by the tiny Mesozoic mammals which, he claimed, could withstand high ambient temperatures better because their larger surface-to-volume ratio enabled them to cool themselves more quickly (by radiation, see Crompton, 1968: 146). This, surely, cannot be true. When the ambient temperature is higher than that of the body, the animal will gain heat from its surroundings; the only way in which heat can be lost against the temperature gradient is by evaporative cooling, which, in an arid environment, uses far too much water. The experiments of Colbert, Cowles & Bogert (1946) on American alligators clearly demonstrated that a small animal is more sensitive than a large one to temperature fluctuations in either direction. It is therefore more likely that the small mammals were better fitted than their larger therapsid cousins, not to endure high ambient temperatures, but to *avoid* them by hiding in crevices and burrows and by adopting nocturnal habits. Meanwhile the large archosaurs, especially the dinosaurs, were better able to *withstand* high ambient temperatures – with no external insulation other than (perhaps) subcutaneous fat – simply by virtue of their much greater size; cf. the discussion below (p. 623) of "inertial homoiothermy".

However, Crompton (1968) seems later to have changed his mind on two points (Crompton, Taylor & Jagger, 1978). First, he felt (Crompton *et al.*, 1978: 336) that "the mammal-like reptiles may well have been ectotherms", in which case the whole of his 1968 explanation would collapse. Secondly, although he did not explicitly recant his earlier statement, he made clear that he had come to realize (Crompton *et al.*, 1978: 333–334) the immense difficulties encountered by a small homoiotherm in trying to cool itself when the ambient temperature is appreciably higher than the preferred body temperature. The easiest "solution" to the problem is to prevent the problem arising, by the simple expedient of preferring a body temperature so high that the ambient temperature is unlikely to rise above it.

Crompton's 1968 explanation, like others in the same "physiological" category, is in any case open to three more criticisms of a fundamental nature. First, the explanation rests upon premises concerning the animals' respective thermal physiologies which are highly conjectural and far from universally accepted – no longer even by Crompton himself. Second, it rests also upon climatic changes which, though substantiated to some degree, are certainly not proven and were certainly too limited (see below, pp. 620–622). Third, the explanation does not accord with the chronological sequence of the actual events. The alleged increase in world-wide temperatures in the Late Triassic could hardly have helped bring about a faunal replacement which reached its climax at the *beginning* of the Late Triassic; indeed, the archosaurs were already evolving rapidly towards a dominant position very much earlier, in the Middle Triassic, when therapsids were still flourishing and world temperatures must still have been relatively cool.

Geist's explanation
Another variation of the "thermal physiology" theory was that of Geist (1972), who seems, quite independently, to have evolved an explanation of the Triassic faunal replacement which is very similar to Crompton's (1968). It differs in two particulars:

(1) The ectothermic grouping, as considered by him, comprised not merely the archosaurs on their own but included *all* the diapsids.

(2) He seemed to be little concerned with the adverse effects on the therapsids of over-heating and the consequent need to restrict endogenous heat production by reducing metabolic activity. Rather did he emphasize his belief that, in the high temperatures of the Late Triassic, the therapsids were at a disadvantage in requiring an enormously greater amount of energy for the unnecessary maintenance of a high body temperature.

All the criticisms made above of Crompton's 1968 explanation apply equally strongly to that of Geist (1972). The climatic differences that he postulated, though they probably existed, were unlikely to have been as clear-cut or as world-wide as Geist would have had us believe. The disadvantages which are allegedly suffered by homoiotherms in a hot climate are not apparent today; surely thermoregulation cannot require the expenditure of any extra energy if the ambient temperature is as high as that of the body, and, in any case, a survey of modern tropical faunas does not suggest that homoiotherms are any less successful than poikilotherms, although some very large tropical homoiotherms do require special cooling adaptations (e.g. the loss of hair and the flapping of ears by elephants). In this connection it is impossible to agree with the assumption of Geist (1972) that the number of extant species in a group is a measure of its superiority or its success. The suggestion that modern mammals are less successful than modern reptiles because they have rather fewer species is untenable; the diversity of form and habit, of geographical and ecological range, is obviously greater in mammals, and the mammalian biomass, although difficult to estimate, is probably far greater too. In fact a cogent case could be argued for the opposite belief, that the existence of a large number of essentially similar species indicates a restricted geographical and/or ecological range for each and therefore poor adaptability. Consider the range of the leopard (*Panthera pardus*) or, to take an extreme case, *Homo sapiens*! Each is highly successful, yet neither is divided into more than one species.

Thermoregulation/archosaur endothermy: Bakker's explanation

Bakker is one of the main proponents of the idea that dinosaurs, and some thecodontians too, were endotherms. Of the two editors and 16 authors represented in the symposium volume *A cold look at the warm-blooded dinosaurs* (Thomas & Olson, 1980) Bakker was the only unequivocal advocate of true endothermy in all dinosaurs.

Several of Bakker's papers refer explicitly to the competitive nature of the therapsid-archosaur replacement and consider its causes. Initially (1968) Bakker thought it due entirely to the superior locomotor system of the archosaurs, reflecting the earlier views of Charig (p. 605); but subsequently (1971, 1972) Bakker ascribed an increasingly important rôle to their postulated endothermal physiology. Indeed, his 1975 paper mentions only endothermy (Bakker, 1975: 77–78), without any clear reference to locomotion. However, his most recent article on the subject (Bakker, 1980) gives equal prominence to locomotion *and* to endothermy as alleged causal factors.

It should be noted that some of these works (particularly Bakker, 1972 and 1980) refer specifically to dinosaurs rather than to archosaurs as a whole; and, as pointed out above, the protagonists in the replacement drama were not the dinosaurs themselves but their thecodontian forbears. However, Bakker's other works suggest endothermy in at least the later thecodontians and the thecodontians' general superiority to therapsids; this shows that he often used the word 'dinosaurs' rather loosely, to mean 'dinosaurs and their kin'.

The question arises, why did Bakker consider "dinosaur" endothermy to be better than therapsid endothermy? He wrote (1975: 77–78) "The success of the dinosaurs . . . can now be seen as the predictable result of the superiority of their high heat production, high aerobic exercise metabolism and insulation." And in 1980 he stated that "Dinosaurs probably had higher and more continuous levels of heat production than therapsids" (1980: 352), and that archosaur success was achieved through "a higher level of endothermy" and a high "level of performance in thermoregulation and locomotion" (1980: 462). On the other hand, he had written earlier (1971: 656) that "possibly the lack of efficient heat-loss mechanisms in therapsids caused their extinction", and the following year (1972: 85) he had opined that "dinosaurs had an advantage over mammal-like reptiles in mobility and the capacity to unload high endogenous heat production", an explanation very similar to Crompton's (1968). Such a fate for the therapsids would be quite independent, of course, of what was happening to the archosaurs. But it does mean that, if archosaurs were endothermic too, then we must suppose that they did possess efficient heat-loss mechanisms! Perhaps the archosaurs scored over the therapsids in being without external insulation, although that would be contrary to Bakker's belief (1975: 58–59) that they were covered in some sort of "feathers".

Bakker excluded large fresh-water predatory archosaurs from his hypothesis. He claimed (1975: 70) that such forms possessed an ectothermal physiology, which would have allowed them to make longer dives and, as in modern ecosystems, given them a competitive advantage over endotherms.

All this might convince those who believe that archosaurs were highly developed endotherms. It is not within the scope of this article to discuss the likelihood of the correctness of such views; nevertheless Bakker's followers are few, which means that his ideas on the causes of faunal replacement are likewise without much support. In any case, they are based on speculation rather than fact.

Nitrogen excretion and high-temperature tolerance: Robinson's explanation

Summary
An entirely original and thought-provoking approach to the subject of the Triassic faunal replacement was that of Robinson in her Twelfth Annual Address to the Palaeontological Association (delivered 1969, published 1971). She pointed out (Robinson, 1971: 138) that most modern "sauropsids", i.e. squamates and birds, excrete their waste nitrogen as the relatively insoluble uric acid, which requires little water for its removal; this "uricotelic" excretion is water-conserving and presumably advantageous in desert conditions. Modern mammals, on the other hand, excrete their nitrogen as the very soluble urea; this "ureotelic" excretion requires more than 10 times as much water. Robinson then postulated (1971: 141) that the Permo-Triassic "sauropsids", which included the archosaurs, were also uricotelic; they were therefore better fitted to withstand the increasingly arid conditions which prevailed in certain regions during the later Triassic than were the

contemporary "theropsids", i.e. mammal-like reptiles and mammals, which she presumed to have been ureotelic like modern mammals. (It should be noted that Reig (1970: 265) had made the contrary suggestion, namely that early archosaurs were ureotelic.) Robinson (1971) stated also that squamates and birds prefer higher body core temperatures than do modern mammals and are therefore better able to endure higher ambient temperatures. Again she supposed a similar difference between the sauropsids and theropsids of the Permo-Triassic and, in consequence, believed that the sauropsids enjoyed yet another advantage in Triassic conditions.

Analysis
These theories clearly rest upon three fundamental premises:

(1) Modern squamates and birds are better adapted to life in hot arid conditions than are mammals, partly because of their physiological mechanisms.

(2) Such physiological character-states are more dependent on phylogenetic relationships than on the external conditions with which the animal has to cope, or, indeed, on any other factor. From this it follows that the physiological character-states of Permo-Triassic reptiles may be inferred from those of their nearest modern relatives. The physiology of therapsids must therefore have been like that of their living descendants, the mammals; and the physiology of the Permo-Triassic sauropsids must have been like that of lizards and snakes (living sauropsids) and of birds (sauropsid derivatives) but not necessarily like that of crocodilians, which, though sauropsids, have been adapted to a semi-aquatic existence since their origin.

(3) During the Triassic (as at some other times) most of the continents were drifting northwards across the parallels of palaeolatitude. Their distribution relative to each other and to the palaeo-equator during the latter part of the period resulted in a particular pattern of climate, characterized by an unusually wide distribution of hot, semi-arid or arid conditions which were quite unlike the warm but moderately well-watered conditions generally prevalent in the Late Permian.

If these three premises are correct, it follows logically that the sauropsids were pre-adapted to the hot, arid conditions which were gradually prevailing over large areas of the continents in Triassic times. The theropsids were not pre-adapted to those conditions, hence the sauropsids' success in the competition.

Adaptations to hot arid conditions [*comments on premise (1) above*]
Robinson's (1971) paper, though admirable in many respects, seems to contain some basic fallacies; her fundamental premises are unacceptable.

Consider first whether modern squamates and birds are really better adapted to life in low-latitude deserts than are mammals. The respective distributions of the three groups in question do not suggest that this is true. (It is true that mammals, being endothermic, are better adapted to cold or fluctuating conditions than are reptiles, but the converse cannot be presumed.) Robinson's text-figure 4 (1971) purports to illustrate in diagrammatic fashion her statement (1971: 140–141) that mammals are less numerous than squamates or birds in arid or semi-arid conditions, but she made no attempt to substantiate that statement; indeed, it is contradicted by Bartholomew's statement (1968: 333) that "very large populations of herbivorous mammals and granivorous birds exist in desert regions and form the base for a flourishing fauna that includes many predatory mammals, birds, and reptiles." The larger species, without doubt, are mostly mammals. Two characters are specifically mentioned by Robinson (1971) as contributing towards the alleged success of the sauropsids under those conditions: their uricotely and their higher body core temperatures.

On the question of uricotely and water conservation, Robinson's comment (1971: 138) "Urea is soluble, and requires much more water for its excretion than uric acid", may seem paradoxical to the uninitiated. How can water loss be reduced by excreting nitrogen as the comparatively insoluble uric acid instead of the highly soluble urea? Surely much *more* water is required to dissolve a solid of low solubility? Surprisingly, however, she is quite correct. It is much easier to resorb water from a dilute solution of uric acid than from a more concentrated solution of urea because of the much lower osmotic pressure of the former; the resorption takes place in the cloaca and the uric acid is thereby precipitated in crystalline form, to be excreted as a solid. Gordon (1968: 275) wrote "Because of uric acid's low solubility and easy precipitation, it is not necessary for reptilian urine to contain much water to keep nitrogenous wastes in solution at low enough concentrations not to cause osmotic problems." Uric acid undoubtedly does help water conservation.

Against this, however, it would be difficult to prove that the presence of uricotely can be correlated with life in an arid environment. True, many modern reptiles live in arid environments, but this is chiefly because arid conditions are generally less favourable than moister ones, and there is consequently less intense competition from birds and mammals. As for birds, they frequent habitats of all degrees of aridity; their speedy locomotion enables them to travel to the nearest water-supply quickly and easily and they can thus survive in the driest of deserts. On the other hand, many uricotelic squamates and birds live in conditions which are not at all arid; and many ureotelic mammals live very successfully in arid conditions, having developed – as Robinson herself pointed out (1971: 139) – alternative (and equally effective) morphological, physiological or behavioural adaptations for water conservation. Small mammals can thrive as nocturnal foragers, living in burrows or crevices during the heat of the day; large mammals can do very well because of their much lower surface area:volume ratio and because they can travel greater distances in search of their requirements (especially when equipped with cursorial adaptations). Indeed, considering the animal as a whole, mammals seem to be rather better adapted to this life-style than are squamates or birds. (If the Triassic sauropsids enjoyed the advantages of uricotely, might not the therapsids of the time have benefited from the same alternative adaptations as are found in present-day mammals?) Probably the most one can say in favour of Robinson's (1971) thesis is that sauropsids *may* have an innate potential to develop a greater degree of uricotely than non-sauropsids and may therefore be more capable of coping with intermittent dry conditions.

Robinson considered also (1971: 138–139) the relationship between an animal's preferred body core temperature and the ambient temperature. The former, on average, is 4°C higher in birds than in fully endothermic mammals; and in squamates, of course, it is largely dependent on the ambient temperature to which the animal is acclimatized. Now, it is more difficult for a homoiothermic animal to cool itself down than to keep itself warm; this is especially true in arid conditions, when it loses water by evaporation too quickly and cannot replace it, or in saturated conditions, when it is unable to lose water by evaporation at all. The squamate or the bird is therefore at an advantage over the mammal when the ambient temperature is higher than the mammal's body core temperature, typically in the 36–38°C range. The squamate moreover possesses an additional advantage in that it has little endogenous heat to lose. Once again the mammal has alternative physical, physiological or behavioural adaptations which offset its disadvantages; for example, it may be very large and therefore less susceptible to changes in the ambient temperature, or it may be nocturnal and/or crevice-living or fossorial in its habits.

As for the lethal effects of very high temperatures, Robinson stated (1971: 140) that some modern lizards and snakes show astonishingly high values of preferred body temperature (up to 44°C) that would be lethal to any mammal. But the published data on lethal temperatures vary greatly and are difficult to assess. The lethal temperature is sometimes confused with the critical maximal temperature (the thermal point at which locomotor activity becomes so disorganized that the animal can no longer escape from the conditions that will probably lead to its death); it may even be confused with the highest body temperature that the animal normally tolerates. Further, different workers quote different results for experiments on the same species – possibly because some measure the body temperature at the moment of death and others the ambient temperature. In any case, Robinson's observation (1971) simply cannot be true if the published data are correct. Lethal temperatures quoted (Prosser, 1950; Cloudsley-Thompson, 1971) are in the 37.5–49°C range for a variety of reptiles (both extreme values being for lizards, i.e. sauropsids), in the 43.5–47°C range for birds and in the 40–48°C range for most mammals. A rat endured 45°C for 79 min before it died, a guinea-pig 48°C for 100 min, two cats 50°C for over 8 h, and – surprisingly – a calf 52°C for 5½ h (Heilbrunn, Harris, Le Fevre, Wilson & Woodward, 1946: 406).

There are, in fact, several different mechanisms causing heat death in vertebrates. The lethal temperature depends on, *inter alia*, the following:

(i) The degree of humidity to which the animal has been accustomed.
(ii) The ambient temperature to which it has been accustomed.
(iii) The rate at which the temperature has been raised.
(iv) The duration of the exposure to the elevated temperature.

Thus the lethal temperature depends to a considerable extent upon the nature of the animal's normal environment. It does not depend on phylogenetic position but can vary widely within a group. There is no reason to suppose that reptiles as a class can, in general, endure higher temperatures than can mammals.

Physiology and phylogeny [*comments on premise (2) above*]
With regard to certain basic physiological functions – in particular nitrogen excretion and high-temperature tolerance – Robinson stated (1971: 137) that nearly all modern sauropsids (i.e. squamates and birds) share one condition and nearly all modern theropsids (i.e. mammals) share another, the two being markedly different. From this she argued that the differences "must therefore have arisen at an early date", implying that the Permo-Triassic sauropsids and theropsids were respectively characterized by the same character-states that distinguish their present-day relatives. In short, she claimed a strong correlation between physiology and phylogeny.

That, however, is manifestly not true of *every* aspect of physiology. Careful observation shows that physiology in general is not especially dependent on either environment or phylogeny and it may differ widely between closely related forms. Thus the ectothermic, ammoniotelic crocodilians are phylogenetically more closely related to the endothermic, uricotelic birds than they are to any other creatures alive today; yet birds and mammals, quite unrelated phylogenetically, share not only endothermy but also certain features of the excretory system which are absent in modern reptiles (e.g. the presence of a thin loop of Henle and the ability to excrete a hypertonic urine; see Prosser, 1950: 58). Indeed, Robinson herself ignored her own precept, for she explicitly stated her belief (1971: 141) that the Permo-Triassic theropsids – be they therapsids or mammals – resembled the contemporary sauropsids in possessing an ectothermic physiology and were thus quite different from *modern* mammals. Bearing in mind that the preferred body core temperature must surely depend to a very large extent upon whether the animal is an ectotherm or an endotherm, it is difficult to understand why Robinson's arguments (see p. 615) employed the relationship between the preferred body core temperature in therapsids and the ambient temperature as though we knew it to be similar to the corresponding relationship in modern mammals – which, of course, we do not. (Even if we, unlike Robinson, accepted the majority view that the Triassic *mammals* were at least *incipient* endotherms, it is hard to see how the characters of the mammals in the Late Triassic could have so influenced the course of tetrapod evolution as to lead to – or at least to stimulate – a faunal replacement which, by the time the mammals appeared, had already taken place. To make this even more unlikely, the "Triassic" mammals are now considered to be of Early Jurassic age – see Olsen & Galton, 1977.)

All that apart, the reaction provoked by Robinson's arguments is that they are curiously inconsistent. She deliberately excluded crocodilians from consideration as sauropsids in this connection because they "have been semi-aquatic in habits since Lower Jurassic times, and this has had very considerable effects on their physiology" (Robinson, 1971: 137); in other words, physiology depends on environment. She continued "They can no longer be regarded as truly terrestrial vertebrates physiologically" and therefore considered that, for the purposes of her argument, "living sauropsids" included only squamates – lizards and snakes – and, as indubitable sauropsid derivatives, the birds. Then, however, she deduced that the physiology of the dinosaurs must have been similar to that of squamates and birds because of their common sauropsid origin; in other words, physiology depends on phylogeny!

Despite the apparent contradiction, Robinson's two 'statements' are both perfectly correct in so far as they relate to nitrogen excretion; that function does seem to depend both on environment and on phylogeny. It cannot be denied that modern squamates and birds are predominantly uricotelic or that mammals are chiefly ureotelic. On the other hand,

Robinson greatly over-simplified the picture of the distribution of ureotely and uricotely among living tetrapods; most reptiles are partly ureotelic, and, conversely, some mammals excrete a proportion of their waste nitrogen as uric acid. The numerous exceptions to the broad general rule demonstrate that the chemical form in which waste nitrogen is excreted is a labile character, related to the stress of water supply in both embryo and adult and therefore dependent to a large degree upon the habitat. Prosser wrote (1950: 203) "... which [enzymatic] route and which [waste nitrogen] product become dominant in a given species depend more on its immediate osmotic needs than on its ancestry." Again (Prosser, 1950: 191) "The correlation between [nitrogenous] excretory products and habitat of adult or embryo is so good that nitrogen excretion appears to be an important adaptive character." The nature of the product can be greatly affected by minor genetic variations. Further, the form of nitrogen excreted can change during ontogeny (e.g. from ammonia to urea at anuran metamorphosis; see Campbell, 1973: 305) and even to suit variations in external water relationships (e.g. terrestrial tortoises, normally excreting both urea and uric acid, excrete mainly the latter during dehydration; see Khalil & Haggag, 1955). It could well be that the above generalization, namely that modern squamates and birds are predominantly uricotelic and mammals ureotelic, is due not to the common phylogenetic heritage shared by the members of each category but to a common adaptation to a common method of reproduction – oviparity in squamates and birds, viviparity in mammals. A high concentration of urea in the cleidoic egg would certainly be toxic, whereas the deposition of uric acid is relatively harmless. Who knows how the therapsids of the Triassic reproduced themselves?

It is abundantly evident (from a study of biochemistry and embryology as well as from the present-day distribution of nitrogenous waste-products within the Animal Kingdom) that the evolution of nitrogen excretion has led either from ammoniotely through ureotely to uricotely or sometimes directly from ammoniotely to uricotely (Campbell, 1973). Ammonia will be the waste product where there is plenty of water and easy diffusion, urea (or trimethylamine oxide) where there is some stress of water supply, and uric acid where there is extreme need of water retention and the kidneys are unable to cope. Uricotely may be regarded as a negative character-state in so far as it generally results from the absence of specific enzymes (uricases) found in the more primitive ureotelic forms or, less commonly, from the inability to resorb uric acid in the kidney for conversion to allantoin in the liver; it can therefore develop homoplasously in different groups (even at the

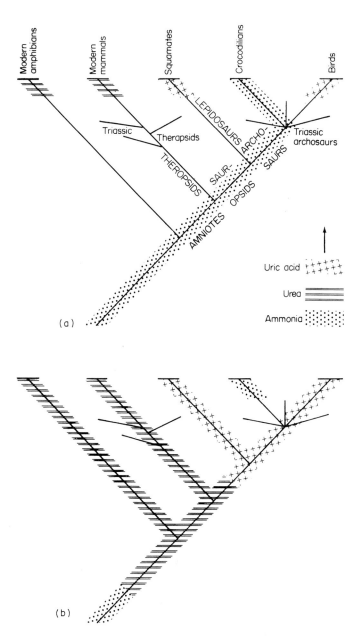

FIG. 3. Alternative cladograms showing two different evolutionary pathways, either of which would explain the origin of the different modes of nitrogen excretion in modern tetrapods. These cladograms incorporate a time-dimension. The stippling, striping and crosses indicate the nature of the main nitrogenous waste-product, as known in the Recent forms and presumed in their ancestors. For the sake of simplicity the testudinates have been omitted from these diagrams. This is because:
  (i) The position of their dichotomy from the main line leading to birds, relative to the position of the theropsid dichotomy, is not at all clear.
 (ii) They produce both ammonia and urea in significant quantities, in some species uric acid too.
(iii) Their inclusion would not be helpful in any way.

intraspecific level, e.g. in the Dalmatian dog; see Friedman & Byers, 1948). Prosser wrote (1950: 201) "uric acid excretion became dominant in several unrelated groups and represents convergent evolution."

If the ammoniotely of the crocodilians is indeed a primitive character-state, not derived by a reversion from a more highly evolved system of nitrogen excretion, then, in the light of the generally accepted phylogeny, the uricotely of squamates and of birds must have been derived homoplasously in the two lineages [Fig. 3(a)]. Some at least of the Triassic archosaurs must then have been ammoniotelic and others may have been ureotelic or uricotelic or even something different that is quite unknown today. On the other hand, it seems just as likely that uricotely is a truly synapomorphous character-state uniting lepidosaurs and archosaurs as sister-groups within a higher taxon Diapsida, itself part of the Sauropsida [Fig. 3(b)]; in which case the ammoniotely of the crocodilians, within the archosaurs, must have been derived by a reversion from uricotely as an evolutionary response to the requirements of the fresh-water habitat. If the latter alternative were correct we should be forced to admit that such changes could happen in any group, so that Triassic therapsids living in hot arid conditions might also have developed uricotely without our knowing about it. Either way we cannot contrast Triassic theropsids and sauropsids as ureotelic and uricotelic respectively, and the main plank of Robinson's thesis (1971) collapses.

In summary, the legitimacy of Robinson's arguments is questioned, for phylogenetic relationships cannot be considered a sufficiently reliable guide to thermal physiology, high-temperature tolerance, water relations or nitrogen excretion.

Changes in Triassic climates [*comments on premise (3) above*]
One might also question whether the supposed general increase in temperature and aridity of the Triassic climate could indeed have played a major part in bringing about the faunal replacement. Such an increase, if it had occurred, would certainly have resulted in a drastic reduction in the area of the globe over which prevailed terrestrial conditions favourable to tetrapod vertebrates; it might thus have led to an increase in the pressures of natural selection. But, as has already been shown, there is no good reason to suppose that such an increase in selection pressure would have favoured the sauropsids at the expense of the other elements in the fauna; and, in any case, it seems very doubtful whether the climatic change was really as significant in this connection as Robinson (1971) would have us believe. Evidence adduced by Robinson herself suggests that Late Triassic climates were not especially arid, being characterized more by seasonal droughts than by a true

general aridity; she wrote (1971: 148) "environments with high year-round temperatures, and at least a seasonal drought . . . gradually increased in frequency on the world's lands during the Triassic, in the middle and low latitudes, and became especially widespread in Upper [sic] Triassic times." This is borne out by a study of Upper Triassic sedimentology and faunas; for example, the lower part of the Elliot Formation in the "Red Beds" of South Africa and Lesotho contains what is generally believed to be a swamp-dwelling fauna, while the upper part (which in any case is now believed by some to be of Early Jurassic age; see Olsen & Galton, 1977) shows cyclic deposition of a type which indicates occasional torrents. Robinson herself discussed the environments of three widely-separated and differing continental formations of Late Triassic age, and made it clear that none of them was especially arid. She believed (1971: 146) that:

"Evidently the Maleri formation [of the Deccan of India] represents a monsoon-type climate, with fairly high year-round temperatures, and a dry season alternating with a season of abundant rainfall. That this was well-watered country, at least seasonally, is suggested by the locally abundant unionids, and by the presence of three aquatic and two semi-aquatic members of the vertebrate fauna. Probably these members of the Maleri fauna passed through the dry season in the more permanent bodies of water in the region, such as the deeper pools along watercourses."

On the same page she wrote of the Lossiemouth and Findrassie Sandstones of Elgin in Scotland "Probably this was a terrain of water-courses with very dry interchannel areas", and referred several times to the river or its banks, near which latter various members of the fauna made their homes. She described (1971: 147) the Ischigualasto Formation of western Argentina as partly "possibly river channel deposits" but mostly "suggestive of lake-bottom deposits", and further stated that "Cursory examination of the sediments suggests that this was a region with more sustained rainfall in Upper Triassic times, lacking the dry season . . ."

Another point worth considering is that most of the Late Triassic herbivores were dinosaurs with fairly simple teeth, not suitable for dealing with tough xerophilic vegetation. This could well indicate the prevalence of reasonably moist conditions over a good part of the land surface of the globe.

Robinson also suggested (1971: 143) that the droughts tended to be located in the more western areas of the supercontinental land-masses. Again, it may be presumed that any climatic changes took place in a fairly gradual manner – over many thousands, if not millions, of years – so that the theropsid stock would have had plenty of time either to adapt itself to the changing conditions or to alter its pattern of geographical distribution accordingly. Finally, it is again difficult to see how a change which (so it was claimed) arose fairly late in the Triassic could have been a causal agent – let alone *the* causal agent – of a major change in the vertebrate fauna when the more important of the two phases of that faunal replacement had already been completed before the Late Triassic had even begun. In short, the supposed general aridity of the Late Triassic climates – though it may well have had some effect upon the pattern of evolution – could hardly have been a causal factor of the sauropsid (more precisely archosaur) take-over. It was too seasonal, too regional, too gradual, and certainly much too late.

The present author is not competent to criticize Robinson's beliefs on the causes of the changes in climate between Permian and Late Triassic times. In any case, it is sufficient for present purposes merely to consider whether or not those changes actually took place – as manifested by the nature of the sediments – for their causes are irrelevant to the arguments concerning the changes in the fauna.

Combination Theories

Various authors have proposed explanations of the therapsid-archosaur replacement that include both "anatomical" and "physiological" components (see Fig. 2, p. 605). The main anatomical element consists always of the improvements to the locomotor apparatus.

Locomotion/ectothermy: Benton's explanation

Benton (1979b) cited as the causes of the faunal replacement the archosaurs' locomotor improvements together with their alleged ectothermy; like the present author, he recognized that the replacement and radiation had occurred in two "waves", each with a different causal agent. The first wave was among the carnivores, effected by the archosaurs' improved gait and jaw action; the second was among the herbivores, effected by the ectothermal archosaurs' superior abilities to survive on reduced quantities of plant food and water during the hotter, more arid Late Triassic when the areas of lush lowland vegetation were diminished. Those alleged causes of the second wave are somewhat similar to the causes envisaged by Geist (1972). Benton's ideas, of course, are subject to the same critical comments on adaptations to desert life and on the supposed climatic changes as those made above (pp. 613–616, 620–622) in relation to Robinson's theories (1971). They are certainly not the same as the ideas of the present author which rely neither on alleged climatic changes nor on alleged differences in thermal physiology; on the contrary, the scenario presented on pp. 606–607 is nothing more or less than the most parsimonious interpretation of the stratigraphical distribution of the various groups as observed in the fossil record (see Fig. 1).

Locomotion/endothermy: Bakker's/Desmond's explanation

As stated above, most of Bakker's works cite the archosaurs' locomotor improvements together with their alleged endothermy (in thecodontians as well as dinosaurs). Desmond (1975) followed Bakker in this, his locomotor element including the alleged incipient bipedality of even the early archosaurs.

Locomotion/uricotely: Hotton's explanation

Yet another combination was proposed by Hotton (1980), who believed that the major factors in the archosaurs' acquisition of dominance were improved locomotion (hence their high vagility) and uricotely. Neither he nor Bakker seems to have known of Robinson's article (1971); yet, like Robinson, Hotton stressed (1980: 316–317) the effectiveness of uricotely in water conservation, the derived nature of uricotely, and the probability that dinosaurs were uricotelic and therapsids ureotelic. Unlike Robinson, he did not mention tolerance of high temperatures, but he agreed that dinosaurs were "inertial homoiotherms" (see immediately below).

Other Explanations

Inertial homoiothermy: explanation by Colbert et al.

Colbert, Cowles & Bogert (1946, 1947) suggested that the dinosaurs were what we now call "inertial homoiotherms" (McNab & Auffenberg, 1976); this means that they were able to maintain a fairly constant, fairly high body temperature without endothermy by virtue of their large size and the stable, warm climatic conditions in which they lived. We may therefore presume that they were able to enjoy the advantages of homoiothermy, including the ability to remain active at all times, without having to suffer the disadvantages of endothermy (such as a high requirement of food, the difficulty of losing excess endogenous heat and the consequent need to restrict activity or to endure heat stress). These ideas were accepted by Cox (1967), who, contrary to what was stated by Benton (1979b: 991), did not postulate that dinosaurs were endotherms. A similar line of thought was followed by Spotila, Lommen, Bakken & Gates (1973).

All the publications cited refer only to the dinosaurs, in competition with the mammals. The fossil record, however, makes it abundantly clear that the competitors were on one side the thecodontians (and perhaps, towards the end, the earliest dinosaurs) and on the other the therapsids; the mammals were certainly not involved. By the time that dinosaurs and mammals were co-existing (Early Jurassic) the two lineages had come to occupy completely different adaptive zones. It is therefore manifest that these suggestions cannot explain the therapsid-archosaur replacement of the Middle to Late Triassic. In any case, many dinosaurs (especially juveniles) were not particularly large and could not have been inertial homoiotherms.

Insectivory: Watson's explanation

Watson (1957: 390–391) took an opposite point of view. He attributed the success of the sauropsids, which of course included the archosaurs, to their *small* size; their dimensions fitted them to eat insects, which were likely to have been the main herbivorous creatures in Triassic times and therefore freely available. At the same time he believed that those small sauropsids would not have had an opportunity of spreading to other, more difficult modes of life until the exploitation of all land habitats by the theropsids had been reduced in scale. Such a scenario is totally at variance with the facts as presented by the fossil record and does not merit further consideration.

CONCLUSIONS

It cannot be disputed that the archosaurs were the victors in the Triassic faunal competition. But how far, one might ask, do the various explanations proposed for their success satisfy the general requirements for interpretations and scenarios laid down on p. 601? Quite simply, none of the "physiological" explanations is sufficiently well founded on unimpeachable fact. All rely upon an alleged difference between therapsids and archosaurs in their type of thermal physiology, in the chemical composition of their main nitrogenous excretory product or in their degree of tolerance of potentially lethal high temperatures. But in no case have we any real evidence of those hypothesized characteristics, generally presumed – where other evidence is lacking – to be the same as in those present-day animals to which the extinct forms are supposed to be related. First, the extrapolation of highly adaptable physiological characteristics from modern creatures to their distant relatives (directly ancestral or collateral) of more than 200 My ago is too dubious to be considered a scientific process; secondly, the phylogenetic relationships concerned are themselves far from certain. The subjective nature of the argument is amply demonstrated by the conflicting conclusions reached by different authors; witness, for example, the disagreement mentioned above (pp. 607–608) between Robinson (1971) and Bakker (1975) as to whether Triassic therapsids were ectotherms or endotherms, the same authors differing likewise with regard to Triassic archosaurs. The "physiological" explanations depend also upon certain extrinsic conditions, e.g. of temperature, aridity, food supply and so on, which are usually stated in too general a manner ("World climate was moderating in the Triassic", Bakker, 1975: 70) and for which the evidence is often far from satisfactory. Thus the explanation ultimately depends upon the conjectured effects of conjectured environmental conditions upon conjectured physiological characteristics. One conjecture might be permissible in a scientific argument, but not three; three take us into a world of pure fantasy, so speculative that meditating upon it can be nothing more than a waste of the scientist's time.

In contrast to these "physiological" explanations we have their "anatomical" counterparts, which depend fundamentally upon (a) the tangible remains of the Triassic animals and the functional interpretation of those remains (concerning which there is generally very little disagreement) and (b) the conjectured functional superiority of one type of skeleton over another in the conditions of the time. Admittedly, we are again unsure of those conditions, but they are far less important here when assessing the relative merits of, say, one locomotor system or one type of dentition against another than when comparing, say, ectothermy with endothermy or ureotely with uricotely. (Neither category of explanation is likely to be affected to any significant extent by disagreements as to the correct correlation of the fossiliferous strata.)

Thus we know, without a shadow of doubt, that most of the Triassic archosaurs were evolving a type of locomotor system that was generally superior to those of other contemporary terrestrial tetrapods. It is true that certain members of the therapsid-mammal lineage were beginning to evolve a somewhat similar type of improved locomotor system; but the evidence indicates that they did so far more slowly, so that in Triassic times the archosaurs as a whole were much better developed in that respect. What we cannot be sure of is how significant was this locomotor improvement, compared with other factors, in the archosaurs' acquisition of dominance. But in the absence of good evidence of other significant factors it is reasonable to suppose that the improved locomotor system, and the concomitant increase in size which it made possible, played important rôles in bringing about the faunal replacement. It is possible that other factors, mainly physiological, played equally important or even more important parts, but without real evidence of them we can do no more than indulge in idle speculation.

ACKNOWLEDGEMENTS

I am grateful to Dr A. W. Crompton (Museum of Comparative Zoology, Harvard University) and to Dr Max K. Hecht (Queen's College, City University of New York) for their critical reading of my typescript.

REFERENCES

Bakker, R. T. (1968). The superiority of dinosaurs. *Discovery, New Haven, Conn.* **3**(2): 11–22.
Bakker, R. T. (1971). Dinosaur physiology and the origin of mammals. *Evolution, Lawrence, Kans.* **25**: 636–658.
Bakker, R. T. (1972). Anatomical and ecological evidence of endothermy in dinosaurs. *Nature, Lond.* **238**: 81–85.
Bakker, R. T. (1975). Dinosaur renaissance. *Scient. Am.* **232**(4): 58–78.
Bakker, R. T. (1980). Dinosaur heresy – dinosaur renaissance: why we need endothermic archosaurs for a comprehensive theory of bioenergetic evolution. In *A cold look at the warm-blooded dinosaurs*: 351–462. Thomas, R. D. K. & Olson, E. C. (Eds). (*Select. Symp. Am. Ass. Advmt Sci.* **28**.) Boulder, Colorado: Westview Press.
Bartholomew, G. A. (1968). Body temperature and energy metabolism. In *Animal function: principles and adaptations*: 290–354. Gordon, M. S. (Ed.). London: Macmillan.
Benton, M. J. (1979a). Ecological succession among Late Palaeozoic and Mesozoic tetrapods. *Palaeogeogr., Palaeoclimat., Palaeoecol.* **26**: 127–150.
Benton, M. J. (1979b). Ectothermy and the success of dinosaurs. *Evolution, Lawrence, Kans.* **33**: 983–997.
Benton, M. J. (1983a). Large-scale replacements in the history of life. *Nature, Lond.* **302**: 16–17.
Benton, M. J. (1983b). Dinosaur success in the Triassic: a noncompetitive ecological model. *Q. Rev. Biol.* **58**: 29–55.
Benton, M. J. (1983c). The age of the rhynchosaur. *New Scient.* **98**: 9–13.
Bonaparte, J. F. (1976). *Pisanosaurus mertii* Casamiquela and the origin of the Ornithischia. *J. Paleont.* **50**: 808–820.
Bonaparte, J. F. (1978). El Mesozoico de America del Sur y sus tetrapodos. *Op. lilloana* **26**: 1–596.
Bonaparte, J. F. (1982). Faunal replacement in the Triassic of South America. *J. vert. Paleont.* **2**: 362–371.
Campbell, J. W. (1973). Nitrogen excretion. In *Comparative animal physiology*, 3rd Edn, 279–316. Prosser, C. L. (Ed.). Philadelphia, London and Toronto: W. B. Saunders Company.
Casamiquela, R. M. (1960). Noticia preliminar sobre dos nuevos estagonolepoideos argentinos. *Ameghiniana* **2**: 3–9.
Casamiquela, R. M. (1967). Un nuevo dinosaurio ornitisquio triásico (*Pisanosaurus mertii*, Ornithopoda) de la Formación Ischigualasto, Argentina. *Ameghiniana* **5**: 47–64.
Charig, A. J. (1965). Stance and gait in the archosaur reptiles. Unpublished preprint of lecture given to Section C (Geology), Br. Ass. Advmt Sci., 7 September, Cambridge (quoted verbatim in Desmond, 1975: 218, with acknowledgement).
Charig, A. J. (1966). Stance and gait in the archosaur reptiles. *Liaison Rep. Commonw. geol. Liaison Off.* **86**: 18–19.
Charig, A. J. (1972). The evolution of the archosaur pelvis and hind-limb: an explanation in functional terms. In *Studies in vertebrate evolution (essays presented to Dr F. R. Parrington, F.R.S.)*: 121–155. Joysey, K. A. & Kemp, T. S. (Eds). Edinburgh: Oliver & Boyd.
Charig, A. J. (1975). Rise of the dinosaurs. In *Before the ark*: 98–113. Charig, A. J. & Horsfield, C. M. B. London: BBC.

Charig, A. J. (1979). *A new look at the dinosaurs*. London: Heinemann/British Museum (Natural History).
Charig, A. J. (1980). Differentiation of lineages among Mesozoic tetrapods. *Mém. Soc. géol. Fr.* (n.s.) **139**: 207–210.
Charig, A. J. (1982). Problems in dinosaur phylogeny: a reasoned approach to their attempted resolution. *Geobios, Mém. spéc.* **6**: 113–126.
Charig, A. J. (In prep.). The Triassic explosion in tetrapod evolution and the origin of dinosaurs.
Cloudsley-Thompson, J. L. (1971). *The temperature and water relations of reptiles*. Watford: Merrow.
Colbert, E. H., Cowles, R. B. & Bogert, C. M. (1946). Temperature tolerances in the American alligator and their bearing on the habits, evolution, and extinction of the dinosaurs. *Bull. Am. Mus. nat. Hist.* **86**: 327–373.
Colbert, E. H., Cowles, R. B. & Bogert, C. M. (1947). Rates of temperature increase in the dinosaurs. *Copeia* **1947**: 141–142.
Cox, C. B. (1967). Changes in terrestrial vertebrate faunas during the Mesozoic. In *The fossil record*: 77–89. Harland, W. B. *et al.* (Eds). London: Geological Society of London.
Crompton, A. W. (1968). The enigma of the evolution of mammals. *Optima* **18**(3): 137–151.
Crompton, A. W., Taylor, C. R. & Jagger, J. A. (1978). Evolution of homeothermy in mammals. *Nature, Lond.* **272**: 333–336.
Desmond, A. J. (1975). *The hot-blooded dinosaurs: a revolution in palaeontology*. London: Blond & Briggs.
Friedman, M. & Byers, S. O. (1948). Uric acid excretion, Dalmatian dog. *J. biol. Chem.* **175**: 727–735.
Gardiner, B. G. (1982). Tetrapod classification. *Zool. J. Linn. Soc.* **74**: 207–232.
Geist, V. (1972). An ecological and behavioural explanation of mammalian characteristics, and their implication to therapsid evolution. *Z. Säugetierk.* **37**: 1–15.
Gordon, M. S. (1968). Water and solute metabolism. In *Animal function: principles and adaptations*: 230–289. Gordon, M. S. (Ed.). London: Macmillan.
Heilbrunn, L. V., Harris, D. L., Le Fevre, P. G., Wilson, W. L. & Woodward, A. A. (1946). Heat death, heat injury, and toxic factor. *Physiol. Zoöl.* **19**(4): 404–429.
Hotton, N., III. (1980). An alternative to dinosaur endothermy: the happy wanderers. In *A cold look at the warm-blooded dinosaurs*: 311–350. Thomas, R. D. K. & Olson, E. C. (Eds). (*Select. Symp. Am. Ass. Advmt Sci.* **28**.) Boulder, Colorado: Westview Press.
Huene, F. von (1936). *Die fossilen Reptilien des südamerikanischen Gondwanalandes*. Munich: Beck'sche Verlag.
Jenkins, F. A. jr. (1970). Cynodont postcranial anatomy and the "prototherian" level of mammalian organization. *Evolution, Lawrence, Kans.* **24**: 230–252.
Jenkins, F. A. jr. (1971a). The postcranial skeleton of African cynodonts: problems in the early evolution of the mammalian postcranial skeleton. *Bull. Peabody Mus. nat. Hist.* **36**: i–x, 1–216.
Jenkins, F. A. jr. (1971b). Limb posture and locomotion in the Virginia opossum (*Didelphis marsupialis*) and in other non-cursorial mammals. *J. Zool., Lond.* **165**: 303–315.
Khalil, F. & Haggag, G. (1955). Ureotelism and uricotelism in tortoises. *J. exp. Zool.* **130**: 423–432.
McNab, B. K. & Auffenberg, W. (1976). The effect of large body size on the temperature regulation of the Komodo dragon, *Varanus komodoensis*. *Comp. Biochem. Physiol.* **55A**: 345–350.
Olsen, P. E. & Galton, P. M. (1977). Triassic-Jurassic tetrapod extinctions: are they real? *Science, N.Y.* **197**: 983–986.

Ostrom, J. H. (1970). Terrestrial vertebrates as indicators of Mesozoic climates. *Proc. N. Am. paleont. Convention* (D) **1969**: 347–376. Lawrence, Kans.: Allen Press Inc.

Patterson, C. (1980). Cladistics. *Biologist* **27**: 234–240.

Prosser, C. L. (1950). Various chapters in *Comparative animal physiology*. Prosser, C. L. (Ed.). Philadelphia and London: W. B. Saunders Company.

Reig, O. A. (1970). The Proterosuchia and the early evolution of the archosaurs; an essay about the origin of a major taxon. *Bull. Mus. comp. Zool. Harv.* **139**: 229–292.

Robinson, P. L. (1971). A problem of faunal replacement on Permo-Triassic continents. *Palaeontology* **14**: 131–153.

Spotila, J. R., Lommen, P. W., Bakken, G. S. & Gates, D. M. (1973). A mathematical model for body temperatures of large reptiles: implications for dinosaur ecology. *Am. Nat.* **107**: 391–404.

Thomas, R. D. K. & Olson, E. C. (Eds) (1980). *A cold look at the warm-blooded dinosaurs. (Select. Symp. Am. Ass. Advmt Sci.* **28**.) Boulder, Colorado: Westview Press.

Thulborn, R. A. (1975). Dinosaur polyphyly and the classification of archosaurs and birds. *Aust. J. Zool.* **23**: 249–270.

Tucker, M. E. & Benton, M. J. (1982). Triassic environments, climates and reptile evolution. *Palaeogeogr., Palaeoclimat., Palaeoecol.* **40**: 361–379.

Watson, D. M. S. (1957). On *Millerosaurus* and the early history of the sauropsid reptiles. *Phil. Trans. R. Soc.* (B) **240**: 325–400.

Zhang, F.-K. (1975). [A new thecodont *Lotosaurus*, from Middle Triassic of Hunan.] *Vertebr. palasiat.* **13**: 144–147. [In Chinese, with English summary.]

The Histology of Dinosaurian Bone, and its possible bearing on Dinosaurian Physiology

R. E. H. REID

Department of Geology, The Queen's University of Belfast, Belfast BT7 1NN, Northern Ireland

SYNOPSIS

Current knowledge of dinosaurian bone histology is based essentially on casual sampling, and much systematic work will be needed before a full picture can be given. In known cases, the primary compact bone is usually of fibrolamellar types; but bone with "growth rings" (zones and annuli) may also occur. Haversian bone was often developed to mammalian levels, but was not as extensively developed as some authors have implied. In some cases, its local development can be related to muscular attachments. Endochondral bone can be found in expected positions, and its formation from cartilage demonstrated. Secondary cancellous bone is widespread, and may show brecciate structure. Compacted coarse-cancellous bone, metaplastic bone and pathological types also occur. Dinosaurs with fibrolamellar bone had a capacity for sustained rapid growth, which is not shown by modern types of reptiles, and was probably not simply due to inertial homoiothermy; but the presence of similar bone in early therapsids is a warning that it should not be assumed to be evidence of endothermy.

INTRODUCTION

Dinosaurian bone has been studied microscopically since Mantell (1850) saw Haversian canals and "bone cells" (osteocyte lacunae) in the sauropod *Pelorosaurus*, and has been used in several studies which made positive contributions to bone histology. Hasse (1878), for instance, used material from *Thecodontosaurus* for some of his early illustrations of endochondral ossification in fossil reptiles; and Seitz (1907) made extensive use of dinosaurs in firmly establishing the distinction between the primary and secondary types of vascular canals. *Plateosaurus* and *Brachiosaurus* were two of the four reptiles used by Gross (1934) to illustrate his concepts of laminar bone and primary osteons; and prosauropod material was used by Currey (1962) in one of his studies on vascularity in laminar bone. Broili (1922) and Moodie (1928) made early studies of ossified tendons, and Moodie (1923) was able to recognize a range of pathological conditions. Dinosaurs are also of interest because their bone is often similar to that of large mammals, and unlike that of most other reptiles (Gross, 1934; Currey, 1960, 1962; de Ricqlès, 1976); and this has led some authors (e.g. Bakker, 1972, 1975; de Ricqlès, 1974, 1976, 1980) to see its structure as evidence of high metabolic rates.

This paper gives an outline of the nature of dinosaurian bone, as known to the present author from literature and personal studies, and also briefly discusses its possible physiological significance. The present author agrees with de Ricqlès (1976, 1980) that the primary compact bone was most commonly of fibrolamellar types, and that the presence of such bone shows that dinosaurs could sustain rapid growth to large sizes. This ability suggests at least some physiological difference between them and modern types of reptiles, which cannot sustain rapid growth. This difference does not appear to have been simply a result of inertial homoiothermy. However, bone with typical "growth rings" (zones and annuli) can also be found in dinosaurs; and fibrolamellar bone is known from early types of therapsids (deinocephalians, titanosuchians) whose status as endotherms is doubtful. Hence the evidence from bone is not thought to show that dinosaurs were endotherms.

EXTENT OF DATA

Although current records (e.g. de Ricqlès, 1980: table 1) allow the recognition of such general features as the widespread occurrence of fibrolamellar and dense Haversian bone, the detailed bone histology of dinosaurs is very inadequately known. To get an overall picture of the bone of any animal, it is necessary to study at least a wide selection of bones, from different parts of the skeleton; to compare different parts of any given bone; and to compare bones from several individuals, of both similar and different known ages. In some bones, serial sectioning may be needed: for example, when structural variations are related to muscular attachments. Also, in any group of animals, enough genera need to be studied to be fully representative. In contrast, most published data on dinosaurs are based on casual sampling of single bones, often figured from single sections, and many records refer to long bones only. In addition, most records are from theropods, sauropodomorphs and ornithopods, with few from other groups, and many genera have never been sampled. Such haphazard data cannot yield a complete or balanced picture, neither are they an adequate basis for firm physiological conclusions, even if some types of bone are supposed to be significant physiologically.

Material available to the present author has also been very limited, with only *Iguanodon* represented by a fair range of different bones (femur, tibia, radius, ulna, humerus, undetermined limb bones, ribs, scapula, ischium, centra and other parts of dorsal and caudal vertebrae, chevron bones, skull fragments), and other genera by a few bones or single bones [samples from limb bones, ribs, pelvic bones and vertebrae of sauropods; ribs, limb bones and a pubis from megalosaurs; ribs of *Tyrannosaurus*; a femur of *Aristosuchus*; femora of *Hypsilophodon* and *Valdosaurus*, and other bones ascribed to hypsilophodontids; limb bones and ribs of *Rhabdodon*; ribs of *Anatosaurus* ("*Trachodon*"); and various other bones, not identified except as dinosaurian]. The surfaces of a number of other bones were also examined, including some from additional genera (e.g. *Ornithomimus*). No material from ankylosaurs, ceratopians, pachycephalosaurs or stegosaurs was available, and no osteoderms or armour were studied. Hence the author's account of dinosaurian bone should be read as an introduction only, dealing simply with types of bone currently known to occur, and allowing discussion of claims that some are significant physiologically. It is not an attempt at a comprehensive study, for which much more information would be needed.

THE BONE OF DINOSAURS

General

Well preserved dinosaurian bones show considerable histological detail, including osteocyte lacunae and canaliculi (Fig. 1), resting and reversal lines, the presence or absence of Sharpey's fibres, and detail of fibrillar organization when studied with crossed polarizers. Other examples show less detail, owing to diagenetic changes, or (Moodie, 1928) to early fungal invasion. Some loss of detail may also have been due to micropetrosis in life. The retention of characteristic birefringence patterns has been ascribed to mineral replacement at submicroscopic level (Cook, Brooks & Ezra-Cohn, 1962), but some bones seem still to contain traces of organic material.[1]*

Predictably, dinosaurian bone includes compact and cancellous types, and bone of endochondral, periosteal and endosteal origins. Metaplastic (= heteromorphic) bone may also occur. Primary (unremodelled) compact bone is most commonly of fibrolamellar types (e.g. Figs 2, 3 & 4); but lamellar-zonal bone may also occur, as surface bone only (Fig. 7) or more extensively (e.g. Figs 10, 12 & 15). Replacement of primary compact bone by dense Haversian bone (Fig. 20) is

* Superscript numbers refer to notes added in proof at the end of the chapter.

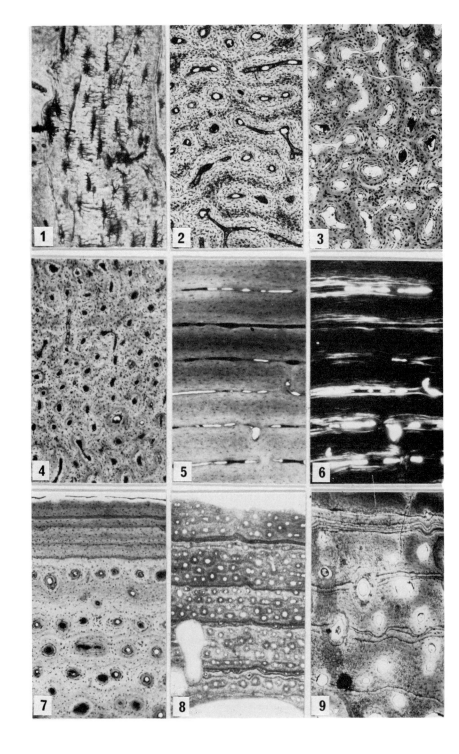

often extensive, and sometimes total; but it may also be restricted to certain parts of bones, or lacking, although scattered secondary osteons may then occur. In some cases, its local development can be related to muscular attachments. Endochondral bone (Figs 27 & 28) can be found in expected situations, and secondary cancellous bone (Figs 29–32) is common. Primary (non-replacive) lamellar bone is seen mainly in the osteons of fibrolamellar bone (Fig. 2), but may also be present in zonal tissues. Secondary (replacive) lamellar bone occurs as compact secondary osteons (or Haversian systems; Figs 18 & 20), in secondary cancellous bone (Fig. 32) or as endosteal bone lining medullary cavities (fig. 35) or large resorption canals. Non-lamellar (fibrous, coarsely bundled, woven) bone occurs as a matrix in fibrolamellar tissues (Fig. 2), and in various other situations. Metaplastic bone is best known from ossified tendons, but may also be present in superficial parts of limb bones (Fig. 37).

In broad terms, various studies (e.g. Gross, 1934; Enlow & Brown, 1956, 1957, 1958; Currey, 1960, 1962; de Ricqlès, 1976, 1980) have shown that dinosaurian bones are often similar to those of mammals histologically, and unlike those of other kinds of reptiles except therapsids and some thecodontians. On the other hand, bone containing typical growth rings was sometimes developed (Figs 8, 9, 10 & 11); and, in material from *Rhabdodon* figured here (Figs 12, 13 & 14), the primary compact bone was similar to that usually found in crocodiles. Such conditions may have been commoner than is currently known.

FIGS 1–9. Osteocytes; fibrolamellar bone; lamellar-zonal bone.
(1) Osteocyte lacunae and canaliculi, as seen when infilled with dark minerals; × 125. From lamellar endosteal bone, lining the medullary cavity of an *Iguanodon* femur. Author's collection.
(2) Laminar fibrolamellar bone, showing the woven periosteal framework (as dark trabeculae) and primary osteons; × 50. Sauropod (*"Cetiosaurus"*) limb bone indet. OUM J29835/p2.
(3) Reticulate fibrolamellar bone, × 50. *Valdosaurus*, femur. B.M.(N.H.) R.8421.
(4) Fibrolamellar bone with mainly parallel primary osteons; × 50. *Tyrannosaurus*, rib. B.M.(N.H.) R.7995.
(5) Primary compact bone with circumferential vascular networks as in laminar bone, but no true woven framework, which is replaced by parallel-fibred bone; × 50. *Iguanodon*, femur. Author's collection.
(6) The bone shown in Fig. 5, as seen with crossed polarizers, showing the presence of an osteon system.
(7) Avascular lamellar-zonal bone (at top) developed as surface bone, outside fibrolamellar bone with parallel primary osteons; × 50. Most of the primary canals also show slight reconstruction. *Iguanodon*, basal part of a chevron bone. B.M.(N.H.) R.5331.
(8) Lamellar-zonal bone with zones and annuli, developed locally in an *Iguanodon* vertebra, in other parts of which the cortex is laminar (Fig. 30); × 25. Large spaces at the bottom of the figure are resorption lacunae, produced at the start of conversion of compact bone into cancellous bone. B.M.(N.H.) R.6311.
(9) Lamellar-zonal bone with zones and annuli, seen in the superficial parts of a megalosaurid rib; × 50. Sandown Museum, I.O.W. 5309.
(B.M.(N.H.) = British Museum (Natural History); OUM = Oxford University Museum.)

Compact Bone

The compact bone of dinosaurs is typically highly vascular (e.g. Figs 2, 10 & 20), and may show a higher vascularity than in mammals of comparable size (Currey, 1962). As in large mammals with similar bone, this high vascularity was fundamentally a primary feature, and not a result of reconstruction as Ostrom (1981: 26) implies.

For the detailed description of the primary compact bone of dinosaurs, it is convenient to follow de Ricqlès (1974, 1975) in distinguishing two principal types.

Fibrolamellar Bone

This term (de Ricqlès, 1974: 53; 1975: 88–92) designates bone of the types containing primary osteons described by Smith (1960: 331–333), including laminar bone *sensu* Gross (1934), and tissues which are similar histologically but have the osteons differently arranged. A finely cancellous framework of non-lamellar (typically woven) bone is first laid down periosteally, enclosing networks of small blood vessels which run through the cancelli but do not fill them. Internally deposited lamellar bone then grows inward progressively towards the blood vessels, to form the primary osteons, which may form the major part of the bone when its growth is complete (e.g. Fig. 2). When they surround single vascular canals, the primary osteons resemble typical Haversian systems; but their formation involves no resorption of pre-existing bone. As a result, they show no peripheral reversal lines, and never encroach on one another like the secondary osteons of dense Haversian bone (e.g. Fig. 18). The arrangement of the osteons is determined by the form of the spaces which are enclosed initially, which may form an irregular labyrinth, or be longitudinal tunnels, or radial or circumferential clefts or both together. The main patterns which result are distinguished by de Ricqlès (1980: table 1) as reticular, parallel-osteoned, radiate, laminar and plexiform, although intermediates also occur.

Fibrolamellar bone is also typically fast-growing, and formed without cyclical interruption by resting lines or annuli [i.e. thin sheets of the bone described by Smith (1960) as surface bone], although irregular interruptions may occur. Laminar bone *sensu* Gross (1934) (= plexiform bone, Enlow & Brown, 1956, 1957, 1958; laminar and plexiform bone *sensu* de Ricqlès, 1974), as developed in mammals, often shows a special form of cyclicity in the presence of "bright lines" between the laminae (see Currey, 1960: figs 2 & 3); but these are not a general feature of fibrolamellar tissues (cf. Smith, 1960), and do not appear to be related to the annuli of lamellar-zonal bone (de Ricqlès, 1975).

Fibrolamellar bone was first recognized as a feature of dinosaurs by Gross (1934), who pointed out laminar bone from *Plateosaurus* and *Brachiosaurus* as unlike the bone of most other reptiles, but similar to bone from a young mammoth and "numerous ungulates". The resemblance of the laminar bone of dinosaurs to that of artiodactyls was also emphasized by Enlow & Brown (1957), who called it plexiform bone, and by Currey (1960, 1962), who confirmed (1960) Gross's view that the lamellar bone forming the osteons is not replacive. De Ricqlès (1980: table 1) has listed published records which he sees as referring to various fibrolamellar types, from genera including prosauropods, sauropods, theropods, ornithopods and a stegosaur; and typical examples are figured here from a sauropod (Fig. 2), a theropod (Fig. 4) and an ornithopod (Fig. 3). It does not appear to be recorded from ankylosaurs or ceratopians, although its absence in these groups is not implied.

Nominally laminar bone from dinosaurs may, however, depart in two ways from the pattern described by Currey (1960). Although some bone with the characteristic circumferential vascular networks shows a well developed primary framework (Fig. 2), in other cases this may be limited to single lines of unflattened osteocyte lacunae, or be apparently absent (Fig. 5). Bone between the osteon systems then has the characteristics of parallel-fibred bone (*sensu* Weidenreich), although use of crossed polarizers shows that an osteon system is still present (Fig. 6). In a further case (Currey, 1962), no distinction can be made between framework and osteons, and the bone shows only a mixture of lamellar and non-lamellar tissues. Hence caution is needed in interpreting figures as showing true fibrolamellar bone, unless the relevant histological details can be recognized. A different type of departure from the pattern described by Currey (1960) is the absence of "bright lines" (e.g. Fig. 2), so that the pauses in growth which they mark do not seem to have occurred.

Lamellar-zonal bone

This term (de Ricqlès, 1974: 53; 1975: 84–88) designates bone of a variety of types, ranging structurally from simple to complex, which are formed partly or wholly from concentrically lamellated bone of periosteal origin. Some examples are as vascular as fibrolamellar bone (e.g. Fig. 8), while others are moderately or sparsely vascular only (e.g. Fig. 13), and some are avascular. Primary osteons may be present or absent when the bone is vascular, and are inevitably absent when it is avascular. The lamellated bone may be formed from lamellar bone *sensu stricto* or parallel-fibred bone, or show both formed alternately, or show intercalated layers of more coarsely fibrous tissue. The most typical zonal bone is divided into "growth rings" or zones, through the cyclical formation of resting lines marking pauses in growth (Fig. 12), or of thin sheets of lamellar bone termed annuli (e.g. Fig. 9) which mark periods of slowed growth. Sometimes both features are present, with the resting lines seen most commonly outside the annuli, but sometimes under them. These "growth rings" have been regarded as annual by some authors (e.g. Peabody, 1961), and are genuinely so in some cases (e.g. in *Alligator mississippiensis*: Ferguson, Honig, Bringas & Slavkin, 1982; Ferguson, this volume). In the most advanced types, bone in the zones shows primary osteons in a matrix of woven bone (cf. Enlow, 1969: 55–57), and thus has the same type of structure as fibrolamellar bone, although classified as zonal because of being formed cyclically.

Lamellar-zonal bone is less well known as a feature of dinosaurs than fibrolamellar bone, but has been figured by several authors (e.g. Gross, 1934, from *Brachiosaurus*; Enlow, 1969, from *Triceratops*) and mentioned by others (e.g. Nopcsa, 1933, from *Rhabdodon*). From data available to the present author, it seems usually to have made only minor contributions to the skeleton, when present at all; but it also appears to have been commoner than has hitherto been recognized, and it sometimes took the place of fibrolamellar bone. Examples in the present author's material show a number of contrasting developments.

(1) In an *Iguanodon* chevron bone (Fig. 7), thin sheets of finely lamellated bone containing few or no vascular canals were developed as surface bone (cf. Smith, 1960: 334), coating fibrolamellar bone with parallel primary osteons. A similar development of surface bone is known from limb bones of *Brachiosaurus* (Gross, 1934: figs 14, 15, upper parts).

(2) The thin cortex of the centrum in *Iguanodon* vertebrae was usually formed by laminar or plexiform fibrolamellar bone, or locally by bone with parallel osteons; but one dorsal vetebra examined had the outer part of the cortex formed by bone with vascular zones and typical annuli (Fig. 8). The outer parts of a megalosaurid rib also showed zones and annuli (Fig. 9), although deeper parts showed traces of older fibrolamellar bone.

(3) In a sauropod pubis (Reid, 1981), the only primary compacta found was typical zonal bone with well developed annuli (Figs 10 & 11). Whether any fibrolamellar bone was formed initially could not be determined, owing to internal reconstruction; but a section through the shaft showed at least 17 zonal increments, in a part where, judged from their average thickness, the maximum possible would be only 28. Thus, any formation of fibrolamellar bone must have been limited to early life only, if it took place at all.

(4) In the iguanodontid *Rhabdodon*, as represented by the material of Nopcsa (1925), the primary compact bone of the long bones (limb bones, ribs) was typical lamellar-zonal bone, with conspicuous "growth rings" (Figs 12 & 13). As noted by Nopcsa (1933: 221), these may only be seen near the surface; but in material examined (nine bones) their inward disappearance is due either to Haversian reconstruction or to diagenetic degradation. In the best preserved material (e.g. Fig. 12), the pattern can be traced to the endosteal margin. In the outer parts, the bone is formed mainly from parallel-fibred material, with fine lamellation, and all osteocytes flattened circumferentially (Fig. 13). Annuli are not conspicuous with plain lighting, but can be seen with crossed polarizers (Fig. 14). Many primary canals simply interrupt the periosteal bone, without enclosure within primary osteons. In the deeper parts, however, primary osteons may be well developed, and parts of the zones may show a matrix of woven bone.

(5) At mid-diaphysis, a small theropod femur, ascribed to *Aristosuchus* (Galton, 1973), has compacta in the form of a simple finely lamellated tissue without evident "growth rings" (Fig. 15). Most osteocyte lacunae are flattened and aligned circumferentially, and no primary osteons are present. The vascular canals are thinly lined with lamellar bone, but peripheral cementing lines (see p. 640 below) show that this is a result of slight internal reconstruction. Whether only bone of this type was produced could not be determined, because of the presence of a large medullary cavity; but, while some of the osteocytes near its margin were not aligned circumferentially, there is no sign of an inward transition into fibrolamellar bone.

(6) Typical zonal bone, whose original extent could not be assessed, was also found in the superficial parts of a pubis of *Megalosaurus* (Fig. 16), outside dense Haversian bone, and in a caudal centrum of *Iguanodon*.

How frequent such occurrences are in dinosaurs is not currently known, but the discovery of these examples in only limited material suggests that they may be fairly common. More data are needed especially if bone is supposed to have a bearing on dinosaurian physiology. For example, the crocodile-like pattern seen in *Rhabdodon* shows that the primary compact bone of dinosaurs was not always like that of large mammals, even in the limb bones, and other examples may exist among genera which have not been investigated.

Periosteal bone and growth rates

Although two main categories of primary compact bone are distinguished above, it has also been noted that intermediates between them may occur. Apparently fibrolamellar bone, for instance, may lack a true

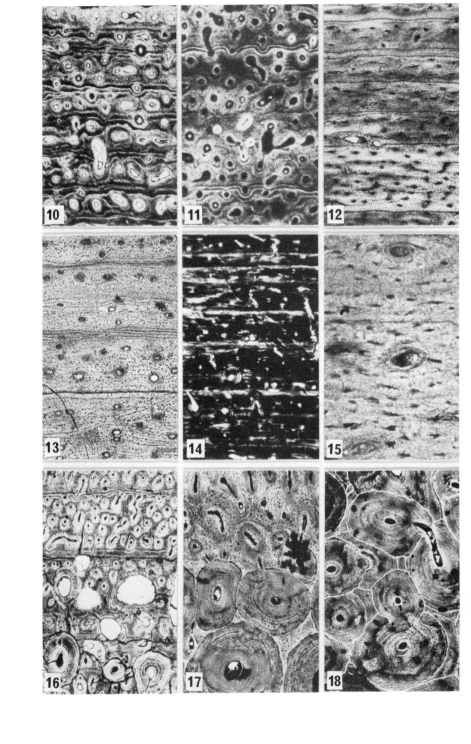

woven matrix, or even show no differentiation into framework and osteons. Conversely, bone from the zones of typical zonal bone may show primary osteons in a matrix of woven bone, and thus may have the same constitution as fibrolamellar bone. Compact bone formed mainly from typical fibrolamellar bone may also show resting lines or annuli in parts near the periosteal surface, marking pauses before growth finally ceased, so that it locally takes on the character of advanced types of zonal bone.

These transitional conditions can occur because the two main types of primary compact bone distinguished by de Ricqlès (1974, 1975) are essentially convenient subdivisions of a continuous spectrum of intergrading histological patterns, and do not differ fundamentally in the same way as, for example, primary and secondary tissues. Variations in the periosteal matrix are related to growth rates, which, in modern forms (Enlow, 1969: 49–50, 55–57), are lowest when the bone which is laid down periosteally is lamellar bone, and highest when it is woven bone. Resting lines mark temporary pauses in growth, while alternating annuli and zones imply a cyclically fluctuating growth rate. The woven

FIGS 10–18. Lamellar-zonal bone; Haversian reconstruction
(10) Lamellar-zonal bone with zones and annuli from the shaft of a sauropod pubis, as developed where growth was slowest; × 50. Undescribed sauropod, distinct from *Cetiosaurus* (A. J. Charig, pers. comm.) B.M.(N.H.) R.9472.
(11) Lamellar-zonal bone from the same pubis, as developed where growth was fastest; × 50. Note that the difference between this bone and that seen in Fig. 10, from the opposite (inward) side of the shaft, is entirely due to different growth rates, and does not reflect lateral drifting.
(12) Lamellar-zonal bone from a femur of *Rhabdodon*, showing zones defined by strong resting lines; × 25. The bone is only moderately vascular, and no primary osteons are present (cf. Fig. 13). Two secondary osteons are developed across the second resting line from the bottom. B.M.(N.H.) R.3839.
(13) Another example of similar bone, showing fine lamellations and the absence of primary osteons; × 50. *Rhabdodon*; limb bone indet. B.M.(N.H.) R.3809.
(14) Similar bone as seen with crossed polarizers, with annuli showing as bright transverse features; × 50. Very similar patterns can be found in bone from crocodiles. *Rhabdodon*; limb bone indet. B.M.(N.H.) R.3816.
(15) Finely lamellated bone with most osteocytes aligned circumferentially, from a femur of *Aristosuchus*; × 250. The thin rings of lamellar bone surrounding the vascular canals are defined externally mainly by reversal lines, and hence mark slight reconstruction. B.M.(N.H.) R.5194.
(16) Haversian reconstruction in zonal bone from a pubis of *Megalosaurus*; × 25. The upper part shows unaltered zonal bone, with many small primary osteons. The central part shows resorption lacunae (or, cross sections of resorption canals), some of which have been lined with secondary lamellar bone. In the lower part, several fully formed Haversian systems (= cross-sections of secondary osteons) are seen. OUM J29797/p1.
(17) Primary and secondary compact bone in a rib of *Tyrannosaurus*; × 50. The primary (fibrolamellar) tissue (cf. Fig. 4) is seen unaltered in the upper part, and as interstitial traces between Haversian systems in the lower part. B.M.(N.H.) R.7995.
(18) Dense Haversian bone from *Tyrannosaurus*, showing bright peripheral cementing lines, and interstitial remnants of older secondary osteons between the last-formed ones; × 50. B.M.(N.H.) R.7995.

matrix of fibrolamellar bone implies rapid growth, and rapid continuous growth when no "bright lines" are present; while its formation as a fine-cancellous framework can be seen as a further adaptation to fast growth (Currey, 1960). Thus, the division of primary compact bone into two major categories is essentially arbitrary; and the principal convenience of those used here is to separate zonal bone with "growth rings" from fibrolamellar types without them.

Although this distinction has no special histological significance, it is relevant if bone is cited in discussion of dinosaurian physiology. In terms of growth rates, a major difference between dinosaurs and modern types of reptiles is the implied ability of those with fibrolamellar bone to sustain rapid growth to large sizes. This has been seen by de Ricqlès (e.g. 1974, 1980) as evidence of high metabolic rates. On the other hand, the bone seen in *Rhabdodon* implies slower and intermittent growth, with a pattern like that now seen in crocodiles.

Haversian bone

In its widest usage (e.g. Bouvier, 1977), this term takes in all endosteally formed lamellar bone which has replaced a pre-existing bone tissue, of endochondral, periosteal or Haversian origin. As used by de Ricqlès (e.g. 1974, 1980) and Bakker, (e.g. 1972), however, it applies to bone formed as typical Haversian systems, and especially to dense Haversian bone (Enlow & Brown, 1956: 410), which de Ricqlès again sees as physiologically significant.

The "typical Haversian systems" seen in sections are cross-sections of cylinders of replacive lamellar bone, known as secondary osteons (Figs 17, 18, 20 & 21), which are produced by a remodelling process known as Haversian reconstruction, replacement or substitution. When this begins, bone surrounding primary vascular canals is resorbed by cells termed osteoclasts to produce enlarged resorption canals, seen in sections as resorption lacunae (Fig. 16). Resorption then ceases and is replaced by deposition of new lamellar bone, which grows inwards to form a cylinder of bone with a vascular canal at its centre. These vascular canals are termed secondary canals, to distinguish them from the primary canals which are formed when blood vessels are enclosed during periosteal growth. Each Haversian system is bounded by a cementing line (e.g. Fig. 18), which is also a reversal line since it marks the outward limit of resorption. Bone left between Haversian systems is called interstitial bone, and initially consists of the unremodelled remnants of the primary compacta (e.g. Fig. 17). Once started, however, the process of reconstruction often continues, so that secondary osteons are formed in successive generations, and replace one another as well as the primary bone. The ultimate result is the tissue called dense Haversian bone (Figs 18, 20 & 21), in which all the interstitial bone between the youngest Haversian systems consists of remnants of older ones.

Haversian bone is recorded from most dinosaurs whose bone has been investigated [cf. de Ricqlès, 1980: table 1; but note that some records ascribed to Nopcsa (1933) are based on a misreading of his text]. The characteristic stages of Haversian reconstruction can often be recognized (Fig. 16); and bodies originally figured as blood cells of *Iguanodon* (Seitz, 1907; fig. 61) were considered to be probable osteoclasts by Moodie (1923) and Swinton (1934). The secondary osteons were usually larger than the primary osteons, as in modern forms, and sometimes they were much larger. For example, in materials from a rib of *Tyrannosaurus* (Fig. 4), no osteocyte in the primary compacta was more than about 0.05 mm from a blood supply; but in the Haversian tissue (Fig. 18) this distance could be up to 0.35 mm.

Haversian reconstruction in dinosaurs was often as extensive as it may be in large mammals, but it was also not as general as some authors have implied. In particular, Enlow & Brown (1957: 202–203) thought that all of the primary canals underwent reconstruction; and Ostrom (1980: 41; 1981: 26) treats all highly vascular bone seen in mammals, birds and dinosaurs as reconstructed. But Enlow & Brown were writing before it was generally recognized that the osteons of fibrolamellar bone are not replacive structures; and high vascularity does not imply that the bone which shows it has been reconstructed. On the contrary, the high vascularity which is usual in dinosaurs is fundamentally a primary vascularity, as in mammals with similar bone, and typically the result of the presence of fibrolamellar bone, which is characteristically highly vascular (e.g. Figs 2–4).

In practice, the amount of reconstruction seen in any given sample can vary considerably, as it does in large mammals. In the present author's material, some samples (e.g. from a rib of *Anatosaurus*, and an ischium of *Iguanodon*) showed only dense Haversian compacta; but most showed both primary and dense Haversian tissue, with either predominant, and some had scattered Haversian systems only. No Haversian systems were seen in a section from a femur of *Aristosuchus*, although this does not mean that they were absent from its other parts. In *Iguanodon*, which provided the widest range of samples, variations were found between different bones, different specimens of a given bone, and different parts of single bones. Diaphyseal sections of limb bones up to 12 cm in diameter showed primary laminar bone predominant, with dense Haversian bone restricted to parts near the medullary cavity or to segments under muscular attachments, or even absent. In contrast, sections through metaphyses, in which the cortex is thin and a medullary spongiosa is present, showed compact bone which was mainly or all Haversian. Ribs showed varying conditions, up to more or less total reconstruction. In vertebrae, only slight reconstruction was found in the cortical parts of centra, but extensive reconstruction was seen in neural spines and other processes. A chevron bone still showed considerable primary compacta in a section near its basal arch, but extensive reconstruction in more distal sections.

Few data are available on the factors which produced such variations, but some surmises are possible. Age was probably relevant, since Gross (1934), for example, found reconstruction more complete in the large Berlin *Brachiosaurus* than in a young individual. Heavy weights due to large sizes could be specially relevant in limb bones, if reconstruction is thought to increase strength (e.g. Ostrom, 1980: 42); but the limited reconstruction seen in *Iguanodon* diaphyses (cf. above), including those of bones (femur, tibia) from the weight-supporting hind-limbs, suggests that weight should not be over-emphasized. The influence of osteocyte necrosis (Enlow, 1962a) is probable but difficult to prove, because of similar appearances produced by diagenetic changes. In bones or parts of bones with a medullary spongiosa, the outward expansion of the marrow may have been relevant, since secondary spongiosa often grades outwards into typical Haversian bone. Differences between diaphyses and metaphyses, also noted by Gross (1934) in *Brachiosaurus*, suggest reconstruction triggered by metaphyseal remodelling. In the present author's material, however, the most surely demonstrable relationship was to muscular attachments in two femora.

This relationship was most clearly seen in a series of sections from the mid-part of a femur of *Iguanodon*, although a similar pattern was found in a section from *Hypsilophodon*. Most sections of *Iguanodon* diaphyses showed little histological evidence of muscular attachments, apart from local concentrations of Sharpey's fibres; but the femur showed dense Haversian bone developed in two conspicuous radial segments (Fig. 19). One of these segments runs radially inward from the tip of the large fourth trochanter, and the other obliquely inward from a flattened and roughened lateral surface. The margins of these segments of Haversian bone are strikingly abrupt (Fig. 22), and the trochanteric segment disappears below the tip of the trochanter, between sections only 6 mm apart. Only scattered Haversian systems are seen in the intervening segments, and most of these are near the endosteal margin. In the trochanteric segment, many secondary osteons show an oblique radial arrangement, suggesting alignment with the pull of the caudifemoralis muscles which are usually attached here. In the lateral segment, by contrast, they run obliquely downward, and interstitial bone between

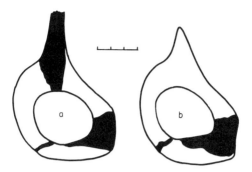

FIG. 19. Distribution of dense Haversian bone (shown black) in two mid-diaphyseal sections of an *Iguanodon* femur. (a) Section through the tip of the fourth trochanter. (b) Section 6 mm lower. Scale line 3 cm long. Remaining bone fibrolamellar, mainly laminar.

them is densely packed with Sharpey's fibres with a similar alignment. These features confirm a relationship to a muscular attachment, which, from its position, should have been that of a femorotibialis muscle. In addition, the most distal sections showed the Haversian bone underlain by an expanding wedge of primary laminar bone; this is consistent with a downward extension of the area of attachment during growth.

In this light, it can probably be assumed that at least some Haversian replacement in metaphyses was related to the maintenance of muscular attachments during growth and external remodelling (Enlow, 1962a: 275, 277–278; 1969: 59, 74). No relationships to specific muscles were demonstrable in the present author's material; but several metaphyseal sections showed superficial or intersitital bone densely packed with Sharpey's fibres (Fig. 36), and partly Haversian tissue with a matrix of this sort was found extending across incised resorption surfaces.

The generally high level of Haversian reconstruction in dinosaurs, compared with most other reptiles, has been seen by de Ricqlès (e.g. 1974, 1980) as due to endothermy; but others (e.g. McNab, 1978; Hotton, 1980) relate it to inertial homoiothermy, and the influence of high primary vascularity is another possibility.

Cancellous Bone

Cancellous bone may be formed (i) by endochondral ossification of cartilage, (ii) by replacement of endochondral bone, or any type of compact bone, in a process essentially similar to Haversian reconstruction, and (iii) in reptiles (Enlow, 1969), by internal resorption of compact bone, without further reconstruction. Only the first two types appear to have been usual in dinosaurs.

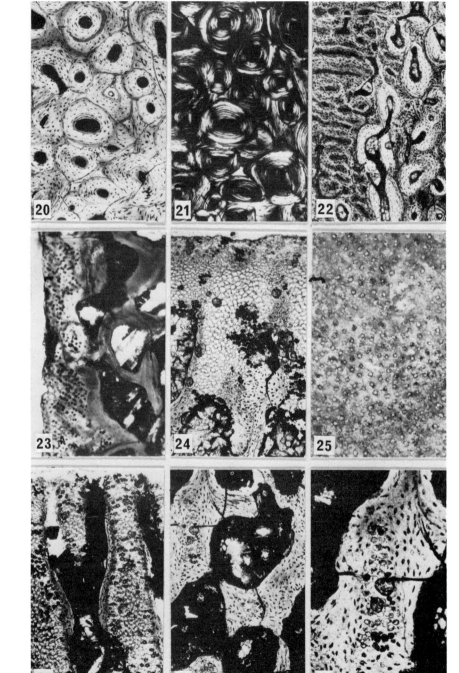

Endochondral bone

This type of bone is called endochondral because its formation involves the replacement of cartilage. In early development, it is the first type of bone formed in all bones which ossify from cartilaginous prototypes. In later life, it is typically formed at epiphyses and in similar situations, e.g. synchondroses. Cartilage near the zone of bone formation becomes hypertrophic and calcified, and shows large globular chondrocytes which are often arranged in columns. Osteoclast-like cells in the marrow attack this cartilage from below, and excavate cavities which marrow processes occupy. The first endochondral bone is then laid down on the walls of these cavities. As growth proceeds, some of this bone is then usually resorbed, while more bone is deposited, to produce a final pattern of trabeculae. Calcified cartilage may persist as small "islands" within the endochondral trabeculae, or, less commonly, core them continuously. Such inclusions identify cancellous bone as endochondral, since they do not occur in secondary (fully reconstructed) trabeculae. The endochondral trabeculae may persist for a considerable distance from the growth zone, or be more rapidly destroyed by reconstruction or by the extension of a medullary cavity.

In the present author's material, the formation of endochondral bone was seen at the articulatory surfaces of *Iguanodon* centra, and at the epiphyses of various limb bones from *Iguanodon*, *Hypsilophodon* and

FIGS 20–28. Dense Haversian bone; endochondral bone; calcified cartilage.
(20) Dense Haversian bone, showing the arrangement of the osteocyte lacunae in the secondary osteons; × 50. Indet. limb bone (? sauropod). Author's collection.
(21) The bone shown in Fig. 20, as seen with crossed polarisers; × 50. The fine light and dark lamellations identify the bone in the osteons as lamellar bone. The dark diagonal "axial crosses" are an optical effect, marking axes of total extinction.
(22) An abrupt radial junction between primary laminar bone (on left) and dense Haversian bone (on right), from the left hand side of the trochanteric wedge of dense Haversian bone shown in Fig. 19A; × 50. *Iguanodon*, femur. Author's collection.
(23) Calcified cartilage (on left) and endochondral cancellous bone (on right) from the centrum of an *Iguanodon* vertebra; × 50. The chondrocyte lacunae (some marked by dark mineral infilling) are arranged in rows inclined outwards towards the margin of the articulatory face. Author's collection.
(24) Calcified cartilage with densely packed chondrocyte lacunae, which are not arranged in columns; × 50. Some endochondral bone is also seen in the lower part (cf. Figs 26 & 27). *Iguanodon*, indet. limb bone. B.M.(N.H.) R.9952.
(25) Epiphyseal surface of the same bone, showing numerous small tubules of bone surrounded by calcified cartilage; × 5. The section seen in Figs 24 and 26–28 are cut longitudinally, at right-angles to this surface. B.M.(N.H.) R.9952.
(26) Longitudinal section through part of the same epiphysis, showing radial cavities (mainly black, but partly clear) once occupied by marrow processes, on the walls of which endochondral bone has started to form; × 50. Calcified cartilage still cores the partition between the cavities. B.M.(N.H.) R.9952.
(27) Typical cancellous endochondral bone, seen about 1 cm below the tissue shown in Fig. 26; × 50. Several small cartilage islands are present. B.M.(N.H.) R.9952.
(28) Bone seen at top left in Fig. 27, showing a cartilage island (chondrocyte lacunae round, mostly unfilled) surrounded by endochondral bone; × 100. B.M.(N.H.) R.9952.

Valdosaurus. In the centra, calcified cartilage can often be recognized without sectioning. It forms a thin coating of dense finely porous tissue seen at the ends of the centra.[2] In section, it is readily identified by the presence of large globular chondrocytes (Fig. 23). In the central parts, they tend to be arranged without order, but farther out they are usually in regular columns, as figured, which are inclined obliquely outwards. The epiphyses, perhaps slightly eroded, show a different pattern, in which calcified cartilage (Fig. 24) surrounds numerous bone-lined hollow tubules which run radially inwards (Figs 25 & 26). A few millimetres under the surface, these tubules merge into a labyrinth of bone-lined tubular tunnels, and this tissue passes rapidly down into typical endochondral trabeculae (Fig. 27). Some of the trabeculae are cored by calcified cartilage for short distances, and small cartilage islands (Fig. 28) can be found well into the metaphyses. The style of ossification was thus similar to that seen in *Crocodylus niloticus*, as described by Haines (1938: 330–333) whose fig. 8b shows the nature of the radial tubules. The cartilage was, however, rather different from that of *Crocodylus*, since the densely packed chondrocyte lacunae show little or no sign of even an irregular columnar arrangement (cf. Haines, 1938).[3]

Under these epiphyseal surfaces, trabeculae which run longitudinally show the radiating pattern found by Haines (1938) in tortoises, crocodiles and dicynodonts. The same pattern was also found in bones from other genera, whose epiphyses were too eroded to show cartilage. The pattern of ossification described above confirms (cf. e.g. Moodie, 1908) that growth in length of the limb bones took place at the articulatory cartilages, as for example in crocodiles, and not at a separate epiphyseal plate as in mammals. Otherwise, epiphyseal cartilage was only seen in an *Ornithomimus* metatarsal, in which it formed a smooth surface with occasional small pores but no perforating tubules. Such surfaces should not, incidentally, be pictured as formed from bone to which cartilage was attached; they represent an internal interface between calcified and uncalcified cartilage, like that shown in Haines (1938: fig. 6).

Some fossil reptiles and *Dermochelys* have cones of endochondral bone extending inwards from the epiphyses to the centres of the limb bones; but this pattern is not known from dinosaurs, in which the diaphyses of the limb bones seem typically to show either a secondary (reconstructed) medullary spongiosa or a hollow medullary cavity.

Secondary cancellous bone

Secondary deposits of lamellar bone, which do not replace cartilage, are sometimes formed simply on the surface of primary endochondral trabeculae, but typically they replace bone removed in the course of internal remodelling. The bone replaced may be endochondral bone, periosteal or Haversian compact bone, or older endosteal cancellous bone.

In dinosaurs, medullary cancellous bone (or spongiosa) can easily be recognized as secondary, if it either shows obvious encroachment on surrounding compact bone (Fig. 29), or passes gradationally outwards into dense Haversian tissue (Fig. 31). In the first stages of replacement, seen best when the junction is abrupt (Figs 29 & 30), excavation of large resorption spaces in compact bone was followed by deposition of thin internal linings of endosteal lamellar bone. Sometimes partitions of primary bone were left between the resorption spaces, and are then seen in the cores of the trabeculae; but other spaces seem to have expanded until only soft tissue was left between them, so that even the first formed trabeculae are entirely endosteal. In addition, once formed the trabeculae were usually subjected to further reconstruction, shown by more or less rapid development of brecciate structure (Fig. 32) with cross-cutting layers of lamellar bone. As a further complication, thick trabeculae, or bone at the junctions of trabeculae, were sometimes subjected to internal Haversian reconstruction, marked by typical Haversian systems cutting through bone which was formed at trabecular surfaces.

Cancellous reconstruction in dinosaurs was thus essentially identical in character with the Haversian reconstruction of compact bone, except in the form and the scale of the features produced, and the endosteal lamellar bone produced was Haversian in the broadest sense (cf. p. 640). This relationship can also be seen in the occurrence of intermediates between compact Haversian systems and lined cancellous spaces, both where transition from primary compact bone to secondary cancellous bone is abrupt (Fig. 29) and also where it is gradational (Fig. 31). The replacement of endochondral bone is less immediately obvious, because it is cancellous already, but is implied whenever secondary cancellous bone is seen where endochondral bone would have been present originally. In the present author's material, most cancellous bone seen in limb bones and vertebral centra was secondary, except in parts near articulatory surfaces. Exceptions were found, however, in the small genus *Valdosaurus*, in which the diaphyses of a femur and a tibia still showed unreconstructed endochondral bone.[4]

As in modern forms, resorption of a bone at the medullary limits would have been necessary for growth of the bone marrow, while weight reduction through replacement of compact bone by cancellous bone could also have been important in large animals. Markedly linear alignments of secondary trabeculae are likely to mark adjustment of

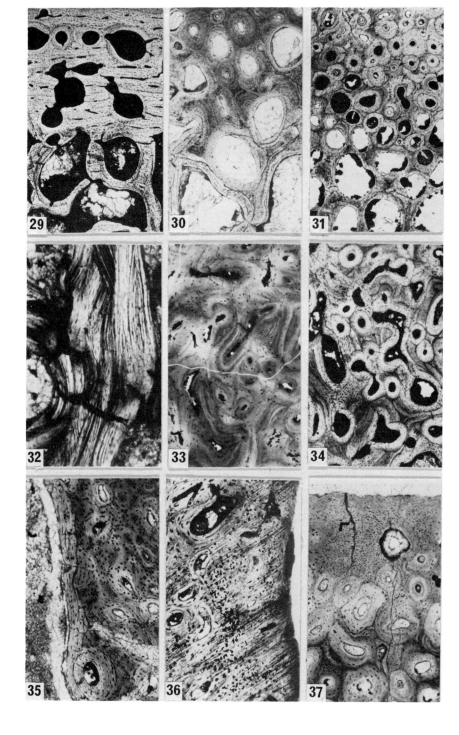

structure to stress fields (cf. Milch, 1940; Smith, 1962). The repeated reconstruction of trabeculae is, however, more usual in mammals than in reptiles (Enlow, 1969; 53–54), and again has been seen by de Ricqlès (e.g. 1976, 1980) as evidence of high metabolic rates.

Compacted Coarse-cancellous Bone

This type of bone is produced by endosteal infilling in inter-trabecular spaces with lamellar bone, and is most typically developed where bones are subjected to external resorption in the course of metaphyseal remodelling or lateral drifting (Enlow, 1962b, 1969). This allows them to maintain a compact external cortex, if the original compact bone is entirely removed. Quite large parts of the surfaces of mammalian limb bones may be formed by such compacted bone. It may also become covered by periosteal bone, which has spread outwards from the diaphyses, and then form an internal stratum on the inside of a major reversal line; or it may form an external or median stratum if internal reversal occurs, with the formation of endosteal compact bone (cf. Enlow, 1962b: figs 21, 22, 26–29).

Bone of this sort (Fig. 33) is not well known from dinosaurs, and little was seen in the present author's material. A typical internal compacted zone was seen in small (juvenile ?) bones from *Valdosaurus*; but, in

FIGS 29–37. Secondary cancellous bone; compacted coarse-cancellous bone; medullary endosteal bone; supposed heteromorphic bone.
(29) First stage in the replacement of laminar primary bone by cancellous bone, as seen in the cortex of an *Iguanodon* centrum; × 25. Some of the resorption spaces seen in the upper part have received thin linings of secondary lamellar bone, which is also seen forming the surfaces of the trabeculae below. B.M.(N.H.) R.6311.
(30) The same stage in the replacement of dense Haversian bone by secondary cancellous bone, as seen in a megalosaurid rib; × 50. Sandown Museum, 5309.
(31) Secondary cancellous bone intergrading with dense Haversian bone in a rib of *Iguanodon;* × 25. Sandown Museum 3907.
(32) Repeated reconstruction in secondary trabeculae, in the megalosaurid rib also shown in Fig. 30, as seen with crossed polarisers; × 100. Sandown Museum, 5309.
(33) Compacted coarse-cancellous bone, showing the "swirling" patterns produced by endosteal compaction; × 25. A few Haversian systems are also present. *Iguanodon*, femur. Author's collection.
(34) Partially compacted cancellous bone, invaded by Haversian systems; × 25. *Iguanodon*, long bone indet., metaphysis. B.M.(N.H.) R.9952.
(35) Endosteal bone (on left) thinly lining the medullary cavity of an *Iguanodon* femur; × 50. Bone seen to the right of this lining is mainly primary, with some intruding Haversian systems. Author's collection.
(36) Densely fibrous bone of supposedly metaplastic origin, intruded by Haversian systems; × 100. Owing to iron staining, the Haversian systems appear black. *Iguanodon*, humerus, distal metaphysis. B.M.(N.H.) 6439.
(37) Densely fibrous bone seen outside and between Haversian systems, in the lateral Haversian segment shown in Fig. 19A. The fibrous inclusions are cut transversely. *Iguanodon*, femur. Author's collection.

sections from other forms, compacted bone was either absent, or seen mainly as interstitial remnants in Haversian areas. In part, this may have been because most sections from large bones were from diaphyses, in which the oldest bone had been either resorbed or reconstructed; but sections through several metaphyses of medium sized *Iguanodon* bones showed only minor compaction of cancellous bone, even when they were cut across areas of obvious external resorption (Fig. 34). Instead, these sections showed a thin cortex formed mainly from Haversian bone, and locally showed interstitial remnants suggesting that the surface was compacted by metaplastic accretion (see below) instead of endosteal deposition. In addition, no sections from diaphyses showed evidence of lateral drifting; nor was this seen in the shaft of a sauropod pubis, in which it might have been expected, but which instead showed only strong asymmetry in periosteal growth. Although this evidence is meagre, it suggests that compaction and drifting were less important in dinosaurs than they often are in mammals.

Endosteal Compact Bone

After the initial enlargement of a growing medullary cavity, growth reversal may occur, with deposition of endosteal lamellar bone on its walls. The deposit may be thin, or form a substantial part of the wall of the medullary cavity. Thin deposits of bone of this sort (Fig. 35) were found lining the medullary cavity in some bones of *Iguanodon* and *Rhabdodon*, but not in others. In some, deposition had been followed by partial resorption, and some *Iguanodon* bones showed traces of several deposits before the final one. No thick endosteal zones were seen.

Metaplastic Bone

Metaplastic bone (Haines & Mohuiddin, 1968), also called heteromorphic bone, is produced by the ossification of tendons and ligaments. In some cases at least, ossification is preceded by chondrification (Badi, 1972).

The best known metaplastic structures in dinosaurs are the ossified tendons of e.g. *Iguanodon* (Dollo, 1886) and *Anatosaurus* ("*Trachodon*"; Broili, 1922; Moodie, 1928). In the latter, the tendons were first converted into bone with many fibrous inclusions, which was then replaced more or less completely by longitudinal secondary osteons. The outermost parts of these osteons may also contain fibrous inclusions, presumably derived form the bone which they replaced (Broili, 1922; figs 1–4).[5]

In addition, some material studied showed evidence of metaplastic accretion at the surfaces of limb bones. For example, in the more distal parts of the lateral muscular attachment described above (p. 642) from an *Iguanodon* femur, where the secondary osteons meet the surface obliquely, interstitial bone seen locally between them is so densely fibrous as to suggest ossification from tendonous material. In more proximal parts, in which the outermost osteons run downwards just under the surface, they were developed in a sparsely vascular tissue (Fig. 37) with numerous longitudinal inclusions, unlike the laminar bone which forms the surface in the unreconstructed segments. Further, densely fibrous bone was also found in several metaphyseal sections (Fig. 36), and has been noted above as extending across resorption areas. It seems reasonably likely that these tissues indicate metaplastic accretion. This could also have been involved in the formation of features like the strongly pendent fourth trochanter of *Hysilophodon*, from which Galton (1969) shows a tendon running down to the tibia.[6]

Pathological Bone

Various pathological conditions in dinosaurian bone are reported by Moodie (1923) and other authors. As summarized by Swinton (1970), these include normal fracture callus; effects of suppurative periostitis after fracture; hypertrophic growth after fracture; various other hyperostoses; osteomyelitis; necrotic sinuses due to osteomyelitis or osteitis; spondylitis deformans; a condition related to avian (not mammalian) osteopetrosis; and various tumours, including an homologous osteoma and a multiple myeloma. Some identifications of such conditions are doubtful, however, because of the absence of soft parts. For example, the condition now seen (Campbell, 1966) as related to viral (avian) osteopetrosis was originally interpreted by Swinton (1934) as a periosteal sarcoma.

Campbell (1966) also hinted that the ossified tendons seen in forms like *Anatosaurus* (Broili, 1922, as "*Trachodon*") should be seen as pathological, and not normal. If so, attitudes shown in some current restorations of dinosaurs are themselves pathological.

BONE AND DINOSAURIAN PHYSIOLOGY

General

As noted already, the mammal-like aspects of dinosaurian bone histology have been interpreted by Bakker (1972) and de Ricqlès (1974, 1976, 1978a, b, 1980) as implying that dinosaurs were endotherms. The principal worker in this field has been de Ricqlès (cf. 1974, etc.), but the concept was first publicized by Bakker (1972), who saw the high vascularity and often Haversian character of the compact bone as implying high levels of calcium and phosphate exchange between bone and body fluids. His views are sometimes represented as based on Haversian bone only (e.g. Bouvier, 1977: Ostrom, 1980), but he also referred to the primary laminar bone described by Currey (1962) as more vascular in prosauropods than in artiodactyls. De Ricqlès (1974 etc.) has drawn a similar inference from the commonly mammal-like levels of Haversian and trabecular reconstruction; but he has also based his central argument on the prevalence of fibrolamellar tissues as primary compact bone. Since the only large modern forms which show similar bone are all endotherms (various large mammals and birds), he has inferred that the sustained rapid growth which its presence implies must require high metabolic rates. However, these views have not been widely accepted, and some authors have offered different opinions. Enlow (1969), for instance, hinted that mammal-like levels of Haversian reconstruction could reflect physiological similarity; but he also pointed out that other factors (Enlow, 1962a, b) affect it in mammals, and that unknown ones may have done so in dinosaurs. Halstead (1975) saw high vascularity in dinosaurian bones as necessary simply because of their size, while McNab (1978) and Hotton (1980) have argued that the features relied on by de Ricqlès are results of inertial homoiothermy.

For further discussion, it needs first to be recognized that the principal difference between dinosaurs and most other reptiles is the widespread occurrence of fibrolamellar bone. In adults, this is otherwise seen mainly in therapsids and some thecodontians. Ostrom (1980, 1981) has explained the similarity of dinosaurian bone to that of mammals and birds as though it were due simply to occurrences of highly vascular reconstructed bone; but this view is untenable since Currey (1960) confirmed that the osteons of the laminar bone of Gross (1934) are not replacive structures. It remains true that Haversian reconstruction was often extensive in dinosaurs; but this is not the main difference between them and modern types of reptiles, and their bone may show high vascularity without showing reconstruction. In addition, is should also be realized that not all mammals and birds have bone like that of typical dinosaurs, and that small forms, in particular, may show bone similar to that of small lizards and amphibians. Such bone is not fibrolamellar, often shows no Haversian reconstruction, and may be entirely avascular.

Arguments for Endothermy

In more detail, Bakker's grounds (1972) for seeing dinosaurian bone as showing evidence of endothermy were based on the function of the skeleton as a reserve in calcium phosphate metabolism. In his view, endothermy must require higher levels of calcium phosphate exchange, because of both more variable energy demands and more precise metabolic regulation. Endotherms show extensive Haversian reconstruction and high vascularity because labile reserves are concentrated in the last formed Haversian systems (i.e. secondary osteons), and because access to reserves depends on the degree of vascularity. Hence the extensive reconstruction and high vascularity seen in dinosaurs implies that they were also endothermic. De Ricqlès (e.g. 1974) also uses the concept of the skeleton as a calcium reserve, though in a rather different sense (cf. below), with cancellous bone also being involved.

In various papers (1974 etc.), de Ricqlès has used a number of arguments, of which the main ones may be outlined as follows.

(i) Although the fibrolamellar bone of dinosaurs is primarily only evidence of rapid growth, the only large modern forms which show it in their long bones are all endotherms. Large ectotherms, in contrast, always show some form of zonal bone; and they also have to grow throughout long lives to reach large sizes, instead of growing rapidly in their early years like endotherms. This implies that, to sustain rapid growth to large sizes, in the manner of large mammals and birds, it would be necessary for dinosaurs to have comparably high metabolic rates.

(ii) Although genuinely annual "growth rings" do occur in some mammals (e.g. Klevezal & Kleinenberg, 1969), they are typically restricted to certain bones (e.g. jaw bones) which grow slowly, and are not seen in the long bones. They are also seen mainly (a) in small animals from cool or cold regions, and especially in forms which hibernate, and (b) in marine forms which experience cold conditions or seasonal temperature changes. These restricted occurrences of "growth rings" do not compromise the observation that their presence or absence in the long bones is typical of ectotherms and endotherms respectively.

(iii) If the resorption of bone in Haversian and trabecular reconstruction is seen as a means by which calcium phosphate reserves are released, these processes should occur most extensively in animals with high metabolic rates. This is confirmed by their being more extensively developed in endotherms than in ectotherms. Hence their extensive development in dinosaurs points to high metabolic rates.

(iv) Although comparative histological data supposed to imply endothermy can be interpreted alternatively as functionally linked with large sizes, it would be necessary for large dinosaurs to grow rapidly, in order to reach such sizes within reasonable life-spans. They would only be able to do this if they had high metabolic rates.

(v) Even if Haversian reconstruction is initially triggered by factors other than thermal physiology, its rate of development will still be controlled by metabolic rates. Its extensive development in dinosaurs implies rapid reconstruction, and hence high metabolic rates.

Here, a difference between Bakker and de Ricqlès should be noted. Bakker (1971) specifically claimed that dinosaurs had an avian type of physiology, with homoiothermy achieved through endothermy. De Ricqlès (e.g. 1974, 1980), in contrast, has emphasized that dinosaurs were probably not exactly like any modern forms physiologically, and may have had some unique intermediate type of physiology. His views should thus not be equated with Bakker's extreme view; and facts which do not fit the notion that dinosaurs were typical endotherms may not be contrary to the concept of an intermediate type of physiology.

Arguments Against Endothermy

Authors who do not accept these arguments for dinosaurs being endotherms have mostly argued from different evidence (e.g. Auffenberg & McNab, 1976; McNab, 1978; Hotton, 1980; Spotila, 1980); but a number of alternative arguments can be based on bone histology itself.

(i) Although the smallest endotherms (e.g. shrews, bats, small birds) have generally the highest metabolic rates, they may show simple avascular bone with no Haversian systems, not distinguishable histologically from that of small lizards. It follows that no strict causal connection can exist between endothermy and (a) high vascularity, (b) fibrolamellar structure, or (c) Haversian reconstruction. The same is shown for fibrolamellar bone by its occurrence in young crocodiles (Enlow, 1969, as plexiform bone), and by occurrences of lamellar-zonal bone in slowly growing bones of large endotherms.

(ii) If there were a causal connection between endothermy and Haversian reconstruction, it would not occur in ectotherms, and would affect all primary canals in endotherms. In fact it occurs in both ectotherms and endotherms (cf. Bouvier, 1977), and large segments of primary bone may remain unreconstructed in endotherms.

(iii) The only sure significance of the high vascularity of fibrolamellar bone, whose occurrence is the primary basis of high vascularity in most dinosaurs, is that rapid deposition of compact bone requires a rich blood supply.

(iv) Although extensive Haversian reconstruction is genuinely found mainly in endotherms, at least at present, it is also seen mainly in bones with high primary vascularities. This suggests that high primary vascularity is a factor which can predispose bone to high levels of Haversian reconstruction, if this process is triggered by factors of the sorts reviewed by Enlow (1962a).

(v) Although de Ricqlès (1980: 124) states that dense Haversian bone is now only found in endotherms, the present author found the cortex of a femur of *Testudo triserrata* to be almost entirely dense Haversian (B.M. (N.H.) R.9730).

(vi) As de Ricqlès (1980) notes, there are few experimental data on how temperature affects bone histology. M. W. J. Ferguson (pers. comm.), however, has found that American alligators (*A. mississippiensis*) farmed at temperatures between 23.5 and 27.0°C (Joanen & McNease, 1976) show increased production of Haversian bone compared with wild examples of similar size. This suggests that inertial homoiothermy, at the level expected in large dinosaurs (cf. Spotila, Lommen, Bakken & Gates, 1973), could have a similar effect.

(vii) As Enlow (1969) notes, the vascular zones of zonal bone in, for example, crocodiles may show primary osteons in a matrix of woven bone. Bone of this sort would take on the character of nominal fibrolamellar bone if inertial homoiothermy led to the suppression of the annuli.

(viii) The changes suggested in (vi) and (vii) above could still reflect an upward shift in metabolic rates, resulting from high steady temperatures achieved through thermal inertia. Such a rise in metabolic rates would have no connection with endothermy, in which high metabolic rates are the basis of homoiothermy and not a result of it.

(ix) If some dinosaurs were "simply too big to have a high metabolic rate" (Halstead, 1975), which could only have been maintained by more food than they could eat, their possession of fibrolamellar bone is not evidence of endothermy. This seems likely to apply to at least the large sauropods.

(x) If the metabolic rates of large dinosaurs declined to ectothermic levels when they reached a size permitting inertial homoiothermy (Halstead, 1975; de Ricqlès, 1980), their continued production of fibrolamellar bone up to the largest size known (in *Brachiosaurus*: cf. Gross, 1934) can again not have depended on high metabolic rates.

(xi) Typical reptilian "growth rings" (zones and annuli) are now known from several dinosaurs (cf. pp. 635–637), and may be commoner than has so far been shown. Some were only formed late in life; but this does not apply to those known from a sauropod pubis (Reid, 1981), or from the limb bones of *Rhabdodon* (e.g. Fig. 12). At the very least, their presence suggests that dinosaurs were closer to ectothermic reptiles than Bakker (e.g. 1972) or de Ricqlès (e.g. 1980) have supposed.

(xii) The physiological inferences which de Ricqlès (e.g. 1974, 1976, 1980) has drawn from fibrolamellar bone depend on the assumption that its present distribution can be used as a basis for assessing its significance in fossils. An alternative view is that what in fact needs to be known is the thermal physiology of the earliest large forms to show it; because even if the presence of such bone in large animals does mark progress towards endothermy, it cannot be claimed to show progress beyond the level at which it first appeared. In addition, without this information, there is no way of knowing whether data from modern forms show that sustained rapid growth is only possible for endotherms, or whether occurrences in fossils may show it to be possible for ectotherms.

(xiii) In therapsids, the earliest appearances of fibrolamellar bone did not coincide with the approach to mammalian organization seen in cynodonts, but took place in deinocephalians and eotheriodonts. This suggests strongly that it first appeared before the emergence of endothermy, and in animals still closer to normal ectotherms than to endotherms.

DISCUSSION

First, it seems appropriate to comment on the term "high metabolic rates". If metabolic rates are measured in terms of the standard metabolic rate (SMR) for oxygen metabolism, those of modern endotherms are on average six or more times higher than those of ectotherms of similar size; but this relationship does not apply between animals of different sizes, because the SMR decreases as size increases (e.g. McFarland, Pough, Cade & Heiser, 1979: figs 7–21). As a result, a small (e.g. 0.1 kg) lizard may have a higher metabolic rate in this sense than a large (e.g. 5 tonne) mammal. Hence, assertions that fossil animals had "high metabolic rates" should be read as implying "higher than would be expected in an ectotherm of similar size".

Further, if a small lizard with lamellar-zonal compact bone can in fact have a higher metabolic rate than large mammals with fibrolamellar and dense Haversian tissues, it can clearly be argued that no true correlation can exist between bone types and specific metabolic levels. None the less, while the bones of small endotherms may be indistinguishable histologically from those of small ectotherms, it remains true that the differences emphasized by de Ricqlès (e.g. 1974,

1976, 1980) are usual in the fastest growing bones of examples of medium to large sizes. This in turn seems to be due to a genuine difference in the maximum growth rates which ectotherms and endotherms can sustain if they grow beyond small sizes.

Next, it seems to the present author that the most important feature of dinosaurs, if their bone is supposed to have a physiological significance, is the widespread occurrences of fibrolamellar bone. Their Haversian bone could also be important, as Bakker (1972) and de Ricqlès (1974, 1976, 1980) believe; but the factors which controlled its development seem currently so uncertain (cf. p. 642 above and de Ricqlès, 1980: 122–124) that its significance is best seen as doubtful. Their fibrolamellar bone, in contrast, has a clear implication of ability to sustain rapid growth to large sizes, which is now shown by mammals and birds but not by any modern reptiles.

The first question to be asked is whether this ability can be simply a result of inertial homoiothermy, as supposed by McNab (1978: 7, reference to non-lamellar bone) in the case of therapsids. This is possible, but does not seem likely. Spotila *et al.* (1973) showed that dinosaurs of 1 m body diameter or over would have been effectively homoiothermic without endothermy; but their model implies much larger animals than small forms like *Deinonychus* or *Hypsilophodon* which show fibrolamellar bone. Spotila *et al.* (1973: fig. 3) also show that animals of the size they envisaged would still have been affected by seasonal temperature variations, and so likely to form "growth rings" if they were simply typical ectotherms. Further, fibrolamellar bone is not seen in the Komodo monitor (max. weight 100 kg; Auffenberg & McNab, 1976), on which McNab (1978) based his conclusions, nor in comparably large pelycosaurs (e.g. *Dimetrodon*) nor still larger cotylosaurs like *Bradysaurus* (250 kg or over). This makes it difficult to see its presence in, for example, small specimens of *Hypsilophodon* as due simply to bulk homoiothermy.

The alternative is then to suppose that both dinosaurs and therapsids, which have fibrolamellar bone, must have taken some step forward in physiological evolution, which has not been taken in monitors, and which had not been taken in *Dimetrodon* or *Bradysaurus*. A rise in metabolic rates, above those seen in any modern reptiles, is an obvious possibility. The question then becomes whether this involved the achievement of endothermy, or some close approach to it, or only some lesser advance, which would leave the animals concerned still essentially ectothermic. Again it is relevant to note that the metabolic rates of modern endotherms are on average six or more times higher than those of ectotherms of comparable size. In the two stocks which did achieve endothermy (i.e. those leading to mammals and birds) there must once have been intermediate forms, in which metabolic rates had not risen above various intervening levels.

No certain answer can be given to this question, but a probable answer can be inferred from the level of morphological evolution at which fibrolamellar bone first appears in therapsids. These animals are critically important in this context, because they allow the supposed association between fibrolamellar bone and endothermy (de Ricqlès, e.g. 1974) to be compared with other evidence of the level at which mammalian endothermy originated. This evidence, from gross anatomy and functional studies, has recently been summarized by Kemp (1982). It has long been suspected that mammalian endothermy first emerged in the more advanced cynodonts; but fibrolamellar bone is first seen in early genera (deinocephalians, titanosuchians), and even, in incipient form, in the primitive *Biarmosuchus*. Bakker (1975) pictured early therapsids from the *Tapinocephalus* Zone fauna of southern Africa as needing to be insulated endotherms, to survive winter snows; but they need not in fact have been more endothermic than large pareiasaurs like *Bradysaurus*, which also occur in this fauna, and whose bone is typical of ectotherms. McNab (1978) saw mammalian endothermy as first emerging in cynodonts, and claimed that it could only have originated through a trend to small sizes, after early forms had first achieved inertial homoiothermy. Kemp (1982: 98) is prepared to see some rise in metabolic rates as involved from the start, but still concludes (1982: 251) that most forms earlier than cynodonts were probably inertial homoiotherms. His picture of gradual progress through small correlated changes in all evolving systems (Kemp, 1982: 296–313) also seems fundamentally more probable than a sudden mutational "jump" to endothermy in early forms, for which bone histology is the only evidence. If this view is correct, the ability to sustain rapid growth, implied by fibrolamellar bone in large animals, was first evolved long before endothermy, and need not imply even a close approach to it. On the contrary, the first forms to show it may well have been closer physiologically to modern ectothermic reptiles than to endotherms.

On this basis, the frequent occurrence of fibrolamellar bone in dinosaurs should not be cited as evidence of endothermy, and again need not imply close approach to it. The present author prefers this conclusion to supposing that such bone must imply endothermy. This does not, of course, mean that no dinosaurs can have been endotherms; but, if they were, the evidence to show this will need to come from other sources.

A possible scenario for dinosaurs can then be based on the notion that ability to sustain rapid growth marks some step in physiological evolution, which has to be taken before animals can evolve into endotherms, but is not itself more than a necessary preliminary. This step can be pictured as involving some rise in metabolic rates, to above the levels seen in modern reptiles, but not to an extent where the animals concerned would cease to be basically ectothermic. Early therapsids and dinosaurs (or their ancestors) would then be groups which took this step independently, with different results because of different later histories. Therapsids would be forms in which external insulation, first evolved as an aid to homoiothermy, allowed a trend to small sizes which in turn led to endothermy. Dinosaurs, in contrast, do not seem to have evolved insulation, at least at this stage, and mostly followed an opposite trend to large sizes, which would lead to inertial homoiothermy but not endothermy. In part at least, this could have been due to their evolving in warmer conditions than those in which therapsids first appeared (cf. Fedducia, 1973). Some further progress towards higher metabolic rates could have been made in the forms described by Hopson (1976) as designed to process large amounts of food; but most of these at least would be too large to evolve into endotherms. Little or no progress would occur in forms like sauropods, which seem unlikely to have eaten enough to have high metabolic rates. The most active small theropods (e.g. deinonychosaurs) *might* have become endothermic; but, unless they had external insulation, this can still be thought unlikely. Thus, if endothermy was ever developed in dinosaurs, it could have been limited to very small forms ancestral to birds, if this is really where birds came from.

This picture is, of course, hypothetical; but it seems clear that all that is certainly implied by the fibrolamellar bone of dinosaurs is an ability to sustain rapid growth. What this means in terms of physiology is not currently known, and will not be understood, if it ever is, until much more is known about the factors which control the growth of bone. We also need much more information about the bone of all kinds of dinosaurs. None the less, it still appears likely that most dinosaurs at least could only become homoiothermic by growing to large sizes. Such animals cannot properly be considered endothermic, even if they had relatively high metabolic rates, since endothermy involves an ability to maintain homoiothermy at small and very small sizes. We need to know that dinosaurs possessed that ability before we start to call them endotherms.

ACKNOWLEDGEMENTS

Thanks are due first to Dr H. W. Ball, Dr A. J. Charig, Dr Alan Insole and Mr Phillip Powell for permission to use material in their charge, and to Mr Nick Chase for the use of a specimen from his personal collection. They are also due to Dr A. C. Milner, Mr C. A. Walker and Mr S. Hutt for assistance in locating suitable material, and to Mr Walker for helpful discussion of various topics. I must also acknowledge the kind encouragement of Professor T. J. Harrison and Professor A. D. Wright, and especially the kindness and advice I have received from Dr M. W. J. Ferguson. Sections cut in my Department were made by Mr W. G. Allingham and Miss E. A. Lawson, and Mr S. Watters processed my photographs. Payment of various expenses by The Queen's University of Belfast must also be gratefully acknowledged. Last, but not least, I would like to mention a debt to my old teachers, R. W. Haines and Tom Barnard.

REFERENCES

Auffenberg, W. & McNab, B. K. (1976). The effect of large body size on the temperature regulation of the Komodo dragon, *Varanus komodoensis*. *Comp. Biochem. Physiol.* **55A**: 345–350.

Badi, M. H. (1972). Calcification and ossification of fibrocartilage in the attachment of the patellar ligament in the rat. *J. Anat.* **112**: 415–421.

Bakker, R. T. (1971). Dinosaur physiology and the origin of mammals. *Evolution, Lawrence, Kans.* **25**: 636–658.

Bakker, R. T. (1972). Anatomical and ecological evidence of endothermy in dinosaurs. *Nature, Lond.* **238**: 81–85.

Bakker, R. T. (1975). Dinosaur renaissance. *Scient. Am.* **232**: 58–78.

Bouvier, M. (1977). Dinosaur Haversian bone and endothermy. *Evolution, Lawrence, Kans.* **31**: 449–450.

Broili, F. (1922). Über den feineren Bau der "Verknöcherten Sehnen" (= verknöcherten Muskeln) von *Trachodon*. *Anat. Anz.* **55**: 465–475.

Campbell, J. G. (1966). A dinosaur bone lesion resembling avian osteopetrosis with some remarks on the mode of development of the lesions. *Jl R. microsc. Soc.* **85**: 163–174.

Cook, S. F., Brooks. S. T. & Ezra-Cohn, H. E. (1962). Histological studies on fossil bone. *J. Paleont.* **36**: 483–494.

Currey, J. D. (1960). Differences in the blood-supply of bone of different histological types. *Q. Jl. microsc. Sci.* **101**: 351–370.

Currey, J. D. (1962). The histology of the bone of a prosauropod dinosaur. *Palaeontology* **5**: 238–246.

Dollo, L. (1886). Note sur les ligaments ossifiés des dinosauriens de Bernissart. *Archs Biol.* **7**: 249–264.

Enlow, D. H. (1962a). Functions of the Haversian system. *Am. J. Anat.* **110**: 268–306.

Enlow, D. H. (1962b). A study of the post-natal growth and remodelling of bone. *Am. J. Anat.* **110**: 79–102.

Enlow, D. H. (1969). The bone of reptiles. In *Biology of the Reptilia* **1**: 45–80. Gans, C. & Bellairs, A. d'A. (Eds). London and New York: Academic Press.

Enlow, D. H. & Brown, S. O. (1956). A comparative histological study of fossil and Recent bone tissue. Part I. *Tex. J. Sci.* **8**: 405–443.

Enlow, D. H. & Brown, S. O. (1957). A comparative histological study of fossil and Recent bone tissue. Part II. *Tex. J. Sci.* **9**: 186–214.
Enlow, D. H. & Brown, S. O. (1958). A comparative histological study of fossil and Recent bone tissue. Part III. *Tex. J. Sci.* **10**: 187–230.
Feduccia, A. (1973). Dinosaurs as reptiles. *Evolution, Lawrence, Kans.* **27**: 166–169.
Ferguson, M. W. J., Honig, L. S., Bringas Jr, P. & Slavkin, H. C. (1982). In vivo and in vitro development of first branchial arch derivatives in *Alligator mississippiensis*. In *Factors and mechanisms influencing bone growth*: 275–286. Dixon, A. D. & Sarnat, B. (Eds). New York: Alan R. Liss, Inc.
Galton, P. M. (1969). The pelvic musculature of the dinosaur *Hypsilophodon* (Reptilia, Ornithischia). *Postilla* **131**: 1–64.
Galton, P. M. (1973). A femur of a small theropod dinosaur from the Lower Cretaceous of England. *J. Paleont.* **47**: 996–997.
Gross, W. (1934). Die Typen des mikroskopischen Knochenbaues bei fossilen Stegocephalen und Reptilien. *Z. Anat.* **103**: 731–764.
Haines, R. W. (1938). The primitive form of epiphysis in the long bones of tetrapods. *J. Anat.* **72**: 323–343.
Haines, R. W. & Mohuiddin, A. (1968). Metaplastic bone. *J. Anat.* **103**: 527–538.
Halstead, L. B. (1975). *The evolution and ecology of the dinosaurs*. London: Peter Lowe.
Hasse, C. (1878). Die fossilen Wirbel. Morphologische Studien. Die Histologie fossiler Reptilwirbel. *Morph. Jb.* **4**: 480–502.
Hopson, J. A. (1976). Hot-, cold-, or lukewarm-blooded dinosaurs? *Paleobiology* **2**: 271–275.
Hotton, N. (1980). An alternative to dinosaur endothermy. The happy wanderers. In *A cold look at the warm-blooded dinosaurs*: 311–350. Thomas, R. D. K. & Olson, E. C. (Eds). (*Select. Symp. Am. Ass. Advmt Sci.* **28**.) Boulder, Colorado: Westview Press.
Joanen, T. & McNease, L. (1976). Culture of immature American alligators in controlled environmental chambers. *Proc. ann. Meet. Wld Maricult. Soc.* **7**: 201–211.
Kemp, T. (1982). *Mammal-like reptiles and the origin of mammals*. London: Academic Press.
Klevezal, G. A. & Kleinenberg, S. E. (1969). *Age determination of mammals from layered structures in teeth and bone*. Jerusalem: Israel program for scientific translations.
Mantell, G. A. (1850). On the *Pelorosaurus*, an undescribed gigantic terrestrial reptile, whose remains are associated with those of *Iguanodon* and other saurians in the strata of Tilgate Forest, Sussex. *Phil. Trans. R. Soc.* **140**: 379–390.
McFarland, W. N., Pough, F. H., Cade, T. J. & Heiser, J. B. (1979). *Vertebrate life*. New York: McMillan Publishing Co., Inc.
McNab, B. K. (1978). The evolution of endothermy in the phylogeny of mammals. *Am. Nat.* **112**: 1–21.
Milch, H. (1940). Photoelastic studies of bone form. *J. Bone Jt Surg.* **22**: 621–626.
Moodie, R. L. (1908). Reptilian epiphyses. *Am. J. Anat.* **7**: 442–467.
Moodie, R. L. (1923). *Palaeopathology*. Urbana: University of Illinois Press.
Moodie, R. L. (1928). The histological nature of ossified tendons found in dinosaurs. *Am. Mus. Novit.* No. 311: 1–15.
Nopcsa, F. (1925). Dinosaurierreste aus Siebenbürgen, IV: Die Wirbelsäule von *Rhabdodon* und *Orthomerus*. *Paleont. hung.* **1**: 273–288.
Nopcsa, F. (1933). On the histology of the ribs in immature and half-grown trachodont dinosaur. *Proc. zool. Soc. Lond.* **1933**: 221–223.
Ostrom, J. (1980). The evidence for endothermy in dinosaurs. In *A cold look at the warm-blooded dinosaurs*: 15–54. Thomas, R. D. K. & Olson, E. C. (Eds). (*Select. Symp. Am. Ass. Advmt Sci.* **28**.) Boulder, Colorado: Westview Press.
Ostrom, J. (1981). *Dinosaurs*. (*Carolina Biol. Readers* No. 98.) Burlington, N. Carolina: Carolina Biological Supply Co.

Pawlicki, R. (1977). Histochemical reactions for mucopolysaccharides in the dinosaur bone. Studies on Épon- and methacrylate-embedded semithin sections as well as on isolated osteocytes and ground sections of bone. *Acta histochem.* **58**: 75–78.
Peabody, F. E. (1961). Annual growth zones in living and fossil vertebrates. *J. Morph.* **108**: 11–62.
Reid, R. E. H. (1981). Lamellar-zonal bone with zones and annuli in the pelvis of a sauropod dinosaur. *Nature, Lond.* **292**: 49–51.
Ricqlès, A.J. de (1974). Evolution of endothermy: histological evidence. *Evolut. Theory* **1**: 51–80.
Ricqlès, A.J. de (1975). Recherches paléohistologiques sur les os longs des tétrapodes. VII. Sur la classification, la signification fonctionnelle et l'histoire des tissus osseux des tétrapodes. Première partie: structures. *Annls Paléont. (Vert.)* **61**: 49–129.
Ricqlès, A.J. de (1976). On bone histology of fossil and living reptiles, with comments on its functional and evolutionary significance. In *Morphology and biology of reptiles*: 123–150. Bellairs, A. d'A. & Cox, C. B. (Eds). (*Linn. Soc. Lond. Symp.* **3**) London: Academic Press.
Ricqlès, A.J. de (1978a). Recherches paléohistologiques sur les os longs des tétrapodes. VII. Sur la classification, la signification fonctionnelle et l'histoire des tissus osseux des tétrapodes. Troisième partie: évolution. *Annls Paléont. (Vert.)* **64**: 85–111.
Ricqlès, A.J. de (1978b). Recherches paléohistologiques sur les os longs des tétrapodes. VII. Sur la classification, la signification fonctionnelle et l'histoire des tissus osseux des tétrapodes. Troisième partie: évolution (fin). *Annls Paléont. (Vert.)* **64**: 153–184.
Ricqlès, A.J. de (1980). Tissue structure of dinosaur bone. Functional significance and possible relation to dinosaur physiology. In *A cold look at the warm-blooded dinosaurs*: 103–139. Thomas, R. D. K. & Olson, E. C. (Eds). (*Select. Symp. Am. Ass. Advmt Sci.* **28**.) Boulder, Colorado: Westview Press.
Seitz, A. (1907). Vergleichende Studien über den mikroskopischen Knochenbau fossiler und rezenter Reptilien und dessen Bedeutung für das Wachstum und Umbildung des Knochengewebes im Allgemeinen. *Nova Acta Leop.* **87**: 230–270.
Smith, J. W. (1960). Collagen fibre patterns in mammalian bone. *J. Anat., Lond.* **94**: 329–344.
Smith J. W. (1962). The structure and stress relations of fibrous epiphyseal plates. *J. Anat. Lond.* **96**: 209–225.
Spotila, J. R. (1980). Constraints of body size and environment on the temperature regulation of dinosaurs. In *A cold look at the warm-blooded dinosaurs*: 235–252. Thomas, R. D. K. & Olson, E. C. (Eds). (*Select. Symp. Am. Ass. Advmt Sci.* **28**.) Boulder, Colorado: Westview Press.
Spotila, J. R., Lommen, P. W., Bakken, G. S. & Gates, D. M. (1973). A mathematical model for body temperatures of large reptiles: implications for dinosaur ecology. *Am. Nat.* **107**: 391–404.
Swinton, W. E. (1934). *The dinosaurs. A short history of a great group of extinct reptiles.* London: Thomas Murby.
Swinton, W. E. (1970). *The dinosaurs.* New York: Wiley-Interscience.

NOTES ADDED IN PROOF

1. Pawlicki (1977) records unidentified mucopolysaccharides from bone of the tyrannosaurid *Tarbosaurus*.

2. In an unidentified sacral vertebra, a similar tissue was seen lining cavernous spaces exposed where sacral ribs had been detached.

Histology of Dinosaurian Bone   663

3. A similar pattern can be found in the epiphyses of ostriches, again where ossification takes place from the articular cartilage.

4. The centra of dinosaurian vertebrae may show either normal coarse cancellous bone, or cavernous tissues in which thin partitions separate spaces up to several centimetres wide. In the author's material, several vertebrae from a megalosaurid and one sauropod caudal showed cavernous conditions, produced by progressive enlargement of normal inter-trabecular cancelli; but one large sauropod caudal had the whole centrum built from thin sheets of compact bone, 1–3 mm thick, which formed a thin external shell and an internal plexus of longitudinal partitions. The sheets of bone either showed a thin central Haversian stratum between external layers of lamellated bone, or were locally formed entirely from lamellated tissue. Flattened osteocytes in some lamellae showed a plumose arrangement, suggesting lateral growth. In the absence of external pleurocoels, this pattern appears to have resulted from a specialized reconstruction process, with normal vertebral structure entirely replaced. A practical result would be very substantial lightening of the centrum.

5. In *Iguanodon*, the central parts of ossified tendons may show conversion into secondary cancellous bone.

6. In the *Iguanodon* femur, apparently metaplastic tissue was also found on the inward facing side of the large fourth trochanter (Fig. 19a), with fibres aligned as expected at the insertion of a caudifemoralis brevis muscle. With occurrences in metaphyses (p. 650 and Fig. 36), this suggests that Haversian reconstruction was sometimes specially developed in parts in which primary ossification took place from tendonous material instead of from normal periosteum. This could be why other evident muscular attachments showed no special development of Haversian bone, but only Sharpey's fibres inserted in otherwise normal fibrolamellar bone. Why reconstruction should affect such parts specially is uncertain; but one of its effects would be to raise vascularity to nearer the level seen in parts formed from fibrolamellar bone (cf. Figs 22, 37).

Closing Address
With comments on the organ of Jacobson and the evolution of Squamata, and on the intermandibular connection in Squamata

A. d'A. BELLAIRS*

Department of Anatomy, St Mary's Hospital Medical School, London W 2, England

INTRODUCTORY REMARKS

I am greatly touched that so many of my friends and colleagues have been able to come and talk at this symposium. It is also a great pleasure to meet so many others who have come from far away, whose names have long been familiar to me, but whom I have not previously had the chance of meeting. And, of course, my very best thanks are due to Mark Ferguson, who conceived the idea of this symposium (besides putting the alligator on the map as an experimental embryo! – see Ferguson, 1981). I also greatly appreciate the fact that this meeting has been held here in the rooms of the Zoological Society of London and has been supported also by the Anatomical Society of Great Britain and Ireland and by the British Herpetological Society – three learned societies with which I have had the happiest associations throughout most of my adult life.

The contributions which have been presented at this symposium cover the widest variety of topics, from molecular biology to palaeoecology. This is surely a testimony to the impressive way in which herpetology spans what could perhaps be thought of as the "two cultures" in modern biology. I mean on the one hand the prestigious culture of developmental, cellular and molecular biology, and on the other, what one might call the "natural history culture": systematics, morphology, palaeontology, ecology and ethology. I suspect that this latter culture is at present somewhat under-valued by the scientific estalishment in Britain. Sometimes these two cultures seem almost as far apart as C. P. Snow (1959) supposed his "two cultures" of the arts and sciences to be. But both of them have been well represented in the talks which we have heard here during the last two days.

In this context I would particularly like to pay tribute to the work of

---

*Honorary Herpetologist, Zoological Society of London.

Carl Gans, who was unfortunately not able to come to this meeting, although he has contributed a paper for publication. Quite apart from his original contribution to herpetology, Carl Gans has done us all an immense service by his able editorship and promotion of *Biology of the Reptilia*, a work which seems to provide an admirable synthesis of the two cultures in so far as these are relevant to a single class of animals.

It would be appropriate for me to mention three men who gave me great help at an early stage in my career and who became lifelong friends; alas, they are no longer with us. One was Malcolm Smith, who began to study reptiles when he was Physician to the Court of Siam, and who later worked for many years at the British Museum (Natural History). He produced not only his classical volumes on reptiles in the *Fauna of British India* series (Smith, 1931–43), but also much the best book which exists on the British herpetofauna (Smith, 1973). There was also Rex Parrington, who gave the most delightful and stimulating lectures on vertebrates at Cambridge in the 1930s, and had a quite remarkable talent for clothing old bones with flesh. And there was Dixon Boyd, who was not really a herpetologist at all, but a human anatomist. He appointed me to the first job in which I was actually able to earn money while studying reptiles. He loved natural history and collaborated with me in my first work on Jacobson's organ in reptiles (see Bellairs & Boyd, 1950). Above all, he taught me the immense value of micro-anatomy, a technique which can reveal so much about the relationship between structure and lifestyle in animals.

I should now like to touch on two quite different problems in herpetology, to which, as it seems to me, micro-anatomical techniques can make a significant contribution.

THE ORGAN OF JACOBSON AND THE EVOLUTION OF THE SQUAMATA

It is said that people never forget their first loves, and I should like to return to two of mine and to try to relate them: the organ of Jacobson, which so much impressed me when I first saw it on a microscope slide over 40 years ago, and the problem of the origin of snakes. This last is not perhaps the kind of problem which would appeal to those who follow Medawar's description (1969) of research as "the art of the soluble". None the less, it has intrigued me for many years, and I know that Garth Underwood has also fallen under its spell. Indeed, he has coined the delightful phrase "lower and higher snakes" (Underwood, 1967), the latter having strayed the furthest from the Garden of Eden by completely discarding their limb vestiges and other obsolescent saurian features.

I should add that palaeontology can, as yet, make little direct contribution to either of these topics. The fragile and partly cartilaginous skeleton which surrounds the organ of Jacobson is seldom, if ever, preserved in fossils. Snakes are unknown in the geological record before the Lower Cretaceous, although their origin most probably occurred at a much earlier date (see Rage, 1982a). Of the other animals which concern us here, the Eosuchia is an ill-defined group of primitive reptiles which flourished in Permo-Triassic times and probably contained the ancestors of all other lepidosaurs, and perhaps of archosaurs as well (Evans, 1980; Benton, 1982). *Sphenodon* and its extinct allies are customarily placed in the order Rhynchocephalia, but it is likely that this order does not form a natural group, and Evans prefers to classify the Sphenodontidae as an independent family within the Eosuchia. True lizards may well have appeared during the Triassic or even earlier, but the dating is problematical since the classification of many of the earliest lizard-like forms is open to question (Evans, 1980; see Estes, 1983). Lizards of modern type (i.e. which can be referred to some of the extant infra-orders) occur in the late Jurassic (Hoffstetter, 1968).

Sixty years ago and more, most herpetologists had no difficulty in believing that the snakes were descended from lizards, probably from lizards of varanoid or at least of anguimorph type. Thus, Camp (1923: 359) regarded the snakes as "highly modified anguimorphine lizards near the platynotid stock", and believed that the range of differences between lizards and snakes was not much beyond the range of differences between different types of snake-like lizards. A rather similar view was expressed by Dowling (1959). The difficulty of distinguishing between the less typical members of the two suborders was illustrated by the idea of McDowell & Bogert (1954) that the Typhlopidae are not really snakes, but should either be placed in a suborder of their own, or perhaps regarded as aberrant burrowing lizards.

More recently the pendulum has swung in the opposite direction. Hoffstetter (1962) wrote that it is not at all certain that the snakes were derived from lizards. The differences between lizards and snakes have been increasingly emphasized, and Underwood (1967) has been able to demonstrate 16 characters in which all (or nearly all) snakes differ from all known lizards. This trend dates from 1942, when Walls cogently demonstrated the striking anatomical contrast between the eyes of lizards and those of snakes, which led him to put forward his now famous theory that the snakes had originated from burrowing ancestors (see also Bellairs & Underwood, 1951; Bellairs, 1972). Walls' findings (1942) have in the main been confirmed and have been extended by

Underwood (1967, 1970), who has shown that the basic duplex pattern of the ophidian retina is actually more primitive than that in modern lizards, from which the original types of rod have been discarded. Underwood (1970) concluded that although "the relationship between lizards and snakes is not in doubt", the "division of the Squamata into snakes and lizards antedates the origins of any of the groups of Recent lizards".

However, Hoffstetter (1968) suggested a rather more radical interpretation of ophidian phylogeny, stating that the common ancestor of lizards and snakes was a very ancient form, "probably an unknown protosaurian, or even a pre-saurian, probably very different from modern lizards". This view has now been carried further by Rage (1982a) in his admirable review of the history of snakes. He suggests that the latter are not descended from lizards, but that the two groups arose from a common ancestor which must be looked for among the Eosuchia.

Let us consider the bearing of the organ of Jacobson on these questions. The facts which I am going to mention are well known, but they seem to have implications which can bear re-statement. The paired organs of Jacobson or vomeronasal organs were described in mammals by L. L. Jacobson, an anatomist and physician in Copenhagen and later a doctor in the French army (Dobson, 1946; see also Negus, 1956). His paper was transmitted to the Memoirs of the Natural History Museum of Paris by Cuvier (1811).

Jacobson's organ is basically a specialized region of the nose and is concerned with a chemical sense akin to smell. It was doubtless a part of the reptilian heritage from amphibian ancestors and is present in some form in all three groups of Recent Amphibia (see Parsons, 1967; Bertmar, 1981). Although present in Chelonia, its structure, like that of certain other features of the nose, is so different from that in other reptiles that it suggests a very early divergence of the turtles from the primitive amniote stock (Parsons, 1970). The organ was retained in the lepidosaurs, becoming much elaborated in Squamata. In the archosaurian lineage, on the other hand, it has apparently been lost, since in crocodilians and birds it is represented, if at all, only by transient and dubious embryonic vestiges (Parsons, 1967, 1970). Jacobson's organ also doubtless occurred in the ancestors of mammals (Duvall, King & Graves, 1983) as it does in most mammals today, playing an important part in the sexual behaviour of at least some forms (Powers & Winans, 1975).

Sphenodon is as near a living eosuchian as one can get, and also shows many affinities with lizards (Carroll, 1977). Its organ of Jacobson has been described by several authors (see Parsons, 1970; Gabe & Saint

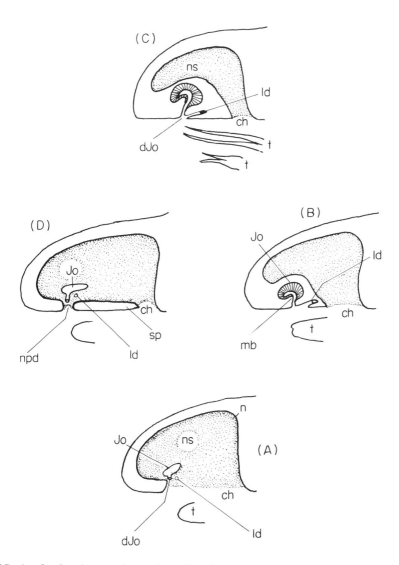

FIG. 1. Jacobson's organ in amniotes. The diagrams show the snout cut in longitudinal section, seen from the left side. Jacobson's organ is shown intact in (A) and (D), and the approximate position of the opening of the lachrymal duct on the *lateral* side of the nose is also indicated. The organ is shown sectioned in (B) and (C). (A) Primitive reptilian, *Sphenodon*-like condition. (B) Lizard with slightly notched tongue tip, as in geckos and some skinks. (C) Snake or lizard with forked tongue (two alternative tongues are shown). (D) Mammal. Abbreviations: ch, choana, dJo, duct of Jacobson's organ; Jo, Jacobson's organ; ld, lachrymal duct; mb, mushroom body; n, nasal sac; npd, nasopalatine duct; ns, nasal septum; sp, secondary palate; t, tongue.

Girons, 1976). As in both Squamata and mammals the organ arises in the embryo as a ventromedial diverticulum from the nasal sac. It becomes a relatively simple, cylindrical structure, devoid of a mushroom body; it communicates simultaneously with both the nose and the mouth, opening into the anterior part of the extensive choana [Fig. 1 (A)]. As Broom (1906) observed, its morphology in *Sphenodon* is in many ways more like that in mammals than in lizards. In many mammals it opens into or near the nasopalatine duct, which seems to represent a vestige of the embryonic choanal opening before the secondary palate develops [see Bellairs, 1951; Negus, 1956; Fig. 1 (D) here].

In almost all Squamata, on the other hand, the organ has a more complicated structure [Fig. 1(B), (C); Fig. 2]. It is rounded rather than cylindrical and has a high dorsal dome lined with sensory epithelium

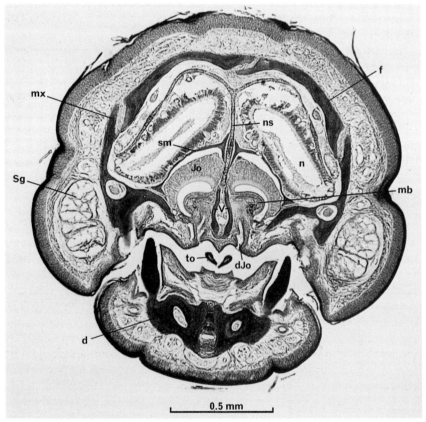

FIG. 2. *Loveridgea ionidesi* (Amphisbaenia). Transverse section through snout showing Jacobson's organs. Abbreviations: d, dentary; dJo, duct of Jacobson's organ; f, frontal; Jo, Jacobson's organ; mb, mushroom body of Jo; mx, maxilla; n, nasal sac; ns, nasal septum; sg, supralabial gland; sm, septomaxilla; to, tongue; v, vomer.

resembling, though not identical with, that of the nose (Wang & Halpern, 1980a). Projecting into its ventral aspect is a mushroom body covered with ciliated epithelium and nearly always supported by a knob of cartilage, the vomeronasal concha; in some lizards and snakes there is a bony knob derived from the vomer as well. Each organ loses its primitive connection with the nasal sac during embryonic life and comes to open on to the front of the palate through a narrow duct. Thus the ducts of the two organs, one on each side of the mid-line, come into a functional relationship with the anterior part of the tongue. Secretions of the eye glands (the Harderian, and when present, the lachrymal) are carried to the vicinity of the duct of Jacobson's organ by the lachrymal duct and may play some part in the transport of odorous substances into its lumen (Bellairs & Boyd, 1950; see Saint Girons, 1982). In both *Sphenodon* and mammals the lachrymal duct has a rather different relationship, discharging into the lateral wall of the anterior part of the nasal sac.

Although there are quantitative variations in the relative size of Jacobson's organ and the extent of its sensory epithelium among different Squamata (Gabe & Saint Girons, 1976), its morphology is on the whole remarkably constant, more so in fact that that of the nose proper. Only in a few lizards such as chameleons and certain iguanids is the organ absent or significantly small (see Armstrong, Gamble & Goldby, 1953; Parsons, 1970). In these lizards the sensory epithelium of the nose is also poorly developed and this reduction of the chemical sense organs may be regarded as a secondary feature associated with arboreal life.

Important recent studies on the fine structure of Jacobson's organ in Squamata and on the role which it plays in certain types of behaviour such as prey-finding, sexual interactions and aggregation have been made (Halpern, 1980, 1983; Wang & Halpern, 1980a, b; Heller & Halpern, 1982; Andrén, 1982; Halpern & Kubie, 1980, 1983; Reformato, Kirschenbaum & Halpern, 1983; Duvall, King & Graves, 1983). In many of these animals, at least, the organ of Jacobson seems to have taken up the role of an alternative organ of smell which complements rather than supplements the function of the nose proper; apparently the vomeronasal and main olfactory nerves have different central connections with the brain. It has been suggested that the organ of Jacobson is concerned with fine discrimination between different chemical stimuli in the close vicinity, while the nose proper responds less selectively to more distant stimuli (see Duvall, 1981; Halpern, 1983). In some Squamata, such as *Varanus* and snakes, the sensory cells in the organ of Jacobson are relatively more numerous than those in the nose (Gabe & Saint Girons, 1976), and there can be no question that throughout the

Squamata generally the organ of Jacobson is of great functional significance.

One can envisage, perhaps in early Triassic times, a population of *Sphenodon*-like reptiles which differed from their nearest relatives in showing at least two unique innovatory trends: the elaboration of Jacobson's organ beyond the relatively primitive stage in which it exists in *Sphenodon*, and the evolution of paired copulatory organs which are absent in the male *Sphenodon* (though it does have certain cloacal structures which may represent hemipenial precursors; E. N. Arnold, pers. comm.). Such animals would probably also have been acquiring the streptostylic quadrate which is so characteristic of Squamata, though Evans (1980) has shown that this trend was also evident in a variety of eosuchians which were not directly ancestral to Squamata. However, one might suggest that it was a combination of these critical adaptations which gave the early Squamata "the edge" over their eosuchian, sphenodontid ancestors. One might reasonably claim that those reptiles which possessed them, even in a nascent form, could properly be called lizards. It is easier to believe that the elaborate organ of Jacobson with its mushroom body, and the hemipenes, were evolved once rather than two or three times in the origin of the main squamate groups, the Sauria, the Amphisbaenia and the Serpentes.

It seems likely that in the early lizards, as in *Sphenodon*, the tongue was broad and fleshy; it probably served as a useful instrument for "manipulating" food in the mouth, and for lapping fluid. It may also have possessed taste buds, such as occur in some modern lizards, but there is little evidence that taste is important in reptiles (see Duvall, 1981). At some stage, however, the tongue acquired a further function, the transport of scents to the vicinity of the organ of Jacobson. This stage is perhaps exemplified in the iguanids and agamids, which retain the broad, manipulative tongue and in which the organ of Jacobson seems in general to be rather less well developed than in the majority of other lizards. Nevertheless, some of these lizards, at least, have established the habit of exploratory tongue protrusion (Gove, 1979), so that odorous substances (e.g. pheromones produced by conspecifics: Duvall, 1981) can be picked up by the tongue tip. Geckos also have a broad fleshy tongue, though it may be notched at the tip as in many skinks [Fig. 1(B)]. However, the organ of Jacobson is well developed in geckos (Gabe & Saint Girons, 1976), perhaps in association with the various cutaneous gland-like structures which probably produce odoriferous secretions (Maderson, 1970).

In many other lizards, such as lacertids, teiids and anguids, the tongue has become narrower and in some cases is definitely deeply forked at its tip [Fig. 1(C)]. It may retain some of its primitive manipulatory function, but has also become more highly adapted to acting in conjunction with Jacobson's organs. The tongue tips are particularly fine in the monitor lizards (*Varanus*) and in snakes, squamates in which the organ is especially well developed (Gabe & Saint Girons, 1976) and perhaps has a more important function than the nose proper. It is significant that snakes have evolved a more complex pattern of exploratory tongue movement than is shown by many lizards (Gove, 1979).

The precise mechanism of scent transport from the tongue to the organ of Jacobson is not yet fully understood. It has been generally supposed that odorous substances are wiped off the tongue on to the palate, into or against the openings of the ducts of Jacobson's organs, or into palatal grooves, lubricated by saliva and the secretions of the orbital glands, which in many lizards lead to the vicinity of the ducts. The dissolved particles are then carried to the lumen of each organ by a current set up by the cilia on the mushroom body. It has been suggested that the fine tongue tips of varanid lizards and snakes can be actually inserted into the lumina of Jacobson's organs, but this is now thought very doubtful (Oelofsen & Van Den Heever, 1979). Another view, which does not necessarily exclude the first, is that dissolved particles are sucked into the organ by the release of pressure on its floor, previously exerted by the tongue, or by material in the mouth. The slender tongues of snakes, however, would not be suitable for this purpose (see Pratt, 1948; Parsons, 1970; Bertmar, 1981). A third and new hypothesis, supported by some experimental evidence, has been advanced by Gillingham & Clark (1981). They believe that in snakes the odorous particles are wiped off the *ventral* surface of the tongue on to a pair of ridges or pads ("anterior processes") which project from the floor of the mouth. These ridges with their adherent particles are elevated just after the tongue has been retracted and pressed into the openings of the ducts of Jacobson's organs, which lie just above them.

It is interesting that the romantic naturalist-writer W. H. Hudson devoted a whole chapter of one of his books [1919] to the snake's tongue. Disillusioned with the academic zoology of his day, he believed that the reason for the tongue's flickering activity could only be elucidated by observation in the wild. This may, indeed, be partly true. Yet a single transverse section through the snout of a snake showing Jacobson's organs and their ducts, with the tongue tips situated beneath, suggests some association between these organs (see Fig. 2 of a similar section of an amphisbaenian). One might therefore suggest that on the basis of the structure of Jacobson's organ (admittedly a single organ), the most primitive members of the Squamata were indeed lizards, as has been traditionally believed. Also that both the amphisbaenians and the snakes diverged independently from the saurian stock *after* the organ of Jacobson had become elaborated and the tongue had become more or less deeply forked. Rage (1982b) has suggested that in some respects the amphisbaenians are closer to the snakes than to the lizards. However, many peculiar features of these animals, such as the morphology of the anterior part of the brain-case (Bellairs & Gans, 1983) seem to indicate an independent origin from snakes. The organ of Jacobson is well developed in amphisbaenians (Fig. 2), as one might expect in burrowers, but the tongue (as in "micro-teiids") is broad and fleshy at the base and quite finely forked at the tip (Bogert, 1964). Perhaps this type of tongue best combines both the main lingual functions found among squamates: manoeuvring food in the mouth, and the transport of odorous substances to the organ of Jacobson (Underwood, 1971).

I would finally like to comment on one significant difference between typical lizards and typical snakes which has been cited as evidence that the latter could not readily have been derived from the former (Hoffstetter, 1968). In lizards the chondrocranium is tropitrabic, with the trabeculae fused at mid-orbital level and usually participating in the interorbital septum; in snakes the trabeculae are characteristically unfused at this level (platytrabic) and there is no cartilaginous interorbital septum. However, the distinction is not so clear-cut as has been suggested, since in some Typhlopidae (as in some amphisbaenians), the condition, though technically tropitrabic, seems to be intermediate between the more typical kinds of tropitraby and platytraby (see Bellairs & Kamal, 1981). A shift between the two, accompanied by the loss of the orbital cartilages (as is seen in the pygopodid lizard *Aprasia repens*), might have occurred as an adaptation to burrowing life involving changes in the relative proportions of the skull and the principal sense organs. Such changes could have occurred in saurian, as opposed to eosuchian, ancestors.

THE INTERMANDIBULAR CONNECTION IN SQUAMATA

My second topic, which also calls for micro-anatomical methods of study, is less well known than the previous one; in fact, very little has been written about it. This is the anterior connection between the two sides of the lower jaw in reptiles. In many tetrapods the tips of the two dentaries articulate at a symphysial joint which allows varying degrees of intermandibular mobility. Such a joint is generally present in lizards (Gans, 1961), in crocodilians, and in many, probably the majority of mammals (see Scapino, 1965; Moore, 1981); in higher Primates including man the joint is lost through fusion of the mandibular symphysis, beginning shortly after birth. Fusion also occurs in birds (see Hogg, 1983) and in chelonians, here perhaps in association with the development of a continuous horny beak.

In *Lacerta, Anguis* and *Chamaeleo* the tips of the dentaries and of the persistent Meckel's cartilages are joined by dense connective tissue which in places contains fibrocartilage. This last tissue is less evident in a Weigert-stained preparation of *Varanus griseus*, where numerous transverse elastic fibres can be seen in the symphysial region; possibly this feature is correlated with the considerable mobility of the varanid jaw. In the amphisbaenian *Loveridgea* the fibrocartilaginous component of the symphysis is well developed [Fig. 3(A)]; Meckel's cartilage seems to have disappeared.

Snakes are remarkable for the fact that there is typically no symphysial joint between the tips of the dentaries, which are quite separate from each other. This is, of course, an adaptation to the swallowing of relatively large prey, and allows independent movement of the two sides of the lower jaw, laterally, vertically and in the longitudinal plane (Gans, 1961; Kardong, 1977; Cundall & Gans, 1979). Owen (1866: 151) stated that in snakes "the front of the lower jaw is united to that of the opposite ramus by an elastic ligament". Similar remarks were made by Huxley (1871: 202) and appear in more recent texts (e.g. Smith, 1943; Orr, 1976). However, the actual conditions have been little studied despite the number of recent papers dealing with the jaw muscles and kinesis of snakes (e.g. Albright & Nelson, 1959a, b; Kardong, 1977). In this context the account of Groombridge (1979) is most relevant.

In the Typhlopidae the mandibular connection is exceptional among snakes in being relatively rigid, the dentary tips being joined by a cartilaginous nodule which is continuous with the front of Meckel's cartilage on either side (Bellairs & Kamal, 1981). In other primitive snakes of which sectioned material is available (*Anilius, Rhinophis, Uropeltis* and newborn *Boa* and *Eryx*, the dentary tips appear in transverse section to be connected by fibrous tissue which may be condensed in the mid-line to form a median band [see Fig. 3 (B) of *Anilius*], here labelled the intermandibular ligament. Variations among these primitive snakes undoubtedly occur (see Groombridge, 1979) and might profitably be studied by histological methods.

In some, possibly in most or all caenophidian snakes, a more striking intermandibular structure can be seen. It appears much more conspicuous in histological preparations than in dissections, which is probably why it has attracted so little attention. Its presence has, however, been noted by some previous workers [e.g. Groombridge

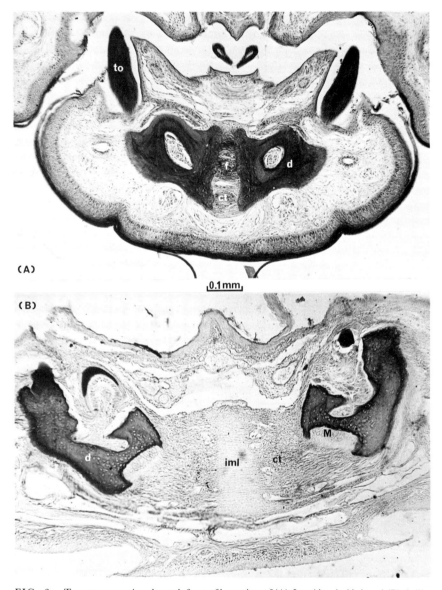

FIG. 3. Transverse section through front of lower jaw of (A) *Loveridgea ionidesi*, and (B) *Anilius scytale*, subadult, showing intermandibular connection. Abbreviations: ct, dense connective tissue; d, dentary; f, fibrocartilage; iml, intermandibular ligament; M, Meckel's cartilage; to, tooth.

(1979) in *Xenochrophis piscator*; Bellairs & Kamal (1981: 162)]; it can also be seen in sections of newborn *Vipera berus*. A short paper has been specifically written about it by Kiran (1981) in *Oligodon arnensis*; she terms it the corpus musculo-fibrosus. However, the term intermandibular nodule seems preferable and affords an adequate distinction from the less well differentiated intermandibular ligament of primitive snakes.

In hatchling *Natrix natrix* the general shape of this nodule, with its expanded head and tapering body, is shown in Figs 4 and 5. It lies in the mid-line, beneath the mucosa of the floor of the mouth, anterior to the

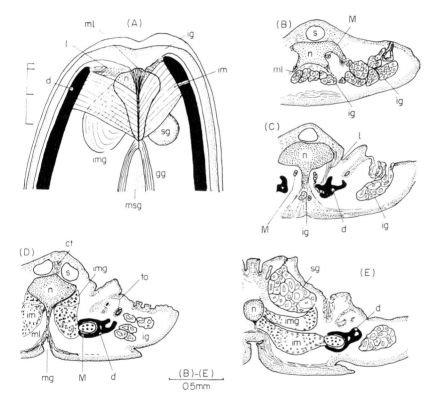

FIG. 4. (A) Diagram showing intermandibular nodule and associated structures in caenophidian snake in ventral view; structures spread out laterally. In this view the nodule lies beneath the muscles. The pars posterior of the intermandibularis anterior muscle, and the protractor laryngeus are not shown. (B)–(E) *Natrix natrix*, hatchling. Transverse section through lower jaw at progressively posterior levels approximately indicated by horizontal lines on left of (A). Abbreviations: ct, connective tissue; d, dentary; gg, genioglossus muscle; ig, infralabial gland; im, intermandibularis anterior muscle, pars anterior; img, gland constrictor part of im; l, lateral ligament; M, Meckel's cartilage; mg, mental groove; ml, median ligament; msg, median sublingual gland; n, intermandibular nodule; s, blood sinus; sg, lateral sublingual gland; to, tooth and dental lamina.

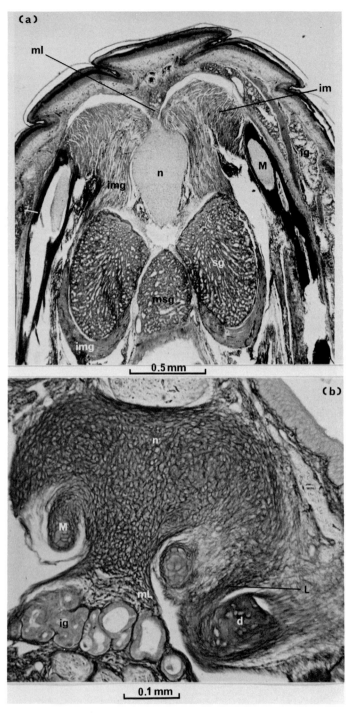

FIG. 5. *Natrix natrix*, hatchling. (A) Horizontal section through lower jaw showing intermandibular nodule and associated structures. (B) Transverse section through nodule at level between sections (B) and (C) in Fig. 4. Abbreviations: d, dentary; ig, infralabial gland; im, pars anterior of intermandibularis anterior muscle; img, gland constrictor part of im; l, lateral ligament; M, Meckel's cartilage; ml, part of median ligament; msg, median sublingual gland; n, intermandibular nodule; sg, lateral sublingual gland.

tongue sheath and just ventral to a large blood sinus which bifurcates posteriorly. It is composed of very dense fibrous tissue which seems to resemble the "fibrovesicular tissue" of certain unchondrified and unossified sesamoid bodies described by Haines (1969) in reptilian limbs. It also resembles procartilage, and looks as if it might become chondrified in maturity, though I have found no evidence for this.

Anteriorly, the intermandibular nodule of *Natrix* is attached to the infralabial gland which is continuous across the front of the lower jaw, while in the ventral mid-line it is attached by a big median ligament to the subcutaneous tissue of the mental groove, between the infralabial scales. Dorsally, it is attached to the oral mucosa by connective tissue. The head of the nodule is also connected by a lateral ligament with the tip of the dentary on each side. A Weigert preparation indicates that this ligament contains some elastic tissue, and one would expect it to become stretched when the snake was engulfing prey.

The pars anterior of the intermandibularis anterior muscle originates from each side of the nodule and from the ventral median ligament [Fig. 4(A), (D), (E)]. The fibres of the ventral part of this muscle are inserted on to the front part of the dentary and on to Meckel's cartilage as it lies in its groove in the bone. The more dorsal fibres of this muscle become increasingly distinct as one passes backwards, and surround the lateral sublingual or mandibular gland (see Smith & Bellairs, 1947; Kochva, 1978); similar conditions have been described in other snakes by Groombridge (1979) and previous workers. The protractor laryngeus and the medial head of the genioglossus muscle on each side are also attached to the posterior tip of the nodule, and other muscles such as the pars posterior of the intermandibularis anterior may be attached here also. This last muscle appears to lie posterior to the nodule in *Natrix natrix* and is not shown in Fig. 2. The nodule is deeply forked posteriorly in acrochordid snakes, perhaps in association with specializations of the anterior mandibular region (Groombridge, 1979).

Kiran (1981) supposed that the intermandibular nodule "works almost like an angler's reel and can deal out lengths of musculotendinous fibres in accordance with the degree of distension imposed on the throat in feeding". A somewhat different explanation is suggested here. The complex movements made by the snake's mandible during feeding take place at both the jaw joint proper and at the intramandibular hinge between the dentary and the more posterior bones. They include protraction and retraction, and abduction and adduction of the dentary tips. The latter movement is produced mainly by contraction of the pars anterior of the intermandibularis anterior; abduction is mainly though not entirely a passive movement produced by the distensive effect of prey in the mouth (Albright & Nelson,

1959b). The ligamentous connections of the intermandibular nodule would tend to hold it in or near the mid-line, providing an anchorage for the origin of the pars anterior muscle. This would seem to facilitate independent movement of the two "mandibles" by preventing the tension exerted by one contracting pars anterior muscle from being transmitted across to the other side of the jaw. Such an effect might operate when one dentary tip was adducted while the other one remained relatively abducted, as shown by Cundall & Gans (1979: fig. 3d) and in Fig. 4(A) here. It is true that these authors found that with the mandibles in this position, as in the last phase of jaw closure during a swallowing cycle (in *Nerodia rhombifera* and *N. fasciata*), the anterior intermandibular muscles were active on both sides. Possibly, however, the muscle on the abducted side was active while being lengthened. Additionally, the nodule provides a stable origin for the gland constrictor fibres of the intermandibularis anterior, and attachments for certain other muscles as well.

It also seems likely that by virtue of its position in the mid-line of the floor of the mouth, the dense nodule with its ligamentous attachments will assist the muscles, integument and other tissues in preventing over-distension and possible damage caused by struggling prey.

It seems likely that the nodule has been evolved from a simpler type of longitudinal intermandibular ligament such as is found in some primitive snakes, and that as B. C. Groombridge (pers. comm) suggests, its development is associated with improved, or at least more sophisticated methods of swallowing prey.

ACKNOWLEDGEMENTS

I am grateful to Dr E. N. Arnold, Prof. C. L. Foster, Prof. C. Gans, Dr B. C. Groombridge, Dr D. R. Kershaw and Dr G. Underwood for their helpful comments, and to Mr M. Gross for taking the photomicrographs.

REFERENCES

Albright, R. G. & Nelson, E. M. (1959a). Cranial kinetics of the generalised colubrid snake *Elaphe obsoleta quadrivittata*. 1. Descriptive morphology. *J. Morph.* **105**: 193–240.

Albright, R. G. & Nelson, E. M. (1959b). Cranial kinetics of the generalised colubrid snake *Elaphe obsoleta quadrivittata*. 2. Functional morphology. *J. Morph.* **105**: 241–292.

Andrén, C. (1982). The role of the vomeronasal organs in the reproductive behavior of the adder *Vipera berus*. *Copeia* **1982**: 148–157.
Armstrong, J. A., Gamble, H. J. & Goldby, F. (1953). Observations on the olfactory apparatus and the telencephalon of *Anolis*, a microsmatic lizard. *J. Anat.* **87**: 288–307.
Bellairs, A. d'A. (1951). Observations on the incisive canaliculi and nasopalatine ducts. *Br. dent. J.* **91**: 281–291.
Bellairs, A. d'A. (1972). Comments on the evolution and affinities of snakes. In *Studies in vertebrate evolution. Essays presented to Dr F. R. Parrington, FRS*: 157–172. Joysey, K. A. & Kemp, T. S. (Eds). Edinburgh: Oliver & Boyd.
Bellairs, A. d'A. & Boyd, J. D. (1950). The lachrymal apparatus in lizards and snakes. II. The anterior part of the lachrymal duct and its relationship with the palate and with the nasal and vomeronasal organs. *Proc. zool. Soc. Lond.* **120**: 269–310.
Bellairs, A. d'A. & Gans, C. (1982). A reinterpretation of the amphisbaenian orbitosphenoid. *Nature, Lond.* **302**: 243–244.
Bellairs, A. d'A. & Kamal, A. M. (1981). The chondrocranium and the development of the skull in Recent reptiles. In *Biology of the Reptilia*, **11**: 1–263. Gans, C. & Parsons, T. S. (Eds). London: Academic Press.
Bellairs, A. d'A. & Underwood, G. (1951). The origin of snakes. *Biol. Rev.* **26**: 193–237.
Benton, M. J. (1982). The Diapsida: revolution in reptile relationships. *Nature, Lond.* **296**: 306–307.
Bertmar, G. (1981). Evolution of vomeronasal organs in vertebrates. *Evolution, Lawrence, Kans.* **35**: 359–366.
Bogert, C. M. (1964). Amphisbaenids are a taxonomic enigma. *Nat. Hist., N.Y.* **73**: 16–25.
Broom, R. (1906). On the organ of Jacobson in *Sphenodon*. *J. Linn. Soc. (Zool.)* **29**: 414–420.
Camp, C. L. (1923). Classification of the lizards. *Bull. Am. Mus. nat. Hist.* **48**: 289–481.
Carroll, R. L. (1977). The origin of lizards. In *Problems in vertebrate evolution.* (Linnean Society Symposium Series No. 4: 359–396.) Andrews, S. Mahala, Miles, R. S. & Walker, A. D. (Eds). London: Academic Press.
Cundall, D. & Gans, C. (1979). Feeding in water snakes: an electromyographic study. *J. exp. Zool.* **209**: 189–208.
Cuvier [G.L.C.F.D. Baron] (1811). Fait à l'Institut, sur un mémoire de M. Jacobson, intitulé: Description anatomique d'un organe observé dans les mammifères. *Annls Mus. Hist. nat. Paris* **18**: 412–424. [Apparently a transmission of L. L. Jacobson's paper.]
Dobson, J. (1946). *Anatomical eponyms*. London: Baillière, Tindall & Cox.
Dowling, H. G. (1959). Classification of the Serpentes. *Copeia* **1959**: 38–52.
Duvall, D. (1981). Western fence lizard *(Sceloporus occidentalis)* chemical signals. II. A replication with naturally breeding adults and a test of the Cowles and Phelan hypothesis of rattlesnake olfaction. *J. exp. Zool.* **218**: 351–361.
Duvall, D., King, M. B. & Graves, B. M. (1983). Fossil and comparative evidence for possible chemical signaling in the mammal-like reptiles. In *Chemical signals in vertebrates*, **3**: 25–44. Müller-Schwarze, D. & Silverstein, R. M. (Eds). New York: Plenum.
Estes, R. (1983). The fossil record and early distribution of lizards. In *Advances in herpetology and evolutionary biology. Essays in honor of Ernest E. Williams*: 365–398. Rhodin, A. G. J. & Miyata, K. (Eds). Cambridge, Mass.: Museum of Comparative Zoology, Harvard University.
Evans, S. E. (1980). The skull of a new eosuchian reptile from the Lower Jurassic of South Wales. *Zool. J. Linn. Soc.* **70**: 203–264.

Ferguson, M. W. J. (1981). The value of the American Alligator (*Alligator mississippiensis*) as a model for research in craniofacial development. *J. Craniofacial Genet. Devel. Biol.* **1**: 123–144.
Gabe, M. & Saint Girons, H. (1976). Contribution à la morphologie comparée des fosses nasales et de leurs annexes chez les lépidosoriens. *Mém. Mus. natn. Hist. nat. Paris* (A) **98**: 1–87.
Gans, C. (1961). The feeding mechanism of snakes and its possible evolution. *Am. Zool.* **1**: 217–227.
Gillingham, J. C. & Clark, D. L. (1981). Snake tongue-flicking: transfer mechanics to Jacobson's organ. *Can. J. Zool.* **59**: 1651–1657.
Gove, D. (1979). A comparative study of snake and lizard tongue-flicking, with an evolutionary hypothesis. *Z. Tierpsychol.* **51**: 58–76.
Groombridge, B. C. (1979). Comments on the intermandibular muscles of snakes. *J. nat. Hist.* **13**: 477–498.
Haines, R. W. (1969). Epiphyses and sesamoids. In *Biology of the Reptilia*, **1**: 81–115. Gans, C., Bellairs, A. d'A. & Parsons, T. S. (Eds). London: Academic Press.
Halpern, M. (1980). Chemical ecology of terrestrial vertebrates. In *Animals and environmental fitness*: 263–282. Gilles, R. (Ed.). Oxford: Pergamon Press.
Halpern, M. (1983). Nasal chemical senses in snakes. In *Advances in vertebrate neuroethology*: 141–176. Ewert, J. P., Capranica, R. R. & Ingle, D. J. (Eds). New York: Plenum.
Halpern, M. & Kubie, J. L. (1980). Chemical access to the vomeronasal organs of garter snakes. *Physiology Behav.* **24**: 367–371.
Halpern, M. & Kubie, J. L. (1983). Snake tongue flicking behavior: clues to vomeronasal system functions. In *Chemical signals in vertebrates*, **3**: 45–72. Silverstein, R. M. & Muller-Schwarze, D. (Eds). New York: Plenum.
Heller, S. B. & Halpern, M. (1982). Laboratory observations of aggregative behaviour of garter snakes, *Thamnophis sirtalis*: roles of the visual, olfactory and vomeronasal senses. *J. comp. Physiol. Psychol.* **96**: 984–999.
Hoffstetter, R. (1962). Revue des récentes acquisitions concernant l'histoire et la systématique des squamates. *Colloq. int. Cent. nat. Rech. Sci.* No. 104: 243–279.
Hoffstetter, R. (1968). [Review of] *A contribution to the classification of snakes* [by Garth Underwood]. *Copeia* **1968**: 201–213.
Hogg, D. A. (1983). Fusions within the mandible of the domestic fowl (*Gallus gallus domesticus*). *J. Anat.* **136**: 535–541.
Hudson, W. H. (1919). *The book of a naturalist*. London: Nelson.
Huxley, T. H. (1871). *A manual of the anatomy of vertebrated animals*. London: J. & A. Churchill.
Kardong, K. V. (1977). Kinesis of the jaw apparatus during swallowing in the cottonmouth snake, *Agkistrodon piscivorus*. *Copeia* **1977**: 338–348.
Kiran, U. (1981). A new structure in the lower jaw of colubrid snakes. *Snake* **13**: 131–133.
Kochva, E. (1978). Oral glands of the Reptilia. In *Biology of the Reptilia*, **8**: 43–162. Gans, C. & Gans, K. A. (Eds). London: Academic Press.
Maderson, P. F. A. (1970). Lizard glands and lizard hands: models for evolutionary study. *Forma et Functio* **3**: 179–204.
McDowell, S. B. & Bogert, C. M. (1954). The systematic position of *Lanthanotus* and the affinities of the anguinomorphan lizards. *Bull. Am. Mus. nat. Hist.* **105**: 1–142.
Medawar, P. B. (1969). *The art of the soluble*. Harmondsworth, Middlesex: Penguin Books.
Moore, W. J. (1981). *The mammalian skull*. Cambridge University Press.
Negus, V. E. (1956). The organ of Jacobson. *J. Anat.* **90**: 515–519.

Oelofsen, B. W. & Van Den Heever, J. A. (1979). Role of the tongue during olfaction in varanids and snakes. *S. Afr. J. Sci.* **75**: 365–366.
Orr, R. T. (1976). *Vertebrate biology*. 4th Edn. Philadelphia: Saunders.
Owen, R. (1866). *On the anatomy of vertebrates*. I. *Fishes and reptiles*. London: Longmans, Green.
Parsons, T. S. (1967). Evolution of the nasal structure in the lower tetrapods. *Am. Zool.* **7**: 397–413.
Parsons, T. S. (1970). The nose and Jacobson's organ. In *Biology of the Reptilia*, **2**: 99–191. Gans, C. & Parsons, T. S. (Eds). London: Academic Press.
Powers, J. B. & Winans, S. S. (1975). Vomeronasal organ: critical role in mediating sexual behavior of the male hamster. *Science, Wash.* **187**: 961–963.
Pratt, C. W. McE. (1948). The morphology of the ethmoidal region of *Sphenodon* and lizards. *Proc. zool. Soc. Lond.* **118**: 171–201.
Rage, J-C. (1982a). L'histoire des serpents. *Pour la Science* 1982: 16–27.
Rage, J-C. (1982b). La phylogénie des lépidosauriens (Reptilia): une approche cladistique. *C. r. Séanc Acad. Sci, Paris* **294**: 563–566.
Reformato, L. S., Kirschenbaum, D. M. & Halpern, M. (1983). Preliminary characterisation of response-eliciting components of earthworm extract. *Pharmacol. Biochem. Behav.* **18**: 247–254.
Saint Girons, H. (1982). Histologie comparée des glandes orbitaires des Lépidosauriens. *Annls Sci. nat.* (Zool.) (13) **4**: 171–191.
Scapino, R. P. (1965). The third joint of the canine jaw. *J. Morph.* **116**: 23–50.
Smith, M. A. (1931–1943). *The fauna of British India, Ceylon and Burma* (3 *including the whole of the Indo-Chinese sub-region*). *Reptilia and Amphibia*, 3 vols. London: Taylor & Francis.
Smith, M. [A.] (1973). *The British amphibians and reptiles*, 5th Edn, revised by A. d'A. Bellairs & J. F. D. Frazer. London: Collins.
Smith, M. [A.] & Bellairs A. d'A. (1947). The head glands of snakes, with remarks on the evolution of the parotid gland and teeth of the Opisthoglypha. *J. Linn. Soc. (Zool).* **41**: 351–368.
Snow, C. P. (1959). *The two cultures and the scientific revolution*. (Rede Lectures.) Cambridge: Cambridge University Press.
Underwood, G. (1967). *A contribution to the classification of snakes*. London: Trustees of the British Museum (Natural History).
Underwood, G. (1970). The eye. In *Biology of the Reptilia*, **2**: 1–97. Gans, C. & Parsons, T. S. (Eds). London: Academic Press.
Underwood, G. L. (1971). A modern appreciation of Camp's "Classification of the lizards". In *Facsimile reprint of C. L. Camp: Classification of the lizards*: vii–xvii. Lawrence, Kans.: Society for the study of amphibians and reptiles.
Walls, G. L. (1942). The vertebrate eye and its adaptive radiation. *Bull. Cranbrook Inst. Sci., Michigan* No. 19: 1–785.
Wang, R. T. & Halpern, M. (1980a). Scanning electron microscopic studies of the surface morphology of the vomeronasal epithelium and olfactory epithelium of garter snakes. *Am. J. Anat.* **157**: 399–428.
Wang, R. T. & Halpern, M. (1980b). Light and electron microscopic observations on the normal structure of the vomeronasal organ of garter snakes. *J. Morph.* **164**: 47–67.

Subject Index

Numbers in italics refer to figures

Ablepharus, 499
Abrictosaurus, 542
Abronia deppii, 82
Acanthodactylus, 67, 72, 499
Acerosodontosaurus, *578*, 581, 582
Acontias, 68, 509
 A. plumbeus, 82
Acontias sp., *508*
Acontophiops, 509
Acrochordus granulatus, 368, *369*
Acrodont ankylosis, dentitions, 558
Adolfus, 72
Aetosauroides, 600
Agama
 A. flavimaculata, 81
 A. tuberculata, 366
 A. yemenensis, 81
Agamidae, 59, 74
Age structure analysis, Australian crocodile, 321, 328, 343
Aging technique, alligators, 252
Agkistrodon, 14, 474
 A. piscivorus, 367, *369*
Algyroides, 513
 A. fitzingeri, 508, *508*, *509*, 513
Allelic exclusion, regulatory gene function, 312
Alligator, 458
 A. mississippiensis, 169, 223–273, 282, 285, 292, 323–355, 357, *374*, 443, 444, 464, 465, 636, 655
Alligator limb buds, *214*
Alocodon, 522, 542
Alveolar bone, tooth attachment, 570
Amblyrhynchus, 56, *57*, 540
 A. cristatus, 81
Ambystoma, 180, 181
Ameiva, 66, 366
 A. ameiva, 82, *508*
Amelogenesis, 275–304
American alligator, 282
Amia, 89
 A. calva, 129

Amino acids
 enamel matrix proteins, vertebrates, *283*
 porcine amelogenins, *284*
Amniote limb bud, *198*
Amniote limbs, 197–221
Amphibolurus muricatus, 81
Amphisbaenians, 69, 77
Anabas, 396
Anadia, 66
 A. brevifrontalis, 82
Anaerobic glycolysis, habits, 425–441
Anatomy
 squamate epidermis, 112
 therapsids and archosaurs, 604
 tympanic membranes, *134*
Anatosaurus, 631, 650, 673
Androgen cycles, snakes, *369*
Angling, turtles, 88
Angolosaurus, 68
 A. skoogi, 82
Anguidae, 62, 75
Anguis, 62, 500, 675
 A. fragilis, 82, *508*, 510, *510*
Anilius, 517, 675
 A. scytale, 676
Anisolepis, 59, 74
 A. grillii, 81
Anniella, 47, 75, 78, 500, 513, 514
Anniella pulchra, 82, *507*, 510, *510*
Anniellidae, 64, 75
Annual reproductive cycle, *Alligator*, 375
Anolis
 A. bimaculatus, 59, 81
 A. biporcatus, 59, 81
 A. carolinensis, 374, 420
 A. chlorocyanus, 59, 81
 A. chrysolepis, 59, 81
 A. coelestinus, 59, 81
 A. cristellatus, 81
 A. cuvieri, 59, 81
 A. cybotes, 59, 81
 A. equestrio, 59, 81

Anolis (cont.)
  A. garmani, 59, 81
  A. lemurinus, 81
  A. lineatopus, *484*
  A. nebulosus, 59, 81
  A. richardi, 59, 81
  A. sagrei, 59, 81
  A. valencienni, 59
Anolis sp., 119
Anterior extremity, anterior limb, *Varanus*, 36
Anterior limb, *Varanus*, 27–45
Anterior plastron, turtles, 102, *103*
Anteroposterior axis, amniote limb, 203
Antigens, enamel proteins, vertebrates, 291, 293
Aprasia, 505, 513, 514
  A. repens, 511, *512*, 674
Aprasia sp., *508*
Aptycholaemus, 59, 74
  A. longicaudus, 81
Archosaurs, 597–627
Aristosuchus, 631, *638*, 641
Askeptosaurus, 580, 587, 588, *588*
Aspidolaemus affinis, 82
Atherostonia, 583
Atractaspis, 478
Australian freshwater crocodile, 319–355
Avian embryonic limb, 200

B

Balaenoptera sp., 129
Basiliscus, 56
  B. vittatus, 81
Basking, metabolic rate, 437
Batrachemys, 95, 104, 105, 108
Bavarisaurus, 583
Behavioural factors, skin shedding, squamates, 120
Bellairs, Professor Angus d'Albini, 3–10
Biarmosuchus, 658
Blomosaurus, 581
Blood flow and habits, 426
Boa, 675
Body size, Nile crocodile, 458
Body temperature
  lizards, *409*
  Nile crocodile, 463, 448
Body twist, squamates, 21

Bone
  development, 155–176
  dinosaurs, 629–663, *632–648*
Boreopricea, 580
Brachial arches, craniofacial development, 226
Brachiosaurus, 629, 635, 636, 642, 655
Brachylophus, 56
  B. fasciatus, 81
  B. vitiensis, 81
Brachyrhinodon, 581, *590*
Bradypodion pumilus, *508*
Bradysaurus, 657, 658
Breeding, gharial, 385–406
Breeding enclosure, design, gharial, 390
Breeding stock, gharial, 397
Bristol fissures, 292
Brookesia, 60, 82
Bufo, 208
Bungarus, 474

C

Caiman crocodilus, 443, 463, 465
Callisaurus, 57, 74, 483–502
  C. draconoides, 81
Callopistes, 64, 72
  C. maculus, 82
Calotes versicolor, 81
Camptosaurus, 528, 529, *543*, 544
Cancellous bone, dinosaurs, 643
Captive breeding, crocodiles, 385–406
Carboniferous reptiles, *584*
Cardiac skeleton, 166
Caretta caretta, 99, *100*
Cartilage, development, 155–176
Casarea dussumieri, 160
Cells, vertebrate limb development, *190*
Cellular basis, signalling, amniote limbs, 205
Cerberus rhynchops, 368
Cercosaura ocellata, 82
Cetiosaurus, *632*, *638*
Chalaradon, 58
  C. madagascariensis, 81
Chalcides ocellatus, *508*
Chamaeleo, 59, 675
  C. africanus, 60, 82
  C. dilepis, 60, 82
  C. johnstonii, 82

Subject Index

Chamaeleonidae, 59, 75
Chamaelinorops, 59
   C. wetmorei, 81
Character analysis
   lizards, 483–502
   scleral ossicles, lizards, 489, 496
Chasmatosaurus, 580, *586*
Chasmatosuchus, 580
Chelidae, 86–110
   long-necked, evolution, 86–110
Chelodina, 87–110, *102*
   C. intergularis, 101
   C. longicollis, 100
   C. novaeguineae, *96*, *97*, 104
   C. oblonga, 108
   C. rugosa, 101
   C. steindachneri, 108
Chelonia mydas, 99, *100*, 360, 363, 365
Chelus
   C. colombianus, 101
   C. fimbriatus, 87–110, *96*, *107*
Chelydra, 87–110
   C. serpentina, 91, 347, 363
Chinemys reevesi, *91*
Chiniquodon, 600
Chitra, 87–110
   C. indica, 92, *92*
Chitracephalus, 87
   C. dumonni, 94
Chlamydosaurus, 59
   C. kingi, 81
Chorda tympani
   lizards, 136
   mammals, 136
Chromosomal hijacking, 311
Chromosomal inactivation
   insects, 314
   mammals, 305–316
   reptiles, 305–316
Chrysemys, 362, 363
   C. picta, 348, 360, *364*, 365
Clade models, evolution, *552*
Clades, evolution, *556*, 563
Clarazia, 587, 588
Clarias, 396
Classification, Diapsida, 576
Classification, Permo-Triassic diapsids, 580
Claudiosaurus, 576, 580, 582, *584*, 585
Cleft lip, alligator development, *262*
Clevosaurus, 581, 585, 592, *592*

Cloacal muscles, lizards, 47–85
Cloacal region, *Sphenodon*, 70
Cnemidophorus, 66
   C. lemniscatus, 366
Coelurosauravus, 576, 580, 583, *584*, 585
Coleonyx, 61
   C. elegans, 82
Colubrifer, 581
Compact bone, dinosaurs, 634
Compacted coarse-cancellous bone, dinosaurs, 649
Competition, therapsids and archosaurs, 597–627
Competitive replacement, therapsids and archosaurs, 601
Conolophus subcristatus, 81
Conservation strategy, crocodiles, 385–406
Cooling rate, Nile crocodile, 456, 464
Cophosaurus, 57, 74, 496
   C. texanus, 81
Cordylidae, 68
Cordylus, 68, 499
   C. cordylus, 82
Corythophanes, 56
   C. percarinatus, 81
Cosesaurus, 580
Cranial architecture, ornithopod dinosaurs, 522
Cranial flexibility, *Iguanodon*, 535
Cranial morphology, ornithopod dinosaurs, 521–547
Craniofacial development, *Alligator mississippiensis*, 223–273
Crocodilia, 374
Crocodilurus lacertinus, 82
Crocodylus
   C. acutus, 389, 443, 444, 464
   C. johnstoni, 223–273, 319–355, *325*, *328*, *334*, *336*, *337*, *342*, 443, 444, 458, *461*, 464
   C. niloticus, 129, 226, 249, 252, 374, 431, 443–467, *456*, 646
   C. novaeguineae, 443, 463
   C. palustris, 385, 386, 398
   C. porosus, 226, 228, 249, 251, 323–355, 385, 389, 391, 398, 443, 464, 466
   C. rhombifer, 389
   C. siamensis, 252, 389
Crotalus, 474
   C. durissus terrificus, 472

C. scutulatus, 476
C. terrificus, 166
C. viridis helleri, 475
C. viridis viridis, 255
Ctenoblepharis, 58, 74
C. adspersus, 81
Ctenosaura similis, 81
Cutaneous water loss, squamate reptiles, 116
Cyclic activity, squamate epidermis, 113
Cyclura, 56
C. nubila, 81
Cylindrophis, 517
Cynognathus, 585, *586*, 587
C. crateronotus, 570
Cyrtodactylus louisiadensis, 82

D

Daedalosaurus, 583
Dasia, 68
D. smaragdina, 82
Deirochelys, 87–110, 99, 657
D. reticularia, 91
Dentition, evolution, 549–574
Dermochelys, 646
D. coriacea, 168
Detrahens mandibulae muscle, lizards and mammals, 142
Development, 133
 alligator, craniofacial, 223–273
 amniote limb, 197
 cartilage and bone, 155–176
 vertebrate appendages, 177–196
Diapsida, 575
Dibamus, 513–520
D. novaeguineae, 504, 505, *515*, *516*
Dibamus sp., *508*
Dicrodon, 66
D. guttulatum, 82
Didelphis virginiana, 129, *134*
Dimetrodon, 657
Diploglossus, 63
D. monotropis, 82
D. sternus, 82, *63*
Diplolaemus, 58
D. bibronii, 81
Diplometopon, 69
D. zàrudnyi, 82
Dipsosaurus, 56

D. dorsalis, 81
Dispersal, Australian crocodile, 322
Dispholidus typus, 478
Diving, metabolic rate, 434
Dorsoventral axes, amniote limbs, 202
Draco, 583
Drosophila, 200, 310
Dryosaurus, 542
Dysalotosaurus, 544

E

Early Jurassic, 592
Early Triassic, 585
Early evolution, Diapsida, 575–596
Ears
 mammals, *131*
 sauropsid, *131*
Echinodon, 522, 542
Echinosaura horrida, 82
Echis carinatus, 15, 23
Ecological factors, skin shedding, squamates, 120
Ecology, lizard growth, 422
Ecpleopus montius, 82
Ectothermy
 archosaurs, 608
 therapsids and archosaurs, 621
Edmontosaurus, 532
E. regalis, 533
Egg handling, gharial breeding, 399
Egg production, Australian crocodiles, *331*
Elaps, 478
Elephantulus myurus, 129
Elongate squamates, 13–26
Elseya, 99, 108
Embryo development, alligators, *327*
 closing palate, 258
 closure zone, 246
 head, *238*, *243*
 palatal shelves, *240*, *244*, *250*
Embryology, tympanic membrane, 130
Embryonic limbs, 211
Embryos, alligators, *227*, *231*
Emydoidea, 99
Emydura, 99, 108
Enamel gene expression, 294
Enamel gene products, development, 275–304

Enamel matrix proteins
  reptiles, 281
  vertebrates, *288*
Endangered crocodilians, 385–406
Endochondral bone, dinosaurs, 645
Endochondral ossification, *159*
Endocrinology, reproduction, 357–383
Endogenous aqueous flux, 118
Endosteal compact bone, dinosaurs, 650
Endothermy
  archosaurs, 610
  dinosaurs, 653
  therapsids and archosaurs, 622
Energy costs, habits, *435*
Energy production, dinosaurs, 439
England, diapsids, 592
Enyalioides, 58
  E. heterolepis, 81
Enyalius, 58
  E. iheringii, 81
Enzymes, snake venoms, *474*
Eozostrodon, 592
Epidermis, squamates, 111–126
Epithelia, response to aqueous flux changes, 117
Eptatretus stoutii, 293
Eremias, 67
Erinaceus europaeus, 129, *131*
Erpetosuchus, 589, *590*
Eruption, tooth attachment, 569
Erythrosuchus, *578*, 580, 585, *586*, 587
Eryx, 675
Eublepharis hardwickii, 82
Eumeces, 77
  E. algeriensis, 68, 82
Eunectes murinus, 255
Euparkeria, *578*, 580, 585, *586*, 587
Eusarkia rotundiformis, 99
Euspondylus manicatus, 82
Evolution, 481
  bone, 155–176
  cartilage, 155–176
  Chelidae, 86–110
  chromosomes, reptiles and mammals, 305–316
  cranial morphology, ornithopod dinosaurs, 521–547
  Diapsida, 575–596
  enamel gene products, reptiles, 275–304
  lizard skull miniaturization, 503–520
  lizard tympanic membrane, 145
  ornithopod dinosaurs, 521
  scleral ossicles, lizards, 497
  sex chromosomes, 305–316
  Squamata, 665–683
  therapsid membrane, 145
  tooth attachment, 549–574
Experimental tests, regeneration and development, 184
External auditory meatus, lizards and mammals, 130, 141

F

Fabrosauroid cranial architecture, 522
Fabrosaurus, 522, 523, 542
Facial clefting, alligator development, 259
Fast, metabolic rate and habits, 431
Faunal evolution, 581
Feeding, metabolic rate and habits, 431
Felis
  F. domestica, 129
  F. domesticus, *133*
Feylinia
  F. currori, 508
  F. elegans, 508
Fibrolamellar bone, dinosaurs, 634
Field nests, Australian crocodile, 326
Food
  breeding in gharial, 396
  growth dynamics, lizards, 417
  requirements, metabolic rate, 437
Forelimb buds, turtles, *213*
Forelimb development, amniotes, *201*
Function, lizard skull miniaturization, 503–520
Functional advantages, tooth attachment, 566
Functional morphology, anterior limb, *Varanus*, 33

G

Galesphyrus, 580
Gallus, 131
  G. domesticus, 129, *160*
Gambelia, 56
  G. wislizeni, 81
Gaping, Nile crocodile, 458, 465

G

Garjainia, 580
Gavialis gangeticus, 385–406
Gehyra, 498
Gekko
  G. gecko, 82, 114, *508*
  G. vittatus, 82
Gekkonidae, 60, 75
Geographic variation, Australian crocodiles, 338
Gephyrosaurus, 581, 592, *593*
Germany, 589
Gerrhonotus, 62, 498, 500
  G. coeruleus, *63*, 82
Gerrhosaurus, 68, 499
Gerrhosaurus sp., 82
Gharial, 385–406
Glossopteris, 583
Glyptops, 87
  G. plicatulus, 91
Gomphosis, dentitions, 559
Gonads, Australian crocodile, *325*
Gonocephalus, 59
  G. modestus, 81
Grafting, vertebrate appendages, *185*
Grafts
  chick wing buds, 212
  polarizing region, amniote limbs, *203*
  retinoic acid, amniote limbs, *207*
  urodeles, *187*
  vertebrate appendages, *186*, *188*, *189*
Graptemys, 212
Graptemys sp., 336
Grenzbitumenzone, 587
Growth dynamics, lizards, 420
Growth parameters, lizards, 419
Growth patterns, lizards, 418
Growth rates, dinosaurs, 637
Growth, lizards, 407–424
Gymnophthalmus pleii, 82

H

Habits, 425–441
Hadrosauroid cranial architecture, 532
Halticosaurus, 591
Haramiya, 592
Hatching, gharial breeding, 400
Hatchlings, Australian crocodile, 326
Haversian bone, dinosaurs, 640
Heating rate, Nile crocodile, 456, 464

Heleosaurus, 576, 577, 580
Heloderma, 500
  H. horridum, 82, *508*
Helodermatidae, 64, 75
Hemidactylus, 498
Hemipenial muscles, lizards, 47–85
Hemipenis
  lizards, 57
  origin, lizards, 72
Hemitheconyx taylori, 82
Hescheleria, 587, 588
Heterodontosaurus, 542
Heteromorphic bone, dinosaurs, 650
Hipposideros sp., 129, *133*
Histogenesis, reptile epidermis, *114*
Histology, dinosaurian bone, 629–663
Holbrookia, 57, 74, 483–502
  H. maculata, 81
  H. propinqua, 494, 496
Homeothermy, 438
Homo, 136
  H. sapiens, 129
Homoeosaurus, 581
Homoplasies, muscles, lizards, 78
Hoplocercus, 58
  H. spinosus, 81
Hoplodactylus, 62
  H. pacificus, 82
Hovasaurus, *578*, 581, 582, 584, *584*
Howesia, *578*, 580, 587
Hydraspis, 108
Hydromedusa, 87–110, *102*
  H. maximiliani, 100, 102
  H. tectifera, *96*, *97*, 97, 100, *106*, 107
Hyperodapedon, *578*, 580, 589, *590*
Hypsilophodon, 525–542, 631, 642, 645, 651, 657
  H. foxii, 526
Hypsilophodontoid cranial architecture, 525

I

Icarosaurus, 580
Ichnotropis, 67
Iguana, 33, 56, *57*
  I. iguana, 81, *608*
Iguanidae, 56, 73
Iguanodon, 528–544, *543*, 631–651, *632–649*
  I. atherfieldensis, *530*, *536*, *537*

Iguanodontoid cranial architecture, 528
Immunohistochemical localization, enamel proteins, 291
Immunological determinants, enamel proteins, 288
Inductive tissue interactions, timing, 168
Inertial homoiothermy, therapsids and archosaurs, 622
Insectivory, theropsids and archosaurs, 623
Insects, chromosomal inactivation, 314
Insulation, Nile crocodile, 457
Intermandibular connection, squamates, 665–683, *676–678*
Intratendinous ossification, 165

J

Jacobson's organ, amniotes, *669*
Jacobson's organ, Amphisbaenia, 670
Jaw adductor musculature, lizard skull, 513

K

Kadimakara, 580
Kannemeyeria, 585
Kentropyx, 66
 K. calcaratus, 82
Kenyasaurus, *578*, 582, 584
Kinesis
 Iguanodon, 537
 ornithopod dinosaurs, 539
Kinosternon sp., 347
Knollenmergel, 589
Kudnu, 581
Kuehneosaurus, 580, *593*
Kuehneosuchus, 580,
Kuehneotherium, 592

L

Labrus bergylta, 558
Lacerta, 66, 499, 675
 L. agilis, 47, 83, *508*
 L. lepida, 72, 255
 L. lilfordi, *508*
 L. muralis, 129
 L. sicula, 373
 L. stirpium, 83
 L. viridis, 372
 L. vivipara, 129, *131*, *133*, *134*, 157, 255, *371*, 407–424, *412–418*
Lacertidae, 66, 76
Lacertulus, 580
Lachesis
 L. alternatus, 166
 L. lanceolatus, 166
 L. neuwiedii, 166
Locomotion, squamates, 13–26
Laemanctus, 56
 L. serratus, 81
Lamellar-zonal bone, dinosaurs, 635
Lanthanotidae, 75
Lanthanotus, 500
 L. borneensis, 82, *508*
Latastia, 66, 67, 72
Late Permian, 581
Late Triassic, 588
Laticauda
 L. colubrina, 368
 L. semifasciata, 473
Leiocephalus, 57
 L. varius, 81
Leiolepis, 59, 75
 L. belliana, 81
 L. muricatus, 81
Leiolopisma, 68, 82
Lepidochelys olivacea, 129
Lepidophyma, 67, 485
 L. flavimaculatum, 82
 L. tuxtlae, 82
Leptopleuron, 589
Lesothosaurus, 522–528
 L. diagnosticus, *523*
Lialis jicari, *508*
Limb development, 167, 197–221
Limbs, amniotes, *199*
Liolaemus, 58, 74
 L. gravenhorsti, 370
 L. multiformis, 81
Lissemys, 93
Lizards, 47, 127–152, 407–424, 483–502, 503–520
Locomotion, 13
 therapsids and archosaurs, 604–606, 621
Long-distance signalling action, amniote limb, 204

Lossiemouth Sandstone Formation, 588
Lotosaurus, 599, 600
Loveridgea, 675
L. ionidesi, 670, *676*
Lower Sakamena Formation, 583
Lower jaw, alligator development, 256, 260
Lycorhinus, 542
Lystrosaurus, 585, *586*

M

Mabuya, 68
 M. brevicollis, 82
 M. wrighti, 82
Macroclemys sp., 347
Macroclemys temminckii, 86–110
Macrocnemus, *578*, 580, 587, *588*
Macropus, 143
Malagasy, 581
Male reproduction, endocrinology, 357–383
Male reptiles, 357–383
Malerisaurus, 576, 580
Mammalian dentitions, 555
Mammals, 127–152, 305–316, 549–574
Maternal age, alligator development, 253
McKinlay River, 321
Meckel's cartilage, *169*
Megalosaurus, 637, *638*
Meroles, 67, 72
Mesalina brevirostris, 66
Mesoclemmys, 109
Mesos granule, garter snake, *121*
Mesosaurus, 574, 575
Mesosuchus, 577, *578*, 580, 585
Metabolic rate, habits, 425–441
Metaplastic bone, 165, 650
Methionine, enamel proteins, vertebrates, *289*
Micro-ELISA assay vertebrate enamel proteins, 293
Middle Triassic, 587
Miniaturization, lizard skull, 503–520
Mixed ankylosis/gomphosis, 562
Mochlodon, 542
Molecular mechanism, retinoids, amniote limbs, 208
Monjurosuchus, 576, 580

Monopeltis, 69
 M. capensis, 82
Morganucodon, 148, 149, 592
Morphology, 11
 anterior limb, *Varanus*, 27–45
 cranial, ornithopod dinosaurs, 521–547
Mouse amelogenins, 290
Movement, Australian crocodile, 322
Mus, 131
 M. musculus, 129
Muscle glycogen resynthesis, 433
Muscle patterns
 lizards, 72
 squamates, *53*
Muscles
 anguid lizards, *63*
 geckoes, *60*
 lizards, 47–85, *49*, *50*
 squamates, *54*
 teiid lizards, *65*
Mustela domestica, 129
Muttaburrasaurus, 528, 529, 544
Myotis sp., 129, *133*

N

Naja naja, 368
Nanosaurus, 522, 542
Nasal development, alligators, *234*
Nasal pits, craniofacial development, alligators, 233
Nasal placodes, craniofacial development, 233
Natrix, 679
 N. natrix, 83, 255, 677, *677*, *678*
Nerodia, 14
 N. fasciata, 367, *369*, 680
 N. rhombifera, 680
 N. sipedon, 373
Nessia layardi, 508
Nesting, Australian crocodile, 323
Netting, turtles, 89
Neural bones, turtles, *98*
Neusticurus
 N. bicarinatus, 82
 N. strangulatus, 82
Nile crocodile, 443–467
Nitrogen excretion, therapsids and archosaurs, 612–618
Non-muscular structures, lizards, 48

Subject Index

Notechis scutatus, 156, 157
Noteosuchus, 576, 580, 585
Notophthalmus, 180, 181
Nucras, 66
Number of teeth, evolution, 554

O

Odontogenesis, American alligator, *286*, *287*
Oedura marmorata, 82
Olgokyphus, 592
Oligodon arnensis, 677
Ontogeny, dentitions, 549–574
Opheodrys aestivus, 367, *369*
Ophiocephalus, 396
Ophiodes, 63, 78
   O. intermedius, 82
Ophiodes sp., *508*
Ophisaurus, 62, 500
   O. gracilis, 82
Ophryoessoides, 57, 58, 74
   O. trachycephalus, 81
Oplurus cuvieri, 81
Organ of Jacobson, squamates, 665–683
Origin
   diapsids, 581
   snakes, 516
Ornithomimus, 631, 646
Ornithopod dinosaurs, 521–547
Ornithorhynchus, *134*
   O. anatinus, 129
Ornithosuchus, 589, *590*
Ouranosaurus, 528, 529, *543*, 544
Ovis aries, 129
Oxirhopus sp., 166

P

Palaeagama, 581, 585
Paliguan, 581, 585, *586*
Parksosaurus, 542
Pathological bone, dinosaurs, 651
Pattern formation, amniote limb development, 197–221
Pattern regulation, limb development, 207
Pelorosaurus, 629
Pelusios sinuatus, 99

Periosteal bone, dinosaurs, 637
Periosteal ossification, 165
Permo-Triassic diapsids, 581
   classification, 580
Permo-Triassic reptiles, *578*
Petrodromus tetradactylus, 129
Petrolacosaurus, *578*, 581, *584*
Petrosaurus, 57, 496
   P. thalassinus, 81
Phrynops, 95, 99
   P. (Batrachemys) nasutus wermuthi, 97
   P. geoffroanus, 104, 105
Phrynosoma, 47, 56, 57, 74, 496
   P. coronatum, 47, 83
   P. orbiculare, 81
Phylogenetic pattern analysis, scleral ossicles, 492
Phylogeny
   dentitions, 549–574
   muscles, lizards, 72
Phymaturus, 57, 74
   P. pallumus, 81
Physignathus, 59
   P. lesueurii, 81
Physiological ecology, 317
Physiology
   dinosaurs, 629–663
   therapsids and archosaurs, 607
Pipistrellus pipistrellus, 129, *133*
Pisanosaurus, 600
Piscivory, turtles, 86–110
Planocephalosaurus, 585, 592, *592*
Plasma testosterone, male alligators, *376*
Platemys, 99, 109
   P. platycephala, 104
Plateosaurus, 591, *591*, 629, 635
Platysaurus sp., 82
Pletholax, 505, 511, 513, 514
   P. gracilis, *508*, *512*
Pleurodeles, 181
Pleurokinesis, *Iguanodon*, 538
Plica, 58
   P. umbra, 81
Podarcis
   P. muralis, 408, 421
   P. sicula, 421
Podocnemis, 88, 104
   P. venezuelensis, 99
Polar co-ordinate model, regeneration, 183
Polarity, muscles, lizards, 72

Polarizing region
  amniote limbs, 203
  reptilian limbs, 211
Polarizing zone model, regeneration and
  development, 182
Polychrus, 58, 63, 77
  P. marmoratus, 81
Polypterus, 129
Population size, Australian crocodile, 321
Post-hatching mortality, Australian
  crocodile, 322
Posterior limb bud, turtles, *216*
Potamogale velox, 129
Power production, *431*
Primary palate, alligator development,
  235
Primitive tetrapods, 127–152
Primitive tetrapod membrane, 139
Prionodactylus, 66
  P. manicatus, 82
Pristurus, 61
  P. carteri, 82
Probactrosaurus, 528, 544
Proctoporus unicolor, 82
Proctotretus, 58, 74
Proganochelys, 591
Progress zone model, regeneration and
  development, 182
Prolacerta, 576, *578*, 580, 585, 585, *586*, 587
Propulsive pattern, squamates, 16
Proterosuchus, *578*, 580, 585, 587
Protorosaurus, 576, 577, *578*, 580
Proximodistal axes, amniote limbs, 202
Pseudemydura, 99
  P. umbrina, 108
Pseudocordylus subviridis, 82
Pterospondylus, 591
Ptyas korros, 120
Ptyodactylus hasselquistii, 82
Pygopodidae, 62, 75
Pygopus, 62
  P. lepidopodus, 82, *508*, 511, *511*

R

Rabbit enamelins, 291
Radiation, Nile crocodile, 456
Rana, 180, 181, 208, 210
  R. pipiens, 129
Rattus norvegicus, 129

Regeneration, vertebrate appendages,
  177–196
Rehabilitation, gharial, 388
Repeated DNA, 308
Replacement, therapsids by archosaurs,
  599
Reproducibility, metabolic rate and
  habits, 431
Reproduction
  Australian crocodile, 323
  endocrinology, 357–383
Respiratory rate, Nile crocodile, 457
Retinoic acid, limb development, 207
Rhabdodon, 631, 636, 637, *637*, 640, 650,
  656
Rhabdophis, 477
Rhacodactylus trachyrhynchus, 82, *508*
Rhinesuchus, 583
Rhinoclemmys, 104
Rhinophis, 675
Rhynchosaurus, *578*, 580
Root morphology, tooth attachment, 599

S

Salmo trutta, 129
Saltopus, 589, *590*, 591
Salvinia, 457, 465
  S. molesta, 445, 446, *449*
Sanctuaries, gharial breeding, 404
Sapheosaurus, 581
Sauromalus, 56
  S. ater, 81
Saurosternon, 581, 585
Scaphonyx, *578*, 580
Sceloporus, 57
  S. torquatus, 81
Sceloporus sp., 408
Scincidae, 68, 77
Scincus, 68, 499
  S. mitranus, 82
  S. scincus, 82
Scleral ossicles, lizards, 483–502
Scleral ring, lizards, *484*
Scleromochlus, 589, *590*
Scoring, scleral ossicles, lizards, 485
Scotland, diapsids, 588
Scutellosaurus, 542
Second visceral arch blastema, lizards,
  135

Second visceral blastema, mammals, 135
Secondary callus cartilage, *160*
Secondary cancellous bone, dinosaurs, 646
Secondary cartilage, 156
Secondary centres, cartilage, 163
Secondary palate, alligator development, 236
Serpentes, 367
Sesamoid bones, 165
Sex chromosomes, evolution, reptiles and mammals, 305–316
Sex determination, 305
  Australian crocodile, 323, 346
  snakes, 307
Sex ratio, Australian crocodile, 319–355
Shansisuchus, 580
Shape gradients, dentitions, 553
Shinisaurus, 500
Signalling, retinoic acid, limbs, 210
Size and habits, 426
Skeletal development, 155
Skeleton, turtles, 106, *107*
Skin shedding, squamates, 111–126
Skull
  diameter reduction, lizards, 505
  lizard, 503–520, *509–518*
  miniaturization, lizards, 503–520
  trionychids, 92
  turtles, 91, 95, 96, 97
Slide-pushing, 13–26
Snakes, 69, 77, 469–480
  lateral undulation, *17*
  movement, *29*
  slide-pushinhg, *15*, *18*
Sorex araneus, 129
South Africa, diapsids, 585
Spear-fishing, turtles, 90
Spermatogenic stages, *363*
Sphaerodactylus, 508, 513
  S. molei, *507*, *508*
Sphenodon, 47, 69, *70*, 71, 72, 163, 483–502, 505, 575, 581, 667, *667*
  S. punctata, 129
  S. punctatus, 82, *508*
Spiracular tympanic membrane, lizards and mammals, 140
Spontaneous malformations, alligators, 251
Sprawling gait, *Varanus*, 27–45
Squalus acanthias, 293

Squamata, 366, 665–683
Squamates, 13–26, 111–126
  transitions, slide-pushing, *24*
Stagonolepis, 589, *590*
Staphylococcus aureus, 293
Stenaulorhynchus, *578*, 580
Stenocercus, 58, 74
  S. roseiventris, 81
  S. simonsii, 81
  S. varius, 81
Steptostyly, *Iguanodon*, 535
Sternotherus, 89, 363
  S. odoratus, 347, 360, *361*, *364*
Strobilurus, 58, 74
  S. torquatus, 81
Survivorship, Australian crocodile, 319–355
Sus scrofa, 129
Sustained effort, metabolic rate, 437
Switzerland, diapsids, 587
Systematics, Ornithopoda, 541

T

Tachyglossus aculeatus, 129
Tadarida mexicana, 129
Tail, lizards, *51*
Takydromus, 66, 72, 76
Tangasaurus, *578*, 581
Tanystropheus, 580, 587, *588*
Tanytrachelos, *578*, 580
Tapinocephalus, 658
Tarentola mauritanica, *507*, *508*
Taxonomy, scleral ossicles, lizards, 494, 497
Tectoseptal processes, alligator development, 236
Teeth, elephant, *571*
Teiidae, 64, 76
Teius, 66
  T. teyou, *65*, 82
Temperature, metabolic rate and habits, 431
Temperature-dependent sex determination, 323, 346
Temperature-dependent survivorship, 326
Temperature tolerance, therapsids and archosaurs, 612
Tenontosaurus, 528, 542

Teratology, alligator development, 255
Teratoscincus, 61
　T. scincus, 82
Testudo
　T. hermanni, 360
　T. triserrata, 655
Tetradactylus, 68
　T. africanus, 82
Thadeosaurus, 578, 581, 582, 584, *584*
Thalattosaurus, 580
Thamnophis, 14, 366
　T. elegans, 255
　T. melanogaster, 368
　T. sirtalis, 115, 119, 120, *121, 369*
　T. sirtalis parietalis, 367
Thamnophis sp., 122
Thecodontosaurus, 592, *593*, 627
Thelotornis kirtlandi, 478
Therapsids, 597–627
Thermoregulation
　growth, lizards, 407–424
　Nile crocodile, 443–467
　therapsids and archosaurs, 607
Theropsids, 612–621
Thescelosaurus, 529, 542
Thomasia, 592
Thrinaxodon, 555, *557, 568*
Ticinosuchus, 587, *588*
Tiliqua, 372, 499
　T. rugosa, 370
Tooth attachment, evolution, 549–574, 556
Tooth initiation sequence, 550
Tooth morphogenesis, vertebrates, *279*
Tongue, craniofacial development, alligators, 245
Toxolothosaurus, 581
Trachodon, 631, 650, 651
Transitions, slide pushing, squamates, 22
Transmission, sound, lizards and mammals, 146
Triassic, 585–588, 597–627
Triceratops, 636
Trichosurus, 143
　T. vulpecula, 129
Tricuspisaurus, 592
Trilophosaurus, 576, 580
Trimucrodon, 522, 542
Trionychidae, 93
Trionyx, 93, 104, 362
　T. sinensis, 359, 360

T. triunguis, 92
Triturus vulgaris, 129
Tropidonotus natrix, 166
Tropidurus, 57, 58, 74
　T. peruvianus, 58, 81
　T. torquatus, 58, 81
Tupinambio, 65
　T. nigropunctatus, 65, 82
　T. teguixin, *508*
Tubo-tympanic recess, lizards and mammals, 141
Turtles, 86–110
Tympanic cavity, 130
Tympanic membrane
　Morguanucodon, *148*
　lizards, 127–152
　mammals, 127–152
　primitive tetrapods, 127–152
Tympanic meatal plate, lizards and mammals, 142
Tympanic region
　lizards, *133*
　mammals, *133*
Typhlosaurus, 505, 509, 514
　T. cregoi, 516
　T. gariepensis, 504
　T. lineatus, 504, 510
Typhlosaurus sp., *508*
Tyrannosaurus, 631, *638*, 641

U

Uma, 57, 74, 483–502
　U. notata, 81, 494
Universality, patterning mechanisms, 178
Uracentron, 58
　U. flaviceps, 81
Uranoscodon, 58, 74
　U. superciliosus, 81
Uricotely, therapsids and archosaurs, 622
Uromastyx, 59, 372
　U. hardwicki, 81, 370
　U. macfadyeri, 81
Uropeltis, 675
Urostrophus, 58
　U. torquatus, 81
Uta, 57, 496
　U. stansburiana, 81

V

Valdosaurus, 544, 631, *632*, 646, 647, 649
Varanidae, 64, 75
Varanus, 27–45, *31, 34, 40*, 164, 485, 498, 500, 671, 673
  V. dumerilii, 64
  V. eremius, 72
  V. exanthematicus, 64, *508*
  V. gilleni, 64
  V. gouldi, 64
  V. griseus, 71, 675
Varanus sp., *82*
Variodens, 592
Venoms, snakes, 469–480
Ventral caudal muscles, lizards, 48
Vertebrate appendages, 177–196
Vertebrates
  amelogenesis, 277
  enamel proteins, 291, 293
  limb development, 197
Vipera
  V. berus, 83, 255, 675
  V. palestinae, 478
Vjushkovia, 580

W

W chromosome heterochromatin, 308

Water supply, breeding in gharial, 394
Weigeltisaurus, 576, 580, 583
Windspeed, Nile crocodile, 456

X

X chromosome inactivation, 310
Xantusia, 485
  X. henshawi, 67
  X. vigilis, 67
Xantusiidae, 67, 76
Xenochrophis piscator, 677
Xenopus, 180, 181
Xenosauridae, 62, 75
Xenosaurus, 62, 75, 500
  X. grandis, 82
Xenosaurus sp., 82

Y

Youngina, 576, *578*, 581, 585

Z

Zahnreihen theory, *551*
Zephyrosaurus, 540, 542
Zonosaurus madagascariensis, 82

DATE DUE			
AUG 11 '86			

PRINTED IN U.S.A.